T0199379

I.G. PETROWSKY
SELECTED WORKS

PART I

Classics of Soviet Mathematics

This book is part of a series. The publisher will accept continuation orders which may be cancelled at any time and which provide for automatic billing and shipping of each title in the series upon publication. Please write for details.

I.G. PETROWSKY
SELECTED WORKS

PART I
Systems of Partial Differential Equations
and
Algebraic Geometry

Edited by

O.A. Oleinik
Department of Mathematics and Mechanics
Moscow State University, Moscow, Russia

Translated from the Russian by

G.A. Yosifan

LONDON AND NEW YORK

First published 1996 by OPA (Overseas Publishers Association)

Published 2019 by Routledge
2 Park Square, Milton Park, Abingdon, Oxon OX14 4RN
52 Vanderbilt Avenue, New York, NY 10017

Routledge is an imprint of the Taylor & Francis Group, an informa business

First issued in paperback 2019

British Library Cataloguing in Publication Data

Petrowsky, I.G.
   Selected Works. – Part 1: Systems of
   Partial Differential Equations and
   Algebraic Geometry. – (Classics of
   Soviet Mathematics Series, ISSN 0743-9199;
   Vol. 5)
   I. Title   II. Yosifan, G.A.
   III. Oleinik, O.A.   IV. Series
   515.353

ISBN 13: 978-0-367-44916-2 (pbk)
ISBN 13: 978-2-88124-978-5 (hbk)(Part I)
ISBN 13: 978-2-88124-980-8 (2 part set)

I.G. PETROWSKY
SELECTED WORKS

PART I
Systems of Partial Differential Equations
and
Algebraic Geometry

Edited by

O.A. Oleinik
Department of Mathematics and Mechanics
Moscow State University, Moscow, Russia

Translated from the Russian by

G.A. Yosifan

LONDON AND NEW YORK

First published 1996 by OPA (Overseas Publishers Association)

Published 2019 by Routledge
2 Park Square, Milton Park, Abingdon, Oxon OX14 4RN
52 Vanderbilt Avenue, New York, NY 10017

Routledge is an imprint of the Taylor & Francis Group, an informa business

First issued in paperback 2019

Publisher's Note
The publisher has gone to great lengths to ensure the quality of this reprint but points out that some imperfections in the original may be apparent.

British Library Cataloguing in Publication Data

Petrowsky, I.G.
   Selected Works. – Part 1: Systems of
   Partial Differential Equations and
   Algebraic Geometry. – (Classics of
   Soviet Mathematics Series, ISSN 0743-9199;
   Vol. 5)
   I. Title   II. Yosifan, G.A.
   III. Oleinik, O.A.   IV. Series
   515.353

ISBN 13: 978-0-367-44916-2 (pbk)
ISBN 13: 978-2-88124-978-5 (hbk)(Part I)
ISBN 13: 978-2-88124-980-8 (2 part set)

CONTENTS

PETROWSKY'S ARTICLES ON PARTIAL DIFFERENTIAL EQUATIONS

FOREWORD

Ivan Georgievich Petrowsky (1901–1973) was a much loved and respected Rector of Moscow State University during a long and difficult time. As a mathematician, Petrowsky was not so well known in the West, partly because his most important papers, some in Russian, were difficult to read and far ahead of their time. But Petrowsky could also treat simpler problems in an exquisite way, as evidenced by his papers on the Dirichlet problem and a delicate boundary value problem for the heat equation. The value of this book is greatly enhanced by the learned commentaries, which in many cases amount to contemporary reviews of the many fields Petrowsky worked in.

Lars Gårding,
Mathematics Department, Lund University, Sweden

Ivan Georgievich Petrowsky
1901–1973

EDITOR'S PREFACE

This book contains the major works of Ivan Georgievich Petrowsky on systems of partial differential equations and on algebraic geometry. These works played a prominent role in the development of modern mathematics; they have become classics and their fundamental character remains manifest to this day.

The collection of Petrowsky's papers printed in the first section of this book focuses on the theory of systems of partial differential equations and is fundamental to the modern general theory of such systems. It was in these works that the classes of elliptic, hyperbolic and parabolic systems were singled out and studied for the first time; in modern mathematics, these systems are referred to as elliptic, hyperbolic and parabolic in the Petrowskian sense. These investigations are at the origin of vast and important disciplines in the theory of partial differential equations itself and in its applications.

The articles written by Petrowsky on algebraic geometry, and also printed here are of crucial importance for the topology of real algebraic manifolds, after the classic works of Harnack and Hilbert written at the close of the 19th century. These investigations of Petrowsky have been the source of intensive development of the theory of real algebraic manifolds in recent decades.

Some of the articles printed in this book were originally written by Petrowsky in languages other than Russian. During the preparation of the Russian edition of Petrowsky's *Selected Works* a number of changes, based on the author's preserved manuscripts and drafts, were introduced into the Russian translation of these articles. The present English edition takes into account all of these changes.

From among the various transliterations of his name, it is "Petrowsky" that has been chosen in the present translation, for it is this spelling that seems to have been preferred by the author himself.

The text of Petrowsky's articles is supplied with editor's notes and explanatory notes written by specialists in the field and printed either in footnotes or at the end of the article. The Appendix to this book contains commentaries on the featured articles. In particular, these commentaries describe the results obtained under the influence of Petrowsky's work and will acquaint the reader with the present state of the various fields. Two supplementary articles are also included in the Appendix: one gives the proof of Petrowsky's theorem about lacunas and is based on the paper by M. Atiyah, R. Bott and L. Gårding; the other is a review of the results on the Cauchy problem, with I.G. Petrowsky as co-author.

Part II of Petrowsky's *Selected Works* contains his major papers on ordinary and partial differential equations, mathematical physics, probability theory, and the theory of functions.

Ivan Georgievich Petrowsky

by A.N. Kolmogorov

Twenty two years of work as Rector of the first and largest university of our great country, and work as a true leader, participating in all projects and initiating of the most diverse undertakings in all directions of the activities of the University – this is quite a unique achievement.

Petrowsky did not belong to those of early flowering talent. His youth was far from easy. At the age of 21, in 1922, he came to the University of Moscow, already equipped with sound mathematical knowledge, which he had acquired by independent reading. He followed the normal five-year course as an undergraduate and three year course as a graduate. It seems no accident that he chose as his supervisor D.F. Egorov, a profound and subtle scholar, but a somewhat stern man, an unwearying toiler, avoiding all ostentation.

Even in his student years, Petrowsky was passionately concerned with community action – he was elected a student's representative on the mathematical syllabus commission and threw himself into this task with all the seriousness that was one of his characteristics. I recall him in 1927, a modest young man wearing a blue blouse, his hair close-cropped, as he was entrusted with the duty of greeting the First Congress of Soviet Mathematicians in the name of the students of the University of Moscow.

In 1926, while still a student, Petrowsky completed his first independent research paper on the Dirichlet problem. The theme arose in Egorov's seminar and was connected with recent work of L.A. Lusternik. But Petrowsky's results are remarkable by their conclusiveness. The very character of his subsequent cycle of papers is remarkable. He unexpectedly became interested in a problem from a completely new field. He studied this field, starting from the first steps of a beginner. And after a year, or several years... a solution emerged of the problem that had interested him at the very beginning. Thus, we have his famous papers on the topology on real algebraic curves, which began to appear in 1933. Just as suddenly, he became interested in the problem of justifying the limit theorems of probability theory by means of differential equations of limit stochastic processes with continuous time, a topic that concerned me closely, and after concentrated efforts, he met with success. The behavior of the integral curves of a system of ordinary differential equations close to a singular point also attracted Petrowsky's attention, and he brought it to very complete results.

The development of his most celebrated work, which started in 1936, on conditions of analyticity and of solvability of Cauchy's problem for systems of partial differential equations followed a similar course. His work on partial differential equations transformed entirely this important field of mathematics. It can even be said that before Petrowsky the subject dealt predominantly either with "analytic theory," using Taylor series and leading to formal results, which conceal the real features of individual types of equation, or else with special types, arising out of individual problems of mathematical physics. In effect, with Petrowsky's work there began a period of predominant interest in the systematic study of systems of partial differential equations of the most general type, with the object of selecting from them natural classes distinguished by some general property (solvability of Cauchy's problem, analyticity of all solutions, etc.). The influence of Petrowsky's work in this direction is growing even now in both Soviet and world science.

The fundamental inspiration of this principal direction of Petrowsky's research lies in freeing a problem from its connection with the chance peculiarities of actual problems that had arisen in particular physical investigations. This is a sensible tendency to abstraction that has led to the creation of large new general mathematical theories. But it does not mean that Petrowsky had no concern for precisely stated specific problems of immediate application. A number of examples could be quoted where Petrowsky made a considerable contribution to their solution. A time may come when something will probably be said, too, about other cases where Petrowsky took part in work on current applications of mathematics, but which are not reflected in his publications. I mention how often in recent years special attention was paid in his seminars to questions of modern theoretical physics, which stressed, once more, his inexhaustible spiritual youth, his capacity to grasp something new again and again.

But, to return to the description of Petrowsky's personality, I would stress once more, that the starting point of his scientific enquiries was usually that he became convinced of the current importance of a problem and the necessity of solving it. A mathematician needs exceptional powers if, by this approach to his work, he is to be so productive and almost free from error in his individual activity as was the case with Petrowsky. For would-be imitators, there is the danger that in becoming involved in the contemplation of problems that strike them as especially attractive, they will restrict themselves to such contemplation. The organized scientific effort of a team consists of distinguishing the most important problems arising and the assignment to each problem of people capable of solving it. But in Petrowsky's case, it often happened that he was attracted to problems in entirely new fields and solved them himself.

For Petrowsky, work as a scientist was always inseparably connected with professorial activity – the training of young scientists. His carefully delivered lecture courses were transformed into standard textbooks. His seminars were always centers of lively scientific thought. He trained his own pupils in the traditions of interest in science and its current problems, and not in their own personal achievements.

In 1940, Petrowsky was elected Dean of the Faculty of Mechanics and Mathematics of the University of Moscow. His time as Dean included the difficult war years, when the main body of the University operated in Ashkhabad and later in Sverdlovsk. The whole Faculty recall Petrowsky's constant concern about material resources and his unswerving efforts to maintain a high potential of academic and scientific work. In 1949–1951 Petrowsky was Academic Secretary of the Department of Physics and Mathematics of the Academy of Sciences of the USSR, devoting, as always, a full measure of his forces to this new work. From it he went on to the post of Rector of the University of Moscow, a position which he filled with all his energy until literally the last moments of his life.

All Petrowsky's close acquaintances, and even people who merely came into contact with him, will carry an indelible impression of this eternal toiler and man of inexhaustible youthful enthusiasm. The word "feat" can be used in relation to his entire life without any exaggeration.

I.G. Petrowsky and Modern Mathematics

by O.A. Oleinik

At the entrance to the main building of Moscow State University there is a memorial bas-relief dedicated to I.G. Petrowsky. The inscription under the sculptural portrait reads: "Ivan Georgievich Petrowsky – eminent scientist and community leader, recipient of the highest State Order, rector of the University, full member of the Academy of Sciences – worked in this building from 1953 to 1973." This concise statement rather conceals the enormity of the work performed by one of the greatest mathematicians of the 20th century, talented organizer of scientific endeavor, notable personality in state and community affairs – I.G. Petrowsky. His rectorship is one of the most remarkable pages in the history of Moscow University.

The studies conducted by Ivan Georgievich belong to the treasury of mathematical science. To him should be credited fundamental achievements in several diverse branches of mathematics: algebraic geometry, theory of probability, theory of ordinary differential equations, mathematical physics and partial differential equations.

The scientific interests of Ivan Georgievich have always been concentrated around great crucial problems of mathematics. His attention was drawn to concrete exactly formulated difficult questions. His choice of research subject inevitably proved to be of vital importance for a particular branch of mathematics. Any chosen problem was solved with finality and perfection; and the profound nature of the results obtained, the pioneering methods, and newly discovered problems made his investigation a starting point of a new direction in mathematics.

Many of Petrowsky's works were decades ahead of their time. Even now, his studies of the '30s and the '40s are the focus of attention of the leading mathematicians all over the world. The mighty advance of mathematical analysis and algebraic topology in recent years allows one to comprehend the profundity of the results and the ideas stored in the works of Ivan Georgievich, and sheds new light on the fundamental facts discovered by him. I.G. Petrowsky is the originator of the theory of systems of partial

4

differential equations, a discipline undergoing intensive development in recent decades, many of its departments being a natural continuation of the investigations started by Ivan Georgievich.

I.G. Petrowsky was born on January 18, 1901, not far from Bryansk, in a small ancient Russian town Sevsk. There, in 1917, he graduated from a Real School. Ivan Georgievich liked to look back upon the days of his childhood, the atmosphere of his grandfather's house where he grew up, and his native town where he had spent seventeen years of his life.

Even now the town of Sevsk keeps memory of its offspring who became a world-famous scientist. The building of the former Real School and the house where Ivan Georgievich was born and grew up are vested with memorial tablets; a school and one of the streets bear his name. Recently his name has been given to the Pedagogical Institute of Bryansk.

In 1917, on graduation from the Real School, I.G. Petrowsky joins the Natural Department of the Faculty of Physics and Mathematics of Moscow State University, in the hope of making a career in chemistry or biology. As a young man, he decided to become a scientist; however, at that time, circumstances prevented him from attending the University. In 1918, Ivan Georgievich, together with his relatives, had to move to the city of Elisabethgrad. For some time he works as a clerk and then enters a College of Mechanics and Tool Design. It was a mere accident that a book by N.E. Zhukovsky on theoretical mechanics got into his hands. At that time, his knowledge of mathematics was not enough to cope with the monograph, and to fill the gap he had to resort to other books. His first book on mathematics was Dirichlet's "Theory of Numbers," which, as he used to remember, amazed him by the beauty of its ideas and results, and turned his mind to mathematics for good.

In 1922, I.G. Petrowsky comes to Moscow with a desire to study at the University of Moscow. It was the same year that he transferred from the Natural to the Mathematical Department of the Faculty of Physics and Mathematics. As a student, he attended few lectures, and his education was mainly based on book study. On his arrival to Moscow in 1922, I.G. Petrowsky had neither accommodation nor any means of subsistence, and he had to spend several nights at a railway station. But he was fortunate enough to see an advertisement which said that an orphanage wanted to employ a yard-keeper and a stoker, board and lodging provided. Having reached the place, Ivan Georgievich found the stoker vacancy occupied by another student and had to resign himself to the yard-keeper's job. Actually, he had to work for the both, since the fellow-student was rather negligent of his duty.

From the fall of 1923, I.G. Petrowsky begins to teach mathematics to worker students at the Superior Art and Technology Workshop and continues until 1930. With some of his students, who later became artists, sculptors and musicians, I.G. Petrowsky kept friendly relations for many years. The teaching job allowed him to spend more time on his studies at the University of Moscow. During the 1923–1924, I.G. Petrowsky also taught mathematics at a secondary school in Moscow.

Even as a student, Ivan Georgievich revealed his talent to organize people and showed natural inclination to social work: he was constantly involved in community action. In particular, he was representing the students in the mathematical syllabus commission at the Faculty of Physics and Mathematics, which shows the high esteem he enjoyed among the students.

In 1927, I.G. Petrowsky greeted the First Congress of Russian Mathematicians in the name of the students of the Faculty of Physics and Mathematics of Moscow University. There is a photograph showing the Council of the Congress and student I.G. Petrowsky addressing the audience from the nearby rostrum. The text of that speech has also been preserved; it contains the idea of unity of theory and practice, which later became the basis of his entire work as a scientist and organizer.

At the University of Moscow, it was professor Dmitri Egorov whose influence had the greatest effect on I.G. Petrowsky and whom he regarded as his teacher. All these years he kept a portrait of D.F. Egorov in his apartment.

In the early period of his studies at the University, I.G. Petrowsky attends a seminar on the number theory, and in 1926 joins the seminar supervised by D.F. Egorov on the problem of Dirichlet. He reviewed O. Perron's works where the method of upper and lower functions was applied for the solution of the Dirichlet problem. At this time, L.A. Lusternik published his results on the solution of the Dirichlet problem by the method of finite differences. Ivan Georgievich decided to compare the two methods and obtained a number of interesting results on the Dirichlet problem for the Laplace equation. This research (see [1]) was the first work of I.G. Petrowsky as an undergraduate student, and he reported it at a meeting of the Moscow Mathematical Society. It was published in "Matematichesky Sbornik" in 1928 and became his graduation paper.

In 1927–1930, I.G. Petrowsky studied as a post-graduate under the supervision of D.F. Egorov. In later years, he recalled an episode of that period with D.F. Egorov. The professor asked him to make a review of two papers, one in Italian and another in English. Ivan Georgievich said that he did not know these languages. Egorov was surprised to hear that: "Does it matter? go and learn them," was his reply. And these two articles were diligently studied, summarized and reviewed at the seminar. It should be mentioned that all his life Ivan Georgievich had an inclination to study foreign languages. He could easily read German, English and French. In order to improve his pronunciation, he took private lessons in English and French, even as Rector of the University of Moscow. In his personal library, lots of foreign books bear traces of his reading. An unknown word he happened to come across while reading he liked to write out on the margin, together with its phonetic transcription.

In his post-graduate years, Ivan Georgievich also attended the seminars of A.Ya. Khinchin and A.N. Kolmogorov on the theory of probability. He was always willing to review articles at these seminars. Even then, A.Ya. Khinchin used to predict: "That young man will become an outstanding analyst." Ivan Georgievich always said that, for him, it was an arduous and time-consuming task to go into details of other people's proofs, and many a time he even referred to himself as "dull-witted." For this reason, while reviewing an article, he had to invent his own proofs, which task he found easier for himself. And it often turned out that he actually found a more general approach to a particular problem and suggested stronger methods, which led him to essentially new results. Thus appeared the famous work of I.G. Petrowsky on the problem of random walks; the original impulse to write this paper was due to the necessity of reviewing a German paper on the subject at the Kolmogorov's seminar. One is amazed to see the vast extent of Petrowsky's scientific interests during his graduate and post-graduate years, and also his extraordinary diligence.

Still preserved are the annual reports written by I.G. Petrowsky, as a post-graduate student, about his work during the years 1927–1929 at the Institute for Research in Mathematics and Mechanics at the University of Moscow. These reports are the evidence of his deep interest in mathematics, his thirst for knowledge, and his creative approach to the study of mathematical science.

In 1929, I.G. Petrowsky enters on academic career at the University of Moscow, and remains committed to it until his last day. From 1933, I.G. Petrowsky is a full professor at the Faculty of Mechanics and Mathematics. The '30s constitute a period of Petrowsky's amazingly intense creative activity, astonishing in the comprehensiveness of its scope and the power of its results. It was in this period that Petrowsky created the works that are nowadays regarded as classical.

Petrowsky's principal works, in spite of their "purely mathematical" character, essentially manifest his attitude to mathematics as an indispensable part of natural science, a basis for the discovery and understanding of qualitative and quantitative laws that constitute the subject of the sciences describing nature. It was his firm conviction that all concrete and deep mathematical facts would eventually find their application.

Next, we give a brief description of Petrowsky's principal works and illustrate the influence they had and continue to have on the development of modern mathematics.

1. Second Order Partial Differential Equations and the Problems of Probability Theory

Petrowsky's studies of boundary value problems for second order elliptic and parabolic equations are closely linked to his works on the theory of probability. The ideas and the methods that he initially had suggested and developed for the examination of second order partial differential equations were applied by him to the problems of probability. His results on the existence of solutions for the first boundary value problem for the heat equation have had immediate application to the theory of random processes. In the aforementioned Petrowsky's first paper [1], written in his student years, among other results we find a proof of the general uniqueness theorem for solutions of the Dirichlet problem for the Laplace equation: a bounded harmonic function is uniquely defined by its values at regular points of the boundary. In 1940, I.G. Petrowsky once again turns to this problem: in [2], he gives an elegant exposition of Perron's method for solving the Dirichlet problem, and in [3] – a new proof of the existence of solutions for the Dirichlet problem for the Laplace equation by the method of finite differences. In order to prove that the solutions of the difference equations are convergent, he used the finite difference counterpart of Bernstein's estimates; and, to examine the behavior of the solutions, barrier functions were applied.

In 1934, I.G. Petrowsky publishes a paper (see [4]) where, on the basis of Perron's method, he gives a solution of the first boundary value problem for the heat equation. He shows this problem to admit a solution in a domain G bounded by two straight lines parallel to the x-axis, and by the curves $x = \varphi_1(t)$ and $x = \varphi_2(t)$, $\varphi_1(t) \leq \varphi_2(t)$, for any continuous boundary value function, provided that $\varphi_1(t)$ and $\varphi_2(t)$ satisfy the following condition: for any t belonging to the domain of $\varphi_1(t)$ and $\varphi_2(t)$, there is a positive monotone function $\rho(h)$ defined for $h < 0$ and such that $\rho(h) \to 0$ as $h \to 0$,

$$\varphi_1(t+h) - \varphi_1(t) \geq -2\sqrt{h \ln \rho(h)} , \quad \varphi_2(t+h) - \varphi_2(t) \leq 2\sqrt{h \ln \rho(h)} ,$$

for all sufficiently small $|h|$ and

$$\int_C^\varepsilon \rho(h) h^{-1} \sqrt{|\ln \rho(h)|} \, dh \to \infty ,$$

as $\varepsilon \to -0$, C being a negative constant. On the other hand, assume that for some t there exists a positive monotone function $\rho(h)$ such that $\rho(h) \to 0$ as $h \to 0$; and let, for sufficiently small $|h|$, at least one of the inequalities

$$\varphi_1(t+h) - \varphi_1(t) \leq -2\sqrt{h \ln \rho(h)} ,$$

$$\varphi_2(t+h) - \varphi_2(t) \geq 2\sqrt{h \ln \rho(h)}$$

be satisfied, provided that the integral

$$\int_C^\varepsilon \rho(h)h^{-1}\sqrt{|\ln \rho(h)|}\,dh$$

is convergent as $\varepsilon \to -0$, $C = $ const < 0; then there exists a continuous function on the boundary such that the corresponding Dirichlet problem has no solution in the domain G. It follows, in particular, that in the domain bounded by the straight line $t = t_0 < 0$ and the curve

$$x^2 = 4(1 + \varepsilon)|t|\,[\ln|\ln|t|\,|]$$

the first boundary value problem for the heat equation always has a solution for $\varepsilon \leq 0$, and has no solution for $\varepsilon > 0$. The latter result was also obtained by Khinchin, who used probabilistic methods; it corresponds to the iterated logarithm law in the theory of probability. A stronger version of this law is of great importance for the study of random processes such as that of Brownian motion. The solvability conditions obtained by Ivan Georgievich for the first boundary value problem for the heat equation made it possible to apply the methods of Kolmogorov for the derivation of the stronger version of Khinchin's iterated logarithm law.

Closely related to [4], as regards analytic methods, is Petrowsky's paper [5] on the theory of probability, where he considers the problem of random walk. This paper has had a great effect on all subsequent development of the theory of random processes. In modern terms, the problem can be formulated as follows. In Euclidean space $\mathbb{R}^n$ consider a Markov chain such that $p(x, A)$ is the probability of a particle moving, in a single step, from the point x into the set A. Let G be a domain in $\mathbb{R}^n$ with a smooth boundary, and let $\pi(x, A)$ be the probability of the event that a particle starting its motion at x happens to enter A at the moment of its first exit from G. The problem is to examine the limit behavior of the integral

$$v(x) = \int_{\mathbb{R}^n \setminus G} \varphi(y)\pi(x, dy)$$

for a given continuous function φ defined in $\mathbb{R}^n \setminus G$, when the transition function $p(x, A)$ varies in such a way that the displacement of the particle tends, in a certain sense, to zero. In [5], Petrowsky proves that $v(x)$ converges to a solution of the Dirichlet problem for a second order elliptic equation, its boundary values being equal to φ.

To solve this probability problem, Petrowsky made use of the methods that he had previously developed with a view to study the Dirichlet problem for the Laplace equation, in particular, the method of upper and lower functions. A detailed exposition of Petrowsky's method and its further developments was later given by Khinchin in his well-known monograph [6]. The relation between the probabilities of a Markov process' exit from the domain and the solution of the Dirichlet problem was established by Petrowsky in limit form. However, further investigations have shown that there is no need for passing to the limit if one considers a Markov process with continuous time. This observation has made probabilistic methods applicable to the study of many problems in the theory of partial differential equations.

2. Theory of Functions

In 1929, Petrowsky published his article [7] on the theory of functions of real variable that was in the process of intensive development at that time by the efforts of the analysts belonging to the so-called Moscow mathematical school headed by D.F. Egorov and N.N. Lusin. One might be interested to know how this paper appeared. Petrowsky wrote it in his post-graduate period, while getting prepared to the examination on the theory of probability. In a report describing his work during the academic year 1928/29, Ivan Georgievich writes: "While studying the theory of probability, I often came across the notion of the Stiltjes integral; this made me read a chapter about the Stiltjes integral in the latest edition of Lebesgue's monograph "Leçons sur l'Integration et la Recherche des Primitives." At the end of his book, Lebesgue raises the question: Let $f(x)$ and $\varphi(x)$ be two continuous functions defined on (a, b); assume that at every point of (a, b) the function $f(x)$ has a finite derivative with respect to $\varphi(x)$, i.e.,

$$\lim_{h \to 0} \frac{f(x+h) - f(x)}{\varphi(x+h) - \varphi(x)}$$

exists and is finite for every $x \in (a, b)$. Is there another continuous function $f_1(x)$ having the same derivative with respect to $\varphi(x)$ at any point of (a, b)?

If $\varphi(x)$ is a function of bounded variation, the answer to this question is known to be negative. I have been able to show that the answer is also negative in the general case; to this end, I proved that if

$$\lim_{h \to 0} \frac{f(x+h) - f(x)}{\varphi(x+h) - \varphi(x)} = 0$$

everywhere on (a, b), then $f(x)$ is constant."

3. Topology of Real Algebraic Manifolds

Petrowsky's famous work [8] on the topology of real algebraic curves appeared in 1933. Its results he reported to the Congress of Soviet Mathematicians in 1934, and gave their detailed exposition in [9].

To describe the mutual position the branches of a plane algebraic curve may occupy is one of the oldest mathematical problems. For the curves of degree 2, it was solved by the ancients (Apollonius, et al.); the curves of degree 3 were studied by Descartes, and the curves of degree 4 by Newton. Fifth order curves can be examined by similar methods. However, the obstacles in the way of studying the curves of degrees larger than 5 had been so strong, that Hilbert was prompted to formulate his 16th problem about the topology of real algebraic curves and surfaces.

As early as 1876, Harnack [10] showed that for a plane algebraic curve of degree n on the projective plane the number of pieces does not exceed $\frac{1}{2}(n-1)(n-2)+1$; he also constructed algebraic curves with this maximal number of components. Since then and up to 1933, when Petrowsky's paper appeared, i.e., for more than half a century, no perceptible advance could be observed in the topology of real algebraic curves, in spite of all the efforts made by Hilbert and his disciples to examine real algebraic manifolds.

The 16th problem posed by Hilbert (see [11]) raises many questions, very simple in form but requiring extreme efforts for their solution. One of these questions is about the mutual position of the 11 ovals forming an algebraic curve of degree 6. In 1891, Hilbert made the conjecture that an algebraic curve of degree 6 with maximal number of components cannot consist of 11 mutually exterior ovals. Once again, in 1900, he expressed this hypothesis while formulating his 16th problem at the International Mathematical Congress held in Paris.

At the beginning of the 20th century, there were several attempts to prove the correctness of Hilbert's conjecture. I.G. Petrowsky should be credited with a deep investigation into the topology of real algebraic curves on the projective plane (see [8], [9]); the results of this investigation imply, as a very special case, the statement of Hilbert about the curves of degree 6. I.G. Petrowsky established the fact that an algebraic curve of even degree n cannot consist of more than $\frac{1}{8}(3n^2-6n)+1$ mutually exterior ovals.

An oval of an algebraic curve $F(x,y)=0$ (where F is a polynomial of degree n with real coefficients) is called positive (resp., negative), if the inequality $F(x,y)>0$ (resp., $F(x,y)<0$) holds for the points enclosed in this oval and belonging to its neighborhood. Petrowsky's theorem about algebraic curves of even degree n can be formulated as follows. Consider a curve of even degree n with p positive ovals and m negative ones; then $|p-m|\leq\frac{1}{8}(3n^2-6n)+1$, and there exist curves for which this upper bound is attained. A similar result has been proved by Petrowsky for curves of odd degree n. It is easy to verify that $p-m$, for n even, is equal to the Euler characteristic of the closure of the set formed by finite points of the projective plane (x,y,z) such that $z^n F\left(\frac{x}{z},\frac{y}{z}\right)\geq 0$. Thus formulated, Petrowsky's theorem can be extended to algebraic surfaces of any degree n in real projective space P^m (see [12]), and also to spatial algebraic curves (see [13]).

Topology of real algebraic surfaces is the subject of joint publications of
I.G. Petrowsky and O.A. Oleinik. Basically, the method used in [8], [9],
[12]–[14] is that of examining how the topological structure of an algebraic
curve or surface changes with the variation of the coefficients in the corresponding algebraic equation. Of crucial importance for this analysis are
the Euler-Jacobi formulas and the results obtained by Morse [15] concerning the changes of topological structure of the level manifold while passing
through critical values of a function.

The results of I.G. Petrowsky, on a par with those of A. Harnack and
D. Hilbert, are fundamental to the theory of real algebraic manifolds.

In 1969, D.A. Gudkov gave a complete solution to the problem of describing mutual positions occupied by the pieces of an algebraic curve of
degree 6. This work was deeply influenced by the ideas and methods of
I.G. Petrowsky.

During the recent 15 years, the topology of real algebraic manifolds has
been greatly enriched by new important ideas and results. A summary of
these new developments is given by V.M. Kharlamov in a separate article
published in this volume (see Appendix).

4. Ordinary Differential Equations

One of the basic problems in the qualitative theory of ordinary differential
equations is the subject of Petrowsky's paper [17], written in 1934, where
he studies the behavior of integral curves near the origin for a system of
ordinary differential equations

$$\frac{\partial x_j}{\partial t} = \sum_{k=1}^{n} a_{jk} x_k + \varphi_j(x), \qquad x = (x_1, \ldots, x_n), \tag{1}$$

where a_{jk} are constant; the functions $\varphi_j(x)$ have continuous derivatives near
the origin, $\varphi_j(0) = 0$, $\mathrm{grad}\, \varphi_j(0) = 0$, $j = 1, \ldots n$.

Initial studies of the behavior of trajectories in a neighborhood of singular points date back to the works of A. Poincaré. This problem is also the
subject of important articles written by O. Perron. The most comprehensive
study on the behavior of the trajectories for system (1) in a neighborhood
of the origin in phase space is due to I.G. Petrowsky [17], who showed
that, under the above conditions on the functions $\varphi_j(x)$, the qualitative
behavior of the trajectories of (1) near the origin is mainly determined by
the matrix $\|a_{jk}\|$, provided that all roots λ of the characteristic equation
$\det(\|a_{jk}\| - \lambda E) = 0$ have non-vanishing real parts; i.e., in this case, the
trajectories of (1) behave near the origin in the same way as the trajectories
of the system with constant coefficients which is obtained from (1) by setting $\varphi_j \equiv 0$, $j = 1, \ldots, n$. If the characteristic equation admits roots with

zero real part, then one can specify a class of trajectories whose behavior near the origin is determined by the matrix $\|a_{jk}\|$.

This paper stimulated further research in the qualitative theory of differential equations (see [18]). It was I.G. Petrowsky who initiated the qualitative analysis of solutions of ordinary differential equations with rational right-hand sides in complex region. In this field, a number of fundamental concepts and ideas owe their existence to Ivan Georgievich. For many years his attention had been drawn to the problem concerning the number of limit cycles for a first order ordinary differential equation with rational right-hand side; and, although his works do not give a complete solution to this problem, the methods indicated by him for its investigation, the ideas and new concepts introduced by him had great impact on the development of a new mathematical discipline – the qualitative theory of ordinary differential equations in complex region – and some related fields.

The article [19], written by I.G. Petrowsky as a survey on the behavior of the solutions of a system of ordinary differential equations near a singular point, was published as Appendix to the monograph by A. Poincaré: "On the Curves Defined by Differential Equations." I.G. Petrowsky also wrote the chapter "Ordinary differential Equations" as part of the book "Mathematics, its Subject, Methods, and Meaning" [20]. This chapter contains an easy exposition of fundamental facts from the theory of ordinary differential equations and is aimed at a wide audience.

5. Systems of Partial Differential equations; Cauchy's Problem

In the 19th and the early part of the 20th century, the attention of analysts was mainly being drawn to concrete problems of mathematical physics, in particular, to some special equations describing certain physical processes. During this period, such special second order equations as the Laplace equation, the heat equation and the wave equation were subject to thorough analysis, together with the properties of their solutions and the related boundary and initial value problems; a large number of separate facts had been gathered, and some methods for the solution of separate problems were developed. The need for a general theory based on all these achievements became evident. And it was I.G. Petrowsky who, in his articles published in 1936–1945, laid the foundation of the general theory of systems of partial differential equations.

Before the works of I.G. Petrowsky, deep and fairly complete results had been mainly obtained for second order equations; among them, the classes of elliptic, hyperbolic, and parabolic equations had been specified, the properties and the boundary value problems characteristic of each class had been examined. Ivan Georgievich singled out and studied the classes of systems which, in their properties, resemble elliptic, hyperbolic, and parabolic equations. His results formed a basis for the general theory of systems of partial differential equations and determined the essential features of its subsequent development.

Starting from 1936, I.G. Petrowsky published a series of papers about the Cauchy problem for systems of partial differential equations. These works were of crucial importance for the development of the theory of partial differential equations as a whole.

In his famous monograph on the Cauchy problem [21], J. Hadamard introduced the notion of a well-posed Cauchy problem for partial differential equations. This notion embraces the existence of solutions of the Cauchy problem, their uniqueness and continuous dependence on the initial functions.

In [22] (see also [23]), I.G. Petrowsky introduces the notion of a hyperbolic system and examines the so-called correct solvability of the Cauchy problem for such systems.

Consider a nonlinear system

$$\Phi_j \equiv D_0^{n_j} u_j - F_j(t, x, u, D^\alpha u) = 0 , \qquad j = 1, \ldots, N , \qquad (2)$$

where $|\alpha| \leq n_k$, $\alpha_0 < n_k$ for the derivatives $D^\alpha u_k$ taken as arguments of the functions F_j. Here

$$\alpha = (\alpha_0, \alpha') = (\alpha_0, \alpha_1, \ldots, \alpha_n) , \quad |\alpha| = \alpha_0 + \alpha_1 + \cdots + \alpha_n , \; \bullet$$

$$|\alpha'| = \alpha_1 + \cdots + \alpha_n , \quad D_0 = \frac{\partial}{\partial t} , \quad D_j = \frac{\partial}{\partial x_j} , \quad j = 1, \ldots, n ,$$

$$D^\alpha = D_0^{\alpha_0} D_1^{\alpha_1} \cdots D_n^{\alpha_n} , \quad u = (u_1, \ldots, u_N) , \quad x = (x_1, \ldots, x_n) .$$

Petrowsky calls system (2) hyperbolic, if the equation

$$\det \left\| \sum_{|\alpha|=n_k} \frac{\partial \Phi_j}{\partial(D^\alpha u_k)} \lambda^{\alpha_0} \xi^{\alpha'} \right\| = 0 \qquad \xi^{\alpha'} = (\xi_1^{\alpha_1} \cdots \xi_n^{\alpha_n}) \qquad (3)$$

with respect to λ has real mutually distinct roots for any real non-zero vector $\xi = (\xi_1, \ldots, \xi_n)$.

I.G. Petrowsky [23] showed that for a nonlinear hyperbolic system (2) the Cauchy problem with initial conditions at $t = 0$ is correctly solvable for sufficiently small t; and for a linear hyperbolic system with variable coefficients it is correctly solvable on any time interval. His proof is based on energy estimates derived for solutions of a first order hyperbolic system with coefficients periodic in x. The solution of the Cauchy problem for system (2) is obtained as a limit of analytic solutions of the Cauchy problems for some approximating systems; these solutions are constructed according to the Kowalevskaya theorem, and for the proof of their convergence the energy estimates and the compactness theorems are used. It might be interesting to note that, initially, energy integrals determined by singular (pseudo-differential) operators appeared in [23].

At the end of the '50s, Calderon [24], Mizohata [25], and Yamaguti [26] once again obtained energy estimates for hyperbolic systems, using, as it were, the method initially proposed by Petrowsky. Being in possession of the technique of pseudo-differential operators, already developed at that time, these authors were able to simplify the formal aspect of Petrowsky's proof. Energy estimates for a single higher order hyperbolic equation were obtained by Leray [27], whose method was different. Gårding [28], using Leray's method, established energy estimates for the conjugate operator and proved the existence theorem for the Cauchy problem on the basis of some theorems from functional analysis.

In recent years, the mixed problem for systems hyperbolic in the sense of Petrowsky has been solved. At present, efforts of many mathematicians are applied to the study of weakly hyperbolic equations and systems, i.e., the case of equation (3) admitting multiple real roots. It is well-known that in order to have a well-posed Cauchy problem in this case, additional conditions should be imposed on the coefficients by the lower order derivatives in these linear equations and systems (the so-called conditions of Lewy). A detailed account of research in this direction is given in a separate article written by V.Ya. Ivrii and L.R. Volevich (see Appendix).

In [29], I.G. Petrowsky studies the question of well-posedness for systems with the Cauchy data prescribed on the entire hypersurface $t = 0$. He considers an arbitrary linear system with smooth coefficients depending only on t; the system has the form

$$D_0^{n_j} u_j = \sum_{k=1}^{N} \sum_{|\alpha| \leq M} A_\alpha^{kj}(t) D^\alpha u_k + f_j(t, x), \qquad j = 1, \ldots, N, \qquad (4)$$

(M is an integer, $\alpha_0 < n_j$), and the initial Cauchy conditions are

$$D_0^s u_j \Big|_{t=0} = \varphi_{js}(x), \quad s = 0, 1, \ldots, n_j - 1, \quad j = 1, \ldots, N, \qquad (5)$$

where the functions $f_j(t, x)$, $\varphi_{js}(x)$ are assumed bounded, together with their derivatives up to a certain order, $x = (x_1, \ldots, x_n)$. I.G. Petrowsky found a necessary and sufficient condition (the famous Condition A) for the Cauchy problem (4), (5) to be well-posed in the class of bounded functions. For systems of the first order in t, Condition A can be expressed as follows: let

$$\frac{dV_j(t, \xi)}{dt} = \sum_{k=1}^{N} \sum_{|\alpha| \leq M} A_\alpha^{kj}(t)(i\xi)^{\alpha'} V_k(t, \xi)$$

be the system of ordinary differential equations obtained by taking the Fourier transform of (4) with respect to x; then the fundamental matrix of solutions of this system must have all its elements growing slower than a certain power of $|\xi|$ as $|\xi| \to \infty$.

In [29], I.G. Petrowsky relies on a systematic application of the Fourier method, which has later become one of the major tools of investigation in the field of partial differential equations and systems. The development of the theory of distributions and their Fourier transform resulted, during the '50 – '60s, in numerous studies of the Cauchy problem for system (4); essentially, these studies continue the research performed by Petrowsky in [29]. The theory of the Fourier transform in spaces of rapidly growing functions has made it possible to specify for the solutions of system (4) the fastest admissible growth rate at infinity, with respect to the variables x, when the Cauchy problem (4), (5) remains well-posed (see, for instance, [30]).

Condition A is fulfilled for hyperbolic systems. I.G. Petrowsky specified another important class of systems satisfying Condition A, namely, parabolic systems, which are now called parabolic in the sense of Petrowsky.

To define a parabolic system, let us associate weight 1 to the differentiation in x_j $(j = 1, \ldots, n)$, and weight $2b$ to the differentiation in t, where b is an integer. Assume that $2b$ is chosen such that the total order of differentiation of any function u_j in (4) does not exceed $2bn_j$. Removing the terms with the derivatives of u_j $(j = 1, \ldots, N)$ of total order less than $2bn_j$, we obtain the so-called principal part of system (4). System (4) is said to be parabolic, if the roots λ of the characteristic equation

$$\det \left\| \sum_{|\alpha'|+2b\alpha_0=2bn_k} A_\alpha^{kj}(t)(i\xi)^{\alpha'} \lambda^{\alpha_0} - \delta_{kj}\lambda^{n_j} \right\| = 0$$

have their real parts smaller than a negative constant, for all ξ such that $|\xi| = 1$.

For such systems, I.G. Petrowsky constructed the Green matrices, and for the solution of the Cauchy problem (4), (5) he found an integral representation similar to the Poisson integral in the case of the heat equation. Extending the Green matrices to the complex region, he proved analyticity with respect to spatial variables x for solutions of the parabolic system (4), assuming the functions $f_j(t, x)$ to be analytic in x. In [29], Ivan Georgievich indicates various possibilities for the generalization of his results, and shows the way of their realization. Subsequently, these generalizations have been done by other mathematicians. In particular, a general theory of boundary value problems has been developed for parabolic systems. The theory of systems parabolic in the sense of Petrowsky is outlined in a commentary written by O.A. Oleinik and V.P. Palamodov on Article 2 of this volume (see Appendix).

6. Diffusion of Waves and Lacunas

Petrowsky's articles [31] and [32] are dedicated to the qualitative theory of hyperbolic equations; the aim is to study how the solution of the Cauchy problem may depend on the initial functions. In these papers, the notion of lacunas is introduced, which has a definite physical meaning, and a necessary and sufficient condition is provided for the existence of lacunas for hyperbolic equations of arbitrary order with constant coefficients. I.G. Petrowsky developed an effective method to reduce the problem of the existence of lacunas to the examination of hyperplane cross-sections of complex algebraic manifolds.

For the wave equation with odd number of spatial variables, it is well-known that the value of a solution of the Cauchy problem at the vertex of the characteristic cone depends on the initial data prescribed on the periphery of the characteristic cone's base on the hyperplane $t = 0$: the propagating wave has a forefront and a rear. In this case, it is common to say that there is a lacuna. If the number of spatial variables is even, the value at the vertex depends on the initial data taken on the entire basis of the characteristic cone: then the propagating wave has only a forefront. In this case, one speaks of diffusion of waves.

I.G. Petrowsky gave definition of a lacuna for hyperbolic equations of order m and obtained a necessary and sufficient condition for the existence of stable lacunas for a homogeneous hyperbolic equation of the form

$$D_0^m u = \sum_{|\alpha|=m,\, \alpha_0 < m} a_\alpha D^\alpha u \qquad (6)$$

with constant coefficients. The existence of a lacuna for such an equation is determined by topological properties of the surface

$$1 - \sum_{|\alpha|=m,\, \alpha_0 < m} a_\alpha z^{\alpha'} = 0, \quad z^{\alpha'} = z_1^{\alpha_1} \cdots z_n^{\alpha_n},$$

in complex space $(z_1, \ldots, z_n)$.

Having in view the lacuna problem, Ivan Georgievich suggested a new method to derive the formulas representing the solution of the Cauchy problem for a homogeneous equation of type (6) with initial conditions at $t = 0$; a different version of this representation had been previously obtained by Herglotz. At present, these expressions for the solutions are known as Herglotz-Petrowsky's formulas. I.G. Petrowsky also solved the lacuna problem for hyperbolic equations with variable coefficients and two independent variables, as well as for some classes of systems with many independent variables.

It should be observed that Petrowsky's works on the diffusion of waves and lacunas are distinguished by an exceptionally high level of ideas and methods, both from the theory of differential equations and from complex algebraic geometry. His investigations were far ahead of their time. The well-known analysts M. Atiyah (UK), R. Bott (USA), L. Gårding (Sweden), who studied these excellent and deep results of Petrowsky, have written an article dedicated to Ivan Georgievich, where they give an alternative exposition of his results (see [33]). For further developments in Petrowsky's theory of the diffusion of waves and lacunas, the reader is referred to the commentary written by A.M. Gabrielov and V.P. Palamodov on Articles 5 and 6 of this volume (see Appendix). Owing to the results of [31], [32], in 1946, I.G. Petrowsky was honored by the State Prize of the USSR for his great contribution to science.

7. Elliptic systems; Analyticity

In 1937, I.G. Petrowsky (see [34], [35]) gave the most comprehensive and, in some sense, exhaustive answer to a question which forms the 19th problem posed by Hilbert. The question concerns the characterization of all differential equations and systems whose sufficiently smooth solutions can only be analytic functions. Analyticity of solutions of differential equations attracted the attention of many well-known mathematicians. Thus, Hilbert wrote: "One of the most remarkable facts in the foundations of the theory of analytic functions is, in my opinion, the existence of partial differential equations whose integrals can only be analytic functions of their arguments; or, in brief, these equations admit only analytic solutions."

I.G. Petrowsky singled out a class of systems, now called 'elliptic in the sense of Petrowsky', and proved that all sufficiently smooth solutions of these systems are analytic functions, provided that the equations forming the system are given by analytic functions of their arguments. He also showed that in the case of systems with constant coefficients the condition of ellipticity is necessary and sufficient for the analyticity of all its solutions (provided that the characteristic matrix is not identically degenerate). It is of interest to note that for the proof of his result I.G. Petrowsky does not resort to any methods or ideas unknown to his predecessors who had studied this problem. By simple means Ivan Georgievich obtains very deep results. In order to prove analyticity of a solution of an elliptic system, he defines the solution for complex values of the argument x_1 by extending to the complex region in x_1 the Fourier coefficients of its expansion with respect to $x_2, \ldots, x_n$, so that the function corresponding to the resulting Fourier coefficients would satisfy the Cauchy-Riemann conditions.

A system of the form

$$F_j(x, u, D^\alpha u) = 0 \, , \qquad j = 1, \ldots, N \, ,$$

(here $x = (x_1, \ldots, x_n)$, $u = (u_1, \ldots, u_N)$, and the order of the derivatives of u_j in the left-hand side does not exceed n_j) is called elliptic in the sense of Petrowsky in a domain where the arguments of F_j take their values, if

$$\det \left\| \sum_{|\alpha|=n_k} \frac{\partial F_j}{\partial (D^\alpha u_k)} \xi_1^{\alpha_1} \cdots \xi_n^{\alpha_n} \right\| \neq 0$$

for all real values of $\xi = (\xi_1, \ldots, \xi_n)$ such that $|\xi| \neq 0$.

For second order elliptic equations with two independent variables, analyticity of solutions was initially proved in 1903 by S.N. Bernstein [36], who used his own method of normal series; this result for elliptic equations of arbitrary order with two independent variables was established in 1927 by H. Lewy [37]. A review of the results pertaining to the analyticity problem for partial differential equations can be found in [38], as well as in the commentary written by O.A. Oleinik and L.R. Volevich on Petrowsky's Article 4 of this volume (see Appendix, and also [39]).

The article [35] has been a starting point for the theory of elliptic systems – a vast and important branch of the theory of partial differential equations. After this work of I.G. Petrowsky, elliptic systems became an object of numerous studies. For systems elliptic in the sense of Petrowsky, a complete general theory of boundary value problems has been constructed (see [40], [41]), the index problem has been solved for such problems (see [42], [43]), and local differentiability properties of solutions have been investigated, in accordance with the conjectures made by I.G. Petrowsky in [44].

8. Problems of Mathematical Physics and Biology

In 1937, A.N. Kolmogorov, I.G. Petrowsky, and N.S. Piskunov published their research performed in connection with a problem in biology – the problem of proliferation of a genus (see [45]). The ideas developed in this work happen to be fruitful in many fields of mathematical physics.

In [45], the authors consider a nonlinear diffusion equation of the form

$$\frac{\partial v}{\partial t} - k \frac{\partial^2 v}{\partial x^2} = F(v) , \qquad (7)$$

where $F(v)$ is a sufficiently smooth function defined for $0 \leq v \leq 1$ and such that

$$F(0) = F(1) = 0 , \quad F(v) > 0 \quad \text{for} \quad 0 < v < 1 , \quad F'(0) = a > 0 ,$$

$$F'(v) < a \quad \text{for} \quad v > 0 , \quad k = \text{const} > 0 .$$

It has been shown that equation (7) admits solutions having the form of travelling waves: $v(x,t) = V(x + \lambda t + C)$, $\lambda, C = $ const, and satisfying the conditions $V(-\infty) = 0$, $V(+\infty) = 1$ for any $\lambda \geq \lambda_0 = 2\sqrt{ak}$. Among these, it is only the solution corresponding to the minimal velocity $\lambda_0 = 2\sqrt{ak}$ that is stable in the following sense: any solution of the Cauchy problem with arbitrary initial data such that

$$v(x,0) \equiv 0 \quad \text{for} \quad x \leq x_0, \qquad 0 < v(x,0) < 1 \quad \text{for} \quad x_0 < x < x_1,$$

$$v(x,0) \equiv 1 \quad \text{for} \quad x \geq x_1,$$

where x_0, x_1 are arbitrary real constants, converges, as $t \to \infty$, to the solution having the form of a travelling wave corresponding to $\lambda = \lambda_0$. It turned out that similar questions arise in various problems of mathematical physics, particularly, in combustion theory. Publications on the mathematical theory of combustion essentially rely on the ideas initially proposed in [45] (see, for instance, [46]). An enormous number of purely mathematical works, as well as those connected with biology and physics, apply, generalize and develop the results of [45].

In 1945, I.G. Petrowsky considered a problem dealing with some concrete questions arising in the theory of elasticity [47]. He examined the propagation speed of discontinuities for displacement derivatives on the surface of an inhomogeneous elastic body free of external forces, under certain assumptions about the surface of the body and boundary conditions; he proved that a discontinuity of a certain n-th order derivative across the line on the surface of the body propagates, at each point M of this line, with such a speed, as if the body were infinite, bounded by a plane, and homogeneous, with the parameters defining its elastic properties equal to their values at M. This problem is related to the propagation of Rayleigh waves.

The nature of the lines and surfaces of discontinuity for solutions of the wave equation was considered by I.G. Petrowsky in [48]. At present, propagation of discontinuities of solutions of partial differential equations is a problem that attracts the attention of many analysts, and it is one of the central problems in the theory of partial differential equations (see, for instance, [49], [50]).

9. Textbooks

Being a professor at the University of Moscow, Ivan Georgievich was actively engaged in the teaching process: he was giving regular and special courses in mathematics, conducting seminars, and supervising graduate students and post-graduates. In 1936–1937, he was reading lectures on the theory of ordinary differential equations, first, at the University of Saratov, and then at the University of Moscow. The records of this course formed the basis of his textbook on ordinary differential equations [51] (initially published in 1939). It was Petrowsky's conviction that a textbook should contain only the main carefully chosen topics, and their exposition should be rigorous and possibly short. The good taste he manifested in his scientific work was also present here.

Another of Petrowsky's textbooks, "Lectures on the Theory of Integral Equations," appeared in 1948. This course is based on the lectures he was reading at the University of Moscow in 1946. It contains an elegant exposition of the theories of Fredholm and Hilbert-Schmidt from the standpoint of functional analysis.

For many years I.G. Petrowsky was lecturing on partial differential equations at the University of Moscow. The first edition of his "Lectures on Partial Differential Equations" appeared in 1950. The deep presentation of the material in this course, which has become classical, raised the teaching of the theory of partial differential equations to a new level in our country, as well as abroad.

The textbooks on the three major university courses have a very moderate size and were written by I.G. Petrowsky under the influence of new ideas characteristic of the so-called Moscow mathematical school. They have been reprinted and re-edited in Russia, and there exist their translations into several foreign languages. These textbooks were granted the State Prize of the USSR.

10. Problems Set by I.G. Petrowsky; His Scientific School

In 1946, I.G. Petrowsky published the article entitled "On some problems in the theory of partial differential equations" [44], where he describes the contemporary state of the theory and indicates the directions of its further development. Now, fifty years after, it can be claimed with certainty that the theory of partial differential equations has been developing mainly along the lines pointed out by I.G. Petrowsky in his remarkable article, which has been a source of ideas and problems for scientists all over the world. Many hypotheses stated by I.G. Petrowsky in that article have found confirmation in subsequent works. The recent four volume edition of the "Analysis of Linear Partial Differential Operators" by L. Hörmander [52] is a reference work on the developments in the general theory of linear partial differential equations during the last decades. These books make evident the great role of Petrowsky's contribution and the importance of the problems posed by him.

Of fundamental importance, according to I.G. Petrowsky, is the problem of the uniqueness of solutions for the Cauchy problem. In recent decades, the attention of many mathematicians was focused on this problem. The results obtained in this direction can be found, for instance, in Hörmander's monographs [41], [52].

The problem of solvability of partial differential equations was also included by I.G. Petrowsky into the class of fundamental problems in the theory of differential equations. He wrote: "For some very simple non-analytic equations, we usually do not know whether they have even one solution. It is very important to examine this problem."

In 1957, H. Lewy published his paper [53], where he constructs a first order linear equation without solutions, while its coefficients are analytic complex valued functions and the right-hand side is infinitely smooth. The work of H. Lewy was followed by a large number of publications studying the conditions of solvability for linear equations and systems (see [41], [52], [54]).

In [44], I.G. Petrowsky stresses the necessity to consider the mixed problem for hyperbolic equations and systems. Substantial progress has been made in the analysis of this problem during recent decades. A detailed account of these investigations is given in the commentary written by V.Ya. Ivrii and L.R. Volevich on Petrowsky's Articles 1 and 3 of this volume (see Appendix).

The need to continue research on the lacunas for hyperbolic equations, also mentioned in [44], was met by the most prominent mathematicians (see Appendix for a commentary by A.M. Gabrielov and V.P. Palamodov on Petrowsky's Articles 5 and 6 published in this volume).

Another of the tasks set by I.G. Petrowsky is to compare the sets of regular boundary points for the Dirichlet problem for the Laplace equation and for a general linear second order equation. For the first time, this problem was solved in [55].

I.G. Petrowsky also mentioned the necessity to study general boundary value problems for elliptic equations and systems; and, in this connection, he expected a lot from the application of the parametrix method. The subsequent development of the theory of elliptic systems justified these expectations, for this method allowed to give a comprehensive answer to the question about general boundary value problems (see [40], [41], [52]).

Full confirmation has been given to Petrowsky's hypothesis that elliptic equations and systems with sufficiently smooth coefficients can have only smooth solutions. This problem is solved in [56], [57], et al.

I.G. Petrowsky marked as really interesting and promising an article published by L. Bers and A. Gelbart in 1944 about solutions of an elliptic system consisting of two equations with two independent variables (see [58]); on the basis of these solutions, the authors construct a theory similar to the theory of analytic functions. Later, these questions were investigated more deeply and resulted in the theory of generalized analytic functions created by the efforts of I.N. Vekua [59], L. Bers [60], and their followers.

Of great importance for the theory of parabolic equations and systems, I.G. Petrowsky regarded the problem of characterizing precise uniqueness classes for solutions of the Cauchy problem. He suggested this problem to O.A. Ladyzhenskaya as a topic for her graduation thesis [61]. A description of further developments in this direction is given in the commentary written by O.A. Oleinik and V.P. Palamodov on Article 2 of this volume (see Appendix).

Another challenging problem indicated by I.G. Petrowsky pertains to solving boundary value problems for second order equations with a non-negative characteristic form, in the case of equations that cannot be reduced to parabolic ones. At present, there is a fairly extensive theory of boundary value problems for second order equations with a non-negative characteristic form; among the questions well-studied for these equations are the existence of solutions, their uniqueness and smoothness, the problem of hypoellipticity (see [62], [63]).

The list of important problems formulated by I.G. Petrowsky in [44] does not end here. It should also be mentioned that many interesting problems were posed by Ivan Georgievich at the sessions of his seminar; these problems were suggested to his graduate and post-graduate students as topics for their theses. Thus, he stressed the need of studying the stability of difference schemes; and, being offered this subject, his students S.K. Godunov, V.S. Ryabenkij, and A.F. Filippov [64], [65] engaged themselves in these studies. Uniqueness of solutions for the Cauchy problem was considered by A.D. Mishkis [66] and E.M. Landis, who were also his students. Discontinuous solutions of quasilinear hyperbolic equations and the development of mathematical theory of shock waves — these problems, being regarded by I.G. Petrowsky as extremely important, called the attention of his students S.K. Godunov [68] and O.A. Oleinik [69].

After the publication by S.L. Sobolev of his paper [70] on what is now known as Sobolev's equations, I.G. Petrowsky realized the need to examine general differential equations and systems that cannot be resolved with respect the highest order derivative in the time variable (these systems are not of Kowalevskaya type). This investigation was undertaken in [71] by S.A. Galpern, and also in a series of subsequent publications.

Ivan Georgievich attributed primary importance to the investigation of problems arising in mathematical physics, mechanics and other natural sciences. In his view, the problems related to natural sciences are capable of showing the correct path for theoretical research: he always advised his pupils to take up applied problems and be in contact with specialists in mechanics and physics.

From 1946 up to the last day of his life, Petrowsky's seminar on differential equations continued its work. All important events in the history of differential equations of that period were reflected in the activities of this seminar. To a great extent, this seminar itself was making that history. Here, many new theories were begotten and grown (theory of discontinuous solutions of nonlinear equations, theory of differential equations with a retarded argument, equations with a small parameter by the highest order derivatives, theory of stability of finite difference schemes, qualitative theory of partial differential equations, etc.). Out of innumerable publications all over the world, I.G. Petrowsky new how to select the most important and promising ones for reviewing at his seminar, called the attention of young mathematicians to these works and formulated new problems arising from them. Actually, the school of I.G. Petrowsky took shape at this seminar, and many of his disciples have become prominent scientists.

Petrowsky's intensive teaching activities in the '30s did not prevent him from carrying on excellent mathematical research described above. It seems that in the history of mathematics there are not too many manifestations of creative power that would allow a single person to solve, in a very short period, most difficult crucial problems from various far removed departments of mathematics.

In 1940, I.G. Petrowsky was elected Dean of the Faculty of Mechanics and Mathematics at the University of Moscow; it was the first time ever that a Dean was appointed by election. During the World War II, when the University was evacuated to another region and its condition was that of extreme hardship, I.G. Petrowsky did not spare the efforts to keep the Faculty going, to provide living necessities for the students, teachers, other employees, and their families. It was an urgent need of the time to intensify the work for defence, to bring up highly qualified specialists the country required then and would require during the post-war years. In May 1943, the Faculty was brought back to Moscow, and its Dean did his best to gather its members and effectively engage them in science and education. In all matters concerning the Faculty I.G. Petrowsky felt great responsibility. He stayed as Dean of the Faculty until the middle of 1944.

From 1950 to the end of his life Ivan Georgievich headed the Department of Differential Equations at the University of Moscow. Under his guidance, the Department became one of the leading scientific teams in the world. I.G. Petrowsky was proud of his Department and its famous scientists; he took great care to make the scientific work in the Department reach the highest possible level, to make available for the students special courses on the most important and modern topics, to make the regular courses provide firm ground in any branch of the theory of differential equations.

During the early post-war years, I.G. Petrowsky took active part in research aimed at the development of new technologies. The culmination of this work was recognized as a great success of Russian scientists.

In May 1951, the Government appointed I.G. Petrowsky to be Rector of the Moscow State University. For almost 22 years he held this office. Prodigious in its scale was the activity of the prominent scientist, who had taken upon himself the hard labor of guiding one of the world's greatest educational institutions. To this cause he was willing to give his skills and talents, the noble qualities of his soul, his energy and wisdom. He showed great concern for scientific research and educational activities at the University, and took equally good care of the people. There were many who sought his advice and, in difficult moments of their lives, even asked for support, knowing that he was ever ready to help; among these were students, technical staff, and even professors. It often happened that his support decided a person's fate.

Ivan Georgievich was a person of vast education and had deep understanding of modern science, together with all its intricate relationships. He possessed keen insight and could look far ahead: he easily recognized promising new directions in science and gave them his support. Indeed, it is a hard task to mention everything I.G. Petrowsky did to bring the University of Moscow to the high position it justly occupies among the leading world's universities, with its standard of education and scientific research. I.G. Petrowsky helped to organize at the University of Moscow the Department of Numerical Analysis, the first one in our country; he took part in the organization of one of the first computer centers in the USSR, the Faculty of Psychology, the Institute for Afro-Asian Studies, the Faculty of Journalism, the Faculty of Cybernetics and Numerical Analysis, the Department of Scientific Information, the Laboratory of Molecular Biology and Biochemistry, etc. He was at the origin of more than 80 educational departments and 200 laboratories. In all his endeavors, I.G. Petrowsky was not inclined to act as administrator, but rather as a scientist with a unique insight into the general system of contemporary knowledge. There is only one precedent of such industry in the history of Russian universities — the nineteen year long rectorship of the great N.I. Lobachevsky at the University of Kazan.

I.G. Petrowsky was Chairman of the Organizing Committee and President of the International Mathematical Congress held in Moscow during the summer of 1966. This congress happened to be the largest by the number of participants and one of the best in organization; 4275 mathematicians from 54 countries attended the Congress. The preparations for this congress required great strength unsparingly spent by I.G. Petrowsky.

A considerable part of Petrowsky's career belonged to the Academy of Sciences of the USSR. In 1943, he was elected its corresponding member and, in 1946, he was granted full membership. For several years he worked at the Steklov Mathematical Institute of the Academy of Sciences, where he held the post of Deputy Director. During 1949–1951, I.G. Petrowsky was Academic Secretary of the Department of Physics and Mathematics of the Academy of Sciences of the USSR; and in 1953 he became member of the Presidium of the Academy. Moreover, for many years he worked as editor-in-chief of two leading mathematical journals: "Matematichesky Sbornik" and "Proceedings of Steklov's Mathematical Institute." Ivan Georgievich was not indifferent to the level of education in secondary schools: at his insistence, the institution called "Upgrading Courses for Secondary School Teachers" was launched; he also helped to organize a correspondence school for studying mathematics and a boarding-school at the University of Moscow for talented teen-agers.

His great general culture and vast education, deep understanding of the needs of his country distinguish I.G. Petrowsky among the community and state leaders. He was elected Deputy of the Supreme Soviet of the USSR and member of its Presidium. In his home country, I.G. Petrowsky had the honor to receive the Highest State Award (the so-called "Hero of Socialist Labor"), 5 orders of Lenin, and 3 orders of Red Banner; he was promoted to the Legion of Honor in France and received decorations and honorary titles in Bulgaria, Germany, Mongolia, and Hungary; he was also made *Doctor Philosophiae Honoris Causa* at the universities of Bucharest and Prague, and at Lund University in Sweden.

Ivan Georgievich Petrowsky belongs to the select few whose works determine the present state of mathematical science.

On January 15, 1973, I.G. Petrowsky passed away, while performing his regular duties, this time, getting prepared for an important report at the Conference of Rectors, which contained many valuable suggestions on the work of educational institutions.

In the main building of Moscow University there is a lecture-hall named after I.G. Petrowsky and intended to symbolize his being constant companion to the team of the Faculty of Mechanics and Mathematics, where he belonged for more than half a century.

Passengers of every bus going along Leninsky Prospect in Moscow hear the words "Petrowsky's Street" announced by the driver, as the bus stops by the street where I.G. Petrowsky lived since 1949. There is a memorial tablet on the front wall of his house. His name has been given to a large ship used by the University of Moscow for the purposes of science; a mountain passage discovered by the University mountaineers in Tien Shan mountains, at their request, has also been given his name. Petrowsky's former office in the old building of the University is preserved as a memorial; it contains his unique personal library donated to the University after his death. The best students of the Faculty of Mechanics and Mathematics are entitled to be paid Petrowsky's grant.

Annually, since 1978, joint sessions of Petrowsky's Seminar and Moscow Mathematical Society are held in January, in commemoration of the great scientist. At these sessions, every year, more than a hundred scientists from all parts of the USSR have the opportunity to communicate and discuss their results, mostly in the fields related to Petrowsky's works. The abstracts of the reports given at these meetings are published in "Russian Mathematical Surveys"; they clearly show the increasing influence of Petrowsky's ideas and results on the development of modern mathematics. Many of the reported studies either directly continue Petrowsky's works or belong to theoretical disciplines originated by him. Annual issues of the "Proceedings of Petrowsky's Seminar," which started in 1975, make public the works of professors, teachers, and post-graduate students from the Faculty of Mechanics and Mathematics of the University of Moscow.

It is beyond doubt that the scientific endeavors of Ivan Georgievich will have an ever increasing effect on the development of mathematics in future. His investigations form an indispensable part in the foundations of modern mathematics.

References

[1] Petrowsky, I.G. Einige Bemerkungen zu den Arbeiten von Herren O. Perron und L.A. Lusternik über das Dirichlet'sche Problem. *Mat. Sbornik* **35:1** (1928) 105–110.

[2] Petrowsky, I.G. Perron's method for solving the Dirichlet problem. *Uspekhi Mat. Nauk* **8** (1941) 107–114.

[3] Petrowsky, I.G. A new proof of the existence of solutions for the Dirichlet problem by the finite difference method. *Uspekhi Mat. Nauk* **8** (1941) 161–170.

[4] Petrowsky, I.G. Zur ersten Randwertaufgabe der Wärmeleitungsgleichung. *Compos. Math.* **1:3** (1935) 383–419.

[5] Petrowsky, I.G. Über das Irrfahrtproblem. *Math. Ann.* **10:3** (1934) 425–444.

[6] Khinchin, A. Ya. *Asymptotic Laws in the Theory of Probability.* Moscow, Gostekhizdat, 1936.

[7] Petrowsky, I.G. Sur les fonctions primitives par rapport à une fonction continue arbitraire. *C. R. Acad. Sci. Paris* **189** (1929) 1242–1244.

[8] Petrowsky, I.G. Sur la topologies des courbes planes réelles et algébriques. *R.R. Acad. Sci. Paris* **197** (1933) 1270–1272.

[9] Petrowsky, I.G. On the topology of real plane algebraic curves. *Ann. Math.* **39:1** (1938) 189–209.

[10] Harnack, A. Über die Vielheitkeit der ebenen algebraischen Kurven. *Math. Ann.* **10** (1876) 189–198.

[11] *Hilbert's Problems: A Collection of Articles.* Moscow, Nauka, 1969.

[12] Petrowsky, I.G.; Oleinik, O.A. On the topology of real algebraic surfaces. *Izvest. Akad. Nauk. SSSR, Ser. Mat.* **13:5** (1949) 389–402.

[13] Oleinik, O.A. On the topology of real algebraic curves on an algebraic surface. *Mat. Sbornik* **29:1** (1951) 133–156.

[14] Oleinik, O.A. Estimates for the Betti numbers of real algebraic surfaces. *Mat. Sbornik* **28** (1951) 635–640.

[15] Morse, M. Critical points. *Trans. Amer. Math. Soc.* **27** (1925) 345–396.

[16] Gudkov, D.A. Topology of real projective algebraic manifolds. *Uspekhi Mat. Nauk* **29:4** (1974) 3–79.

[17] Petrowsky, I.G. Über das Verhalten der Integralkurven eines Systems gewöhnlicher Differentialgleichungen in der Nähe eines singulären Punktes. *Matem. Sbornik* **41:1** (1934) 107–156.

[18] Nemytsky, V.V.; Stepanov, V.V. *Qualitative Theory of Differential Equations.* 2nd Ed. Moscow, Gostekhizdat, 1949.

[19] Petrowsky, I.G. On the behavior of integral curves near a singular point for systems of ordinary differential equations. In: Poincaré, A. *On Curves Specified by Differential equations.* Moscow, Leningrad, Gostekhizdat, 1947, pp. 336–347.

[20] *Mathematics, Its Subject and Meaning.* Moscow, Izd.-vo Akad. Nauk SSSR, 1956, Vol. 2, pp. 3–47.

[21] Hadamard, J. Le Problème de Cauchy et les Équations aux Dérivées Partielles Linéaires Hyperboliques. Paris, Hermann, 1932.

[22] Petrowsky, I.G. Sur le probléme de Cauchy pour un systéme d'équations aux derivées partielles dans le domaine réel. *C.R. Acad. Sci. Paris* **202** (1936) 1010–1012.

[23] Petrowsky, I.G. Über das Cauchysche Problem für Systeme von partiellen Differentialgleichungen. *Mat. Sbornik* **2:5** (1937) 815–870.

[24] Calderon, A.P. *Integrales Singulares y sus Aplicaciones a Ecuaciones Diferenciales Hiperbolicas.* Buenos Aires, Univ. Press, 1960.

[25] Mizohata, S. Note sur le traitement par les operateurs d'integrale singuliere du probléme de Cauchy. *J. Math. Soc. Japan* **11:3** (1959) 234–240.

[26] Yamaguti, M. Sur l'inegalité d'energie pour les systémes hyperboliques. *Proc. Jap. Acad. Sci.* **35** (1959) 37–41.

[27] Leray, J. *Hyperbolic Differential Equations.* Princeton, N.J.: Inst. Adv. Study, 1953.

[28] Gårding, L. Cauchy's Problem for Hyperbolic Equations.

[29] Petrowsky, I.G. On the Cauchy problem for systems of linear partial differential equations in classes of non-analytic functions. *Bull Mosc. Univ. Mat. Mekh.* **1:7** (1938) 1–72.

[30] Gelfand, I.M.; Shilov, G.E. *Some Problems in the Theory of differential equations. 3.* Moscow, Fizmatgiz, 1958.

[31] Petrowsky, I.G. On the diffusion of waves and the lacunas for systems of hyperbolic equations. *Izvest. Akad. Nauk SSSR, Ser. Mat.* **8:3** (1944) 101–106.

[32] Petrowsky, I.G. On the diffusion of waves and the lacunas for hyperbolic equations. *Mat. Sbornik* **17:3** (1945) 289–370.

[33] Atiyah, M.; Bott, R.; Gårding, L. Lacunas for hyperbolic differential operators with constant coefficients, I, II. *Acta Math.* **124** (1970) 110–189; **131** (1973) 145–206.

[34] Petrowsky, I.G. On systems of differential equations that can have only analytic solutions. *Dokl. Akad. Nauk* **17**:7 (1937) 339–342.

[35] Petrowsky, I.G. Sur l'analyticité des solutions des systèmes d'équations différentielles. *Mat. Sbornik* **5**:1 (1939) 3–70.

[36] Bernstein, S.N. Sur la nature analytique des solutions des certaines équations aux dérivées partielles du second ordre. *C.R. Acad. Sci. Paris* **137** (1903) 778–781.

[37] Lewy, H. Über den analytischen Charakter der Lösungen elliptischer Differentialgleichungen. *Nachr. Kgl. Ges. Wiss. Göttingen. Math.-phys.* **K1** (1927) 178–186.

[38] Oleinik, O.A. On the 19th problem of Hilbert. In: *Problems of Hilbert*. Moscow, Nauka, 1969, pp. 216–219.

[39] Oleinik, O.A.; Radkevich, E.V. On the analyticity of solutions for linear partial differential equations. *Mat. Sbornik* **90**:4 (1973) 592–606.

[40] Lopatinsky, Ya. B. A method for the reduction of boundary value problems for an elliptic system of differential equations to regular integral equations. *Ukr. Mat. Zhurn.* **5** (1953) 132–151.

[41] Hörmander, L. *Linear Partial Differential Operators*. Springer-Verlag, Berlin, 1963.

[42] Atiyah, M.; Singer, I. The index of elliptic operators. I. *Ann. Math.* **87** (1968) 484–530.

[43] Palais, R. *Seminar on the Atiyah-Singer Index Theorem*. Princeton, N.J., 1965.

[44] Petrowsky, I.G. On some problems in the theory of partial differential equations. *Uspekhi Mat. Nauk* **1**:4 (1946) 44–70.

[45] Kolmogorov, A.N.; Petrowsky, I.G.; Piskunov, N.S. Studies of the Diffusion with Increasing Quantity of the Substance and its Application to a Biological Problem. *Bull. Mosc. Univ. Mat. Mekh.* **1**:6 (1937) 1–26.

[46] Barenblatt, G.I.; Zeldovich, Ya. B. Intermediate Asymptotics in Mathematical Physics. *Uspekhi Mat. Nauk* **26**:2 (1971) 115–129.

[47] Petrowsky, I.G. On the propagation speed of discontinuities of displacement derivatives on the surfaces of an elastic body of arbitrary shape. *Dokl. Akad. Nauk* **47** (1945) 258–261.

[48] Petrowsky, I.G.; Chudov, L.A. On curves and two-dimensional surfaces of discontinuity for solutions of the wave equation. *Uspekhi Mat. Nauk* **9**:3 (1954) 175–180.

[49] Duistermaat, J.; Hörmander, L. Fourier integral operators. II. *Acta Math.* **128** (1972) 183–269.

[50] Hörmander, L. Uniqueness theorems and wave front sets for solutions of linear differential equations with analytic coefficients. *Comm. Pure Appl. Math.* **24** (1971) 671–704.

[51] Petrowsky, I.G. *Lectures on the Theory of Ordinary Differential Equations,* 7th ed., Moscow, Izd.-vo Mosk. Univ. 1984.

[52] Hörmander, L. *The Analysis of Linear Partial Differential Operators.* Vols. 1–4. Berlin, Springer-Verlag, 1985.

[53] Lewy, H. An example of a smooth linear partial differential equation without solutions. *Ann. Math.* **66:2** (1957) 155–158.

[54] Egorov, Yu.V. *Linear Differential Equations of Principal Type.* Moscow, Nauka, 1984.

[55] Oleinik, O.A. On the Dirichlet problem for elliptic equations. *Mat. Sbornik* **24** (1949) 3–14.

[56] Douglis, A.; Nirenberg, L. Interior estimates for elliptic systems of partial differential equations. *Comm. Pure Appl. Math.* **8** (1955) 503–538.

[57] Agmon, S.; Douglis, A.; Nirenberg, L. Estimates near the boundary for solutions of elliptic partial differential equations satisfying general boundary conditions. *Comm. Pure Appl. Math.* **12** (1959) 623–727.

[58] Bers, L.; Gelbart, A. On a class of functions defined by partial differential equations. *Trans. Amer. Math. Soc.* **56** (1944) 67–93.

[59] Vekua, I.N. *Generalized Analytic Functions.* Moscow, Fizmatgiz, 1959.

[60] Bers, L.; John, F.; Schechter, M. *Partial Diffeential Equations.* N.Y., Wiley, 1964.

[61] Ladyzhenskaya, O.A. On the uniqueness of solutions for the Cauchy problem for a linear parabolic equation. *Mat. Sbornik* **27:2** (1950) 175–184.

[62] Oleinik, O.A.; Radkevich, E.V. *Second Order Equations with Non-Negative Characteristic Form.* N.Y. AMS, Plenum Press, 1973.

[63] Oleinik, O.A.; Radkevich, E.V. On local smoothness of weak solutions and hypoellipticity of second order differential equations. *Uspekhi Mat. Nauk* **26:2** (1971) 265–281.

[64] Ryabenkij, V.S.; Filippov, V.S. *Stability of Finite Difference Equations.* Moscow, Gostekhizdat, 1956.

[65] Godunov, S.K.; Ryabenkij, V.S. *An Introduction to the Theory of Difference Schemes.* Moscow, Fizmatgiz, 1962.

[66] Mishkis, A.D. Uniqueness of solutions for the Cauchy Problem. *Uspekhi Mat. Nauk* **3:2** (1948) 3–46.

[67] Landis, E.M. On the uniqueness of solutions for the Cauchy problem for parabolic equations. *Dokl. Akad. Nauk SSSR* **83** (1952) 345–348.

[68] Godunov, S.K. A finite difference method for numerical calculation of discontinuous solutions for the equations of fluid dynamics. *Mat. Sbornik* **47:3** (1959) 271–306.

[69] Oleinik, O.A. Discontinuous solutions of nonlinear differential equations. *Uspekhi Mat. Nauk* **12:3** (1957) 3–73.

[70] Sobolev, S.L. The Cauchy problem for a special system that does not belong to the Kowalevskaya type. *Dokl. Akad. Nauk SSSR* **82:2** (1952) 205–208.

[71] Galpern, S.A. The Cauchy problem for general systems of linear partial differential equations. *Trudy. Mosc. Mat. Ob.-va* **9** (1960) 401–423.

Ivan Georgievich Petrowsky and Partial Differential Equations

Lars Gårding

The title of my talk is perhaps somewhat misleading. I do not intend to review the entire work of Ivan Georgievich Petrowsky in the domain of partial differential equations. Instead, I will restrict myself to the part of it which made a large impact on me and determined the direction of my work and the work of people which I have influenced.

Three papers form the main subject of my talk: the article from 1937 on Cauchy's problem for hyperbolic systems, the one from 1939 on the analyticity of solutions of elliptic systems, and the one from 1945 on the lacunas for solutions of hyperbolic systems. In all three cases, the papers presented new, complete results, which it took 20 years to understand and reprove. By now Petrowsky's frontal attacks on his problems have been replaced by more human procedures. To tell the truth, Petrowsky's original proofs are extremely difficult to understand, although every detail seems clear. The author somehow managed to conceal the powerful intuition which made it possible for him to marry fire and water, in this case, the Fourier transform and non-linearity. I believe that Petrowsky was the first to use the Fourier transform in the analysis of higher order equations with variable coefficients.

Before going into the details of the papers, I will try to sketch how they appear against the traditions at the time in the field of partial differential equations. With a beginning in the eighteenth century, this field went through a rapid development in the nineteenth. There was a complete theory for first order equations, where not much difference was made between real and complex theory. The theory of second order equations centered on the construction of general solutions by quadrature, but lost momentum faced with difficulties which proved in the end that the goal was too ambitious.

31

Instead, the initiative went to mathematicians who worked with boundary value problems for the equations of mathematical physics. There was much to choose from: wave propagation, hydrodynamics, propagation of heat, optics, potential theory and electrodynamics. In the twenties the situation was summed up in the influential book *Methoden in der Mathematischen Physik* by Courant and Hilbert. There were three basic models: Laplace's equation of the potential theory, the wave equation and the heat equation for the elliptic, hyperbolic and parabolic equations, respectively. All three cases had produced beautiful and refined mathematical theory.

Although I learned some mathematical physics from Courant-Hilbert and am grateful for it, I found the book too much *ad hoc* from the mathematical point of view. The radically different properties of the solutions of the three main classes of differential equations somehow came from and were explained by physics. For many people this is enough and even desirable, for others, with more mathematical minds, the same fact could be disturbing. Mathematics is a subject in itself and there ought to be a mathematical explanation. Another disturbing fact was the absence of any kind of theory or classification of higher order equations.

My first encounters with Ivan Georgievich came in 1946–47, which I spent at the Mathematics Department of Princeton University. At that time I had behind me a study of two differential operators of higher order associated with the determinants

$$\det \left(\varepsilon_{jk} \frac{\partial}{\partial x_{jk}} \right) , \qquad \varepsilon_{jk} = \frac{1}{2}(1 + \delta_{jk}) ,$$

where $j, k = 1, \ldots, n$, and the matrix $x = x_{jk}$ is either real or complex and Hermitian symmetric. They appear in natural extensions of the socalled fractional integrals of my teacher Marcel Riesz, which he had used to simplify Hadamard's method of *partie finie* for the wave operator. For these operators I had found what is now called fundamental solutions with support in the cones of positive matrices of rank 2 and 1, respectively. For large n, this meant that the support is much smaller than the entire cone of positive matrices of full rank and there was an obvious connection with Huyghens's principle for the propagation of light.

Only afterwards did I discover that my equations were totally hyperbolic in the words of Courant-Hilbert. This purely *ad hoc* notion was not a very satisfactory description and said nothing about the possible intrinsic properties. Here Ivan Georgievich came to my aid. The redoubtable librarian of the Department, miss Shields, held a treasure for me, namely, the unbound sections of the first volume of the international series of the Bulletin of Moscow University. And there was a long article by Petrowsky on Cauchy's problem for non-analytic functions. It was exactly what I wanted for my further education, namely, a systematic exploration of Hadamard's notion of correctly posed problems. Since miss Shields was very strict with limits for the loans, I had to borrow the Bulletin again and again.

Petrowsky's necessary and sufficient criterion had to do with, at the most, logarithmic growth of the real parts of some eigenvalues. For constant coefficients I could prove a conjecture stated in a footnote that boundedness was sufficient. As an act of admiration for the master mixed with satisfaction of myself, I wrote a letter to Petrowsky about my little discovery. After some time I got an answer posted in New York and probably sent by diplomatic mail.

At about the same time I mentioned to my friend Irving Segal that I had dreams about a general theory of Huyghens' principle which would explain this mysterious phenomenon. He answered to me: "Why of course, there is one already; just look in the last issue of Sbornik." I looked almost all night and to my dismay, I understood nothing. Repeated efforts produced no results. I was simply totally ignorant of algebraic integrals and topology. But I ordered a photocopy of the article and had it bound.

During my year in Princeton, other portentous things happened to me. Jean Delsarte arrived from France and gave lectures on Schwartz's theory of distributions. The reception was not enthusiastic from a department dominated by topology. Delsarte had two listeners and I was one of them. In these lectures I learned about infinitely differentiable functions with compact supports, and this notion gave me what I had been looking for: an intrinsic definition of hyperbolic differential operators with constant coefficients, which combines finite propagation velocity with a certain continuity condition. It was equivalent to the classical condition for the characteristic polynomials. There remained, however, the case of variable coefficients and the solution of Cauchy's problem. By now I knew that Petrowsky had done all this in his big paper from 1937, but the proofs seemed impenetrable.

In the fall of 1947 I returned home to Lund University. By then Petrowsky's paper on Cauchy's problem for hyperbolic systems was constantly on my mind and his paper on lacunas loomed at the horizon. When Laurent Schwartz visited Lund and lectured on distributions, he lived in my small apartment. I declared to him: "Petrowsky is God and I am his prophet." I believe that this was the origin of Schwartz's and later also Leray's interest in Petrowsky's work, more precisely, the 1937 article on Cauchy's problem for hyperbolic systems. My efforts with my friends at the Courant Institute were less successful. Kurt Friedrichs told me that the hyperbolic systems in physics were always symmetric, which gives an immediate energy integral. The main result of Petrowsky's 1937 paper is indeed the existence of such an energy integral for a first order linear system

$$\frac{\partial}{\partial t} = A\left(t, x, \frac{\partial}{\partial x}\right) u + B(t, x)u + f(t, x)\,, \tag{1}$$

where $u(t, x) \in C^n$, $x \in \mathbb{R}^n$, the coefficients are sufficiently smooth and the system is hyperbolic in the sense that all eigenvalues of

$$A(t, x, \xi)$$

are real and uniformly separate. Under the hypothesis that everything in sight is periodic, Petrowsky proves that

$$\int |u(t, x)|^2 \, dx$$

can serve as an energy. It is bounded by its value for $t = 0$ and by

$$\int_0^t \int |f(s, x)|^2 \, dx \, ds$$

with bounds exponential in t. For second order equations, the result is included in the classical inequalities by Friedrichs and Lewy. The only trouble here is that the proof consists of 39 pages of inequalities and close reasoning with very little explanation otherwise. In this article, Petrowsky goes so far as to prove the existence of the solution of Cauchy's problem for hyperbolic quasi-linear systems of the Cauchy-Kowalevskaya type.

Laurent Schwartz, who read the article, admitted that he could not penetrate it. But the result cried for a simpler proof. The first simplification appeared in Leray's Princeton lectures of 1953, and relied on what became known later as Gårding's inequality. He also stressed the fundamental property of finite propagation velocity. The theory for systems was derived from that of a single equation. As Petrowsky had done, Leray constructed the solution by approximation from the analytic case.

A slight intermezzo was Leray's discovery that the use of a certain partition of unity in Petrowsky's paper implicitly supposes that the spheres used are parallelizable. This created a stir, but the situation was clarified by Petrowsky himself.

The next step was a solution in 1956 of Cauchy's problem by functional analysis alone. The year before Lars Hörmander had published his brilliant Lund thesis which laid the foundations of a general theory of partial differential operators. It starts with a reference to Petrowsky's 1946 review of problems in the field. My claim to fame in this connection is to have introduced Hörmander both to the field and to Petrowsky's papers. The great progress made in the field since 1950 can be seen in Hörmander's four volumes 1983–85 *The Analysis of Linear Partial Differential Operators*. In 1968, when Lund University celebrated its 300 years of existence, we had the pleasure of seeing Petrowsky receive the insignia of a *Doctor Philosophiae Honoris Causa*.

The 1937 article on hyperbolic equations was followed in 1939 by an equally monumental article on the analyticity of solutions of elliptic systems.

As before, this was a big step from the second order case. The analyticity was one of Hilbert's problems and had been solved successively by Serge Bernstein and Hans Lewy in special cases.

In this new problem, Petrowsky uses his mastery of the Fourier transform. For instance, let us consider system (1), again with the real variables t, x. We assume that the system is elliptic in the sense that the matrix

$$\tau I + A(t, x, \xi)$$

is non-singular for all real $\tau, \xi \neq 0$. When all coefficients are analytic in t, Petrowsky shows that a solution $u(t, x)$ is also analytic in t, provided that it is sufficiently differentiable with respect to all variables. The idea is to devise a system satisfied by a function $v(t, \varphi(t, x)\sigma, x)$ which plays the role of $u(t + i\sigma, x)$ when t, x are small. The function $\varphi(t, x)$ is supposed to be sufficiently differentiable, supported in a small neighborhood of the origin and equal to 1 in another neighborhood. This construction permits the use of the Fourier series with respect to x. Here again, the method is that of a frontal attack. The proof is very difficult to read and it is difficult to see where ellipticity is used.

In contrast to this, the proof at the end that the ellipticity condition is necessary is very enlightening and inspired Hörmander's characterization, given in his thesis, of hypoelliptic operators and the existence of null solutions.

One had to wait 19 years for a simpler proof of sufficiency. It was done by Charles Morrey in 1957. The basic tool is the construction of a fundamental solution $E(x)$ of a single elliptic homogeneous operator $A(\partial/i\partial x)$ of order m with constant coefficients. It is given by the formula

$$E(x) = (2\pi i)^{-n} \int_{\rho=1} A(\xi)^{-1} \chi_{m-n}(x \cdot \xi + i\theta)\omega(\xi) ,$$

where $x\xi = x \cdot \xi$, and $\chi_k(t + i0)$ is obtained from $\log(t + i0)$ by integration k times for $k > 0$ and differentiation $-k$ times in the opposite case. The function $\rho(\xi) > 0$ is supposed to be homogeneous of degree 1, and

$$\omega(\xi) = \sum (-1)^{k-1} \xi_k \, d\xi_1 \ldots d\xi_n ,$$

with $d\xi_k$ missing in the product of the differentials. With this construction, the differential of the integrand is a polynomial of degree $m - n$ in x, which permits a shift of the region of integration into a complex direction η such that $x\eta > 0$ for a given x, and this proves the analyticity. Morrey's proof uses the fact known from the potential theory that $E * f$ is of class $C^{m+\mu}$ when $f \in C^\mu$, i.e., f is Hölder continuous of order $0 < \mu < 1$, and turns this into an integral equation, from which the analyticity follows.

Finally, let me turn to Petrowsky's lacuna paper from 1945. As I said, I had tried to read it in 1946 during one night in the mathematics library of Princeton University. But I understood nothing. The explanation was very simple: I was ignorant of topology and algebraic geometry. I tried to approach the local topologists, but this made me not much wiser, perhaps, because I was not able to express myself clearly. I now regret that I did not approach the redoubtable Chairman of the Department, Solomon Lefschetz, whose classical work of 1924 on the topology of algebraic manifolds had been Petrowsky's main source for his paper. In this way, I probably also robbed Lefschetz of the pleasure of seeing his work so thoroughly understood. Lefschetz's method was in the classical tradition of Riemann surfaces. In the forties, this method was already out of fashion and Lefschetz himself and the crew of topologists around him were busy laying strict foundations of algebraic topology.

Let me now give a brief description of the lacuna paper. Suppose that $A(\xi)$ is a polynomial in n variables, which is homogeneous of degree m and hyperbolic with respect to some direction θ in the sense that $A(\theta + i\xi) \neq 0$ for all real ξ. The hyperbolicity subsists with respect to all η in the convex hyperbolicity cone $\Gamma(A, \theta)$ of A which is the component of the complement of the real hypersurface $\operatorname{Re} A$: $A(\xi) = 0$ containing θ. The dual cone $C = C(A, \theta)$ of all x such that $x\Gamma(A\theta) \geq 0$ is called the propagation cone of A, for a reason which will become apparent shortly.

If we allow distributions in formula (2), we can write a fundamental solution $E = E(A, \theta, x)$ of the operator $A(\partial/i\partial)$ with support in the propagation cone $C(A, \theta)$. The formula is

$$E(x) = (2\pi i)^{-n} \int \chi_{m-n}(x\xi + i0) A(\xi - i\theta)^{-1} \omega(\xi) \tag{3}$$

with integration as before. When $m < n$, $E(x)$ is a distribution.

The wave front surface $W(A, \theta)$, which consists of all x such that the real hyperplane $\operatorname{Re} X$: $x\xi = 0$ touches $\operatorname{Re} A$, is contained in the propagation cone $C(A, \theta)$, and $E(x)$ is holomorphic there outside the wave front surface. A component of $C(A, \theta)$ outside the wave front surface where $E(x)$ is a homogeneous polynomial of degree $m - n$ is called a lacuna for E. The exterior of $C(A, \theta)$ is a trivial lacuna, since $E(x)$ vanishes there.

When x is outside the wave front surface, we can move the region of integration in (3) into the complex domain and perform various residue operations. If we do this, we arrive at Abelian integrals over certain cycles depending on the component of $C(A, \theta) \setminus W(A, \theta)$ where x is. If the cycles are suitably homologous to zero, x is in a lacuna. Petrowsky's big feat is a proof of the converse in a generic situation.

In papers from 1926 and 1928, Herglotz had managed to construct the proper fundamental solutions and to express them as Abelian integrals in a special case. Similar cases had been studied before by Fredholm in 1900 and by Zeilon in 1911 and 1919–1921. There were several difficulties with the general case. One is that the fundamental solution is a distribution when $m < n$. Petrowsky circumvents this difficulty by going to a fundamental solution for a hyperbolic operator AB and differentiating afterwards. He found that the critical cycles, here called Petrowsky cycles, are

$$C_{\text{real}}(x) \quad \text{and} \quad C_{\text{imag}}(x)$$

homologically locally independent of x and associated with n even and odd, respectively. To describe them, let A, X be the complex hypersurfaces $A(\xi) = 0$ and $x\xi = 0$. The first cycle above is simply the projective intersection of $\text{Re}\,A$ and $\text{Re}\,X$, the second one is a kind of complex intersection best described as a cycle whose surrounding tube is homologous to $2\text{Re}\,X$ symmetrically detached from $\text{Re}\,A \cap \text{Re}\,X$ into X. The Petrowsky condition for a lacuna at x is that $C_{\text{real}}(x)$ be homologous to zero in the complex projective intersection of A and X, and $C_{\text{imag}}(x)$ homologous to zero in the complex projective X outside A.

To prove his essential point that the condition is necessary for a lacuna, Petrowsky turned to the general situation of Abelian integrals over cycles of middle dimension $n-1$ of an algebraic non-singular projective hypersurface $H(x) = 0$ in complex projective n-space. The corresponding homology had been studied by Lefschetz using his theory of vanishing cycles. These are cycles which contract to a point of degeneration of the surface when it varies in the space of all surfaces. If a non-vanishing Abelian integral is extended over a vanishing cycle, it will exhibit a singularity as the surface degenerates. This behavior and the fact proved by Lefschetz that practically all the homology in the middle dimension is spanned by vanishing cycles forms the backbone of Petrowsky's proof. As usual with the master, the proof is a massive frontal attack, which leaves the reader in despair. What is difficult with Lefschetz's pioneering paper returns here with a vengeance. Around 1960, Jean Leray read the paper but hesitated to go into the details. Instead, he used the Petrowsky cycles for an extension of the original Laplace transform to several variables (1962).

What Leray hesitated to do was for me an impossibility. But I had Petrowsky's lacuna paper constantly on my mind. The opportunity came to me in 1968, during a visit to Oxford, where I was invited by Michael Atiyah. I suggested to him and to Raoul Bott that we should avail ourselves of the copies of Petrowsky's paper and try to understand the topology. By that time I had also found how to deal with singular hyperbolic polynomials. The result of the most delightful collaboration of my life was a two-part paper in *Acta Mathematica*, dedicated to Petrowsky, where Atiyah, Bott and I extended Petrowsky's theory to the singular case and in some other directions. The determination of the homology in the middle direction of a projective algebraic hypersurface is replaced by an algebraic de Rham theorem which implies that the cohomology of the complement of the hypersurface is spanned by rational differential forms with poles on the hypersurface (see Atiyah, Bott & Gårding, II, pp. 146–147). The proof of this, in turn, uses the technique of sheaf cohomology, some vanishing theorems and Hironaka's major theorem about the resolution of singularities.

I will close with short indications how (3) can lead to Abelian integrals. We assume that Re X and Re A do not touch. Let $\xi \to v(\xi) \neq 0$ be a small real vector field, homogeneous of degree 1 in ξ and chosen so that $v(\xi)$ points out from Re A on the same side as $-\theta$, and $v(\xi) \in \text{Re } X$. The cycle

$$\xi \to \xi - iv(\xi)$$

oriented by $x\xi\omega(\xi) > 0$ will be denoted by $\alpha(x)$. It is a cycle in $(Z - A)$ mod X, where $Z = \mathbb{C}^n$. The homology class of its image in projective space belongs to

$$H^{n-1}\left(Z^* - A^*, X^*\right),$$

where the star denotes image in projective space. Its boundary $\partial\alpha^*(x)$ lies in $X^* - A^* \cap X^*$ and is a tube around the original Petrowsky cycles, and the Petrowsky condition in the present context reads

$$\partial\alpha^*(x) \sim 0.$$

It guarantees that $E(x) = E(A, x)$ is a polynomial of degree $m - n$ in the component of $C(A, \theta) = W(A, \theta)$ which contains x. More generally, it applies also to fundamental solutions $E\left(A^k, x\right)$ of powers of A, if the degree of the polynomial is changed to $mk - n$. For large enough k, the condition is also necessary.

When I look back on my work with hyperbolic equations, the work I initiated on lacunas and my activities as a propagandist, I believe I can say with a good conscience: Petrowsky is God and I am his prophet.

References

Petrowsky, I.G.

[1937] Über das Cauchysche Problem für Systeme von Partiellen Differentialgleichungen. *Mat. Sbornik* **2(44)** (1937) 815–870.

[1938] Über das Cauchysche Problem für ein Systeme linearer Differentialgleichungen im Gebeit der nichtanalytischen Funktionen. Bull. Mosk. Univ. Mat. Mekh. **7** (1938) 1–74.

[1939] Sur l'analyticité des systèmes d'équations différentielles. *Mat. Sbornik* **5(47):1** (1939) 1–70.

[1945] On the diffusion of waves and the lacunas for hyperbolic equations. *Mat. Sbornik* **17(50)** (1945) 289–370.

[1946] On some problems of the theory of partial differential equations. *Uspekhi Mat. Nauk* **1:3-4** (1946).

Atiyah, M.; Bott, R.; Gårding, L.
[1970] Lacunas for hyperbolic differential operators with constant coefficients. I,II. *Acta Math.* **124** (1970) 110–189; **131** (1973) 145–206.

Fredholm, I.
[1900] Sur les équations de l'équilibre d'un corps solide élastique. *Acta Math.* **23** (1900) 1–42.

Gårding, L.
[1953] Dirichlet's problem for linear partial differential equations. *Math. Scand.* bf 1 (1953) 55–72.
[1956] Solution directe du problème de Cauchy pour les équations hyperboliques. *Coll. Int. CNRS Nancy*, 1956, 71–90.

Herglotz, G.
[1926-28] Über die Integration linearer partieller Differentialgleichungen mit Konstanten Koeffizienten. I,II,III. *Ber. Sächs. Akad. Wiss.* **78** (1926) 93–126, 287–318; **80** (1928) 69–114.

Hörmander, L.
[1955] On the theory of general partial differential operators. *Acta Math.* **94** (1955) 161–248.
[1983-85] *The Analysis of Linear Partial Differential Operators. I,II,III,IV.* Berlin, Springer-Verlag, 1983–1985.

Lefschetz, S.
[1924] *L'analysis situs et la géométrie algébrique.* Coll. Borel. Gauthier-Villars et C., Paris, 1924.

Leray, J.
[1953] *Hyperbolic Differential Equations.* Princeton, Inst. Adv. Study, 1953; reprinted 1955.
[1962] Un prolongement de la transformation de Laplace...(Probleme de Cauchy). *Bull. Soc. Math. France* **90** (1962) 39–156.

Morrey, C.B.
[1958] On the analyticity of the solutions of analytic nonlinear elliptic systems of partial differential equations. I,II. *Amer. J. Math.* **80** (1958) 198–218, 219–237.

Zeilon, N.
[1911] Das Fundamentalintegral der allgemeinen partiellen Differentialgleichung mit konstanten Koeffizienten. *Ark. Mat. Fys. Astr.* **6:38** (1911) 1–32.
[1919-21] Sur les équations aux dérivées partielles et le problème optique des milieux birefringents. I,II. *Nova Acta Reg. Soc. Sc. Uppsal. IV* **5:3** (1919) 1–59; **5:4** (1921) 1–131.

PETROWSKY'S ARTICLES ON
PARTIAL DIFFERENTIAL EQUATIONS

1

On the Cauchy Problem for Systems of Partial Differential Equations[*][1]

Introduction

1. We consider the following system of equations[2]

$$\frac{\partial^{n_i} u_i}{\partial t^{n_i}} = F_i \quad (i = 1, 2, \ldots, N),\tag{1}$$

where all the variables are real; t, x_k $(k = 1, \ldots, n)$ are independent variables; F_i depend on $t, x_1 \ldots, x_n$, and the unknown functions u_j $(j = 1, \ldots, N)$, as well as their partial derivatives with respect to x_k and t up to the order n_j; among the arguments of F_i, the order of the derivatives of u_j in t cannot exceed $n_j - 1$. We follow here the ideas of J. Hadamard as expounded in his monograph [3]. Consider the initial functions

$$\varphi_i^{(j)}(x, \ldots, x_n) \quad (i = 1, \ldots, N, \quad j = 1, \ldots, n_i - 1),$$

which are supposed to be bounded and continuous, together with all their partial derivatives up to a finite order L. We say that the Cauchy problem[3]

---

[*]Original title: "Über das Cauchysche Problem für Systeme von partiellen Differentialgleichungen." *Matem. Sbornik* **2:5** (1937) 815–870. See Appendix for a commentary by V.Ya. Ivrii and L.R. Volevich.

[1]The proofs in Petrowsky's original paper have several gaps, as pointed out by J. Leray and J. Schauder. In a subsequent publication ("Some remarks concerning my papers on the Cauchy problem." *Matem. Sbornik* **39** (1956) 267–272), I.G. Petrowsky gave an indication of how these gaps might be filled. The present translation takes into account all suggestions made by the author. Since the author's remarks make use of a shorter multi-index notation for the Fourier coefficients, it has been deemed possible to use this type of notation in some lengthy formulas of the translated version. The bibliographical references in footnotes of the original paper have been replaced by a more common reference system with a list of references at the end of each article. – *Ed. note.*

[2]See my notes [1], [2].

[3]sometimes called *the initial value problem.*

42

with initial functions $\varphi_i^{(j)}$, is *well-posed*[4] in a bounded domain G adjacent to the hyperplane[5] $t = 0$ in the space of the variables $t, x_1, \ldots, x_n$, if the following two conditions are satisfied:

I. Let $\bar{\varphi}_i^{(j)}(x_1, \ldots, x_n)$ be arbitrary functions which, together with their (continuous) partial derivatives in x_k up to a given finite order L, differ from the respective functions $\varphi_i^{(j)}$ (this shortened notation will be occasionally employed for $\varphi_i^{(j)}(x_1, \ldots, x_n)$) and their partial derivatives by less than a positive constant ε. Then there exists, in the domain G, one and only one system of functions $\bar{u}_i$ which, together with their partial derivatives with respect to all x_k and t up to a certain finite order, are bounded and continuous in G, satisfy equations (1), and, for $t = 0$, take the initial values $\bar{\varphi}_i^j$, together with their partial derivatives in t up to the order $(n_i - 1)$.

II. For each positive η there is a $\delta > 0$ such that the functions $\bar{u}_i(t, x)$ and $u_i(t, x)$ differ in the domain G less than by η, provided that the functions $\bar{\varphi}_i^{(j)}$ and $\varphi_i^{(j)}$, together with all their partial derivatives up to the order L, differ less than by δ.

2. A system of equations of the form

$$\frac{\partial^{n_i} u_i}{\partial t^{n_i}} = \sum_{k,j} A_{ij}^{(k_0, k_1, \ldots, k_n)} \frac{\partial^k u_j}{\partial t^{k_0} \partial x_1^{k_1} \cdots \partial x_n^{k_n}} + F_i(t, x_1, \ldots, x_n) \qquad (2)$$

has been studied in my paper [4] quoted above. Here the index k implies that the sum is taken over all non-negative k_s whose sum is not greater than n_j, $k = \sum_{s=0}^n k_s$, $k_0 < n_j$. In the present paper we are going to prove the following result which, in some sense, is a generalization of a theorem previously established[6] for system (2).

Consider a system of type (1). This system will be denoted by (I), if it satisfies the following conditions:

Whatever values of the arguments be considered, the functions F_i possess continuous bounded derivatives up to the order $(4n + 4)$ with respect to all their arguments. Moreover, all functions u_i, together with all their derivatives entering the right-hand sides, have absolute values bounded by a positive constant M. Let G be a domain adjacent to the hyperplane $t = 0$ in the $(t, x_1, \ldots, x_n)$-space. Then the matrix

---

[4] See my paper "On the Cauchy problem for systems of linear partial differential equations in classes of non-analytic functions," to be published shortly. (The author refers to the paper [4], translated as Article 2 of this volume. – *Ed. note.*)

[5] It is quite easy to reduce to this situation the general case of the initial conditions for u_i and their normal derivatives prescribed on a curved portion of a hypersurface adjacent to G.

[6] See [4, Part 3].

$$\left\| \sum_{(k)} \frac{\partial F_i}{\partial \left\{ \frac{\partial^{n_j} u_j}{\partial t^{k_0} \partial x_1^{k_1} \dots \partial x_n^{k_n}} \right\}} \lambda^{k_0} \alpha_1^{k_1} \dots \alpha_n^{k_n} \right\| - \left\| \begin{matrix} \lambda^{n_1} & & & \\ & \lambda^{n_2} & & \\ & & \ddots & \\ & & & \lambda^{n_N} \end{matrix} \right\| \tag{3}$$

has the form

$$\left\| \begin{matrix} \boxed{M_1} & & & \\ & \boxed{M_2} & & \\ & & \ddots & \\ & & & \boxed{M_l} \end{matrix} \right\|, \tag{4}$$

where all elements lying outside each of the square matrices M_r are identically equal to zero. The sum in (3) is taken over $k_s \geq 0$ such that $\sum k_s = n_j$. The determinant of each matrix M_s can have only real roots λ for all real values of α_s such that $\sum \alpha_k^2 = 1$, all the roots being different from one another, the difference of each pair of roots of the same determinant having an absolute value larger than a positive δ.

System (1) which possesses only the last property is said to be *hyperbolic* in a neighborhood of the system of functions $u_i = 0$. It will be shown that:

The Cauchy problem for system (1), with identically vanishing initial values prescribed at $t = 0$, is well-posed in a domain $G_1 \subset G$.[7]

If, for $t = 0$, the real parameters α_s can be chosen such that the characteristic determinant for the matrix (4) possesses at least one complex root, then, in the case of a linear system with coefficients depending on t only, the Cauchy problem cannot be well-posed in any domain adjacent to the plane $t = 0$.[8] In the case when the determinant of each matrix M_r has only real roots, but some of these determinants have multiple roots for at least one choice of real α_k, the question whether the Cauchy problem is well-posed can be decided, in general, by knowing, apart from the properties of the coefficients by the highest order derivatives of u_i, the behavior of the other coefficients as well.

In order to emphasize the distinction between various difficult points and also to clarify the main ideas of the proof, we start with the justification of our theorem in the case of the following first order system

$$\frac{\partial u_i}{\partial t} = F_i\left(t, x, u_1, \dots, u_N, \frac{\partial u_1}{\partial x_1}, \dots, \frac{\partial u_1}{\partial x_n}, \dots, \frac{\partial u_N}{\partial x_1}, \dots, \frac{\partial u_N}{\partial x_n} \right), \tag{5}$$

where $i = 1, \dots, N$; $x = (x_1, \dots, x_n)$.

---

[7] This result seems to be known only in the case $n = 1$ (see [5]).
[8] See [4, Part 3, §3].

In Part II, while proving the theorem in its general form, we shall limit ourselves to those considerations only which differ essentially from the corresponding arguments of Part I. In the case of $n = 1$, the general theorem follows directly from the special result for the first order system of partial differential equations (5). Actually, in this case the hyperbolic system (1) can be easily reduced to a system of type (5), which turns out to be hyperbolic, too. In the general case, however, it is yet to be shown that system (1) reduces to system (5); for some α_k, the elementary divisors of the matrix (3) for this system may turn out to be of an order higher than the first. For this reason, hyperbolic systems (1) require some special considerations, which are the main subject of Part II of the present paper.

PART I

Hyperbolic Systems of First Order Partial Differential Equations

§1. Reduction of the General First Order System of Hyperbolic Type to a Quasilinear System

Each system of type (5) can be reduced to a system of the form

$$\frac{\partial u_i}{\partial t} = \sum_{j,k} a_{ij}^{(k)}(t, x, u_1, \ldots, u_{N_0}) \frac{\partial u_j}{\partial x_k} + b_i(t, x, u_1, \ldots, u_{N_1}) \qquad (6)$$

$$(i, j = 1, 2, \ldots, N, \ldots, N_0, \ldots, N_1) ,$$

where all derivatives of the functions $u_1, \ldots, u_{N_0}$, up to the order l with respect to x_k, coincide with the functions $u_{N_0+1}, u_{N_0+2}, \ldots, u_{N_1}$. The values of N_0 and N_1 in system (6) are, in general, greater than N in (5).

As a first step, the above theorem will be proved for $l = 0$. To this end we differentiate each equation in (5) with respect to each x_p. Setting

$$\frac{\partial u_j}{\partial x_k} = u_j^{(k)} , \qquad (7)$$

we can rewrite the original system (5) in the form

$$\frac{\partial u_i}{\partial t} = F_i\left(t, \ldots, x_k, \ldots, u_j, \ldots, \ldots, u_j^{(k)}, \ldots\right) . \tag{8}$$

The result of differentiation in x_p can be written in the form

$$\frac{\partial u_i^{(p)}}{\partial t} = \sum_{j,k} \frac{\partial F_i}{\partial u_j^{(k)}} \frac{\partial u_j^{(p)}}{\partial x_k} + \sum_j \frac{\partial F_i}{\partial u_j} u_j^{(p)} + \frac{\partial F_i}{\partial x_p} . \tag{9_p}$$

Thus, it has been shown that every sufficiently smooth solution of system (5) is also a solution of system (8), (9)[9], which satisfies the assumptions of our theorem for $l = 0$.

In general, the converse statement is not true. However, if the initial conditions for $t = 0$ be chosen so as to satisfy the relations (7), then (7) will also hold for all $t > 0$ (as well as for $t < 0$). Therefore, knowing beforehand any particular solution of system (8), (9) to satisfy (7) for $t = 0$, we get a solution of the original system.

In order to prove our statement, we employ the following procedure. First, let us differentiate equation (8) with respect to x_p and subtract the result from (9_p). Setting

$$u_j^{(p)} - \frac{\partial u_j}{\partial x_p} = \bar{u}_j^{(p)} , \tag{10}$$

$$\frac{\partial u_j^{(p)}}{\partial x_k} - \frac{\partial u_j^{(k)}}{\partial x_p} = \bar{u}_j^{(p,k)} , \tag{10'}$$

we obtain after the subtraction

$$\frac{\partial \bar{u}_j^{(p)}}{\partial t} = \sum_{j,k} \frac{\partial F_i}{\partial u_j^{(k)}} \bar{u}_j^{(p,k)} + \sum_j \frac{\partial F_i}{\partial u_j} \bar{u}_j^{(p)} . \tag{11}$$

Secondly, we differentiate equation (9_p) in x_r, and equation (9_r) in x_p, and consider the difference of the expressions obtained. We get

$$\frac{\partial \bar{u}_i^{(p,r)}}{\partial t} = \sum_{j,k} \frac{\partial F_i}{\partial u_j^{(k)}} \frac{\partial \bar{u}_j^{(p,r)}}{\partial x_k} + \cdots , \tag{12}$$

where the terms implied by the dots have the form of a product of partial derivatives of F and $\partial u_j/\partial x_k$, $u_j^{(m)}$, $\partial u_j^{(m)}/\partial x_k$. In every one of these terms we replace all $\partial u_j/\partial x_k$ and all $\partial u_j^{(k)}/\partial x_p$ with $k > p$ by their expressions

---

[9]System (9) consists of equations (9_1), (9_2), ..., (9_n).

obtained from (10), (10′). We claim that after this, among the terms implied by the dots, only those will remain which contain at least one of the functions

$$\bar{u}_j^{(m)}, \bar{u}_j^{(m,s)} \quad (m, s = 1, \ldots, n)$$

as a factor.

Indeed, let us substitute arbitrary functions v_i in the original equations (5) instead of u_i. Denote the difference between the left-hand side and the right-hand side of the i-th equation after this substitution by $f_i(t, x_1, \ldots, x_n)$. Then the functions v_i satisfy the following system

$$\frac{\partial v_i}{\partial t} = F_i + f_i \quad (i = 1, \ldots, N). \tag{13}$$

Let us subject system (13) to the operations just applied to system (5). Then we obtain a system formed by the counterparts of equations (11), (12), where the terms due to f_i are absent and the remaining equalities differ from (11) and (12) only in that they have u_i replaced by v_i. On the other hand, it is obvious that all the right-hand sides of this system are equal to zero, as well as each function $\bar{v}_j^{(m)}$ and $\bar{v}_j^{(m,s)}$. Therefore, the sums of all those terms which do not contain as a factor any of the functions $\bar{v}_j^{(m)}$, $\bar{v}_j^{(m,s)}$, nor any of their derivatives, must be equal to zero. Since the functions v_i were chosen arbitrary, these sums must vanish identically for all v_j, $v_j^{(m)}$, and $\partial v_j^{(m)}/\partial x_k$. Hence follows the above statement.

Now let us assume all the functions u_j, $u_j^{(p)}$ and their derivatives entering equations (11), (12) to be known, and consider $\bar{u}_j^{(p)}$ and $\bar{u}_j^{(p,k)}$ as the unknown functions. Then the latter functions will be known to vanish for $t = 0$, and to satisfy system (11), (12) for $t > 0$. This system is linear and of hyperbolic type, since we have assumed the original equations to be hyperbolic. In the sequel, for systems of this type we prove a uniqueness theorem which is independent of all the preceding considerations. This theorem will imply that for $t > 0$ all $\bar{u}_j^{(p)}$, $\bar{u}_j^{(p,k)}$ vanish identically, i.e., $u_j^{(p)} \equiv \partial u_j / \partial x_p$.

Now let us subject the system consisting of equations (8) and (9) to the operations performed over system (5). Then, on the basis of similar considerations we obtain, in the case of $l = 1$, a proof of the theorem formulated at the beginning of this section.

Repeating all the preceding arguments with respect to the newly obtained system, we get a proof of the theorem for $l = 2$, etc.

Thus, we ultimately arrive at a quasilinear system of the type described in the main theorem of the present section. If the original system is hyperbolic, then the system obtained as a result of the procedures described in this section will also be hyperbolic, since its characteristic matrix has the form

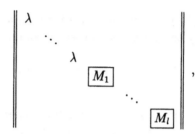

where the matrices M_r are the same as in (3).

§2. Main Theorem on the Growth Rate of an Integral; Special Case

Consider a linear hyperbolic system

$$\frac{\partial u_i}{\partial t} = \sum_{j,k} a_{ij}^{(k)} \frac{\partial u_j}{\partial x_k} + \sum_j b_{ij} u_j + f_i . \tag{14}$$

Here every $a_{ij}^{(k)}$ is a function of $t, x_1, \ldots, x_n$ admitting an l-periodic exten-sion with respect to all x_k which is continuous and bounded, together with its partial derivatives up to the order 3n. The functions $\partial a_{ij}^{(k)}/\partial t$ and b_{ij} have l-periodic extensions with continuous bounded derivatives[10] in all x_k up to the order 2n. The functions u_j are assumed to have continuous l-periodic extensions with respect to every x_k.

Denote by Q_l the cube

$$0 \le x_k \le l \qquad (k = 1, 2, \ldots, n) .$$

Let $\alpha_1, \ldots, \alpha_n$ be real numbers that do not vanish simultaneously. Let us rotate the system of coordinates in the $(x_1, \ldots, x_n)$-space, so that the new axis Ox_1' be orthogonal to the plane $\sum \alpha_k x_k = 0$, and the rest of the axes, $Ox_2', \ldots, Ox_n'$, lie in that plane. Since system (14) is hyperbolic, it can be reduced to the form (see [6])

$$\frac{\partial u_i'}{\partial t} = k_i' \frac{\partial u_i'}{\partial x_1'} + \sum_{\substack{k=2 \\ j}}^{k=n} a_{ij}'^{(k)} \frac{\partial u_j'}{\partial x_k'} + \sum_{j=1}^{n} b_{ij}' u_j' + f_i'(t, x_1', \ldots, x_n') . \tag{15}$$

It is assumed here that

---

[10]The derivatives $\partial a_{ij}^{(k)}/\partial t$ may be allowed to have points of discontinuity on a finite number of planes $t = \text{const}$; at these points, instead of the differentiability of $a_{ij}^{(k)}$ with respect to t, one can assume the existence of derivative numbers ("nombres dérivées").

$$u_i' = u_i'(t, x_1, \ldots, x_n; \alpha_1, \ldots, \alpha_n) = \sum_{j=1}^{N} k_{ij}(t, x, \alpha) u_j =$$

$$= u_i'(t, x_1, \ldots, x_n; \alpha_1', \ldots, \alpha_n') = \sum_{j=1}^{N} k_{ij}(t, x, \alpha') u_j, \qquad (16)$$

where

$$\alpha_i' = \frac{\alpha_i}{|\alpha|}, \qquad |\alpha| = \left(\sum_{i=1}^{n} \alpha_i^2 \right)^{\frac{1}{2}}$$

For the sake of brevity, we shall say that the transformation (16) reduces the matrix

$$\left\| \sum_k a_{ij}^{(k)} \alpha_k \right\| - \lambda E$$

to canonical form.

It follows from the definition of hyperbolic systems that the functions k_{ij} admit periodic extensions with continuous partial derivatives in x_k and t of the same order as the functions $a_{ij}^{(k)}$, and continuous partial derivatives of any order with respect to α_k'.

Indeed, each matrix M_r in (4) can be reduced to canonical form independently of the other matrices. It is well-known that the coefficients $k_{ij}(t, x, \alpha)$ of any of these matrices satisfy the respective systems of homogeneous linear equations

$$k_i' k_{ij} = \sum_{s,k} k_{is} a_{sj}^{(k)} \alpha_k'. \qquad (*)$$

Fixing a value of i, we obtain for the coefficients k_{ij} a system whose solutions coincide with minor determinants $M_{sj}^{(i)}$ for, say, the j-th column of the matrix

$$\left\| \begin{matrix} \sum_k a_{11}^{(k)} \alpha_k' - k_i' & \sum_k a_{12}^{(k)} \alpha_k' & \cdots \\ \sum_k a_{21}^{(k)} \alpha_k' & \sum_k a_{22}^{(k)} \alpha_k' - k_i' & \cdots \\ \cdots & \cdots & \cdots \end{matrix} \right\| .$$

Since all roots of the determinant of each matrix M_r are different, the rank of the matrix obtained is equal to $N_r - 1$, where N_r is the order of the matrix M_r. Setting

$$k_{is} = \frac{M_{s1}^{(i)}}{\pm \sqrt{\sum_p M_{p1}^{(i)2}}} = \frac{M_{s2}^{(i)}}{\pm \sqrt{\sum_p M_{p2}^{(i)2}}} = \cdots = \frac{M_{sN_r}^{(i)}}{\pm \sqrt{\sum_p M_{pN_r}^{(i)2}}} \qquad (**)$$

(with suitably chosen signs in the denominator), we obtain solutions of systems $(*)$ such that

$$\sum_s k_{i_s}^2 = 1 .$$

These solutions are differentiable with respect to t, x_k as many times as the coefficients a_{ij}^k, since at least one of the sums

$$\sum_p M_{p1}^{(i)2} , \dots , \sum_p M_{pN_r}^{(i)2}$$

is always larger than a positive constant. Using the solution $(**)$, we can easily find a transformation (15) whose determinant is equal to 1. To this end it suffices to multiply the coefficients k_{ij}, obtained from $(*)$, by

$$\frac{\pm 1}{|k_{ij}|^{1/(N_r-1)}} ,$$

where $|k_{ij}|$ is the determinant of the matrix with elements k_{ij}. The absolute value of this determinant will always be larger that a positive constant, provided that the system under consideration is hyperbolic and $\sum \alpha_k^2 = 1$.

Let $|\alpha| > 0$. Then the functions $u_i'(t, x_1, \dots, x_n, \alpha_1 \dots, \alpha_n)$ possess the following property: $u_i'(t, x, \alpha_1, \dots, \alpha_n)$ *are single-valued functions of* α_i' *having continuous derivatives at all points of the sphere*

$$\alpha_1'^2 + \cdots + \alpha_n'^2 = 1 . \qquad (***)$$

In order to prove this result, it suffices to show that each root λ of the determinant of each matrix M_s is a single-valued function defined on the sphere $(***)$. If the last statement were untrue in the case $n = 2$, then going round the circle $(***)$ once, we would arrive at a value of λ different from the original one. On the other hand, it is obvious that going round the circle several times, we inevitably arrive at the original value of λ. Therefore the circle $(***)$ must contain a point at which two roots λ of the same determinant $|M_s|$ coincide, which is incompatible with our assumption. Assume now that there is a root λ which is not a single-valued function on the sphere $(***)$ for $n > 2$. Then, the sphere must contain a closed curve K which, being followed, would lead us to a value of λ other than the initial one. Since λ is a single-valued holomorphic function in a neighborhood of each point on the sphere $(***)$, it can be uniquely extended beyond the curve K. Therefore, contracting the curve K on the sphere $(***)$ to a point, we obtain a branch point for λ, which is impossible.

Since the coefficients $k_{ij}(t, x, \alpha)$ are normalized, we see that for each i_1 there is one and only one i_2 such that

$$k_{i_1}'(t, x, \alpha) = -k_{i_2}'(t, x, -\alpha) \quad \text{and} \quad u_{i_1}'(t, x, \alpha) = u_{i_2}'(t, x, -\alpha)\delta(i_2) ,$$

where

$$\delta(i_1) = \delta(i_2) = \pm 1 .$$

In the case of $\sum_{k=1}^{n} \alpha_k^2 = 0$ we assume that

$$k_{ij}(t, x, 0, \ldots, 0) = 0 \quad \text{for} \quad i \neq j , \qquad k_{ii}(t, x, \ldots, 0) = 1 ,$$

and

$$x_i = x_i' \quad (i = 1, \ldots, n) .$$

Let us multiply (15) by

$$e^{-\frac{2\pi i}{l}(\alpha, x)} \equiv e^{-\frac{2\pi i}{l} x_1' |\alpha|}$$

where $(\alpha, x) = \sum \alpha_k x_k$, $|\alpha| = \sqrt{\sum \alpha_k^2}$ and integrate over Q_l. Then for $k > 1$ we find that

$$\int_{Q_l} a_{ij}^{\prime(k)} \frac{\partial u_j'}{\partial x_k'} e^{-\frac{2\pi i}{l} x_1' |\alpha|} \, dx' = -\int_{Q_l} \frac{\partial a_{ij}^{\prime(k)}}{\partial x_k'} u_j' e^{-\frac{2\pi i}{l} x_1' |\alpha|} \, dx' +$$

$$+ \int \delta a_{ij}^{\prime(k)} u_j' e^{-\frac{2\pi i}{l} x_1' |\alpha|} \, dx_1' \ldots dx_{k-1}' \, dx_{k+1}' \ldots dx_n' .^{*}$$

In the last integral, we have $\delta = \pm 1$, and the integration is over the surface of cube Q_l. Because of the assumed l-periodicity of the functions $a_{ij}^{(k)}$ and u_i, this integral is equal to zero. Thus, for $k > 1$ we have

$$\int_{Q_l} a_{ij}^{\prime(k)} \frac{\partial u_j'}{\partial x_k'} e^{-\frac{2\pi i}{l} x_1' |\alpha|} \, dx' = -\int_{Q_l} \frac{\partial a_{ij}^{\prime(k)}}{\partial x_k'} u_j' e^{-\frac{2\pi i}{l} x_1' |\alpha|} \, dx' .$$

In exactly the same way it can be shown that

$$\int_{Q_l} k_i' \frac{\partial u_i'}{\partial x_1'} e^{-\frac{2\pi i}{l}(\alpha, x)} \, dx = \int_{Q_l} k_i' \frac{\partial u_i'}{\partial x_1'} e^{-\frac{2\pi i}{l} x_1' |\alpha|} \, dx' =$$

$$= -\int_{Q_l} \frac{\partial k_i'}{\partial x_1'} u_i' e^{-\frac{2\pi i}{l} x_1' |\alpha|} \, dx' + \frac{2\pi i}{l} |\alpha| \int_{Q_l} k_i' u_i' e^{-\frac{2\pi i}{l} x_1' |\alpha|} \, dx' .$$

Let us introduce the notation

$$\frac{1}{l^n} \int_{Q_l} u_i'(t, x, \alpha) e^{-\frac{2\pi i}{l}(\beta, x)} \, dx = a_{i(\alpha_1, \ldots, \alpha_n)}^{\prime(\beta_1, \ldots, \beta_n)} = a_{i(\alpha)}^{\prime(\beta)} .$$

Then the result obtained after the integration of the right-hand side of (15) multiplied by $e^{-(2\pi i/l)(\alpha, x)}$ can be written as

---

*Here and in what follows, for the sake of brevity, dx and dx' stand for $dx_1 \ldots dx_n$ and $dx_1' \ldots dx_n'$, respectively. – Ed. note.

$$l^n \, \frac{da_{i(\alpha)}^{\prime(\alpha)}}{dt} = \frac{2\pi i}{l} |\alpha| \int_{Q_l} k_i' u_i' e^{-\frac{2\pi i}{l}(\alpha,x)} \, dx + \sum_{j=1}^{N} \int_{Q_l} a_{ij}' u_j' e^{-\frac{2\pi i}{l}(\alpha,x)} \, dx +$$

$$+ \int_{Q_l} f_i' e^{-\frac{2\pi i}{l}(\alpha,x)} \, dx \qquad (i = 1, 2, \ldots, N) \, .$$

Here a_{ij}' stand for linear combinations (with constant coefficients) of functions b_{ij}' and the first derivatives of $a_{ij}^{\prime(k)}$ and k_i.

Let us multiply the above equation by $a_{i(\alpha)}^{\prime(-\alpha)}$ and sum the resulting equations with respect to all α_k and i. Taking into account the relations

$$\frac{1}{2} \frac{d}{dt} \sum_{i,\alpha} |a_{i(\alpha)}^{\prime(\alpha)}|^2 = \frac{1}{2} \frac{d}{dt} \sum_{i,\alpha} a_{i(\alpha)}^{\prime(\alpha)} a_{i(\alpha)}^{\prime(-\alpha)} =$$

$$= \frac{1}{2} \sum_{i,\alpha} a_{i(\alpha)}^{\prime(-\alpha)} \frac{da_{i(\alpha)}^{\prime(\alpha)}}{dt} + \frac{1}{2} \sum_{i_1,\alpha} a_{i_1(\alpha)}^{\prime(\alpha)} \frac{da_{i_1(\alpha)}^{\prime(-\alpha)}}{dt} =$$

$$= \frac{1}{2} \sum_{i,\alpha} a_{i(\alpha)}^{\prime(-\alpha)} \frac{da_{i(\alpha)}^{\prime(\alpha)}}{dt} + \frac{1}{2} \sum_{i_1,\alpha} a_{i_1(-\alpha)}^{\prime(-\alpha)} \frac{da_{i_1(-\alpha)}^{\prime(\alpha)}}{dt} =$$

$$= \frac{1}{2} \sum_{i,\alpha} a_{i(\alpha)}^{\prime(-\alpha)} \frac{da_{i(\alpha)}^{\prime(\alpha)}}{dt} + \frac{1}{2} \sum_{i_2,\alpha} a_{i_2(\alpha)}^{\prime(-\alpha)} \frac{da_{i_2(\alpha)}^{\prime(\alpha)}}{dt} = \sum_{i,\alpha} a_{i(\alpha)}^{\prime(-\alpha)} \frac{da_{i(\alpha)}^{\prime(\alpha)}}{dt} \, ,$$

we obtain

$$\frac{1}{2} l^{2n} \frac{d}{dt} \sum_{i,\alpha} |a_{i(\alpha)}^{\prime(\alpha)}|^2 = \sum_{i,\alpha} \frac{2\pi i}{l} |\alpha| \int_{Q_l} k_i' u_i' e^{-\frac{2\pi i}{l}(\alpha,x)} \, dx \int_{Q_l} u_i' e^{\frac{2\pi i}{l}(\alpha,x)} \, dx +$$

$$+ \sum_{i,j,\alpha} \int_{Q_l} a_{ij}' u_j' e^{-\frac{2\pi i}{l}(\alpha,x)} \, dx \int_{Q_l} u_i' e^{\frac{2\pi i}{l}(\alpha,x)} \, dx +$$

$$+ \sum_{i,\alpha} \int_{Q_l} f_i' e^{-\frac{2\pi i}{l}(\alpha,x)} \, dx \int_{Q_l} u_i' e^{\frac{2\pi i}{l}(\alpha,x)} \, dx \, . \qquad (17)$$

Using (16) we can rewrite (17) in the following, sometimes more convenient, form

$$\frac{1}{2} l^{2n} \frac{d}{dt} \sum_{i,\alpha} |a_{i(\alpha)}^{\prime(\alpha)}|^2 =$$

$$= \frac{2\pi i}{l} \sum_{i,r,s,\alpha} |\alpha| \int_{Q_l} k_i' k_{ir} u_r e^{-\frac{2\pi i}{l}(\alpha,x)} \, dx \int_{Q_l} k_{is} u_s e^{\frac{2\pi i}{l}(\alpha,x)} \, dx +$$

$$+ \sum_{i,r,s,\alpha} \int_{Q_l} a_{ir}'' u_r e^{-\frac{2\pi i}{l}(\alpha,x)} \, dx \int_{Q_l} k_{is} u_s e^{\frac{2\pi i}{l}(\alpha,x)} \, dx +$$

$$+ \sum_{i,r,s,\alpha} \int_{Q_l} k_{ir} f_r e^{-\frac{2\pi i}{l}(\alpha,x)} \, dx \int_{Q_l} k_{is} u_s e^{-\frac{2\pi i}{l}(\alpha,x)} \, dx . \qquad (17')$$

Here $a_{ir}'' = \sum_j a_{ij}' k_{jr}$. The convergence of the series in (17) and (17') has not yet been proved. Therefore, for the time being, all these preliminary calculations should be considered as formal. However, their validity will be established shortly.

To begin with, consider the first sum in the right-hand side of (17). We introduce the following notations:

$$b_{i(\alpha)}^{(\beta)} = \frac{1}{l^n} \int_{Q_l} k_i'(t, x, \alpha) e^{-\frac{2\pi i}{l}(\beta, x)} \, dx ,$$

$$c_{i(\alpha)}^{(\beta)} = \frac{1}{l^n} \int_{Q_l} k_i'(t, x, \alpha) u_i'(t, x, \alpha) e^{-\frac{2\pi i}{l}(\beta, x)} \, dx .$$

Then the sum under consideration can be rewritten in the form

$$\frac{2\pi i}{l} \sum_{i,\alpha} |\alpha| c_{i(\alpha)}^{(\alpha)} a_{i(\alpha)}'^{(-\alpha)} = \frac{2\pi i}{l} \sum_{i,\alpha,\beta} |\alpha| a_{i(\alpha)}'^{(-\beta)} a_{i(\alpha)}'^{(-\alpha)} b_{i(\beta)}^{(\alpha+\beta)} , \qquad (18')$$

provided that the sum is absolutely convergent. Making this assumption, we proceed to transform the sum as follows:

$$\frac{2\pi i}{l} \sum_{i,\alpha,\beta} |\alpha| a_{i(\alpha)}'^{(-\beta)} a_{i(\alpha)}'^{(-\alpha)} b_{i(\alpha)}^{(\alpha+\beta)} =$$

$$= \frac{\pi i}{l} \sum_{i_1,\alpha,\beta} |\alpha| a_{i_1(\alpha)}'^{(-\beta)} a_{i_1(\alpha)}'^{(-\alpha)} b_{i_1(\alpha)}^{(\alpha+\beta)} + \frac{\pi}{l} \sum_{i_1,\alpha,\beta} |\beta| a_{i_1(\beta}'^{(-\alpha)} a_{i_1(\beta)}'^{(-\beta)} b_{i_1(\beta)}^{(\alpha+\beta)} =$$

$$= \frac{\pi i}{l} \sum_{i_1,\alpha,\beta} |\beta| a_{i_1(\alpha)}'^{(-\beta)} a_{i_1(\alpha)}'^{(-\alpha)} b_{i_1(\alpha)}^{(\alpha+\beta)} - \frac{\pi i}{l} \sum_{i_2,\alpha,\beta} |\beta| a_{i_2(-\beta)}'^{(-\alpha)} a_{i_2(-\beta)}'^{(-\alpha)} b_{i_2(-\beta)}^{(\alpha+\beta)} =$$

$$= \frac{\pi i}{l} \sum_{i,\alpha,\beta} \left[|\alpha| a_{i(\alpha)}'^{(-\beta)} a_{i(\alpha)}'^{(-\alpha)} b_{i(\alpha)}^{(\alpha+\beta)} - |\beta| a_{i(-\beta)}'^{(-\alpha)} a_{i(-\beta)}'^{(-\beta)} b_{i(-\beta)}^{(\alpha+\beta)} \right] =$$

$$= \sum_{i,\alpha,\beta} (|\alpha| - |\beta|) \, a_{i(\alpha)}'^{(-\beta)} a_{i(\alpha)}'^{(-\alpha)} b_{i(\alpha)}^{(\alpha+\beta)} +$$

$$+ \frac{\pi i}{l} \sum_{i,\alpha,\beta} |\beta| \left[b_{i(\alpha)}^{(\alpha+\beta)} - b_{i(-\beta)}^{(\alpha+\beta)} \right] a_{i(-\beta)}'^{(-\alpha)} a_{i(-\beta)}'^{(-\beta)} +$$

$$+ \frac{\pi i}{l} \sum_{i,\alpha,\beta} |\beta| \left[a_{i(\alpha)}'^{(-\alpha)} - a_{i(-\beta)}'^{(-\alpha)} \right] a_{i(-\beta)}'^{(-\beta)} b_{i(\alpha)}^{(\alpha+\beta)} +$$

$$+ \frac{\pi i}{l} \sum_{i,\alpha,\beta} |\beta| \left[a_{i(\alpha)}'^{(-\beta)} - a_{i(-\beta)}'^{(-\beta)} \right] a_{i(\alpha)}'^{(-\alpha)} b_{i(\alpha)}^{(\alpha+\beta)} . \qquad (18)$$

Let us estimate the absolute value of the sum in the right-hand side of (18). We can assume that the sum here is taken over the combinations of α_k and β_k which satisfy the inequality $\sum \alpha_k^2 + \sum \beta_k^2 > 0$.

We assume in advance that the coefficients $a_{ij}^{(k)}$ of system (14), and therefore also the coefficients k_{ij} and k_i', possess bounded derivatives with respect to all x_k up to the order $3n$. Hence we obtain

$$|b_{i(\alpha)}^{(\alpha+\beta)}| \leq \frac{M_0^{(3m)}}{|\alpha_{i_1} + \beta_{i_1}|^3 \ldots |\alpha_{i_m} + \beta_{i_m}|^3 (\frac{2\pi}{l})^{3m}} , \qquad (19')$$

where $M_0^{(k)}$ is the upper bound for all k-th order derivatives of the functions k_i'. The denominator in the right-hand side contains only the terms with non-vanishing $\alpha_k + \beta_k$. Denote the number of such terms by m.

We can also write a stronger version of the above inequality:

$$|b_{i(\alpha)}^{(\alpha+\beta)}| \leq \frac{M_0}{[|\alpha_1 + \beta_1| + 1]^3 \ldots [|\alpha_n + \beta_n| + 1]^3} , \qquad (19)$$

where the denominator in the right-hand side contains all $[|\alpha_k + \beta_k| + 1]$, and M_0 is the largest among $M_0^{(0)}, M_0^{(1)} l, \ldots, M_0^{(3n)} l^{3n}$.

Setting

$$k_{ir(\alpha)}^{(\beta)} = \frac{1}{l^n} \int_{Q_l} k_{ir}(x, \alpha) e^{-\frac{2\pi i}{l}(\beta, x)} \, dx ,$$

$$a_i^{(\beta)} = \frac{1}{l^n} \int_{Q_l} u_i(x_1, \ldots, x_n) e^{-\frac{2\pi i}{l}(\beta, x)} \, dx ,$$

we find by virtue of (16) that

$$a_{i(\alpha)}^{'(\beta)} = \sum_{r,\gamma} a_r^{(\gamma)} k_{ir(\alpha)}^{(\beta-\gamma)} . \qquad (20)$$

Therefore

$$\left| \sum_{i,\alpha,\beta} (|\alpha| - |\beta|) a_{i(\alpha)}^{'(-\beta)} a_{i(-\alpha)}^{'(-\alpha)} b_{i(\alpha)}^{(\alpha+\beta)} \right| \leq$$

$$\leq \sum_{i,r,s,\alpha,\beta,\gamma,\delta} \left| \frac{\sum(\alpha_p + \beta_p)(\alpha_p - \beta_p)}{|\alpha| + |\beta|} a_r^{(\gamma)} k_{ir(\alpha)}^{(-\beta-\gamma)} a_s^{(\delta)} k_{is(-\alpha)}^{(-\alpha-\delta)} b_{i(\alpha)}^{(\alpha+\beta)} \right| \leq$$

$$\leq \frac{1}{2} \sum_{i,p,r,s,\alpha,\beta,\gamma,\delta} |\alpha_p + \beta_p| \cdot |a_r^{(\gamma)}|^2 \cdot |k_{ir(\alpha)}^{(-\beta-\gamma)}| \cdot |k_{is(-\alpha)}^{(-\alpha-\delta)}| \cdot |b_{i(\alpha)}^{(\alpha+\beta)}| +$$

$$+ \frac{1}{2} \sum_{i,p,r,s,\alpha,\beta,\gamma,\delta} |\alpha_p + \beta_p| \cdot |a_s^{(\delta)}|^2 \cdot |k_{ir(\alpha)}^{(-\beta-\gamma)}| \cdot |k_{is(-\alpha)}^{(-\alpha-\delta)}| \cdot |b_{i(\alpha)}^{(\alpha+\beta)}| . \qquad (21)$$

For $|k_{ir(\alpha)}^{(\beta)}|$ we obtain an estimate similar to (19):

$$|k_{ir(\alpha)}^{(\beta)}| \leq \frac{M_1}{[|\beta_1| + 1]^2 \ldots [|\beta_n| + 1]^2} , \qquad (22)$$

where M_1 is the largest among $M_1^{(0)}, M_1^{(1)}l, \ldots, M_1^{(2n)}l^{2n}$, and $M_1^{(k)}$ is the upper bound for the absolute values of all k-th order derivatives of the functions k_{ir}. Therefore

$$\sum_{i,p,r,s,\alpha,\beta,\gamma,\delta} |\alpha_p + \beta_p| \cdot |a_r^{(\gamma)}|^2 \cdot |k_{ir(\alpha)}^{(-\beta-\gamma)}| \cdot |k_{is(-\alpha)}^{(-\alpha-\delta)}| \cdot |b_{i(\alpha)}^{(\alpha+\beta)}| \leq$$

$$\leq M_1^2 M_0 \sum_{i,p,r,s,\alpha,\beta,\gamma,\delta} \frac{|\alpha_p + \beta_p| \cdot |a_r^{(\gamma)}|^2}{\prod_k [|\beta_k + \gamma_k| + 1]^2 \cdot [|\alpha_k + \delta_k| + 1]^2 \cdot [|\alpha_k + \beta_k| + 1]^3} \leq$$

$$\leq M_1^2 M_0 n \sum_{i,r,s,\alpha,\beta,\gamma,\delta} \frac{|a_r^{(\gamma)}|^2}{\prod_k [|\beta_k + \gamma_k| + 1]^2 \cdot [|\alpha_k + \delta_k| + 1]^2 \cdot [|\alpha_k + \beta_k| + 1]^2} =$$

$$= M_1^2 M_0 C^3 N^2 n \sum_{r,\gamma} |a_r^{(\gamma)}|^2 . \qquad (23)$$

Here

$$C = \sum_\alpha [|\alpha_1| + 1]^{-2} \cdots [|\alpha_n| + 1]^{-2} ,$$

and the sum is taken over all $\alpha_1, \ldots, \alpha_n$.

In a similar fashion we find that

$$\sum_{i,p,r,s,\alpha,\beta,\gamma,\delta} |\alpha_p + \beta_p| \cdot |a_s^{(\delta)}|^2 \cdot |k_{ir(\alpha)}^{(-\beta-\gamma)}| \cdot |k_{is(\alpha)}^{(-\alpha-\delta)}| \cdot |b_{i(\alpha)}^{(\alpha+\beta)}| \leq$$

$$\leq M_1^2 M_0 C^3 N^2 n \sum_{s,\delta} |a_s^{(\delta)}|^2 = M_1^2 M_0 C^3 N^2 n \sum_{r,\gamma} |a_r^{(\gamma)}|^2 . \qquad (23')$$

Next we obtain an estimate for the absolute value of the second sum in the right-hand side of (18). First of all, we observe that the coefficients of the transformation (16) do not change if all α_k are multiplied by the same positive number. Therefore, if $\sum \alpha_k^2 > 0$ and $\sum \beta_k^2 > 0$, then

$$b_{i(\alpha)}^{(\alpha+\beta)} - b_{i(-\beta)}^{(\alpha+\beta)} = b_{i(\alpha/|\alpha|)}^{(\alpha+\beta)} - b_{i(-\beta/|\beta|)}^{(\alpha+\beta)} =$$

$$= \frac{1}{l^n} \int_{Q_l} \left[k_i' \left(t, x, \frac{\alpha}{|\alpha|} \right) - k_i' \left(t, x, -\frac{\beta}{|\beta|} \right) \right] e^{-\frac{2\pi i}{l}(\alpha+\beta,x)} \, dx . \qquad (24)$$

If among $(\alpha_1 + \beta_1), \ldots, (\alpha_n + \beta_n)$ there are m non-vanishing numbers, then we integrate by parts $3m$ times, as previously. Let us consider the expression

$$k_i'\left(t, x, \frac{\alpha_1}{|\alpha|}, \ldots, \frac{\alpha_n}{|\alpha|}\right) - k_i'\left(t, x, \frac{-\beta_1}{|\beta|}, \ldots, \frac{-\beta_1}{|\beta|}\right)$$

and apply the following procedure to estimate its derivative (D^{3m}) of order $3m$ with respect to any of the variables $x_1, \ldots, x_n$, using the formula of finite increments. We move from the values $\alpha_s/|\alpha| = \alpha_s'$ $(s = 1, \ldots, n)$ to the values $-\beta_s/|\beta| = -\beta_s'$ $(s = 1, \ldots, n)$ along the arc of the great circle on the sphere $\sum_{s=1}^n \gamma_s^2 = 1$. All γ_s can be considered as functions of $\rho(\alpha', \gamma)$, which is the distance along the geodesic from the point $\gamma = (\gamma_1, \ldots, \gamma_n)$ to the point $\alpha' = (\alpha_s', \ldots, \alpha_s')$. Therefore

$$\left| D^{3m} k_i'\left(t, x, \frac{\alpha}{|\alpha|}\right) - D^{3m} k_i'\left(t, x, -\frac{\beta}{|\beta|}\right) \right| =$$

$$= \left| \left[\frac{d}{d\rho} D^{3m} k_i'(t, x, \alpha'(\rho)) \right]_{\rho=\tilde\rho} \rho(\alpha', -\beta') \right| \le$$

$$\le \pi \left| \sum_s \frac{\partial}{\partial \alpha_s'} D^{3m} k_i'(t, x, \alpha'(\rho)) \frac{d\alpha_s'}{d\rho} \right|_{\rho=\tilde\rho} |\alpha' + \beta'| \le$$

$$\le \pi \sum_s \left| \frac{\partial}{\partial \alpha_s'} D^{3m} k_i'(t, x, \alpha'(\rho)) \right|_{\rho=\tilde\rho} \sum_k |\alpha_k' + \beta_k'|$$

$$(0 \le \tilde\rho \le \rho(\alpha', -\beta')) .$$

Let $M_2^{(k)}/l^k$ be the upper bound for the absolute values of k-th order derivatives with respect to various combinations of $x_1, \ldots, x_n$, taken of the first order derivatives of the functions

$$k_i'\left(t, x_1, \ldots, x_n, \frac{\alpha_1}{|\alpha|}, \ldots, \frac{\alpha_n}{|\alpha|}\right)$$

with respect to any of the variables $\alpha_j/|\alpha|$. By M_2 we denote the largest among the constants $M_2^{(0)}, M_2^{(1)}, \ldots, M_2^{3n}, M_0$. Using the same arguments as before, we find that

$$|b_{i(\alpha)}^{(\alpha+\beta)} - b_{i(-\beta)}^{(\alpha+\beta)}| \le \frac{M_2 n \pi \sum_k \left| \frac{\alpha_k}{|\alpha|} + \frac{\beta_k}{|\beta|} \right|}{[|\alpha_1 + \beta_1| + 1]^3 \cdots [|\alpha_n + \beta_n| + 1]^3} .$$

On the other hand, we have

$$\frac{\alpha_k}{|\alpha|} + \frac{\beta_k}{|\beta|} = \frac{\alpha_k|\beta| + \beta_k|\alpha|}{|\alpha||\beta|} = \frac{\alpha_k + \beta_k}{|\beta|} + \alpha_k \frac{|\beta| - |\alpha|}{|\alpha||\beta|} =$$

$$= \frac{\alpha_k + \beta_k}{|\beta|} + \alpha_k \frac{|\beta| - |\alpha|}{|\alpha||\beta|(|\alpha| + |\beta|)} =$$

$$= \frac{\alpha_k + \beta_k}{|\beta|} + \alpha_k \frac{\sum(\alpha_p + \beta_p)(\beta_p - \alpha_p)}{|\alpha||\beta|(|\alpha| + |\beta|)} . \tag{25}$$

Hence

$$\left| |\beta| \left[b_{i(\alpha)}^{\alpha+\beta)} - b_{i(-\beta)}^{(\alpha+\beta)} \right] \right| \le$$

$$\le \sum_k \frac{M_2 |\beta||\alpha_k + \beta_k|n\pi}{[|\alpha_1 + \beta_1| + 1]^3 \cdots [|\alpha_n + \beta_n| + 1]^3 |\beta|} +$$

$$+ \sum_k \frac{M_2 |\beta||\alpha_k| \sum_p |\alpha_p + \beta_p| |\alpha_p - \beta_p|n\pi}{[|\alpha_1 + \beta_1| + 1]^3 \cdots [|\alpha_n + \beta_n| + 1]^3 |\alpha||\beta|(|\alpha| + |\beta|)} \le$$

$$\le \sum_k \frac{M_2 n\pi}{[|\alpha_1 + \beta_1| + 1]^2 \cdots [|\alpha_n + \beta_n| + 1]^2} +$$

$$+ n \sum_k \frac{M_2 n\pi}{[|\alpha_1 + \beta_1| + 1]^2 \cdots [|\alpha_n + \beta_n| + 1]^2} . \tag{26}$$

The last estimate is also valid if

$$\sum \alpha_k^2 > 0 , \quad \sum \beta_k^2 = 0 \quad \text{or} \quad \sum \alpha_k^2 = 0 , \quad \sum \beta_k^2 > 0 .$$

In the first case the estimate is obvious. In the second case we have

$$\left| |\beta| \left[b_{i(\alpha)}^{(\alpha+\beta)} - b_{i(-\beta)}^{(\alpha+\beta)} \right] \right| = |\beta| \left| b_{i(0)}^{(\beta)} - b_{i(-\beta)}^{(\beta)} \right| ,$$

and $b_{i(0)}^{(\beta)}$, $b_{i(-\beta)}^{(\beta)}$ satisfy the following inequalities:

$$|b_{i(0)}^{(\beta)}| \le \frac{M_0}{[|\beta_1| + 1]^3 \cdots [|\beta_n| + 1]^3} \le \frac{M_2}{|\beta| [|\beta_1| + 1]^2 \cdots [|\beta_n| + 1]^2} ,$$

$$|b_{i(-\beta)}^{(\beta)}| \le \frac{M_0}{[|\beta_1| + 1]^3 \cdots [|\beta_n| + 1]^3} \le \frac{M_2}{|\beta|[|\beta_1| + 1]^2 \cdots [|\beta_n| + 1]^2} .$$

Hence we find that

$$\left| \sum_{i,\alpha,\beta} |\beta| a_{i(-\beta)}^{\prime(-\alpha)} a_{i(-\beta)}^{\prime(-\beta)} \left[b_{i(\alpha)}^{(\alpha+\beta)} - b_{i(-\beta)}^{(\alpha+\beta)} \right] \right| \le$$

$$\le \sum_{i,\alpha,\beta} a_{i(-\beta)}^{\prime(-\alpha)} a_{i(-\beta)}^{\prime(-\beta)} \frac{M_2 n^2(n+1)\pi}{[|\alpha_1 + \beta_1| + 1]^2 \cdots [|\alpha_n + \beta_n| + 1]^2} \le$$

$$\le \sum_{i,r,s,\alpha,\beta,\gamma,\delta} \frac{\pi M_2 n^2(n+1)|a_r^{(\gamma)}| |k_{ir(-\beta)}^{(-\alpha-\gamma)}|}{[|\alpha_1 + \beta_1| + 1]^2 \cdots [|\alpha_n + \beta_n| + 1]^2} |a_s^{(\delta)}| \cdot |k_{is(-\beta)}^{(-\beta-\delta)}| \le {}^{11}$$

---

[11] cf. (20)

$$\leq \frac{1}{2} M_2 n^2 (n+1) \pi \times$$

$$\times \sum_{i,r,s,\alpha,\beta,\gamma,\delta} \frac{\left[|a_r^{(\gamma)}|^2 + |a_s^{(\delta)}|^2 \right] M_1^2}{\prod_k [|\alpha_k + \beta_k| + 1]^2 [|\alpha_k + \delta_k| + 1]^2 [|\beta_k + \gamma_k| + 1]^2} = {}^{12}$$

$$= \pi M_1^2 M_2^2 n^2 (n+1) C^3 N^2 \sum_{r,\gamma} |a_r^{(\gamma)}|^2 . \tag{27}$$

As before, C denotes the sum of the series

$$\sum_{\alpha} \frac{1}{[|\alpha_1| + 1]^2 \cdots [|\alpha_1| + 1]^2} , \tag{28}$$

where the summation is carried over all possible α_k.

Now we turn to the proof of an estimate for the absolute value of the third sum in the right-hand side of (18). First of all, we note that

$$a_{i(\alpha)}^{'(\beta)} = \frac{1}{l^n} \int_{Q_l} \sum_j k_{ij}(t, x, \alpha) u_j(x) e^{-\frac{2\pi i}{l}(\beta, x)} \, dx .$$

If $|\alpha| > 0$ and $|\beta| > 0$, the above formula implies that

$$a_{i(\alpha)}^{'(-\alpha)} - a_{i(-\beta)}^{'(-\alpha)} = \frac{1}{l^n} \int \sum_s [k_{is}(t, x, \alpha) - k_{is}(t, x, -\beta)] u_s e^{\frac{2\pi i}{l}(\alpha, x)} \, dx =$$

$$= \sum_{s,\gamma} a_{(s)}^{(\gamma)} \left[k_{is(\alpha)}^{(\alpha-\gamma)} - k_{is(-\beta)}^{(-\alpha-\gamma)} \right] . \tag{29}$$

In order to estimate the absolute value of the difference in square brackets, we use the same method as that used to estimate the left-hand side of (24). Denote by $M_3^{(k)}/l^k$ the upper bound for the absolute values of various derivatives of order k with respect to $x_1, \ldots, x_n$ taken from first order derivatives of the coefficients $k_{is}\left(t, x, \frac{\alpha_1}{|\alpha|}, \ldots, \frac{\alpha_n}{|\alpha|}\right)$ with respect to any of the variables $\alpha_j/|\alpha|$. Let M_3 be the largest constant among $M_3^{(0)}, M_3^{(1)}, \ldots, M_3^{(2n)}, M_1$. We then obtain

$$\left| |\beta| \left[k_{is(\alpha)}^{(-\alpha-\gamma)} - k_{is(-\beta)}^{(-\alpha-\gamma)} \right] \right| \leq \sum_k \frac{M_3 |\beta| |\alpha_k + \beta_k| n\pi}{|\beta| [|\alpha_1 + \gamma_1| + 1]^2 \cdots [|\alpha_n + \gamma_n| + 1]^2} +$$

$$+ \sum_k \frac{M_2 |\beta| |\alpha_k| \sum_p |\alpha_p + \beta_p| \cdot |\alpha_p - \beta_p| \cdot n\pi}{[|\alpha_1 + \gamma_1| + 1]^2 \cdots [|\alpha_n + \gamma_n| + 1]^2 |\alpha| |\beta| (|\alpha| + |\beta|)} .$$

Hence

---
[12]by virtue of (22)

$$\left| |\beta| \left[k_{is(\alpha)}^{(-\alpha-\gamma)} - k_{is(-\beta)}^{(-\alpha-\gamma)} \right] \right| \leq$$

$$\leq M_3(n+1) \sum_k \frac{|\alpha_k + \beta_k| n\pi}{[|\alpha_1 + \gamma_1| + 1]^2 \cdots [|\alpha_n + \gamma_n| + 1]^2} . \quad (30)$$

This estimate is also valid if $\sum \beta_k^2 = 0$ or $\sum \alpha_k^2 = 0$. In the first case the estimate is trivial. In the second case,

$$\left| |\beta| \left[k_{is(\alpha)}^{(-\alpha-\gamma)} - k_{is(-\beta)}^{(-\alpha-\gamma)} \right] \right| = |\beta| \left| k_{is(0)}^{(-\gamma)} - k_{is(-\beta)}^{(-\gamma)} \right| \leq$$

$$\leq \frac{2M_1 |\beta|}{[|\gamma_1| + 1]^2 \cdots [|\gamma_n| + 1]^2} \leq \frac{2M_3 \sum_k |\beta_k|}{[|\gamma_1| + 1]^2 \cdots [|\gamma_n| + 1]^2} .$$

Hence, using the estimate (19) and equation (20), we get

$$\left| \sum_{i,\alpha,\beta} |\beta| a_{i(-\beta)}^{'(-\beta)} \left[a_{i(\alpha)}^{'(-\alpha)} - a_{i(-\beta}^{'(-\alpha)} \right] b_{i(\beta)}^{(\alpha+\beta)} \right| \leq$$

$$\leq \sum_{i,k,r,s,\alpha,\beta,\gamma,\delta} \frac{n\pi M_3(n+1)|\alpha_k + \beta_k| \cdot |a_s^{(\gamma)}| M_0 |a_r^{(\delta)}| k_{is(-\beta)}^{(-\beta-\delta)}}{\prod_p [|\alpha_p + \gamma_p| + 1]^2 [|\alpha_p + \beta_p| + 1]^2} \leq$$

$$\leq \frac{M_0 M_1 M_3}{2} \times$$

$$\times \sum_{i,k,r,s,\alpha,\beta,\gamma,\delta} \frac{n\pi(n+1) \left[|a_r^{(\delta)}|^2 + |a_s^{(\gamma)}|^2 \right]}{\prod_p [|\alpha_p + \gamma_p| + 1]^2 [|\alpha_p + \beta_p| + 1]^2 [|\beta_p + \delta_p| + 1]^2} =$$

$$= \pi M_0 M_1 M_3(n+1)n^2 N^2 C^3 \sum_{r,\delta} |a_r^{(\delta)}|^2 , \quad (31)$$

where C, as usual, stands for the sum of the series (28).

The absolute value of the last sum estimates the penultimate sum in the right-hand side of (18), as well as the last one; therefore, the same estimate is obtained for both.

Hence we finally get the following estimate for the absolute value of the left-hand side of (18'):

$$\left| \frac{2\pi i}{l} \sum_{i,\alpha} |\alpha| c_{i(\alpha)}^{(\alpha)} a_{i(\alpha)}^{'(-\alpha)} \right| \leq \frac{4\pi^2}{l} n^2(n+1) N^2 C^3 M^3 \sum_{r,\gamma} |a_r^{(\gamma)}|^2 , \quad (32)$$

where M is the largest among M_0, M_1, M_2, M_3 in (23), (23'), (27) and (31).

Next we establish an estimate for the absolute value of the second sum in the right-hand side of (17). It will be convenient to write this sum in the form (17'). Set

$$d_{ij(\alpha)}^{(\beta)} = \frac{1}{l^n} \int_{Q_l} a_{ij}''(t, x, \alpha_1, \ldots, \alpha_n) e^{-\frac{2\pi i}{l}(\beta, x)} \, dx \ .$$

Then we have

$$\frac{1}{l^n} \int_{Q_l} k_{is}(t, x, \alpha) u_s e^{\frac{2\pi i}{l}(\alpha, x)} \, dx = \sum_\beta k_{is(\alpha)}^{(-\alpha-\beta)} a_s^{(\beta)}$$

and

$$\frac{1}{l^n} \int_{Q_l} a_{ir}''(t, x, \alpha) u_r e^{-\frac{2\pi i}{l}(\alpha, x)} \, dx = \sum_\gamma d_{ir(\alpha)}^{(\alpha-\gamma)} a_r^{(\gamma)} \ .$$

Hence we obtain

$$\left| \sum_{i,j,\alpha} \frac{1}{l^n} \int_{Q_l} u_i'(t, x, \alpha) e^{\frac{2\pi i}{l}(\alpha, x)} \, dx \; \frac{1}{l^n} \int_{Q_l} a_{ij}'(t, x, \alpha) u_j'(t, x, \alpha) e^{-\frac{2\pi i}{l}(\alpha, x)} \, dx \right| \leq$$

$$\leq \sum_{i,r,s,\alpha,\beta,\gamma} |k_{is(\alpha)}^{(-\alpha-\beta)}| \cdot |a_s^{(\beta)}| \cdot |d_{ir(\alpha)}^{(\alpha-\gamma)}| \cdot |a_r^{(\gamma)}| \leq$$

$$\leq \frac{M_1 M_4}{2} \sum_{i,r,s,\alpha,\beta,\gamma} \frac{|a_s^{(\beta)}|^2 + |a_r^{(\gamma)}|^2}{\prod_k [|\alpha_k + \beta_k| + 1]^2 [|\alpha_k - \gamma_k| + 1]^2} =$$

$$= M_1 M_4 N^2 C^2 \sum_{s,\beta} |a_s^{(\beta)}|^2 \ . \tag{33}$$

where M_4 is the largest constant among $M_4^{(0)}, M_4^{(1)}, \ldots, M_4^{(2n)}$, and $M_4^{(k)}/l^k$ is the upper bound for the absolute values of the k-th order derivatives, with respect to various combinations of $x_1, \ldots, x_n$, of the functions a_{ij}'', for all possible real values of α_k. Here a_{ij}'' are polynomials of the variables

$$b_{ij}(t, x) \ , \quad k_{ij}(t, x, \alpha) \ , \quad \frac{\partial k_{ij}}{\partial t} \ , \quad \frac{\partial k_{ij}}{\partial x_k} \ , \quad a_{ij}^{(k)}(t, x) \ , \quad \frac{\partial a_{ij}^{(k)}}{\partial t} \ , \quad \frac{\partial a_{ij}^{(k)}}{\partial x_p} \ .$$

It follows from our considerations, that all sums in the inequality (33) are absolutely convergent. With the help of similar arguments it can be shown that the right-hand side of (18') is also absolutely convergent. To this end, we should only take into account the following relation

$$\frac{2\pi}{l} \left| |\alpha| a_{i(-\alpha)}'^{(-\alpha)} \right| = \frac{1}{l^n} \left| \int_{Q_l} \frac{\partial u_i'}{\partial x_1'} e^{-\frac{2\pi i}{l} |\alpha| x_1'} \, dx \right|$$

and repeat the above arguments replacing the factor u_i' by $\partial u_i'/\partial x_1'$ in all terms of the sum to be estimated.

Finally, we obtain an estimate for the last sum in the right-hand side of (17'):

$$\frac{1}{l^{2n}} \left| \sum_{i,r,s,\alpha} \int_{Q_l} k_{is}(t,x,\alpha) u_s e^{\frac{2\pi i}{T}(\alpha,x)} \, dx \int_{Q_l} k_{ir}(t,x,\alpha) f_r(t,x) e^{-\frac{2\pi i}{T}(\alpha,x)} \, dx \right| \leq$$

$$\leq \frac{1}{2l^{2n}} \sum_{i,r,s,\alpha} \left| \int_{Q_l} k_{is}(t,x,\alpha) u_s e^{\frac{2\pi i}{T}(\alpha,x)} \, dx \right|^2 +$$

$$+ \frac{1}{2l^{2n}} \sum_{i,r,s,\alpha} \left| \int_{Q_l} k_{is}(t,x,\alpha) f_r(t,x) e^{-\frac{2\pi i}{T}(\alpha,x)} \, dx \right|^2 . \tag{34}$$

Each term in the first sum can be represented in general form as

$$\int_{Q_l} k_{is}(t,x,\alpha) u_s e^{-\frac{2\pi i}{T}(\alpha,x)} \, dx \int_{Q_l} k_{is}(t,x,\alpha) u_s e^{\frac{2\pi i}{T}(\alpha,x)} \, dx =$$

$$= l^{2n} \sum_{\beta,\gamma} k_{is(\alpha)}^{(\alpha-\beta)} a_s^{(\beta)} k_{is(\alpha)}^{(-\alpha-\gamma)} a_s^{(\gamma)} .$$

Hence, taking into account the estimate (22), we get

$$\sum_{i,r,s,\alpha} \frac{1}{l^{2n}} \left| \int_{Q_l} k_{is}(t,x,\alpha) u_s(t,x) e^{\frac{2\pi i}{T}(\alpha,x)} \, dx \right|^2 \leq$$

$$\leq \sum_{i,r,s,\alpha,\beta,\gamma} \frac{|a_r^{(\beta)}| \cdot |a_s^{(\gamma)}| M_1^2}{\prod_k [|\alpha_k - \beta_k| + 1]^2 [|\alpha_k + \gamma_k| + 1]^2} \leq$$

$$\leq \frac{M_1}{2} \sum_{i,r,s,\alpha,\beta,\gamma} \frac{|a_r^{(\beta)}|^2 + |a_s^{(\gamma)}|^2}{\prod_k [|\alpha_k - \beta_k| + 1]^2 [|\alpha_k + \gamma_k| + 1]^2} =$$

$$= M_1^2 C^2 N^2 \sum_{s,\gamma} |a_s^{(\gamma)}|^2 . \tag{35}$$

In exactly the same way we find that the absolute value of second sum in the right-hand side of (34) does not exceed

$$M_1^2 C^2 N^2 \sum_{s,\gamma} |f_s^{(\gamma)}|^2 ,$$

where

$$f_s^{(\gamma)} = \frac{1}{l^n} \int_{Q_l} f_s(t,x) e^{-\frac{2\pi i}{T}(\gamma,x)} \, dx .$$

Now, combining all estimates obtained above, we deduce that

$$\frac{d\sum_{i,\alpha} |a_{i(\alpha)}'^{(\alpha)}|^2}{dt} \leq \psi_1 \sum_{r,\gamma} |a_r^{(\gamma)}|^2 + \psi_2 \sum_{r,\gamma} |f_r^{(\gamma)}|^2 , \tag{36}$$

where the functions ψ_1 and ψ_2 depend only on n, N, and M_i.

Now let us assume that

$$M_5^{*2} C^2 N^2 + M_5^{(0)} M_5^* C N^2 < 1 . \tag{37}$$

Here M_5^* is a positive constant whose exact value will be determined later on; it depends only on l and the coefficients of the system under consideration. *We show that in this case*

$$\sum_{r,\gamma} |a_r^{(\gamma)}|^2 < \psi_3 \sum |a_{r(\gamma)}^{'(\gamma)}|^2 , \tag{38}$$

where ψ_3 is a constant which depends only on the coefficients of the system under consideration. The exact value for ψ_3 will be indicated at the end of the proof.

Proof. We have

$$S = \sum_{i,\alpha} |a_i^{(\alpha)}|^2 = \sum_{i,\alpha} a_i^{(\alpha)} a_i^{(-\alpha)} ,$$

$$a_i^{(\alpha)} = \frac{1}{l^n} \int_{Q_l} u_i(t,x) e^{-\frac{2\pi i}{l}(\alpha,x)} \, dx =$$

$$= \frac{1}{l^n} \int_{Q_l} \sum_j k_{ij}'(t,x,\alpha) u_j'(t,x,\alpha) e^{-\frac{2\pi i}{l}(\alpha,x)} \, dx . \tag{39}$$

The formula

$$u_i(t,x) = \sum_j k_{ij}'(t,x,\alpha) u_j'(t,x,\alpha)$$

defines the inverse transformation of (16). Setting

$$k_{ij(\alpha)}'^{(\beta)} = \frac{1}{l^n} \int_{Q_l} k_{ij}'(t,x,\alpha) e^{-\frac{2\pi i}{l}(\beta,x)} \, dx ,$$

we can represent S in the form

$$S = \sum_{i,j,r,\alpha,\beta,\gamma} k_{ij(\alpha)}'^{(\alpha-\beta)} a_{j(\alpha)}'^{(\beta)} k_{(-\alpha)}'^{(-\alpha+\gamma)} a_{r(-\alpha)}'^{(-\gamma)} \le$$

$$\le \frac{1}{2} \sum_{i,j,r,s} |k_{ij(\alpha)}'^{(0)} k_{ir(-\alpha)}'^{(0)}| \left[|a_{j(\alpha)}'^{(\alpha)}|^2 + |a_{r(-\alpha)}'^{(-\alpha)}|^2 \right] +$$

$$+ \frac{1}{2} \sum_{i,j,r,\alpha,\beta\neq\alpha} |k_{ij(\alpha)}'^{(\alpha-\beta)} k_{ir(-\alpha)}'^{(0)}| \left[|a_{j(\alpha)}'^{(\beta)}|^2 + |a_{r(-\alpha)}'^{(-\alpha)}|^2 \right] +$$

$$+ \frac{1}{2} \sum_{i,j,r,\alpha,\gamma\neq\alpha} |k_{ij(\alpha)}'^{(0)} k_{ir(-\alpha)}'^{(-\alpha+\gamma)}| \left[|a_{j(\alpha)}'^{(\alpha)}|^2 + |a_{r(-\alpha)}'^{(-\gamma)}|^2 \right] +$$

$$+ \frac{1}{2} \sum_{i,j,r,\alpha,\beta\neq\alpha,\gamma\neq\alpha} |k_{ij(\alpha)}'^{(\alpha-\beta)} k_{ir(-\alpha)}'^{(-\alpha+\gamma)}| \left[|a_{j(\alpha)}'^{(\beta)}|^2 + |a_{r(-\alpha)}'^{(-\gamma)}|^2 \right] . \tag{40}$$

Denote by $M_1^{'(0)}$ the upper bound for the absolute values of $k_{ij}'(t, x, \alpha)$. Then

$$|k_{ij(\alpha)}^{'(0)}| = \frac{1}{l^n}\left|\int_{Q_l} k_{ij}'(t, x, \alpha)\, dx\right| \leq M_1^{'(0)}. \tag{41}$$

Denote by $M_1^{'*}$ the upper bound for the absolute values of the k-th order derivatives in x_p of the coefficients $k_{ij}'(t, x, \alpha)$ multiplied by l^k ($k = 1, \ldots, 2n$). Assuming that $\beta_k \neq 0$ for at least one k, we obtain

$$|k_{ij(\alpha)}^{'(\beta)}| = \frac{1}{l^n}\left|\int_{Q_l} k_{ij}'(t, x, \alpha)e^{-\frac{2\pi i}{l}(\beta, x)}\, dx\right| =$$

$$= \frac{l^{2m}}{\beta_{i_1}^2 \cdots \beta_{i_m}^2 (2\pi)^{2m} l^n}\left|\int_{Q_l} k_{ij}^{'(2m)}(t, x, \alpha)e^{\frac{-2\pi i}{l}(\beta, x)}\, dx\right| \leq$$

$$\leq \frac{M_1^{'*}}{\beta_{i_1}^2 \cdots \beta_{i_m}^2 (2\pi)^{2m}} \leq \frac{M_1^*}{[|\beta_1| + 1]^2 \cdots [|\beta_n| + 1]^2}. \tag{42}$$

Here $\beta_{i_1}, \ldots, \beta_{i_n}$ are those, from among β_k, that are not equal to zero. Hence, by virtue of (40) and (41), we find that

$$S \leq N^2 M_1^{'(0)2} \sum_{i,\alpha} |a_{j(\alpha)}^{'(\alpha)}|^2 + N^2 M_1^{'(0)} M_1^{'*} \sum_{j,\alpha,\gamma} \frac{|a_{j(\alpha)}^{'(\alpha)}|^2}{\prod_k [|\alpha_k - \gamma_k| + 1]^2} +$$

$$+ N^2 M_1^{'(0)} M_1^{'*} \sum_{j,\alpha,\beta} \frac{|a_{j(\alpha)}^{'(\beta)}|^2}{\prod_k [|\alpha_k - \beta_k| + 1]^2} +$$

$$+ \sum_{i,j,r,\alpha,\beta,\gamma} \frac{M_1^{'*2}\left[|a_{j(\alpha)}^{'(\beta)}|^2 + |a_{r(\alpha)}^{'(\gamma)}|^2\right]}{2\prod_k [|\alpha_k - \beta_k| + 1]^2 [|\alpha_k - \gamma_k| + 1]^2} \leq$$

$$\leq \left[N^2 M_1^{'(0)2} + N^2 M_1^{'(0)} M_1^{'*} C\right] \sum_{j,\alpha} |a_{j(\alpha)}^{'(\alpha)}|^2 +$$

$$+ \left[N^2 M_1^{'(0)} M_1^{'*} + M_1^{'*2} C N^2\right] \sum_{j,\alpha,\beta} \frac{|a_{j(\alpha)}^{'(\beta)}| \cdot |a_{j(-\alpha)}^{'(-\beta)}|}{\prod_k [|\alpha_k - \beta_k| + 1]^2}. \tag{43}$$

Note that

$$a_{i(\alpha)}^{'(\beta)} = \frac{1}{l^n}\int_{Q_l} u_i'(t, x, \alpha)e^{-\frac{2\pi i}{l}(\beta, x)}\, dx =$$

$$= \frac{1}{l^n}\int_{Q_l} \sum_j k_{ij}(t, x, \alpha, \beta)u_j'(t, x, \beta)e^{-\frac{2\pi i}{l}(\beta, x)}\, dx = \sum_{j,\gamma} k_{ij(\alpha,\beta)}^{(\beta-\gamma)}a_{j(\beta)}^{'(\gamma)},$$

where

$$k^{(\gamma)}_{ij(\alpha,\beta)} = \frac{1}{l^n} \int_{Q_l} k_{ij}(t,x,\alpha,\beta) e^{-\frac{2\pi i}{l}(\gamma,x)}\, dx \ .$$

The functions $k_{ij}(t,x,\alpha,\beta)$ coincide with the coefficients in the linear transformations by which $u'_j(t,x,\beta)$ $(j = 1,\ldots,N)$ and $u'_j(t,x,\alpha)$ $(j = 1,\ldots,N)$ are related. Therefore

$$\sum_{j,\alpha,\beta} \frac{|a'^{(\beta)}_{j(\alpha)}| \cdot |a'^{(-\beta)}_{j(-\alpha)}|}{\prod_k [|\alpha_k - \beta_k| + 1]^2} \le$$

$$\le \sum_{j,r,s,\alpha,\beta,\gamma,\delta} \frac{|k^{(\beta-\gamma)}_{jr(\alpha,\beta)}| \cdot |a'^{(\gamma)}_{r(\beta)}| \cdot |k^{(-\beta+\delta)}_{js(-\alpha-\beta)}| \cdot |a'^{(-\delta)}_{s(-\beta)}|}{\prod_k [|\alpha_k - \beta_k| + 1]^2} =$$

$$= \sum_{j,r,s,\alpha,\beta} \frac{|k^{(0)}_{jr(\alpha,\beta)}| \cdot |k^{(0)}_{js(-\alpha,-\beta)}| \cdot |a'^{(\beta)}_{r(\beta)}| \cdot |a'^{(-\beta)}_{s(-\beta)}|}{\prod_k [|\alpha_k - \beta_k| + 1]^2} +$$

$$+ \sum_{j,r,s,\alpha,\beta,\gamma\neq\beta} \frac{|k^{(\beta-\gamma)}_{jr(\alpha,\beta)}| \cdot |a'^{(\gamma)}_{r(\beta)}| \cdot |k^{(0)}_{js(-\alpha,-\beta)}| \cdot |a'^{(-\beta)}_{s(-\beta)}|}{\prod_k [|\alpha_k - \beta_k| + 1]^2} +$$

$$+ \sum_{j,r,s,\alpha,\beta,\delta\neq\beta} \frac{|k^{(0)}_{jr(\alpha,\beta)}| \cdot |a'^{(\beta)}_{r(\beta)}| \cdot |k^{(-\beta+\delta)}_{js(-\alpha,-\beta)}| \cdot |a'^{(-\delta)}_{s(-\beta)}|}{\prod_k [|\alpha_k - \beta_k| + 1]^2} +$$

$$+ \sum_{j,r,s,\alpha,\, zb, \gamma\neq\beta, \delta\neq\beta} \frac{|k^{(\beta-\gamma)}_{jr(\alpha,\beta)}| \cdot |a'^{(\gamma)}_{r(\beta)}| \cdot |k^{(-\beta+\delta)}_{js(-\alpha,-\beta)}| \cdot |a'^{(-\delta)}_{s(-\beta)}|}{\prod_k [|\alpha_k - \beta_k| + 1]^2} \ . \quad (44)$$

Denote by $M^{(k)}_5/l^k$ the upper bound for the absolute values of the k-th order $(k = 0, 1, \ldots)$ derivatives of the functions $k_{ij}(t,x,\alpha,\beta)$ with respect to various combinations of $x_1, \ldots, x_n$, for all possible values of i, j, α, β. Let M^*_5 be the largest constant among $M^{(1)}_5, \ldots, M^{(2n)}_5$. Then

$$|k^{(0)}_{ij(\alpha,\beta)}| = \frac{1}{l^n} \left| \int_{Q_l} k_{ij}(t,x,\alpha,\beta)\, dx \right| \le M^{(0)}_5 \ . \quad (45')$$

If at least one of $\gamma_1, \ldots, \gamma_n$ is different from zero, then

$$|k^{(\gamma)}_{ij(\alpha,\beta)}| = \frac{1}{l^n} \left| \int_{Q_l} k_{ij}(t,x,\alpha,\beta) e^{-\frac{2\pi i}{l}(\gamma,x)}\, dx \right| \le$$

$$\le \frac{l^{2m}}{\gamma^2_{i_1}\gamma^2_{i_2}\cdots\gamma^2_{i_m}(2\pi)^{2m}l^n} \left| \int_{Q_l} k^{(2m)}_{ij}(t,x,\alpha,\beta) e^{-\frac{2\pi i}{l}(\gamma,x)}\, dx \right| \le$$

$$\le \frac{M^{2m}_5}{\gamma^2_{i_1}\gamma^2_{i_2}\cdots\gamma^2_{i_m}(2\pi)^{2m}} \le \frac{M^*_5}{[|\gamma_1| + 1]^2 \cdots [|\gamma_n| + 1]^2} \ . \quad (45)$$

Here $\gamma_{i_1}, \gamma_{i_2}, \ldots, \gamma_{i_m}$ are those, from among $\gamma_1, \ldots, \gamma_n$, which are different from zero. Hence, by virtue of $(44), (45')$ and (45), we find that

$$\sum_{j,\alpha,\beta} \frac{|a'^{(\beta)}_{j(\alpha)}| \cdot |a'^{(-\beta)}_{j(-\alpha)}|}{\prod_{k}[|\alpha_k - \beta_k| + 1]^2} \leq \frac{1}{2} \sum_{j,r,s,\alpha,\beta} \frac{M_5^{(0)2}\left[|a'^{(\beta)}_{r(\beta)}|^2 + |a'^{(\beta)}_{s(\beta)}|^2\right]}{\prod_{k}[|\alpha_k - \beta_k| + 1]^2} +$$

$$+\frac{1}{2}\sum_{j,r,s,\alpha,\beta,\gamma} \frac{M_5^{(0)} M_5^* \left[|a'^{(\gamma)}_{r(\beta)}|^2 + |a'^{(-\beta)}_{s(-\beta)}|^2\right]}{\prod_{k}[|\alpha_k - \beta_k| + 1]^2[|\beta_k - \gamma_k| + 1]^2} +$$

$$+\frac{1}{2}\sum_{j,r,s,\alpha,\beta,\delta} \frac{M_5^{(0)} M_5^* \left[|a'^{(\beta)}_{r(\beta)}|^2 + |a'^{(\delta)}_{s(\beta)}|^2\right]}{\prod_{k}[|\alpha_k - \beta_k| + 1]^2[|\beta_k - \delta_k| + 1]^2} +$$

$$+\frac{1}{2}\sum_{j,r,s,\alpha,\beta,\gamma,\delta} \frac{M_5^{*2}\left[|a'^{(\gamma)}_{r(\beta)}|^2 + |a'^{(\delta)}_{s(\beta)}|^2\right]}{\prod_{k}[|\beta_k - \gamma_k| + 1]^2[|-\beta_k + \delta_k| + 1]^2[|\alpha_k - \beta_k| + 1]^2} \leq$$

$$\leq \left[M_5^{(0)2}CN^2 + M_5^{(0)}M_5^*C^2N^2\right]\sum_{j,\alpha}|a'^{(\alpha)}_{j(\alpha)}| +$$

$$+\left[M_5^{(0)2}M_5^*CN^2 + M_5^{*2}C^2N^2\right]\sum_{j,\alpha,\beta}\frac{|a'^{(\beta)}_{j(\alpha)}|^2}{\prod_{k}[|\alpha_k - \beta_k| + 1]^2} \,. \tag{46}$$

If the condition (37) is satisfied, then

$$\sum_{j,\alpha,\beta}\frac{|a'^{(\beta)}_{j(\alpha)}|^2}{\prod_{k}[|\alpha_k - \beta_k| + 1]^2} \leq$$

$$\leq \frac{M_5^{(0)} M_5^* C^2 N^2 + M_5^{(0)2}CN^2}{1 - M_5^{(0)}M_5^*CN^2 - M_5^{*2}C^2N^2}\sum_{j,\alpha}|a'^{(\alpha)}_{j(\alpha)}|^2 \,. \tag{47}$$

It follows from $(43), (47)$, and (37) that

$$\sum_{i,\alpha}|a_i^{(\alpha)}|^2 \leq \psi_3(N, M_1^{(0)}, M_1'^{*}, M_5^{(0)}, M_5^*)\sum_{j,\alpha}|a'^{(\alpha)}_{j(\alpha)}|^2 \,,$$

and thereby our proposition is proved.

Remark. Since each matrix M_r can be reduced to canonical form independently of the other matrices, the inequality (38) can also be proved in the case when, instead of (37), only the following condition is satisfied:

$$M_5^{*2}C^2N'^2 + M_5'^{(0)}M_5^*CN'^2 < 1 \,. \tag{37'}$$

Here N' is the largest order of the matrices $M_1, M_2, \ldots, M_l$. To prove this statement, we should restrict the summation in $(39), (40), (43), (44), (46), (47)$ to those i, j, r, s which correspond the one of the matrices $M_1, M_2, \ldots, M_l$.

By virtue of (36), we find from (38) that

$$\frac{d\sum_{i,\alpha} |a_{i(\alpha)}'^{(\alpha)}|^2}{dt} \leq \psi_4 \sum_{i,\alpha} |a_{i(\alpha)}'^{(\alpha)}|^2 + \psi_2 \sum_{i,\alpha} |f_i^{(\alpha)}|^2 , \tag{48}$$

where ψ_i are, as usual, positive constants depending only on l and the maximum of the absolute values of the coefficients of the system under consideration and their derivatives up to the order $3n$. We obtain from (48)

$$\sum_{i,\alpha} |a_{i(t,\alpha)}'^{(\alpha)}|^2 \leq e^{\psi_4 t} \sum_{i,\alpha} |a_{i(0,\alpha)}'^{(\alpha)}|^2 + \psi_2 \int_0^t \sum_{i,\alpha} |f_{i(\tau)}^{(\alpha)}|^2 e^{\psi_4(t-\tau)} \, d\tau \leq$$

$$\leq e^{\psi_4 t} \left[\sum_{i,\alpha} |a_{i(0,\alpha)}'^{(\alpha)}|^2 + \psi_2 \int_0^t \sum_{i,\alpha} |f_{i(\tau)}^{(\alpha)}|^2 \, d\tau \right].$$

Now we replace the values of $\sum_{i,\alpha} |a_i'(\alpha)|^2$ for $t = 0$ and t by their expressions in terms of $\sum_{i,\alpha} |a_i(\alpha)|^2$. To estimate the left-hand side we use the general estimate (38), and for the right-hand side we use the estimate (35). We get

$$\sum_{i,\alpha} |a_{i(t)}^{(\alpha)}|^2 \leq e^{\psi_4 t} \left[\psi_5 \sum_{i,\alpha} |a_{i(0)}^{(\alpha)}|^2 + \psi_6 \int_0^t \sum_{i,\alpha} |f_{i(\tau)}^{(\alpha)}|^2 \, d\tau \right],$$

where ψ_5 and ψ_6 depend on the same quantities as the former ψ_i. The application of Parseval's relation allows us to rewrite the last estimate as follows:

$$\int_{Q_l} \sum_i u_i^2(t, x) \, dx \leq$$

$$\leq e^{\psi_4 t} \left[\psi_5 \int_{Q_l} \sum_i u_i^2(0, x) \, dx + \psi_6 \int_0^t \int_{Q_l} \sum_i f_i^2(\tau, x) \, d\tau \, dx \right] \tag{49}$$

It should be stressed once again[13] that *the above inequality holds under the condition that (37) is satisfied and, in the domain under consideration, the constants ψ_i depend only on the coefficients of the given system, partial derivatives of $a_{ij}^{(k)}$ up to the order $3n$ with respect to various x_k, and partial derivatives of the functions $b_{ij}, \partial a_{ij}/\partial t$ up to the order $2n$ with respect to $x_1, \ldots, x_n$.*

---

[13]Cf. remark in footnote 10.

Lemma. *Consider an arbitrary point P on the plane $t = 0$ and let Q_l be a cube such that*

1) *P coincides with the center of cube Q_l;*

2) *the faces of Q_l are parallel to the coordinate hyperplanes;*

3) *the edges of Q_l have length l (to be specified later on).*

Assume that the origin of the coordinate system coincides with a vertex of cube Q_l, and the orientation of the coordinate axes is such that Q_l belongs to the first quarter.

   Denote by $\zeta(x)$ a function with the following properties:

1) *$\zeta(x) = 1$ for $\eta \leq x \leq 1 - \eta$ $(0 < \eta < 0.1)$;*

2) *$\zeta(0) = \zeta'(0) = \cdots = \zeta^{(m)}(0) = 0,$ $m \geq 3n$;*

3) *$\zeta(x) = \zeta(1 - x)$,*

4) *$\zeta(x)$ is a monotone function on the interval $(-\infty, 1/2)$, with continuous derivatives up to the order m.*

For any function $f(t, x_1, \ldots, x_n)$, set

$$\bar{f}(t, x_1, \ldots, x_n) = f\left(t, \frac{l}{2}, \ldots, \frac{l}{2}\right) +$$

$$+ \left[f(t, x_1, \ldots, x_n) - f\left(t, \frac{l}{2}, \ldots, \frac{l}{2}\right)\right] \zeta\left(\frac{x_1}{l}\right) \cdots \zeta\left(\frac{x_n}{l}\right) . \qquad (50)$$

   Let f be a continuous function possessing bounded derivatives with respect $x_1, \ldots, x_n$ in a bounded domain of the $(t, x_1, \ldots, x_n)$-space containing the point P. Then, for $l \to 0$, the least upper bound for the absolute value of any k-th order $(0 < k \leq m)$ partial derivative of $\bar{f}$ multiplied by l^k converges to zero.

Proof. Each k-th order derivative of the second term in the right-hand side of (50), with respect to a combination of $x_1, \ldots, x_n$, is equal to a sum of products, each product consisting of a k_1-th order derivative of

$$f(t, x_1, \ldots, x_n) - f\left(t, \frac{l}{2}, \ldots, \frac{l}{2}\right) \qquad (51)$$

and a $(k - k_1)$-th order derivative of $\zeta(x_1/l) \cdots \zeta(x_n/l)$, where $0 \leq k_1 \leq k$. In each product, we multiply the first factor by l^{k_1} and the second one by l^{k-k_1}, so that the product itself is multiplied by l^k. For $k_2 = k - k_1$, any

k_2-th order derivative (with respect to a combination of x_p) of the function $\zeta(x_1/l)\cdots\zeta(x_n/l)$ multiplied by l^{k_2} always remains bounded. A product of l^{k_1} by any of the k_1-th order derivatives (with respect to a combination of x_p) of the function (51) tends to zero as $l \to 0$, for $k_1 = 0$, as well as for $k_1 > 0$. In the case $k_1 = 0$, this is due to the fact that, because of the continuity of $f(t, x_1, \ldots, x_n)$, the maximum of the absolute value of (51) converges to zero. In the case $k_1 > 0$, we see that the derivatives of $f(t, x_1, \ldots, x_n)$ with respect to various x_p are independent of l, and therefore, being multiplied by l^{k_1}, tend to zero. Thereby our last assertion is proved.

On the basis of this lemma, we can construct, in a neighborhood of every point P of the domain where the given system is hyperbolic, a new system which coincides with the given one near that point, and also satisfies the conditions that have been shown above to guarantee the estimate (49). To this end, letting P be the center of cube Q_l, we construct in Q_l the functions $\bar{a}_{ij}^{(k)}$, $\bar{b}_{ij}$, $\bar{f}$ corresponding to $a_{ij}^{(k)}$, b_{ij}, f, as indicated in the above lemma. Denote by $(\overline{14})$ the system obtained from (14), if $a_{ij}^{(k)}$, b_{ij}, f are replaced with $\bar{a}_{ij}^{(k)}$, $\bar{b}_{ij}$, $\bar{f}$. Since the original system is hyperbolic, system $(\overline{14})$ will also be hyperbolic for sufficiently small l. Indeed, for sufficiently small l, the non-vanishing coefficients in $(\overline{14})$ are close to the corresponding coefficients of system (14), whereas the identically vanishing coefficients of (14) have zero counterparts in $(\overline{14})$. Therefore the roots of the determinant of each matrix M_r (see (4)) for equations $(\overline{14})$ are close to the roots of the determinant of the corresponding matrix M_r associated with system (14). Thus, for every such determinant in $(\overline{14})$, the roots are mutually different if this is the case for the roots of the determinants in (14). It follows that for sufficiently small l, the denominators of the fractional expressions for the functions $k_{is}(t, x, \alpha)$, as well as for $k_{is}(t, x, \alpha, \beta)$ in the case of system $(\overline{14})$, are close to the denominators of the fractions representing $k_{is}(t, x, \alpha)$ and $k_{is}(t, x, \alpha, \beta)$, respectively, in the case of system (14) (cf. formulas (**)). Therefore, we can assume all these denominators to be larger than a positive constant. The numerators of these fractions are polynomials with arguments $\bar{a}_{ij}^{(k)}$. Thus, according to our lemma, conditions (37) will be satisfied for system $(\overline{14})$, if l is sufficiently small. Therefore, in a cylinder going along the t-axis and having Q_l as its base, all the solutions of $(\overline{14})$ allowing for a continuous l-periodic extension satisfy the inequality (49). It is obvious that the condition of l-periodicity of the coefficients and their derivatives in the cylinder, under which the estimate (49) has been proved, is also valid for system $(\overline{14})$.

§3. An Auxiliary Result from the Analytic Theory of Differential Equations[14]

Consider the following system of equations

$$\frac{\partial u_i}{\partial t} = \sum_{j,k} a_{ij}^{(k)} \frac{\partial u_j}{\partial x_k} + \sum_j b_{ij} u_j + f_i \qquad (i,j = 1,2,\dots,N) . \tag{52}$$

Assume that the coefficients and every f_i are entire analytic functions[15] with respect to all arguments t, x_k.

Proposition. *Let R be a rectilinear cylinder whose top and bottom belong to the planes $t = T$ and $t = 0$, respectively. Then for any plane E ($t = $ const) having a non-empty intersection with R and any entire analytic initial functions, the Cauchy problem for (52) admits an analytic solution in a layer of width δ around E, where $\delta > 0$ is a constant depending only on $a_{ij}^{(k)}, b_{ij}$ and the cylinder R.*

Proof. As usual, we reduce our problem to that with zero initial functions; as a result, we get the same system with modified f_i. Let

$$G\left(x_1 + \cdots + x_n + \frac{t}{\rho} \right)$$

be a common majorant for $a_{ij}^{(k)}$ and b_{ij} in cylinder R; and let

$$H\left(x_1 + \cdots + x_n + \frac{t}{\rho} \right)$$

be a majorant for f_i. Assume that the radius of convergence for $G(h)$ and $H(h)$ is larger than 1. It is easy to see that $G(h)$ can be chosen such as to be independent of the position of the origin inside R. Then the system

$$\frac{\partial v_i}{\partial t} = \left(\sum_{j,k} \frac{\partial v_j}{\partial x_k} + \sum_j v_j \right) G + H \qquad (i,j = 1,2\dots,N) \tag{53}$$

can be shown to majorize system (52).

It has been shown elsewhere that in order to prove, in a neighborhood $\mathcal{G}$ of the origin, the existence of a solution of the Cauchy problem for system (52) with zero initial functions, it suffices to establish for system (53) the

---

[14]This result, in its formulation and proof, bears almost complete resemblance to a theorem concerning second order hyperbolic equations proved by Schauder [7], pp. 229–231].

[15]As in the case considered by J. Schauder, this result can be proved only under the assumption that these functions are holomorphic in a sufficiently large domain.

existence of an analytic solution which can be expanded in a series with non-negative coefficients. In particular, it suffices to construct a solution which depends on $h = x_1 + x_2 + \cdots + x_n + t/\rho$ only, and consists of functions identical to one another. Each of the latter functions must satisfy the equation

$$\frac{1}{\rho}\frac{dv}{dh} = GnN\frac{dv}{dh} + GNv + H \ ,$$

or

$$\left(\frac{1}{\rho} - GnN\right)\frac{dv}{dh} = GNv + H \ . \tag{54}$$

Set

$$\mu = \max_{|\xi| \le 1}|G(\xi)|, \qquad \rho = \frac{1}{2}\mu nN \ .$$

We limit ourselves to the case of h with $|h| \le 1$. Then

$$\frac{dv}{dh} = F_1(h)v + F_2(h) \ , \tag{55}$$

where the functions $F_1(h)$ and $F_2(h)$, for $|h| < 1$, can be expanded in absolutely convergent series of powers of h with positive coefficients. Hence, it is easy to conclude that the solution of (55) that vanishes for $h = 0$ can also be expanded in a series with positive coefficients by the powers of h, the series being absolutely convergent for $|h| < 1$.

This implies the existence of a solution of the Cauchy problem for system (52) in any domain which belongs to the interior of R and whose points satisfy the inequality

$$\sum_{i=1}^{n}|x_i - x_i^{(0)}| + \rho^{-1}|t - t^0| < 1 \ ,$$

for any $(t^0, x_1^0, \ldots, x_n^0)$ belonging to the intersection of the cylinder R with the plane E. Therefore, an analytic solution exists in a layer of width ρ around the plane E. This proves our statement.

Remark. Let the coefficients of equations (52), the functions f_i, and the initial functions be l-periodic with respect to all x_k. Then the uniqueness of an analytic solution of the Cauchy problem under consideration implies that the functions u_i constructed above are also l-periodic with respect to all x_k.

§4. Proof of the Existence of a Global Solution in Cylinder R for the Cauchy Problem in the Case of a Linear System with Periodic Entire Coefficients

In this section we consider the following linear system

$$\frac{\partial u_i}{\partial t} = \sum_{j,k} a_{ij}^{(k)} \frac{\partial u_j}{\partial x_k} + \sum_j b_{ij} u_j + f_i \qquad (i = 1, 2, \ldots, N) \qquad (56)$$

with periodic entire coefficients and infinitely differentiable periodic initial functions.

We assume that $a_{ij}^{(k)}$, b_{ij}, and f_i are defined in a cylinder

$$R : 0 \leq x_k \leq l, \ \ 0 \leq t \leq T,$$

and can be represented in the form of trigonometric polynomials of the arguments $2\pi x_k/l$, the coefficients of the polynomials being entire analytic functions of t. Moreover, the given system is assumed to satisfy the condition (37') in the entire R. We also assume that the initial functions possess all derivatives everywhere in Q_l (the base of cylinder R), and that the initial functions and their derivatives admit continuous l-periodic extension with respect to all x_k. According to the result of §3, for all initial functions u_i having the form of trigonometric polynomials of arguments $2\pi x_k/l$, there exist solutions of system (56) analytic in a layer R_1 of width δ adjacent to the cylinder's bottom on the plane $t = 0$. In particular, for a sequence of polynomials $M_i^{(m)}$, consider the sequence of corresponding solutions $u_i^{(m)}$ for this system. Assume that these polynomials, together with their derivatives up to the order $(n + M + 1)$, converge to the given initial functions u_i and their respective derivatives, as $m \to \infty$. According to the remark at the end of §3, each function $u_i^{(m)}$ admits a periodic extension with respect to all x_k, the extension being continuous, together with all its partial derivatives. Using the inequality (49), we find that all integrals

$$\int_{R_1} \sum_i u_i^{(m)2} \, dt \, dx$$

are uniformly bounded.

Let us differentiate each equation in (56) k times ($k \leq n + M + 1$) with respect to some of the variables $t, x_1, \ldots, x_n$. Using the same symbol D to denote various derivatives, we can express the result of this differentiation in the form

$$\frac{\partial D u_i}{\partial t} = \sum_{j,k} a_{ij}^{(k)} \frac{\partial D u_j}{\partial x_k} + \sum_{j,k} D a_{ij}^{(k)} D u_j + \sum_{j,k} D a_{ij}^{(k)} u_j +$$

$$+ \sum_j u_j D b_{ij} + \sum_j b_{ij} D u_j + D f_i. \qquad (57)$$

It is easy to see that equations (57) and (56), considered jointly, again form a hyperbolic system, this time, with respect to the functions u_i and their various derivatives up to the order $(n + M + 1)$. In this case, the λ-matrix associated with the new system is split into separate λ-matrices: some of them coincide with the λ-matrix for the original system, and several other matrices consist of a single element λ. Therefore, the condition $(37')$ is satisfied for the new system, provided that it holds for the original one. Now, if we repeat, with respect to the new system, the considerations used for system (56), we find that the property of being uniformly bounded applies not only to the integrals over the layer R_1 of the sums of squared functions $u_i^{(m)}$, but also to the integrals of the sums of all their squared derivatives up to the order $(n + M + 1)$. Hence, on the basis of the theorem established by Courant, Friedrichs & Lewy [8], we deduce that the sequence $u_i^{(m)}$ ($i = 1, \ldots, N$, $m = 1, 2, \ldots$) contains a subsequence formed by functions which, together with their partial derivatives up to the order M, uniformly converge in R_1.

It is clear that the limit functions $\bar{u}_i$ satisfy system (56) in R_1, together with the functions forming the said subsequence; the limit functions also admit continuous l-periodic extensions with respect to all x_k. Moreover, for $t = 0$ they satisfy the given initial conditions. The fact that M can be chosen arbitrarily large implies that the functions $\bar{u}_i$ constructed above possess derivatives of any order. The functions $\bar{u}_i$ corresponding to different M must coincide on R_1. Indeed, assuming the contrary, we obtain a homogeneous system (56) with a non-trivial solution which is l-periodic in all x_k and satisfies zero initial conditions. But this is impossible, since otherwise the inequality (49) would take the form

$$\int_{Q_l} \sum_i u_i^2(t, x)\, \mathrm{d}x \le e^{\psi_1 t} \psi_5 \int_{Q_l} \sum_i u_i^2(0, x)\, \mathrm{d}x \,,$$

which, together with the relation

$$\int_{Q_l} \sum_i u_i^2(0, x)\, \mathrm{d}x = 0 \,,$$

implies that for all t we have

$$\int_{Q_l} \sum_i u_i^2(t, x)\, \mathrm{d}x = 0 \,.$$

Having thus obtained the values of the functions u_i at the top of the cylinder, let us apply the arguments used for R_1 to the cylinder $R_2 : 0 \le x_k \le l$, $\delta \le t \le 2t$ in order to find the values of u_i in R_2. We can proceed in this way to find u_i in R_3, etc., until we arrive at the top end of the cylinder R.

§5. Proof of the Existence of a Solution for the Cauchy Problem in a Sufficiently Small Domain for an Arbitrary Linear Hyperbolic System with Sufficiently Smooth Coefficients

In this section we consider the following linear hyperbolic system:

$$\frac{\partial u_i}{\partial t} = \sum_{j,k} a_{ij}^{(k)} \frac{\partial u_j}{\partial x_k} + \sum_j b_{ij} u_j + f_i \qquad (i = 1, 2, \ldots, N), \qquad (58)$$

assuming the coefficients and the initial functions to be sufficiently smooth.

The above system is given in a cylinder R with top and bottom belonging to the planes $t = T > 0$ and $t = 0$, respectively, and generatrix parallel to the t-axis. We assume that the coefficients $a_{ij}^{(k)}$ and b_{ij} have continuous partial derivatives up to the order $(3n + 2)$ with respect to all x_k and[16] t, and the functions f_i have continuous partial derivatives up to the order $(n + 2)$. Let P be an interior point of the domain forming the bottom of R. We construct a cube Q_l with center at P whose side length l is so small that the system

$$\frac{\partial \bar{u}_i}{\partial t} = \sum_{j,k} \bar{a}_{ij}^{(k)} \frac{\partial \bar{u}_j}{\partial x_k} + \sum_j \bar{b}_{ij} \bar{u}_j + \bar{f}_i \qquad (i, j = 1, 2, \ldots, N) \qquad (\overline{58})$$

satisfies the condition (37′) in R_l, where the functions $\bar{a}_{ij}^{(k)}, \bar{b}_{ij}, \bar{f}_i$ are constructed from $a_{ij}^{(k)}, b_{ij}, f_i$ in the same way as $\bar{f}(t, x_1, \ldots, x_n)$ was constructed from $f(t, x_1, \ldots, x_n)$ in the Lemma of §2; R_l is the cylinder having Q_l as its bottom and the length of its generatrix parallel to the t-axis equal to T. In exactly the same way, we construct $\bar{\varphi}_i(x_1, \ldots, x_n)$ from the initial values φ_i of the functions u_i. For system $(\overline{58})$ obtained in this fashion, we prove the existence of a solution for the corresponding Cauchy problem. The solutions $\bar{u}_i$ of this problem satisfy equations (58) in a cylinder R' inside R_l wherein systems $(\overline{58})$ and (58) coincide.

Consider a sequence of systems $(58_1), (58_2), \ldots, (58_p) \ldots$, of the same type as system $(\overline{58})$[17], but having as their coefficients $a_{ij,p}^{(k)}, b_{ij,p}$ and functions $f_{i,p}$ trigonometric polynomials with arguments $2\pi x_k/l$. The coefficients of these polynomials are analytic in t and, together with their derivatives in x_k and t up to the order $(3n + 2)$ and $(n + 2)$, respectively, uniformly converge, as $p \to \infty$, to the respective functions $\bar{a}_{ij}^{(k)}, \bar{b}_{ij}, \bar{f}_i$ (associated with system $(\overline{58})$) and their derivatives. Moreover, consider a sequence $\varphi_{i,p}(x_1, \ldots, x_n)$ $(i = 1, 2, \ldots, n; p = 1, 2, \ldots)$ of trigonometric polynomials in $2\pi x_k/l$ which, together with their derivatives up to the order $(n + 2)$,

---

[16]It suffices to assume that there is only one differentiation with respect to t.

[17]i.e., the characteristic matrices for these systems are split into matrices M_r, same in number and of the same order as for system (58).

uniformly converge to the functions $\bar{\varphi}_i(x_1, \ldots, x_n)$ and their partial derivatives, respectively (for instance, we can consider the sequence of Fejer's polynomials constructed for the functions $\varphi_i(x_1, \ldots, x_n)$).

For all sufficiently large p, systems (58_p) satisfy the conditions $(37')$. For all l-periodic solutions of system (58_p) the inequalities (49) hold *with the same constants*. Indeed, these constants depend on the values of the coefficients $a_{ij,p}^{(k)}$, on the coefficients $k_{ij,p}(t, x, \alpha)$ defining the transformation (16), and their derivatives with respect to α_s, and on the derivatives of all these functions up to the order $3n$ with respect various combinations of x_k. Moreover, these constants depend on $\partial a_{ij,p}^{(k)}/\partial t$, $\partial k_{ij,p}/\partial t$, $b_{ij,p}(t, x)$, the coefficients $k_{ij,p}(t, x, \alpha, \beta)$ defined above (see §2), and the derivatives of all these functions up to the order $2n$ with respect to various combinations of x_k. By construction, the coefficients of systems (58_p), together with all their derivatives under consideration, uniformly converge, as $p \to \infty$, to the respective coefficients of system $(\overline{58})$ and their derivatives. The coefficients $k_{ij,p}(t, x, \alpha, \beta)$, expressed in terms of α, can be explicitly written out as linear combinations of the coefficients of the transformation (16); whereas, expressed in terms of β, they can be represented as linear combinations of the coefficients of the inverse transformation to (16). However, the relations $(**)$ (see §2) imply that all coefficients $k_{ij,p}(t, x, \alpha)$, and therefore $k_{ij,p}(t, x, \alpha, \beta)$ and all their derivatives under consideration, uniformly converge, as $p \to \infty$, to the respective $k_{ij}(t, x, \alpha)$, $k_{ij}(t, x, \alpha, \beta)$ and their derivatives associated with system $(\overline{58})$. It follows that, for all sufficiently large p, all l-periodic solutions of system (58_p) satisfy the inequalities of type (49) with the same constants.

The results of the preceding sections imply that for every sufficiently large p there exists a solution $u_{i,p}$ ($i = 1, \ldots, N$; $p = 1, 2, \ldots$) of system (58_p) which is infinitely differentiable in the entire cylinder R_l, and has the polynomials $\varphi_{i,p}$ as its initial values at $t = 0$. By assumption, the coefficients of system (58_p) and the functions $\varphi_{i,p}$, together with their derivatives, converge to the respective coefficients of system $(\overline{58})$ and the functions $\bar{\varphi}_i$. This fact allows us to make the conclusion (in exactly the same way as in the preceding section) that the inequalities (49) are valid not only for the functions $u_{i,p}$ (i.e., the solutions of the Cauchy problem for system (58_p)), but also for all their derivatives in x_k and t up to the order $(n + 2)$. Therefore, from the sequence of the systems of functions $u_{i,p}$ we can extract a subsequence which converges uniformly on R_l, together with their first order partial derivatives. The limit functions turn out to represent a solution of the Cauchy problem for system $(\overline{58})$.

Remark I. In order to show that the set of functions $u_{i,p}$ contains a uniformly convergent sequence, it suffices to verify the following condition: all squared functions $u_{i,p}$ themselves and their squared derivatives up to the order $(n+1)$ of the form

$$\frac{\partial^k u_{i,p}}{\partial x_{i_1} \dots \partial x_{i_k}} \quad \text{and} \quad \frac{\partial^{k+1} u_{i,p}}{\partial t \partial x_{i_1} \dots \partial x_{i_k}},$$

where $i_1, i_2, \dots, i_k$ are different from one another, have uniformly bounded integrals (see Courant, Friedrichs & Lewy [8, pp. 51–52]).

This remark is essential for what follows. On its basis, even in this section, we could assume the coefficients of the system to have a moderate number of derivatives with respect to t.

Remark II. If the coefficients of system (58), the functions f_i, and the initial values possess continuous derivatives up to the order $(3n + 2 + M)$ and $(n + 2 + M)$, respectively, then the solution constructed above can be claimed to have continuous partial derivatives up to the order M. The proof of this result is almost identical to the proof of a similar statement in §4.

Remark III. Quite recently, Sobolev [9] has shown that a set of functions of n variables contains a uniformly convergent sequence, if the integrals of the squares of these functions and their partial derivatives up to the order $\left[\frac{n}{2}\right] + 1$ are uniformly bounded. This fact allows us to obtain the results of the present section assuming only that the coefficients $a_{ij}^{(k)}$ have continuous partial derivatives up to the order $\max\left\{3n, 2n + 3 + \left[\frac{n+1}{2}\right]\right\}$, the coefficients b_{ij} are continuously differentiable up to the order $2n + \left[\frac{n+1}{2}\right] + 2$, while f_i and φ_i up to the order $\left[\frac{n+1}{2}\right] + 2$.

§6. Uniqueness

A uniqueness theorem for the Cauchy problem (in the space of real variables) for a linear system of the form (58) with analytic coefficients was first proved by Holmgren [10] in 1901. For a linear system with coefficients continuously differentiable up to any order, but not analytic ones, this theorem has remained without proof until the present day. As shown by Hadamard [11, Remark 1], the problem of establishing the uniqueness theorem for nonlinear equations can be reduced to the case of *linear* equations with non-analytic coefficients. For linear hyperbolic systems this problem can be solved on the basis of the method used by Holmgren. The proof of this result is given below. For the sake of simplicity, we assume that $n = 1$, although the same method is applicable in the case of any positive integer n. Moreover, it is clear that we can limit our proof to the case of a homogeneous system.

Let us write the homogeneous hyperbolic system in the form

$$\frac{\partial u_i}{\partial t} - \sum_{j=1}^{N} a_{ij} \frac{\partial u_j}{\partial x} - \sum_{j=1}^{N} b_{ij} u_j = 0 \qquad (i = 1, 2, \dots, N) . \qquad (59)$$

The coefficients of this system are assumed to be defined in a domain containing a segment AB on the Ox-axis, with the origin at the middle point of AB. Assume that all functions u_i vanish on this segment, and that a_{ij} and b_{ij} possess continuous derivatives up to the orders

$$\max \left\{ 3n, 2n + \left[\frac{n+1}{2} \right] + 3 \right\} \quad \text{and} \quad 2n + \left[\frac{n+1}{2} \right] + 2 ,$$

respectively.

Let us change the variables, setting

$$x_1 = x ,$$
$$t_1 = t + x^2 .$$

Then the segment AB will be transformed into the arc $A_1 B_1$ of a parabola in the (t_1, x_1)-plane. In a neighborhood of the new origin, the system obtained from (59) after the above transformation can also be resolved with respect to $\partial u_i / \partial t$. Let the new system have the form

$$F_i(u) = \frac{\partial u_i}{\partial t_1} - \sum_j a_{ij}^* \frac{\partial u_j}{\partial x_1} - \sum_j b_{ij}^* u_j = 0 . \qquad (59')$$

Clearly, the new system is hyperbolic in the same domain as the original one. Indeed, its matrix (3) splits into the same matrices M_r as the matrix (3) for the original system. Each matrix M_r for the transformed system (59') is obtained from the corresponding M_r for the original system by a linear transformation of the variables α, λ.

Let v_i ($i = 1, \dots, N$) be arbitrary differentiable functions. Integrating the sum $\sum_i v_i F_i(u)$ over an arbitrary domain $\mathcal{G}$, we obtain

$$\iint_{\mathcal{G}} \sum_i v_i F_i(u) \, dt_1 \, dx_1 = - \iint_{\mathcal{G}} \sum_i u_i G_i(v) \, dt_1 \, dx_1 +$$

$$+ \int_L \left(\sum_i u_i v_i \, dx_1 - \sum_{i,j} a_{ij}^* u_j v_i \, dt_1 \right) , \qquad (60)$$

where the curve L is the boundary of the domain $\mathcal{G}$, and

$$G_i(v) \equiv \frac{\partial v_i}{\partial t_1} - \sum_j a_{ij}^* \frac{\partial v_i}{\partial x_1} - \sum_j \left(\frac{\partial a_{ij}^*}{\partial x_1} - b_{ij}^* \right) v_j . \qquad (61)$$

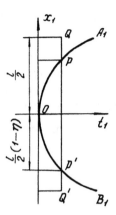

Fig. 1.

Let us take a segment of length l on the Ox_1-axis, having the origin as its middle point. In a rectangular domain R_l lying in the half-plane $t_1 > 0$ and having the said segment as its base, consider the following system

$$\frac{\partial v_i}{\partial t} = \sum_j \bar{a}^*_{ij} \frac{\partial v_j}{\partial x_1} + \sum_j \left(\frac{\partial \bar{a}^*_{ij}}{\partial x_1} - \bar{b}^*_{ij} \right) v_j , \qquad (62)$$

where the new coefficients are obtained from the respective coefficients a^*_{ij}, b^*_{ij} by the procedure described in the Lemma of §2. If l is small enough, then system (62) is also of hyperbolic type and satisfies the conditions (37) in the entire rectangular domain R_l. Let us take as $\mathcal{G}$ in (60) a domain bounded by the arc $A_1 B_1$ of the parabola $t_1 = x_1^2$ and a segment PP' parallel to the Ox_1-axis and lying so close to it that the absolute values of the x_1-coordinates of the points P and P' are not larger than $l(1 - \eta)/2$ (here η is the parameter involved in the operations described in the Lemma of §2; see Fig. 1). According to the results established in the preceding section, there exists, in the rectangle R_l, a solution of system (62) which takes any given sufficiently smooth values on the segment QQ'. These values, together with their partial derivatives in x_1 of a sufficiently high order, can be extended as l-periodic functions of x_1. Let us take as v_i the solution of system (62) which, on the segment PP', has its values sufficiently close to those of the corresponding functions u_i. Then (60) takes the form

$$0 = \int_{PP'} \sum_i u_i v_i \, dx_1 ,$$

which is impossible because of the closeness of v_i to u_i, unless u_i vanish identically on the segment PP'. By virtue of the arbitrary choice of the segment PP' in the vicinity of the Ox_1-axis, we have $u_i \equiv 0$ in the domain $\mathcal{G}$, q.e.d.

The above construction with respect to the middle point of the segment AB can be reproduced in relation to any of its interior points. Thereby we get a domain in which the values of the functions u_i satisfying system (59) are determined by their values on the segment AB.

Now we turn to the proof of the uniqueness of solutions of the Cauchy problem for nonlinear equations. We can limit ourselves to the case of quasi-linear systems, since, as we have shown in §1, the Cauchy problem for any given nonlinear system can be reduced to the problem for a quasilinear one. Assume that for $t \geq 0$ the functions $\bar{u}_i$ and $\bar{\bar{u}}_i$ $(i = 1, 2, \ldots, N)$ satisfy the following system

$$\frac{\partial u_i}{\partial t} = \sum_{j,k} a_{ij}^{(k)}(t, x, u) \frac{\partial u_j}{\partial x_k} + f_i(t, x, u) \qquad (i = 1, 2, \ldots, N) \qquad (63)$$

and take equal initial values at $t = 0$. Let us replace u_i in (63), first by $\bar{\bar{u}}_i$, then by $\bar{u}_i$, and subtract the equations of the second system from the corresponding equations of the first one. Setting $\bar{\bar{u}}_i - \bar{u}_i = u_i$, we get

$$\frac{\partial u_i}{\partial t} = \sum_{j,k} a_{ij}^{(k)}(t, x, \bar{\bar{u}}) \frac{\partial u_j}{\partial x_k} + \sum_{j,k} \left[a_{ij}^{(k)}(t, x, \bar{\bar{u}}) - a_{ij}^{(k)}(t, x, \bar{u}) \right] \frac{\partial \bar{u}_j}{\partial x_k} +$$

$$+ \left[f_i(t, x, \bar{\bar{u}}) - f_i(t, x, \bar{u}) \right] \qquad (i, j = 1, 2, \ldots N) .$$

J. Hadamard, in his remark quoted above, has shown that

$$F(\bar{\bar{z}}_1, \ldots, \bar{\bar{z}}_n) - F(\bar{z}_1, \ldots, \bar{z}_n) =$$

$$= \sum_{k=1}^{n} \bar{\Phi}_k(\bar{\bar{z}}_1, \ldots, \bar{\bar{z}}_n, \bar{z}_1, \ldots, \bar{z}_n)(\bar{\bar{z}}_k - \bar{z}_k) ,$$

where the functions $\bar{\Phi}_k$ possess continuous derivatives up to the order $(p-1)$, if $F(z_1, \ldots, z_n)$ has continuous derivatives up to the order p with respect to all its arguments. This implies that the functions u_i with zero initial values at $t = 0$ satisfy, for $t > 0$, a homogeneous linear system of the form

$$\frac{\partial u_i}{\partial t} = \sum_{j,k} a_{ij}^{(k)}(t, x, \bar{\bar{u}}) \frac{\partial u_j}{\partial x_k} + \sum_{j} b_{ij} \left(t, x, \bar{u}, \bar{\bar{u}}, \frac{\partial \bar{u}}{\partial x_k} \right) u_j .$$

This system is hyperbolic, provided that system (63) is hyperbolic, and therefore the question of the uniqueness of a solution of the Cauchy problem for this system has already got its answer. Hence $\bar{\bar{u}}_i \equiv \bar{u}_i$ in a neighborhood of any domain on the plane $t = 0$ where $\bar{\bar{u}}_i = u_i$; the functions $a_{ij}^{(k)}, f_i, \bar{u}_i, \bar{\bar{u}}_i$ being differentiable in all their arguments

$$\max\left\{3n, 2n + \left[\tfrac{n+1}{2}\right] + 3\right\}, \quad 2n + \left[\tfrac{n+1}{2}\right] + 3, \quad \max\left\{3n, 2n + \left[\tfrac{n+1}{2}\right] + 3\right\}$$

times, respectively (see Remark III of §5). If merely the result of Courant, Friedrichs & Lewy [8] were used here, then we should have written n instead of $[(n + 1)/2]$.

§7. Proof of the Existence of a Solution of the Cauchy Problem for a Quasilinear System with Polynomial Coefficients and Polynomial Initial Functions

In this section we consider a system of the form

$$\frac{\partial u_i}{\partial t} = \sum_{j,k} a_{ij}^{(k)} \frac{\partial u_j}{\partial x_k} + f_i \qquad (i, j = 1, 2, \dots, N), \tag{64}$$

where $a_{ij}^{(k)}$ and f_i are trigonometric polynomials with arguments $2\pi x_k/l$ and coefficients having the form of polynomials of the variables t and u_i. We take the initial functions equal to zero.

It is assumed that system (64) remains hyperbolic if the functions u_i, in their absolute value, remain bounded by a positive constant M. Moreover, conditions (37′) should be satisfied for this system, provided that the functions u_i and their derivatives in all x_k up to the order $3n$ have absolute values bounded by M, and these functions admit l-periodic extension in all x_k. When finding the derivatives of $a_{ij}^{(k)}$ in x_p, we should take into account not only their explicit dependence on x_p but also the indirect dependence on x_p through the functions u_j.

Consider a cylinder R of height T, $0 \le t \le T$, having $Q_l : 0 \le x_k \le l$ as its bottom. Let us divide R into cylindrical layers $R_1^m, R_2^m, \dots$ by the planes $E_p : t = \frac{p}{m}, p = 1, 2, \dots, \left[\frac{T}{m}\right]$. Let us substitute for the functions u_i in $a_{ij}^{(k)}$ and f_i their given initial values (in our case, zeroes). Next, for the system obtained in this way we solve the Cauchy problem in cylinder R_1^m with the initial values prescribed for u_i at $t = 0$. Thus we get the values of these functions on the top of the cylinder R_1^m.

In cylinder R_2^m, we replace u_i in $a_{ij}^{(k)}$ and f_i by the functions

$$0 + \frac{t - \frac{1}{m}}{\frac{1}{m}} \left[u_i \left(\frac{1}{m}, x_1, \dots, x_n \right) - 0 \right].$$

For the linear system obtained in this way we solve the Cauchy problem, taking as the initial functions for u_i on the bottom of R_2^m the functions obtained on the top of R_1^m. We thus find the values of u_i on the top of R_2^m.

In cylinder R_3^m, we replace u_i in $a_{ij}^{(k)}$ and f_i by the functions

$$u_i \left(\frac{1}{m}, x \right) + \frac{t - \frac{2}{m}}{\frac{1}{m}} \left[u_i \left(\frac{2}{m}, x \right) - u_i \left(\frac{1}{m}, x \right) \right] .$$

For the system thus obtained we solve the Cauchy problem, taking as the initial values of u_i on the bottom of R_3^m the values of u_i obtained previously at the top of R_2^m, etc. We proceed in this manner until we either fill up the entire cylinder R, or find ourselves outside the domain where the system is hyperbolic and the condition (37') is satisfied. Denote the resulting functions by $u_i^{(m)}$, and the system they satisfy by (64_m). It is clear that every $u_i^{(m)}$ possesses derivatives of any order with respect to all its arguments[18]; all these functions and their derivatives admit continuous l-periodic extensions in all x_k.

It will be shown that all $u_i^{(m)}$ are defined in a certain fixed part of cylinder R; and, as $m \to \infty$, the functions $u_i^{(m)}$ $(i = 1, 2, \ldots, N)$ uniformly converge on that part of R, together with their first order partial derivatives, to functions u_i which possess derivatives of any order and satisfy the given system (64) and the prescribed initial conditions.

Let us apply, $(4n + 2)$ times, the procedure described in §1 to system (64_m). For the derivatives of $u_i^{(m)}$ in x_k up to the order $(3n+1)$, we introduce the following notations:

$$u_{N+1}^{(m)}, \ u_{N+2}^{(m)}, \ \ldots, \ u_{N_1}^{(m)} \ ;$$

and we denote the rest of the derivatives of $u_i^{(m)}$ in x_k by $v_1^{(m)}, \ldots, v_{N_2}^{(m)}$. Denote the system obtained in this way by $(64_m^{(4n+2)})$. The values to be substituted into $a_{ij}^{(k)}$ and f_i for u_i are denoted by $u_i^{*(m)}$ $(i = 1, \ldots, N)$. Their derivatives up to the $(3n + 1)$-th order and from the $(3n + 2)$-th to the $(4n+2)$-th order, with respect to various combinations of x_k, we denote by $u_i^{*(m)}$ $(i = N + 1, \ldots, N_1)$ and $v_i^{*(m)}$ $(i = 1, 2, \ldots, N_2)$, respectively. The coefficients by $\partial v_j^{(m)}/\partial x_k$ in $(64_m^{(4n+2)})$ are equal to $a_{ij}^{(k)}$ of system (64_m). The coefficients by $v_j^{(m)}$ and $v_j^{*(m)}$ in system $(64_m^{(4n+2)})$ depend on the derivatives of the coefficients of (64_m) up to the order $(n + 1)$. Indeed, every equation of system $(64_m^{(4n+2)})$ is obtained from an equation (64_m) by differentiating the latter $(4n + 2)$ times, while $v_j^{(m)}$ and $v_j^{*(m)}$ coincide with derivatives of $u_j^{(m)}$ and $u_j^{*(m)}$ $(j \leq N)$ of an order higher than $(3n + 1)$.

In cylinder R_1 we have

$$\frac{\partial u_i^{*(m)}}{\partial t} = 0 .$$

In cylinder R_p^m, for $p > 1$, we have

---

[18]With the exception of the planes $t = p/m$, where only the first derivatives in t should exist.

$$\frac{\partial u_i^{*(m)}}{\partial t} = \frac{u_i^{(m)}\left(\frac{(p-1)}{m}, x_1, \ldots, x_n\right) - u_i^{(m)}\left(\frac{(p-2)}{m}, x_1, \ldots, x_n\right)}{\frac{1}{m}}.$$

For $i \leq N$, the functions $\partial u_i^{(m)}/\partial t$ and their derivatives (with respect to various combinations of x_k) of order less than $2n$, depend on t, x_k and $u_i^{(m)}$ ($i = 1, \ldots, N_1$); therefore, if $|u_i^{(m)}| < M$ ($i = 1, \ldots, N_1$), their absolute values remain smaller than a positive constant C which depends on M only. It follows that for $i \leq N$ the absolute values of $\partial u_i^{*(m)}/\partial t$ and of all their partial derivatives (with respect to various combinations of x_k) up to the order $2n$ also remain bounded by C.

So long as we remain within the domain where all $u_i^{(m)}$ and $u_i^{*(m)}$ ($i = 1, \ldots, N_1$) do not exceed the constant M in their absolute value, an inequality of type (49) will hold for system $(64_m^{(4n+1)})$; and in our case this inequality can be written in the form

$$\int_{Q_l} \sum_{i=1}^{N} v_i^{(m)2}(t, x)\, dx \leq e^{C_1 T} \int_0^T \int_{Q_l} \left[C_2 + C_3 \sum_{i=1}^{N_2} v_i^{*(m)2}(\tau, x) \right] dx\, d\tau . \quad (65)$$

Indeed, because of our choice of the initial functions, we have

$$\sum_{i=1}^{N_1} u_i^{(m)2}(0, x) + \sum_{i=1}^{N_2} v_i^{(m)2}(0, x) \equiv 0 ,$$

and C_1, C_2, C_3 in (65) are positive constants depending only on M, i.e., on the maximum of $|u_i^{*(m)}|$ ($i = 1, 2, \ldots, N_1$). By the definition of $u_i^{*(m)}$ ($i \leq N$) on the interval $I_{p+1} : \frac{p}{m} \leq t \leq \frac{(p+1)}{m}$, we have

$$v_i^{*(m)}(t, x) = v_i^{(m)}\left(\frac{p-1}{m}, x\right) + \left[v_i^{(m)}\left(\frac{p}{m}, x\right) - v_i^{(m)}\left(\frac{p-1}{m}, x\right) \right] \frac{t - \frac{p}{m}}{\frac{1}{m}} =$$

$$= v_i^{(m)}\left(\frac{p-1}{m}, x\right) \frac{\frac{(p+1)}{m} - t}{\frac{1}{m}} + v_i^{(m)}\left(\frac{p}{m}, x\right) \frac{t - \frac{p}{m}}{\frac{1}{m}} .$$

We assume here that $v_i^{(m)}\left(-\frac{1}{m}, x\right) \equiv v_i^{(m)}(0, x) \equiv 0$. It follows that

$$\left[v_i^{*(m)}(t, x) \right]^2 \leq \left[v_i^{(m)}\left(\frac{p-1}{m}, x\right) \right]^2 + \left[v_i^{(m)}\left(\frac{p}{m}, x\right) \right]^2 .$$

Set

$$\int_{Q_l} \sum_{i=1}^{N_1} \left[v_i^{(m)}(t, x) \right]^2 dx = y(t) ,$$

$$y^*(t) = y\left(\frac{p}{m}\right) \quad \text{for} \quad \frac{p}{m} \leq t \leq \frac{p+1}{m} ,$$

$$y^{**} = y\left(\frac{p-1}{m}\right) \quad \text{for} \quad \frac{p}{m} \leq t \leq \frac{p+1}{m} .$$

Then the inequality (65) can be written in the form

$$y(t) \le e^{C_1 T} \int_0^t \{C_2 + C_3[y^*(\tau) + y^{**}(\tau)]\} \, d\tau \ .$$

Consider a function $z(t)$ which satisfies the equation

$$z(t) = e^{C_1 T} \int_2^t \{C_2 + C_3[z^*(\tau) + z^{**}(\tau)]\} \, d\tau \ ,$$

where $z^*(t)$ and $z^{**}(t)$ are obtained from $z(t)$ in exactly the same way as $y^*(t)$ and $y^{**}(t)$ from $y(t)$. We also observe that

$$y^*\left(-\frac{1}{m}\right) = y^*(0) = 0 \ .$$

Then we obviously have $z(t) \ge y(t)$. Since $C_2 > 0$ and $C_3 > 0$, we see that $z(t)$ is an increasing function of t. Thus

$$z(t) \le e^{C_1 T} \int_0^t \{C_2 + 2C_3 z(\tau)\} \, d\tau \ .$$

Consider now a function $w(t)$ which satisfies the following equation for $t \ge 0$:

$$w(t) = e^{C_1 T} \int_0^t \{C_2 + 2C_3 w(\tau)\} \, d\tau \ ,$$

with the condition $w(0) = 0$. Clearly $w(t) \ge z(t)$. For $t \le T$, we have

$$w(t) = C_2 e^{C_1 T} \int_0^t e^{2C_3(t-\tau)e^{C_1 T}} \, d\tau \le C_2 T e^{C_1 T + 2C_2 e^{C_1 T}} \ .$$

Thus, for $t \le T$, the inequality

$$\int_{Q_l} \sum_{i=1}^{N_1} \left[v_i^{(m)}(t, x) \right]^2 \, dx \le C_2 T e^{C_1 T + 2C_3 e^{C_1 T}} \tag{66}$$

is satisfied as long as absolute values of all $u_i^{(m)}$ $(i = 1, 2, \ldots, N)$ and all their partial derivatives (with respect to x_k) up to the order $(3n+1)$ remain smaller than M. In what follows, the latter condition is referred to as the *M-condition*.

Let us square both sides of every equation $(64_m^{(4n+1)})$. Then, taking into account that the product of the functions $v_i^{(m)}$ and $v_i^{*(m)}$, in absolute value, does not exceed the sum of their squares, and observing that the absolute values of all the coefficients by $v_i^{(m)}$ and $v_i^{*(m)}$ are less than a constant C, if the M-condition is satisfied, we find that the integral over Q_l, and therefore over R, of the sum of squared derivatives in t of the derivatives of $u_i^{(m)}$ $(i = 1, 2, \ldots, N)$ up to the order $(4n + 1)$ (with respect to various combinations of x_k) remains bounded.

It follows that we can fix T_1 such that for all $t < T_1$ the M-condition holds in cylinder R for all functions

$$u_i^{(m)} \qquad (i = 1, 2, \ldots, N; \quad m = 1, 2, \ldots)$$

uniformly with respect to m. Indeed, assume that for an arbitrarily small t there exist a point of cylinder R and a function $u_i^{(m)}$ ($i \leq N$) such that this function itself, or any of its derivatives (in combinations of x_k) of an order $\leq (3n + 1)$, has absolute value at this point greater than M. Then, in cylinder R, we can find an arbitrarily narrow layer adjacent to its bottom where the M-condition is satisfied and one of the functions $u_i^{(m)}$, or any of its derivatives (with respect to a combination of x_k) of an order $\leq (3n + 1)$, has absolute value larger than $M/2$. However, this is incompatible with the fact that the integral over this layer of the sum of squared functions $u_i^{(m)}$ ($i \leq N$) and their derivatives up to the order $(4n + 2)$ (only one differentiation in t may be involved here) remains bounded in absolute value (cf. Remark I, §5).

Thus, there exists a layer R_1 in cylinder R adjacent to its bottom, where all functions $u_i^{(m)}$ ($i \leq N$), together with their partial derivatives in x_k up to the order $(3n + 1)$, are equicontinuous. Therefore, from the sequence $u_i^{(m)}$ ($i \leq N$) constructed above we can extract a subsequence consisting of functions which uniformly converge in R_1 ($0 \leq t \leq T_1$), together with their partial derivatives in x_k up to the order $(3n + 1)$, the limits being continuous functions, and the initial values of the derivatives being the derivatives of the initial values. Taking into account the fact that $\partial u_i^{(m)}/\partial t$ ($i \leq N$) stay always equal to the right-hand sides of equations (64_m), we find that $\partial u_i^{(m)}/\partial t$ uniformly converge in R_1. It follows that $u_i^{(m)}$ ($i \leq N$) uniformly converge in R_1 to a solution of the Cauchy problem. This solution possesses continuous partial derivatives with respect to all x_k up to the order $(3n + 1)$. Differentiating system (64) with respect to t and x_k, we find that the solution obtained possesses continuous derivatives up to the order $(3n+1)$ with respect to t, as well as x_k. The above arguments regarding the functions u_i can also be applied to their derivatives of any order, which shows that u_i are infinitely differentiable.

§8. Existence of Solutions of the Cauchy Problem for a General Quasilinear System

In this section we consider a quasilinear hyperbolic system of the form

$$\frac{\partial u_i}{\partial t} = \sum_{j,k} a_{ij}^{(k)} \frac{\partial u_j}{\partial x_k} + f_i \qquad (i = 1, \ldots, N), \tag{67}$$

making only the following two assumptions:

1) $a_{ij}^{(k)}$ and f_i possess continuous partial derivatives up to the order $(4n+3)$ with respect to all their arguments, i.e., u_j, x_k and t.

2) For all values of u_i whose modulus does not exceed a constant M, system (67) remains hyperbolic in cylinder R such that its generatrix is parallel to the t-axis, its top and bottom belong to the planes $t = T$ and $t = 0$, respectively. The initial functions are equal to zero.

Consider a point P on the bottom of cylinder R. Let us construct a cylinder R_l for this point, as it has been done in §5. Let the bottom Q_l of cylinder R_l be so small that the coefficients $\bar{a}_{ij}^{(k)}$ of the system

$$\frac{\partial u_i}{\partial t} = \sum_{j,k} \bar{a}_{ij}^{(k)} \frac{\partial u_j}{\partial x_k} + \bar{f}_i \qquad (i = 1, 2, \ldots, N) \tag{$\overline{67}$}$$

satisfy the condition (37′) in R_l for all functions u_i which, together with their partial derivatives in x_k up to the order $3n$, admit l-periodic extension with respect to every x_k and have their absolute values $\le M$. The functions $\bar{a}_{ij}^{(k)}$ and $\bar{f}_i$ in ($\overline{67}$) are constructed from $a_{ij}^{(k)}$ and f_i in exactly the same way as $\bar{f}$ was constructed from f in the Lemma of §2, where u_i and t are regarded as parameters independent of x_k.

Let us approximate system ($\overline{67}$) on R_l by systems of type[19] (64), so that the coefficients in the systems of type (64), and their derivatives referred to in the above condition 1), would uniformly approximate the respective coefficients in ($\overline{67}$) and their derivatives, provided that absolute values of u_i are less than M. We denote these systems by $(67_1), (67_2) \ldots, (67_p)$. For sufficiently large p, all these systems are hyperbolic on R. Subjecting these systems to a finite number of operations described in §1, we obtain new systems, for which the conditions (37′) hold uniformly[20], provided that the functions u_i $(i \le N)$ and their partial derivatives in x_k up to the order $3n$, in their absolute value, do not exceed M. Then, according to the preceding section, in cylinder R_1 next to its bottom there is a layer R_{1l} in which all equations (67_p) (for p large enough) admit solutions satisfying the conditions (49) with the same constants. Therefore, the sequence of solutions of the Cauchy problem for systems (67_p) contains a subsequence formed by functions which uniformly converge, together with their derivatives in x_k up to the order $(3n+1)$. Obviously, the limit of this subsequence is a solution of system (67) possessing continuous derivatives in x_k and[21] t up to the order $(3n+1)$. We can also deduce from this that there exists a solution of system (67) in a cylinder R' which belongs to the interior of R. This solution has continuous derivatives up to the order $(3n+1)$. In order to prove its

---

[19]See footnote 17.
[20]since the left-hand side in (37′) remains bounded by a constant smaller than 1.
[21]Cf. the end of §7.

existence, it suffices to assume that $a_{ij}^{(k)}$ and f_i have derivatives up to the order $(4n + 2)$. Under the assumption that $a_{ij}^{(k)}$ and f_i possess continuous partial derivatives up to the order $\max\{4n + 2,\ 3n + [(n + 1)/2] + 4\}$ with respect to all their arguments, we find for system $(\overline{67})$, and consequently, for system (67), a solution with continuous derivatives in t and x_k up to the order $\max\{3n+1,\ 2n+[(n+1)/2]+3$. Next, with a view to establishing the uniqueness of the solution of the Cauchy problem for (67), one can apply the arguments of §6.

Note that for $n > 1$ we have

$$3n + \left[\frac{n+1}{2}\right] + 4 \le 4n + 3 \ .$$

However, in the case $n = 1$, one can prove the uniqueness theorem, assuming that $a_{ij}^{(k)}$ and f_i have continuous derivatives up to the sixth order[22] only. Here, we can merely say that our proof of the uniqueness result is valid if $a_{ij}^{(k)}$ and f_i have continuous derivatives up to the order $(4n + 3)$.

§9. Proof of Continuous Dependence of the Solution of the Cauchy Problem for Hyperbolic Systems on the Equations Themselves and the Initial Functions

We limit ourselves to the proof of continuous dependence of the solution on the right-hand sides of the equations. The analysis of its dependence on the initial functions can be reduced to the case considered here.

Let

$$\frac{\partial u_i}{\partial t} = F_i\left(t, x_1, \ldots, x_n; u_1, \ldots, u_N; \frac{\partial u_1}{\partial x_1}, \ldots, \frac{\partial u_N}{\partial x_n}\right) \tag{68}$$

$$(i = 1, \ldots, N)$$

be a given hyperbolic system. Consider also a system, denoted by (68_ε), which is related to (68) by the following conditions:

1) The matrices (3) for both systems have similar structure.

2) The right-hand sides $F_{i,\varepsilon}$ of system (68_ε), together with their partial derivatives up to the order $(4n + 4)$ with respect to all arguments, differ from the right-hand sides of system (68) and their respective derivatives less than by ε, provided that the point $(t, x_1, \ldots, x_n)$ varies within a domain G adjacent to the plane $t = 0$, and absolute values of u_j and $\partial u_j / \partial x_k$ are bounded by $M > 0$.

---

[22]Cf. [6, p. 550 ff]

Let us apply (once only) the procedure described in §1 to systems (68) and (68_ε). For all sufficiently small ε, we obtain quasilinear hyperbolic systems amenable to the considerations of §8. Therefore, in a domain G' independent of ε, each of the systems (68), (68_ε) has a solution with zero initial conditions at $t = 0$.

We claim that for sufficiently small ε the solutions of these systems are arbitrarily close to each other in the domain G'.

Indeed, the very method used for the construction of the solution of system (68_ε) implies that, in the domain G', all derivatives of the solution up to the order $(3n + 1)$ are bounded by M. Therefore, by virtue of the Arzelà theorem, from each sequence of these solutions (as $\varepsilon \to 0$) we can extract a subsequence consisting of functions which uniformly converge in G', together with their derivatives in all x_k, and therefore, in t. Obviously, the limit functions satisfy system (68). Hence we obtain the required statement.

§10. A Remark on Generalization of the above Results to a Wider Class of Differential Equations

In our proof of the existence results for the Cauchy problem for hyperbolic systems, only the following properties have been used:

I. For any point P of the domain on the plane $t = 0$ where the initial values for the functions u_i are prescribed, we construct a cube Q_l with its edges parallel to the coordinate axes. In cylinder R_l whose generatrix is parallel to the t-axis, Q_l being the bottom of R_l, we can define functions $\overline{F}_i$ depending on the same arguments as F_i and possessing the following properties:

  1. Functions $\overline{F}_i$ coincide with F_i at the points of a cylinder lying inside R_l and having point P on its bottom[23].

  2. Functions $\overline{F}_i$, as well as their partial derivatives up to the order[24] $(4n + 4)$ with respect to all variables, admit l-periodic extension in all x_k.

II. For each $\overline{F}_i$ we can define functions $\overline{\overline{F}}_i$, each $\overline{\overline{F}}_i$ being entire analytic with respect to $t, x_k, u_j, \partial u_j / \partial x_k$ and l-periodic in x_k. For $|u_j|, |\partial u_j / \partial x_k| < M$, the functions $\overline{\overline{F}}_i$ approximate, uniformly on R_l and with any given accuracy, the functions $\overline{F}_i$, as well as their partial derivatives (in all arguments) up to the order $(4n + 4)$; moreover, the matrices

---

[23]Here and in what follows, u_j and $\partial u_j / \partial x_k$ are regarded as parameters whose absolute values do not exceed M

[24]Our arguments can be applied to the case $n = 1$ if we replace $(4n + 4)$ by $(4n + 5)$.

$$\sum_k \alpha_k \left\| \frac{\partial \overline{\overline{F}}_i}{\partial \dfrac{\partial u_j}{\partial x_k}} \right\| - \lambda E$$

satisfy the following conditions: they can be reduced to canonical form by close transformations whose coefficients k_{ij}, together with their first derivatives in all α_k (resp., in t), possess continuous uniformly bounded derivatives in all $x_k, u_j, \partial u_j/\partial x_k$ up to the order $3n$ (resp., $2n$). If $k_{ij}^{(m)}$ is a partial derivative of k_{ij} of m-th order with respect to a combination of x_k, then for $l \to 0$ and $k > 0$ we have

$$k_{ij}^{(m)} l^m \to 0 \ .$$

In particular, this is the case if, everywhere in the domain under consideration, the matrix

$$\sum_k \alpha_k \left\| \frac{\partial \bar{F}_i}{\partial \dfrac{\partial u_j}{\partial x_k}} \right\| - \lambda E \ , \quad \sum \alpha_k^2 = 1 \ ,$$

and the matrices

$$\sum_k \alpha_k \left\| \frac{\partial \overline{\overline{F}}_i}{\partial \dfrac{\partial u_j}{\partial x_k}} \right\| - \lambda E \ , \quad \sum \alpha_k^2 = 1 \ ,$$

have all their elementary divisors real and of degree 1, the elementary divisors for each matrix can be divided into groups of elements in such a way that the number of equal elements in each group for all $\overline{\overline{F}}_i$ and $\bar{F}_i$ is an absolute constant (cf. Ch.III, §1 of our article [4] concerning the Cauchy problem for linear systems).

We say that a system is *hyperbolic in generalized sense*, if it satisfies the conditions I and II (which implies that our proof of the existence of solutions for the Cauchy problem can be applied to such systems).

It should be mentioned that some systems of equations in mathematical physics are likely to have this form in the case when the variables x_p and t enter the functions F_i exclusively through the agency of some known sufficiently smooth functions $\psi_k(t, x_1, \ldots, x_n)$, and F_i are analytic entire[25] functions with respect to $u_j, \partial u_j/\partial x_k, \psi_k$ and t, but the coefficients of the transformation (16), which reduces the matrix (3) to canonical

---

[25]On the basis of footnote 15, the same can be proved under the assumption that these functions are holomorphic in a sufficiently large domain, although not necessarily entire.

form, together with their derivatives in all α_k, possess continuous bounded derivatives in all ψ_k, u_j and $\partial u_j / \partial x_k$ up to the order $3n$. With a view to constructing $\overline{F}_i$ and $\overline{\overline{F}}_i$ for the right-hand sides F_i of such a system, the functions ψ_k should be subjected to the transformations similar to those performed over F_i in §5 and §8.

The solution of the Cauchy problem for a generalized hyperbolic system cannot be claimed to possess continuity (in a neighborhood of the domain where initial conditions are prescribed) with respect to continuously varied right-hand sides of the equations and their derivatives. However, it does depend in a continuous way on the initial functions u_i, together with their partial derivatives in x_k of a sufficiently high order.

PART II

Hyperbolic Systems of Higher Order

§1. Reduction of a General Hyperbolic System to a Quasilinear System of Special Type

Let

$$\frac{\partial^{n_i} u_i}{\partial t^{n_i}} = F_i(t, x_1, \ldots, x_n; u_1, \ldots, u_N, \ldots) \qquad (i = 1, 2, \ldots, N) . \qquad (69)$$

be a given hyperbolic system. In the right-hand sides we have not written out explicitly the derivatives of u_j $(j = 1, \ldots, N)$ up to the order n_j, with the highest order of differentiation in t equal to $(n_j - 1)$.

Our aim is to show that

Any hyperbolic system of type (69) can be reduced to the form

$$\frac{\partial^{n_i} u_i}{\partial t^{n_i}} = \sum_{j,k} A_{ij}^{(k_0, k_1, \ldots, k_n)} \frac{\partial^{n_j} u_j}{\partial t^{k_0} \partial x_1^{k_1} \cdots \partial x_n^{k_n}} + \Phi_i \qquad (70)$$

$$(i, j = 1, 2, \ldots, N, \ldots, N_0, \ldots, N_1) ,$$

*provided that all the functions involved admit differentiation of a suffi-
ciently high order. The summation in the right-hand side is carried over
all non-negative k_j whose sum is equal to n_j, with $k_0 < n_j$. The coefficients
$A_{ij}^{(k_0,k_1,\ldots,k_n)}$ depend on t, x_k, u_j $(j = 1, 2, \ldots, N, \ldots, N_0)$ and their derivatives up to the order $(n_j - 1)$, where all the derivatives of u_j $(j = 1, \ldots, N_0)$
in x_k up to the order l are identified with the variables $u_{N_0+1}, \ldots, u_{N_1}$. The
functions Φ_i depend on t, x_k, u_j $(j = 1, 2, \ldots, N, \ldots, N_0, \ldots, N_1)$ and their
derivatives up to the order $(n_j - 1)$.*

First, we prove the above theorem in the case $l = 0$. To this end we set

$$\frac{\partial u_i}{\partial x_k} = u_i^{(k)} . \tag{71}$$

Then the original system can be written in the form

$$\frac{\partial^{n_i} u_i}{\partial t^{n_i}} = F_i^* \qquad (i = 1, 2, \ldots, N) , \tag{72}$$

where F_i^* depend on $u_j, u_j^{(k)}$ $(j = 1, \ldots, N)$ and their derivatives up to the
order $(n_j - 1)$. Differentiating (72) with respect to x_p, we obtain

$$\frac{\partial^{n_i} u_i^{(p)}}{\partial t^{n_i}} = \sum_{D,j,k} \frac{\partial F_i^*}{\partial (D^{n_j-1} u_j^{(k)})} D^{n_j} u_j^{(p)} + \sum_{D,j,m,k} \frac{\partial F_i^*}{\partial (D^{m-1} u_j^{(k)})} D^m u_j^{(p)} +$$

$$+ \sum_j \frac{\partial F_i^*}{\partial u_j} u_j^{(p)} + \frac{\partial F_i^*}{\partial x_p} \tag{73_p}$$

$$(i, j = 1, \ldots, N; \quad k, p = 1, 2, \ldots, n, \quad 0 < m < n_j) .$$

The first sum is taken over all j, k and all possible combinations of derivatives D^{n_j} of order n_j. The second sum has the same structure as the first
one. Here $D^0 u = u$.

Thereby it has been shown that differentiation and introduction of new
variables allow us to reduce each system of type (69) to the form specified
in the above proposition with $l = 0$. Naturally, we cannot expect that all
functions that satisfy the system composed of equations (72) and (73_p) $(p =
1, \ldots, n)$ will also satisfy equations (69), (71). However, if the initial values
for the functions u_i and $u_i^{(p)}$ are chosen so as to satisfy the relations (71),
then these relations will hold for all t. In order to prove this statement, we
perform the following operations.

First, without introducing any new notations, we differentiate equation
(72) with respect to x_p, and subtract the result from (73_p). We denote the
equations obtained in this way by (74). Further, we differentiate equations

(73_p) in x_r, and equations (73_r) in x_p, still without resorting to new notations. While performing these operations, we always differentiate under the sign of D^s or $\partial^{n_i}/\partial t^{n_i}$. Let us consider the difference of the respective resulting equations. The equations obtained in this way are denoted by (75). Let us introduce the notations

$$u_j^{(p)} - \frac{\partial u_j}{\partial x_p} = \bar{u}_j^{(p)},$$

$$\frac{\partial u_j^{(p)}}{\partial x_k} - \frac{\partial u_j^{(k)}}{\partial x_p} = \bar{u}_j^{(p,k)}.$$

(76)

Everywhere in (74) and (75), except in the expressions entering partial derivatives of F_i, we replace the terms $\partial u_j/\partial x_k$ and $\partial u_j^{(p)}/\partial x_q$, for $p < q$, with their expressions provided by (76). Next, we show that in every equation thus obtained the sum of the terms containing neither the factors of the form $\bar{u}_j^{(p)}, \bar{u}_j^{(p,k)}$, nor $D^s \bar{u}_j^{(p)}, D^s \bar{u}_j^{(p,k)}$, is equal to zero. Indeed, let us replace the functions u_i in the given equations (69) by arbitrary functions v_i. After this substitution, the difference between the left-hand side of the i-th equation and its right-hand side becomes equal to $f_i(t, x_1, \ldots, x_n)$. Therefore, v_i satisfy the following system

$$\frac{\partial^{n_i} v_i}{\partial t^{n_i}} = F_i + f_i(t, x) \qquad (i = 1, \ldots, N),$$

(77)

where the functions F_i are exactly the same as those in (69). In relation to system (77), we perform the same operations as those just performed over system (69). The terms resulting from the summands f_i will disappear from the equations that are the counterparts of (74), (75) (we denote these by $(74^*), (75^*)$). The equations (74), (75) differ from their counterparts only in that u is replaced by v. On the other hand, it is clear that the right-hand sides of equations $(74^*), (75^*)$ must vanish identically, as well as the functions $\bar{v}_j^{(p)}, \bar{v}_j^{(p,k)}$, taken separately. Consequently, the terms involving neither the functions $\bar{v}_j^{(p)}, \bar{v}_j^{(p,k)}$, nor their derivatives as factors, must also vanish. Since all the functions $D^{(m)} v_j^{(p)}$ and $\partial D^{(m)} v_j^{(p)}/\partial x_q$ entering these terms are arbitrary (if $p > q$), it follows that they are also identically equal to zero. Hence we obtain the desired conclusion.

Thus equations (74), (75) form a linear homogeneous system with respect to the functions $\bar{u}_j^{(p)}$ and $\bar{u}_j^{(p,k)}$. The right-hand sides of equations (74) contain derivatives of $\bar{u}_j^{(p)}$ and $\bar{u}_j^{(p,k)}$ up to the order $(n_j - 1)$. The terms of the highest order in equations (75) have the form

$$\frac{\partial^{n_i} \bar{u}_i^{(r,p)}}{\partial t^{n_i}} = \sum \frac{\partial F_i}{\partial (D^{n_j} u_j)} D^{n_i} \bar{u}_j^{(r,p)} + \cdots \qquad (i = 1, 2, \ldots, N).$$

Therefore, regarding $\bar{u}_j^{(p)}$ and $\bar{u}_j^{(r,p)}$ as the only unknown functions in equations (74), (75), we see that (74), (75) form a linear hyperbolic system with respect to the functions $\bar{u}_j^{(p)}$ and $\bar{u}_j^{(r,p)}$. In the sequel, we establish a uniqueness theorem for the Cauchy problem for such equations, its proof being independent of the considerations of this section. Thus, if $\bar{u}_j^{(p)}$ and $\bar{u}_j^{(r,p)}$ vanish at $t = 0$, then they must vanish identically. It follows that the functions u_i satisfy system (69), q.e.d.

Thereby the theorem formulated at the beginning of this section has been proved for $l = 0$. If we subject system (72), (73) to the operations performed over system (69) to derive (72), (73), we obtain a proof of our theorem for $l = 1$. Proceeding in this way, we can establish the theorem for arbitrary l.

§2. Estimate for an Integral of the Sum of Squared Unknown Functions and Their Derivatives

1. Special case. Let R_l be a cylinder of height T, whose bottom coincides with the cube Q_l ($0 \le x_k \le 1$) on the plane $t = 0$. In R_l we consider the following hyperbolic system:

$$\frac{\partial^{n_i} u_i}{\partial t^{n_i}} = \sum_{j,k} a_{ij}^{(k_0,\ldots,k_n)} \frac{\partial^k u_j}{\partial t^{k_0} \partial x_1^{k_1} \cdots \partial x_n^{k_n}} + f_i(t, x_1, \ldots, x_n) \qquad (78)$$

$$(i = 1, 2, \ldots, N),$$

where $a_{ij}^{(k)}$ are functions of $t, x_1, \ldots, x_n$; $k = \sum k_s \le n_j$, $k_s \ge 0$, $k_0 < n_j$.

It is assumed that for each t the following functions admit bounded continuous l-periodic extension with respect to all variables x_k:

1) the coefficients $a_{ij}^{(k_0,\ldots,k_n)}$ with $\sum_{s=0}^n k_s = n_j$, and all their derivatives in x_k up to the order $3n$;

2) the functions $\dfrac{\partial a_{ij}^{(k_0,\ldots,k_n)}}{\partial t}$ for $\sum k_s = n_j$, and the coefficients $a_{ij}^{(k_0,\ldots,k_n)}$ for $\sum k_s < n_j$, as well as the derivatives (in x_k) of all these functions up to the order $2n$;

3) the functions u_i and all their derivatives up to the order $(n_j - 1)$.

The $(x_1, \ldots, x_n)$-space is assumed to be Euclidean, and the axes $Ox_1, \ldots,$ Ox_n mutually orthogonal. Let us rotate this coordinate system so that the new coordinate plane $x_1' = 0$ coincide with the plane

$$\sum \alpha_k x_k = 0 \ .$$

Set

$$\alpha_k = \alpha \alpha_k' \ , \quad \alpha = \sqrt{\sum \alpha_k^2} = |\alpha| \ .$$

Then

$$x_1' = \sum_k \alpha_k' x_k \ .$$

In new variables, system (78) takes the form

$$\frac{\partial^{n_i} u_i}{\partial t^{n_i}} = \sum_{j,k} a_{ij}'^{(k_0,\ldots,k_n)} \frac{\partial^k u_j}{\partial t^{k_0} \partial x_1'^{k_1} \cdot \partial x_n'^{k_n}} + f_i(t,x') \tag{79}$$

$$(i,j = 1,\ldots,N) \ .$$

It is clear that for all α_k' the coefficients $a_{ij}'^{(k_0,\ldots,k_n)}$ remain bounded and have the form of linear combinations of $a_{ij}^{(k_0,\ldots,k_n)}$ with bounded coefficients depending on α_s' and the direction cosines of the other coordinate axes Ox_k' $(k > 1)$.

For $\sum_{s=0}^n p_s < n_j$, we set

$$u_j'^{(p_0\ldots,p_n)} = \frac{\partial^p u_j}{\partial t^{p_0} \partial x_1'^{p_1} \cdots \partial x_n'^{p_n}} \ ,$$

$$u_j^{(p_0\ldots,p_n)} = \frac{\partial^p u_j}{\partial t^{p_0} \partial x_1^{p_1} \cdots \partial x_n^{p_n}} \ .$$

Then the functions $u_i'^{(p_0\ldots,p_n)}$ satisfy the following system

$$\frac{\partial u_i'^{(n_i-1,0,\ldots,0)}(\alpha')}{\partial t} = \sum_{j,\sum k_s = n_j,\, l} a_{ij}'^{(k_0,\ldots,k_n)} \frac{\partial u_j'^{(k_0,\ldots,k_{l-1},k_l-1,k_{l+1},\ldots,k_n)}(\alpha')}{\partial x_l'} +$$

$$+ \sum_{j,\sum k_s \le n_j-1} a_{ij}'^{(k_0,\ldots,k_n)} u_j'^{(k_0,\ldots,k_n)}(\alpha') + f_i(t,x') \ ; \tag{$79_1'$}$$

$$\frac{\partial u_i'^{(k_0,k_1,\ldots,k_{l-1},k_l,k_{l+1}\ldots,k_n)}(\alpha')}{\partial t} = \frac{\partial u_i'^{(k_0+1,k_1,\ldots,k_{l-1},k_l-1,k_{l+1},\ldots,k_n)}(\alpha')}{\partial x_l'} \tag{$79_2'$}$$

for $\sum k_s = n_i - 1$, $k_0 < n_i - 1$;

$$\frac{\partial u_i'^{(k_0,\ldots,k_n)}(\alpha')}{\partial t} = u_i'^{(k_0+1,k_1,\ldots,k_n)} \tag{$79_3'$}$$

for $\sum k_s < n_i - 1$.

The subscript l in $(79'_1)$, $(79'_2)$ is positive and equal to the largest subscript of k_s other than zero. In particular, $l = 1$ only if $k_s = 0$ for all $s \geq 2$.

We choose the new coordinate axes $Ox'_2, \ldots, Ox'_n$ in such a way that their direction cosines, in relation to the original axes $Ox_1, \ldots, Ox_n$, are functions of $\alpha'_1, \ldots, \alpha'_n$ possessing continuous first order derivatives. Then

$$u_i'^{(k_0,\ldots,k_n)}(\alpha') = \sum_{p,j} c_{ij}^{(p_0,\ldots,p_n)}(\alpha') u_j^{(p_0,\ldots,p_n)} , \tag{80}$$

and

$$u_i^{(k_0,\ldots,k_n)} = \sum_{p,j} c_{ij}^{*(p_0,\ldots,p_n)}(\alpha') u_j'^{(p_0,\ldots,p_n)} , \tag{80*}$$

where $c_{ij}^{(p_0,\ldots,p_n)}(\alpha')$ and $c_{ij}^{*(p_0,\ldots,p_n)}(\alpha')$, together with all their first order derivatives in α_k, are bounded continuous functions.

Let us construct the λ-matrix $\mathcal{M}$ for system $(79')$, replacing therein $\partial/\partial t$ by λ, $\partial/\partial x'_1$ by α, and all the rest of $\partial/\partial x'_k$ by 0. It is clear that each matrix M_r, from among those into which the matrix (4) for system (78) is split, corresponds to one of the matrices, say M'_r, which compose $\mathcal{M}$. Each matrix M'_r, in its turn, can be split into a matrix M''_r and several other matrices consisting of a single element λ on the main diagonal. *The determinant of the λ-matrix M''_r is equal to the determinant of the corresponding λ-matrix M_r.* This statement can be proved as follows: First of all, we notice that the construction of the above determinants does not require the coefficients $A_{ij}^{(k_0,\ldots,k_n)}$ with $\sum k_s < n_j$ to be taken into consideration, nor does this construction in any way rely on the dependence of $A_{ij}^{(k_0,\ldots,k_n)}$ on $t, x_1, \ldots, x_n$. Therefore, the relation between the said determinants in the general case is the same as in the case of constant coefficients $A_{ij}^{(k_0,\ldots,k_n)}$ equal to zero for $\sum k_s < n_j$. In the latter case, the fact that the determinant of the matrix M_r vanishes means that the equations corresponding to this matrix (we denote these equations by A) have a non-trivial solution

$$u_i = C_i e^{\lambda t + (\alpha,x)} , \quad \text{where} \quad (\alpha, x) = \sum \alpha_k x_k . \tag{81}$$

The vanishing determinant of M''_r shows that the equations associated with this matrix (we denote them by B) have a solution of the form

$$C_i e^{\lambda t + \alpha x'_1} . \tag{82}$$

Each solution of system A having the form (81) corresponds to a solution (82) of system B. Consequently, each root of the determinant of the matrix M_r is a root of the determinant of M''_r. Since all roots of the determinant of M_r are different from one another, and the determinants of both M_r and M''_r have 1 as the coefficient by the highest degree of λ, these two determinants coincide.

By assumption (according to the definition of hyperbolic systems), all roots of the λ-matrices M_r are real and the difference between any two roots is larger than a positive δ, if $\sum \alpha_k^2 = 1$. Therefore, there exists a linear transformation

$$u_i^{*(k_0,\ldots,k_n)}(\alpha') = \sum_{j,p} \tau_{ij}^{(p_0,\ldots,p_n)}(\alpha') u_j^{'(p_0,\ldots,p_n)}(\alpha') \tag{83}$$

which reduces the matrix $\mathcal{M}$ to canonical form (i.e., the resulting matrix has non-zero elements on the main diagonal only), its determinant being equal to 1.

Combining (80) and (83), we obtain the following transformation

$$u_i^{*(k_0,\ldots,k_n)}(\alpha') = \sum_{j,p} \tau_{ij}^{*(p_0,\ldots,p_n)}(\alpha') u_j^{(p_0,\ldots,p_n)}, \tag{84}$$

which reduces system (79) to the form

$$\frac{\partial u_i^*(\alpha')}{\partial t} = k_i(t, x, \alpha') \frac{\partial u_i^*(\alpha')}{\partial x_1'} + \sum_{j,k>1} A_{ij}^{*(k)}(t, x, \alpha') \frac{\partial u_j^*(\alpha')}{\partial x_k'} +$$

$$+ \sum_j B_{ij}^*(t, x, \alpha') u_j^*(\alpha') + f_i^*(t, x, \alpha') \tag{85}$$

$$(i, j = 1, \ldots, N_1 > N) .$$

The transformation (84) possesses the following properties:

1^0. Its determinant is equal to 1.

2^0. The coefficients of the transformations, corresponding to the non-trivial cells M_r'' of the matrix $\mathcal{M}$, and their derivatives in α_k' possess continuous derivatives (with respect to x_k and t) of the same order as the coefficients $a_{ij}^{(k_0,\ldots,k_n)}$, $\sum k_s = n_j$, of the original hyperbolic system.

Note that the λ-matrix $\mathcal{M}$ for system (79) does not depend on the choice of the axes $Ox_2', \ldots, Ox_n'$. Indeed, for the construction of the matrix $\mathcal{M}$, only those equations in $(79_1')$ and $(79_2')$ have any importance which contain differentiation with respect to x_1' in the right-hand side. Because of the procedure adopted here for the derivation of equations $(79_1')$ and $(79_2')$, the terms with $\partial/\partial x_1'$ involve the functions $u_i^{'(k_0,k_1,0,\ldots,0)}$ that coincide with derivatives of u_i' with respect to t, x_1'. But the coefficients of the transformation (80) for the functions $u_i^{'(k_0,k_1,0,\ldots,0)}$ do not depend on the directions of the axes $Ox_2', \ldots, Ox_n'$. Every non-trivial cell M_r'' corresponds to a "part" of the transformation (84). It is easy to see that the part of the transformation (83) corresponding to M_r'' depends only on the structure of M_r'' and is, therefore, independent of the choice of the axes $Ox'_2, \ldots, Ox'_n$. On the other hand, as noted above, the coefficients of the transformation (80) corresponding to non-trivial cells of the matrix $\mathcal{M}$ are also independent of the directions of $Ox'_2, \ldots, Ox'_n$. Thus we finally find the cell M''_r to be associated with a part of the transformation (84) independent of the directions of the axes $Ox'_2, \ldots, Ox'_n$. The coefficients τ^*_{ij} entering that part of the transformation are uniquely determined to within a normalizing factor, and, being suitably normalized, become continuously differentiable with respect to α'_k on the entire sphere $\sum_{i=1}^n \alpha_i'^2 = 1$. It follows that the coefficients $k_i(t, x, \alpha')$ of system (85) are also independent of the directions of $Ox'_2, \ldots, Ox'_n$, and are continuously differentiable in α'_k on the entire sphere $\sum_{i=1}^n \alpha_i'^2 = 1$.

As regards the rest of the coefficients τ^*_{ij} (these coefficients correspond to $k_i(t, x, \alpha') \equiv 0$), we cannot claim their differentiability in α'_k, nor even their continuity in α'_k. It is only needed that these coefficients be uniformly bounded, together with their derivatives in x_k up to a suitable order, and this is obviously the case.

Let us multiply every equation in (85) by

$$\frac{1}{l^n} e^{-\frac{2\pi i}{l}(\alpha, x)} = \frac{1}{l^n} e^{-\frac{2\pi i}{l}|\alpha|x'_1}$$

and integrate the resulting equality over the cube Q_l, assuming that t is constant. Since all the functions under consideration admit l-periodic extension, we can apply integration by parts (with respect to x'_k) to every term in the right-hand side of every equation; then, under the sign of the integral, the derivatives of $u^*_i(\alpha')$ in x'_k will disappear, and there will be no new terms depending on the values of $u^*_i(\alpha')$ on the faces of cube Q_l. Setting

$$a_{i(\alpha)}^{*(\beta)} = \frac{1}{l^n} \int_{Q_l} u^*_i(\alpha') e^{-\frac{2\pi i}{l}(\beta, x)} \, dx = \frac{1}{l^n} \int_{Q_l} u^*_i(\alpha') e^{-\frac{2\pi i}{l}|\beta|x'_1} \, dx' ,$$

we obtain for $a_{i(\alpha)}^{*(\alpha)}$ the following system of ordinary differential equations:

$$\frac{da_{i(\alpha)}^{*(\alpha)}}{dt} = \frac{2\pi i}{l} \cdot \frac{|\alpha|}{l^n} \int_{Q_l} k_i(t, x, \alpha') u^*_i(t, x, \alpha') e^{-\frac{2\pi i}{l}(\alpha, x)} \, dx +$$

$$+ \sum \int_{Q_l} A^*_{ij}(t, x, \alpha') u^*_i(t, x, \alpha') e^{-\frac{2\pi i}{l}(\alpha, x)} \, dx + \int_{Q_l} f^*_i(t, x, \alpha') e^{-\frac{2\pi i}{l}(\alpha, x)} \, dx .$$

(86)

Here the number of equations, N_1, is the same as that of the unknown functions u^*_i; A^*_{ij} are linear combinations of B^*_{ij} and the first order derivatives in x'_k of the functions $A^{*(k)}_{ij}$ and k_i.

Assuming the derivatives of the coefficients in (78) to be sufficiently small, one can prove an inequality similar to (46) by the same method as that used in §2 of Part I for a system of first order equations. This inequality has the form

$$\int_{Q_l} \sum_{k,i} \left[u_i^{(k_0,k_1,\ldots,k_n)}(t,x) \right]^2 dx \le e^{C_1 t} \left\{ C_2 \int_{Q_l} \left[\sum_{k,i} u_i^{(k_0,k_1,\ldots,k_n)}(0,x) \right]^2 dx + \right.$$

$$\left. + C_3 \int_0^t \int_{Q_l} \sum_i f_i^2(\tau,x)\, dx\, d\tau \right\}, \tag{87}$$

where $\sum_{k,i}$ implies summation over all i from 1 to N and all non-negative k_s whose sum does not exceed $n_i - 1$; C_1, C_2 and C_3 are constants depending on the maximum of the absolute values of the coefficients of the original system, and, of course, their derivatives. Moreover, they depend on the structure of the characteristic λ-matrix for that system and on the smallest difference between the roots of each λ-matrix M_r composing the entire λ-matrix if $\sum \alpha_k^2 = 1$.

2. The inequality (87) admits generalization to the case of systems whose coefficients, or their derivatives, cannot be l-periodically extended, and also systems whose coefficients do not have small enough derivatives, so that no condition of type (37′) holds. This generalization can be obtained in exactly the same way as in §2 of Part I, and therefore its details are not reproduced here.

§3. A Lemma from the Analytic Theory of Linear Differential Equations

Proposition. *Let*

$$\frac{\partial^{n_i} u_i}{\partial t^{n_i}} = \sum_{j,k} a_{ij}^{(k_0,k_1,\ldots,k_n)} \frac{\partial^k u_j}{\partial t^{k_0} \partial x_1^{k_1} \cdots \partial x_n^{k_n}} + f_i \quad (i,j = 1,\ldots,N). \tag{88}$$

be a given system, where $a_{ij}^{(k_0,k_1,\ldots,k_n)}$ and f_i are entire[26] analytic functions of x_k and t. Let R be a cylinder of finite height, with generatrix parallel to the axis Ot. Then there exists $\delta > 0$, depending only on $a_{ij}^{(k_0,k_1,\ldots,k_n)}$ and R, such that for any plane E ($t =$ const) having a non-empty intersection with R, and any entire[27] analytic initial values prescribed on E for the functions u_i and their derivatives in t up to the order $(n_i - 1)$, the Cauchy problem has a solution in the layer of width δ enclosing E.

---

[26]See footnote 15.

[27]See *ibid.*

Proof. As in the preceding section, we set

$$u_i^{(k_0,k_1,\ldots,k_n)} = \frac{\partial^k u_i}{\partial t^{k_0} \partial x_1^{k_1} \cdots \partial x_n^{k_n}} \qquad \left(\sum k_s < n_i\right).$$

If the functions u_i satisfy system (88), then $u_i^{(k_0,k_1,\ldots,k_n)}$ satisfy the following system

$$\frac{\partial u_i^{(n_i-1,0,\ldots,0)}}{\partial t} = \sum_{j,\sum k_s = n_j-1,\,l} a_{ij}^{(k_0,k_1,\ldots,k_n)} \frac{\partial u_j^{(k_0,k_1,\ldots,k_{l-1},k_l-1,k_{l+1},\ldots,k_n)}}{\partial x_l} +$$

$$+ \sum_{j,\sum k_s < n_j} a_{ij}^{(k_0,k_1,\ldots,k_n)} u_j^{(k_0,k_1,\ldots,k_n)} + f_i \,, \tag{89_1}$$

$$\frac{\partial u_i^{(k_0,\ldots,k_{l-1},k_l,k_{l+1},\ldots,k_n)}}{\partial t} = \frac{\partial u_i^{(k_0+1,k_1,\ldots,k_{l-1},k_l-1,k_{l+1},\ldots,k_n)}}{\partial x_l} \,, \tag{89_2}$$

$$k_0 < n_i - 1, \quad \sum k_s \leq n_i - 1 \,,$$

$$\frac{\partial u_i^{(k_0,0,\ldots,0)}}{\partial t} = u_i^{(k_0+1,0,\ldots,0)} \,, \quad k_0 < n_i - 1 \,. \tag{89_3}$$

The integer $l > 0$ in (89_1), (89_2) is equal to the largest subscript of the indices k_s other than zero.

This system satisfies all the conditions required by the Theorem in Part I, §3, if it satisfies the conditions of the preceding sections. Under the assumptions of the present theorem, we can immediately conclude that a solution of system (89) exists in a layer of fixed width δ which depends neither on the initial conditions, nor on the height of the plane E ($t =$ const), on which the initial conditions are prescribed. The only result to be proved is as follows: if the initial conditions for system (89) are chosen in such a way that for $t = t_0$ we have

$$u_i^{(k_0,k_1,\ldots,k_n)} = \frac{\partial^k u_i}{\partial t^{k_0} \partial x_1^{k_1} \cdots \partial x_n^{k_n}} \,, \tag{90}$$

then these relations hold also for the other values of t, and therefore we obtain a solution of system (88). This statement can be proved in the following way: First of all, equations (89_2) show that the above statement is always valid if $\sum_{s=1}^n k_s = 0$. The right-hand sides of the equations (89_2) with $\sum_s^n k_s = 1$ (thus, among all k_s with $s > 0$, only one, say k_l, is different from 0) can be written as

$$\frac{\partial u_i^{(k_0,0,\ldots,0,k_l,0,\ldots,0)}}{\partial t} = \frac{\partial^2 u_i^{(k_0,0,\ldots,0,0,0,\ldots,0)}}{\partial x_l \partial t}$$

or

$$\frac{\partial}{\partial t}\left[u_i^{(k_0,0,\ldots,0,k_l,0,\ldots,0)} - \frac{\partial u_i^{(k_0,0,\ldots,0,0,0,\ldots,0)}}{\partial x_l}\right] = 0 , \qquad k_l = 1 .$$

Hence it is easy to see that the relation (90) with $\sum_1^n k_s = 1$ holds for any t, if it holds for $t = t_0$. Using the equations (89$_2$) written out for $\sum_1^n k_s = 2$, we can prove, in exactly the same way, that for $\sum_1^n k_s = 2$ the relations (90) are valid for any t, if this is the case for $t = t_0$, etc.

§4. Remarks on Further Exposition as Compared to Part I

1. Theorems similar to those proved in Part I, §4 and §5, can now be established in exactly the same manner. However, from the very beginning, we should estimate not only the sum of squared functions approximating u_i (we denote them by $u_{i,\varepsilon}$), but also of their squared derivatives up to the order $(n_i - 1)$. To this end the estimate (87) should be used. Next, the approximating system S_ε should be subjected, $(n + 2)$ times, to the operations described in §1 of Part II. Applying the estimate (87) to the resulting system $S_\varepsilon^{(n+2)}$, we obtain an estimate for the integral of the sum of squared functions $u_{i,\varepsilon}^{(k_0,k_1,\ldots,k_n)}$ $\left(\sum_{s=0}^n k_s \leq n_i - 1\right)$ and their derivatives in all x_k up to the order $(n + 2)$. Making use of system $S_\varepsilon^{(n+1)}$, we thence obtain an estimate for the integral of the sum of squared derivatives of the functions $u_{i,\varepsilon}^{(k_0,k_1,\ldots,k_n)}$, up to the order $(n + 2)$. Hence we can establish the differentiability in t. This leads us to the uniform convergence of the functions $u_{i,\varepsilon}^{(k_0,k_1,\ldots,k_n)}$ and their first derivatives in x_k. Therefore, we have uniform convergence of all first order derivatives, since the system S_ε sets a relation between the first order derivatives of the functions $u_{i,\varepsilon}^{(k_0,k_1,\ldots,k_n)}$ with respect to t and the first order derivatives of these functions with respect to x_k.

2. While proving the uniqueness theorem, we can proceed as follows. Assume that the homogeneous system (78) admits a non-trivial solution which satisfies the conditions

$$u_i \equiv \frac{\partial u_i}{\partial t} \equiv \cdots \equiv \frac{\partial^{n_i-1} u_i}{\partial t^{n_i-1}} \equiv 0$$

in a domain on the plane $t = 0$.

Let us change the independent variables, in analogy to §6 of Part I. In new variables, the system will have the same form as (78). The new system is hyperbolic, and all the derivatives of every u_i up to the order $(n_i - 1)$ must vanish on a paraboloid in a neighborhood of its vertex (Fig. 1 pertains to the case $n = 1$). In order to avoid new notations, we assume that system (78) already possesses all these properties. Let us construct the sum

$$\sum v_i F_i(u) \,,$$

where v_i are arbitrary sufficiently smooth functions and

$$F_i(u) \equiv \frac{\partial^{n_i} u_i}{\partial t^{n_i}} - \sum_{j,k} a_{ij}^{(k_0,k_1,\ldots,k_n)} \frac{\partial^k u_j}{\partial t^{k_0} \partial x_1^{k_1} \cdots \partial x_n^{k_n}} \,.$$

We integrate the above sum over a segment $\mathcal{G}$ of the paraboloid, $\mathcal{G}$ being cut off by a plane parallel to $t = 0$ (see Fig. 1). Since

$$F_i(u) = 0 \quad \text{in the interior of} \quad \mathcal{G}\,,$$

and all the derivatives of the functions u_i up to the order $(n_i - 1)$ vanish on the curved portion of the boundary of $\mathcal{G}$, we obtain

$$0 \equiv \int_{\mathcal{G}} \sum u_i G_i(v) \, dt \, dx + \int_{PP'} \left\{ \left[\sum_i v_i \frac{\partial^{n_i-1} u_i}{\partial t^{n_i-1}} - \frac{\partial v_i}{\partial t} \frac{\partial^{n_i-2} u_i}{\partial t^{n_i-2}} + \cdots \right. \right.$$

$$\cdots + (-1)^{n_i-1} \frac{\partial^{n_i-1} v_i}{\partial t^{n_i-1}} u_i \bigg] - \sum_{i,j,k_0} \left[v_i a_{ij}^{(k_0,0,\ldots,0)} \frac{\partial^{k_0-1} u_j}{\partial t^{k_0-1}} - \right.$$

$$\left. - \frac{\partial(v_i a_{ij}^{(k_0,0,\ldots,0)})}{\partial t} \frac{\partial^{k_0-2} u_j}{\partial t^{k_0-2}} + \cdots \pm \frac{\partial^{k_0-1}(v_i a_{ij}^{(k_0,0,\ldots,0)})}{\partial t^{k_0-1}} u_j \right] \bigg\} \, dx \,.(91)$$

Here PP' stands for the plane part of the boundary of the segment $\mathcal{G}$, and

$$G_i(v) \equiv \frac{\partial^{n_i} v_i}{\partial(-t)^{n_i}} - \sum_{j,k} \frac{\partial^k \left(a_{ji}^{(k_0,k_1,\ldots,k_n)} v_j \right)}{[\partial(-t)]^{k_0} [\partial(-x_1)]^{k_1} \cdots [\partial(-x_n)]^{k_n}} \,.$$

Obviously, the system of equations

$$G_i(v) = 0 \tag{92}$$

is hyperbolic, if the original system is hyperbolic. Indeed, its characteristic matrix is obtained from that for system (78) by replacing α_k with $-\alpha_k$, and λ with $-\lambda$, and mutually transposing the columns and the rows. The functions v_i are chosen as follows. In $\mathcal{G}$, these functions must satisfy system (92). All the derivatives of v_i up to the order $(n_i - 2)$ with respect to t must vanish on PP'. The derivatives of v_i of order $(n_i - 1)$ in t must take values close enough to $(-1)^{n_i-1} u_i$. Since system (92) is hyperbolic, all these conditions can be fulfilled (cf., Part I, §6), provided that PP' lies close enough to the plane $t = 0$. Since the value of k_0 in $a_{ij}^{(k_0,0,\ldots,0)}$ does not exceed $n_j - 1$, (91) takes the form

$$0 = \int_{PP'} \sum_i (-1)^{n_i-1} \frac{\partial^{n_i-1} v_i}{\partial t^{n_i-1}} u_i \, dx \,.$$

Because of the arbitrary choice of PP' near the vertex and our assumptions concerning $\partial^{n_i-1} v / \partial t^{n_i-1}$, the above equality is possible only if $u_i \equiv 0$ in a neighborhood of $\mathcal{G}$, q.e.d.

3. We say that system S is *quasilinear*, if its coefficients depend on the derivatives of u_i up to the order $(n_i - 1)$. The same arguments as those used in Part I can be applied to prove the existence of a solution of the Cauchy problem for such systems, the proof being similar to that in Part I, §7. In this case, the estimate (87) should be used in relation to the systems resulting from the approximation of the given system and the operation from §1, Part II, applied to them $(4n + 2)$ times.

In this way, we obtain a sequence of functions

$$u_{i,m}^{(k_0, k_1, \ldots, k_n)} \quad \left(\sum k_s \le n_i - 1, \ i = 1, 2, \ldots, N, \ m \to \infty \right)$$

which uniformly converge, together with their first order derivatives in all x_k and t. The limit functions $u_i^{(0,\ldots,0)} = u_i$ satisfy the given system.

Hence it is easy to see that the proof of the existence of the unique solution for system (1) vanishing at $t = 0$ requires that the functions F_i possess continuous derivatives up to the order $(4n + 4)$ with respect to all their arguments. In order to reduce the general case of arbitrary initial values for u_i and $\partial^k u_i / \partial t^k$ $(k = 1, \ldots, n_i - 1)$ (i.e., arbitrary functions $\varphi_i^{(k)}$) to the case considered above, we should assume that the functions $\varphi_i^{(k)}$ have continuous derivatives up to the order $(n_i + 4n + 4)$.

4. In complete analogy to what we have defined, for first order equations, as the class of systems hyperbolic in generalized sense, we can define the class of systems hyperbolic in generalized sense for equations of higher order. Our proof of the existence theorem for the Cauchy problem can be extended to the latter class.

References

[1] Petrowsky, I.G. Sur le problème de Cauchy pour un système d'équations aux dérivées partielles dans le domain réel. *C. R. Acad. Sci. Paris* **202** (1936) 1010–1012.

[2] Petrowsky, I.G. Sur le problème de Cauchy pour un système linéaire d'équations aux dérivées partielles dans un domain réel. *C. R. Acad. Sci. Paris* **202** (1936) 1246–1248.

[3] Hadamard, J. *Le problème de Cauchy et les équations aux dérivées partielles linéaires hyperboliques.* Hermann, Paris, 1932.

[4] Petrowsky, I.G. On the Cauchy problem for systems of linear partial differential equations in a class of non-analytic functions. *Bulletin of Moscow State Univ. Math., Mech.* **1:7** (1938) 1–72.

[5] Friedrichs, K; Lewy, H. Das Anfangswertproblem einer beliebigen nichtlinearen hyperbolischen Differentialgleichung beliebiger Ordnung in zwei Variablen. Existenz, Eidentigkeit und Abhangingkeitsbericht der Losung. *Math. Ann.* **99** (1928) 200–221.

[6] Perron, O. Über Existenz und Nichtexistenz von Integralen partiellen Differentialgleichungssysteme in reelen Gebiet. *Math. Zeitschr.* **27:4** (1928) 550–564.

[7] Schauder, J. Das Anfangswertproblem einer quasilinearen hyperbolischen Differentialgleichung zweiter Ordnung in beliebiger Anzahl von unabhängiger Veränderlichen. *Fund. Math.* **24** (1935) 213–246.

[8] Courant, R; Friedrichs, K; Lewy, H. Über die partiellen Differenzengleichungen der mathematischen Physik. *Math. Ann.* **100** (1928) 32–74.

[9] Sobolev, S.L. Some estimates pertaining to families of functions with square summable derivatives. *Doklady Akad. Nauk SSSR* **1:7** (1936) 267–270. (See Sobolev, S.L. *Some Applications of Functional Analysis in Mathematical Physics.* AMS Translation, 1991. – *Ed. note.*)

[10] Holmgren, E. Über Systeme von linearen partiellen Differentialgleichungen. *Ovfersigt. Kgl. Vetenskaps. Akad. Förhandl.* **58** (1901) 91–103.

[11] Hadamard, J. *Leçons sur la Propagation des Ondes et les Équations de l'Hydrodynamique.* Hermann, Paris, 1903.

On the Cauchy Problem for Linear Systems of Partial Differential Equations in Classes of Non-Analytic Functions*

Introduction

The systems considered in this paper have the form

$$\frac{\partial^{n_i} u_i}{\partial t^{n_i}} = \sum_{j=1}^{N} \sum_{(k_s)} A_{ij}^{(k_0,k_1,\ldots,k_n)}(t) \frac{\partial^{k_0+k_1+\cdots+k_n} u_j}{\partial t^{k_0} \partial x_1^{k_1} \cdots \partial x_n^{k_n}} + f_i(t, x_1, \ldots, x_n) \quad (1)$$

$$(i = 1, \ldots N) ,$$

where $\sum_{(k_s)}$ implies summation over all integer $k_s \geq 0$ such that their sum does not exceed a constant M, and $k_0 < n_j$, $n_j > 0$. It is assumed that $A_{ij}^{(k_0,k_1,\ldots,k_n)}$, f_i, u_i are complex-valued functions of real arguments, the coefficients $A_{ij}^{(k_0,k_1,\ldots,k_n)}$ depend on t only, while f_i may depend on $t, x_1, \ldots, x_n$; the coefficients and the functions f_i are defined for all real values of x_k and $0 \leq t \leq T$. The solutions u_i are, in general, complex-valued, and we examine these solutions in the same domain.

According to a concept developed by J. Hadamard in his book [1], we say that the Cauchy problem for system (1) is *uniformly well-posed* on the interval[1] $(0, T)$, if the following two conditions are satisfied:

1. *Let $\varphi_i^{(k)}(x_1, \ldots, x_n)$ $(i = 1, \ldots, N; k = 0, 1, \ldots, n_i - 1)$ be an arbitrary system of functions which, together with their partial derivatives up to a certain finite order L, are bounded over the entire $(x_1, \ldots, x_n)$-plane.*

---

*Originally published in: *Bulletin of Moscow Univ. Math., Mech.* **1**:7 (1938) 1–72. Notes [1-12] by V.P. Palamodov. See Appendix for a commentary by O.A. Oleinik & V.P. Palamodov, and a supplementary article by I.M. Gelfand, I.G. Petrowsky & G.E. Shilov.

[1]In what follows, any interval (a, b) is assumed closed, i.e., containing its end-points.

Then there exists one and only one system of functions u_i such that u_i are bounded, together with all their partial derivatives of sufficiently high orders; u_i satisfy system (1) for any t from the interval $0 \leq t_0 < t \leq T$, and for $t = t_0$ take, together with their derivatives in t up to the order $(n_i - 1)$, the values $\varphi_i^{(k)}$, respectively.

2. *For each t_0 from the interval $(0,T)$ and any given $\varepsilon > 0$, there exists a positive η, independent of t_0, such that any variation of the functions $\varphi_i^{(k)}$ and of their partial derivatives up to the order L by a quantity less than η, may only cause a variation of the functions u_i by a quantity less than ε, for all t such that $t_0 \leq t \leq T$* [2].

If for some t_0 and any $T > t_0$ the latter condition is violated, then (as noted by J. Hadamard) a solution of the Cauchy problem for system (1), even in the case of its existence, would hardly be of any consequence for physical problems, since however small the measurement errors for the values of $\varphi_i^{(k)}(x_1, \ldots, x_n)$, or even their derivatives of any order, the distortion of the functions u_i satisfying system (1) might be considerable, even for the values of time arbitrarily close to t_0 [2].

Assuming the coefficients $A_{ij}^{(k_0, k_1, \ldots, k_n)}(t)$ and functions f_i to have bounded derivatives (with respect to all their arguments) of a sufficiently high order, we find, in what follows, a necessary and sufficient condition (**Condition A**) for the Cauchy problem for system (1) to be uniformly well-posed on the interval $(0,T)$. In order to formulate this condition, we reduce our system to another one with $n_i = 1$ for every i. This reduction of the general system can be easily done if we introduce the notations

$$u_{N+1}, \ u_{N+2}, \ \cdots, \qquad \text{for} \qquad \frac{\partial u_i}{\partial t}, \ \frac{\partial^2 u_i}{\partial t^2}, \cdots, \ \frac{\partial^{n_i-1} u_i}{\partial t^{n_i-1}}.$$

If every $n_i = 1$, then every $k_0 = 0$, and therefore, we can drop the latter index. Now **Condition A** can be expressed as follows:

Condition A. *Let the functions $v_i^{(l)}$ satisfy the equations*

$$\frac{dv_i^{(l)}}{dt} = \sum_{j=1}^{N} \sum_{(k_s)} A_{ij}^{(k_1, \ldots, k_n)}(t)(i\alpha_1)^{k_1} \cdots (i\alpha_n)^{k_n} v_j^{(l)} \qquad (2)$$

on the interval (t_0, T), together with the initial conditions at $t = t_0$:

$$v_i^{(l)}(t_0) = \begin{cases} 0 & \text{if } i \neq l, \\ 1 & \text{if } i = l. \end{cases}$$

---

[2]Cf. my notes [2], [3]. It would be interesting to find out whether the first condition always implies the second [1].

Then the quantities $\max \left| v_i^{(l)} \right|$ *should not grow (uniformly in t_0) faster than a certain fixed power of* $\max_s |\alpha_s|$, *for all* α_s *taking only real values and* $\max |\alpha_s| \to \infty$ $(s = 1, \ldots, n)$.

The question whether Condition A is fulfilled can be easily solved in the case of constant coefficients $A_{ij}^{(k_0, k_1, \ldots, k_n)}$. *In a more general case of* $A_{ij}^{(k_0, k_1, \ldots, k_n)}$ *depending on t, Condition A holds for systems that we call "hyperbolic" and "parabolic", under the assumption that the coefficients are sufficiently smooth.*

In order to give the definition of a *parabolic system*, let us choose a positive p such that

$$k_0 p + \sum_{s=1}^{n} k_s \le n_j p , \quad j = 1, \ldots, N , \tag{3}$$

and consider the matrix

$$\left\| \sum_{((k_s))_p} A_{ij}^{(k_0, k_1, \ldots, k_n)}(t) \lambda^{k_0} (i\alpha_1)^{k_1} \cdots (i\alpha_n)^{k_n} \right\| - \left\| \begin{matrix} \lambda^{n_1} & & & 0 \\ & \lambda^{n_2} & & \\ & & \ddots & \\ 0 & & & \lambda^{n_N} \end{matrix} \right\| , \tag{4}$$

where $\sum_{((k_s))_p}$ implies summation over all non-negative integers k_s such that

$$k_0 p + \sum_{s=1}^{n} k_s$$

has its largest value $n_j p$. When writing the conditions (3), we take into account only such combinations of k_s that are attached to the coefficients $A_{ij}^{(k_0, k_1, \ldots, k_n)}(t) \not\equiv 0$.

System (1) *is called p-parabolic, or simply, parabolic on an interval* $(0, T)$, *if, for all real α_k such that $\sum \alpha_k^2 = 1$, all roots of the determinant of the matrix* (4) *have real parts less than a negative constant* $-\delta$.

This condition will be referred to as *Condition B*. For parabolic systems one can establish the existence and the uniqueness of a solution for the Cauchy problem on the interval $t_0 \le t \le T$, assuming only that the functions $f_i(t, x_1, \ldots, x_n)$ and $\varphi_i^{(k)}(x_1, \ldots, x_n)$, together with their partial derivatives up to the order $(2M - p + \Delta)$, are continuous and bounded at all points; here $\Delta = 0$ if $M = p$, and $\Delta = 1$ if $M > p$. Solutions of the Cauchy problem for parabolic systems can be easily represented in terms of the "Green matrix"; the values of these solutions at each point $(t, x_1, \ldots, x_n)$ turn out to be determined by the values of the functions

$\varphi_i^{(k)}(x_1, \ldots, x_n)$ *on the entire hyperplane* $t = t_0$. One can also show that if on the interval (t_0, T) the functions $f_i(t, x_1, \ldots, x_n)$ are analytic in all x_k, then all sufficiently smooth solutions $u_i(t, x_1, \ldots, x_n)$ are also analytic in all x_k for $t_0 < t \leq T$, even if the coefficients $A_{ij}^{(k_0, k_1, \ldots, k_n)}$ or the functions f_i are non-analytic with respect to t.

It should be observed that system (1) can be parabolic only if p in the definition of p-parabolic systems is an even integer. Indeed, the structure of the matrix (4) is such that if every α_k is multiplied by the same factor α, then all the roots of the determinant of this matrix are multiplied by α^p, where α^p may take any real, as well as any complex value; in particular, setting $\alpha = \pm 1$, we come to the conclusion that every root of the determinant of (4) multiplied by $(\pm 1)^p$ must have negative real part, which is possible only for even values of p. We can also formulate a converse statement:

Let p *be a positive integer satisfying the conditions* (3). *Assume that there exists a value of* t *from the interval* $(0, T)$, *as well as real values of* α_k, *such that one of the roots of the determinant of the matrix* (4) *has positive real part. Then the Cauchy problem for system* (1) *on the interval* $(0, T)$ *does not have the property of being well-posed.*

This condition will sometimes be referred to as *Condition* $(-B)$, in contrast to Condition B used as a definition of parabolic systems.

We say that *system* (1) *is hyperbolic, if the following conditions are fulfilled:*

1. $\sum\limits_{s=0}^{n} k_s \leq n_j$, $k_0 < n_j$, $n_j > 0$.[3]

2. *The matrix*

$$\left\| \sum_{((k_s))} A_{ij}^{(k_0, k_1, \ldots, k_n)}(t) \lambda^{k_0} \alpha_1^{k_1} \cdots \alpha_n^{k_n} \right\| - \left\| \begin{matrix} \lambda^{n_1} & & & 0 \\ & \lambda^{n_2} & & \\ & & \ddots & \\ 0 & & & \lambda^{n_N} \end{matrix} \right\|, \quad (5)$$

where $\sum_{((k_s))}$ *implies summation over all* k_s *such that* $\sum_{s=0}^{n} k_s = n_j$, *has the form*

$$\left\| \begin{matrix} M_1 & & & \\ & M_2 & & \\ & & \ddots & \\ & & & M_l \end{matrix} \right\|,$$

_____

[3]Systems which satisfy only this condition we call *systems of Kowalevskaya type.*

the elements outside all square matrices M_s being identically equal to zero. (It is possible that a single square matrix M_s entirely fills up the above matrix).

3. For $0 \leq t \leq T$ and all real values of α_k such that $\sum \alpha_k^2 = 1$, the determinant of every matrix M_s can have only real mutually different roots.

Conversely, if there exist t and real values of α_s such that the determinant of at least one M_s has a non-real root, then the Cauchy problem for system (1) on the interval $(0, T)$ has neither the property of being uniformly well-posed, nor that of being simply well-posed (the definition of well-posedness for the Cauchy problem is given in Ch. I, §7).

If for all real α_s the determinant of each M_s has only real roots, but there exists a matrix M_s whose determinant has multiple roots for a particular choice of real α_s, then the Cauchy problem for system (1) may happen to be well-posed or not. Some examples of such systems are given later on. We also formulate a criterion for the Cauchy problem for system (1) to be uniformly well-posed in this situation.

In the case of linear hyperbolic systems, the solution of the Cauchy problem at a point $(t, x_1, \ldots, x_n)$ is known (see [4], and also [5], [2]) to be completely determined by the values of the functions $\varphi_i^{(k)}(x_1, \ldots, x_n)$ at the points of a certain bounded domain on the hyperplane $t = t_0$.

The restriction imposed here on the coefficients $A_{ij}^{(k_0, \ldots, k_n)}(t)$ (viz., their dependence on t only) appears to be caused, in most cases, by the method employed throughout the paper, namely, the Fourier method. We have tried here to examine only the main features of the Cauchy problem, as part of the theory of partial differential equations. However, we are confident that most of the theorems proved in this paper hold for more general linear, or even nonlinear, systems.

A generalization of the theorems proved in this paper for hyperbolic systems has already been made by the author in [2]. A similar generalization seems also possible for parabolic systems. For this purpose, the "Green matrices", constructed by the author for some special parabolic systems, are likely to have application, in analogy with what has been done for a single parabolic equation (see [6]).

Some other, still open, questions arising in connection with this paper are mentioned in footnotes.

CHAPTER I

Condition A for Systems with Constant Coefficients; Some examples

§1. Necessity of Condition A

First we consider the case of $n_i = 1$ for all i. Assume that Condition A does not hold. This means that for arbitrarily large constants C and L and some t_0, $t_0 + \tau$ ($\tau > 0$) from the interval $(0, T)$, there exists a system of real

$$\alpha_1^*, \alpha_2^*, \ldots, \alpha_n^*$$

such that for some i_0 and l_0 the following inequality is satisfied

$$|v_{i_0}^{(l_0)}(t_0 + \tau)| > C[\max |\alpha_s^*| + 1]^L . \tag{6}$$

Here and in what follows we adopt the notations used in Condition A (see the Introduction).

If the Cauchy problem for system (1) admits no solution for any initial values of the functions

$$u_i(t, x_1, \ldots, x_n) \quad \text{at} \quad t = t_0 \ (0 \le t_0 < T) ,$$

then condition 1 in the definition of a uniformly well-posed Cauchy problem on the interval $(0, T)$ is violated.

Assume that there is at least one system of functions $\varphi_i(x_1, \ldots, x_n)$ such that the Cauchy problem for system (1), with these functions as the initial values of $u_i(t, x_1, \ldots, x_n)$ at $t = t_0$, has a solution. Let us add to φ_{l_0} a function of the form

$$\varphi(x_1, \ldots, x_n) = \varepsilon_1 e^{i(\alpha^*, x)} , \quad \alpha^* = (\alpha_1^*, \ldots, \alpha_n^*) ,$$

where ε_1 is an arbitrary real constant. We obviously have

$$|D^{(k)} \varphi(x_1, \ldots, x_n)| \le \varepsilon_1 \max |\alpha_s^*|^k , \tag{7}$$

where $D^{(k)} \varphi$ stands for a k-th order partial derivative of φ. Accordingly, to $u_i(t, x_1, \ldots, x_n)$ the functions $u_i^*(t, x_1, \ldots, x_n)$ will be added, which satisfy the following homogeneous system

$$\frac{\partial u_i^*}{\partial t} = \sum_{j=1}^{N} \sum_{(k_s)} A_{ij}^{(k_1, \ldots, k_n)}(t) \frac{\partial^{k_1 + k_2 + \cdots + k_n} u_j^*}{\partial x_1^{k_1} \partial x_2^{k_2} \cdots \partial x_n^{k_n}}$$

for $t_0 < t \le T$; and for $t = t_0$ every u_i^* vanishes, except for the function $u_{l_0}^*$, which becomes equal to φ. Obviously

$$u_i^*(t, x_1, \ldots, x_n) = \varepsilon_1 v_i^{(l_0)}(t, t_0, \alpha_1^*, \ldots, \alpha_n^*) e^{i(\alpha^*, x)} .$$

By virtue of (6) we have

$$|u_{i_0}^*(t_0 + \tau, x_1, \ldots, x_n)| > \varepsilon_1 C [\max |\alpha_s^*| + 1]^L .$$

Since the constant C can be arbitrarily large, the above inequality and (7) imply that ε_1 can be chosen in such a way that the function $\varphi(x_1, \ldots, x_n)$, together with all its partial derivatives up to the order L, is arbitrarily small, while the absolute value $|u_{i_0}^*(t_0 + \tau, x_1, \ldots, x_n)|$ is arbitrarily large. Therefore, the second condition in the definition of a uniformly well-posed Cauchy problem is not satisfied.

Next we consider the case of $n_i > 1$ for some i. As mentioned in Introduction, this case can always be reduced to the case of all $n_i = 1$, if we consider additional unknown functions u_i ($i = N+1, N+2, \ldots, N_1$) which stand for the derivatives

$$\frac{\partial u_i}{\partial t}, \frac{\partial^2 u_i}{\partial t^2}, \ldots, \frac{\partial^{n_i-1} u_i}{\partial t^{n_i-1}} .$$

Let us denote the new system by (1^*). As shown above, the fact that the Cauchy problem for system (1^*) is uniformly well-posed implies that Condition A holds for that system. Assume now that Condition A is not valid for system (1^*). Then, according to what has been just proved, the homogeneous system (1^*) admits solutions of the form

$$u_i(t, t_0, x_1, \ldots, x_n) = v_i(t, t_0) e^{i(\alpha, x)} \qquad (i = 1, \ldots, N_1)$$

such that

$$|v_i(t_0, t_0)| \le \varepsilon , \tag{8}$$

where ε is an arbitrary positive constant; however, for no constants C, L the inequalities

$$|v_i(t, t_0)| \le C[1 + \alpha_M]^L \qquad (\alpha_M = \max |\alpha_k|) \tag{9}$$

can be satisfied for all $i = 1, \ldots, N_1$ and all t ($t_0 \le t \le T$).

But this is not enough to guarantee the lack of uniform well-posedness of the Cauchy problem for the homogeneous system (1), since it is still possible that the latter inequalities are satisfied for some C, L and $i \le N$, and are not valid only for some $i > N$. Let us show that actually *there is no such possibility.* To this end, we have to assume that *the coefficients $A_{ij}^{(k_0,k_1,\ldots,k_n)}$ possess continuous derivatives up to the order*[4] $(n_i - 1)$.

---

[4]Nowhere else in the present chapter this restriction is needed. It would be interesting to find out whether the mere assumption of continuity is sufficient here [[9]].

Proof. If the functions $u_i(t, t_0, x_1, \ldots, x_n)$ $(i = 1, \ldots, N_1)$ satisfy the homogeneous system (1^*), then the functions

$$u_i(t, t_0, x_1, \ldots, x_n) \qquad (i = 1, \ldots, N)$$

satisfy the homogeneous system (1), and therefore $v_i(t, t_0)$ satisfy the system

$$\frac{d^{n_i} v_i}{dt^{n_i}} = \sum_{j=1}^{N} \sum_{(k_s)} A_{ij}^{(k_0, k_1, \ldots, k_n)}(t)(i\alpha_1)^{k_1} \cdots (i\alpha_n)^{k_n} \frac{d^{k_0} v_j}{dt^{k_0}}. \qquad (10)$$

Let us integrate each of the above equations $(n_i - 1)$ times with respect to t, each time the lower limit of integration being 0. When integrating the right-hand side, we use the integration by parts so as to exclude the derivatives of the functions v_i. It follows that if the inequalities (8) hold for $i = 1, \ldots, N_1$, and the inequalities (9) hold for $i = 1, \ldots N$, then the first derivatives of the functions v_i $(i = 1, \ldots, N)$ in t satisfy the inequalities of type (9). Likewise, integrating equations (10), $(n_i - 2)$ times with respect to t, we obtain similar estimates for the second derivatives in t, etc.

§2. Auxiliary Results

Lemma 1. *Consider the following two systems of ordinary differential equations:*

$$\frac{dv_i}{dt} = \sum_{j=1}^{N} C_{ij}(t, \alpha_1, \ldots, \alpha_n) v_j \qquad (i = 1, \ldots, N) \qquad (11)$$

and

$$\frac{dv_i^*}{dt} = \sum_{j=1}^{N} C_{ij}(t, \alpha_1, \ldots, \alpha_n) v_j^* + F_i(t, \alpha_1, \ldots, \alpha_n) \qquad (11^*)$$

$$(i = 1, \ldots, N),$$

where v_i, C_{ij}, F_i are complex valued functions of real variables $t, \alpha_1, \ldots, \alpha_n$, defined for $0 \leq t \leq T$ and all α_k. Assume that the dependence of C_{ij} on t is continuous and

$$|F_i(t, \alpha_1 \ldots, \alpha_n)| \leq C_1[1 + \alpha_M]^{p_1} \qquad (i = 1, \ldots, N),$$

where C_1 is constant, and p_1 can be either positive or negative. Denote by $v_i^{(l)}(t, t_0, \alpha_1 \ldots, \alpha_n)$ and $v_i^*(t, t_0, \alpha_1 \ldots, \alpha_n)$ (or shortly, $v_i^{(l)}(t, t_0, \alpha)$ and $v_i^*(t, t_0, \alpha)$) the functions which satisfy systems (11) and (11^*), respectively, on the interval (t_0, T), $t_0 \geq 0$, together with the following initial conditions at $t = t_0$:

$$v_l^{(l)}(t_0, t_0, \alpha) = 1, \quad v_i^{(l)}(t_0, t_0, \alpha) = 0 \quad \text{for} \quad i \neq l,$$

$$v_i^*(t_0, t_0, \alpha) = 0 \quad (i = 1, \ldots, N).$$

Assume that for all t such that $t_0 \leq t \leq T$ we have

$$|v_i^{(l)}(t, t_0, \alpha)| \leq C_2[1 + \alpha_M]^{p_2}, \tag{12}$$

where p_2 and C_2 are some non-negative constants. Then there exists a constant C such that

$$|v_i^*(t, t_0, \alpha)| \leq C[1 + \alpha_M]^{p_1 + p_2} \tag{13}$$

on the interval (t_0, T).

Proof. The functions $v_i^*(t, t_0, \alpha)$ are given by the formulas

$$v_i^*(t, t_0, \alpha) = \int_{t_0}^{t} \sum_{j=1}^{N} v_i^{(j)}(t, \tau, \alpha) F_i(\tau, \alpha) \, d\tau \tag{14}$$

(see [2], Vol. 2, Part 2, §418). Hence we obtain the required statement.

Lemma 2. *Assume that all conditions of Lemma 1 are satisfied, and moreover:*

1) *the functions $C_{ij}(t, \alpha_1, \ldots, \alpha_n)$ are polynomials of a degree $\leq M$ with arguments $\alpha_1, \ldots, \alpha_n$; $M \geq 1$;*

2) *for all $k_s \geq 0$ whose sum is $\leq m$, we have*

$$\left| \frac{\partial^{k_1 + \cdots + k_n} F_i}{\partial \alpha_1^{k_1} \cdots \partial \alpha_n^{k_n}} \right| \leq C_1 [\alpha_M + 1]^{p_1}, \tag{15}$$

where C_1 and p_1 are constants, p_1 can be either positive or negative, and $p_2 > 0$.

Then there exists a constant $C \geq 0$ such that

$$\left| \frac{\partial^{k_1 + \cdots + k_n} v_i^*}{\partial \alpha_1^{k_1} \cdots \partial \alpha_n^{k_n}} \right| \leq C[\alpha_M + 1]^{p_1 + p_2 + \sum k_s [p_2 + M - 1]}, \tag{16}$$

for all t from the interval (t_0, T) and all k_s whose sum is $\leq m$.

Proof. Assume first that only one index among $k_1, \ldots, k_n$, say k_1, is equal to 1, while the rest of k_i are zeroes. Set

$$\frac{\partial v_i^*}{\partial \alpha_1} = \tilde{v}_i(t, \alpha_1, \ldots, \alpha_n).$$

Then the functions $\tilde{v}_i^{(l)}$ satisfy the following system

$$\frac{d\tilde{v}_i}{dt} = \sum_{j=1}^{N} C_{ij}(t, \alpha_1, \dots, \alpha_n)\tilde{v}_j + \sum_{j=1}^{N} C'_{ij}v_j^* + \frac{\partial F_i}{\partial \alpha_1} , \qquad (17)$$

where $C'_{ij} = \partial C_{ij}/\partial \alpha_1$. Obviously $\tilde{v}_i^{(l)}(t_0, \alpha_1, \dots, \alpha_n) = 0$.

Using once again formulas (14), as in the proof of Lemma 1, and estimating v_j^* by (13), we obtain

$$|\tilde{v}_i(t, \alpha_1, \dots, \alpha_n)| \le C'_3[\alpha_M + 1]^{p_2 + (M-1) + p_1 + p_2} + C'_4[\alpha_M + 1]^{p_1 + p_2} ,$$

where C'_3 and C'_4 are constant. Hence

$$|\tilde{v}_i(t, \alpha_1, \dots, \alpha_n)| \le C_3^*[\alpha_M + 1]^{p_1 + 2p_2 + M - 1} .^5 \qquad (18)$$

Differentiating (17) with respect to α_s and setting

$$\frac{\partial \tilde{v}_i}{\partial \alpha_s} = \tilde{\tilde{v}}_i(t, \alpha_1, \dots, \alpha_n) ,$$

we get

$$\frac{d\tilde{\tilde{v}}_i}{dt} = \sum_{j=1}^{N} C_{ij}\tilde{\tilde{v}}_j + 2\sum_{j=1}^{N} C'_{ij}\tilde{v}_j + \sum_{j=1}^{N} C''_{ij}v_j^* + \frac{\partial^2 F_i}{\partial \alpha_1 \partial \alpha_s} . \qquad (19)$$

Here C''_{ij} stands for the derivative of C'_{ij} with respect to α_s, and therefore has the form of a polynomial in $\alpha_1, \dots, \alpha_n$ of a degree not greater than $(M - 2)$.

It is easy to see that

$$\tilde{\tilde{v}}_i(t_0, \alpha_1, \dots, \alpha_n) = 0 .$$

The exponent in the right-hand side of (18) is not smaller than that in the right-hand sides of (13) and (15), provided that $M \ge 1$. The degree of C''_{ij} with respect to α_s is smaller than that of C_{ij} with respect to α_s; if $M = 1$, then $C''_{ij} = 0$. Therefore, applying the estimates (13), (15) and (18), we obtain the largest exponent in the estimate for $\sum C'_{ij}\tilde{v}_j$. Hence, using (14), we find that

$$\left|\tilde{\tilde{v}}_i(t, \alpha_1, \dots, \alpha_n)\right| \le C_3^{**}[\alpha_M + 1]^{(p_1 + 2p_2 + M - 1) + M - 1 + p_2} \le$$

$$\le C_3^{**}[\alpha_M + 1]^{p_1 + 3p_2 + 2(M-1)} .$$

Differentiating (19) several more times with respect to α_s, and applying the arguments just used, we arrive at the estimate (16).

---

[5] If $M = 0$, then $C'_3 = 0$, and we obtain for $\tilde{v}_i$ the same estimate as for v_i^* in Lemma 1.

Corollary. *Under the above assumptions, for every t from the interval* (t_0, T) *and all* $k_s \geq 0$ *whose sum is positive, we have*

$$\left| \frac{\partial^{k_1 + \cdots + k_n} v_i^{(l)}}{\partial \alpha_1^{k_1} \cdots \partial \alpha_n^{k_n}} \right| \leq C_4 [\alpha_M + 1]^{p_2 + \sum k_s [p_2 + M - 1]} . \qquad (16')$$

This inequality follows directly from the preceding proof.

Lemma 3. *According to Bochner (see* [8]*), set*

$$E_f(\alpha_1, \ldots, \alpha_n) = \frac{1}{(2\pi)^n} \int e^{-i(\alpha, x)} f(x_1, \ldots, x_n) \, dx , \qquad (20)$$

for any real α_s, *where the integral is taken over the real* $(x_1, \ldots, x_n)$-*space. Assume that the function* $f(x_1, \ldots, x_n)$ *vanishes outside the cube* Q_l:

$$|x_s| \leq a \qquad (s = 1, \ldots, n) ,$$

and possesses continuous partial derivatives up to the order p *with respect to various combinations of* x_k. *Then*

$$|E_f(\alpha_1, \ldots, \alpha_n)| \leq \frac{C}{[|\alpha_1| + 1]^{k_1} \cdots [|\alpha_n| + 1]^{k_n}} , \qquad (21)$$

where $k_s \geq 0$, $\sum k_s \leq p$, *and the constant* C *depends only on* a *and* M, M *being the maximum of the absolute values of the function* f *and its derivatives up to the order* p. *This implies, in particular, that*

$$|E_f(\alpha_1, \ldots, \alpha_n)| \leq C[\alpha_M + 1]^{-p} .$$

Proof. In the case under consideration, the integrals defining $E_f(\alpha_1, \ldots, \alpha_n)$ can be taken over the cube Q_a only. Assume that among α_s only $\alpha_{s_1}, \ldots, \alpha_{s_l}$ have absolute values ≥ 1. Let us then integrate by parts k_{s_1} times in x_{s_1}, k_{s_2} times in x_{s_2}, etc. Since the function f and all its derivatives up to the order p vanish on the boundary of Q_a, we get

$$|E_f(\alpha_1, \ldots, \alpha_n)| \leq \frac{2^n a^n M}{|\alpha_{s_1}|^{k_{s_1}} \cdots |\alpha_{s_l}|^{k_{s_l}} (2\pi)^n} \leq$$

$$\leq \frac{a^n M \cdot 2^q}{[|\alpha_1| + 1]^{k_1} \cdots [|\alpha_n| + 1]^{k_n} \pi^n} ,$$

where $q = \sum k_s$. Lemma 3 is proved.

Lemma 4. *Let*

$$f(x_1,\ldots,x_n) = \int E(\alpha_1,\ldots,\alpha_n)\, e^{i(\alpha,x)}\, d\alpha_1 \ldots d\alpha_n , \qquad (22)$$

where $E(\alpha_1,\ldots,\alpha_n)$ is a complex valued function of real variables $\alpha_1,\ldots,\alpha_n$, and the integral is taken over the entire $(\alpha_1,\ldots,\alpha_n)$-space. Assume that the absolute values of the function $E(\alpha_1,\ldots,\alpha_n)$ and its partial derivatives with respect to various combinations of α_s, up to the order k, are not greater than

$$\frac{C}{[|\alpha_1|+1]^{2+p_1}\cdots[|\alpha_n|+1]^{2+p_n}} , \qquad (23)$$

where C is constant, and p_s are arbitrary non-negative integers whose sum does not exceed p. Then the absolute values of the derivatives of $f(x_1,\ldots,x_n)$ up to the order p do not exceed

$$\frac{C_1}{[|x_1|+1]^{k_1}\cdots[|x_n|+1]^{k_n}} , \qquad (24)$$

where C_1 is constant, and $\sum k_s \le k$. (As usual, any derivative of "zero" order is assumed here to coincide with the function itself.)

Proof. First, we prove Lemma 4 in the case $p = 0$. Let Q_a be a cube in $(\alpha_1,\ldots,\alpha_n)$-space defined by the inequalities

$$|\alpha_s| \le a . \qquad (25)$$

Assume that among x_s only $x_{s_1},\ldots,x_{s_l}$ have absolute values ≥ 1. In the integral

$$\int_{Q_a} E(\alpha_1,\ldots,\alpha_n) e^{i(\alpha,x)}\, d\alpha_1 \cdots d\alpha_n , \qquad (22')$$

we integrate by parts k_{s_1} times with respect to α_{s_1}, k_{s_2} times with respect to α_{s_2}, etc. Because of the estimates (23), the resulting integrals over the boundary converge to 0 as $a \to \infty$. Indeed, an integral of this type taken, for instance, over the faces $|\alpha_1| = a$ does not exceed

$$\int \frac{2C\, d\alpha_2 \ldots d\alpha_n}{[a+1]^2[|\alpha_2|+1]^2\cdots[|\alpha_n|+1]^2} ,$$

where the domain of integration is the entire $(\alpha_1,\ldots,\alpha_n)$-space. The last integral is not larger than

$$\frac{2C L^{n-1}}{[a+1]^2} ,$$

where

$$L = \int_{-\infty}^{+\infty} \frac{d\alpha}{[|\alpha|+1]^2} .$$

Therefore, the said integral converges to 0, as $a \to \infty$. Moreover, the integral (22') converges to $f(x_1, \dots, x_n)$ as $a \to \infty$. Therefore

$$|f(x_1, \dots, x_n)| \leq \frac{CL^n}{|x_{s_1}|^{k_{s_1}} \cdots |x_{s_l}|^{k_{s_l}}} \leq \frac{CL^n \cdot 2^q}{[|x_1| + 1]^{k_1} \cdots [|x_n| + 1]^{k_n}},$$

where $q = \sum k_s$. Now, with a view to estimating the derivatives of $f(x_1, \dots, x_n)$, it should be noticed, first of all, that we can differentiate in x_s under the sign of the integral, since, by virtue of (23), the resulting integrals are uniformly convergent, if the differentiation is of an order not greater than p. Therefore, denoting by $D^{(m)}$ any partial derivative of f of an order not greater than p, we find that

$$D^{(m)} f(x_1, \dots, x_n) = \int E_m(\alpha_1, \dots, \alpha_n) e^{i(\alpha, x)} \, d\alpha_1 \cdots d\alpha_n,$$

where $E_m(\alpha_1, \dots, \alpha_n)$ satisfies the estimates of type (23) for some non-negative p_s. Therefore, the arguments used in the case $p = 0$ are also valid in this case.

§3. Proof of the Existence of Solutions of the Cauchy Problem for a System Satisfying Condition A

First we prove the existence result under the assumption that *the initial functions and $f_i(t, x_1, \dots, x_n)$ may differ from 0 only for $(x_1, \dots, x_n)$ inside the cube Q_l defined by the inequalities*

$$|x_s| \leq a.$$

In order to simplify the notations, we assume that the initial functions are given at $t = 0$, although all our arguments can be reproduced literally in the case of the initial conditions being given at any other moment t_0 from the interval $(0, T)$. All estimates obtained here for $t_0 = 0$ remain valid in the general case, with the same constants.

As mentioned in the Introduction, it suffices to consider the case of $n_i = 1$ for all i. Then system (1) has the form

$$\frac{\partial u_i}{\partial t} = \sum_{j=1}^{N} \sum_{(k_s)} A_{ij}^{(k_1, \dots, k_n)}(t) \frac{\partial^{k_1 + \cdots + k_n} u_j}{\partial x_1^{k_1} \cdots \partial x_n^{k_n}} + f_i(t, x_1, \dots, x_n) \qquad (1')$$

$$(i = 1, \dots, N).$$

Assume that Condition A is satisfied. This implies that there exist positive constants C and p such that

$$|v_i^{(l)}(t, t_0, \alpha_1, \ldots, \alpha_n)| < C[\alpha_M + 1]^p , \qquad (26)$$

where α_M is the largest among $|\alpha_s|$. Here we retain the notations adopted in the Introduction. Assume that the initial functions $\varphi_i(x_1, \ldots, x_n)$, as well as $f_i(t, x_1, \ldots, x_n)$, possess continuous derivatives with respect to various combinations of x_s up to the order $(2n + 1)(p + M)$. Then these functions can be represented by the Fourier integrals:

$$\varphi_i(x_1, \ldots, x_n) = \int E_{\varphi_i}(\alpha_1, \ldots, \alpha_n) e^{i(\alpha, x)} \, d\alpha_1 \cdots d\alpha_n ,$$

$$f_i(t, x_1, \ldots, x_n) = \int E_{f_i}(t, \alpha_1, \ldots, \alpha_n) e^{i(\alpha, x)} \, d\alpha_1 \cdots d\alpha_n .$$

Denote by $v_i^*(t, \alpha_1, \ldots, \alpha_n)$ a solution of the following system

$$\frac{dv_i^*}{dt} = \sum_{j=1}^{N} \sum_{(k_s)} A_{ij}^{(k_1, \ldots, k_n)}(t)(i\alpha_1)^{k_1} \cdots (i\alpha_n)^{k_n} v_j^* + E_{f_i}(t, \alpha_1, \ldots, \alpha_n) , \quad (27)$$

satisfying the initial conditions

$$v_i^*(0, \alpha_1, \ldots, \alpha_n) = 0 \qquad (i = 1, \ldots, n) . \qquad (28)$$

We claim that *the solution of the Cauchy problem formulated above is given by*

$$u_i(t, x) = \int \sum_l E_{\varphi_l}(\alpha) v_i^{(l)}(t, \alpha) \, e^{i(\alpha, x)} \, d\alpha + \int v_i^*(t, \alpha) \, e^{i(\alpha, x)} \, d\alpha , \qquad (29)$$

where the functions $v_i^{(l)}(t, \alpha_1, \ldots, \alpha_n)$ *are equal to* $v_i^{(l)}(t, 0, \alpha_1, \ldots, \alpha_n)$ *and the domain of integration is the entire real* $(\alpha_1, \ldots, \alpha_n)$-*space.*

We also claim that *the functions* u_i *and all their partial derivatives, with respect to various combinations of* x_k *up to the order* M, *and therefore their first order derivatives in* t, *have absolute values bounded by*

$$\frac{C}{[|x_1| + 1]^2 \cdots [|x_n| + 1]^2} , \qquad (30)$$

where C *is a constant.*

Proof. First we prove that the integrals occupying the first position in the right-hand side of (29) satisfy the homogeneous system obtained from (1') by deleting $f_i(t, x_1, \ldots, x_n)$; and for $t = 0$ these integrals take the values $\varphi_i(x_1, \ldots, x_n)$, respectively. According to Lemma 3, we have

$$|E_{\varphi_i}(\alpha_1, \ldots, \alpha_n)| \le \frac{C_1}{[|\alpha_1| + 1]^{k_1} \cdots [|\alpha_n| + 1]^{k_n}} ,$$

where C_1 is constant, and k_s are arbitrary non-negative integers whose sum does not exceed $2n + p + M$.

By (26) we have

$$|v_i^{(l)}(t, \alpha_1, \ldots, \alpha_n)| \leq C[\alpha_M + 1]^p .$$

Therefore, the integrals

$$\int \sum_i E_{\varphi_l}(\alpha) v_i^{(l)}(t, \alpha) \, e^{i(\alpha, x)} \, d\alpha , \tag{31}$$

as well as their formal derivatives in x_k up to the order M, are uniformly convergent. In exactly the same way we can verify for these integrals uniform convergence of their formal first order derivatives in t. For this purpose, it suffices to notice that the functions $v_i^{(l)}(t, \alpha_1, \ldots, \alpha_n)$ satisfy system (2). Since the functions $v_i^{(l)}$ satisfy the homogeneous system (2), it follows that the integrals (31) satisfy the homogeneous system (1'). The fact that for $t = 0$ these integrals take the values $\varphi_i(x_1, \ldots, x_n)$ is a direct consequence of the initial conditions for $v_i^{(l)}$, viz.: $v_l^{(l)}(0) = 1$, $v_i^{(l)} = 0$ if $i \neq l$.

Now let us prove that the integrals occupying the second position in the right-hand sides of (29) satisfy the nonhomogeneous system (1') and vanish at $t = 0$. To this end, we first note that, under the above assumptions concerning $f_i(t, x_1, \ldots, x_n)$, Lemma 3 yields

$$|E_{f_i}(t, \alpha_1, \ldots, \alpha_n)| \leq \frac{C_3}{[|\alpha_1| + 1]^{k_1} \cdots [|\alpha_n| + 1]^{k_n}} ,$$

where C_3 is constant, and k_s are non-negative integers whose sum does not exceed $2n + M + p$. Therefore, by virtue of Lemma 1, we have

$$|v_i^*(t, \alpha_1, \ldots, \alpha_n)| \leq \frac{C_4}{[|\alpha_1| + 1]^{k_1} \cdots [|\alpha_n| + 1]^{k_n}} ,$$

where k_s are arbitrary non-negative integers whose sum is not greater than $2n + M$. It follows that the integrals

$$\int v_i^*(t, \alpha) \, e^{i(\alpha, x)} \, d\alpha \tag{32}$$

themselves, as well as their formal partial derivatives with respect to various combinations of x_k up to the order M, are uniformly convergent, and thereby the differentiation under the sign of the integral is justified. Taking into account that the functions v_i^* satisfy system (27), one can easily justify the differentiation with respect to t under the sign of the integral. Hence it can be easily seen that the integrals (32) satisfy system (1') and vanish at $t = 0$.

Now let us show that the estimates (30) hold for the functions (29) and their partial derivatives with respect to various combinations of x_k up to the order M, and therefore, hold for their first derivatives in t. To this end, because of Lemma 4, it suffices to show that the absolute values of the functions

$$E_{\varphi_l}(\alpha_1,\ldots,\alpha_n)v_i^{(l)}(t,\alpha_1,\ldots,\alpha_n),\qquad v_i^*(t,\alpha_1,\ldots,\alpha_n)$$

and of all their partial derivatives with respect to various combinations of α_k up to the order $2n$ are bounded by

$$\frac{C_5}{[|\alpha_1|+1]^{p_1+2}\cdots[|\alpha_n|+1]^{p_n+2}}, \tag{33}$$

where C_5 is a constant, and p_s are arbitrary non-negative integers whose sum is not greater than M. In order to prove this statement, it suffices to make the following observation:

The formulas

$$E_{\varphi_i}(\alpha_1,\ldots,\alpha_n) = \frac{1}{(2\pi)^n}\int_{Q_a} \varphi_i(x)\,e^{-i(\alpha,x)}\,dx$$

directly imply that the absolute values of these functions and of all their derivatives with respect to various combinations of α_s up to the order m are bounded by

$$\frac{C_6}{[|\alpha_1|+1]^{k_1}\cdots[|\alpha_n|+1]^{k_n}},$$

where C_6 is a constant, and k_s are non-negative integers whose sum does not exceed $(2n+1)(p+M)$. According to the Corollary of Lemma 2, the functions $v_i^{(l)}(t,\alpha)$ and their partial derivatives with respect to various combinations of α_s up to the order m $(2n \geq m \geq 0)$ have absolute values bounded by

$$C_7[\alpha_M+1]^{p+m(p+M-1)}.$$

Therefore, the estimate (33) holds for the absolute values of all products

$$E_{\varphi_l}(\alpha_1,\ldots,\alpha_n)v_i^{(l)}(t,\alpha_1,\ldots,\alpha_n)$$

and of all their partial derivatives with respect to various combinations of α_s up to the order $2n$.

The formulas

$$E_{f_i}(t,\alpha) = \frac{1}{(2\pi)^n}\int_{Q_a} f_i(t,x)\,e^{-i(\alpha,x)}\,dx$$

imply that these functions, together with all their partial derivatives with respect to various combinations of α_s up to the order $2n$, have absolute values bounded by

$$C_7[\alpha_M + 1]^{-(2n+1)(p+M)} \, .$$

Applying Lemma 2, we find that the functions $v_i^*(t, \alpha_1, \ldots, \alpha_n)$, as well as all their partial derivatives with respect to various combinations of α_s up to the order $2n$, have absolute values bounded by

$$C_8[\alpha_M + 1]^{-2n-M} \, ,$$

where C_8 is a constant. Hence, the estimate (33) is obtained for

$$v_i^*(t, \alpha_1, \ldots, \alpha_n) \, .$$

Now, in order to get rid of the restriction imposed on the functions f_i and φ_i (namely, that they may differ from 0 only inside Q_a) we proceed as follows. Let us construct a function

$$\Phi_a(x_1, \ldots, x_n)$$

with the following properties:

1) $0 \le \Phi_a \le 1$;

2) Φ_a possesses continuous derivatives with respect to all combinations of $x_1, \ldots, x_n$ up to the order $(M + p)(2n + 1)$;

3) $\Phi_a = 1$ in the cube $Q_a : |x_s| \le a$;

4) $\Phi_a = 0$ outside $Q_{\frac{3}{2}a}$.

Consider a function $f(x_1, \ldots, x_n)$ defined everywhere in the $(x_1, \ldots, x_n)$-space. We divide this space into cubes using the planes

$$x_s = n_s \, ,$$

where n_s are arbitrary odd numbers, positive as well as negative. Let us number the centers of these cubes in some order, so that

$$(x_1^{(i)}, \ldots, x_n^{(i)})$$

is the i-th center. Set

$$f^{(1)}(x_1, \ldots, x_n) = f(x_1, \ldots, x_n)\Phi_1(x_1, \ldots, x_n)$$

and

$$f^{(k)}(x) = \left[f(x) - \sum_{m=1}^{k-1} f^{(m)}(x) \right] \Phi_1(x - x^{(k)}) \, ,$$

where $x = (x_1, \ldots, x_n)$, $x^{(k)} = (x_1^{(k)} \ldots x_n^{(k)})$. It is easy to see that

$$f(x_1, \ldots, x_n) = \sum_{k=1}^{\infty} f^{(k)}(x_1, \ldots, x_n) \,,$$

and at each point $(x_1, \ldots, x_n)$ no more than 2^n functions from among $f^{(k)}(x_1, \ldots, x_n)$ can be different from 0.

Assume that *in the entire $(x_1, \ldots, x_n)$-space the functions $\varphi_i(x_1, \ldots, x_n)$ and $f_i(t, x_1, \ldots, x_n)$ possess bounded continuous derivatives with respect to various combinations of x_s up to the order $(M + p)(2n + 1)$*. For these functions we construct $\varphi_i^{(k)}$ and $f_i^{(k)}$ in the same way as $f^{(k)}$ has been constructed for f. Denote by $u_i^{(k)}$ the functions taking the respective initial values $\varphi_i^{(k)}$ for $t = 0$, and for $t \geq 0$ satisfying system $(1'^{(k)})$ obtained from $(1')$ by the replacement of f_i with $f_i^{(k)}$. As shown above, such functions $u_i^{(k)}$ exist and, together with their partial derivatives entering system $(1')$, have absolute values bounded by

$$\frac{C}{[|x_1 - x_1^{(k)}| + 1]^2 \cdots [|x_n - x_n^{(k)}| + 1]^2} \,. \tag{34}$$

Therefore the functions

$$u_i = \sum_{k=1}^{\infty} u_i^{(k)} \tag{35}$$

satisfy system $(1')$ for $t \geq 0$, and take the values φ_i for $t = 0$.

In conclusion, it should be pointed out that the solutions $u_i(x_1, \ldots, x_n)$, constructed above, depend continuously on the initial functions and their derivatives up to the order $(M + p)(2n + 1)$. This statement follows from the fact that the difference of any two solutions of system $(1')$ satisfies the homogeneous system obtained from $(1')$ after deleting f_i, and for $t = 0$ is equal to the difference of the corresponding initial functions. If the initial functions $\varphi_i(x_1, \ldots, x_n)$ and all their partial derivatives up to the order $(M + p)(2n + 1)$ are small enough, and the functions $f_i(x_1, \ldots, x_n)$ are identical to 0, then the previous arguments clearly show that the constant C in the estimates (30) and (34) can be made arbitrarily small. Therefore, the functions (35) and all their derivatives entering system $(1')$ will also be small in absolute value.

§4. Proof of the Uniqueness

Obviously, we can limit ourselves to the case of $n_i = 1$ for all i. In order to prove the uniqueness, let us assume the contrary, i.e., that there exist two different solutions

$$\bar{u}_i(t, x_1, \ldots, x_n) \quad \text{and} \quad \bar{\bar{u}}_i(t, x_1, \ldots, x_n)$$

of system $(1')$ for $t_0 \leq t \leq T$, with the same initial values at $t = t_0$.

Assume that the functions $\bar{u}_i$ and $\bar{\bar{u}}_i$, together with all their derivatives in x_k up to the order $(M-1)$ (we denote the latter derivatives by $D^* u_i$), have bounded absolute values; the derivatives in x_s up to the order M and the first order derivatives in t are assumed to be continuous for $t_0 \leq t \leq T$. Set $\bar{u}_i - \bar{\bar{u}}_i = u_i$. Obviously, the functions $u_i(t, x_1, \ldots, x_n)$ vanish at $t = t_0$, are bounded, together with their derivatives $D^* u_i$, and satisfy, for $t_0 \leq t \leq T$, the homogeneous system obtained from $(1')$ by deleting f_i. The latter system will be denoted by $(1'')$. In order to simplify the notations, we assume, as in the preceding section, that $t_0 = 0$. Let $(\tau, x_1^0, \ldots, x_n^0)$, $\tau > t_0$, be a point where at least one of the functions u_i, say u_1, does not vanish. Then we multiply the left-hand sides of the equations

$$\frac{\partial u_i}{\partial t} - \sum_{j=1}^{N} \sum_{(k_s)} A_{ij}^{(k_1,\ldots,k_n)}(t) \frac{\partial^{k_1 + \cdots + k_n} u_j}{\partial x_1^{k_1} \cdots \partial x_n^{k_n}} = 0 \qquad (i = 1, \ldots, N) \qquad (36)$$

by v_i, respectively (at this stage, we only assume the functions v_i to be sufficiently smooth, and later on, they are specified with greater accuracy). For the sake of brevity, denote the left-hand side of the i-th equation in (36) by $L_i(u)$. Let us integrate the sum

$$\sum_{i=1}^{N} v_i L_i(u)$$

over the cylinder R_a defined by the inequalities

$$|x_s| \leq a, \quad 0 \leq t \leq T .$$

Each term of the sum should be integrated by parts in such a way that, under the sign of the integral over R_a, only expressions containing no derivatives of u_j remain. These expressions can be written in the form

$$\sum_{i=1}^{N} \left\{ \frac{\partial v_i}{\partial(-t)} u_i - \sum_{j=1}^{N} \sum_{(k_s)} A_{ij}^{(k_1,\ldots,k_n)}(t) \frac{\partial^{k_1 + \cdots + k_n} v_i}{[\partial(-x_1)]^{k_1} \cdots [\partial(-x_n)]^{k_n}} u_j \right\} =$$

$$= \sum_{i=1}^{N} u_i \left\{ \frac{\partial v_i}{\partial(-t)} - \sum_{j=1}^{N} \sum_{(k_s)} A_{ji}^{(k_1,\ldots,k_n)}(t) \frac{\partial^{k_1 + \cdots + k_n} v_j}{[\partial(-x_1)]^{k_1} \cdots [\partial(-x_n)]^{k_n}} \right\} .$$

The expressions integrated over the lateral surface of R_a can be written as sums with the terms having the form

$$A_{ij}^{(k_1,\ldots,k_n)}(t) \frac{\partial^{k_1 + \cdots + k_n} v_i}{\partial x_1^{k_1} \cdots \partial x_n^{k_n}} \frac{\partial^{p_1 + \cdots + p_n} u_j}{\partial x_1^{p_1} \cdots \partial x_n^{p_n}} ,$$

where $\sum k_s$ and $\sum p_s$ do not exceed $M - 1$, M being the highest order of the derivatives in system $(1')$. The integral over the bottom $(t = 0)$ of the cylinder R_a is equal to zero, since $u_i = 0$ for $t = 0$. The integral over the top of R_a is equal to

$$\int \sum_{i=1}^{N} u_i v_i \, dx_1 \cdots dx_n .$$

Now, let us choose functions v_i which satisfy the following conditions:

1. For t from the interval $(0, \tau)$ and all real values of x_s, the functions v_i satisfy the equations

$$\frac{\partial v_i}{\partial(-t)} = \sum_{j=1}^{N} \sum_{(k_s)} A_{ij}^{(k_1, \ldots, k_n)}(t) \frac{\partial^{k_1 + \cdots + k_n} v_j}{[\partial(-x_1)]^{k_1} \cdots [\partial(-x_n)]^{k_n}} \qquad (37)$$

$$(i = 1, \ldots, N) .$$

This system is called *conjugate* to system (1').

2. The functions $v_i(t, x_1, \ldots, x_n)$, as well as their partial derivatives with respect to various combinations of x_s up to the order $(M - 1)$, satisfy the estimates of the form

$$|v_i(t, x_1, \ldots, x_n))| \leq \frac{C}{[|x_1| + 1]^2 \cdots [|x_n| + 1]^2} , \qquad (38)$$

where C is constant.

3. For $t = \tau$, all v_i, $i \neq 1$, vanish identically, and $v_1(\tau, x_1, \ldots, x_n)$ is defined as follows. Outside a small neighborhood of the point $(\tau, x_1^0, \ldots, x_n^0)$ where $|u_1|$ is larger than a positive constant ε,

$$v_1(\tau, x_1, \ldots, x_n) = 0 ,$$

and v_1 is positive within that neighborhood. For $0 \leq \tau \leq t_0$, the functions $v_1(\tau, x_1, \ldots, x_n)$ possess continuous derivatives in x_s up to the order M.

Now, let us pass to the limit in the identity

$$\int_{R_a} v_i L_i(u) \, dt \, dx_1 \cdots dx_n = 0 ,$$

as $a \to +\infty$, after transforming its left-hand side as indicated above. Taking into account the estimates (38) and the assumption that the functions u_i and their partial derivatives $D^* u_i$ are bounded, we see that the integrals over the lateral surface of R_a tend to 0 as $a \to \infty$. The limit expression has the form

$$\int u_1(\tau, x) v_1(\tau, x) \, dx = 0 ,$$

where the domain of integration is the entire hyperplane $t = \tau = \text{const.}$
Obviously, this is impossible[6].

It remains to show that functions $v_i(\tau, x)$ with the above properties
can actually be constructed. For this purpose, it suffices to establish the
following

Lemma. *Assume that Condition A holds for system* (1'). *Let* $w_i^{(l)}(t, t_0,$
$\alpha_1, \ldots, \alpha_n)$ *be functions which satisfy the equations*

$$\frac{dw_i}{d(-t)} = \sum_{j=1}^{N} \sum_{(k_s)} A_{ji}^{(k_1, \ldots, k_n)}(t)(i\alpha_1)^{k_1} \cdots (i\alpha_n)^{k_n} w_j \qquad (i = 1, \ldots, N), \quad (39)$$

on the interval $(0, t_0)$, $t_0 \leq T$, *and for* $t = t_0$ *satisfy the initial conditions*

$$w_i^{(l)}(t_0, t_0, \alpha_1, \ldots, \alpha_n) = 0 \quad \text{if} \quad i \neq l \,, \tag{40}$$

$$w_i^{(i)}(t_0, t_0, \alpha_1, \ldots, \alpha_n) = 1 \,.$$

Then, there exist constants C_1^* *and* p_1^*, *independent of* t_0, α_s *and such that*

$$|w_i^{(l)}(t, t_0)| \leq C_1^*[\alpha_M + 1]^{p_1^*}$$

on the interval $(0, t_0)$.

This Lemma implies that system (37) satisfies Condition A on the interval $(0, T)$, and therefore the Cauchy problem for (37) is uniformly well-
posed, provided that the initial conditions are prescribed at $t = t_0$ and the
solution is sought for on the interval $(0, t_0)$. Moreover, we can also conclude
that the estimates (38) hold for the solutions of this problem, if the initial
functions are sufficiently smooth and vanish identically outside a certain
bounded domain. Therefore, this Lemma implies the existence of functions
v_i with the above properties.

Proof. System (39) is the conjugate one to (2). Any two solutions $v_i(t)$ and
$w_i(t)$ of conjugate systems (2) and (39), respectively, are known (see [2],
Vol. 2, Part 2, §419) to be related by

$$\sum_{i=1}^{N} v_i(t) w_i(t) = C \,,$$

where C is a constant independent of t. In particular, let us write the
above relation for the solution $v_i^{(l)}(t, t_1)$ of system (2) and any fixed solution

---

[6]In order to drop the condition that the derivatives of u_i in x_s be continuous at the
points of the plane $t = 0$, we can integrate the sum $\sum v_i L_i(u)$ over a thin layer in R_a
between the planes $t = \varepsilon > 0$ and $t = \tau$, and then make ε tend to 0.

$w_i^{(m)}(t, t_0)$ $(i = 1, \ldots, N)$ of system (39), assuming that $0 \leq t_1 \leq t_0$. We thus obtain the following linear system for $w_i^{(m)}(t, t_0)$:

$$\sum_{i=1}^{N} v_i^{(l)}(t, t_1) w_i^{(m)}(t, t_0) = C_{lm} \qquad (l = 1, \ldots, N) . \tag{41}$$

In order to estimate the constants C_{lm}, set $t = t_0$ in the above equations. Then, by virtue of (40), we find:

$$v_m^{(l)}(t_0, t_1) = C_{lm} ,$$

and therefore,

$$|C_{lm}| \leq C[\alpha_M + 1]^p ,$$

since Condition A is satisfied for system (1') and $t_1 \leq t_0$.

Setting $t = t_1$ in (41), we obtain

$$w_l^{(m)}(t_1, t_0) = C_{lm} ,$$

and therefore

$$|w_l^{(m)}(t_1, t_0)| \leq C[\alpha_M + 1]^p ,$$

since t_1 is an arbitrary point of the interval $(0, t_0)$. The Lemma is proved.[7]

§5. Systems with Constant Coefficients

Consider a system with constant coefficients:

$$\frac{\partial^{n_i} u_i}{\partial t^{n_i}} = \sum_{j=1}^{N} \sum_{(k_s)} A_{ij}^{(k_0, k_1, \ldots, k_n)} \frac{\partial^{k_0 + k_1 + \cdots + k_n} u_j}{\partial t^{k_0} \partial x_1^{k_1} \cdots \partial x_n^{k_n}} + f_i(t, x_1, \ldots, x_n) \tag{42}$$

$$(i = 1, \ldots, N) .$$

We can formulate the following *necessary and sufficient condition for the Cauchy problem for system (42) to be uniformly well-posed: for all real α_s, the determinant of the matrix*

$$\left\| \sum_{(k_s)} A_{ij}^{(k_0, k_1, \ldots, k_n)} \lambda^{k_0} (i\alpha_1)^{k_1} \cdots (i\alpha_n)^{k_n} \right\| - \left\| \begin{matrix} \lambda^{n_1} & & & 0 \\ & \lambda^{n_2} & & \\ & & \ddots & \\ 0 & & & \lambda^{n_N} \end{matrix} \right\| , \tag{43}$$

---

[7]It would be interesting to find out whether a uniqueness result similar to the theorem just proved holds for the Cauchy problem in the case of systems that do not satisfy Condition A [4].

Another interesting question is whether the uniqueness theorem for the Cauchy problem is still valid, if merely the functions $u_i(t, x_1, \ldots, x_n)$ themselves are required to be bounded and satisfy system (1) only for $t > t_0$ [5].

where $\sum_{(k_s)}$ implies summation over all non-negative integers k_s, can only have roots λ with real part $\leq p\ln[\alpha_M+1]+\ln C$, where p and C are positive constants; α_M is, as usual, the largest amongst $|\alpha_s|$.[8]

First we establish the following

Lemma 5. *Let*

$$\frac{dv_i}{dt} = \sum_{j=1}^{N} a_{ij}v_j \qquad (i=1,\ldots,N) \tag{44}$$

be a given system with constant coefficients. Let

$$|a_{ij}| \leq K \ .$$

Then there exists a linear transformation

$$w_i = \sum_{j=1}^{N} C_{ij}v_j \qquad (i=1,\ldots,N) \tag{45}$$

with constant coefficients which reduces system (44) to the form

$$\frac{dw_i}{dt} = \sum_{j=1}^{i} a_{ij}^* w_j \qquad (i=1,\ldots,N) \ , \tag{46}$$

where $|a_{ij}^*|$, for $j < i$, can be made smaller than any given positive ε (we assume that $\varepsilon \leq (N-1)! \cdot 2^N K$), and $a_{ii}^* = \lambda_i$, λ_i being the roots of the determinant of the matrix

$$\|a_{ij}\| - \lambda E \ . \tag{47}$$

The absolute value of the determinant corresponding to system (45) is equal to

$$\left[(N-1)! \cdot 2^N \frac{K}{\varepsilon}\right]^{\frac{N(N-1)}{2}} ,$$

and

$$|C_{ij}| \leq \left[(N-1)! \cdot 2^N \frac{K}{\varepsilon}\right]^{N-1} .[9]$$

Proof. It is well-known that there always exist constants b_i ($b_i \neq 0$ for some i) such that the function

$$w = \sum_{i=1}^{N} b_i v_i$$

---

[8]It seems possible that this condition is equivalent to the following one: every root of the determinant of the matrix (43) has its real part bounded from above, for all real values of α_s [6].

[9]These assertions, although somewhat stronger than those needed in this section, are proved here with a view to further applications.

satisfies the equation

$$\frac{dw}{dt} = \lambda_1 w \, ,$$

where λ_1 is a root of the characteristic equation for system (44), i.e., a root of the determinant of the matrix (47).

Obviously, the above condition is insufficient to determine the coefficients b_i uniquely: for instance, they can all be multiplied by an arbitrary constant. Let us choose this constant in such a way that the largest of $|b_i|$ becomes equal to 1. For definiteness, let $|b_1| = 1$. Next, we make the following linear transformation

$$v_1^{(1)} = \sum_{j=1}^{N} b_j v_j \, , \quad v_i^{(1)} = v_i \quad \text{for} \quad i > 1 \, . \tag{48}$$

The absolute value of the determinant of this transformation is equal to 1, and the absolute values of its coefficients are ≤ 1. Let us find v_i from (48) and substitute them into (44). We get

$$\frac{dv_1^{(1)}}{dt} = \lambda_1 v_1^{(1)} \, , \tag{49}$$

$$\frac{dv_i^{(1)}}{dt} = a_{i1}^{(1)} v_1^{(1)} + \sum_{j=2}^{N} a_{ij}^{(1)} v_j^{(1)} \quad (i = 2, \dots, N) \, .$$

It is obvious that

$$|a_{ij}^{(1)}| \leq 2K \, .$$

The roots of the characteristic equation for system (49) are known to coincide with those of the characteristic equation for system (44). It follows that the roots of the determinant of the matrix

$$\left\| a_{ij}^{(1)} \right\| - \lambda E \, ,$$

where $i, j = 2, \dots, N$, and E is the identity matrix of size $N - 1$, are the same as the roots of the determinant of (47), with the exception of λ_1.

Let us apply to the latter $N - 1$ equations in (49) the arguments similar to those used for system (44), assuming the function $v_1^{(1)}$ to be known. Then we find a linear transformation such that the absolute value of the corresponding determinant is equal to 1, the absolute values of the coefficients are bounded by 1, and after that transformation, system (49) becomes

$$\frac{dv_1^{(2)}}{dt} = \lambda_1 v_1^{(2)} \, ,$$

$$\frac{dv_2^{(2)}}{dt} = a_{21}^{(2)} v_1^{(2)} + \lambda_2 v_2^{(2)} \, ,$$

$$\frac{dv_i^{(2)}}{dt} = a_{i1}^{(2)} v_1^{(2)} + a_{i2}^{(2)} v_2^{(2)} + \sum_{j=3}^{N} a_{ij}^{(2)} v_j^{(2)} .$$

Here $v_1^{(1)} = v_1^{(2)}$, λ_2 is the second root of the determinant (47),

$$|a_{21}^{(2)}| \leq (N-1)2K \quad \text{and} \quad |a_{ij}^{(2)}| \leq 2^2 K \quad \text{for} \quad i > 2 .$$

Combining these two inequalities, we can write

$$|a_{ij}^{(2)}| \leq (N-1) \cdot 2^2 K .$$

Proceeding in like manner, we arrive at the following system

$$\frac{dv_i^{(N)}}{dt} = \sum_{j=1}^{i} a_{ij}^{(N)} v_j^{(N)} \qquad (i = 1, \dots, N) . \tag{50}$$

Here $a_{ii}^{(N)} = \lambda_i$. As regards the rest of $a_{ij}^{(N)}$, it can only be claimed that their absolute values are bounded by

$$(N-1)! \cdot 2^N K .$$

Now we set

$$w_i = v_i^{(N)} \left(\frac{(N-1)! \cdot 2^N K}{\varepsilon} \right)^{N-i} . \tag{51}$$

Thus we obtain a system for which all assertions of this lemma are valid. The determinant of the transformation reducing (44) to (50) has absolute value equal to 1. The determinant of the transformation (51) is equal to

$$\left((N-1)! \cdot 2^N \frac{K}{\varepsilon} \right)^{\frac{N(N-1)}{2}} .$$

Consequently, the determinant of the transformation reducing (44) to (46) has the same absolute value.

Next, we turn to the proof of the Theorem formulated at the beginning of this section.

Proof of the Theorem in the case of $n_i = 1$ for all i. In order to find out whether Condition A holds for system (42) in this case, let us reduce the system

$$\frac{dv_i}{dt} = \sum_{j=1}^{N} \sum_{(k_s)} A_{ij}^{(k_1, \dots, k_n)} (i\alpha_1)^{k_1} \cdots (i\alpha_n)^{k_n} v_j$$

to the form (46), making use of the preceding lemma and setting $\varepsilon = (N-1)! \cdot 2^N K$. The equations forming system (46) can be integrated in a successive order starting from the first equation. If the determinant of the matrix (43) has all its roots subject to the inequality

$$\Re\{\lambda_i\} < p\ln[\alpha_M + 1] + \ln C , \qquad (52)$$

then system (42) satisfies Condition A. Indeed, in this case, integrating the equations of system (46) in successive order and applying Lemma 1, we find that the functions $w_j^{(l)}(t, t_0)$[10], which are bounded for $t = t_0$, satisfy the inequalities

$$|w_j^{(l)}(t, t_0)| < C[\alpha_M + 1]^{p_1 T} \quad (t \geq T) ,$$

and therefore, by virtue of (45), similar inequalities hold for the functions $v_i^{(l)}(t, t_0)$.

Now let us show that if, on the other hand, there exist no such constants p and C that the inequality (52) is satisfied, then Condition A does not hold for system (42) with all $n_i = 1$. We enumerate the roots of the determinant of the matrix (43) in such a way that it is for λ_1 that the inequality (52) cannot hold with any of the constants p and C. Then the function $w_1(t)$ will grow faster than any polynomial of $\alpha_M + 1$. Since the determinant of the transformation (45) is equal to 1 (it is assumed again that $\varepsilon = (N-1)! \cdot 2^N K$), and the coefficients are bounded, it follows that at least one of the functions $v_i(t)$ has to grow faster than any polynomial of $\alpha_M + 1$, and therefore Condition A is indeed violated for system (42).

Proof of the Theorem in the case of $n_i > 1$ for some i. As indicated above, this case can always be reduced to the case of all $n_i = 1$, if

$$\frac{\partial u_i}{\partial t} , \frac{\partial^2 u_i}{\partial t^2} , \ldots, \frac{\partial^{n_i-1} u_i}{\partial t^{n_i-1}}$$

are regarded as new unknown functions u_j ($j > N$). The resulting system, denoted by (42'), is obviously equivalent to the old one. Now it only remains to find out the relation between the determinants of the characteristic matrices for these two systems. To this end, the following observation can be used. The condition that the characteristic matrices of systems (42) and (42') have zero determinants is necessary and sufficient for homogeneous systems (42) and (42') to have non-trivial solutions of the form

$$u_i = C_i e^{\lambda t + i(\alpha, x)} .$$

Obviously, if one of these systems admits a solution of this type, the other

---

[10]The functions $w_i^{(l)}(t, t_0)$ are related to $v_i^{(l)}(t, t_0)$ by (45).

must also admit solutions of this type. Therefore, the roots of the determinants for both systems must be the same. It follows that if these determinants possess multiple roots, the multiplicities of these roots must be equal for both determinants, since the case of multiple roots may be considered as the limit case of different roots. It should be noticed that the terms of the highest order with respect to λ are the same in both determinants. Therefore, the determinants for both systems coincide [7]. Now, if we also take into account the results of §1, we may consider our purpose achieved.

§6. Single Equation with the First Order Partial Derivative in t.

To give an instance of easy verification of Condition A, consider an equation of the form

$$\frac{\partial u}{\partial t} = \sum_{(k_s)} A^{(k_1,k_2,\ldots,k_n)}(t) \frac{\partial^{k_1+k_2+\cdots+k_n} u}{\partial x_1^{k_1} \partial x_2^{k_2} \cdots \partial x_n^{k_n}} + f(t, x_1, \ldots, x_n) , \qquad (53)$$

where the summation is over all non-negative integers k_s whose sum does not exceed M; the coefficients $A^{(k_1,\ldots,k_n)}(t)$ are real-valued functions. The corresponding system (2) takes the form

$$\frac{dv}{dt} = \sum_{(k_s)} A^{(k_1,\ldots,k_n)}(t)(i\alpha_1)^{k_1} \cdots (i\alpha_n)^{k_n} v$$

and can be easily integrated:

$$v(t, t_0) = \exp\left\{ \int_{t_0}^{t} \sum_{(k_s)} A^{(k_1,\ldots,k_n)}(\tau)(i\alpha_1)^{k_1} \cdots (i\alpha_n)^{k_n} d\tau \right\} .$$

Hence it is easy to see that the coefficients by the odd order derivatives of u with respect to x_k have no effect on the well-posedness of the Cauchy problem for equation (53), even if the highest order of the derivatives is odd. In the case of this equation, the necessary and sufficient condition for the Cauchy problem on the interval $(0, T)$ to be uniformly well-posed reads as follows:

For all t, t_0 such that $0 \leq t_0 \leq t \leq T$, the real part of the expression

$$\sum_{(k_s)} \int_{t_0}^{t} A^{(k_1,\ldots,k_n)}(\tau) d\tau (i\alpha_1)^{k_1} \cdots (i\alpha_n)^{k_n} \qquad (54)$$

does not exceed

$$p \ln[\alpha_M + 1] + \ln C , {}^{11}$$

where p and C are constants. In particular, this condition holds, if for all t $(0 \le t \le T)$ the form

$$\sum_{(k_s)} A^{(k_1,\ldots,k_n)}(t)(i\alpha_1)^{k_1} \cdots (i\alpha_n)^{k_n} \tag{55}$$

is negative definite (the summation here is over all non-negative integers k_s whose sum is equal to M for M even, and equal to $M-1$ for M odd). Conversely, this condition is violated, if the form (55) takes positive values for at least one choice of real α_s.

§7. Example of a Non-Uniformly Well-Posed Problem

We say that the Cauchy problem for system (1) on the interval $(0, T)$ is *simply well-posed*, if the following conditions are satisfied:

1. Let

$$\varphi_i^{(k)}(x_1, \ldots, x_n) \quad (i = 1, \ldots, N ; \quad k = 0, 1, \ldots, n_i - 1)$$

be a system of functions bounded over the entire $(x_1, \ldots, x_n)$-hyperplane, together with all their partial derivatives up to a finite order L. Then there exists one and only one system of functions u_i such that u_i are bounded, together with all their partial derivatives of sufficiently high orders, satisfy system (1) on $(0, T)$, and for $t = 0$ take, together with their derivatives in t up to the order $(n_i - 1)$, the values $\varphi_i^{(k)}$, respectively.

2. For every given $\varepsilon > 0$ there exists $\eta > 0$ such that a variation of the functions $\varphi_i^{(k)}$, together with their partial derivatives up to the order L, by a quantity less than η causes a variation of the functions u_i by a quantity less than ε, for all $t : 0 \le t \le T$.

The following equation is an example of a system, for which the Cauchy problem on the interval $(0, 2)$ is well-posed, but *not* uniformly well-posed:

$$\frac{\partial u}{\partial t} = f(t) \frac{\partial^2 u}{\partial x^2} , \tag{56}$$

where the function $f(t)$ possesses the following properties:

---

[11]It seems possible that, in the case of sufficiently smooth coefficients $A^{(k_1,\ldots,k_n)}(t)$, this condition is equivalent to the following one: for all t $(0 \le t \le T)$, the real part of the polynomial

$$\sum_{(k_s)} A^{(k_1,\ldots,k_n)}(t)(i\alpha_1)^{k_1} \cdots (i\alpha_n)^{k_n}$$

is bounded from above [8].

1) $f(t)$ is a continuous function defined on $(0,2)$;

2) $f(t) \geq 0$ on $(0,1)$, $f(t) \leq 0$ on $(1,2)$; $f(t) = 0$ only at $t = 1$;

3) $\int_0^2 f(t)\, dt = \tau_1 > 0$.

The non-uniform well-posedness of the Cauchy problem for equation (56) on the interval $(0,2)$ is due to the fact that

$$\int_1^2 f(t)\, dt = -\tau_2 < 0$$

(see the preceding section).

In order to show that the Cauchy problem for this equation on the interval $(0,2)$ is well-posed, we proceed as follows. Equation (56) can be rewritten as

$$\frac{\partial u}{\partial \tau} = \frac{\partial^2 u}{\partial x^2}, \tag{57}$$

where

$$\tau = \int_0^t f(\xi)\, d\xi .$$

For t varying from 0 to 2 the value of τ, first, increases from 0 up to $\tau_1 + \tau_2$, and then decreases from $\tau_1 + \tau_2$ to τ_1. The Cauchy problem for equation (57) on the interval $(0, \tau_1 + \tau_2)$ is uniformly well-posed, and its solution $u(\tau, x)$ yields the solution

$$u\left(\int_0^t f(\xi)\, d\xi\, , x\right)$$

of the Cauchy problem for equation (56) on the interval $(0,2)$, the latter solution being continuously dependent on the initial functions. It remains to prove that this solution is unique. To this end we note that, because of the uniform well-posedness of the Cauchy problem for (56) on $(0,1)$, its solution is unique on $(0,1)$. On the other hand, if we assume the existence on $(1,2)$ of a bounded function $u(t, x)$ that vanishes at $t = 1$ but does not vanish identically, this would imply the existence of a bounded solution of equation (57) that vanishes at $t = \tau_1 + \tau_2$, and therefore, vanishes for $t > \tau_1 + \tau_2$, but does not vanish identically on $(\tau_1, \tau_1 + \tau_2)$. However, no such solution is possible. Indeed, the solution admits a representation by Poisson's integral, which implies its analyticity with respect to x and t for $t > \tau_1$, and therefore its identity to 0, since it vanishes identically for $t \geq \tau_1 + \tau_2$.

CHAPTER II

Parabolic Systems

§1. Sufficiency of Condition B

Theorem. *Assume that the coefficients $A_{ij}^{(k_1,...,k_n)}(t)$ are continuous functions. Let Condition B (see Introduction) be satisfied on the interval $(0, T)$. Then the Cauchy problem is uniformly well-posed on $(0, T)$.*

Proof. Let Condition B hold on the interval $(0, T)$. Then we can choose a positive integer p such that conditions (3) are satisfied, and all roots of the determinant of the matrix (4), for $\sum \alpha_s^2 = 1$, $0 \leq t \leq T$, have real parts less than a negative constant $-\delta$. We have introduced the term p-parabolic, or simply, parabolic, to designate systems which satisfy these conditions (see Introduction). In order verify Condition A for such systems, we start, as we have done many times before, with reducing system (1) to the form (1′) with $n_i = 1$ for all i. To this end, we denote by u_i $(i = N+1, N+2, \ldots, N_1)$ the functions

$$\frac{\partial u_i}{\partial t}, \frac{\partial^2 u_i}{\partial t^2}, \ldots, \frac{\partial^{n_i-1} u_i}{\partial t^{n_i-1}}.$$

Next, we introduce the following modifications in the system of ordinary differential equations (2) corresponding to system (1′). First, we set

$$\tau = t|\alpha|^p, \quad \text{where} \quad |\alpha| = \sqrt{\sum \alpha_s^2}; \tag{58}$$

secondly, if

$$\frac{d^{k_j} v_i}{dt^{k_j}} = v_j, \quad i \leq N, \quad k_j \geq 0,$$

we set

$$v_j = v_j^* |\alpha|^{(k_j - n_i)p}, \tag{59}$$

or (since n_i depends on j)

$$v_j = v_j^* |\alpha|^{(k_j - n(j))p}. \tag{59′}$$

Let us substitute for the functions v_i in (2) their expressions in terms of v_i^*. We then find that the functions v_i^* satisfy a system of equations whose coefficients remain bounded as $|\alpha| \to \infty$; the coefficients that do not enter the matrix (4) are uniformly convergent to 0 on the interval $(0, T)$ as $|\alpha| \to \infty$. Let this system have the form

$$\frac{dv_i^*}{d\tau} = \sum_{j=1}^{N_1} a_{ij}\left(\frac{\tau}{|\alpha|^p}, \alpha_1, \ldots, \alpha_n\right) v_j^* + \sum_{j=1}^{N_1} b_{ij}\left(\frac{\tau}{|\alpha|^p}, \alpha_1, \ldots, \alpha_n\right) v_j^* \quad (2^*)$$

$$(i = 1, \ldots, N_1) .$$

Here $b_{ij}\left(\frac{\tau}{|\alpha|^p}, \alpha_1, \ldots, \alpha_n\right)$ are the coefficients tending to 0 as $|\alpha| \to \infty$.

Let us show that *the determinant of the matrix*

$$\|a_{ij}(t, \alpha_1, \ldots, \alpha_n)\| - \lambda E \tag{60}$$

coincides with that of the matrix (4) *having* α_s *replaced by* $\alpha_s/|\alpha|$.

Fix a point $\tau_0^* = t_0^*|\alpha|^p$ from the interval $(0, T|\alpha|^p)$ and consider the following system

$$\frac{dv_i^*}{d\tau} = \sum_{j=1}^{N_1} a_{ij}\left(\frac{\tau_0^*}{|\alpha|^p}, \alpha_1, \ldots, \alpha_n\right) v_j^* \quad (i = 1, \ldots, N_1) . \tag{61}$$

The very method used for the construction of system (2^*) shows that system (61) is equivalent to the following one:

$$\frac{d^{n_i} v_i^*}{d\tau^{n_i}} = \sum_{j=1}^{N} \sum_{((k_s))_p} A_{ij}^{(k_0, k_1, \ldots, k_n)}(t_0^*) \left(\frac{i\alpha_1}{|\alpha|}\right)^{k_1} \cdots \left(\frac{i\alpha_n}{|\alpha|}\right)^{k_n} \frac{d^{k_0} v_j^*}{d\tau^{k_0}} , \tag{62}$$

where $\sum_{((k_s))_p}$ is used in the same sense as in Introduction.

Therefore, according to the remark at the end of §5, Ch. I, the determinants of the characteristic matrices are the same for both systems. Thereby our statement is proved.

Now let us fix a point $t_0^* = \tau_0^*/|\alpha|^p$ from the interval $(0, T)$ and apply to system (2^*) the linear transformation (45), given by a triangular matrix (see Lemma 5, §5, Ch. I), which reduces system (61) to the form

$$\frac{dw_i}{d\tau} = \sum_{j=1}^{i} a_{ij}^*(t_0^*, \alpha_1, \ldots, \alpha_n) w_j \quad (i = 1, \ldots, N_1) . \tag{63}$$

Here a_{ii}^* are equal to the roots of the determinant of the matrix (60) with real α_s such that $|\alpha| = 1$. Therefore, the real parts of a_{ii}^* always remain smaller than a negative $-\delta$. According to Lemma 5, the transformation (45) can be chosen in such a way as to make the absolute values of the coefficients a_{ij}^*, for $i \neq j$, arbitrarily small. The system obtained from (2^*) is denoted by $(63')$. Observe that: 1) for all t from the interval $(0, T)$, the coefficients $b_{ij}(t, \alpha_1, \ldots, \alpha_n)$ are arbitrarily small as compared to δ, for sufficiently large $|\alpha|$; 2) on the closed set: $0 \leq t \leq T$, $|\alpha| = 1$, the coefficients $a_{ij}(t, \alpha_1, \ldots, \alpha_n)$ are continuous functions of t and $\alpha_s/|\alpha|$. These two facts allow us to conclude that there exist positive constants ε, M_0 depending only on η and such that for sufficiently large $|\alpha|$ and any t_0^* from the open interval $(0, T)$ the following three properties hold for τ varying within the interval[12] $[(t_0^* - \varepsilon)|\alpha|^p, (t_0^* + \varepsilon)|\alpha|^p]$:

1) all coefficients of system (63') outside the main diagonal have absolute values smaller than an arbitrary given $\eta > 0$;

2) the coefficients on the main diagonal differ from $a_{ii}^*(t_0^*, \alpha_1, \ldots, \alpha_n)$ less than by η;

3) the determinant of the linear transformation reducing system (2*) to the form (63') has its absolute value between 1 and M_0; and the absolute values of the coefficients do not exceed M_0.

Then on the interval $[(t_0^* - \varepsilon)|\alpha|^p, (t_0^* + \varepsilon)|\alpha|^p]$ we have

$$\frac{d}{d\tau} \sum_i |w_i|^2 = \frac{d}{d\tau} \sum_i w_i \bar{w}_i =$$

$$= 2 \sum_i \Re\{a_{ii}^*(t_0^*, \alpha_1, \ldots, \alpha_n)\}|w_i|^2 + \sum_{ij} a_{ij}^{**}(t, \alpha_1, \ldots, \alpha_n) w_i \bar{w}_i ,$$

where

$$|a_{ij}^{**}(t, \alpha_1, \ldots, \alpha_n)| < 2\eta .$$

It follows that

$$\frac{d}{d\tau} \sum_i |w_i|^2 \leq [-2\delta + 2N_1\eta] \sum_i |w_i|^2 .$$

Let us take η so small that

$$\eta < \frac{\delta}{2N_1} .$$

Then

$$\frac{d}{d\tau} \sum_i |w_i|^2 \leq -\delta \sum_i |w_i|^2 ,$$

and therefore

$$\sum_i |w_i(\tau, t_0^*, \alpha_1, \ldots, \alpha_n)|^2 \leq$$

$$\leq \sum_i |w_i((t_0^* - \varepsilon)|\alpha|^p, t_0^*, \alpha_1, \ldots, \alpha_n)|^2 e^{-\delta(\tau - (t_0^* - \varepsilon)|\alpha|^p)} . \quad (64)$$

---

[12]If this interval happens to stretch outside $(0, T|\alpha|^p)$, then its portion belonging to $(0, T|\alpha|^p)$ should be taken.

Now let us go back to the functions $v_i^*(\tau, \alpha_1, \ldots, \alpha_n)$. Since the determinant of the transformation (45) has its absolute value between 1 and M_0, we find from (64) that on the interval $[(t_0^* - \varepsilon)|\alpha|^p, (t_0^* + \varepsilon)|\alpha|^p]$ the following inequality is satisfied:

$$\sum_i \left| v_i^* \left(\frac{\tau}{|\alpha|^p}, \alpha_1, \ldots, \alpha_n \right) \right|^2 \le$$

$$\le M_1 \sum_i |v_i^*((t_0^* - \varepsilon), \alpha_1, \ldots, \alpha_n)|^2 \, e^{-\delta(\tau - (t_0^* - \varepsilon)|\alpha|^p)} \, ,$$

where M_1 is a function of M_0. Applying the above inequality, in successive order, to the intervals

$$[t_0|\alpha|^p, (t_0 + \varepsilon)|\alpha|^p] \, , \quad [(t_0 + \varepsilon)|\alpha|^p, (t_0 + 3\varepsilon)|\alpha|^p] \, , \ldots,$$

which cover the entire interval $(t_0|\alpha|^p, T|\alpha|^p)$, their number being not larger than[13] $\left[\frac{T}{2\varepsilon} \right] + 2 = m$, we find that

$$\sum_i \left| v_i^* \left(\frac{\tau}{|\alpha|^p}, \alpha_1, \ldots, \alpha_n \right) \right|^2 \le$$

$$\le M_1^m \sum_i |v_i^*(t_0, \alpha_1, \ldots, \alpha_n)|^2 \, e^{-\delta(\tau - t_0|\alpha|^p)} \, . \qquad (64')$$

Set

$$v_j^{(l)} = v_j^{(l)*} |\alpha|^{(k_j - n_i)p} \, , \qquad (65)$$

where the functions $v_j^{(l)}$ are such that

$$v_j^{(l)}(t_0, \alpha_1, \ldots, \alpha_n) = \begin{cases} 0 & \text{if } j \ne l, \\ 1 & \text{if } j = l, \end{cases}$$

and k_j is the same as in (59). Then, for $|\alpha| \ge 1$ we have

$$\sum_i |v_i^{(l)*}(t, \alpha_1, \ldots, \alpha_n)|^2 \le |\alpha|^{2M} \, .$$

Therefore

$$\sum_i |v_i^{(l)*}(t, \alpha_1, \ldots, \alpha_n)|^2 \le M_1^m \, e^{-\delta |\alpha|^p (t - t_0)} |\alpha|^{2M} \, ;$$

hence

$$|v_i^{(l)*}(t, \alpha_1, \ldots, \alpha_n)| \le C^* e^{-\frac{\delta |\alpha|^p}{2}(t - t_0)} |\alpha|^M \le C^*(|\alpha| + 1)^M \, e^{-\frac{\delta |\alpha|^p}{2}(t - t_0)} , (66)$$

---

[13]Here $[x]$ stands for the entire part of x.

$$|v_i^{(l)}(t, \alpha_1, \ldots, \alpha_n)| \leq C(|\alpha| + 1)^{M-p} e^{-\frac{\delta|\alpha|^p}{2}(t-t_0)} , \qquad (67)$$

where

$$C^* = C = M_1^{\frac{m}{2}} . \qquad (68)$$

Therefore, Condition A is satisfied. Note also that an estimate of type (67) (possibly with another constant C) holds not only for large values of $|\alpha|$, but for arbitrary $|\alpha|$ as well, because the functions $v_i^{(l)}$ are bounded for bounded $|\alpha|$.

§2. Construction of the Green Matrix

1. It is henceforth assumed in this chapter, except in its last section, that the system we are dealing with is parabolic on the interval $(0, T)$, and $n_i = 1$ for all i. First we consider the case of a *homogeneous* system. As shown in §3, Ch. I, the assumptions at the beginning of §3 guarantee that the solution of the Cauchy problem with initial conditions prescribed at $t = t_0$ is given by

$$u_i(t, x) = \int \sum_l E_{\varphi_l}(\alpha) v_i^{(l)}(t, t_0, \alpha) e^{i(\alpha, x)} \, d\alpha , \qquad (69)$$

where $x = (x_1, \ldots, x_n)$, $\alpha = (\alpha_1, \ldots, \alpha_n)$,

$$E_{\varphi_l}(\alpha) = \frac{1}{(2\pi)^n} \int e^{-(\alpha, x)} \varphi_l(x) \, dx .$$

If we regard (69) as a multiple integral with respect to

$$x_1, \ldots, x_n, \alpha_1, \ldots, \alpha_n ,$$

then, changing the order of integration, we naturally arrive at the following system of functions

$$G_i^{(l)}(t_1, t_0, x, \xi) = \frac{1}{(2\pi)^n} \int e^{i(\alpha, x-\xi)} v_i^{(l)}(t, t_0, \alpha) \, d\alpha \qquad (70)$$

$$(i, l = 1, \ldots, N_1) ,$$

called the *Green matrix*. The estimates (67) imply that the integrals (70) are convergent for $t > t_0$.

Theorem. *The solution of the Cauchy problem in the layer (t_0, T) is given by the formulas*

$$u_i(t, x) = \int \sum_l G_i^{(l)}(t, t_0, x, \xi) \varphi_l(\xi) \, d\xi , \qquad (71)$$

valid under the assumption that the functions $\varphi_l(x_1, \ldots, x_n)$ are bounded and continuous, together with their partial derivatives up to the order $(M - p + + \Delta)$. Here $\Delta = 0$ if $M = p$; $\Delta = 1$ if $M > p$.

2. Lemma. *Assume that Condition B holds for system* (2). *Then the functions $v_i^{(l)}(t, t_0, \alpha_1, \ldots, \alpha_n)$ constructed for system* (2) *satisfy the following estimates*

$$\left| \frac{\partial^k v_i^{(l)}(t, t_0, \alpha_1, \ldots, \alpha_n)}{\partial \alpha_1^{k_1} \cdots \partial \alpha_n^{k_n}} \right| \leq$$

$$\leq C_k(|\alpha| + 1)^{M-p-k} \sum_{\substack{s > 0 \\ s \geq k - M}} (t - t_0)^s (|\alpha| + 1)^{ps} e^{-\frac{\delta |\alpha|^p}{2}(t - t_0)}, \quad (72)$$

where C_k is a constant independent of t, t_0, α_s;

$$|\alpha| = \sqrt{\sum \alpha_s^2}, \quad k = \sum k_s ;$$

for $k = 0$, it is assumed that $\sum\limits_{s>0}^{k} (t - t_0)^s (|\alpha| + 1)^{ps} = 1$.

Proof. In order to prove (72), we first show that for $k \geq 0$ and sufficiently large $|\alpha|$ the following inequality is valid:

$$\left| |\alpha|^{(n(i) - k_i)p} \frac{\partial^k v_i^{(l)}(t, t_0, \alpha_1, \ldots, \alpha_n)}{\partial \alpha_1^{k_1} \cdots \partial \alpha_n^{k_n}} \right| \leq$$

$$\leq C_k^*(|\alpha| + 1)^{M-k} \sum_{\substack{s > 0 \\ s \geq k - M}} (t - t_0)^s (|\alpha| + 1)^{ps} e^{-\frac{\delta |\alpha|^p}{2}(t - t_0)}, \quad (73)$$

For $k = 0$, the inequality (73) is obvious. In order to establish (73) for $k > 0$, we make the following observation. The functions $v_i^{(l)}$ satisfy the equations

$$\frac{dv_i}{dt} = \sum_j C_{ij}(t, \alpha_1, \ldots, \alpha_n) v_j , \quad (74)$$

where each $C_{ij}(t, \alpha_1, \ldots, \alpha_n)$ is a polynomial with arguments $\alpha_1, \ldots, \alpha_n$, whose degree is not larger than $(n(j) - j)p$, and the coefficients depend on t only. Assume now that the estimate (73) holds for $k = 0, 1, 2, \ldots, q$. Let us show that it holds for $k = q + 1$. To this end we differentiate system (74), $q + 1$ times with respect to a combination of $\alpha_1, \ldots, \alpha_n$ and denote the $(q + 1)$-th order derivative of v_i' by $\tilde{v}_i^{(l)}$. Then

$$\frac{d\tilde{v}_i^{(l)}}{dt} = \sum_j C_{ij}(t, \alpha_1, \ldots, \alpha_n)\tilde{v}_i^{(l)} +$$

$$+ \sum_j \sum_{\substack{s \geq 0 \\ s \geq q+1-M}} C_s^{(q)} D^{(q+1-s)} C_{ij}(t, \alpha_1, \ldots, \alpha_n) D^{(s)} v_j^{(l)} . \quad (75)$$

The symbol $D^{(s)}$ preceding a function is used here to denote an s-th order derivative of that function with respect to a combination of $\alpha_1, \ldots, \alpha_n$; $C_s^{(q)}$ are integers depending on s and q only.

We obviously have

$$D^{(k)} C_{ij}(t, \alpha_1, \ldots, \alpha_n) \equiv 0 \quad \text{for} \quad k > M .$$

It is for this reason that the summation in the right-hand side of (75) starts from $q + 1 - M$, if $q + 1 - M > 0$, and not from 0. For $i, l = 1, 2, \ldots, N$ and $s = 0, 1, \ldots, q$, we set

$$D^{(s)} v_i^{(l)} |\alpha|^{(n(i)-k_i)p} = D_i^s v_i^{(l)} ,$$

$$D^{(q+1)} v_i^{(l)} |\alpha|^{(n(i)-k_i)p} = \tilde{v}_i^{(l)} .$$

Then, multiplying the i-th equation in (75) by $|\alpha|^{(n(i)-k_i)}$, we easily find that

$$\frac{d\tilde{v}_i^{(l)}}{dt} = |\alpha|^p \sum_j C_{ij}^*(t, \alpha)\tilde{v}_j^{(l)} + \sum_j \sum_{\substack{s \geq 0 \\ s \geq q+1-M}}^q K_{ij}^{(p-q-1+s)}(t, \alpha) D_j^s v_j^{(l)} \quad (76)$$

$$(i, j = 1, 2, \ldots, N_1) .$$

Here $C_{ij}^*(t, \alpha_1, \ldots, \alpha_n)$ coincide with the coefficients by v_j^* in the right-hand sides of (2^*); and $K_{ij}^{(r)}(t, \alpha_1, \ldots, \alpha_n)$ is the ratio of a polynomial in $\alpha_1, \ldots, \alpha_n$ with coefficients depending only on t divided by a power of $|\alpha|$; moreover, for $|\alpha| \geq \alpha_0 = \text{const} > 0$, the following inequalities are satisfied

$$|K_{ij}^{(r)}(t, \alpha_1, \ldots, \alpha_n)| \leq C_1 (1 + |\alpha|)^r .$$

Here and in what follows, the indexed letter C is used to denote constants.

Let us introduce the functions $w_i^{(l)}(t, t_0, \alpha_1, \ldots, \alpha_n)$ which satisfy the equations

$$\frac{dw_i}{dt} = |\alpha|^p \sum_j C_{ij}^*(t, \alpha_1, \ldots, \alpha_n) w_i \quad (i, j = 1, \ldots, N_1) ,$$

for $t > t_0$, and for $t = t_0$, the initial conditions

$$w_i^{(l)}(t, t_0, \alpha_1, \ldots, \alpha_n) = \begin{cases} 1 & \text{if } i = l, \\ 0 & \text{if } i \neq l. \end{cases}$$

By (64′) we have

$$|w_i^{(l)}(t, t_0, \alpha_1, \ldots, \alpha_n)| \le C_2 \, e^{-\frac{\delta|\alpha|^p}{2}(t-t_0)}.$$

Therefore, applying (14) to system (76) and taking into account the above estimates for $w_i^{(l)}$ and $K_{ij}^{(r)}$, in combination with the estimates (73) for $k < q + 1$, we find that

$$\left| \tilde{\tilde{v}}_i^{(l)}(t, t_0, \alpha_1, \ldots, \alpha_n) \right| =$$

$$= \left| \int_{t_0}^t \sum_{j,k} w_i^{(k)}(t, \tau, \alpha) \sum_{\substack{s \ge 0 \\ s \ge q+1-M}}^{q} K_{kj}^{(p-q-1+s)}(\tau, \alpha) D_j^s v_j^{(l)} \, d\tau \right| \le$$

$$\le \left| \int_{t_0}^t N_1^2 C_2 \, e^{-\frac{\delta|\alpha|^p}{2}(\tau-t_0)} \sum_{\substack{s \ge 0 \\ s \ge q+1-M}}^{q} C_1 (1 + |\alpha|)^{p-q-1+s} \, e^{-\frac{\delta|\alpha|^p}{2}(\tau-t_0)} \times \right.$$

$$\left. \times (1 + |\alpha|)^{M-s} \sum_{\substack{r > 0 \\ r \ge -M+s}}^{s} (\tau - t_0)^r (1 + |\alpha|)^{pr} \, d\tau \right| \le$$

$$\le C_{q+1}^* \, e^{-\frac{\delta|\alpha|^p}{2}(t-t_0)} (1 + |\alpha|)^{M-q-1} \sum_{\substack{s > 0 \\ s \ge -M+q+1}}^{q+1} (t - t_0)^s (1 + |\alpha|)^{ps}. \tag{77}$$

Thereby (73) is proved, and consequently, (72) holds for $k = q+1$, $|\alpha| \ge \alpha_0$. In order to derive the estimates (72) with $|\alpha| \le \alpha_0$, the constants C_k^* should be increased, if necessary. To see this, it suffices to apply directly to system (75) the arguments used for the proof of (77).

3. Estimates for the functions $G_i^{(l)}(t, t_0, x, \xi)$.

Let us change the integration variables in (70) as follows:

$$\alpha_i (t - t_0)^{\frac{1}{p}} = \tilde{\alpha}_i, \qquad |\alpha|(t - t_0)^{\frac{1}{p}} = |\tilde{\alpha}|.$$

Then we get

$$G_i^{(l)}(t, t_0, x, \xi) = \tag{78}$$

$$= \frac{1}{(2\pi)^n (t - t_0)^{\frac{n}{p}}} \int \exp\left(i \frac{\tilde{\alpha}}{(t - t_0)^{\frac{1}{p}}}, x - \xi\right) v_i^{(l)}\left(t, t_0, \frac{\tilde{\alpha}}{(t - t_0)^{\frac{1}{p}}}\right) d\tilde{\alpha} \, .$$

Hence, using the inequality (67), we obtain

$$\left| G_i^{(l)}(t, t_0, x, \xi) \right| \leq \frac{1}{(t - t_0)^{n/p}} \int C \left(1 + \frac{|\tilde{\alpha}|}{(t - t_0)^{1/p}}\right)^{M - p} e^{-\frac{\delta |\tilde{\alpha}|^p}{2}} d\tilde{\alpha} \leq$$

$$\leq \frac{1}{(t - t_0)^{(n + M - p)/p}} \int \tilde{C} (T^{1/p} + |\tilde{\alpha}|)^{M - p} e^{-\frac{\delta |\tilde{\alpha}|^p}{2}} d\tilde{\alpha} =$$

$$= \frac{C_0^{**}}{(t - t_0)^{(n + M - p)/p}} \, , \tag{79}$$

where C_0^{**} is constant.

We can also obtain another estimate for $G_i^{(l)}$, which yields much better results for large values of $|x - \xi|/(t - t_0)^{1/p}$. To this end, in the right-hand side of (78) we perform integration by parts k times with respect to some $\tilde{\alpha}_s$, each time regarding the expression $e^{\cdots} d\tilde{\alpha}$ as a differential. We thus obtain

$$G_i^{(l)}(t, t_0, x, \xi) = \frac{1}{i^k (2\pi)^n (t - t_0)^{n/p} \prod_j (x_j - \xi_j)^{k_j}} \times$$

$$\times \int \exp\left(i \frac{\tilde{\alpha}}{(t - t_0)^{1/p}}, x - \xi\right) D^{(k)} v_i^{(l)}\left(t, t_0, \frac{\tilde{\alpha}}{(t - t_0)^{1/p}}\right) d\tilde{\alpha} \, .$$

Here $k = \sum k_j$, $k_j \geq 0$, $D^{(k)} v_i^{(l)}$ stands for any k-th order derivative with respect to a combination of the variables

$$\frac{\tilde{\alpha}_1}{(t - t_0)^{1/p}}, \ldots, \frac{\tilde{\alpha}_n}{(t - t_0)^{1/p}} \, .$$

According to the Lemma just proved, we have

$$\left| D^{(k)} v_i^{(l)}\left(t, t_0, \frac{\tilde{\alpha}}{(t - t_0)^{1/p}}\right) \right| \leq$$

$$\leq C_k e^{-\frac{\delta |\tilde{\alpha}|^p}{2}} \sum_{\substack{r > 0 \\ r \geq k - M}}^{k} (t - t_0)^r \left(1 + \frac{|\tilde{\alpha}|}{(t - t_0)^{1/p}}\right)^{rp - k + M - p} \leq$$

$$\leq C_k\, e^{-\frac{\delta|\tilde{\alpha}|^p}{2}} \sum_{\substack{r>0 \\ r\geq k-M}} (T^{1/p}+|\tilde{\alpha}|)^{rp-k+M-p}(t-t_0)^{(k-M+p)/p}\,.^{14}$$

Consequently, for $k = \sum k_j > 0$, $k \neq M+1$, we have

$$\left|G_i^{(l)}(t,t_0,x,\xi)\right| \leq \frac{C(k)(t-t_0)^{(k-n-M+p)/p}}{\prod_j |x_j - \xi_j|^{k_j}},$$

where $C(k)$ is a constant depending only on k. Since k_j comprising k are arbitrary, therefore, in order to obtain a better estimate, we can choose all k_s equal to 0, except for k_s corresponding to the largest of $|x_s - \xi_s|$. Therefore, taking into account that

$$\max\{|x_1 - \xi_1|,\ldots,|x_n - \xi_n|\} \geq \frac{|x-\xi|}{\sqrt{n}},$$

we can rewrite the preceding estimate in the form

$$\left|G_i^{(l)}(t,t_0,x,\xi)\right| \leq \frac{C_k^{**}}{(t-t_0)^{(n+M-p)/p}}\left(\frac{(t-t_0)^{1/p}}{|x-\xi|}\right)^k, \qquad (80)$$

where C_k^{**} are new constants. The estimates (79) can be considered as a special case of (80) with $k = 0$.

4. Estimates for the derivatives of $G^{(l)}(t,t_o,x,\xi)$.
The estimates (67) immediately imply that, for $t > t_0$, we can differentiate with respect to the variables x_s under the sign of the integral in (70) as many times as required. The resulting integrals can be estimated in exactly the same way as the integrals (70). Thus we obtain the following estimate

$$\left|\frac{\partial^m G_i^{(l)}(t,t_0,x,\xi)}{\partial x_1^{m_1}\cdots\partial x_n^{m_n}}\right| \leq \frac{C(k,m)}{(t-t_0)^{(n+m+M-p)/p}}\left(\frac{(t-t_0)^{1/p}}{|x-\xi|}\right)^k. \qquad (80')$$

Here $m = \sum m_s$, $C(k,m)$ are constants, $k \geq 0$.

Taking into account that the functions $v_i^{(l)}$ satisfy system (2), we find that in (70) differentiation under the sign of the integral with respect to t is also justified, and therefore we obtain for $\partial G_i^{(l)}/\partial t$ estimates of type (80′) with $m = M$.

The admissibility of differentiation under the sign of the integral in (70), with respect to t and x_s, implies that the functions $G_i^{(l)}(t,t_0,x,\xi)$ satisfy the homogeneous system (1) with respect to t and x_s.

---

[14]The last inequality can be justified only if for all positive $r \geq k-M$ the condition $rp - k + M - p \geq 0$ is always satisfied. If $p \geq 2$ (which is our assumption), then this condition does not hold only for $k = M+1$. In order to obtain the needed estimate in this case, it suffices to replace the constant M in the preceding calculations by $M+1$, which is quite admissible in this situation. In all subsequent arguments we can assume that $k \neq M+1$.

5. Proof of the Theorem. First of all, it should be observed that for bounded functions $\varphi_l(\xi_1, \ldots, \xi_n)$, *the integrals* (71) *make sense for any* $t > t_0$. This fact is a direct consequence of the estimates (80) for $G_i^{(l)}$, if $|x - \xi|$ is large and $k = 2n$, and the estimates (79) for the other values of $|x - \xi|$. The estimates (80') imply that these integrals can be differentiated in x_s and t under the sign of the integral. Therefore, the integrals (71) satisfy our system, since this is the case for the functions $G_i(t, t_0, x, \xi)$.

Let us prove now that *for* $t \to t_0$, *the functions* $u_i(t, x)$ *defined by* (71) *take the values of the respective given functions* $\varphi_i(x)$ *at every interior point* $(x_1^0, \ldots, x_n^0)$ *of any domain* G *where all these functions, together with their derivatives up to the order* $(M - p + \Delta)$, *are continuous.*

In order to show that the functions $u_i(t, x_1, \ldots, x_n)$ take the values $\varphi_i(x_1^0, \ldots, x_n^0)$ when $(t, x_1, \ldots, x_n)$ approaches an interior point $(x_1^0, \ldots, x_n^0)$ of the domain G along an arbitrary path, it suffices to verify for any interior point of the domain G the following uniform convergence

$$u_i(t, x_1^0, \ldots, x_n^0) \to \varphi_i(x_1^0, \ldots, x_n^0), \quad \text{as} \quad t \to t_0 \tag{81}$$

in a neighborhood of that point, since the functions $\varphi_i(x_1, \ldots, x_n)$ have been assumed continuous in G. In order to prove (81), we observe, first of all, that the functions u_i defined by (69) indeed take the values φ_i at all points of the plane $t = t_0$, provided that the initial functions satisfy the conditions described at the beginning of §3, Ch. I. In fact, it has been found that in this case the solution of the Cauchy problem is given by the first terms in the right-hand sides of the relations (29). Taking into account that here the functions $\varphi_i(\xi_1, \ldots, \xi_n)$ are bounded and have support within the cube

$$|x_s^0 - \xi_s| \le a \,,$$

and the functions $v_i^{(l)}(t, \alpha) = v_i^{(l)}(t, t_0, \alpha)$ satisfy the inequalities (67), we find that for any $t > t_0$ the functions $u_i(t, x)$ can be written in the form of the following absolutely convergent integrals

$$\sum_l \int \varphi_l(\xi) v_i^{(l)}(t, t_0, \alpha) \, e^{i(\alpha, x - \xi)} \, d\alpha \, d\xi \,.$$

Therefore, integration can be performed here with respect to $\alpha_1, \ldots, \alpha_n$, $\xi_1, \ldots, \xi_n$ in any order; in particular, we can first integrate with respect to α_s and then ξ_s. Thereby we obtain (71).

Let us go back to the general case. We construct the functions

$$\varphi_i^{(0)}(x_1, \ldots, x_n) \quad (i = 1, \ldots, N_1)$$

which, together with all their partial derivatives up to the order $(M - p + \Delta)$ at the point $(x_1^0, \ldots, x_n^0)$, take the same values as the given functions $\varphi_i(x_1, \ldots, x_n)$ and satisfy, in addition, all the requirements of §3, Ch. I. Set

$$\varphi_i^*(x_1, \ldots, x_n) = \varphi_i(x_1, \ldots, x_n) - \varphi_i^0(x_1, \ldots, x_n) . \tag{82}$$

The functions φ^* and all their partial derivatives up to the order $(M-p+\Delta)$ are continuous at $(x_1^0, \ldots, x_n^0)$ and vanish at this point. Let us show that the integrals

$$u_i^*(t, x_1^0, \ldots, x_n^0) = \int \sum_l G_i^{(l)}(t, t_0, x^0, \xi)\varphi_l^*(\xi) \, d\xi \tag{83}$$

$$(i = 1, \ldots, N_1)$$

converge to 0 as $t \to t_0$. Let the $(\xi_1, \ldots, \xi_n)$-space be divided into two parts G_1 and G_2. We define the domain G_1 by the inequality

$$|x^0 - \xi| \leq \eta = C(t - t_0)^{\frac{n+M-p}{p(n+M-p+\Delta)}} , \tag{84}$$

where C is an arbitrary positive constant. In the domain G_1, we estimate $G_i^{(l)}$ using the inequalities (79). Since the functions φ_i^* are continuous at the point $\xi = x^0$, together with their derivatives up to the order $(M - p + \Delta)$, and vanish at that point, therefore, in G_1 we have

$$|\varphi_i^*(\xi_1, \ldots, \xi_n)| \leq \varepsilon_1(\eta)|x^0 - \xi|^{M-p+\Delta} , \tag{85}$$

where $\varepsilon_1(\eta) \to 0$ as $\eta \to 0$. It follows that

$$\left| \int_{G_1} \sum_l G_i^{(l)}(t, t_0, x^0, \xi)\varphi_i^*(\xi) \, d\xi \right| \leq$$

$$\leq \int_{G_1} \frac{C_0^{**} N_1}{(t - t_0)^{(M+n-p)/p}} \varepsilon_1(p)|x^0 - \xi|^{M-p+\Delta} \, d\xi \leq$$

$$\leq C_0^{***} N_1 \varepsilon_1(\eta) C^{n+M-p+\Delta} . \tag{86}$$

The last expression in the right-hand side, for every fixed C and t close enough to t_0, can be made arbitrarily small.

In the domain G_2, we estimate $G_i^{(l)}$ with the help of the inequality (80), setting therein k so large that

$$k > 2n \quad \text{and} \quad \frac{\Delta}{n + M - p + \Delta} \geq \frac{M - p}{k - n} . \tag{87}$$

Then, for $t - t_0 \leq 1$ we have

$$\left| \int_{G_2} \sum_l G_i^{(l)}(t, t_0, x, \xi)\varphi_i^*(\xi) \, d\xi \right| \leq$$

$$\leq \int_{G_2} \frac{C_k^{**} N_1}{(t - t_0)^{(n+M-p)/p}} \left(\frac{(t - t_0)^{1/p}}{|x - \xi|} \right)^k K \, d\xi \leq$$

$$\leq KC_k^{**}N_1 \int \left(\frac{(t-t_0)^{\frac{1}{p}\left(1-\frac{M-p}{k-n}\right)}}{|x-\xi|}\right)^k d\left(\frac{\xi_1-x_1}{(t-t_0)^{\frac{1}{p}\left(1-\frac{M-p}{k-n}\right)}}\right)\cdots$$

$$\cdots d\left(\frac{\xi_n-x_n}{(t-t_0)^{\frac{1}{p}\left(1-\frac{M-p}{k-n}\right)}}\right),$$

where the domain of integration is defined by the inequality

$$\sum_s \left(\frac{\xi_s-x_s}{(t-t_0)^{\frac{1}{p}\left(1-\frac{M-p}{k-n}\right)}}\right)^2 \geq C^2,$$

and $K = \sup |\varphi_i^*(\xi)|$. By virtue of (87) this domain contains G_2 for $t-t_0 \leq 1$. Therefore

$$\left|\int_{G_2} \sum_l G_i^{(l)}(t,t_0,x,\xi)\varphi_l^*(\xi)\,d\xi\right| \leq \frac{C_1}{C^{k-n}}, \qquad (88)$$

where C is the same as in (84), and the constant C_1 does not depend on C.

Hence, it is easy to see that for sufficiently large C the left-hand side of (88) is arbitrarily small. Thereby the proof of the above statement concerning u_i^* is complete. As regards the uniformity of the convergence of u_i to φ_i, see Sect. 6, d) below.

6. In the sequel, the following observations will be used:

a) The right-hand sides of the inequalities (86) and (88) do not depend on t_0.

b) These right-hand sides contain only the quantities K and $\varepsilon(\eta)$ depending on the functions φ_i^*.

c) If the functions $\varphi_i(x_1,\ldots,x_n)$ and all their derivatives up to the order $(M-p+\Delta)$ have absolute values bounded by a constant M_0, then the functions $\varphi_i^0(x)$ can be chosen in such a way that

1) these functions themselves and all their partial derivatives up to the order $(2n+1)(p+M)$ (where p is the same as in (26)) have absolute values bounded by the same constant M independent of the chosen point x^0 and depending only on M_0;

2) it is only inside the cube $|x_s - x_s^0| \leq a$, $s = 1,\ldots,n$, that these functions may differ from 0, where a is an absolute constant, say, 1.

Then, for any given $\varepsilon_1 > 0$ there exists an ε_2 such that for $t - t_0 < \varepsilon_2$ the following inequality is satisfied

$$\left|\int \sum_l G_i^{(l)}(t,t_0,x,\xi)\varphi_l^0(\xi)\,d\xi - \varphi_i^0(x)\right| < \varepsilon_1,$$

where φ_l^0 are the functions constructed above for an arbitrary point x^0. Here ε_2 depends only on ε_1 and $\mathcal{M}$, which, in its turn, depends only on $\mathcal{M}_0$. Indeed, the estimates given in §3, Ch. I, allow us to make the following conclusions. Consider the first integrals in the right-hand side of (29) as consisting of two parts: one taken over the outside of the sphere

$$\sum \alpha_s^2 = C^2 ,$$

and another, over the inside of that sphere. The part over the outside, for sufficiently large C depending only on $\mathcal{M}$, can be made arbitrarily small for any fixed $\mathcal{M}$ and any t from the interval $(0, T)$. The part over the inside for $t = t_0$ differs from that for $t = t_0 + \varepsilon_2$ by an arbitrarily small quantity if $\varepsilon_2 > 0$ is sufficiently small, since the absolute value

$$\left| v_i^{(l)}(t_0 + \varepsilon_2, t_0, \alpha_1, \ldots, \alpha_n) - v_i^{(l)}(t_0, \alpha_1, \ldots, \alpha_n) \right|$$

inside the sphere is arbitrarily small, uniformly with respect to α_s.

d) If, in addition, the functions $\varphi_i(x_1, \ldots, x_n)$ and all their partial derivatives up to the order $(M - p + \Delta)$ are uniformly continuous, then the function $\varepsilon_1(\eta)$ in (85) can be chosen independent of the point x^0. Hence, taking into account the observation c) above, we obtain uniform convergence (with respect to all x_s) of the functions $u_i(t, x_1, \ldots, x_n)$, defined by (71), to $\varphi_i(x_1, \ldots, x_n)$. In other words, for each positive ε_1 we can find ε_2 such that

$$\left| \int \sum_l G_i^{(l)}(t, t_0, x, \xi) \varphi_l(\xi) \, d\xi - \varphi_i(x) \right| < \varepsilon_1 , \tag{89}$$

for $t - t_0 < \varepsilon_2$. Here ε_2 can be chosen completely independent of t_0, and depending on the functions $\varphi_l(\xi)$ only through the agency of $\varepsilon_1(\eta)$ and $\mathcal{M}_0$.

e) The integrals (71) have absolute values bounded by a constant which can be influenced by the functions $\varphi_i(x)$ only through the medium of the least upper bound $\mathcal{M}_0$ for the absolute values of these functions and their derivatives up to the order $(M - p + \Delta)$. In order to see this, it suffices to replace the inequality (85) by

$$|\varphi_i^*(\xi)| \leq \mathcal{M}_0 \, (n|x - \xi|)^{M-p+\Delta} ,$$

while estimating the integrals (83).

The integrals

$$\int \sum_l G_i^{(l)}(t, t_0, x, \xi) \varphi_l^0(\xi) \, d\xi$$

are estimated in the same way as in Sect. 6, c).

f) Assume that for all x the functions $\varphi_i(x)$ possess continuous derivatives up to the order $(M - p + \Delta + l)$, bounded by a constant $\mathcal{M}^*$; then, using the estimates (80′), instead of (79), (80), we can show that for $t \to t_0$ all derivatives of the functions $u_i(t, x)$ (defined by (71)) up to the order l, with respect to various combinations of x_k, converge to the corresponding derivatives of the functions $\varphi_i(x)$ and have absolute values bounded by constants depending on $\varphi_i(x)$ only in terms of $\mathcal{M}^*$. To this end, we should replace in all preceding formulas $M - p$ by $M - p + l$. Making use of the fact that the functions $u_i(t, x)$ satisfy the homogeneous system (1), we can show in this case that the derivatives in t of the functions v_i up to the order $m \leq l/M$ have bounded derivatives with respect to various combinations of x_k up to the order $(l - mM)$.

In particular, to guarantee the uniqueness of the solution on the basis of the results of §4, Ch. I, it suffices to assume that[15] $l = M$.

§3. Using the Green Matrix to Solve the Cauchy Problem for a Non-Homogeneous System

Every solution of system (1) can be regarded as the sum of a solution of the given system with zero initial conditions at $t = t_0$ and a solution of the corresponding homogeneous system with the given initial conditions. Therefore, now we can limit ourselves to solutions of non-homogeneous systems vanishing at $t = t_0$. For the sake of brevity, the term "solutions of the system" is used here in the sense of "functions satisfying the system".

Under the assumptions on $f_i(t, x_1, \ldots, x_n)$ made at the beginning of §3, Ch. I, the solution is given by the second terms in the right-hand sides of (29). Substituting for $v_i^*(t, \alpha)$ in (29) their expressions found from (14), we get

$$u_i(t, x) = \int e^{i(\alpha, x)} \left(\int_{t_0}^{t} \sum_l v_i^{(l)}(t, \tau, \alpha) E_{f_i}(\tau, \alpha) \, d\tau \right) d\alpha .$$

We can change the order of integration here, assuming that the functions $f_i(t, x_1, \ldots, x_n)$ may be other than zero only for the values of x_k such that $|x_k - x_k^0| \leq a$ for some fixed values x_k^0, and the functions $v_i^{(l)}(t, \tau, \alpha)$ satisfy the inequalities (67). Then we find that

$$u_i(t, x) = \int_{t_0}^{t} \left[\int \sum_l G_i^{(l)}(t, \tau, x, \xi) f_l(\tau, \xi) \, d\xi \right] d\tau , \qquad (90)$$

where the symbol $\int$ implies integration over the entire real $(\xi_1, \cdots, \xi_n)$-plane.

---

[15] Since, obviously, for $t > t_0$ the integrals (71) have continuous derivatives in x_k of any order, therefore, to ensure the uniqueness it suffices to assume that $l = M - 1$ (see footnote 6).

It will be shown that the functions (90) *satisfy system* (1) *for* $t > t_0$
and vanish at $t = t_0$, *provided that the functions* $f_l(\tau, \xi)$, *together with their
partial derivatives in* ξ_k *up to the order* $(M + \Delta - p + l)$, *are bounded and
continuous* $(l \geq M)$. *Here* Δ *is the same as in the Theorem of Sect. 1,
§2. Moreover, we also show that the functions* (90), *together with all their
partial derivatives in* x_k *up to the order* l, *are bounded, and therefore, their
partial derivatives in* t *up to the order* $\left[\frac{l}{M}\right]$ *are bounded, too.*

Proof. 1) First of all, let us show that under the above assumptions concerning $f_l(t, x)$ the integrals (90), as well as their formal partial derivatives
in x_k up to the order l, are absolutely convergent and bounded in the layer
$0 < t \leq T$, and tend to 0 as $t \to t_0$.

Indeed, the integrals

$$\int \sum_l G_i^{(l)}(t, \tau, x, \xi) f_l(\tau, \xi) \, d\xi \tag{91}$$

have the same structure as those in (71). Therefore, on the basis of the
remarks made in Sect. 6 e),f) of §2, we can see that these integrals, together
with all their partial derivatives with respect to various combinations of x_k
up to the order l, are bounded by the same constant. Hence we derive the
required statement. The fact that the derivatives of the integrals (91) with
respect to x_k are bounded, allows us to differentiate u_i in x_k under the sign
of the integral $\int_{t_0}^t$ (cf. [9], §575).

2) Let us prove now that for $t \geq t_0$ the functions (90) satisfy system
(1). It has already been shown that these functions can be differentiated
under the sign of the integral, with respect to various combinations of x_k
up to the order l. It follows that they can also be differentiated $\left[\frac{l}{M}\right]$ times
with respect to the variable t entering through the functions $G_i^{(l)}$, since the
latter functions satisfy the homogeneous system (1). Therefore, in order to
prove that the functions (90) satisfy system (1), it suffices to show that
their derivatives with respect to the variable t, which is the upper limit of
integration, are equal to $f_i(t, x)$, respectively. This fact, in its turn, can be
proved as follows: According to Sect. 6, d) of §2, for each given positive ε_1
there exists an ε_2 such that

$$\left| \sum_l G_i^{(l)}(t, \tau, x, \xi) f_l(\tau, \xi) \, d\xi - f_i(\tau, x) \right| < \varepsilon_1 \,,$$

provided that $t - \tau < \varepsilon_2$. Therefore

$$\lim_{\Delta t \to +0} \frac{1}{\Delta t} \int_t^{t+\Delta t} d\tau \int \sum_l G_i^{(l)}(t + \Delta t, \tau, x, \xi) f_l(\tau, \xi) \, d\xi = f_i(t, x) \,.$$

§4. Theorem on the Analyticity of Solutions of Homogeneous Parabolic Systems

Theorem. *All solutions of a homogeneous parabolic system which are bounded, together with their partial derivatives in x_k up to the order $(2M - p + \Delta)$, for any real x_k and any t from the interval $(0, T)$, are analytic with respect to all x_k $(n_i = 1)$, for $t > 0$.*

Proof (Cf. [10], p. 239). First of all, it should be observed that, according to Sect. 6 f), §2 of this chapter, all the solutions in question are given by (71) with $t_0 = 0$ and

$$\varphi_i(x_1, \ldots, x_n) = u_i(0, x_1, \ldots, x_n) .$$

In order to establish analyticity of the integrals in (71) with respect to all x_k at any particular real point $P(t, x_1, \ldots, x_n)$, $t > 0$, it suffices to verify uniform convergence of these integrals in a complex (with respect to all x_k) neighborhood of that point on the hyperplane $t =$ const. Without loss of generality, we can assume now that the said point has all its coordinates $x_k = 0$, since we can always shift the origin in the hyperplane parallel to $t = 0$. After this shift system (1) remains unchanged. Thus, we are going to prove uniform convergence of the integrals (71) in a complex neighborhood

$$\sum |x_k|^2 \leq R_0^2 \tag{92}$$

of the origin on the hyperplane $t =$ const > 0; here $R_0 > 0$ is an arbitrary fixed constant.

Before turning to the proof proper, let us examine the behavior of the functions $G_i^{(l)}(t, t_0, x, \xi)$ within the ball (92), for any real values of ξ_k and fixed $t > t_0$. Clearly, in any bounded domain formed by the complex values of $x_k - \xi_k$, the functions $G_i^{(l)}$ are bounded for any fixed $t > t_0$. Indeed, using (67), we find that in the domain

$$|x_k - \xi_k| \leq R \qquad (k = 1, \ldots, n) \tag{93}$$

the following inequality is satisfied

$$\left| e^{i(\alpha, x - \xi)} v_i^{(l)}(t, t_0, \alpha_1, \ldots, \alpha_n) \right| \leq C(|\alpha| + 1)^{M - p} e^{nR|\alpha| - \frac{\delta|\alpha|^p}{2}(t - t_0)} .$$

With a view to estimating $G_i^{(l)}$ when some of the differences $x_k - \xi_k$ may become unbounded, we perform integration in (70) in consecutive order with respect to various α_k, each integration being performed under the sign of the preceding integral. Let

$$x_k - \xi_k = R_k e^{i\varphi_k} ,$$

where R_k and ξ_k are of the same sign.

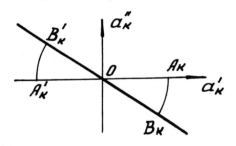

Fig. 1.

If the point $(x_1, \ldots, x_n)$ belongs to the ball (92) and ξ_k takes only real values, then, for ξ_k sufficiently large in absolute value, the absolute value of φ_k becomes arbitrarily small. Assume that for

$$|\xi_k| \geq R_0 , \tag{94}$$

we have

$$|\varphi_k| \leq \theta , \tag{95}$$

where R_0 and θ are positive constants; θ is more accurately defined later on. If the conditions (94), (95) are satisfied, then, instead of the integration along the real axis in the plane of the complex variable α_k, we integrate along the straight line

$$\alpha_k = \alpha_k' + i\alpha_k'' = \rho_k e^{-i\varphi_k} , \tag{96}$$

where φ_k is fixed, and ρ_k varies from $-\infty$ to $+\infty$. Next we show that the result of integration is the same, provided that φ_k is sufficiently small.

Indeed, on the line (96) consider the segment OB_k starting at the origin of the α_k-plane, its length being equal to that of the segment OA_k on the real axis (see Fig. 1). The integrand in (70) is a holomorphic function of α_k, and therefore, the result of its integration along the closed curve OA_kB_kO is equal to zero. In order to prove that the integral along OB_k becomes close to that along OA_k, when the point A_k goes to infinity, we must estimate the integral along the ark A_kB_k.

According to the definition of parabolic systems, for any real α_k such that $\sum \alpha_k^2 = 1$, all roots of the determinant of the matrix (4) have real parts smaller than a negative constant $-\delta$. On the other hand, every root of this determinant is a uniformly continuous function of real, as well as complex, values of α_k on the sphere $\sum |\alpha_s|^2 = 1$. Consequently, there exists a positive constant θ such that all roots of the determinant of the matrix (4) have real parts smaller than $-\delta/2$, provided that (95) holds for all t from the interval $(0, T)$ and all α_s given by (96) and satisfying the condition $\sum |\alpha_s|^2 = 1$. Then, for all complex α_k with arguments satisfying the condition (95), the estimates of type (67) will be satisfied with the following modifications: δ should be replaced by $\delta/2$; the constant C, possibly, by another one; and $|\alpha|$ stands for $\sqrt{\sum |\alpha_k|^2}$, instead of simply $\sqrt{\sum \alpha_k^2}$. For brevity, this estimate is denoted by $(67\frac{1}{2})$. On the other hand, for the points of the ark $A_k B_k$ of the circle of radius ρ_k with center at the origin, we have

$$\left| e^{i\alpha_k (x_k - \xi_k)} \right| \leq e^{\rho_k R_k} .$$

Therefore

$$\left| \int_{\widetilde{A_k B_k}} e^{i\alpha_k (x_k - \xi_k)} v_i^{(l)}(t, t_0, \alpha_1, \ldots, \alpha_n) \, d\alpha_k \right| \leq$$

$$\leq \left| \int_{\widetilde{A_k B_k}} C'(|\alpha| + 1)^{M-p} \exp\left(\rho_k R_k - \frac{\delta |\alpha|^p}{4}(t - t_0) \right) d\alpha_k \right| \leq$$

$$\leq 2\pi C'(\rho_k + 1)^{M-p+1} \exp\left(\rho_k R_k - \frac{\delta \rho_k^p}{4}(t - t_0) \right) \tag{97}$$

and, consequently, the integral in the left-hand side converges to 0 as $\rho_k \to \infty$. Here $\int_{\widetilde{A_k B_k}}$ implies integration along the ark $A_k B_k$ of the circle of radius ρ_k having its center at the origin and lying on the plane of the complex variable α_k. In exactly the same manner we can show that the integral along the ark $A_k' B_k'$, symmetric to $A_k B_k$ with respect to the origin, is also convergent to 0 as $\rho_k \to \infty$. Thus we have justified the possibility of replacing the integration along the real axis of the plane of the complex variable α_k by the integration along the straight line $B_k' B_k$, provided that $|\varphi_k| < \theta$.

Now it remains to estimate the integrals (70) over the hyperplane H, which is the topological product of the straight lines $B_k' B_k$, such that their respective φ_k satisfy condition (95), and the real axes of those complex planes α_k whose respective x_k satisfy condition (93). It has been pointed out above that at the points of this hyperplane the estimates $(67\frac{1}{2})$ are valid, as well as the estimates of type (72), which will be referred to as $(72\frac{1}{2})$. At the points of the straight lines $B_k' B_k$ we have

$$\left| e^{i\alpha_k (x_k - \xi_k)} \right| = 1 .$$

Therefore, the arguments used in the case of real x_k to obtain the estimates (79), (80), (80') will lead us to similar estimates in the case of

$(x_1, \ldots, x_n)$ belonging to the ball (92). The latter estimates are denoted by $(79\frac{1}{2})$, $(80\frac{1}{2})$, $(80\frac{1}{2}')$, respectively; they differ from (79), (80), (80') in that δ is replaced by $\delta/2$, and the constants C by new ones.

   Using these estimates, we can easily prove uniform convergence of the integrals (71) in the ball (92).

   Note that the results established in Sect. 6 a), f), §2 of the present chapter remain valid if the point $(x_1, \ldots, x_n)$ belongs to the ball (92).

§5. Theorem on the Analyticity of Solutions of Nonhomogeneous Parabolic Systems

Theorem. *For a real point* $(t^0, x_1^0, \ldots, x_n^0)$, *consider its neighborhood* G_1 *consisting of points* $(t, x_1, \ldots, x_n)$ *such that the variables*

$$x_{k_1}, x_{k_2}, \ldots, x_{k_m} \tag{98}$$

take complex values, while the rest of the independent variables are real valued. Assume that the functions $f_i(t, x_1, \ldots, x_n)$ $(i = 1, \ldots, N_1)$ *are holomorphic with respect to* $x_{k_1}, x_{k_2}, \ldots, x_{k_m}$. *Let the functions* $u_i(t, x_1, \ldots, x_n)$ *satisfy the nonhomogeneous parabolic system* (1) *in a real neighborhood* G_0 *of* $(t^0, x_1^0, \ldots, x_n^0)$ *and possess, together with their first order derivatives in* t, *continuous derivatives in* x_k *up to the order* $(4M - 2p + 2\Delta)$. *Then* u_i *can be extended to a domain* $G_2 \subset G_1$ *as holomorphic functions of the variables* (98) $(n_i = 1)$.

Proof. First of all, we note that the functions $u_i(t, x_1, \ldots, x_n)$ satisfying system (1) in the domain G_0 can always be extended to the whole infinite real layer $0 \leq t \leq T$ in such a way that the extensions, denoted by $u_i^*(t, x_1, \ldots, x_n)$, satisfy the following conditions:

   1) The extended functions coincide with the given ones in a neighborhood $G_{00} \subset G_0$ of the point $(t^0, x_1^0, \ldots, x_n^0)$.
   2) In the layer $0 \leq t \leq T$, the new functions themselves, their first order derivatives in t, and all their derivatives in x_k up to the order $(4M - 2p + 2\Delta)$ are bounded.
   3) The functions $u_i^*(t, x_1, \ldots, x_n)$ satisfy a parabolic system of type (1) in the entire layer $0 \leq t \leq T$, with $f_i(t, x_1, \ldots, x_n)$ replaced by bounded functions $f_i^*(t, x_1, \ldots, x_n)$ possessing continuous bounded derivatives in x_k up to the order $(3M - 2p + 2\Delta)$.
   4) $u_i^*(0, x) = 0$; $i = 1, \ldots, N_1$ (we assume that $t^0 > 0$).

   In particular, such extensions can be constructed by multiplying the functions defined in G_0 by a function of the form $f(r)f(|t - t^0|)$, where

$$r = \sqrt{\sum (x_s - x_s^0)^2}\,,$$

and f has the following properties:

a) $f(r) = 1$ for $0 \le r \le \varepsilon$, where ε is so small that the entire cylinder
$Z(2\varepsilon) : r \le 2\varepsilon,\ |t - t^0| \le 2\varepsilon$ lies inside G_0;

b) $f(r) = 0$ for $r \ge 2\varepsilon$;

c) $f(r)$ is sufficiently smooth.

According to §4 of Ch. I, and §3 of the present chapter, the functions
$u_i^*(t, x)$, at the points of the layer $0 \le t \le T$, can be represented in the form

$$u_i^*(t, x) = \int_0^t \int \sum_l G_i^{(l)}(t, \tau, x, \xi) f_l^*(\tau, \xi)\, d\xi\, d\tau \quad (i = 1, \ldots, N_1)\,. \tag{99}$$

For $T \ge t > t_0$, the integrals

$$u_i^{**}(t, x) = \int_0^{t_0} \int \sum_l G_i^{(l)}(t, \tau, x, \xi) f_l^*(\tau, \xi)\, d\xi\, d\tau \tag{100}$$

represent solutions of the homogeneous system (1) which, together with
their derivatives in x_k up to the order $(2M - p + \Delta) \ge M$, are bounded
functions (cf. §3 of the present chapter)[16]. Therefore, according to the preceding section, these solutions are analytic in that layer with respect to all
x_k at any point of the complex $(x_1, \ldots, x_n)$-hyperplane. It follows that in
order to prove our Theorem, it suffices to show that for $t_0 \to t$ the integrals
(100) are uniformly convergent in a domain G_2 of the type described in the
Theorem. Thus, we need to verify uniform convergence to 0 for the integrals

$$u_i^*(t, x) - u_i^{**}(t, x) = \int_{t_0}^t \int \sum_l G_i^{(l)}(t, \tau, x, \xi) f_l^*(\tau, \xi)\, d\xi_1 \cdots d\xi_n\, d\tau\,, \tag{101}$$

as $t_0 \to t$. To this end, let us perform integration with respect to each
variable in consecutive order. The integration in the variables corresponding
to (98) should be performed not along the real axis, but along a broken
line of the type shown in Fig. 2 as $ABCDE$, with the angles CBD and
CDB smaller than θ in (95), the points B and D belonging to a fixed
segment $B_0 D_0$. The domain G_2 is chosen such that while the variables
(98) range within their respective lines BCD, and the other independent
variables range within some intervals of the real axis, the point $(x_1, \ldots, x_n)$

---

[16]One can also prove that u_i^{**}, $t > t_0$, are holomorphic functions, without referring to
§3 but applying directly to the integrals (100) the arguments similar to those of §4. In
that case, it would be necessary to assume that f_l^* and u_i possess continuous bounded
derivatives in x_k up to the order $(2M + \Delta - p)$ and $(3M + \Delta - p)$, respectively.

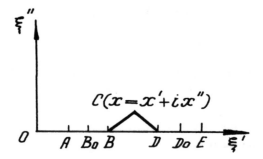

Fig. 2.

remains within a domain G_3 belonging to the interior of $G_1 \cap Z(\varepsilon)$. Then the functions $f_i^*(\tau, \xi)$ and their derivatives in ξ_k up to the order $(3M - 2p + +2\Delta)$, which enter the integrals (101), will have bounded absolute values. According to what has been proved in the preceding section, the functions $G_i^{(l)}$ can be estimated by $(79\frac{1}{2})$ and $(80\frac{1}{2})$ with constants independent of τ, since the argument of each $x_k - \xi_k$ does not exceed θ in absolute value. This allows us to argue in exactly the same way as in Sect. 6 e) of §2 to prove that the integrals

$$\int \sum_l G_i^{(l)}(t, \tau, x, \xi) f_i^*(\tau, \xi) \, d\xi$$

are uniformly bounded on G_2, and therefore, the integrals (101) are uniformly convergent to 0 on G_2, as $t_0 \to t$ [9].

Remark. If we assume that the functions $u_i(t, x)$ satisfy system (1) in the entire layer $0 \le t \le T$ and are bounded, together with their partial derivatives in x_k up to the order $(2M - p + \Delta) \ge M$, then they can be represented as sums of integrals having the form (83) and (99) (cf. §4, Ch. I, and §§3,2, Ch. II), provided that $f_i(t, x)$ are assumed bounded, together with their partial derivatives in x_k up to the order $(3M + 2\Delta - 2p)$. The integrals (83) have already been shown to represent functions analytic in all x_k. The considerations of the present section show that if all $f_i(t, x)$ are holomorphic functions in a neighborhood G_1 of any given real point $(t^0, x_1^0, \ldots, x_n^0)$, then the functions given by the integrals (99) are holomorphic in a neighborhood G_2 of the same point (here the fact of G_1 and G_2 being neighborhoods is understood in the same sense as at the beginning of this section). To sum up, if we impose the condition that the functions $u_i(t, x)$ satisfy system (1) in the entire layer $0 \le t \le T$, then the main theorem of this section can be proved under weaker assumptions.

§6. Condition (−B)

Theorem. *Assume that there exist real $\alpha_1, \ldots, \alpha_k$ and $p > 0$ such that for $t = 0$ the determinant of the matrix (4) possesses at least one root with positive real part. Then, on any interval $(0, T)$, $T > 0$, the Cauchy problem for system (1) is ill-posed.*

Proof. It follows from our assumptions that there exist real values of $\alpha_1, \ldots, \alpha_n, \sum \alpha_k^2 = 1$, and $p > 0$ such that for $t = 0$ the number of roots of the determinant of the matrix (4) with positive real parts is equal to $N_0 > 0$. Let δ be the smallest among the real parts of these roots. Let us subject system (2) (which consists of ordinary differential equations corresponding to the original system (1)) to transformations similar to those performed over this system in §1 of the present chapter. The equations obtained for the functions $w_i(\tau, \alpha_1, \ldots, \alpha_n)$ can be written in concise form

$$\frac{dw_i}{d\tau} = \lambda_i w_i + \sum_j a_{ij}^{**} w_j,$$

where λ_i are the roots of the determinant of the matrix (4) for $t = 0$, and

$$\left| a_{ij}^{**}(\tau, \alpha_1, \ldots, \alpha_n) \right| < \frac{\delta}{4N_1}$$

for $\tau \le t_0 |\alpha|^p$, $t_0 = \text{const} > 0$, and sufficiently large $|\alpha| = \sqrt{\sum \alpha_k^2}$.

Let us assume that the functions w_i are enumerated in such a way that the first N_0 subscripts are attributed to the functions whose respective λ_i have positive real parts. Set

$$w_i^* = w_i\, e^{-\frac{\delta \tau}{2}}. \tag{102}$$

Then

$$\frac{dw_i^*}{dt} = \lambda_i^* w_i^* + \sum_j a_{ij}^{**} w_j^*,$$

where

$$\lambda_i^* = \lambda_i - \frac{\delta}{2}.$$

Therefore, on the interval $(0, t_0 |\alpha|^p)$, for sufficiently large $|\alpha|$, we have

$$\frac{d}{d\tau} \left[\sum_{i=1}^{N_0} |w_i^*|^2 - \sum_{i=N_0+1}^{N_1} |w_i^*|^2 \right] = \sum_{i=1}^{N_0} \left[w_i^* \frac{d\bar{w}_i^*}{d\tau} + \bar{w}_i^* \frac{dw_i^*}{d\tau} \right] -$$

$$- \sum_{i=N_0+1}^{N_1} \left[w_i^* \frac{d\bar{w}_i^*}{d\tau} + \bar{w}_i^* \frac{dw_i^*}{d\tau} \right] = \sum_{i=1}^{N_0} \left[\bar{\lambda}_i^* w_i^* \bar{w}_i^* + \lambda_i^* \bar{w}_i^* w_i^* \right] +$$

$$+\sum_{i=1}^{N_0}\sum_{j=1}^{N_1}\left[\bar{a}_{ij}^{**}\bar{w}_j^*w_i^* + a_{ij}^{**}w_j^*\bar{w}_i^*\right] - \sum_{i=N_0+1}^{N_1}\left[\bar{\lambda}_i^*w_i^*\bar{w}_i^* + \lambda_i^*\bar{w}_i^*w_i^*\right] -$$

$$-\sum_{i=N_0+1}^{N_1}\sum_{j=1}^{N_1}\left[\bar{a}_{ij}^{**}\bar{w}_j^*w_i^* + a_{ij}^{**}w_j^*\bar{w}_i^*\right] \geq \frac{\delta}{2}\sum_{i=1}^{N_0}|w_i^*|^2 .$$

From this inequality, taking into account that

$$\sum_{i=1}^{N_0}|w_i^*|^2 - \sum_{N_0+1}^{N_1}|w_i^*|^2 \leq \sum_{i=1}^{N_0}|w_i^*|^2 ,$$

we obtain

$$\frac{ds}{d\tau} \geq \frac{\delta}{2}s ,$$

where

$$s = \sum_{i=1}^{N_0}|w_i^*|^2 - \sum_{i=N_0+1}^{N_1}|w_i^*|^2 .$$

Consequently (see [11]),

$$s(\tau) \geq s(0)\, e^{\frac{\delta}{2}\tau} .$$

This inequality, together with (102), implies that if the initial values of the functions $v_i(t, \alpha_1, \ldots, \alpha_n)$ are chosen in such a way that $s(0) > 0$ (the possibility of this choice is due to the fact that the linear transformation which maps the functions v_i into w_i is non-degenerate), then, with the increase of $|\alpha|$, the largest of the absolute values of $v_i(t, \alpha_1, \ldots, \alpha_n)$, $t > 0$, will grow faster than $|\alpha|^{-M} \exp\left(\frac{\delta|\alpha|^p}{4}t\right)$, and therefore, faster than any polynomial of $|\alpha|$. On the basis of the arguments of §1, Ch. I, it follows that in this situation the Cauchy problem for the original system is neither uniformly nor even simply well-posed.

CHAPTER III

Systems of Kowalevskaya Type

§1. Basic Theorem on the Reduction of a Matrix to Canonical Form

Theorem. *Consider the following system of ordinary differential equations:*

$$\frac{dv_i}{dt} = \sum_{j=1}^{N} \alpha a_{ij}(\beta_1, \ldots, \beta_n)v_j \qquad (i = 1, \ldots, N),\qquad (103)$$

where α, β_k are real constants. Assume that the coefficients $a_{ij}(\beta_1, \ldots, \beta_n)$ and all their partial derivatives with respect to various combinations of $\beta_1, \ldots, \beta_n$ up to the order L are continuous (and therefore, bounded) functions of the point $P(\beta_1, \ldots, \beta_n)$ belonging to a neighborhood U of the point P_0. These coefficients are allowed to take real, as well as complex, values. Assume that for any point P belonging to U, the matrix

$$\|a_{ij}(P)\| - \lambda E \qquad (104)$$

has elementary divisors of degree 1 only, and the number of equal elementary divisors is independent of P. Then there exists a non-degenerate linear transformation

$$v_i = \sum_{j=1}^{N} C_{ij}w_j \qquad (105)$$

whose coefficients, together with their partial derivatives with respect to $\beta_1, \ldots, \beta_n$ up to the order L, are bounded, continuous in a neighborhood of P_0, and independent of α; moreover, this transformation reduces system (103) to canonical form

$$\frac{dw_i}{dt} = \alpha \lambda_i(P)w_i \qquad (i = 1, \ldots, N),\qquad (106)$$

where $\lambda_i(P)$ are the roots of the determinant of the matrix (104) [10].

Proof. First of all, let us show the roots $\lambda_k(P)$ ($k = 1, \ldots, N$) to have continuous partial derivatives up to the order L at the points of U. Denote the determinant of the matrix (104) by

$$D(\lambda, \beta_1, \ldots, \beta_n) = D(\lambda, P) .$$

Since the coefficient of this determinant by the highest power of λ is equal to 1, all its roots continuously depend on P. Let p_k be the multiplicity of the root $\lambda_k(P)$. By assumption, p_k is independent of P. For $\lambda = \lambda_k(P)$, the determinant $D(\lambda, P)$, and all its derivatives in λ up to the order $(p_k - 1)$, are identically equal to 0 as functions of P; and at no point of U does its derivative in λ of the order p_k vanish. Therefore, differentiating the identity

$$D_\lambda^{(p_k-1)}(\lambda_k(P), \beta_1, \ldots, \beta_n) = 0$$

with respect to any one of the variables $\beta_1, \ldots, \beta_n$, say β_1, we obtain the following equation

$$D_\lambda^{(p_k)}(\lambda_k(P), P) \frac{\partial \lambda_k}{\partial \beta_1} + \frac{\partial D_\lambda^{(p_k-1)}}{\partial \beta_1} = 0$$

with a non-zero coefficient by $\partial \lambda_k / \partial \beta_1$. By $D_\lambda^{(p)}(\lambda, P)$ we denote the p-th order derivative of $D(\lambda, P)$ with respect to λ. The above equation yields

$$\frac{\partial \lambda_k}{\partial \beta_1} = -\frac{1}{D_\lambda^{(p_k)}(\lambda_k(P), P)} \frac{\partial D^{(p_k-1)}}{\partial \beta_1} .$$

In order to find higher order derivatives of $\lambda_k(P)$ with respect to β_i $(i = 1, \ldots, N)$, this equation should be differentiated in β_i several times more. Hence, it can be easily seen that the roots $\lambda_k(P)$ possess, in the neighborhood U, continuous derivatives with respect to various combinations of β_i up to the order L, and thus our statement is proved.

Next, we turn to the proof of the Theorem proper. Consider the following system of homogeneous linear equations

$$\lambda C_i = \sum_{j=1}^{N} a_{ij} C_j \qquad (i = 1, \ldots, N) . \tag{107}$$

This system admits a non-trivial solution, if λ coincides with a root of the determinant $D(\lambda, P)$. Set, for instance, $\lambda = \lambda_1$. Then, there exists a system of functions

$$C_1(P), C_2(P), \ldots, C_N(P) ,$$

defined in a neighborhood of P_0 and having the following properties:

1) these functions satisfy equations (107);

2) they do not vanish simultaneously at any one point;

3) they are continuous, together with all their derivatives in $\beta_1, \ldots, \beta_n$ up to the order L.

Indeed, the matrix (104) has been assumed to possess elementary divisors of degree 1 only, and therefore, if $\lambda = \lambda_1$, all its minor determinants of order $N - p_1 + 1$ vanish identically. On the other hand, there exists at least one minor determinant M of order $N - p_1$ which does not vanish in a neighborhood of P_0. Therefore, we can find functions $C_1(P), \ldots, C_N(P)$ satisfying (107) by making an arbitrary choice of as many as p_1 of these functions which correspond to the rows of the matrix (104) containing no elements involved in M. We take all the latter p_1 functions, but one, equal to 0, and the remaining function set equal to 1. Then, all the rest of the $N - p_1$ functions $C_i(P)$ will have the form of fractions with denominator M and numerators involving minor determinants (other than M) of the matrix (104). Obviously, the system of functions thus obtained satisfies the conditions 1) – 3), and there exist as many as p_1 linearly independent systems of such functions. Arguing likewise in relation to every root $\lambda = \lambda_1, \lambda_2, \ldots$ of the determinant $D(\lambda, P)$, we obtain a basis formed by eigenvectors of the matrix $\|a_{ij}(P)\|$:

$$C_{1i}(P), \ldots, C_{Ni}(P) \qquad (i = 1, \ldots, N) . \qquad (108)$$

Here different i correspond to different eigenvectors, the respective eigenvalues are denoted by λ_i. Therefore, we have

$$\lambda_1 = \cdots = \lambda_{p_1}, \quad \lambda_{p_1+1} = \cdots = \lambda_{p_1+p_2} , \quad \text{etc.}$$

In this basis, $\|a_{ij}\|$ becomes a diagonal matrix. Let us show that linear transformation (105) reduces system (103) to the form (106). Indeed, by virtue of (103), we have

$$\sum_k C_{ik} \frac{dw_k}{dt} = \frac{dv_i}{dt} = \alpha \sum_j a_{ij} v_j = \alpha \sum_j \sum_k a_{ij} C_{jk} w_k =$$

$$= \alpha \sum_k C_{ik} \lambda_k w_k \qquad (i = 1, \ldots, N) . \qquad (109)$$

The matrix $\|C_{ik}\|$ is invertible, since (108) is a basis. Applying the inverse matrix to both sides of (109), we obtain (106). On the other hand, (103) can be obtained from (106) in a similar way. The Theorem is proved.

Remark. If the coefficients of system (103) and all roots of the determinant of the matrix (104) take only real values in the neighborhood of P_0, then the coefficients of the transformation (105) are also real valued functions. This fact is due to the formulas expressing the coefficients C_{ij} in terms of the minor determinants of the matrix (104).

§2. Hyperbolic Systems

1. In this section we establish sufficient conditions for the Cauchy problem to be uniformly well-posed, in the case of some systems of Kowalevskaya type. In order to verify whether Condition A holds for a system of this type, we first reduce it, as usual, to a system involving only first order partial derivatives in t, and write out the corresponding system of ordinary differential equations (2).

In the latter system we set

$$v_j = v_j^* |\alpha|^{k-n_i} ,$$

where $|\alpha| = \sqrt{\sum \alpha_s^2}$, $N_1 \geq j \geq 0$, $n_i - 1 \geq k \geq 0$, provided that

$$\frac{d^k v_i}{dt^k} = v_j .$$

We find that the functions v_i^* $(i = 1, \ldots, N_1)$ satisfy a system of the form

$$\frac{dv_i^*}{dt} = |\alpha| \sum_{j=1}^{N_1} a_{ij} \left(t, \frac{\alpha_1}{|\alpha|}, \ldots, \frac{\alpha_n}{|\alpha|} \right) v_j^* + \sum_{j=1}^{N_1} b_{ij} \left(t, \frac{\alpha_1}{|\alpha|}, \ldots, \frac{\alpha_n}{|\alpha|} \right) v_j^* \quad (115)$$

$$(i = 1, \ldots, N_1) ,$$

where a_{ij}, b_{ij} are bounded functions, and a_{ij} depend only on $A_{ij}^{(k_0, \ldots, k_n)}$ with $\sum k_s = n_j$. If, instead of t, a new independent variable τ be introduced by

$$|\alpha| t = \tau ,$$

then all the transformations we are dealing with now would coincide with those introduced at the beginning of §1, Ch. II, had we set $p = 1$ therein. Likewise, we could show that the determinant of the matrix

$$\|a_{ij}\| - \lambda E \tag{116}$$

is obtained from the determinant of the matrix (5) by the replacement of α_s with $i\alpha_s$, provided that $\sum \alpha_s^2 = 1$. Therefore, the roots of these determinants differ only by the factor i.

Theorem. *Assume that there exists a linear transformation*

$$v_i = \sum C_{ij} \left(t_0, \frac{\alpha_1}{|\alpha|}, \ldots, \frac{\alpha_n}{|\alpha|} \right) w_j \tag{117}$$

satisfying the following conditions: its coefficients, as well as their first order derivatives in t_0, are bounded and continuous; the absolute value of the *corresponding determinant is larger than a positive constant; for all real* α_s *and* t_0 *from the interval* $0 \le t_0 \le T$, *the above linear transformation reduces the system*

$$\frac{dv_i}{dt} = |\alpha| \sum_{j=1}^{N_1} a_{ij}\left(t_0, \frac{\alpha_1}{|\alpha|}, \dots, \frac{\alpha_n}{|\alpha|}\right) v_j \qquad (i = 1, \dots, N_1) \qquad (118)$$

to canonical form

$$\frac{dw_k}{dt} = i|\alpha|\lambda_k w_k, \qquad k = 1, \dots, N_1, \qquad (119)$$

where all λ_k *are real. Systems satisfying this condition will be called hyperbolic in generalized sense. The Cauchy problem for such systems on the interval* $(0, T)$ *is uniformly well-posed.*

Proof. Let us apply the transformation

$$v_i^* = \sum C_{ij}\left(t, \frac{\alpha_1}{|\alpha|}, \dots, \frac{\alpha_n}{|\alpha|}\right) w_j^*$$

to system (115). Since the functions C_{ij} are assumed to possess continuous bounded derivatives in t, the above transformation results in the following system

$$\frac{dw_k^*}{dt} = i|\alpha|\lambda_k w_k^* + \sum_{j=1}^{N_1} b_{kj}^*\left(t, \frac{\alpha_1}{|\alpha|}, \dots, \frac{\alpha_n}{|\alpha|}\right) w_j^*, \qquad (120)$$

where b_{kj}^* are bounded continuous functions. Hence we find

$$\frac{d}{dt} \sum_k |w_k^*|^2 = \sum_k \bar{w}_k^* \frac{dw_k^*}{dt} + \sum_k w_k^* \frac{d\bar{w}_k^*}{dt} =$$

$$= 2\sum_k \sum_j \Re\{b_{kj}^* w_j^* \bar{w}_k^*\} \le N_1 B \sum_k |w_k^*|^2,$$

where B is the least upper bound for $2|b_{kj}^*|$. It follows that

$$\sum_k |w_k^*(t)|^2 \le \sum_k |w_k^*(t_0)|^2 e^{N_1 B(t-t_0)},$$

and therefore

$$\sum_k |v_k^*(t)|^2 \le M \sum_k |v_k^*(t_0)|^2, \qquad (121)$$

where M is an absolute constant. For large values of $|\alpha|$ (which case is the only one that matters in Condition A), we have

$$|\alpha|^{-m}|v_j^*| \le |v_j| \le |\alpha|^{-1}|v_j^*|,$$

where m is the largest of n_j. Hence, by virtue of (121), we obtain the following estimate

$$\sum_k |v_k(t)|^2 \le M|\alpha|^{2(m-1)} \sum_k |v_k(t_0)|^2 .$$

Therefore, Condition A holds for our system.

2. Special Cases of Systems Hyperbolic in Generalized Sense.

I. According to the Theorem of the preceding section, and the Heine–Borel Lemma, a system of Kowalevskaya type is hyperbolic in generalized sense if the following conditions are satisfied:

1) the coefficients $A_{ij}^{(k_0,\dots,k_n)}(t)$, with $\sum k_s = n_j$, possess continuous derivatives on the closed interval $(0,T)$;

2) for all real α_s, $\sum \alpha_s^2 = 1$, and all t from the interval $(0,T)$, the determinant of the matrix (5) can have only real roots, or equivalently, the determinant of the matrix

$$\left\| \sum_{((k_s))} A_{ij}^{(k_0,\dots,k_n)}(t)\lambda^{k_0}(i\alpha_1)^{k_1}\cdots(i\alpha_n)^{k_n} - \begin{Vmatrix} \lambda^{n_1} & & & 0 \\ & \lambda^{n_2} & & \\ & & \ddots & \\ 0 & & & \lambda^{n_{N_1}} \end{Vmatrix} \right\| \quad (122)$$

can have only imaginary roots;

3) the matrix (116) has elementary divisors of degree 1 only. The number of equal elementary divisors of these matrices is independent of t and α_s.

II. The matrix (5) has the form

$$\begin{Vmatrix} \boxed{M_1} & & & \\ & \boxed{M_2} & & \\ & & \ddots & \\ & & & \boxed{M_l} \end{Vmatrix}, \quad (123)$$

where every element whose position is outside all matrices M_k is identically equal to zero. Every matrix M_k and its respective matrix of the form (116) possess all the properties described in the preceding case. In the present situation, the conditions of generalized hyperbolicity are satisfied, since the matrix (116) associated with the system under consideration is split into separate matrices $\mathcal{M}_1,\dots,\mathcal{M}_l$, in complete analogy with the matrix (123) corresponding to our system. Every one of these matrices can be independently reduced to canonical form. Every matrix $\mathcal{M}_p$ is constructed from its counterpart M_p in exactly the same way as the matrix (116) has been constructed from (5).

III. The matrix (5) has the form (123). The determinant of each matrix M_p has only real roots, different from one another. In this situation, our system is said to be simply *hyperbolic*.

§3. A Condition for the Lack of Well-Posedness of the Cauchy Problem for Systems of Kowalevskaya Type

Theorem. *Assume that there exist real* $\alpha_1, \ldots, \alpha_n$ *and* t *from the interval* $(0, T)$ *such that at least one root of the determinant of the matrix (5) is strictly complex, or equivalently, at least one root of the determinant of (122) is not purely imaginary. Then the Cauchy problem for the Kowalevskaya system under consideration has no property of being uniformly well-posed.*

Proof. Let the determinant of (122) have a root that is not purely imaginary for some $t = t_0$ and some real $\alpha_1, \ldots, \alpha_n$; then, changing the sign of every α_s, if necessary, we can always find a system of real α_s such that the determinant has a root with positive real part. In this situation, our statement follows directly from the Theorem proved in §6, Ch. II.

Remark. If the assumption of the above theorem is valid for $t = 0$, then the Cauchy problem for the system under consideration is ill posed on the interval $(0, T)$.

§4. Example of an Ill-Posed Cauchy Problem for a a System of Kowalevskaya Type

The Cauchy problem for the following system of Kowalevskaya type is an instance of the absence of well-posedness in the situation when all elementary divisors of the matrix (116) have degree 1 and the determinant of the matrix (5) has only real roots. The system has the form

$$\frac{\partial u_1}{\partial t} = \frac{\partial u_1}{\partial x_1} - \frac{\partial u_2}{\partial x_2} ;$$

$$\frac{\partial u_2}{\partial t} = -\frac{\partial u_1}{\partial x_2} - \frac{\partial u_3}{\partial x_2} ; \qquad (124)$$

$$\frac{\partial u_3}{\partial t} = -u_2 .$$

The matrix (5) corresponding to the above system (under the assumption that $\sum \alpha_s^2 = 1$) has the form

$$\begin{Vmatrix} -\lambda + \alpha_1 & -\alpha_0 & 0 \\ -\alpha_2 & -\lambda & -\alpha_2 \\ 0 & 0 & -\lambda \end{Vmatrix} . \tag{125}$$

Its determinant is equal to

$$-\lambda(\lambda^2 - \alpha_1 \lambda - \alpha_2^2) .$$

For real α_1, α_2 such that $\alpha_1^2 + \alpha_2^2 = 1$, the above determinant has only real roots; two of its roots coincide only if $\alpha_2 = 0$. For $\alpha_2 = 0$, the matrix (125) can obviously have elementary divisors of degree 1 only. Let us show that, nevertheless, the Cauchy problem for system (124) is ill-posed. By virtue of §5, Ch. I, a necessary and sufficient condition for the well-posedness of the Cauchy problem for system (124) can be stated as follows: *the determinant*

$$D(\lambda) = \begin{vmatrix} \lambda - i\alpha_1 & i\alpha_2 & 0 \\ i\alpha_2 & \lambda & i\alpha_2 \\ 0 & 1 & \lambda \end{vmatrix}$$

has no roots with real parts tending to $+\infty$ *faster than* $\ln |\alpha|$ *as* $|\alpha| = (\sum \alpha_s^2)^{1/2} \to \infty$. Let us show that this condition does not hold if $\alpha_1 = \alpha_2^2$. For this purpose we have to solve the equation

$$D(\lambda) \equiv \lambda^3 - i\alpha_1 \lambda^2 + (\alpha_2^2 - i\alpha_2)\lambda - \alpha_1 \alpha_2 = 0 .$$

Setting

$$\lambda = \mu + \frac{i\alpha_1}{3} ,$$

we find that

$$\mu^3 + \left(\frac{\alpha_1^2}{3} + \alpha_2^2 - i\alpha_2 \right) \mu + \frac{2}{27} i\alpha_1^3 + i\frac{1}{3}\alpha_1 \alpha_2^2 - \frac{2}{3}\alpha_1 \alpha_2 = 0 . \tag{126}$$

Recall that the solutions of the equation

$$\mu^3 + p\mu + q = 0$$

are given by the Cardano formulas

$$\mu_1 = \frac{1}{3}(\rho_1 + \rho_2) ,$$

$$\mu_2 = \frac{1}{3}(\rho^2 \rho_1 + \rho \rho_2) ,$$

$$\mu_3 = \frac{1}{3}(\rho \rho_1 + \rho^2 \rho_2) ,$$

where

$$\rho_1 = \left(-\frac{27}{2}q + \frac{3}{2}\sqrt{-3D}\right)^{\frac{1}{3}},$$

$$\rho_2 = \left(-\frac{27}{2}q - \frac{3}{2}\sqrt{-3D}\right)^{\frac{1}{3}},$$

$$\rho = -\frac{1}{2} + \frac{1}{2}\sqrt{-3},$$

and $D = -4p^3 - 27q^2$.

For equation (126) we have

$$D = 4\left(i\alpha_1^4\alpha_2 - \frac{1}{4}\alpha_1^2\alpha_2^4 + 5i\alpha_1^2\alpha_3^3 - 2\alpha_1^2\alpha_2^2 - \alpha_2^6 + 3i\alpha_2^5 + 3\alpha_2^4 - i\alpha_2^3\right).$$

Hence, it is easy to see that

$$D = 4i\alpha_2^9(1 + O(\alpha_2^{-1})), \quad q = \frac{2i}{27}\alpha_2^6(1 + O(\alpha_2^{-2})),$$

as $\alpha_1 = \alpha_2^2 \to \infty$. Therefore

$$\rho_1 = \left[-i\alpha_2^6(1 + O(\alpha_2^{-2})) + 3\sqrt{-3i\alpha_2}\,\alpha_2^4(1 + O(\alpha_2^{-1}))\right]^{\frac{1}{3}} =$$

$$= \left[-i\alpha_2^6 + 3\sqrt{-3i\alpha_2}\,\alpha_2^4 + O(\alpha_2^4)\right]^{\frac{1}{3}} = i\alpha_2^2 + \sqrt{-3i\alpha_2} + O(1).$$

In exactly the same way, we find that

$$\rho_2 = i\alpha_2^2 - \sqrt{-3i\alpha_2} + O(1)\ [{}^{11}].$$

Hence

$$\mu_2 = \frac{1}{3}\left(\frac{-1 - \sqrt{-3}}{2}\rho_1 + \frac{-1 + \sqrt{-3}}{2}\rho_2\right) =$$

$$= -\frac{1}{6}\left(\rho_1 + \rho_2 + \sqrt{-3}\,(\rho_1 - \rho_2)\right) = -\frac{i}{3}\alpha_2^2 + \frac{1}{2}\sqrt{\alpha_2} +$$

$$+ \frac{i}{2}\sqrt{\alpha_2} + O(1).$$

It follows that the growth of $\Re\mu_2$, as $\alpha_2 \to \infty$, is similar to that of $\sqrt{\alpha_2}$, and therefore, faster than $\ln|\alpha|$, q.e.d.

Remark. The above example also implies that the matrix (125) cannot be reduced to canonical form by any linear transformation with bounded coefficients and the determinant having its absolute value larger than a positive constant. Otherwise, the Cauchy problem for system (124) would be well-posed.

§5. The Case of the Matrix (116) with Purely Imaginary Elementary Divisors of Degree Greater than 1

1. As shown by examples, the question whether the Cauchy problem in this case is well-posed cannot be answered merely on the basis of information about the coefficients by the highest order derivatives of the unknown functions; the behavior of the other coefficients is also important in this case.

The Cauchy problem for the following system serves as an example of a uniformly well-posed problem:

$$\frac{\partial u_1}{\partial t} = a(t) \frac{\partial u_1}{\partial x} + b_1(t) \frac{\partial u_1}{\partial x} \ ;$$

$$\dotfill \tag{127}$$

$$\frac{\partial u_{n-1}}{\partial t} = a(t) \frac{\partial u_{n-1}}{\partial x} + b_{n-1} \frac{\partial u_n}{\partial x} \ ;$$

$$\frac{\partial u_n}{\partial t} = a(t) \frac{\partial u_n}{\partial x} \ .$$

This system can be solved if we start from the last equation, find u_n and substitute this function into the penultimate equation, then find u_{n-1} and substitute it into the $(n-2)$-th equation, etc.

An example of a system, for which the Cauchy problem is ill-posed, can be obtained from system (127), if we assume the coefficients a and b_i to be constant and add certain linear combinations of u_i (with constant coefficients) to the right-hand sides. As shown in §5, Ch. I, the question of well-posedness for such a system can be decided by the examination of the roots of a certain determinant for large values of $|\alpha| = \sqrt{\sum \alpha_k^2}$. The coefficients by u_i in the right-hand sides of the new system can be chosen such that the determinant has a root whose real part tends to $+\infty$ faster than $\ln |\alpha|$, as evidenced by the following example:

$$\frac{\partial u_1}{\partial t} = \frac{\partial u_2}{\partial x} \ ,$$

$$\frac{\partial u_2}{\partial t} = u_1 \ .$$

2. For a system of Kowalevskaya type consisting of two first order partial differential equations with constant coefficients, the following necessary condition can be given for the Cauchy problem to be uniformly well-posed:

Set

$$\alpha_k = |\alpha|\beta_k \ , \quad \text{where} \quad |\alpha| = \sqrt{\sum \alpha_k^2} \ .$$

Then, with the help of a non-degenerate linear transformation (whose coefficients are bounded and continuous, together with their first derivatives in t), the corresponding system of ordinary differential equations (2) can be reduced to the form

$$
\begin{aligned}
\frac{dv_1}{dt} &= i|\alpha|a_{11}v_1 + i|\alpha|a_{12}v_2 + b_{11}v_1 + b_{12}v_2 , \\
\frac{dv_2}{dt} &= \phantom{i|\alpha|a_{11}v_1 +} i|\alpha|a_{22}v_2 + b_{21}v_1 + b_{22}v_2 .
\end{aligned}
\tag{128}
$$

Here the coefficients a_{ij} and b_{ij}, in general, may depend on t and β_k, but are independent of $|\alpha|$. These coefficients are real valued bounded continuous functions of t and β_k. Assume now that, for some $\beta_1, \ldots, \beta_n$ and all t from an interval I, the characteristic equation for our system has purely imaginary roots, and the characteristic matrix possesses an elementary divisor of degree 2. Consequently, we have $a_{11} = a_{22}$ on this interval, the coefficients a_{11}, a_{22} are real valued and a_{12} does not vanish at any point. Then, the necessary condition for the Cauchy problem to be uniformly well-posed on an interval containing I can be formulated as follows:

$$
b_{21} = 0 \quad \text{everywhere on} \quad I .
\tag{129}
$$

Indeed, setting

$$
v_1 = \exp\left\{ i|\alpha| \int_{t_0}^{t} a_{11}(\tau)\, d\tau \right\} |\alpha|^{\frac{1}{2}} \tilde{v}_1 ,
$$

$$
v_2 = \exp\left\{ i|\alpha| \int_{t_0}^{t} a_{22}(\tau)\, d\tau \right\} \tilde{v}_2 ,
$$

we reduce system (128) to the form

$$
\begin{aligned}
\frac{d\tilde{v}_1}{dt} &= i\sqrt{|\alpha|}\, a_{12}\tilde{v}_2 + b_{11}\tilde{v}_1 + \frac{1}{\sqrt{|\alpha|}} b_{12}\tilde{v}_2 , \\
\frac{d\tilde{v}_2}{dt} &= \sqrt{|\alpha|}\, b_{21}\tilde{v}_1 + b_{22}\tilde{v}_2 .
\end{aligned}
$$

The characteristic equation for this system can be written in the form

$$
\begin{vmatrix}
-\lambda & i\sqrt{|\alpha|}\, a_{12} \\
\sqrt{|\alpha|}\, b_{21} & -\lambda
\end{vmatrix}
= 0 ,
$$

or equivalently,

$$
\lambda^2 = i|\alpha| b_{21} a_{12} .
$$

If $b_{21} \neq 0$ for some t from the interval I, then the above equation has a root whose real part tends to $+\infty$ at the same rate as $\sqrt{|\alpha|}$ for this value of t and $|\alpha| \to +\infty$. Therefore, in complete analogy to §6, Ch. II, it is possible to show that $|\tilde{v}_i|$, and consequently $|v_i|$, can grow faster than any power of $|\alpha|$. Thus, if the condition (129) is violated on the interval I, then the Cauchy problem for our system has no property of being uniformly well-posed.

§6. On the Domain where the Solution of the Cauchy Problem for a System of Kowalevskaya Type is Uniquely Determined by the Cauchy Data

If (1) is not a system of Kowalevskaya type, then, in general, the values of u_i at a point $(t, x_1, \ldots, x_n)$ $(t > t_0)$ depend on the initial conditions for these functions on the entire hyperplane $t = t_0$. For instance, this fact is well-known in the case of the heat equation

$$\frac{\partial u}{\partial t} = \frac{\partial^2 u}{\partial x^2}.$$

It follows from (71) that this statement can also be made in relation to every parabolic system [^{12}]. For linear Kowalevskaya systems with analytic coefficients, Holmgren [4] (see also [5]) proved in 1901 that the initial values of the functions u_i prescribed in a bounded domain G_0 on the hyperplane $t = t_0$ completely determine these functions in a domain G belonging to the $(t, x_1, \ldots, x_n)$-space and adjacent to G_0. The only assumption here is that the functions u_i, together with their derivatives up to the order n_i, respectively, are continuous and bounded. This result has also been proved by the author (see [12]) for linear systems (whose coefficients may lack analyticity) hyperbolic in generalized sense, and even for nonlinear systems hyperbolic in generalized sense.

Remark. Therefore, if it is in the domain G only that we examine the behavior of the solution of the Cauchy problem for a system of Kowalevskaya type with constant coefficients, it is quite unimportant what initial conditions for the functions $u_i(t, x_1, \ldots, x_n)$ be prescribed at the points of the hyperplane $t = t_0$ outside G_0. For instance, we can extend the initial values prescribed on G_0 to the entire hyperplane $t = t_0$ as sufficiently smooth periodic functions with respect to every x_k. Then the Fourier series could have been used in our previous arguments, instead of the Fourier integrals.

§7. Proof of the Impossibility for the Cauchy Problem to be Non-Uniformly Well-Posed in the Case of Kowalevskaya Systems

Assume that the Cauchy problem for a given Kowalevskaya system of the type considered throughout the paper is well-posed on the interval $(0, T)$. Let $v_i^{(l)**}(t, 0, \alpha_1, \dots, \alpha_n)$ be the functions which satisfy system (115) on the interval $(0, T)$ and are defined by the following initial conditions

$$v_i^{(l)**}(0, 0, \alpha_1, \dots, \alpha_n) = \begin{cases} 0 & \text{if } i \neq l, \\ 1 & \text{if } i = l. \end{cases}$$

Then, on the basis of §1, Ch. I, it is easy to see that for the given system and the interval $(0, T)$ there exist positive constants L and C such that

$$\left| v_i^{(l)**}(t, 0, \alpha_1, \dots, \alpha_n) \right| \le C(|\alpha| + 1)^L \quad \text{on} \quad (0, T), \tag{130}$$

where $|\alpha| = \sqrt{\sum \alpha_k^2}$, as usual. On the other hand, if the Cauchy problem is not uniformly well-posed on the interval $(0, T)$, then for every given C_1 and L_1 there exist $\alpha_1, \dots, \alpha_n, t_1, t_2 \ (0 \le t_1 < t_2 \le T), i_0, l_0$ such that

$$\left| v_{i_0}^{(l_0)**}(t_2, t_1, \alpha_1, \dots, \alpha_n) \right| > C_1(|\alpha| + 1)^{L_1}. \tag{131}$$

Here $v_i^{(l)**}(t, t_1, \alpha_1, \dots, \alpha_n)$ stand for the functions which satisfy system (115) on the interval (t_1, T), with the following initial conditions at $t = t_1$:

$$v_i^{(l)**}(t_1, t_1, \alpha_1, \dots, \alpha_n) = \begin{cases} 0 & \text{if } i \neq l, \\ 1 & \text{if } i = l. \end{cases}$$

Let $w(t)$ be the Wronski determinant for the system of functions $v_i^{(l)**}(t, 0, \alpha_1, \dots, \alpha_n)$. By the Liouville theorem we obtain

$$w(t) = w(0) \exp \left\{ - \int_0^t \left[|\alpha| \sum_{i=1}^{N_1} \left(a_{ii} \left(\tau, |\alpha|^{-1} \alpha \right) + b_{ii} \left(\tau, |\alpha|^{-1} \alpha \right) \right) \right] d\tau \right\}.$$

If the functions $v_i^{(l)**}(t, 0, \alpha_1, \dots, \alpha_n)$ satisfy the conditions (130), then

$$\left| \exp \left\{ - \int_0^t |\alpha| \sum_{i=1}^{N_1} a_{ii} \, d\tau \right\} \right| \ge \frac{\exp \left\{ - \int_0^t \sum_{i=0}^{N_1} |b_{ii}| \, d\tau \right\}}{N_1! C^{N_1} (|\alpha| + 1)^{N_1 L}}. \tag{132}$$

Indeed, assuming the contrary and taking $-|\alpha|$ (resp., $-\alpha_s$) instead of $|\alpha|$ (resp., α_s) in the left-hand side of (132), we would have

$$|w(t)| > N_1! C^{N_1} (|\alpha| + 1)^{N_1 L} w(0) = N_1! C^{N_1} (|\alpha| + 1)^{N_1 L},$$

which is incompatible with (130).

Now, let us represent the functions $v_i^{(l)**}(t, t_1, \alpha_1, \ldots, \alpha_n)$ in the form of linear combinations of $v_i^{(l)**}(t, 0, \alpha_1, \ldots, \alpha_n)$, setting

$$v_i^{(l)**}(t, t_1, \alpha_1, \ldots, \alpha_n) = \sum_k C_k^{(l)} v_i^{(k)**}(t, 0, \alpha_1, \ldots, \alpha_n). \qquad (133)$$

The constants $C_k^{(l)}$ can be found from the following system of linear equations

$$\sum_k C_k^{(l)} v_i^{(k)**}(t_1, 0, \alpha_1, \ldots, \alpha_n) = v_i^{(l)**}(t_1, t_1, \alpha_1, \ldots, \alpha_n). \qquad (134)$$

Let us fix l in (134). Then the constants $C_k^{(l)}$ are defined as fractions with $w(t_1)$ in the denominator and minor determinants of the matrix associated with the Wronski determinant in the numerator. Therefore, expressing $C_k^{(l)}$ from (134), we find, because of (130) and (132), that

$$|C_k^{(l)}| \le N_1! C^{N_1}(|\alpha| + 1)^{N_1 L}(N_1 - 1)! C^{N_1 - 1}(|\alpha| + 1)^{(N_1 - 1)L} \times$$

$$\times \exp\left\{ 2 \int_0^T \sum |b_{ii}| \, dt \right\} = C_2(|\alpha| + 1)^{(2N_1 - 1)L}.$$

Hence, by virtue of (133), (130), we find that

$$\left| v_i^{(l)**}(t_2, t_1, \alpha) \right| \le N_1 C_2(|\alpha| + 1)^{(2N_1 - 1)L} C(|\alpha| + 1)^L = C_3(|\alpha| + 1)^{2N_1 L}.$$

This estimate is in contradiction with (131), provided that C_1 and L_1 are large enough.

§8. Reversibility Theorem

Theorem. *Assume that for a given system of Kowalevskaya type the Cauchy problem, in its usual setting adopted here, is uniformly well-posed on the interval $(0, T)$. Then another Cauchy problem for the same system on the same interval is also well-posed, namely, the problem with the values of the unknown functions being prescribed at the end t_1 of the interval $(0, t_1)$ $(0 < t_1 \le T)$, and the solutions being sought for on the interval $(0, t_1)$.*

The Cauchy problem of the latter type is called *reverse* with respect to the former.

Proof. Denote by $\tilde{v}_i^{(l)**}(t_1, t, \alpha_1, \ldots, \alpha_n)$ the functions of t satisfying system (115) on the interval $(0, t_1)$, and the following conditions at $t = t_1$:

$$\tilde{v}_i^{(l)**}(t_1, t_1, \alpha_1, \ldots, \alpha_n) = \begin{cases} 0 & \text{if } i \neq l, \\ 1 & \text{if } i = l. \end{cases}$$

It is then obvious that the reverse Cauchy problem for our system on the interval $(0, T)$ is uniformly well-posed if and only if there exist positive constants C_2 and L_2 such that for any t from the interval $(0, t_1)$ the following inequalities are satisfied:

$$\left| \tilde{v}_i^{(l)**}(t_1, t, \alpha_1, \ldots, \alpha_n) \right| \leq C_2(|\alpha| + 1)^{L_2}. \tag{135}$$

On the other hand, because of the assumed uniform well-posedness of the usual Cauchy problem for our system, the estimates (130) are valid. Therefore, our aim is to show that (130) implies (135). To this end, let us represent the functions $\tilde{v}_i^{(l)**}(t_1, t, \alpha_1, \ldots, \alpha_n)$ in the form of linear combinations of $v_i^{(l)**}(t, 0, \alpha_1, \ldots, \alpha_n)$, setting

$$\tilde{v}_i^{(l)**}(t_1, t, \alpha_1, \ldots, \alpha_n) = \sum_k C_k^{(l)} v_i^{(k)**}(t, 0, \alpha_1, \ldots, \alpha_n).$$

In order to find the constants $C_k^{(l)}$, we set $t = t_1$ in these equations. Then we get

$$\tilde{v}_i^{(l)**}(t_1, t_1, \alpha_1, \ldots, \alpha_n) = \sum_k C_k^{(l)} v_i^{(k)**}(t_1, 0, \alpha_1, \ldots, \alpha_n).$$

Hence, in complete analogy to the preceding section, we obtain estimates for $C_k^{(l)}$ and then easily establish (135). Note that for hyperbolic systems it is easy to show that $L_2 = L = 0$.

References

[1] Hadamard, J. *Le Problème de Cauchy et les Équations aux Dérivées Partielles Linéaires Hyperboliques.* Paris: Hermann, 1932.,

[2] Petrowsky, I.G. Sur le problème de Cauchy pour un système d'équations aux dérivées partielles dans le domaine réel. *C.R. Acad. Sci. Paris* **202** (1936) 1010–1012.

[3] Petrowsky, I.G. Sur le problème de Cauchy pour un système linéaire d'équations aux dérivées partielles dans un domaine réel. *C.R. Acad. Sci. Paris* **202** (1936) 1246–1248.

[4] Holmgren, E. Über Systeme von linearen partiellen Differentialgleichungen. *Ovfersigt. Kgl. Vetenskaps. Akad. Förhandl.* **58** (1901) 91–103.

[5] Hadamard, J. *Leçons sur la Propagation des Ondes et les Équations de l'Hydrodynamique.* Paris: Hermann, 1903. Note 1.

[6] Gevrey, M. Sur les équations aux dérivées partielles du type parabolique, I, II. *J. Math. Pures et Appl.* (1913) 405–471; (1914) 105–143.

[7] Goursat, E. *Cours d'Analyse Mathématique.* Tome II, Paris, 1923.

[8] Bochner, S. *Vorlesungen über Fouriersche Integrale.* Leipzig, 1932.

[9] Carathéodory, C. *Vorlesungen über Reelle Funktionen.* Leipzig-Berlin, 1927.

[10] Levi, E.E. Sull'equazione del calore. *Ann. di Matematica, Ser. 3,* **14** (1908) 187–264.

[11] Perron, O. *Math. Ztschr.* **29** (1929) 139–140.

[12] Petrowsky, I.G. Über das Cauchysche Problem für Systeme von partiellen Differentialgleichungen. *Matem. Sbornik* **2:5** (1937) 815–870.

NOTES

(by V.P. Palamodov)

[1] (Introduction). Indeed, the first condition implies the second. In order to prove this statement, consider a mapping $A : U \mapsto \Phi$ of Banach spaces, where U consists of vector valued functions $u(t_0, t, x)$ defined in the domain $0 \leq t_0 \leq t \leq T$, $x \in \mathbb{R}^n$, satisfying system (1) for each t_0, and bounded, together with their derivatives up to "a sufficiently high" order l, each vector valued function $u(t_0, t_0, x)$ being bounded, together with its derivatives in x up to the order L. The space Φ is formed by vector valued functions $\varphi(t_0, x)$ defined in the layer $0 \leq t_0 \leq T$, $x \in \mathbb{R}^n$, and bounded, together with all their derivatives up to the order L. The operator A is defined by: $u(t_0, t, x) \to u(t_0, t_0, x)$, and is easily seen to be continuous. According to the first condition, the operator A is a linear isomorphism. On account of the Banach theorem, A is an open mapping, and therefore the second condition is satisfied.

[2] (Introduction). This statement seems to be much too categorical. The words of Hadamard [1] in this connection are more cautious: "Physical processes occur in such a way that the values of the Cauchy data do not determine the unknown function." It would be erroneous to identify the physical significance of a particular mathematical problem, for instance, the Cauchy problem, with the possibility of the "determination" of its solution as understood by J. Hadamard, i.e., the calculation of the solution on the basis of the initial data available only in approximation. Several examples of physically meaningful problems which, however, are not well-posed are provided by fluid dynamics (of this type, in particular, is the problem describing the appearance and development of turbulent flows). The simplest problem that is ill-posed in the sense of Hadamard consists in restoring a holomorphic function from its values on the real axis. The possibility of solving this problem has been repeatedly put to use in theoretical physics.

However, it is beyond doubt that the class of well-posed problems, defined and characterized by I.G. Petrowsky in this paper, constitutes a part of the linear theory, important from the physical, as well as mathematical, standpoint. The processes described by well-posed and ill-posed Cauchy problems are physically distinct.

[⁹] (Ch. I, §1). The assumption of continuity of the coefficients $A_{ij}^{(k_0,k_1,...,k_n)}(t)$ is meant here. In order to prove that this assumption is sufficient, we show that if a solution

$$\{v_l(t,\alpha) \equiv v_l(t,t_0,\alpha_1,\ldots,\alpha_n),\ 1 \le l \le N\}$$

of system (10) satisfies the inequalities

$$|v_l(t,\alpha)| \le C(1+|\alpha|)^L \qquad (l=1,\ldots,N;\ |\alpha|=\alpha_M),\tag{I}$$

then, for any l and $j < n_l$, there exist constants C_1 and L_1 such that

$$\left|\frac{\partial^j v_l(t,\alpha)}{\partial t^j}\right| \le C_1(1+|\alpha|)^{L_1},\tag{II}$$

Assume the contrary, viz.: there exist l and j such that (II) can hold for no C_1, L_1. Set

$$V_{r,i}(\alpha) = \max_{t_0 \le t \le T}\left|\frac{\partial^i v_r(t,\alpha)}{\partial t^i}\right|,\qquad i \le n_r,\quad r=1,\ldots,N\,.$$

By assumption, there exit l, $j < n_l$, and a sequence of points α_p, $|\alpha_p| \to \infty$, such that $V_{l,j}(\alpha_p)$ tends to infinity faster than any power of $|\alpha_p|$ and

$$V_{l,j}(\alpha_p) = \max_{i<n_r} V_{r,i}(\alpha_p)\,.\tag{III}$$

For any positive functions f and g defined at the points of the sequence $\{\alpha_p\}$, we write $f \prec g$, if for any positive integer s

$$f(\alpha_p) = o(|\alpha_p|^{-s} g(\alpha_p)),\quad p \to \infty\,.$$

Let us take "a chain of maximal length" consisting of functions $V_{l,i}$, $0 \le i \le n_l$, which satisfy the relations

$$V_{l,k} \prec V_{l,k+1} \prec \cdots \prec V_{l,m-1} \prec V_{l,m}\,.\tag{IV}$$

This set contains at least two elements, since $V_{l,0} \prec V_{l,j}$ by virtue of (I). We also have $m < n_l$, since (10) and (III) imply that

$$V_{l,n_l} \le C(1+|\alpha|)^K \max_{i<n_r} V_{r,i} = C(1+|\alpha|)^K V_{l,j}\,,$$

where K is the largest value taken by the sum $k_1 + \cdots + k_n$ formed by the indices of the non-zero coefficients of system (10). Since the chain (IV) is maximal, it follows that there is an infinite subsequence $\{\alpha_p\}$ such that

$$V_{l,m+1}(\alpha_p) \leq C_0(1 + |\alpha_p|)^{s_0} V_{l,m}(\alpha_p) , \qquad (V)$$

where C_0, s_0 are constants. From now on, we consider only this subsequence, retaining the same notation $\{\alpha_p\}$. Let t_p be a point at which the function $|\partial^m v_l(t, \alpha_p)/\partial t^m|$ attains its largest value equal to $V_{l,m}(\alpha_p)$. Set

$$\varepsilon = [2C_0(1 + |\alpha_p|)^{s_0}]^{-1}$$

and consider the neighborhood $[t_p - \varepsilon, t_p + \varepsilon]$ of that point. At least one of the halves of this interval belongs to the interval $[t_0, T]$, on which our solution is defined. It follows from (V) that at the points of this semi-neighborhood we have

$$\left| \frac{\partial^m v_l(t, \alpha_p)}{\partial t^m} \right| \geq \frac{1}{2} V_{l,m}(\alpha_p) .$$

Therefore the oscillation of the function $\partial^{m-1} v_l(t, \alpha_p)/\partial t^{m-1}$ on this semi-neighborhood is less than $\frac{\varepsilon}{2} V_{l,m}(\alpha_p)$, and consequently,

$$V_{l,m-1}(\alpha_p) \geq \frac{\varepsilon}{4} V_{l,m}(\alpha_p) = \frac{1}{8C_0}(1 + |\alpha_p|)^{-s_0} V_{l,m}(\alpha_p) .$$

The above inequality is in contradiction with the relation $V_{l,m-1} \prec V_{l,m}$ in (IV). Thereby the estimate (II) is proved.

[4] (Ch. I, §4). The works of I.G. Petrowsky have stimulated the studies on the uniqueness of solutions of the Cauchy problems, in particular, of the characteristic Cauchy problem (see the Commentary).

[5] (Ch. I, §4). The uniqueness theorem for the Cauchy problem remains valid if the functions $u_i(t, x)$ are only assumed to be bounded. In order to prove this result, it suffices to consider the convolution $u_i * \varphi$, where φ is a smooth compactly supported function of spatial variables. The fact that u_i is bounded implies that all partial derivatives of the convolution with respect to x_s are bounded, too. The functions $u_i * \varphi$ $(i = 1, \ldots, N)$ satisfy system (1) (with $n_i = 1$), and, according to what has been proved in the paper, are identically equal to 0. Hence, $u_i \equiv 0$ $(i = 1, \ldots, N)$.

[6] (Ch. I, §5). The conjecture of I.G. Petrowsky is correct. Its justification can be obtained from a theorem proved by Seidenberg (1954) (see, e.g., E.A. Gorin, Uspekhi Mat. Nauk, 16:1 (1961) 91–118).

[7] (Ch. I, §5). These arguments can be easily transformed into a rigorous proof.

[8] (Ch. I, §6). This is not the case. For instance, the said condition is not satisfied by the binomial $\alpha - t\alpha^2$, $\alpha \in \mathbb{R}^1$ for $t = 0$; however, the integral

$$\int_0^1 (\alpha - t\alpha^2)\, dt = \alpha - \frac{1}{2}\alpha^2$$

is bounded from above.

[9] (Ch. II, §5). The proof of the theorem about analyticity, in its final part, where the author is referring to the arguments of §2, requires more delicate analysis. It is of no avail to single out compactly supported summands in the right-hand sides, because it is their analyticity that is needed in order to estimate the integral (101). However, one can use the decomposition of the form $f_i^* = f_i^0 + f_i^{**}$, where the functions f_i^0 are analytic with respect to the variables $x_{k_1}, \ldots, x_{k_m}$ in G_1, and decay rapidly, as $|x| \to \infty$, together with their derivatives whose order depends on the applicability of the arguments of §3, Ch I. The functions f_i^{**} and their derivatives up to the order $M - p + \Delta$ vanish at the point $(x_1^0, \ldots, x_n^0)$. Next, one can repeat the arguments of §2, making use of the estimates for the derivatives of f_i^0, uniform with respect to the parameters $x_1^0, \ldots, x_n^0$.

[10] (Ch. III, §1). In the text of the original, the Theorem of §1, Ch. III, was formulated in a somewhat different manner: the possibility of reducing system (103) to the form (106) was claimed in the case of an arbitrary closed domain. As noticed by L.S. Pontryagin, this reduction is impossible in the general situation because of the required continuity of the coefficients of the linear transformation used for this reduction. The present version of the Theorem and its proof incorporate the corrections suggested by I.G. Petrowsky.

[11] (Ch. III, §4). The value of the cubic root in the definition of ρ_2 is chosen in accordance with the relation $p = -\frac{1}{3}\rho_1\rho_2$.

[12] (Ch. III, §6). It is claimed here that every parabolic system (1) possesses the following property (*): for any point (t^0, x^0), $t^0 > 0$, $x^0 = (x_1^0, \ldots, x_n^0)$, and any domain U in the $(x_1, \ldots, x_n)$-space, there exist functions $\varphi_i(x_1, \ldots, x_n)$, with a compact support in U, such that the solution $u = (u_1, \ldots, u_N)$ of the homogeneous system (1) with the initial data φ_i at $t = 0$ does not vanish at the point (t^0, x^0). Indeed, this result can be derived from (71) if we take into account the impossibility of all the functions $G_i^{(l)}(t^0, x^0, \xi)$ being identically equal to zero for $\xi = (\xi_1, \ldots, \xi_n) \in U$. Let us justify the latter assertion. The functions $G_i^{(l)}(t^0, x, \xi)$ satisfy system (1) with respect to the variables t^0, x, and therefore, are analytic in x by virtue of the Theorem of §4, Ch. II. Since the coefficients of system (1) are independent of x, the functions $G_i^{(l)}$ depend only on the difference $x - \xi = (x_1 - \xi_1, \ldots, x_n - \xi_n)$, i.e.,

$$G_i^{(l)}(t^0, x, \xi) \equiv g_i^{(l)}(t^0, x - \xi) .$$

If we assume that $G_i^{(l)}(t^0, x^0, \xi) \equiv 0$ in U, then $g_i(t^0, x - \xi) = 0$ for every x and ξ. According to (71), it follows that $u_i(t^0, x) \equiv 0$ for any initial functions $\psi_1, \ldots, \psi_N$. If ψ_i are chosen constant, then the solution u will be independent of x, because of the uniqueness theorem of Ch. I, and therefore, will satisfy a system of ordinary differential equations resolved with respect to the highest order derivatives. By the classical uniqueness theorem, this solution can vanish at no point t^0, unless all constants ψ_i are equal to zero. Thereby the identity $g_i^{(l)}(t^0, x - \xi) \equiv 0$ is shown to be impossible, and thus (∗) is proved.

On the Cauchy Problem in Classes of Non-Analytic Functions*

In our joint paper with S.L. Sobolev about the works of J. Hadamard on partial differential equations (see [1]), we have discussed in detail the fundamental difference between solutions of the Cauchy problem for partial differential equations in classes of sufficiently smooth functions (i.e., functions possessing a sufficiently large number of partial derivatives) and solutions in the class of analytic functions. Many questions, which are of essential importance when solving this problem in classes of non-analytic functions, become quite meaningless in relation to the class of analytic functions. This type of question, for instance, arises in connection with the size of the domain where a solution of the Cauchy problem is completely determined by the values of the unknown functions on a bounded portion of a surface. With respect to classes of non-analytic functions, this question of size is very important, whereas it has no meaning at all if we consider only analytic functions, since an analytic function is completely defined by its values in an arbitrarily small domain.

Following the ideas of J. Hadamard, we introduce the notion of a well-posed Cauchy problem (see [1], pp. 84–85). The exact definition is as follows: In a real domain, consider the system of equations

$$\frac{\partial^{n_i} u_i}{\partial t^{n_i}} = F_i \qquad (i = 1, 2, \ldots, N) , \tag{1}$$

where F_i are functions of independent variables $t, x_1, \ldots, x_n$, of the unknown functions $u_1, \ldots, u_N$ (with arguments $t, x_1, \ldots, x_n$), and their partial derivatives

$$\frac{\partial^{k_0 + k_1 + \cdots + k_n} u_j}{\partial t^{k_0} \partial x_1^{k_1} \cdots \partial x_n^{k_n}} , \qquad j = 1, \ldots, N ; \quad \sum_{s=0}^{n} k_s \leq m_j$$

---

*Originally published in *Uspekhi Mat. Nauk* **3** (1937) 234–238. See Appendix for a commentary by V.Ya. Ivrii and L.R. Volevich.

$(k_s \geq 0;\ k_0 < n_j;\ n_j > 0,\ m_j$ are positive integers). We say that the Cauchy problem for the above system is well-posed in a domain G belonging to the $(t, x_1, \ldots, x_n)$-space and adjacent to the hyperplane $t = 0$, with the initial values $\varphi_i^{(j)}(x_1, \ldots, x_n)$ $(i = 1, \ldots, N;\ j = 1, \ldots, n_i - 1;\ \varphi_i^{(j)}$ are continuous and bounded, together with all their partial derivatives up to a certain finite order L) for the unknown functions u_i and their partial derivatives in t up to the order $(n_i - 1)$, respectively, if the following conditions are satisfied:

1. In the domain G, there exists a unique solution of the Cauchy problem for system (1) with any initial values $\bar\varphi_i^{(j)}$ such that the functions $\bar\varphi_i^{(j)}$, together with all their partial derivatives up to the order L, differ from the respective $\varphi_i^{(j)}$ and their partial derivatives by less than a constant $c > 0$.

2. For any given positive ε, there is a positive η such that the condition that the functions $\bar\varphi_i^{(j)}$ and all their partial derivatives up to the order L differ from the respective $\varphi_i^{(j)}$ and their partial derivatives by less than η implies that in the entire domain G, for every i, the following inequalities are satisfied:

$$|u_i - \bar u_i| < \varepsilon\,,$$

where u_i (resp., $\bar u_i$) are solutions of the Cauchy problem with the initial functions $\varphi_i^{(j)}$ (resp., $\bar\varphi_i^{(j)}$).

If $m_j = n_j$ for every j, then we say that (1) is a *system of Kowalevskaya type*. The Cauchy problems arising in various applications are usually for systems of this type.

The results obtained so far concerning the Cauchy problem for Kowalevskaya systems in classes of non-analytic functions can be summarized as follows: In the case of a single first order partial differential equation with one unknown function, the Cauchy problem was proved to be well-posed quite a long time ago. When considered in ordinary text-books, this equation is assumed to have the right-hand side F holomorphic with respect to all its arguments (see, e.g., [2], Part 2). However, it is easy to see that this assumption is unessential.

At the beginning of this century, Volterra, in collaboration with his students, proved well-posedness for the Cauchy problem in the case of a linear second order hyperbolic equation with constant coefficients, a single unknown function, and arbitrarily many independent variables. In 1904–1908, using some fairly complicated constructions, J. Hadamard (see [1], p.86) proved well-posedness for this problem in the general case of a linear second order hyperbolic equation. In 1928 a simpler proof (by the method of finite differences) of this result was suggested by Courant, Friedrichs & Lewy [3]; and in 1933 another method was found by Sobolev [4]. In 1935, Schauder [5] established well-posedness for the Cauchy problem for a quasilinear second order equation of hyperbolic type (the term "quasilinear second order equation" is applied to an equation that is linear with respect to second derivatives).

The methods employed to study second order hyperbolic equations proved incapable (to the best of my knowledge) of being adjusted to the case of higher order equations; and no results have been known pertaining to the well-posedness of the Cauchy problem for such equations with an arbitrary number of independent variables.

Another group of results deals with systems of Kowalevskaya type containing partial derivatives in two independent variables only; in this situation, it does not matter at all whether the system contains merely first order derivatives or those of a higher order, since the latter case can be easily reduced to the former one. First of all, the paper of Holmgren [6] should be mentioned in this connection, where well-posedness of the Cauchy problem is proved for a class of linear equations which, in analogy with second order hyperbolic equations, it is natural to call hyperbolic systems. In 1927–1931, Friedrichs & Lewy [7] obtained similar results for a system of nonlinear equations, using the method of finite differences. However, all these constructions proved inapplicable to systems with partial derivatives in more than two independent variables. The greater the number of independent variables, the more difficult appears the proof of well-posedness for the Cauchy problem. In the case of a single independent variable, we have systems of ordinary differential equations, for which well-posedness of the Cauchy problem was proved many years ago. When the number of independent variables becomes 3, we have to face all the difficulties arising in the case of many dimensions.

In relation to these two groups of studies, a somewhat separate position, in our opinion, belongs to an extremely important work by Holmgren [8] about the uniqueness of solutions of the Cauchy problem for linear Kowalevskaya systems with analytic coefficients, in a class of sufficiently smooth but not necessarily analytic functions[1]. So far, no version of this theorem in the case of linear systems with non-analytic coefficients has been found. The great importance of such an extension is due to the fact, established by Hadamard[2], that it is to this case that one can reduce the proof of the uniqueness of solutions for the Cauchy problem for nonlinear systems in classes of non-analytic functions.

My own studies (see [10], [11]) also concern the question of well-posedness for the Cauchy problem in the case of the most general systems of Kowalevskaya type with any number of independent variables and unknown functions, as well as arbitrary orders of partial derivatives involved.

---

[1] An exposition of Holmgren's results can also be found in [9], Note 1, pp. 348–354.

[2] See the note of Hadamard referred to in the previous footnote.

The basic theorem, established by the author for such systems, reads as follows:

The Cauchy problem for system (1) *with zero initial values is well-posed in a domain[3] $G' \subset G$, the domain G being adjacent to the hyperplane $t = 0$, if the following two conditions are satisfied in the domain G, for all functions u_j which, together with their partial derivatives entering F_i, have absolute values bounded by a positive constant c_1.*

1. *The functions F_i possess continuous bounded derivatives up to the order[4] $(6n + 4)$ with respect to all their arguments.*

2. *For any real $\alpha_1, \ldots, \alpha_n$, $\sum \alpha_k^2 = 1$, the matrix*

$$
\left\| \sum_{k_0, k_1, \ldots, k_n} \frac{\partial F_i}{\partial \frac{\partial^{k_0 + k_1 + \cdots + k_n} u_j}{\partial t^{k_0} \partial x_1^{k_1} \cdots \partial x_n^{k_n}}} \lambda^{k_0} \alpha_1^{k_1} \cdots \alpha_n^{k_n} \right\| - \left\| \begin{matrix} \lambda^{n_1} & 0 & & 0 \\ 0 & \lambda^{n_2} & & \\ & & \ddots & \\ 0 & & & \lambda^{n_N} \end{matrix} \right\|, \quad (2)
$$

where $\sum_{s=0}^{n} k_s = n_j$, $k_s \geq 0$, has the form

$$
\left\| \begin{matrix} \boxed{M_1} & & & \\ & \boxed{M_2} & & \\ & & \ddots & \\ & & & \boxed{M_l} \end{matrix} \right\|,
$$

where every element outside the squares M_k is identically equal to zero; moreover, the determinant of each matrix M_k can have only real roots λ, the difference of each pair of roots of the same determinant has absolute value larger than a positive constant δ.

It is natural to refer to systems satisfying the above two conditions as *hyperbolic.*

The proof of this result, as suggested by the author, is essentially based on estimates for the integrals of squared solutions (and their derivatives) of systems approximating the given one, in analogy with what has been done by Courant, Friedrichs, and Lewy.

On the other hand, for linear systems of Kowalevskaya type with coefficients depending on t only, the following result can be proved: *The Cauchy*

---

[3] Probably, my estimates for the size of the domain G' can be substantially improved. Recently, this question, in the case of nonlinear second order hyperbolic equations, has been studied by F.I. Frankl and S.A. Khristianovich.

[4] In analogy with the results for second order hyperbolic equations, this number can be expected to allow a reduction, to about n in the case of nonlinear systems, and to about $n/2$ in the linear case. This question for nonlinear hyperbolic second order equations has also been studied by F.I. Frankl and S.A. Khristianovich.

problem is ill-posed if for $t = 0$ there exist real values of $\alpha_1 \ldots, \alpha_n$, such that the determinant of at least one matrix M_k has a complex root.

If the determinants of all M_k, for all real α_k, have only real roots, but some of these determinants allow multiple roots, then the Cauchy problem can be either well-posed or not, as shown by examples. In the case of systems with constant coefficients, a necessary and sufficient condition for the Cauchy problem to be well-posed can be stated as follows: *No root of the determinant of the matrix* (2) *can have its real part tending to $+\infty$ faster than* $\ln \max |\alpha_s|$, *as the largest among* $|\alpha_1|, |\alpha_2| \ldots, |\alpha_n|$ *tends to $+\infty$*; here it is assumed that for the construction of the matrix (2) the summation should be extended to all non-negative integers k_s, and not be restricted to those only whose sum is equal to n_j.

The author has also examined the question of well-posedness for systems which, in some sense, are of a more general type than Kowalevskaya systems, namely, such that the condition $\sum_{s=0}^{n} k_s \leq n_j$ may be absent. However, within this class, only systems whose coefficients depend on t only have been considered. It should be mentioned that in order that a solution of the Cauchy problem for such a system be unique, the initial values of the unknown functions should, in general, be prescribed on the entire hyperplane $t = 0$, and not only on a bounded portion of that hyperplane, as is the case with hyperbolic systems.

References

[1] Petrowsky, I.G; Sobolev, S.L. On the works of J. Hadamard in the theory of partial differential equations. *Uspekhi Mat. Nauk* **2** (1936) 82–91.

[2] Goursat, E. *Cours d'Analyse Mathématique.* Tome II, Paris, 1923.

[3] Courant, R; Friedrichs, K; Lewy, H. Über die partiellen Differenzengleichungen der mathematischen Physik. *Math. Ann.* **100** (1928) 32–74

[4] Sobolev, S.L. On a generalization of the Kirchhoff formula. *Doklady Akad. Nauk SSSR* **6** (1933) 256–262.

[5] Schauder, J. Das Anfangswertproblem einer quasilinearen hyperbolischen Differentialgleichung zweiter Ordnung in beliebiger Anzahl von unabhängiger Veränderlichen. *Fund. Math.* **24** (1935) 213–246.

[6] Holmgren, E. Sur les systemes linéaires aux derivées partielles du premier ordre à deux variables independantes à caracteristique réelles et distinctes. *Ark. Mat. Astron. Fys.* **5** (1909) 1–13.

[7] Friedrichs, K; Lewy, H. *Math. Ann.* Bd. **97, 99, 101, 104.**

[8] Holmgren, E. Über Systeme von linearen partiellen Differentialgleichungen. *Ovfersigt. Kgl. Vetenskaps. Akad. Förhandl.* **58** (1901) 91–103.

[9] Hadamard, J. *Leçons sur la Propagation des Ondes et les Équations de l'Hydrodynamique.* Paris: Hermann, 1903.

[10] Petrowsky, I.G. Sur le problme de Cauchy pour un système d'équations aux dérivées partielles dans le domaine réel. *C.R. Acad. Sci. Paris* **202** (1936) 1010–1012.

[11] Petrowsky, I.G. Sur le problme de Cauchy pour un système linéaire d'équations aux dérivées partielles dans un domaine réel. *C.R. Acad. Sci. Paris* **202** (1936) 1246–1248.

4

On the Analyticity of Solutions of Systems of Partial Differential Equations[*1]

Introduction

In 1903, S.N Bernstein [2] solved one of the problems posed by D. Hilbert at the International Mathematical Congress held in 1900. He proved analyticity in x and y of every solution of an elliptic partial differential equation of the form

$$F\left(x, y, u, \frac{\partial u}{\partial x}, \frac{\partial u}{\partial y}, \frac{\partial^2 u}{\partial x^2}, \frac{\partial^2 u}{\partial x \partial y}, \frac{\partial^2 u}{\partial y^2}\right) = 0,$$

where F is an analytic function of all its arguments, and the solution is assumed to be real valued and continuous, together with its partial derivatives up to the third order. In 1928 this result was proved once again by H. Lewy [3], who assumed the function u to possess continuous partial derivatives up to the fourth order. The method suggested by H. Lewy is essentially different from that of S.N. Bernstein. The problem of the analyticity of u with respect to x and y was reduced by H. Lewy to establishing the existence in the $(x + ix^*, y + iy^*)$-space of a function

$$u(x, x^*, y, y^*) = u_{\text{real}}(x, x^*, y, y^*) + u_{\text{imag}}(x, x^*, y, y^*)$$

satisfying the Cauchy–Riemann conditions with respect to x and x^*, y and y^*, as well as the conditions

---

[*]Original title: "Sur l'analyticité des solutions des systèmes d'équations différentielles." *Matem. Sbornik* **5**:1 (1939) 3–70. Notes [1-6] at the end of the article by L.R. Volevich. See Appendix for a commentary by O.A. Oleinik and L.R. Volevich. The original version contains a number of footnotes, at times of more than a page length. For the convenience of the reader, we have incorporated the more lengthy footnotes into the main text in the form of remarks. Bibliographical references in footnotes have been replaced by a more common system of references. – *Ed. note.*

[1]See also Petrowsky [1].

$$u_{\text{real}}(x, 0, y, 0) = \Re\{u(x, y)\}, \qquad u_{\text{imag}}(x, 0, y, 0) = \Im\{u(x, y)\}.$$

The latter problem was reduced by H. Lewy to the Cauchy problem for a hyperbolic system of differential equations. The essential point in the Lewy's method is that the equation under consideration contains partial derivatives only with respect to *two* independent variables; this method cannot be applied to other equations.

The following theorem has been proved by the author. Consider a system of the form

$$F_i(x_0, x_1, \ldots, x_n; \; u_1, \ldots, u_N; \ldots) = 0, \qquad i = 1, 2, \ldots, N. \qquad (1')$$

The partial derivatives of the functions u_j ($j = 1, \ldots, N$) up to the order n_j are not explicitly written in the left-hand sides. We assume that in a domain G (where the arguments in $(1')$ take their values) formed by complex values of x_0, complex values of the functions u_i and their partial derivatives, and real values of the variables $x_1, \ldots, x_n$, the left-hand sides of equations $(1')$ are analytic (holomorphic) with respect to x_0, all u_j and their partial derivatives, and possess continuous derivatives in $x_1, \ldots, x_n$ up to the order

$$2n + 2\left[\frac{n+1}{2}\right] + n^* + 7, \qquad (2)$$

where n^* is the largest of n_i ($i = 1, \ldots, N$), and $[x]$, as usual, denotes the entire part of x. Further, it is assumed that for all real α_k, $\sum \alpha_k^2 = 1$, the characteristic matrix

$$\left\|\sum_{(k)} \frac{\partial F_i}{\partial\left\{\dfrac{\partial^{k_0 + k_1 + \cdots + k_n} u_j}{\partial x_0^{k_0} \partial x_1^{k_1} \cdots x_n^{k_n}}\right\}} \alpha_0^{k_0} \cdots \alpha_n^{k_n}\right\|, \qquad (3)$$

(here the sum is taken over all non-negative integers $k_0, k_1, \ldots, k_n$ such that $\sum_{s=0}^{n} k_s = n_j$) has the form

$$\left\|\begin{array}{cccc} \boxed{M_1} & & & \\ & \boxed{M_2} & & \\ & & \ddots & \\ & & & \boxed{M_r} \end{array}\right\|. \qquad (3')$$

Here all elements outside the matrices M_s are identically equal to zero. Assume further that the left-hand sides F_i ($i = 1, \ldots, N$) of system $(1')$ contain only those derivatives of the functions u_j corresponding to the matrix M_s which are taken with respect to the variables

$$x_{0s} = x_0, x_{1s}, \ldots, x_{ls} . \tag{4}$$

It is then assumed that the determinant of M_s does not vanish for any real values of

$$\alpha_{0s} = \alpha_0, \alpha_{1s}, \ldots, \alpha_{ls}$$

such that $\alpha_0^2 + \alpha_{1s}^2 + \cdots + \alpha_{ls}^2 = 1$. The absolute values of the determinants of the above matrices are continuous functions of their arguments; therefore, these absolute values are bounded by a positive constant independent of $x_0, x_1, \ldots, x_n, u_1, \ldots, u_N$, or their derivatives. (It should be observed that throughout the paper we are dealing with functions defined on closed sets.) We do not exclude the possibility of any matrix M_s being formed by a single element $\alpha_0^{n_i}$.

Remark 1. The subsequent arguments can also be applied to systems of equations containing, instead of the derivatives of u_i (the functions corresponding to the matrix M_s) with respect to the variables (4), the derivatives of these functions with respect to the variables $y_{0s} = x_0, y_{1s}, \ldots, y_{ls}$ having the form of linear combinations of $x_1, \ldots, x_n$. In this situation, one should consider the matrix

$$\left\| \sum_{(k)} \frac{\partial F_i}{\partial \left\{ \dfrac{\partial^{k_0 + k_1 + \cdots k_n} u_j}{\partial x_0^{k_0} \partial y_{1s}^{k_1} \cdots \partial y_{ls}^{k_n}} \right\}} \alpha_0^{k_0} \alpha_1^{k_1} \cdots \alpha_{ls}^{k_l} \right\|$$

and assume that its determinant does not vanish for any real values of $\alpha_0, \alpha_{1s} \ldots, \alpha_{ls}$ such that $\alpha_0^2 + \alpha_{1s}^2 + \cdots \alpha_{ls}^2 = 1$.

Under the above assumptions, we are going to prove the following result.

Theorem. *In the domain G, every solution of our system is analytic with respect to x_0, provided that this solution possesses continuous derivatives up to the order*

$$2n + 2 \left[\frac{n+1}{2} \right] + n_i + 6 + n^* . \tag{5}$$

Remark 2. For the sake of brevity, we say that "a solution of system $(1')$ is analytic in x_0", "a solution is bounded", etc., to mean that the functions u_i $(i = 1, \ldots, N)$, which actually constitute the solution, are analytic in x_0, are bounded, etc. A solution of system $(1')$ is conceived of as any set of functions u_i defined for $\Im(x_0) = \text{const}$, taking, in general, complex values, and satisfying equations $(1')$, where the derivatives in x_0 are understood as partial derivatives with respect to the real part of x_0.

Remark 3. There is an misprint in my note [1], which should be corrected as follows. The expression

$$\max \left\{ 2n + \left[\frac{n+1}{2}\right], n^* \right\} + \left[\frac{n+1}{2}\right] + 7$$

in (2) should be replaced by

$$\max \left\{ 2n + \left[\frac{n+1}{2}\right], n^* \right\} + \left[\frac{n+1}{2}\right] + 7 + n^* ;$$

and the expression

$$\max \left\{ 2n + \left[\frac{n+1}{2}\right], n^* \right\} + \left[\frac{n+1}{2}\right] + 6 + n^* ;$$

in (5) should read as

$$\max \left\{ 2n + \left[\frac{n+1}{2}\right], n^* \right\} + \left[\frac{n+1}{2}\right] + 6 + n^* + n_i . \qquad \Box$$

In order to prove our theorem, we immerse (following H. Lewy) the $(x_0, x_1, \ldots, x_n)$-space (with x_0 real) into the $(x_0^*, x_0, x_1, \ldots, x_n)$-space and consider a $(n+2)$-dimensional neighborhood of a point P, regarding the latter as the origin. In this neighborhood, we define functions

$$u_i = u_{i\,\text{real}} + i u_{i\,\text{imag}} \qquad (i = 1, 2, \ldots, N)$$

such that each pair $u_{i\,\text{real}}$, $u_{i\,\text{imag}}$ satisfies the system of Cauchy–Riemann with respect to the independent variables x_0 and x_0^*, and the functions u_i coincide with the given ones in a $(n+1)$-dimensional neighborhood of the point P. The method suggested here for the construction of such functions is completely different from that adopted by H. Lewy.

On the other hand, we establish the existence of non-analytic (in x_0) solutions of system (1') possessing partial derivatives of any given order[2], under the following assumptions:

1. There exists a system of real $\alpha_1, \ldots, \alpha_n$ such that the determinant of the matrix (3) admits a real non-zero root α_0.

2. This determinant does not vanish identically as a function of α_0, $\ldots, \alpha_n$.

3. The functions F_i are linear with respect to u_j $(j = 1, 2, \ldots, N)$ and their derivatives (the coefficients in the corresponding linear combinations are constant); the terms containing neither u_j nor their derivatives are analytic functions of $x_0, x_1, \ldots, x_n$.

---

[2]See Remark 2.

Imposing additional restrictions, to be specified in the sequel, one can prove a similar statement for nonlinear systems.

The problem of existence of non-analytic solutions has not been fully examined in the case of the determinant of the matrix (3) having the root $\alpha_0 = 0$ for some system of real $\alpha_1, \ldots, \alpha_n$ such that at least one $\alpha_s \neq 0$. It is possible to construct linear systems with constant coefficients which, under the above conditions, admit solutions non-analytic in x_0 and at the same time possessing derivatives of any order. On the other hand, there exist systems of this type with all their solutions analytic in x_0 [1].

We say that system (1') is *elliptic* in the domain G (wherein the arguments of the functions F_i take their values), if the determinant of the matrix (3) does not vanish for any real $\alpha_0, \ldots, \alpha_n$ such that $\sum_{k=0}^{n} \alpha_k^2 = 1$.

Let G be a complex domain with respect to all its coordinates, and let F_i be analytic functions of all their arguments. Assume that system (1') is elliptic in that domain. Then every solution of system (1') having continuous derivatives up to the order

$$2n + 2 \left[\frac{n+1}{2} \right] + n_i + 6$$

is analytic with respect to any linear combination of independent variables with constant coefficients, and in particular, analytic with respect to all x_k $(k = 0, \ldots, n)$ simultaneously.

On the other hand, the theorem formulated above implies that system (1') admits solutions non-analytic with respect to all their arguments, but possessing continuous derivatives of any order, provided that the following conditions are satisfied:

1. There exist real $\alpha_0, \ldots, \alpha_n$ such that at least one α_s is different from zero, and the determinant of the matrix (3) vanishes for these values of α_k.

2. The determinant of the matrix (3) is not identically equal to zero as a function of $\alpha_0, \ldots, \alpha_n$.

3. F_i are linear functions (with constant coefficients) of u_j $(j = 1, \ldots, N)$ and their derivatives.

The condition that the determinant of the matrix (3) does not identically vanish with respect to $\alpha_0, \ldots, \alpha_n$ is essential, as evidenced by the example in the text. Under certain restrictions on F_i (specified below), similar results can be proved for nonlinear systems.

§1. Preliminary Remarks

1. First of all, we note that each solution of system (1′) formed by the functions u_i having continuous derivatives up to the order $k + n^* + n_i + 2$ corresponds to a system of functions u_ρ having continuous derivatives up to the order $k + n_\rho$ and forming a solution of a system which has the form

$$\sum_{j=1}^{N} \sum_{(k)} a_{ij}^{(k_0 \ldots k_n)} \frac{\partial^{n_j} u_j}{\partial x_0^{k_0} \cdots \partial x_n^{k_n}} + b_i = 0 , \qquad i = 1, \ldots, N ; \qquad (1'')$$

here n_ρ is the highest order of the derivatives of the functions u_ρ in equations (1″); the number N for system (1″) is, in general, larger than N in (1′).

Let the square matrices (3) be of the form (3′). Assume also that the left-hand sides F_i of system (1′) contain only those derivatives of the functions u_j corresponding to the square matrix M_s which are taken with respect to the variables (4). Assume that for no system of real

$$\alpha_{0s} = \alpha_0, \alpha_{1s} \ldots, \alpha_{ls} , \quad \text{with} \quad \alpha_0^2 + \alpha_{1s}^2 + \cdots + \alpha_{ls}^2 = 1 ,$$

the determinant M_s may vanish. Then the square matrix

$$\left\| \sum a_{ij}^{(k_0 \ldots k_n)} \alpha_0^{k_0} \cdots \alpha_n^{k_n} \right\|$$

will also have the form

$$\left\| \begin{array}{cccc} \boxed{M_1} & & & \\ & \boxed{M_2} & & \\ & & \ddots & \\ & & & \boxed{M_r} \end{array} \right\| .$$

(The value of r in this matrix is, in general, larger than r for system (1′)). Moreover, if equations (1″) contain only those derivatives of the functions u_j corresponding to the square matrix M_s which are taken with respect the variables

$$x_{0s} = x_0, x_{1s}, \ldots, x_{ls} ,$$

then the determinant M_s does not vanish for any real $\alpha_{0s} = \alpha_0, \alpha_{1s}, \ldots, \alpha_{ls}$ such that $\alpha_0^2 + \alpha_{1s}^2 + \cdots + \alpha_{ls}^2 = 1$.

The coefficients $a_{ij}^{(k_0 \ldots k_n)}$ depend on $x_0 + i x_0^*$, $x_1, \ldots, x_n$, and the functions u_ρ whose corresponding l is equal to n and $n_\rho \geq 2$; they also depend on the derivatives of these functions up to the order $n_\rho - 2$. The coefficients b_i depend, in general, on $x_0 + i x_0^*$, $x_1, \ldots, x_n$, and all functions u_ρ and their derivatives in x_{ks} up to the order $n_\rho - 1$. Moreover, the coefficients $a_{ij}^{(k_0 \ldots k_n)}$ (resp., b_i) are holomorphic with respect to $x_0 + i x_0^*$, u_ρ and all their derivatives, and are continuously differentiable in $x_1, \ldots, x_n$ up to the order $k_1 + n^* + 2$ (resp., k_1), provided that the left-hand sides of equations $(1')$ are holomorphic with respect to $x_0 + ix_0^*$, u_j and their derivatives, and are continuously differentiable in $x_1, \ldots, x_n$, up to the order $k_1 + n^* + 2$.

In order to prove the above statement, we set

$$\frac{\partial u_i}{\partial x_{p_1}} = u_i^{(p_1)}, \qquad \frac{\partial^2 u_i}{\partial x_{p_1} \partial x_{p_2}} = u_i^{(p_1, p_2)},$$

$$\frac{\partial^{n^*+2} u_i}{\partial x_{p_1} \cdots \partial x_{p_{n^*+2}}} = u_i^{(p_1, \ldots, p_{n^*+2})}, \tag{6}$$

$$i = 1, \ldots, N, \qquad p_1, \ldots, p_{n^*+2} = 0, 1, \ldots, n.$$

Then

$$\sum_{k=0}^{n} \frac{\partial^2 u_i}{\partial x_k^2} = \sum_{k=0}^{n} u_i^{(kk)}, \tag{7_0}$$

$$\sum_{k=0}^{n} \frac{\partial^2 u_i^{(p_1)}}{\partial x_k^2} = \sum_{k=0}^{n} u_i^{(p_1 kk)}, \tag{7_{p_1}}$$

$$\cdots \cdots \cdots \cdots \cdots \cdots \cdots \cdots \cdots$$

$$\sum_{k=0}^{n} \frac{\partial^2 u_i^{(p_1 \cdots p_{n^*})}}{\partial x_k^2} = \sum_{k=0}^{n} u_i^{(p_1 \cdots p_{n^*} kk)}, \tag{$7_{p_1 \cdots p_{n^*}}$}$$

$$i = 1, \ldots, N, \quad p_1, p_2, \ldots, p_{n^*} = 0, 1, \ldots, n.$$

Differentiating every equation in $(1')$ with respect to $x_{p_1}, \ldots, x_{p_{n^*+1}}$, and also with respect to $x_{p_1}, \ldots, x_{p_{n^*+2}}$, we find that

$$\sum_j \sum_k \frac{\partial F_i}{\partial \left\{ \dfrac{\partial^{n_j} u_j}{\partial x_0^{k_0} \cdots \partial x_n^{k_n}} \right\}_u} \frac{\partial^{n_j} u_j^{(p_1 \cdots p_{n^*+1})}}{\partial x_0^{k_0} \cdots \partial x_n^{k_n}} + \cdots = 0, \tag{8}$$

$$\sum_j \sum_k \frac{\partial F_i}{\partial \left\{ \dfrac{\partial^{n_j} u_j}{\partial x_0^{k_0} \cdots \partial x_n^{k_n}} \right\}_u} \frac{\partial^{n_j} u_j^{(p_1 \cdots p_{n^*+2})}}{\partial x_0^{k_0} \cdots \partial x_n^{k_n}} + \cdots = 0, \tag{9}$$

$$i = 1, \ldots, N, \quad p_1, p_2, \ldots, p_{n^*+2} = 0, 1, \ldots, n,$$

where the dots in the left-hand sides stand for the terms containing u_j, $u_j^{(p_1)}, \ldots, u_j^{(p_1 \cdots p_{n^*+2})}$, and the derivatives of $u_j^{(p_1 \cdots p_{n^*+2})}$ of the order $\leq n_j - 1$. Here

$$\overline{\partial\left\{\dfrac{\partial F_i}{\dfrac{\partial^{n_j} u_j}{\partial x_0^{k_0} \cdots \partial x_n^{k_n}}}\right\}_u}$$

denotes the expression obtained from

$$\overline{\partial\left\{\dfrac{\partial F_i}{\dfrac{\partial^{n_j} u_j}{\partial x_0^{k_0} \cdots \partial x_n^{k_n}}}\right\}}$$

by replacing the derivatives

$$\frac{\partial u_\nu}{\partial x_{p_1}}, \ldots, \frac{\partial^{n_\nu} u_\nu}{\partial x_{p_1} \cdots \partial x_{p_{n_\nu}}}$$

with $u_\nu^{(p_1)}, \ldots, u_\nu^{(p_1 \cdots p_{n_\nu})}$.

Obviously, the system consisting of equations $(7) - (9)$ possesses the same properties as system $(1'')$.

An elliptic system can be reduced to the form $(1'')$, if we apply the operator

$$\frac{\partial^2}{\partial x_0^2} + \cdots + \frac{\partial^2}{\partial x_n^2}$$

to the functions F_i. In what follows, we limit ourselves to systems of the form $(1'')$ whose coefficients $a_{ij}^{(k_0 \cdots k_n)}$ (resp., b_i) are holomorphic with respect to $x_0 + i x_0^*$, all u_ρ and their derivatives, and are also continuously differentiable in $x_1, \ldots, x_n$ up to the order

$$2n + 2 \left[\frac{n+1}{2}\right] + 6 \quad \left(\text{resp. } 2n + 2 \left[\frac{n+1}{2}\right] + 5\right) .$$

Our aim is to prove that every solution (i.e., the functions u_ρ) of such a system is holomorphic in $x_0 + i x_0^*$, if it is continuously differentiable up to the order

$$2n + 2 \left[\frac{n+1}{2}\right] + n_\rho + 4 .$$

In order to avoid new notations, we assume, in what follows, that systems $(1')$ and $(1'')$ are identical and refer to them as system (1).

2. Consider system (1), assuming that

$$\left|\frac{\partial^k u_j}{\partial x_0^{k_0} \cdots \partial x_n^{k_n}}\right| \le 2M , \quad j = 1, \ldots, N , \quad k = \sum k_\sigma \le n_j , \ k_\sigma \ge 0 , \quad (10)$$

in the parallelepiped $R_{LL_0^*}$ defined by the inequalities

$$|x_k| \le L , \quad k = 0, \ldots, n , \quad |x_0^*| \le L_0^* .$$

Here and in what follows, x_0 stands for the real part of the previously considered variable x_0, and its imaginary part is denoted by ix_0^*.

We assume that, in general, a solution[3] of system (1) is composed of complex valued functions

$$v_i(x_0, x_1, \ldots, x_n), \qquad i = 1, 2, \ldots, N,$$

defined in the cube Q_L:

$$|x_k| \leq L, \quad k = 0, 1, \ldots, n,$$

for $x_0^* = 0$, and that the functions v_i and their partial derivatives up to the order $2n + 2 \left[\frac{n+1}{2}\right] + n_i + 4$ satisfy the inequalities

$$\left| \frac{\partial^k v_i}{\partial x_0^{k_0} \cdots \partial x_n^{k_n}} \right| \leq M_0, \tag{11}$$

where M_0 is a sufficiently small constant whose value is specified below (see $\widetilde{(11)}$ and the end of §5).

It should be observed that no loss of generality arises from the above restrictions imposed on the range of the variables $x_0^*, x_0, x_1, \ldots, x_n, u_1, \ldots, u_N, v_1, \ldots, v_N$, and their derivatives. The general case can be reduced to this one by shifting the coordinate axes in the $(x_0^*, x_0, \ldots, x_n)$-space and introducing new functions

$$\tilde{u}_1, \tilde{u}_2, \ldots, \tilde{u}_N, \quad \tilde{v}_1, \ldots, \tilde{v}_N,$$

related to the old ones by

$$\tilde{u}_i = u_i + M_i(x_0 + ix_0^*, x_1, \ldots, x_n),$$

$$\tilde{v}_i = v_i + M_i(x_0, x_1, \ldots, x_n), \quad i = 1, 2, \ldots, N,$$

where M_i are polynomials. In the sequel, the range of the variables can be made even more narrow, if necessary.

As mentioned in the Introduction, we are going to extend the functions v_i to a neighborhood of the origin in the $(x_0^*, x_0, x_1, \ldots, x_n)$-space, so as to make them satisfy the Cauchy-Riemann equations in that neighborhood with respect to x_0 and x_0^*. To this end, we introduce the functions $u_i^*(x_0^*, x_0, x_1, \ldots, x_n)$, setting

$$u_i^*(x_0^*, x_0, x_1, \ldots, x_n) = u_i\left(\varphi(x_0^*, x_0, x_1, \ldots, x_n), x_0, x_1, \ldots, x_n\right), \tag{12}$$

for $i = 1, 2, \ldots, N$. Here $\varphi(x_0^*, x_0, x_1, \ldots, x_n) = x_0^* \varphi_0(x_0, x_1, \ldots, x_n)$, where the function $\varphi_0(x_0, x_1, \ldots, x_n)$ has the following properties:

---

[3]See Remark 2.

$$\varphi_0(x_0, x_1, \ldots, x_n) = \begin{cases} 1 & \text{in } Q_{0.6L} , \\ 0 & \text{outside } Q_{0.7L} . \end{cases}$$

Moreover, φ_0 takes positive values between 0 and 1 in $Q_{0.7L} \setminus Q_{0.6L}$; φ_0 is continuous in the entire cube Q_L, together with its derivatives up to the order

$$2n + 2 \left[\frac{n+1}{2} \right] + n^* + 5 ,$$

where n^* denotes the largest of n_i $(i = 1, \ldots, N)$. We find from (12) that

$$\frac{\partial u_i^*}{\partial x_0^*} = \frac{\partial u_i}{\partial \varphi} \frac{\partial \varphi}{\partial x_0^*} ,$$

$$\frac{\partial u_i^*}{\partial x_k} = \frac{\partial u_i}{\partial \varphi} \frac{\partial \varphi}{\partial x_k} + \frac{\partial u_i}{\partial x_k} , \qquad k = 0, 1, \ldots, n , \tag{13}$$

where $\partial u_i / \partial x_k$ stands for the partial derivative of u_i with respect to the variable x_k which enters u_i explicitly, and not through the agency of φ. If the function $u_i(x_0^*, x_0, x_1, \ldots, x_n)$ satisfies the Cauchy-Riemann equations with respect to x_0 and x_0^*, then

$$\frac{\partial u_i}{\partial \varphi} = i \frac{\partial u_i}{\partial x_0} . \tag{14}$$

Substituting this expression into the right-hand side of (13), we find that

$$\frac{\partial u_i}{\partial x_k} = \frac{\partial u_i^*}{\partial x_k} - i \frac{\partial u_i}{\partial x_0} \frac{\partial \varphi}{\partial x_k} , \qquad k = 0, 1, \ldots, n . \tag{15}$$

Hence, for $k = 0$, we get

$$\frac{\partial u_i}{\partial x_0} = \frac{\partial u_i^*}{\partial x_0} : \left(1 + i \frac{\partial \varphi}{\partial x_0} \right) , \tag{16}$$

and consequently, for each k,

$$\frac{\partial u_i}{\partial x_k} = \frac{\partial u_i^*}{\partial x_k} - i \frac{\partial u_i^*}{\partial x_0} \frac{\partial \varphi}{\partial x_k} : \left(1 + i \frac{\partial \varphi}{\partial x_0} \right) . \tag{17}$$

Now it can be easily seen that if $|x_0^*|$ is sufficiently small (this can be ensured by choosing a sufficiently small L_0^*), then the coefficient by the second term in the right-hand side of the last equality can be made arbitrarily small. In a similar fashion, one can establish the relations between higher order derivatives of the functions u_i and u_i^* and write out the differential equations for u_i^* corresponding to equations (1). Let these equations be written in the form

$$F_i^* \left(x_0^*, x_0, \ldots, x_n; \ldots, \frac{\partial^k u_j^*}{\partial x_0^{k_0} \cdots \partial x_n^{k_n}}, \ldots \right) \equiv$$

$$\equiv \sum_j \sum_{(k)} a_{ij}^{*(k_0 \ldots k_n)} \frac{\partial^{n_j} u_j^*}{\partial x_0^{k_0} \cdots \partial x_n^{k_n}} + b_i^* = 0 , \qquad (18)$$

$$i, j = 1, 2, \ldots, N , \qquad k = \sum k_\sigma \leq n_j , \qquad k_\sigma \geq 0 .$$

With the help of (18), we construct a new system

$$\widetilde{F}_i^* \left(x_0^*, x_0, \ldots, x_n; \ldots, \frac{\partial^k u_j^*}{\partial x_0^{k_0} \cdots \partial x_n^{k_n}}, \ldots \right) = 0 , \qquad (19)$$

$$i, j = 1, 2, \ldots N , \qquad k = \sum k_\sigma \leq n_j , \qquad k_\sigma \geq 0 ,$$

setting

$$\widetilde{F}_i^* \left(x_0^*, x_0, \ldots, x_n; \ldots, \frac{\partial^k u_j^*}{\partial x_0^{k_0} \cdots \partial x_n^{k_n}}, \ldots \right) =$$

$$= F_i^* \left(x_0^*, 0, \ldots, 0; \ldots, \frac{\partial^k u_j^*}{\partial x_0^{k_0} \cdots \partial x_n^{k_n}}, \ldots \right) +$$

$$+ \left\{ F_i^* \left(x_0^*, x_0, \ldots, x_n; \ldots, \frac{\partial^k u_j^*}{\partial x_0^{k_0} \cdots \partial x_n^{k_n}}, \ldots \right) - \right.$$

$$\left. - F_i^* \left(x_0^*, 0, \ldots, 0; \ldots, \frac{\partial^k u_j^*}{\partial x_0^{k_0} \cdots \partial x_n^{k_n}}, \ldots \right) \right\} \varphi_1(x_0, x_1, \ldots, x_n) \equiv$$

$$\equiv \sum_j \sum_{(k)} \tilde{a}_{ij}^{*(k_0 \ldots k_n)} \frac{\partial^{n_j} u_j}{\partial x_0^{k_0} \cdots \partial x_n^{k_n}} + \tilde{b}_i^* .$$

Here $\varphi_1(x_0, \ldots, x_n)$ is a function with the following properties:

1) $\varphi_1 = 1$ in $Q_{0.8L}$;

2) $\varphi_1 = 0$ outside $Q_{0.9L}$;

3) in Q_L the function φ_1 has continuous derivatives with respect to all its arguments up to the order $2n + 2 \left[\dfrac{n+1}{2} \right] + 6$.

Obviously, system (19) possesses the following properties:

1. The functions $\widetilde{F}_i^*$ coincide with their respective F_i^* in the parallelepiped $R_{0.8LL_0^*}$ for all functions u_j^* and their derivatives under consideration.

2. If L and L_0^* are sufficiently small, then, for all u_j^* and their derivatives up to the order n_j with absolute values bounded by M, the functions $\tilde{a}^*$ (resp., $\tilde{b}^*$) possess, in the entire parallelepiped $R_{LL_0^*}$, continuous derivatives in $x_0, x_1, \ldots, x_n$ up to the order

$$2n + 2 \left[\frac{n+1}{2}\right] + 6, \quad \left(\text{resp.,} \quad 2n + 2 \left[\frac{n+1}{2}\right] + 5\right)$$

and are holomorphic functions with respect to $x_0^*, u_1^*, \ldots, u_N^*$ and their derivatives. Outside the parallelepiped $R_{0.9LL_0^*}$, the functions $\tilde{F}_i^*$ are independent of $x_0, x_1, \ldots, x_n$ for any fixed $u_1^*, \ldots, u_N^*$ and their derivatives.

3. If L and L_0^* are sufficiently small, and the absolute values of the functions u_j^*, as well as of their derivatives up to the order n_j, are bounded by M, then the first derivatives of F_i^* ($i = 1, 2, \ldots, N$) with respect to

$$\frac{\partial^{n_j} u_j^*}{\partial x_0^{k_0} \cdots \partial x_n^{k_n}}$$

(in other words $a_{ij}^{*(k_0 \ldots k_n)}$) differ by an arbitrarily small quantity from the corresponding first derivatives of the functions F_i with respect to

$$\frac{\partial^{n_j} u_j}{\partial x_0^{k_0} \cdots \partial x_n^{k_n}}.$$

4. System (19) contains only the derivatives of u_j^* with respect to the same variables x_k as the derivatives of u_j in system (1).

Because of the above properties, the characteristic matrix for system (19) is split into matrices M_s^*, in analogy with the characteristic matrix for system (1) being split into the matrices M_s. If L_0 and L_0^* are sufficiently small (this is taken as an assumption in what follows), and if the absolute values of all u_j^* and their derivatives up to the order n_j are bounded by M in the parallelepiped $R_{LL_0^*}$, then the determinants of the matrices M_s^* have absolute values larger than a positive constant, provided that the sum of squared real $\alpha_{0s}, \alpha_{1s}, \ldots, \alpha_{ls}$ is equal to 1.

Let us multiply both sides of (14) by $\partial\varphi/\partial x_0^*$. Then, taking into consideration (16) and the first equality in (13), we obtain

$$\frac{\partial u_i^*}{\partial x_0^*} = \psi \frac{\partial u_i^*}{\partial x_0}, \tag{20}$$

where

$$\psi = i \frac{\partial\varphi}{\partial x_0^*} : \left(1 + i \frac{\partial\varphi}{\partial x_0}\right). \tag{21}$$

Hence, $\psi = 0$ outside the parallelepiped $R_{0.7LL_0^*}$. Conversely, it is easy to see that if the function u_i^* satisfies equation (20) in $R_{LL_0^*}$, then the respective $u_i(x_0^*, x_0, \ldots, x_n)$ satisfies the Cauchy–Riemann equations with respect to x_0 and x_0^* at each point of $R_{0.6LL_0^*}$. Therefore, in order to prove analyticity of the functions $v_i(x_0, \ldots, x_n)$ in a neighborhood of the origin, it suffices to construct functions u_i^* which coincide, for $x_0^* = 0$, with the functions

$$\tilde{v}_i(x_0, x_1, \ldots, x_n) = \varphi_1(x_0, x_1, \ldots, x_n)v_i(x_0, x_1, \ldots, x_n) ,$$

$$i = 1, 2, \ldots, N ,$$

and for the rest of x_0^* satisfy equations (20) in $R_{LL_0^*}$.

If the constant M_0 in the right-hand side of (11) is small enough, then the functions $\tilde{v}_i$ and their derivatives up to the order $2n + 2\left[\frac{n+1}{2}\right] + n_i + 4$ satisfy the inequalities

$$\left| \frac{\partial^k \tilde{v}_i}{\partial x_0^{k_0} \cdots \partial x_n^{k_n}} \right| \leq \frac{1}{4}M . \tag{$\widetilde{11}$}$$

For the function u_i corresponding to the matrix M_s (i.e., the derivatives of u_i are among the elements of M_s), we set

$$p_{ik_0 \ldots k_l} = \frac{\partial^k u_i^*}{\partial x_{0s}^{k_0} \cdots \partial x_{ls}^{k_l}} , \quad \text{where} \quad \sum_{\sigma=0}^l k_\sigma \leq n_i - 1 , \tag{22}$$

$$p_{i00\ldots0} = u_i^* , \quad i = 1, 2, \ldots, N .$$

In what follows, it will be convenient to denote the functions $p_{ik_0 \ldots k_l}$, with $\sum k_\sigma > 0$, by the same character u^* marked by a subscript taking integer values from $N + 1$ to, say, N_1. If u_i $(i = 1, \ldots, N)$ satisfy equations (20), then the functions $p_{ik_0 k_1 \ldots k_l}$ must satisfy the following equations

$$\frac{\partial p_{ik_0 \ldots k_l}}{\partial x_0^*} = \frac{\partial^k \left(\psi \dfrac{\partial u_i^*}{\partial x_0} \right)}{\partial x_{0s}^{k_0} \cdots \partial x_{ls}^{k_l}} =$$

$$= \sum_{m_\sigma \leq k_\sigma} C_{k_0 \ldots k_l}^{m_0 \ldots m_l} \frac{\partial^{k_0 - m_0 + \cdots + k_l - m_l} \psi}{\partial x_{0s}^{k_0 - m_0} \cdots \partial x_{ls}^{k_l - m_l}} \frac{\partial^{m_0 + 1 + \cdots + m_l} u_i^*}{\partial x_{0s}^{m_0 + 1} \cdots \partial x_{ls}^{m_l}} . \tag{23}$$

Here $C_{k_0 \ldots k_l}^{m_0 \ldots m_l}$ are positive integers; the sum $\sum_{m_\sigma \leq k_\sigma}$ is taken over all non-negative integers such that $m_\sigma \leq k_\sigma$ $(\sigma = 0, 1, \ldots, l)$. Let us rotate the coordinate axes $Ox_{0s}, Ox_{1s}, \ldots, Ox_{ls}$ of the $(x_{0s}, x_{1s}, \ldots, x_{ls})$-space ($s$ correspond to the function u_i under consideration; see Introduction) in such a way that these axes remain mutually orthogonal. Let this rotation be given by the following linear transformation

$$x'_{js} = \sum_k c^{(k)}_{js} x_{ks} \ . \tag{24}$$

In order to have symmetry of notation, we write x_{0s} instead of x_0. We can take as $c^{(k)}_{0s}$ arbitrary real numbers the sum of whose squares is equal to 1. For a given s, the indices j and k take the values corresponding to the matrix M^*_s. Then we have

$$\frac{\partial^m u_i}{\partial x^{m_0}_{0s} \cdots \partial x^{m_l}_{ls}} = \sum_p C^{*p_0 \cdots p_l}_{m_0 \cdots m_l} \frac{\partial^m u^*_i}{\partial x'^{p_0}_{0s} \cdots \partial x'^{p_l}_{ls}} \ . \tag{25}$$

Here $C^{*p_0 \cdots p_l}_{m_0 \cdots m_l}$ are bounded constants depending on the above rotation of the coordinate axes; $p_0, \ldots p_l$ take all possible non-negative integer values such that $\sum_r p_r = \sum^l_{\sigma=0} m_\sigma$. Let us substitute for the derivatives

$$\frac{\partial^m u^*_i}{\partial x^{m_0}_{0s} \cdots \partial x^{m_l}_{ls}}$$

in the right-hand sides of the equations (23) corresponding to the matrix M_s the expressions of these derivatives in terms of the right-hand sides of (25). We thus obtain the following equations

$$\frac{\partial p_{ik_0 \ldots k_l}}{\partial x^*_0} = \sum_m \psi^{m_0 \ldots m_l}_{ik_0 \ldots k_l} \frac{\partial^m u^*_i}{\partial x'^{m_0}_{0s} \cdots \partial x'^{m_l}_{ls}} \ , \tag{26}$$

where every $\psi^{m_0 \ldots m_l}_{ik_0 \ldots k_l}$ is a linear combination (with constant coefficients) of the derivatives of the function ψ in $x_{0s}, \ldots, x_{ls}$ up to the order $\sum k_\sigma - \sum m_\sigma$.

Let us perform a similar change of independent variables in the equations of system (19) corresponding to the matrix M^*_s. Taking into consideration that the determinant of M^*_s, for $\sum^l_{k=0} \alpha^2_{ks} = 1$, has absolute value larger than a positive constant, provided that the point $(x^*_0, x_0, \ldots, x_n)$ remains within the parallelepiped R_{LL_0}, and observing that all u^*_j and their derivatives up to the order n_j have absolute values bounded by M, we can resolve equations (1) with respect to $\partial^{n_i} u^*_i / \partial x'^{n_i}_{0s}$. Let us substitute the expressions thus obtained into the right-hand side of (26) and write the resulting equations (for every i) in the form

$$\frac{\partial u^*_i}{\partial x^*_0} = \sum^N_{j=1} \sum_{(p)} A'^{(p_0 \cdots p_l)}_{ij} \frac{\partial^{n_j} u^*_j}{\partial x'^{p_0}_{0s} \cdots \partial x'^{p_l}_{ls}} + B'_i \ , \qquad i = 1, \ldots, N \ . \tag{27}$$

Here the sum is taken over all non-negative $p_0, p_1, \ldots p_l$ such that $\sum^l_{r=0} p_r = n_j$ and at least one of $p_1, \ldots, p_l$ is not equal to 0. The coefficients $A'^{(p_0 \cdots p_l)}_{ij}$ have the following properties:

1. They depend on $x^*_0, x_0, \ldots, x_n$ and the functions u^*_ρ ($\rho = 1, \ldots, N$) with $n_\rho \geq 2$, $l = n$, as well as their derivatives up to the order $n_\rho - 2$.

2. For all values of u_ρ^* and their derivatives with absolute values bounded by M, these coefficients are holomorphic with respect to $x_0^*, u_1^*, \ldots, u_N^*$ and their derivatives, and are continuously differentiable in $x_1, \ldots, x_n$ up to the order

$$2n + 2\left[\frac{n+1}{2}\right] + 6 .$$

3. Outside the parallelepiped $R_{0.7LL_0^*}$ these coefficients vanish for all fixed values of the functions u_ρ^* and their derivatives, since ψ enters $A_{ij}^{'(p_0 \cdots p_l)}$ as a factor.

The functions B_i' possess the following properties:

1. They depend on $x_0^*, x_0, x_1, \ldots, x_n$, as well as u_ρ^* $(\rho = 1, \ldots, N)$ and their derivatives up to the order $n_\rho - 1$.

2. For all u_ρ^* and their derivatives up to the order n_ρ having absolute values bounded by M, the functions B_i' are holomorphic with respect to $x_0^*, u_1^*, \ldots, u_N^*$ and their derivatives, and are continuously differentiable in $x_0, x_1, \ldots, x_n$ up to the order

$$2n + 2\left[\frac{n+1}{2}\right] + 5 .$$

3. Outside the parallelepiped $R_{0.7LL_0^*}$ they vanish for all values of the functions u_ρ^* and their derivatives, since each B_i' contains as a factor the function ψ with suitable indices.

§2. Construction of Approximating Functions

Let us divide the parallelepiped $R_{LL_0^*}$ into equal parts, $2m$ in number, by the planes

$$x_0^* = \frac{L_0^*}{2m}p = hp, \qquad p = 0, \pm 1, \pm 2, \ldots, \pm m .$$

Denote by R_p^m the part of the parallelepiped between the planes

$$x_0^* = h(p - 1) \quad \text{and} \quad x_0^* = hp .$$

For each partition of this type, we are going to construct functions $u_i^{*(m)}$ $(i = 1, \ldots, N_1)$ approximating $u_i^*(i = 1, \ldots, N)$ (considered in the preceding section) and their derivatives up to the order $n_i - 1$. Since, for $p \leq 0$, the structure of the approximations in R_p^m is similar to that for $p > 0$, we limit ourselves to the case of positive p.

For $x_0^* = 0$, we take as $u_i^{*(m)}$ $(i = 1, 2, \ldots, N_1)$ the functions $\tilde{v}_i$, defined on Q_L, and their derivatives ($\tilde{v}_i$ were constructed in the preceding section). Set

$$a_i^{(m)}(x_0^*, \alpha_0, \ldots, \alpha_n) =$$

$$= \frac{1}{2^{n+1} L^{n+1}} \int_{Q_L} u_i^{*(m)}(x_0^*, x_0, \ldots, x_n) \, e^{-\frac{\pi i}{L}(\alpha, x)} \, dx_0 \ldots dx_n \, ,$$

$i = 1, 2, \ldots, N_1$, $(\alpha, x) = \sum_{k=0}^n \alpha_k x_k$. We adopt the convention to write these equalities in the form

$$a_i^{(m)}(x_0^*, \alpha_0, \ldots, \alpha_n) = \int u_i^{*(m)}(x_0^*, x_0, \ldots, x_n) \, e^{-\frac{\pi i}{L}(\alpha, x)} \, dx_0 \ldots dx_n \, ,$$

$$i = 1, 2, \ldots, N_1 \, .$$

For the sake of brevity, similar expressions will also be used to denote the Fourier coefficients of other functions.

The functions $u_i^{*(m)}$ $(i = 1, \ldots, N_1)$ have been defined for $x_0^* = 0$ so as to be periodic in $x_0, x_1, \ldots, x_n$ with period[4] $2L$. Assume now that we have already constructed the functions $u_i^{*(m)}(hp, x_0, \ldots, x_n)$ periodic in every x_k with period $2L$, for a non-negative integer p and all values of i from 1 to N_1; let the functions $u_i^{*(m)}$ $(i = N + 1, N + 2, \ldots, N_1)$ be related to $u_i^{*(m)}$ $(i = 1, \ldots, N)$ as described in §1. Let us define the values of $a_i^{*(m)}$, for $\sum_k \alpha_{ks}^2 > 0$ and $hp < x_0^* \leq h(p + 1)$, so as to satisfy the following relations (counterparts of equations (27)):

$$\frac{a_i^{*(m)}(x_0^*, \alpha_0, \ldots, \alpha_n) - a_i^{*(m)}(hp, \alpha_0, \alpha_1, \ldots, \alpha_n)}{x_0^* - hp} =$$

$$= \int \left(\sum_{j=1}^N \sum_{(p)} A_{ij}^{\prime (p_0 \cdots p_l)} \frac{\partial^{n_j} u_j^{*(m)}}{\partial x_{0s}^{\prime p_0} \cdots \partial x_{ls}^{\prime p_l}} + B_i^\prime \right) e^{-\frac{\pi i}{L}(\alpha, x)} \, dx_0 \ldots dx_n \, , \quad (28)$$

$$i = 1, 2, \ldots, N_1 \, .$$

The integration is performed here for $x_0^* = hp$ under the assumption that the corresponding values of $u_j^{*(m)}$ are known; the values of $\alpha_{0s}, \ldots, \alpha_{ls}$ involved in the derivation of equations (28) are proportional to the coefficients $c_{0s}^{(k)}$ $(k = 0, 1, \ldots, l)$ in (24). Therefore, the derivatives of $e^{-\frac{\pi i}{L}(\alpha, x)}$ with respect to the variables $x_{ks}^\prime$ $(k = 1, 2, \ldots, l)$ are always identically equal to zero. Thus, taking into account that $A_{ij}^{\prime (p_0 \cdots p_l)}$ vanish outside $R_{0.7LL_0^*}$, we can apply integration by parts to the terms of the sum in the right-hand side of (28) containing the derivatives of $u_j^{*(m)}$ of order n_j. The resulting integral contains only derivatives of $u_j^{*(m)}$ $(j = 1, 2, \ldots, N)$ whose order is $\leq n_j - 1$. Thus, we obtain

---

[4]i.e., retaining the values of these functions on Q_L, we can extend them to the entire $(x_0, x_1, \ldots, x_n)$-plane as periodic functions continuous in x_k $(k = 0, 1, \ldots, n)$ and having period $2L$ in every x_k.

$$\frac{a_i^{*(m)}(x_0^*, \alpha_0, \ldots, \alpha_n) - a_i^{*(m)}(hp, \alpha_0, \ldots, \alpha_n)}{x_0^* - hp} =$$

$$= \int \left(\sum_{j=1}^{N} \sum_{(p)} - \frac{dA_{ij}'^{(p_0 \cdots p_l)}}{dx_{ks}'} \frac{\partial^{n_j - 1} u_j^{*(m)}}{\partial x_{0s}'^{p_0} \cdots \partial x_{k-1,s}'^{p_k-1} \partial x_{ks}'^{p_k-1} \cdots \partial x_{ls}'^{p_l}} + B_i' \right) \times$$

$$\times e^{-\frac{\pi i}{L}(\alpha, x)} dx_0 \ldots dx_n , \qquad i = 1, 2, \ldots, N_1 . \qquad (29)$$

The symbol $dA_{ij}'^{(p_0 \cdots p_l)}/dx_{ks}'$ denotes the derivatives of $A_{ij}'^{(p_0 \cdots p_l)}$ with respect to x_{ks}', the dependence of $A_{ij}'^{(p_0 \cdots p_l)}$ on x_{ks}' through the agency of the functions $u_i^{*(m)}$ $(i = 1, 2, \ldots, N)$ and their derivatives being taken into account. The coefficients $A_{ij}'^{(p_0 \cdots p_l)}$ depend only on the derivatives, whose order is $\leq n_\rho - 2$, of the functions $u_\rho^{*(m)}$ $(\rho = 1, \ldots, N)$ with $n_\rho \geq 2$, $l = n$; therefore the expressions obtained under the sign of the integral contain only the derivatives of $u_\rho^{*(m)}$ $(\rho = 1, \ldots, N)$ up to the order $n_\rho - 1$. Let us express them in terms of the functions $u_{N+1}^{*(m)}, \ldots, u_{N_1}^{*(m)}$ (obviously, this is always possible). Then we can rewrite (29) as

$$\frac{a_i^{*(m)}(x_0^*, \alpha_0, \ldots, \alpha_n) - a_i^{*(m)}(hp, \alpha_0, \ldots, \alpha_n)}{x_0^* - hp} =$$

$$= \int F_i'(hp, x_0, \ldots, x_n; u_1^{*(m)}, \ldots, u_{N_1}^{*(m)}) e^{-\frac{\pi i}{L}(\alpha, x)} dx_0 \ldots dx_n , \qquad (30)$$

$$i = 1, 2, \ldots, N_1 .$$

The functions F_i' in the above expression possess the following properties:

1) F_i' are analytic with respect to $u_1^{*(m)}, \ldots, u_{N_1}^{*(m)}$, and x_0^*;

2) F_i' have derivatives in $x_0, x_1, \ldots, x_n$, up to the order $2n + 2 \left[\frac{n+1}{2}\right] + 5$;

3) F_i' vanish outside the parallelepiped $R_{0.7LL_0^*}$.

In fact, it is only in those equations in (29) which correspond to $\partial p_{ik_0 \ldots k_l}/\partial x_0^*$ with $\sum k_\sigma = n_i - 1$ that we have to pass to the variables $x_{0s}', x_{1s}', \ldots, x_{is}'$. We have passed to these variables in the other equations, too, merely with a view to having symmetry and more simple notations.

In the case $\alpha_{0s} = \cdots = \alpha_{ls} = 0$, we obtain the following relations, taking into account (23):

$$\frac{a_i^{*(m)}(x_0^*, \alpha_0, \ldots, \alpha_n) - a_i^{*(m)}(hp, \alpha_0, \ldots, \alpha_n)}{x_0^* - hp} =$$

$$= \int \left(\sum_{m_\sigma \leq k_\sigma} C^{m_0 \ldots m_l}_{k_0 \ldots k_l} \frac{\partial^{k_0 - m_0 + \cdots + k_l - m_l} \psi}{\partial x_{0s}^{k_0 - m_0} \cdots \partial x_{ls}^{k_l - m_l}} \frac{\partial^{m_0 + 1 + \cdots + m_l} u_\rho^*}{\partial x_{0s}^{m_0+1} \cdots \partial x_{ls}^{m_l}} \right) \times$$

$$\times \, e^{-\frac{\pi i}{L}(\alpha, x)} \, dx_0 \ldots dx_n , \qquad i = 1, 2, \ldots, N_1 .$$

Then, integrating by parts, with respect to the variable $x_{0s} = x_0$, the terms containing derivatives of u_ρ^* of order n_ρ, we obtain equations of the form (30).

Right now, we cannot claim the operations that have led us to (30) to be fully justified. Direct application of relations (30) allows us to find only

$$a_i^{*(m)}(x_0^*, \alpha_0, \ldots, \alpha_n) \quad (i = 1, \ldots, N_1) \quad \text{for} \quad hp \leq x_0^* \leq h(p+1) .$$

Thereupon, we can find the functions $u_i^{*(m)}$ $(i = 1, \ldots, N_1)$, expressing them in terms of $a_i^{*(m)}(x_0^*, \alpha_0, \ldots, \alpha_n)$. We are going to prove that the functions $u_i^{*(m)}$ $(i = 1, \ldots, N_1)$ thus defined are continuous, together with their derivatives in $x_0, x_1, \ldots, x_n$ of a sufficiently high order. To this end, it is necessary to estimate the sums of the form

$$\sum_\alpha \left| a_i^{*(m)}(x_0^*, \alpha_0, \alpha_1, \ldots, \alpha_n) \right|^2 |\alpha|^{2\rho} ,$$

where $\sum_\alpha$ is taken over all integer values of $\alpha_0, \ldots, \alpha_n$ (positive, negative, and zero).

§3. Some Estimates

It follows from (30) that

$$\left(\frac{\pi i}{L} \right)^{k_0 + \cdots + k_n} \alpha_0^{k_0} \cdots \alpha_n^{k_n} \frac{a_i^{*(m)}(x_0^*, \alpha_0, \ldots, \alpha_n) - a_i^{*(m)}(hp, \alpha_0, \ldots, \alpha_n)}{x_0^* - hp} =$$

$$= \left(\frac{\pi i}{L} \right)^{k_0 + \cdots + k_n} \alpha_0^{k_0} \cdots \alpha_n^{k_n} \times$$

$$\times \int F_i'(hp, x_0, \ldots, x_n; u_1^{*(m)}, \ldots, u_{N_1}^{*(m)}) e^{-\frac{\pi i}{L}(\alpha, x)} \, dx_0 \ldots dx_n ,$$

$$i = 1, 2, \ldots, N_1 . \tag{31}$$

Note that all functions $u_i^{*(m)}$ $(i = 1, \ldots, N_1)$ are continuously differentiable in $x_0, \ldots, x_n$ up to the order

$$\delta = 2n + \left[\frac{n+1}{2} \right] + 4 ,$$

the derivatives being periodic in $x_0, \ldots, x_n$ with period $2L$. Then, integrating by parts in the right-hand side of (31), k_0 times in x_0, ..., k_n times in x_n, where $\sum_{s=0}^{n} k_s \leq \delta$, we obtain

$$\left(\frac{\pi i}{L}\right)^{k_0 + \cdots + k_n} \alpha_0^{k_0} \cdots \alpha_n^{k_n} \frac{a_i^{*(m)}(x_0^*, \alpha_0, \ldots, \alpha_n) - a_i^{*(m)}(hp, \alpha_0, \ldots, \alpha_n)}{x_0^* - hp} =$$

$$= \int \frac{\partial^k F_i'}{\partial x_0^{k_0} \cdots \partial x_n^{k_n}} e^{-\frac{\pi i}{L}(\alpha, x)} dx_0 \ldots dx_n , \qquad i = 1, 2, \ldots, N_1 . \qquad (32)$$

In a similar manner we find that

$$\left(\frac{\pi i}{L}\right)^{k_0 + \cdots + k_n} \alpha_0^{k_0} \cdots \alpha_n^{k_n} a_i^{*(m)} = \int \frac{\partial^{k_0 + \cdots + k_n} u_i^{*(m)}}{\partial x_0^{k_0} \cdots \partial x_n^{k_n}} e^{-\frac{\pi i}{L}(\alpha, x)} dx_0 \ldots dx_n ,$$

if $\sum k_\sigma \leq \delta$; the boundary terms disappear. Denote by

$$u_i^{*(m)} , \qquad i = N_1 + 1, \ldots, N_2 ,$$

the derivatives of the functions $u_i^{*(m)}$ $(i = 1, \ldots, N_1)$ with respect to the variables $x_0, \ldots, x_n$ up to the order $\delta - n - 2$; and denote by

$$u_i^{*(m)} , \qquad i = N_2 + 1, \ldots, N_3 ,$$

the derivatives of $u_i^{*(m)}$ $(i = 1, \ldots, N_1)$ whose order ranges from $\delta - n - 1$ to δ. Let $a_i^{*(m)}$ $(i = 1, \ldots, N_3)$ be the Fourier coefficients of $u_i^{*(m)}$ $(i = 1, \ldots, N_3)$. Then equations (30) and (32) can be rewritten in the form

$$\frac{a_i^{*(m)}(x_0^*, \alpha_0, \ldots, \alpha_n) - a_i^{*(m)}(hp, \alpha_0, \ldots, \alpha_n)}{x_0^* - hp} =$$

$$= \int F_i'(hp, x_0, x_1, \ldots, x_n; u_1^{*(m)}, \ldots, u_{N_3}^{*(m)}) e^{-\frac{\pi i}{L}(\alpha, x)} dx_0 \ldots dx_n , \qquad (33)$$

$$i = 1, 2, \ldots, N_3 .$$

It should be observed that the expressions F_i' have the form of polynomials in $u_i^{*(m)}$ and the derivatives of $\tilde{F}_j^*$ $(j = 1, 2, \ldots, N)$, the latter functions being also dependent on $u_1^{*(m)}, \ldots, u_{N_1}^{*(m)}$. Each term of such a polynomial can contain only one of the functions $u_{N_2+1}^{*(m)}, \ldots, u_{N_3}^{*(m)}$; otherwise, the orders of the derivatives of $u_i^{*(m)}$ $(i = 1, 2, \ldots, N_1)$ entering as factors in such terms would add up to a number larger than δ, which is impossible. Therefore, equation (33) can be rewritten in the form

$$\frac{a_i^{*(m)}(x_0^*, \alpha_0, \dots, \alpha_n) - a_i^{*(m)}(hp, \alpha_0, \dots, \alpha_n)}{x_0^* - hp} =$$

$$= \sum_{j=N_2+1}^{N_3} \int a_{ij} u_j^{*(m)} e^{-\frac{\pi i}{L}(\alpha, x)} dx_0 \dots dx_n + \int b_i e^{-\frac{\pi i}{L}(\alpha, x)} dx_0 \dots dx_n ,$$

$$i = 1, 2, \dots, N_3 . \tag{34}$$

The coefficients a_{ij} contain the derivatives of $u_i^{*(m)}$ whose order does not exceed $\delta - (n+2) - \left[\frac{n+1}{2}\right] - 1$, while b_i contain the functions $u_i^{*(m)}$ which either coincide with $u_s^{*(m)}$ $(s = 1, 2, \dots, N_1)$, or with their derivatives up to the order $\delta - n - 2$. Let us multiply both sides of every equation in (34) by $|\alpha|^\rho = (\sum \alpha_k^2)^{\rho/2}$, where ρ is an arbitrary non-negative integer less or equal to $\left[\frac{n+1}{2}\right] + 1$. Then we get

$$|\alpha|^\rho \frac{a_i^{*(m)}(x_0^*, \alpha_0, \dots, \alpha_n) - a_i^{*(m)}(hp, \alpha_0, \dots, \alpha_n)}{x_0^* - hp} =$$

$$= \sum_{j=N_2+1}^{N_3} |\alpha|^\rho \int a_{ij} u_j^{*(m)} e^{-\frac{\pi i}{L}(\alpha, x)} dx_0 \dots dx_n+$$

$$+ |\alpha|^\rho \int b_i e^{-\frac{\pi i}{L}(\alpha, x)} dx_0 \dots dx_n , \qquad i = 1, 2, \dots, N_3 .$$

Setting $\alpha_m = \max\{|\alpha_0|, \dots, |\alpha_n|\}$, we obviously have

$$\left|\frac{\alpha_m}{|\alpha|}\right| \geq \frac{1}{\sqrt{n+1}} .$$

Integrating by parts with respect to x_m for $|\alpha| \neq 0$, we find that

$$|\alpha|^\rho \int b_i e^{-\frac{\pi i}{L}(\alpha, x)} dx_0 \dots dx_n = \frac{|\alpha|^\rho}{\alpha_m} \frac{L}{\pi i} \int \frac{db_i}{dx_m} e^{-\frac{\pi i}{L}(\alpha, x)} dx_0 \dots dx_n , \tag{35}$$

where db_i/dx_m stands for the derivative of b_i with respect to x_m, the explicit dependence of b_i on x_m, as well as the dependence through the agency of the functions $u_1^{*(m)}, \dots, u_{N_2}^{*(m)}$, being taken into account. Set

$$\frac{db_i}{dx_m} = \sum_{j=N_2+1}^{N_3} a_{ij}^{(1)} u_{ij}^{*(m)} + b_i^{(1)} .$$

The above sum involves only such terms that contain as factors the derivatives of $u_1^{*(m)}, \dots, u_{N_1}^{*(m)}$ whose order does not exceed $\delta - n - 2$; $b_i^{(1)}$ stands for the rest of the terms. The coefficients $a_{ij}^{(1)}$ are polynomials whose arguments are partial derivatives of $u_s^{*(m)}$ $(s = 1, 2, \ldots, N_1)$ of orders $\leq \delta - (n+2) - \left[\frac{n+1}{2}\right]$; the coefficients of these polynomials are functions of $u_s^{*(m)}$ $(s = 1, 2, \ldots, N_1)$ and $x_0^*, x_0, \ldots, x_n$. Consequently,

$$|\alpha|^\rho \int b_i \, e^{-\frac{\pi i}{L}(\alpha, x)} \, dx_0 \ldots dx_n =$$

$$= \sum_{j=N_2+1}^{N_3} \frac{|\alpha|}{\alpha_m} \frac{L}{\pi i} |\alpha|^{\rho-1} \int a_{ij}^{(1)} u_j^{*(m)} \, e^{-\frac{\pi i}{L}(\alpha, x)} \, dx_0 \ldots dx_n +$$

$$+ \frac{|\alpha|}{\alpha_m} \frac{L}{\pi i} |\alpha|^{\rho-1} \int b_i^{(1)} \, e^{-\frac{\pi i}{L}(\alpha, x)} \, dx_0 \ldots dx_n , \qquad i = 1, 2, \ldots, N_3 .$$

Transforming the last integral several times, in the same way as the integral (35), we finally obtain

$$|\alpha|^\rho \frac{a_i^{*(m)}(x_0^*, \alpha_0, \ldots, \alpha_n) - a_i^{*(m)}(hp, \alpha_0, \ldots, \alpha_n)}{x_0^* - hp} =$$

$$= \sum_{j=N_2+1}^{N_3} \sum_{s=0}^{\rho} \left(\frac{|\alpha|L}{i\alpha_m \pi}\right)^s |\alpha|^{\rho-s} \int a_{ij}^{(s)} u_j^{*(m)} \, e^{-\frac{\pi i}{L}(\alpha, x)} \, dx_0 \ldots dx_n +$$

$$+ \left(\frac{|\alpha|L}{i\alpha_m \pi}\right)^\rho \int b_i^{(\rho)} \, e^{-\frac{\pi i}{L}(\alpha, x)} \, dx_0 \ldots dx_n , \qquad i = 1, 2, \ldots, N_3 . \quad (36)$$

Here $a_{ij}^{(s)}$ are polynomials having as arguments partial derivatives of $u_k^{*(m)}$ $(k = 1, 2, \ldots, N_1)$ whose orders do not exceed $\delta - (n+2) - \left[\frac{n+1}{2}\right] + s - 1$; the coefficients of these polynomials depend on $x_0, \ldots, x_n, u_1^{*(m)}, \ldots, u_{N_1}^{*(m)}$. For the sake of symmetry, we write $a_{ij}^{(0)}$ instead of a_{ij}. Using the formulas for the Fourier coefficients of the product of two functions, we find that

$$\int a_{ij}^{(s)} u_j^{*(m)} \, e^{-\frac{\pi i}{L}(\alpha, x)} \, dx_0 \ldots dx_n =$$

$$= \sum_{\beta} a_{ij}^{(s)}(hp, \alpha_0 - \beta_0, \ldots, \alpha_n - \beta_n) \, a_j^{*(m)}(hp, \beta_0, \ldots, \beta_n) . \quad (37)$$

Here we have set

$$a_{ij}^{(s)}(hp, \gamma_0, \ldots, \gamma_n) = \int a_{ij}^{(s)} \, e^{-\frac{\pi i}{L}(\gamma, x)} \, dx_0 \ldots dx_n ; \quad (38)$$

the sum $\sum_\beta$ is taken over all integer values of $\beta_0, \ldots, \beta_n$ (positive, negative, and zero).

Assume that for $x_0^* = hp$ the functions $u_i^{(m)}$ $(i = 1, \ldots, N_1)$ are continuously differentiable in $x_0, \ldots, x_n$ up to the order δ, the partial derivatives having absolute values bounded by M. Then the functions $a_{ij}^{(s)}$ (resp., $b_i^{(\rho)}$) are continuously differentiable in $x_0, \dots, x_n$, up to the order $n + 2 + \left[\frac{n+1}{2}\right] - s + 1$ (resp., $n + 2$), the derivatives being identically equal to zero outside $R_{0.7LL_0^*}$, with absolute values bounded by a constant M_1 depending on M. Then, integrating by parts in the right-hand side of (38) with respect to the variables $x_0, \dots, x_n$, $n + 3 + \left[\frac{n+1}{2}\right] - s$ times at the most, we find that

$$\left| a_{ij}^{(s)}(hp, \gamma_0, \dots, \gamma_n) \right| \leq \frac{M_1 C_1^*}{|\gamma_0|^{r_0} \cdots |\gamma_n|^{r_n}}, \tag{39}$$

where $r_0, \dots, r_n$ are arbitrary non-negative numbers whose sum does not exceed $n + 3 + \left[\frac{n+1}{2}\right] - s$; C_1^* is an absolute constant. From (39) we conclude that

$$\left| a_{ij}^{(s)}(hp, \gamma_0, \dots, \gamma_n) \right| \leq \frac{M_1'}{(1 + |\gamma|)^{n+3+\left[\frac{n+1}{2}\right]-s}}, \tag{40}$$

where

$$M_1' = M_1 C_1^* (2\sqrt{n+1})^{n+3+\left[\frac{n+1}{2}\right]-s}, \qquad |\gamma| = \sqrt{\sum \gamma_s^2}.$$

Remark 4. Here $\left| a_{ij}^{(s)}(hp, \gamma_0, \dots, \gamma_n) \right|$ are estimated by fractions of the form

$$\frac{M}{(1 + |\gamma|)^k} = \frac{M}{(1 + \sqrt{\gamma_1^2 + \cdots + \gamma_n^2})^k}, \qquad \text{where} \quad k \geq n + 2.$$

Previously (see [4]), similar Fourier coefficients have been estimated in terms of fractions of the form

$$\frac{M}{(1 + |\gamma_1|)^2 \cdots (1 + |\gamma_n|)^2} \qquad \text{or} \qquad \frac{M}{(1 + |\gamma_1|)^3 \cdots (1 + |\gamma_n|)^3}.$$

It is for this reason that in the previous article we had to require that the functions under consideration possess more derivatives than necessary. Had we used the estimates similar to those of the present paper, we could have proved the Cauchy problem for the hyperbolic system

$$\frac{\partial^{n_i} u_i}{\partial t^{n_i}} = F_i(t, x_1, \dots, x_n; u_1, \dots, n_N, \dots), \qquad i = 1, 2, \dots, N$$

with zero initial conditions to be well-posed, provided that F_i possess continuous derivatives up to the order

$$n + 2\left[\frac{n+1}{2}\right] + 6,$$

with respect to all their arguments.   □

In a similar fashion we find that

$$\left| \int b_i^{(\rho)} e^{-\frac{\pi i}{L}(\alpha, x)} \, dx_0 \dots dx_n \right| \le \frac{M_1'}{(1 + |\alpha|)^{n+2}} \, . \tag{41}$$

It follows from (40) that

$$|\alpha|^{\rho - s} \left| a_{ij}^{(s)}(hp, \alpha_0 - \beta_0, \dots, \alpha_n - \beta_n) \, a_j^{*(m)}(hp, \beta_0, \dots, \beta_n) \right| \le$$

$$\le \sum_{\sigma=0}^{\rho-s} C_{\rho-s}^{\sigma} \left(\sqrt{\sum_k (\alpha_k - \beta_k)^2} \right)^{\rho - \sigma - s} \left| a_{ij}^{(s)}(hp, \alpha_0 - \beta_0, \dots, \alpha_n - \beta_n) \right| \times$$

$$\times |\beta|^\sigma \left| a_j^{*(m)}(hp, \beta_0, \dots, \beta_n) \right| \le$$

$$\le \sum_{\sigma=0}^{\rho-s} C_{\rho-s}^{\sigma} \frac{M_1' |a_j^{*(m)}(hp, \beta_0, \dots, \beta_n)| \, |\beta|^\sigma}{\left(1 + \sqrt{\sum_k (\alpha_k - \beta_k)^2} \right)^{n+2}} \le$$

$$\le M_2 \sum_{\sigma=0}^{\rho} \frac{\left| a_j^{*(m)}(hp, \beta_0, \dots, \beta_n) \right| \, |\beta|^\sigma}{\left(1 + \sqrt{\sum_k (\alpha_k - \beta_k)^2} \right)^{n+2}} \, . \tag{42}$$

Here M_2 is a constant depending on $C_{\rho-s}^{\sigma}$ and M_1'; $C_{\rho-s}^{\sigma}$ are the binomial coefficients. From (37) and (42) we obtain

$$\sum_{\alpha} |\alpha|^{2(\rho-s)} \left| \int a_{ij}^{(s)} u_j^{*(m)} e^{-\frac{\pi i}{L}(\alpha, x)} \, dx_0 \dots dx_n \right|^2 \le$$

$$\le \sum_{\alpha} M_2 \sum_{\sigma=0}^{\rho} \frac{\left| a_j^{(*m)}(hp, \beta_0, \dots, \beta_n) \right| \, |\beta|^\sigma}{\left(1 + \sqrt{\sum_k (\alpha_k - \beta_k)^2} \right)^{n+2}} \times$$

$$\times M_2 \sum_{s=0}^{\rho} \sum_{\gamma} \frac{\left| a_j^{*(m)}(hp, \gamma_0, \dots, \gamma_n) \right| \, |\gamma|^s}{\left(1 + \sqrt{\sum_k (\alpha_k - \beta_k)^2} \right)^{n+2}} \le$$

$$\le M_2^2 \sum_{\alpha, \beta, \gamma, s, \sigma} \frac{\left| a_j^{*(m)}(hp, \beta_0, \dots, \beta_n) \right|^2 \, |\beta|^{2\sigma}}{\left(1 + \sqrt{\sum_k (\alpha_k - \beta_k)^2} \right)^{n+2} \left(1 + \sqrt{\sum_k (\alpha_k - \gamma_k)^2} \right)^{n+2}} +$$

$$+ M_2^2 \sum_{\alpha, \beta, \gamma, s, \sigma} \frac{\left| a_j^{*(m)}(hp, \gamma_0, \dots, \gamma_n) \right|^2 \, |\gamma|^{2s}}{\left(1 + \sqrt{\sum_k (\alpha_k - \beta_k)^2} \right)^{n+2} \left(1 + \sqrt{\sum_k (\alpha_k - \gamma_k)^2} \right)^{n+2}} \le$$

$$\le M_3 \sum_{r=0}^{\rho} \sum_{\beta} \left| a_j^{*(m)}(hp, \beta_0, \dots, \beta_n) \right|^2 \, |\beta|^{2r} \, . \tag{43}$$

Here M_3 is a constant depending on n and M_2; M_3 is proportional to the squared sum of the series

$$\sum_{\gamma} \frac{1}{\left(1 + \sqrt{\sum \gamma_k^2}\right)^{n+2}} \cdot$$

To obtain the last inequality, we have carried out summation first with respect to γ and then α in the first sum, and, likewise, in the second sum we have first summed with respect to β and then α.

Further, by virtue of (41), we have

$$\sum_{\alpha} \left| \int b_i^{(\rho)} e^{-\frac{\pi i}{L}(\alpha, x)} \, dx_0 \dots dx_n \right| \le \sum \frac{M_1'^2}{\left(1 + \sqrt{\sum \alpha_k^2}\right)^{n+2}} \le M_4 \, . \tag{44}$$

From (36), (43), and (44) we deduce that

$$\sum_{i=1}^{N_3} \sum_{\alpha} |\alpha|^{2\rho} \left| a_i^{*(m)}(x_0^*, \alpha_0, \dots, \alpha_n) \right|^2 \le$$

$$\le \sum_{i=1}^{N_3} \sum_{\alpha} |\alpha|^{2\rho} \left| a_i^{*(m)}(hp, \alpha_0, \dots, \alpha_n) \right|^2 +$$

$$+(x_0^* - hp) \left\{ \sum_{i=1}^{N_3} \sum_{r=0}^{\rho} \sum_{\alpha} |\alpha|^{2r} \left| a_i^{*(m)}(hp, \alpha_0, \dots, \alpha_n) \right|^2 + M_6 \right\} , \tag{45}$$

$$i = 1, 2, \dots, N_3 \, , \qquad 0 \le \rho \le \left[\frac{n+1}{2} \right] + 1 \, ,$$

where M_5 and M_6 are constants depending on M_3, M_4, n, and N.

Set $\rho = 0$ in (45). Then, writing this inequality for

$$x_0^* = h(p+1), \quad p = 0, 1, 2, \dots,$$

we see that the sum

$$\sum_{i=1}^{N_3} \sum_{\alpha} \left| a_i^{*(m)}(hp, \alpha_0, \dots, \alpha_n) \right|^2$$

is bounded from above by a function $f(t)$ such that

$$f(0) = \sum_{i=1}^{N_3} \sum_{\alpha} \left| a_i^{*(m)}(0, \alpha_0, \dots, \alpha_n) \right|^2 \, ,$$

and for $p > 0$ the function $f(t)$ satisfies the following finite difference equation

$$f(h(p+1)) = (1 + M_5 h)f(hp) + M_6 h .$$

This equation yields[5]

$$f(hp) = (1 + M_5 h)^p \sum_{i=1}^{N_3} \sum_\alpha \left| a_i^{*(m)}(0, \alpha_0, \ldots, \alpha_n) \right|^2 + \sum_{s=0}^{p-1} (1 + M_5 h)^s M_6 h \le$$

$$\le e^{L_0^* M_5} \left\{ \sum_{i=1}^{N_3} \sum_\alpha |a_i^*(0, \alpha_0, \ldots, \alpha_n)|^2 + M_6 L_0^* \right\} .$$

Therefore

$$\sum_{i=1}^{N_3} \sum_\alpha \left| a_i^{*(m)}(hp, \alpha_0, \ldots, \alpha_n) \right|^2 \le$$

$$\le e^{L_0^* M_5} \left\{ \sum_{i=1}^{N_3} \sum_\alpha |a_i^*(0, \alpha_0, \ldots, \alpha_n)|^2 + M_6 L_0^* \right\} .$$

Combining this inequality with (45) for $\rho = 1$, we can similarly estimate the sum

$$\sum_{i=1}^{N_3} \sum_\alpha |\alpha|^2 \left| a_i^{*(m)}(hp, \alpha_0, \ldots, \alpha_n) \right|^2 , \quad \text{etc.}$$

Thus, for all integer ρ from 0 to $\left[\frac{n+1}{2}\right] + 2$, we get

$$\sum_{i=1}^{N_3} \sum_\alpha |\alpha|^{2\rho} \left| a_i^{*(m)}(hp, \alpha_0, \ldots, \alpha_n) \right|^2 \le$$

$$\le C_1 \sum_{i=1}^{N_3} \sum_{r=0}^{\rho} \sum_\alpha |\alpha|^{2r} |a_i^*(0, \alpha_0, \ldots, \alpha_n)|^2 + C_2 . \qquad (46)$$

It is easy to see that similar estimates hold for all x_0^* from the interval $(0, L_0^*)$ (not only for x_0^* that is a multiple of h); C_1 and C_2 are constants depending on M, n, and N.

Set

$$\frac{a_i^{*(m)}(x_0^* + \Delta x_0^*, \alpha_0, \ldots, \alpha_n) - a_i^{*(m)}(x_0^*, \alpha_0, \ldots, \alpha_n)}{\Delta x_0^*} =$$

$$= \frac{\Delta a_i^{*(m)}(x_0^*, \alpha_0, \ldots, \alpha_n)}{\Delta x_0^*} .$$

Repeating the arguments used to derive (46) from (34), and taking into account (46), we obtain

---

[5] We write here a_i^* instead of $a_i^{*(m)}$, since $a_i^{*(m)}$ do not depend on m for $p = 0$.

$$\sum_{i=1}^{N_3} \sum_{\alpha} |\alpha|^{2\rho} \left| \frac{\Delta a_i^{*(m)}(x_0^*, \alpha_0, \ldots, \alpha_n)}{\Delta x_0^*} \right| \le$$

$$\le C_3 \sum_{i=1}^{N_3} \sum_{\alpha} \sum_{r=0}^{\rho} |\alpha|^{2r} \left| a_i^{*(m)}(hp, \alpha_0, \ldots, \alpha_n) \right|^2 + C_4 \le$$

$$\le C_5 \sum_{i=1}^{N_3} \sum_{\alpha} \sum_{r=0}^{\rho} |\alpha|^{2r} \left| a_i^*(0, \alpha_0, \ldots, \alpha_n) \right|^2 + C_6 , \qquad (47)$$

where the constants $C_3, C_4, C_5,$ and C_6 depend on M, N and n;

$$\rho \le \left[\frac{n+1}{2} \right] + 1, \quad hp \le x_0^* < h(p+1), \quad hp \le x_0^* + \Delta x_0^* \le h(p+1) .$$

It should be observed that the inequalities (46) and (47) have been established for x_0^* such that

$$\left| u_i^{*(m)}(x_0^* - h, x_0, \ldots, x_n) \right| \le M \quad \text{if} \quad x_0^* = \left[\frac{x_0^*}{h} \right] h , \qquad (48')$$

and

$$\left| u_i^{*(m)} \left(h \left[\frac{x_0^*}{h} \right], x_0, \ldots, x_n \right) \right| \le M \quad \text{if} \quad x_0^* > \left[\frac{x_0^*}{h} \right] h . \qquad (48'')$$

These conditions have been assumed to hold for $x_0^* = 0$, i.e., in a layer within the parallelepiped $R_{LL_0^*}$ adjacent to the plane $x_0^* = 0$.

§4. Proof of Convergence

Next, we are going to prove that the functions $u_i^{*(m)}(x_0^*, x_0, \ldots, x_n)$ defined by

$$u_i^{*(m)}(x_0^*, x_0, \ldots, x_n) = \sum_{\alpha} a_i^{*(m)}(x_0^*, \alpha_0, \ldots, \alpha_n) e^{-\frac{\pi i}{L}(\alpha, x)} , \qquad (49)$$

$$i = 1, 2, \ldots, N_3 ,$$

are continuous in $x_0^*, x_0, \ldots, x_n$, uniformly bounded (with respect to m) in the domain where the conditions (48) are valid, and moreover, that the series (49) is uniformly convergent with respect to $x_0^*, x_0, \ldots, x_n$, and m. This will allow us to conclude that the functions

$$u_i^{*(m)}(x_0^*, x_0, \ldots, x_n) \qquad (i = N_1 + 1, \ldots, N_3)$$

coincide with the derivatives of $u_i^{*(m)}$ $(i = 1, 2, \ldots, N_1)$ in the domain under consideration. The relations

$$p_{ik_0\ldots k_l} = \frac{\partial^{k_0+\cdots+k_l} u_i^*}{\partial x_{0s}^{k_0} \cdots \partial x_{ls}^{k_l}}, \qquad i = 1, 2, \ldots, N,$$

will be proved in the next section. Since every term in (49) is a continuous function of $x_0^*, x_0, \ldots, x_n$, and

$$\left| e^{-\frac{\pi i}{L}(\alpha, x)} \right| = 1,$$

it suffices to show that for all large enough R and any i from 1 to N_3 we have

$$\sum_{\alpha > R} \left| a_i^{*(m)}(x_0^*, \alpha_0, \alpha_1, \ldots, \alpha_n) \right| < \varepsilon(R), \tag{50}$$

where $\sum_{\alpha > R}$ implies summation over the points of the $(\alpha_0, \alpha_1, \ldots, \alpha_n)$-space with integer coordinates such that $|\alpha_s| > R$ for at least one s; $\varepsilon(R)$ is a constant independent of m and converging to 0 as $R \to \infty$.

In order to prove the estimate (50), we note that for all values of x_0^* under consideration (i.e., x_0^* such that its absolute value does not exceed L_0^* and the conditions (48) are satisfied), for all i from 1 to N_3, and all ρ from 1 to $\left[\frac{n+1}{2}\right] + 1$, the sums

$$\sum_{\alpha} |\alpha|^{2\rho} \left| a_i^{*(m)}(x_0^*, \alpha_0, \ldots, \alpha_n) \right|^2,$$

and therefore, the sums

$$\sum_{\alpha} (1 + |\alpha|)^{2\rho} \left| a_i^{*(m)}(x_0^*, \alpha_0, \ldots, \alpha_n) \right|^2$$

are bounded by a constant M^* independent of m. Since

$$2\left(\left[\frac{n+1}{2}\right] + 1\right) \geq n + 2,$$

we have

$$\sum_{\alpha} (1 + |\alpha|)^{n+2} \left| a_i^{*(m)}(x_0^*, \alpha_0, \ldots, \alpha_n) \right|^2 \leq M^*. \tag{51}$$

Hence, using the Schwartz inequality, we find that

$$\sum_{\alpha > R} \left| a_i^{*(m)}(x_0^*, \alpha_0, \ldots, \alpha_n) \right| =$$

$$= \sum_{\alpha > R} (1 + |\alpha|)^{\frac{n+2}{2}} \left| a_i^{*(m)}(x_0^*, \alpha_0, \ldots, \alpha_n) \right| \frac{1}{(1 + |\alpha|)^{\frac{n+2}{2}}} \leq$$

$$\leq \left(\sum_{\alpha > R} (1 + |\alpha|)^{n+2} \left| a_i^{*(m)}(x_0, \alpha_0, \ldots, \alpha_n) \right|^2 \right)^{\frac{1}{2}} \left(\sum_{\alpha > R} \frac{1}{(1 + |\alpha|)^{n+2}} \right)^{\frac{1}{2}}$$

Because of (51), the first factor in the right-hand side does not exceed $\sqrt{M^*}$, while the second one tends to 0 as $R \to \infty$. Thereby (50) is proved.

Remark 5. It should be pointed out that the above arguments allow us to prove the following statement. *Let $a(\alpha_0, \ldots, \alpha_n)$ be the Fourier coefficients for the function $u(x_0, \ldots, x_n)$. Assume that every series of the form*

$$\sum |\alpha|^{2\rho} |a(\alpha_0, \ldots, \alpha_n)|^2 \tag{52}$$

is convergent for $\rho = 0, 1, \ldots, \left[\frac{n+1}{2}\right] + 1$, so that the series

$$\sum (1 + |\alpha|)^{2\rho} |a(\alpha_0, \ldots, \alpha_n)|^2$$

is convergent, too. Then the Fourier series for the function $u(x_0, \ldots, x_n)$ is convergent, the convergence being absolute and uniform.

It is easy to see that this statement remains valid, if the sums (52) are replaced by

$$\sum |\alpha_{n_1}|^2 \cdots |\alpha_{n_l}|^2 |a(\alpha_0, \ldots, \alpha_n)|^2 ,$$

where $n_1, \ldots, n_l$ take integer values $0, 1, \ldots, n$ such that $n_s \neq n_t$ for $s \neq t$; $l \leq \left[\frac{n+1}{2}\right] + 1$. In this case the series of the form

$$\sum_{\alpha} (1 + |\alpha_{n_1}|)^2 \cdots (1 + |\alpha_{n_l}|)^2 |a(\alpha_0, \ldots, \alpha_n)|^2$$

are convergent, and therefore the same can be said of the series

$$\sum (1 + |\alpha_0|)^{1 + \frac{1}{n+1}} \cdots (1 + |\alpha_n|)^{1 + \frac{1}{n+1}} |a(\alpha_0, \ldots, \alpha_n)|^2 .$$

It follows that if the function u and all its derivatives up to the order $\left[\frac{n+1}{2}\right] + 1$, such that u is differentiated in each of the variables $x_0, \ldots, x_n$ once at the most, admit continuous $2L$-periodic extension, then the Fourier series for u is uniformly and absolutely convergent [2].   □

On the basis of (47), one can show in a similar way that the series

$$\sum_{\alpha} \frac{\Delta a_i^{*(m)}(x_0^*, \alpha_0, \ldots, \alpha_n)}{\Delta x_0^*} e^{-\frac{\pi i}{L}(\alpha, x)}$$

are uniformly and absolutely convergent in the domain where the conditions (48) are satisfied; moreover, the sums of the above series have absolute values bounded by a constant M' which depends on M but is independent of m. It follows that the derivative numbers of $u_i^{*(m)}$ $(i = 1, 2, \ldots, N_3)$ with respect to x_0^* have their absolute values bounded by a constant M' independent of m; consequently, there exists a real $l > 0$ independent of m, such that for all m the absolute values of all functions $u_i^{*(m)}$ $(i = 1, 2, \ldots, N_3)$ are bounded by M in the parallelepiped R_{LI}, since the relations $(\widetilde{11})$ of §1 hold for $x_0^* = 0$. Therefore, the conditions (48) are satisfied in the parallelepiped R_{LI}. In what follows only this parallelepiped is dealt with; therefore, in order to avoid new notations, we assume that $L_0^* = l$.

Next, we show that from the infinite sequence (indexed by $m = 1, 2, 3, \ldots$) of systems of functions

$$u_i^{*(m)}(x_0^*, x_0, x_1, \ldots, x_n) \qquad (i = 1, 2, \ldots N_3)$$

we can extract an infinite subsequence (indexed by $m_1, m_2, \ldots$) of uniformly convergent systems[6]. In this connection, it should be pointed out that the estimate (47) (resp., (46)) implies that for all integer $\alpha_0, \alpha_1, \ldots, \alpha_n$ the quantities

$$\frac{\Delta a_i^{*(m)}(x_0^*, \alpha_0, \ldots, \alpha_n)}{\Delta x_0^*}, \qquad i = 1, 2, \ldots, N_3$$

(resp., $a_i^{*(m)}(x_0^*, \alpha_0, \ldots, \alpha_n)$, $i = 1, 2, \ldots, N_3$) are bounded by a constant[7] M_1 independent of $m, \alpha_0, \ldots, \alpha_n, \Delta x_0^*$. Let us enumerate the points having integer coordinates and belonging to the real $(\alpha_0, \alpha_1, \ldots, \alpha_n)$-space in such a way that the index k marks the point with coordinates

$$(\alpha_0^{(k)}, \alpha_1^{(k)}, \ldots, \alpha_n^{(k)}) .$$

Then, by the Arzelà theorem, from the sequence of integers

$$1, 2, 3, \ldots$$

we can extract a subsequence

$$m_1^1, m_2^1, m_3^1, \ldots$$

such that, for every i from 1 to N_3, the infinite sequences

$$a_i^{*(m_k^1)}(x_0^*, \alpha_0^{(1)}, \ldots, \alpha_n^{(1)}), \quad i = 1, 2, \ldots, N_3, \quad k = 1, 2, \ldots,$$

are uniformly convergent with respect to x_0^*. Applying the Arzelà theorem to the latter subsequence, we obtain the following infinite subsequence

$$a_i^{*(m_k^2)}(x_0^*, \alpha_0^{(2)}, \ldots, \alpha_n^{(2)}), \quad i = 1, 2, \ldots, N_3, \quad k = 1, 2, \ldots,$$

which is also uniformly convergent in x_0^*. Proceeding in a similar fashion to apply the Arzelà theorem, we can find for each positive integer B a sequence of integers

---

[6] For the sake of brevity, we say that "this subsequence is uniformly convergent".

[7] M_1 is not the constant considered in the preceding section.

$$m_1^B, m_2^B, m_3^B, \ldots,$$

such that for all $s \leq B$ and all i from 1 to N_3 the infinite sequences

$$a_i^{*(m_k^B)}(x_0^*, \alpha_0^{(s)}, \ldots, \alpha_n^{(s)}), \quad i = 1, \ldots, N_3, \quad k = 1, 2, \ldots,$$

converge uniformly in x_0^*. Now, let us consider the following sequence

$$a_i^{*(m_k^k)}(x_0^*, \alpha_0, \ldots, \alpha_n), \quad i = 1, 2, \ldots, N_3, \quad k = 1, 2, \ldots,$$

which converges uniformly in x_0^* on any set consisting of finitely many points of the $(\alpha_0, \alpha_1, \ldots, \alpha_n)$-space with integer coordinates. We claim that the corresponding sequence of functions defined by

$$u_i^{*(m_k^k)}(x_0^*, x_0, x_1, \ldots, x_n) = \sum_\alpha a_i^{*(m_k^k)}(x_0^*, \alpha_0, \ldots, \alpha_n) e^{-\frac{\pi i}{L}(\alpha, x)}, \quad (53)$$

$$i = 1, 2, \ldots, N_3, \quad k = 1, 2, \ldots,$$

converges uniformly on the parallelepiped $R_{LL_0^*}$. Because of the uniform convergence (with respect to x_0^*) of the coefficients in any partial sum of the series (53), we only have to show that for every i the following inequality is valid:

$$\sum_{\alpha > R} \left| a_i^{*(m_k^k)}(x_0^*, \alpha_0, \ldots, \alpha_n) \right| < \varepsilon(R), \quad (54)$$

where $\varepsilon(R)$ depends neither on i nor on k, and converges to 0 as $R \to \infty$. The estimate (54) can be proved in exactly the same way as (50). Thus the functions (53) have been shown to converge uniformly on $R_{LL_0^*}$ as $k \to \infty$. The limit functions $u_i^*(x_0^*, x_0, \ldots, x_n)$ for the sequences $u_i^{*(m_k^k)}(x_0^*, x_0, \ldots, x_n)$ are continuously differentiable with respect to $x_0, x_1, \ldots, x_n$ on $R_{LL_0^*}$ up to the order δ.

Now, let us show that the functions

$$u_i^*(x_0^*, x_0, x_1, \ldots, x_n) \quad (i = 1, 2, \ldots, N_3)$$

constructed above as limits of the sequences $u_i^{*(m)}(x_0^*, x_0, x_1, \ldots, x_n)$ ($i = 1, \ldots, N_3$) possess continuous derivatives with respect to x_0^*; and moreover, the functions

$$\frac{\Delta u_i^{*(m_k^k)}}{\Delta x_0^*}, \quad i = 1, 2, \ldots, N_3,$$

are uniformly convergent to the respective derivatives, as $k \to \infty$, $\Delta x_0^* \to 0$. Obviously, it suffices to show that, for all values of x_0 such that $|x_0^*| \leq L_0^*$ (and not only for the multiples of h, as in (34)), the sequence of functions

$$v_i^{(k)}(x_0^*, x_0, x_1, \ldots, x_n) =$$

$$= \sum_\alpha e^{-\frac{\pi i}{L}(\alpha, x)} \int \left(\sum_j a_{ij} u_j^{*(m_k^k)} + b_i \right) e^{-\frac{\pi i}{L}(\alpha, \xi)} d\xi_0 \cdots d\xi_n , \qquad (55)$$

$$i = 1, 2, \ldots, N_3 ,$$

is uniformly convergent on the entire parallelepiped $R_{LL_0^*}$ as $k \to \infty$, the limit functions $v_i(x_0^*, x_0, \ldots, x_n)$ being continuous in $x_0^*, x_1, \ldots, x_n$. Clearly, each term of the series (55) is a continuous function of $x_0^*, x_0, \ldots, x_n$ and converges uniformly on $R_{LL_0^*}$ as $k \to \infty$. Therefore, in order to establish uniform convergence of $v_i^{(k)}$ and continuity of the limit functions, we have to show that

$$\sum_{\alpha > R} \left| \int \left(\sum a_{ij} u_j^{*(m_k^k)} + b_i \right) e^{-\frac{\pi i}{L}(\alpha, \xi)} d\xi_0 \cdots d\xi_n \right| < \varepsilon(R) , \qquad (56)$$

where $\varepsilon(R)$ does not depend on k and converges to 0 as $R \to \infty$. To this end, we note that (37), for $s = 0$, yields the following inequality

$$A_{ij}(x_0^*, \alpha_0, \ldots, \alpha_n) = \left| \int a_{ij} u_j^{*(m_k^k)} e^{-\frac{\pi i}{L}(\alpha, \xi)} d\xi_0 \cdots d\xi_n \right| \leq$$

$$\leq \sum_\gamma |a_{ij}(x_0^*, \alpha_0 - \gamma_0, \ldots, \alpha_n - \gamma_n)| \left| a_j^{*(m_k^k)}(x_0^*, \gamma_0, \ldots, \gamma_n) \right| .$$

Using (40) with $s = 0$, we obtain

$$\sum_j \sum_{\alpha > R} A_{ij}(x_0^*, \alpha_0, \alpha_1, \ldots, \alpha_n) \leq$$

$$\leq \sum_j \sum_{\alpha > R} \sum_\gamma |a_{ij}(x_0^*, \alpha_0 - \gamma_0, \ldots, \alpha_n - \gamma_n)| \left| a_j^{*(m_k^k)}(x_0^*, \gamma_0, \ldots, \gamma_n) \right| \leq$$

$$\leq \sum_j \sum_{\alpha > R} \sum_{\gamma > R_1} \frac{M_1'}{\left(1 + \sqrt{\sum_s (\alpha_s - \gamma_s)^2} \right)^{n+2}} \left| a_j^{*(m_k^k)}(x_0^*, \gamma_0, \ldots, \gamma_n) \right| +$$

$$+ \sum_j \sum_{\alpha > R} \sum_{\gamma \leq R_1} \frac{M_1'}{\left(1 + \sqrt{\sum_s (\alpha_s - \gamma_s)^2} \right)^{n+2}} \left| a_j^{*(m_k^k)}(x_0^*, \gamma_0, \ldots, \gamma_n) \right| .$$

Here $\sum_{\gamma \leq R_1}$ implies summation over all integer $\gamma_0, \ldots, \gamma_n$, each γ_s being $\leq R_1$. Denote the first sum in the right-hand side of the last inequality by $\sum_1$, and the second one by $\sum_2$. Then, obviously,

$$\Sigma_1 \leq \sum_j \sum_\alpha \sum_{\gamma > R_1} \frac{M_1'}{\left(1 + \sqrt{\sum_s (\alpha_s - \gamma_s)^2}\right)^{n+2}} \left| a_j^{*(m_k^k)}(x_0^*, \gamma_0, \ldots, \gamma_n) \right| .$$

After performing summation with respect to $\alpha_0, \ldots, \alpha_n$ in the right-hand side, we find that

$$\Sigma_1 \leq M_1' C \sum_j \sum_{\gamma > R} \left| a_j^{*(m_k^k)}(x_0^*, \gamma_0, \ldots, \gamma_n) \right| ,$$

where C stands for the sum of the series

$$C = \sum_\beta \frac{1}{\left(1 + \sqrt{\beta_s^2}\right)^2} . \tag{57}$$

By virtue of (54) we get

$$\Sigma_1 \leq M_2' N_3 C \varepsilon(R_1) ,$$

where $\varepsilon(R_1)$ can be made arbitrarily small, if R_1 is chosen sufficiently large and independent of x_0^*, k.

Now, let us consider Σ_2. Clearly, for each R_1 fixed, the constant R can be chosen so large as to make Σ_2 arbitrarily small.

Finally, by (41) we have

$$\sum_{\alpha > R} \left| \int b_i \, e^{-\frac{\pi i}{L}(\alpha, \xi)} \, d\xi_0 \cdots d\xi_n \right| \leq \sum_{\alpha > R} \frac{M_1'}{(1 + |\alpha|)^{n+2}} ,$$

and therefore, choosing R sufficiently large, we can make the last sum arbitrarily small. Thus (56) is proved.

To sum up, the functions u_i^* ($i = 1, 2, \ldots, N_1$) and their derivatives in $x_0, \ldots, x_n$ up to the order δ have been proved to be continuous and bounded, as well as their first derivatives in x_0^*. For $\partial u_i^* / \partial x_0^*$ ($i = 1, 2, \ldots, N_1$) we obtain the following Fourier series

$$\frac{\partial u_i^*}{\partial x_0^*} = \sum_\alpha d_i(x_0^*, \alpha_0, \ldots, \alpha_n) \, e^{-\frac{\pi i}{L}(\alpha, x)} , \tag{58'}$$

where

$$d_i(x_0^*, \alpha_0, \ldots, \alpha_n) = \frac{\partial}{\partial x_0^*} \int u_i^* \, e^{-\frac{\pi i}{L}(\alpha, x)} \, dx_0 \ldots dx_n =$$

$$= \int F_i'(x_0^*, x_0, \ldots, x_n; u_1^*, \ldots, u_{N_1}^*) \, e^{-\frac{\pi i}{L}(\alpha, x)} \, dx_0 \ldots dx_n . \tag{58''}$$

§5. Proof of Analyticity

1. Assume that we have

$$u_i^* = p_{\rho k_0 \dots k_l} , \tag{59}$$

where $\sum k_\sigma < n_\rho - 1$ and $i = 1, \dots, N'$. Then, (58) and (23) immediately yield

$$\frac{\partial u_i^*}{\partial x_0^*} = \sum_{m_\sigma \leq k_\sigma} \psi_{\rho k_0 \dots k_l}^{m_0 \dots m_l} \, p_{\rho, m_0+1, m_1, \dots, m_l} , \qquad i = 1, \dots, N' , \tag{60}$$

where the functions $\psi_{\rho k_0 \dots k_l}^{m_0 \dots m_l}$ are linear combinations (with constant coefficients) of ψ and its partial derivatives up to the order n_ρ.

If the relations (59) hold for all k_σ such that $\sum_\sigma k_\sigma = n_\rho - 1$, then it follows from (59), (58), and (29) that

$$\frac{\partial}{\partial x_0^*} \int u_i^* \, e^{-\frac{\pi i}{L}(\alpha, x)} \, dx_0 \dots dx_n =$$

$$= \int \Bigg\{ \sum_{\substack{m_\sigma \leq k_\sigma \\ \sum m_\sigma < n_\rho - 1}} \psi_{\rho k_0 \dots k_l}^{m_0 \dots m_l} \, p_{\rho, m_0+1, m_1, \dots, m_l} + \psi \frac{\partial u_i^*}{\partial x_0} +$$

$$+ \sum_j \frac{\alpha_{0s}^{k_0+1} \cdots \alpha_{ls}^{k_l}}{\alpha'^{n_\rho}} \, \psi k_{ij}(\alpha) \widetilde{F}_{ju}^* +$$

$$+ \sum_{j,m,\sigma} \psi_{ijm_0 \dots m_l \sigma}(\alpha) \left(\frac{\partial p_{jm_0 \dots m_\sigma - 1 \dots m_l}}{\partial x_\sigma} - p_{jm_0 \dots m_\sigma \dots m_l} \right) +$$

$$+ \sum_{j,m,\sigma_1,\sigma_2} \psi_{ijm_0 \dots m_l \sigma_1 \sigma_2}(\alpha) \left(\frac{\partial p_{jm_0 \dots m_{\sigma_1}-1 \dots m_{\sigma_2} \dots m_l}}{\partial x_{\sigma_1}} - \right.$$

$$\left. - \frac{\partial p_{jm_0 \dots m_{\sigma_1} \dots m_{\sigma_2}-1 \dots m_l}}{\partial x_{\sigma_2}} \right) \Bigg\} e^{-\frac{\pi i}{L}(\alpha, x)} \, dx_0 \dots dx_n , \tag{61}$$

where

$$\alpha' = \sqrt{\alpha_{0s}^2 + \cdots \alpha_{is}^2} .$$

For the sake of convenience, we set

$$\frac{\alpha_{0s}^{k_0+1} \cdots \alpha_{ls}^{k_l}}{\alpha'^{n_\rho}} = 0 \quad \text{if} \quad \alpha' = 0 .$$

By $\widetilde{F}_{ju}^*$ we have denoted the expressions obtained from $\widetilde{F}_j^*$ with all the derivatives of u_ν^* $(\nu = 1, \dots, N)$ up to the order $n_\nu - 1$ replaced by the respective $p_{\nu m_0 \dots m_l}$, and the derivatives of order n_ν replaced by the suitable derivatives of $p_{\nu m_0 \ldots m_l}$ with respect to one of the variables $x_0, \ldots, x_n$. The derivatives of order n_ν may correspond to various expressions containing $p_{\nu m_0 \ldots m_l}$; however, we make the convention that in every $\tilde{F}^*_{ju}$ these derivatives are expressed in the same form. The functions ψ and k, with the index α in parentheses, depend not only on $x_0^*, x_0, \ldots, x_n; u_1^*, \ldots, u_{N_1}^*$, but also on $\alpha_0, \ldots, \alpha_n$. For all real $\alpha_0, \ldots, \alpha_n$, the functions ψ and k, together with their derivatives up to the order $n + 4$, are continuous and periodic with respect to all variables x_k, the period being equal to $2L$. (A similar property can be attributed to higher order derivatives; this fact, however, will not be used in the sequel).

The expressions in curly braces under the sign of the integral in the right-hand side of (61) are obtained from their counterparts in (29) as follows:

1) First of all, in the expressions corresponding to $dA_{ij}^{\prime(p_0 \ldots p_l)}/dx_{ks}^\prime$ we replace certain $p_{\nu m_0 \ldots m_l}$ by derivatives of the other functions $p_{\nu m_0 \ldots m_l}$ with respect to one of the variables $x_0, \ldots, x_n$, so that the resulting expression coincide with the total derivative in $x_{ks}^\prime$ of the function obtained from $A_{ij}^{\prime(p_0 \ldots p_l)}$ by replacing the derivatives of $u_1^* \ldots, u_N^*$ with $u_{N+1}^*, \ldots, u_{N_1}^*$.

2) Further, in the expressions thus obtained, we perform integration by parts with respect to $x_{1s}^\prime, \ldots, x_{ls}^\prime$, which are the inverse variables to those used while passing from (28) to (29).

3) Some of the derivatives

$$\frac{\partial p_{\nu m_0 \ldots m_{\sigma_1} - 1 \ldots m_{\sigma_2} \ldots m_l}}{\partial x_{\sigma_1 s}}$$

we replace by

$$\frac{\partial p_{\nu m_0 \ldots m_{\sigma_1} \ldots m_{\sigma_2} - 1 \ldots m_l}}{\partial x_{\sigma_2 s}} .$$

In the above transformations, we deal only with the functions $p_{\nu m_0 \ldots m_l}$ considered before and denoted by $u_{N+1}^*, \ldots, u_{N_1}^*$. Transformations of the first and the third types are responsible for the introduction of the third and the fourth sums under the sign of the integral (61). Simplifying the right-hand side of (61), we obtain (58″) again. We have

$$\psi_{ijm_0 \ldots m_l \sigma}(\alpha) \equiv \psi_{ijm_0 \ldots m_l \sigma_1 \sigma_2} \equiv 0 \quad \text{for} \quad \alpha_{0s} = \alpha_{1s} = \cdots = \alpha_{ls} = 0 .$$

If

$$u_i^* = p_{\rho k_0 \ldots k_\sigma \ldots k_l} ,$$

then we set

$$\frac{\partial p_{\rho k_0 \ldots k_\sigma - 1 \ldots k_l}}{\partial x_{k_\sigma s}} - p_{\rho k_0 \ldots k_\sigma \ldots k_l} = P_{i(\sigma)} . \tag{62}$$

If

$$p_{\rho k_0 \dots k_{\sigma_1} - 1 \dots k_{\sigma_2} \dots k_l} = u_{i_1}^* \qquad \text{and} \qquad p_{\rho k_0 \dots k_{\sigma_1} \dots k_{\sigma_2} - 1 \dots k_l} = u_{i_2}^* \,,$$

then we set

$$\frac{\partial p_{\rho k_0 \dots k_{\sigma_1} - 1 \dots k_{\sigma_2} \dots k_l}}{\partial x_{\sigma_1 s}} - \frac{\partial p_{\rho k_0 \dots k_{\sigma_1} \dots k_{\sigma_2} - 1 \dots k_l}}{\partial x_{\sigma_2 s}} = P_{i_1 i_2} \,. \tag{63}$$

Then equation (61) can be rewritten in the form

$$\frac{\partial}{\partial x_0^*} \int u_i^* \, e^{-\frac{\pi i}{L}(\alpha, x)} \, dx_0 \dots dx_n =$$

$$= \int \left\{ \sum_{\substack{m_\sigma \leq k_\sigma \\ \sum m_\sigma < n_\rho - 1}} \psi_{\rho k_0 \dots k_l}^{m_0 \dots m_l} \, p_{\rho m_0 + 1 \dots m_l} + \psi \frac{\partial u_i^*}{\partial x_0} + \right.$$

$$+ \sum_j \frac{\alpha_{0s}^{k_0 + 1} \cdots \alpha_{ls}^{k_l}}{\alpha'^{n_\rho}} \, \psi k_{ij}(\alpha) \tilde{F}_{ju}^* + \sum_{jk} \psi_{ij(k)} P_{j(k)} +$$

$$\left. + \sum_{j_1 j_2} \psi_{i j_1 j_2} P_{j_1 j_2} \right\} e^{-\frac{\pi i}{L}(\alpha, x)} \, dx_0 \dots dx_n \,. \tag{64}$$

2. Let us single out from (60) the equations which correspond to u_i^* satisfying the relations (59) with $\sum k_\nu < n_\rho - 2$. Differentiating these equations with respect to x_{ks}, and subtracting the resulting equations from the other equations of the system, we obtain

$$\frac{\partial P_{i(k)}}{\partial x_0^*} = \sum_{jq} \psi_{ij(kq)}^{(1)} P_{j(q)} \,. \tag{65'}$$

To simplify the formulas, the indices $m_0, \dots, m_l$ of the function ψ have been replaced by j, and (kq) is allowed to take all integer values from 1 to N_1.

Let us differentiate in x_{ks} both sides of the equation (60) for which the corresponding function u_i^* satisfies (59) with $\sum k_\nu = n_\rho - 2$, and then multiply the resulting equation by

$$\frac{1}{2^{n+1} L^{n+1}} \, e^{-\frac{\pi i}{L}(\alpha, x)}$$

and integrate it over the cube Q_L. Subtracting the equation thus obtained from the corresponding equation of system (61), we find that

$$\frac{\partial}{\partial x_0^*} \int P_{i(k)}\, e^{-\frac{\pi i}{L}(\alpha,x)}\, dx_0 \ldots dx_n = \int \Bigg\{ \sum_{jq} \psi_{ij(kq)}^{(1)} P_{j(q)} + $$

$$+ \sum_{j_1 j_2} \psi_{i(k)j_1 j_2} P_{j_1 j_2} + \sum_j \frac{\alpha_{0s}^{k_0+1} \cdots \alpha_{ls}^{k_l}}{\alpha'_{n\rho}} \, \psi k_{ij}(\alpha) \tilde{F}_{ju}^* \Bigg\} e^{-\frac{\pi i}{L}(\alpha,x)}\, dx_0 \ldots dx_n\,,$$

$$i = N' + 1, \ldots, N_1\,. \tag{66}$$

Here and in what follows, the letter ψ equipped with various indices denotes functions continuously differentiable in Q_L with respect to $x_0, \ldots, x_n$, at least up to the order $n+4$, and vanishing outside the parallelepiped $R_{0.7LL_0^*}$. Some of these functions may happen to be identically equal to zero.

3. Let us differentiate, with respect to $x_{\sigma_1 s}$, both sides of the equation in (60) corresponding to

$$u_{i_1}^* = p_{\rho k_0 \ldots k_{\sigma_1}-1\ldots k_l}\,;$$

then we differentiate, with respect to $x_{\sigma_2 s}$, both sides of the equation in (60) corresponding to

$$u_{i_2}^* = p_{\rho k_0 \ldots k_{\sigma_1} \ldots k_{\sigma_2}-1\ldots k_l}\,,$$

and consider the term-by-term difference of the two resulting equations. Then we get

$$\frac{\partial P_{i_1 i_2}}{\partial x_0^*} = \sum_{jk} \psi_{i_1 i_2 j(k)}^{(2)} P_{j(k)} + \sum_{j_1 j_2} \psi_{i_1 i_2 j_1 j_2} P_{j_1 j_2}\,, \quad i_1, i_2 = 1, \ldots, N'\,. \tag{67'}$$

These relations can also be established by direct computation. However, we employ here another method, which allows us to avoid direct computations and happens to be quite useful in the sequel; this method will be referred to as *method M*.

The expression for $\partial P_{i_1 i_2}/\partial x_0^*$ obtained from (60) involves the characters $p_{\rho m_0 \ldots m_l}$ and their derivatives in $x_{\sigma_1 s}$ and $x_{\sigma_2 s}$. If the relations (22) are satisfied, then one and the same derivative of the function u_ρ^* can be represented, in general, by different symbols of the form

$$p_{\rho m_0 \ldots m_l}\,, \qquad \frac{\partial p_{\rho m_0 \ldots m_l}}{\partial x_{\sigma_1 s}}\,, \qquad \frac{\partial p_{\rho m_0 \ldots m_l}}{\partial x_{\sigma_2 s}}\,. \tag{68}$$

We shall refer to one of these symbols as "marked". By virtue of (62) and (63), all symbols of the form (68) can be expressed in terms of the marked ones and $P_{j(\sigma_1)}$, $P_{j(\sigma_2)}$, $P_{j_1 j_2}$. Since the expression for $\partial P_{i_1 i_2}/\partial x_0^*$ has linear structure with respect to the symbols of the form

$$\frac{\partial p}{\partial x_{\sigma_1 s}} \quad \text{and} \quad \frac{\partial p}{\partial x_{\sigma_2 s}}\,,$$

it is clear that after the said substitution, we obtain the terms linear with respect to $P_{j(\sigma_1)}$, $P_{j(\sigma_2)}$, $P_{j_1 j_2}$ (as in the right-hand side of (67')), and an additional expression involving only the marked symbols. Denote the latter expression by $R_{i_1 i_2}$. Now, it remains to show that $R_{i_1 i_2}$ is identically equal to zero. To this end, we note that if the relations (22) are satisfied for all $\rho, k_0, \ldots, k_l$, then for all functions $u_1^*, \ldots u_N^*$ the equality $R_{i_1 i_2} = 0$ holds identically with respect to $x_0, \ldots, x_n$. Therefore, $R_{i_1 i_2}$ are identically equal to zero for all $p_{\rho k_0 \ldots k_l}$ involved, as well as their derivatives. And this is exactly what we wanted to prove.

Next, we differentiate with respect to $x_{\sigma_1 s}$ every equation in (58') that corresponds to

$$u_{i_1}^* = p_{\rho k_0 \ldots k_{\sigma_1}-1 \ldots k_{\sigma_2} \ldots k_l} \quad \text{with} \quad \sum k_\nu = n_\rho \,.$$

The equation in (58') corresponding to $u_{i_2}^* = p_{\rho k_0 \ldots k_{\sigma_1} \ldots k_{\sigma_2}-1 \ldots k_l}$ we differentiate with respect to $x_{\sigma_2 s}$. Let us consider the difference of the resulting equations, subtracting the respective terms from one another. Then, in the right-hand sides of the expressions obtained for the Fourier coefficients, all the terms containing $\widetilde{F}_{ju}^*$ will be mutually reduced, since the only difference between the terms in F_{i_1}' and F_{i_2}' depending on $\widetilde{F}_{ju}^*$ is in the coefficients

$$\frac{\alpha_{0s}^{k_0+1} \cdots \alpha_{\sigma_1 s}^{k_{\sigma_1}-1} \cdots \alpha_{\sigma_2 s}^{k_{\sigma_2}} \cdots \alpha_{ls}^{k_l}}{\alpha'^{n_\rho}} \quad \text{and} \quad \frac{\alpha_{0s}^{k_0+1} \cdots \alpha_{\sigma_1 s}^{k_{\sigma_1}} \cdots \alpha_{\sigma_2 s}^{k_{\sigma_2}-1} \cdots \alpha_{ls}^{k_l}}{\alpha'^{n_\rho}} \,.$$

We claim that the Fourier coefficients, after this reduction, can be written in the form

$$\int \left(\sum_{jk} \psi_{ij(k)}^{(3)} P_{j(k)} + \sum_{j_1 j_2} \psi_{ij_1 j_2}^{(3)} P_{j_1 j_2} \right) e^{-\frac{\pi i}{L}(\alpha, x)} \, dx_0 \ldots dx_n \,. \tag{69}$$

For $\alpha' = 0$ this statement can be proved by direct calculation. With a view to prove it for $\alpha' > 0$, we integrate by parts the remaining terms with respect to $x_{0s}', \ldots, x_{ls}'$, in the order inverse to that used in §2. Moreover, we apply the integration by parts in $x_{\sigma_1 s}$ to the terms which contain $\alpha_{\sigma_1 s}$ as a factor, and the terms containing $\alpha_{\sigma_2 s}$ as a factor we integrate by parts with respect to $x_{\sigma_2 s}$, so that both these factors disappear. As to the terms with second derivatives, the differentiation with respect to $x_{\sigma_1 s}$ and $x_{\sigma_2 s}$ in these terms can be expressed through the derivatives in $x_{0s}', \ldots, x_{ls}'$:

$$\frac{\partial}{\partial x_{\sigma_\nu s}} = \sum_j k_{\sigma_\nu j s} \frac{\partial}{\partial x_{js}'} \,, \quad \text{where} \quad k_{\sigma_\nu 0 s} = \frac{\alpha_{\sigma_\nu s}}{\alpha'} \,, \quad \nu = 1, 2 \,.$$

Then all the terms containing second derivatives can be written in the form

$$\psi \sum_{pm} k_{p_1 p_2}^{m_0 \ldots m_l} \frac{\partial^2}{\partial x_{p_1 s}' \partial x_{p_2 s}'} \left\{ \frac{\partial^{n_\rho - 1} u_p^*}{\partial x_{0s}'^{m_0} \cdots \partial x_{ls}'^{m_l}} \right\}_u \,,$$

where $\left\{ \partial^{n_\rho-1} u_\rho^* / \partial x_{0s}^{\prime m_0} \cdots \partial x_{ls}^{\prime m_l} \right\}_u$ denotes the derivative in curly braces expressed in terms of $p_{\rho m_0 \ldots m_l}$ (these symbols are assumed here to stand for the derivatives of u_ρ^*); the sum $\sum_{pm}$ is taken over all non-negative integers $m_0, \ldots, m_l, p_1, p_2$ such that $m_0 + \cdots + m_l = n_\rho - 1$, $p_1 + p_2 \neq 0$.

Let us add to the expressions obtained above for the Fourier coefficients of u_i^* the following integrals

$$\int \sum_{j=0}^{l} k_{\sigma_1 js} \frac{\alpha_{0s}^{k_0+1} \cdots \alpha_{\sigma_1 s}^{k_{\sigma_1}-1} \cdots \alpha_{\sigma_2 s}^{k_{\sigma_2}} \cdots \alpha_{ls}^{k_l}}{\alpha^{\prime n_\rho}} \frac{\partial}{\partial x_{js}'} \left(\psi \frac{\partial}{\partial x_{0s}'} \left\{ \frac{\partial^{n_\rho-1} u_\rho^*}{\partial x_{0s}^{\prime n_\rho-1}} \right\}_u \right) \times$$

$$\times e^{-\frac{\pi i}{L}(\alpha, x)} \, dx_0 \ldots dx_n \,,$$

which are equal to

$$\int \frac{\alpha_{0s}^{k_0+1} \cdots \alpha_{\sigma_1 s}^{k_{\sigma_1}} \cdots \alpha_{\sigma_2 s}^{k_{\sigma_2}} \cdots \alpha_{ls}^{k_l}}{\alpha^{\prime n_\rho-1}} \frac{\partial}{\partial x_{0s}'} \left(\psi \frac{\partial}{\partial x_{0s}'} \left\{ \frac{\partial^{n_\rho-1} u_\rho^*}{\partial x_{0s}^{\prime n_\rho-1}} \right\}_u \right) \times$$

$$\times e^{-\frac{\pi i}{L}(\alpha, x)} \, dx_0 \ldots dx_n \,,$$

since the integrals of the other terms in $\sum_{j=0}^{l}$ vanish.

The integral

$$\int \sum_{j=0}^{l} k_{\sigma_2 js} \frac{\alpha_{0s}^{k_0+1} \cdots \alpha_{\sigma_1 s}^{k_{\sigma_1}} \cdots \alpha_{\sigma_2 s}^{k_{\sigma_2}-1} \cdots \alpha_{ls}^{k_l}}{\alpha^{\prime n_\rho}} \frac{\partial}{\partial x_{js}'} \left(\psi \frac{\partial}{\partial x_{0s}'} \left\{ \frac{\partial^{n_\rho-1} u_\rho^*}{\partial x_{0s}^{\prime n_\rho-1}} \right\}_u \right) \times$$

$$\times e^{-\frac{\pi i}{L}(\alpha, x)} \, dx_0 \ldots dx_n \,,$$

which, in its turn, is equal to

$$\int \frac{\alpha_{0s}^{k_0+1} \cdots \alpha_{\sigma_1 s}^{k_{\sigma_1}} \cdots \alpha_{\sigma_2 s}^{k_{\sigma_2}} \cdots \alpha_{ls}^{k_l}}{\alpha^{\prime n_\rho-1}} \frac{\partial}{\partial x_{0s}'} \left(\psi \frac{\partial}{\partial x_{0s}'} \left\{ \frac{\partial^{n_\rho-1} u_\rho^*}{\partial x_{0s}^{\prime n_\rho-1}} \right\}_u \right) \times$$

$$\times e^{-\frac{\pi i}{L}(\alpha, x)} \, dx_0 \ldots dx_n \,,$$

should be added to the expressions obtained above for the Fourier coefficients of $u_{i_2}^*$.

If the functions $p_{\rho k_0 \ldots k_l}$ coincide with the derivatives of u_ρ^*, then, whatever the function u_ρ^* may be, all the terms entering the difference under consideration will be mutually reduced, since the terms of this difference are actually the terms of the same expression for

$$\frac{\partial^{n_\rho} \left(\psi \dfrac{\partial u_\rho^*}{\partial x_0} \right)}{\partial x_{0s}^{k_0} \cdots \partial x_{\sigma_1 s}^{k_{\sigma_1}} \cdots \partial x_{\sigma_2 s}^{k_{\sigma_2}} \cdots \partial x_{ls}^{k_l}} \,.$$

Now, it is easy to verify by direct calculation, or by the method M, that the terms containing no derivatives of the functions $p_{\rho m_0 \ldots m_l}$, or containing only their first derivatives, can be written in the same form as the second term under the sign of the integral in (69).

Consider the terms involving second derivatives of $p_{\rho m_0 \ldots m_l}$ with respect to the variables x'_{ks}. Since the function u^*_ρ can be chosen such that the said derivatives take any given values at any given point, therefore the terms of the expression under discussion which contain

$$\frac{\partial^2}{\partial x'_{p_1} \partial x'_{p_2}} \left\{ \frac{\partial^{n_\rho - 1} u^*_\rho}{\partial x'^{k_0}_{0s} \ldots \partial x'^{k_l}_{ls}} \right\}_u \tag{70}$$

will disappear by virtue of the condition that all such expressions corresponding to one and the same derivative of the function u^*_ρ are equal. We see that the terms of the form (70), with $p_1 = p_2 = 0$, are equal to one another; consequently, they become mutually reduced after the subtraction. Set

$$\frac{\partial^2}{\partial x'_{p_1 s} \partial x'_{p_2 s}} \left\{ \frac{\partial^{n_\rho - 1} u^*_\rho}{\partial x'^{m_0}_{0s} \ldots \partial x'^{m_{p_1}-1}_{p_1 s} \ldots \partial x'^{m_{p_2}-1}_{p_2 s} \ldots \partial x'^{m_{p_3}}_{p_3 s} \ldots \partial x'^{m_{p_4}}_{p_4 s} \ldots \partial x'^{m_l}_{ls}} \right\}_u -$$

$$- \frac{\partial^2}{\partial x'_{p_3 s} \partial x'_{p_4 s}} \left\{ \frac{\partial^{n_\rho - 1} u^*_\rho}{\partial x'^{m_0}_{0s} \ldots \partial x'^{m_{p_1}}_{p_1 s} \ldots \partial x'^{m_{p_2}}_{p_2 s} \ldots \partial x'^{m_{p_3}-1}_{p_3 s} \ldots \partial x'^{m_l}_{ls}} \right\}_u =$$

$$= P_{\rho m_0 \ldots m_l p_1 p_2 p_3 p_4} . \tag{71}$$

Let us single out one of the expressions of the form (70) which correspond to the same derivative of u^*_ρ, assuming that either $p_1 \neq 0$ or $p_2 \neq 0$, and $\sum k_\nu = n_\rho - 1$; this expression will be referred to as "marked". Using the relations (71), we can now write every expression (70) with either $p_1 \neq 0$ or $p_2 \neq 0$ in terms of the marked one and $P_{\rho m_0 \ldots m_l p_1 p_2 p_3 p_4}$. Thereupon, all "marked" symbols entering the expression under consideration become mutually reduced after the subtraction. In every remaining term $P_{\rho m_0 \ldots m_l p_1 p_2 p_3 p_4}$, we have either $p_1 \neq 0$ or $p_2 \neq 0$, and also either $p_3 \neq 0$ or $p_4 \neq 0$. Then, let us express $P_{\rho m_0 \ldots p_4}$ as follows:

$$P_{\rho m_0 \ldots m_l p_1 p_2 p_3 p_4} = P_{\rho m_0 \ldots m_l p_1 p_2 p_1 p_3} + P_{\rho m_0 \ldots m_l p_1 p_3 p_3 p_4} =$$

$$= \frac{\partial}{\partial x'_{p_1 s}} \left(\frac{\partial}{\partial x'_{p_2 s}} \left\{ \frac{\partial^{n_\rho - 1} u^*_\rho}{\partial x'^{m_0}_0 \ldots \partial x'^{m_{p_1}-1}_{p_1 s} \ldots \partial x'^{m_{p_2}-1}_{p_2 s} \ldots \partial x'^{m_{p_3}}_{p_3 s} \ldots \partial x'^{m_l}_{ls}} \right\}_u - \right.$$

$$\left. - \frac{\partial}{\partial x'_{p_3 s}} \left\{ \frac{\partial^{n_\rho - 1} u^*_\rho}{\partial x'^{m_0}_0 \ldots \partial x'^{m_{p_1}-1}_{p_1 s} \ldots \partial x'^{m_{p_2}}_{p_2 s} \ldots \partial x'^{m_{p_3}-1}_{p_3 s} \ldots \partial x'^{m_{p_4}}_{p_4 s} \ldots \partial x'^{m_l}_{ls}} \right\}_u \right) +$$

$$+ \frac{\partial}{\partial x'_{p_3 s}} \left(\frac{\partial}{\partial x'_{p_1 s}} \left\{ \frac{\partial^{n_\rho - 1} u_\rho^*}{\partial x_0^{'m_0} \dots \partial x_{p_1 s}^{'m_{p_1} - 1} \dots \partial x_{p_2 s}^{'m_{p_2}} \dots \partial x_{p_3 s}^{'m_{p_3} - 1} \dots \partial x_{l_s}^{'m_l}} \right\}_u - \right.$$

$$\left. - \frac{\partial}{\partial x'_{p_4 s}} \left\{ \frac{\partial^{n_\rho - 1} u_\rho^*}{\partial x_0^{'m_0} \dots \partial x_{p_1 s}^{'m_{p_1}} \dots \partial x_{p_2 s}^{'m_{p_2}} \dots \partial x_{p_3 s}^{'m_{p_3} - 1} \dots \partial x_{p_4 s}^{'m_{p_4} - 1} \dots \partial x_{l_s}^{'m_l}} \right\}_u \right).$$

Consequently, without introducing the factors $\alpha_0, \dots \alpha_l$, we can integrate by parts the terms of the first parenthesized expression, with respect to $x'_{p_1 s}$, and those of the second one, with respect to $x'_{p_3 s}$.

We thus obtain

$$\frac{\partial}{\partial x_0^*} \int P_{i_1 i_2} \, e^{-\frac{\pi i}{L}(\alpha, x)} \, dx_0 \dots dx_n =$$

$$= \int \left(\sum_{jk} \psi^{(3)}_{i_1 i_2 (k)} P_{j(k)} + \sum_{j_1 j_2} \psi^{(3)}_{i_1 i_2 j_1 j_2} P_{j_1 j_2} \right) e^{-\frac{\pi i}{L}(\alpha, x)} \, dx_0 \dots dx_n, \quad (67'')$$

$$i_1, i_2 = N' + 1, \dots N_1,$$

and thereby our statement is proved.

4. If $u_i^* = p_{\rho k_0 \dots k_l}$, then we set

$$\frac{\partial u_i^*}{\partial x_0^*} - \psi \frac{\partial u_i^*}{\partial x_0} \sum_{\substack{m_\sigma \leq k_\sigma \\ \sum m_\sigma < \sum k_\sigma}} \psi^{m_0 \dots m_l}_{\rho k_0 \dots k_l} \, p_{\rho m_0 + 1 \dots m_l} = Q_i, \quad i = 1, \dots, N_1. \quad (72)$$

Then relations (61) can be rewritten in the from

$$\int Q_i \, e^{-\frac{\pi i}{L}(\alpha, x)} \, dx_0 \dots dx_n = \int \left(\sum_{jk} \psi_{ij(k)} P_{j(k)} + \sum_{j_1 j_2} \psi_{i j_1 j_2} P_{j_1 j_2} + \right.$$

$$\left. + \sum_j \frac{\alpha_{0s}^{k_0 + 1} \dots \alpha_{ls}^{k_l}}{\alpha^{'n_\rho}} \, \psi k_{ij}(\alpha) \tilde{F}_{ju}^* \right) e^{-\frac{\pi i}{L}(\alpha, x)} \, dx_0 \dots dx_n, \quad (73)$$

$$i = N' + 1, \dots, N_1.$$

Here the terms containing $\tilde{F}_{ju}^*$ are exactly the same as in (66). Expressing these terms from (66) and substituting them into (73), we get

$$\frac{\partial}{\partial x_0^*} \int P_{i(k)} \, e^{-\frac{\pi i}{L}(\alpha, x)} \, dx_0 \dots dx_n = \int \left(\sum_{jq} \psi^{(4)}_{ij(kq)} P_{j(q)} + \right.$$

$$\left. + \sum_{j_1 j_2} \psi^{(4)}_{i(k) j_1 j_2} P_{j_1 j_2} + Q_i \right) e^{-\frac{\pi i}{L}(\alpha, x)} \, dx_0 \dots dx_n, \quad (65)$$

$$i = N' + 1, \dots, N_1, \quad k = 0, \dots, l.$$

Here the functions $\psi^{(4)}_{ij(kq)}$ and $\psi^{(4)}_{i(k) j_1 j_2}$ are other than those considered previously; however, they are also continuously differentiable up to the order $n + 4$ with respect to all variables x_k, and they vanish outside $R_{0.7LL_0^*}$.

5a. Now, let us find the expressions for $\partial Q_i / \partial x_0^*$. First, we obtain these expressions for $i = 1, \ldots, N'$. To this end we note that

$$\frac{\partial Q_i}{\partial x_0^*} = \frac{\partial^2 u_i^*}{\partial x_0^{*2}} - \psi \frac{\partial^2 u_i^*}{\partial x_0 \partial x_0^*} - \frac{\partial \psi}{\partial x_0^*} \frac{\partial u_i^*}{\partial x_0} - \tag{74'}$$

$$- \sum_{\substack{m_\sigma \leq k_\sigma \\ \sum m_\sigma < \sum k_\sigma}} p_{\rho m_0 + 1 \ldots m_l} \frac{\partial}{\partial x_0^*} \psi_{\rho k_0 \ldots k_l}^{m_0 \ldots m_l} - \sum_{\substack{m_\sigma \leq k_\sigma \\ \sum m_\sigma < \sum k_\sigma}} \psi_{\rho k_0 \ldots k_l}^{m_0 \ldots m_l} \frac{\partial}{\partial x_0^*} p_{\rho m_0 + 1 \ldots m_l} \; .$$

Let us differentiate both sides of (60), first with respect to x_0, and then with respect to x_0. Let us substitute the expressions obtained for $\partial^2 u_i^* / \partial x_0^{*2}$ and $\partial^2 u_i^* / \partial x_0^* \partial x_0$ into the right-hand side of (74'). All the derivatives of $p_{\rho m_0 + 1 \ldots m_l}$, i.e., all the derivatives of u_j^*, with respect to x_0^*, appearing here, we replace by their expressions obtained from (72). We are going to show that, all these transformations having been executed, it will only be linear combinations of Q_j and $P_{j(k)}$ that will remain in the right-hand side of (74'). The proof of this result is based on the application of the method M. After the said transformations, the right-hand side of (74') will contain, in addition to Q_j, the functions $p_{\rho m_0 + 1 \ldots m_l}$ and their first derivatives in x_0. One of the symbols of the form

$$p_{\rho m_0 + 1 \ldots m_l} \quad \text{or} \quad \frac{\partial p_{\rho m_0 \ldots m_l}}{\partial x_0} \tag{75}$$

corresponding to the same derivative of u_ρ^* we regard as marked. Using marked symbols and the functions $P_{j(k)}$, let us express all symbols of the form (75) in the right-hand side of (74'), after having subjected this right-hand side to the transformations described above. Now we can claim that the sum of all the terms in the right-hand side which contain neither $P_{j(k)}$ nor Q_j (we denote this sum by S) is identically equal to zero. Indeed, the sum S consists of combinations of marked symbols and derivatives of ψ. Suppose that all $p_{\rho m_0 + 1 \ldots}$ coincide with derivatives of a function u_ρ^* related by (12) to another function u_ρ which is analytic in $x_0 + i x_0^*$. Then all $P_{j(k)}$ and Q_j are equal to zero. Consequently, their sum is also equal to zero. On the other hand, whatever values the marked symbols may take, we can always chose a function u_ρ^* of the above type, whose derivatives at any given point coincide with any given values of the marked symbols. Therefore, $S \equiv 0$. Thus we have shown that

$$\frac{\partial Q_i}{\partial x_0^*} = \sum_j \psi_{ij}^{(5)} Q_j + \sum_{jk} \psi_{ij(k)}^{(5)} P_{j(k)} \; , \qquad i = 1, \ldots, N' \; , \tag{76}$$

where the indexed ψ, as usual, denotes a function which is continuously differentiable in all variables x_k up to the order $n+4$ and vanishes outside $R_{0.7LL_0^*}$.

5b. Now let us find the expressions for

$$\frac{\partial Q_i}{\partial x_0^*}, \qquad i = N' + 1, \ldots N_1 .$$

In this case, formulas (74') can be used once again; however, in order to find the derivatives $\partial^2 u_i^*/\partial x_0^{*2}$ and $\partial^2 u_i^*/\partial x_0^* \partial x_0$, we have to resort to (61), and therefore, to consider the corresponding Fourier series. Thus we get

$$\frac{\partial}{\partial x_0^*} \int Q_i\, e^{-\frac{\pi i}{L}(\alpha, x)} \, dx_0 \ldots dx_n =$$

$$= \int \left(\frac{\partial^2 u_i^*}{\partial x_0^{*2}} - \psi\, \frac{\partial^2 u_i^*}{\partial x_0 \partial x_0^*} - \frac{\partial \psi}{\partial x_0^*}\, \frac{\partial u_i^*}{\partial x_0} - \right.$$

$$- \sum_{\substack{m_\sigma \leq k_\sigma \\ \sum m_\sigma < \sum k_\sigma}} p_{\rho m_0 + 1 \ldots m_l}\, \frac{\partial}{\partial x_0^*}\, \psi_{\rho k_0 \ldots k_l}^{m_0 \ldots m_l} -$$

$$\left. - \sum_{\substack{m_\sigma \leq k_\sigma \\ \sum m_\sigma < \sum k_\sigma}} \psi_{\rho k_0 \ldots k_l}^{m_0 \ldots m_l}\, \frac{\partial}{\partial x_0^*}\, p_{\rho m_0 + 1 \ldots m_l} \right) e^{-\frac{\pi i}{L}(\alpha, x)} \, dx_0 \ldots dx_n . \quad (74'')$$

5c. Using (64), we eliminate the terms containing second derivatives of u_i^*. To this end, let us differentiate both sides of (64) with respect to x_0^*. In the terms of the right-hand side involving derivatives in any of the variables x_k, we differentiate under the sign of these derivatives. We replace the derivatives of u_j^* in x_0^* thus obtained by their expressions given by (72). Then the derivatives of Q_j will enter the resulting expressions in the same way as the derivatives of u_j^* enter the integrand in (64). Therefore, the derivatives $\partial Q_j/\partial x_{ks}$ can be eliminated, if we subject the terms containing Q_j to the transformations inverse to those which have led us from (28) to (29), in other words, if we reduce the terms of similar structure and integrate by parts several times.

Now, let us express $d(\psi \widetilde{F}_{ju}^*)/dx_0^*$ in the form suitable for our purposes (this is needed only for $\alpha' > 0$). We start with an observation that

$$\frac{\partial \psi}{\partial x_0^*} = i\varphi_0 \frac{-i\,\dfrac{\partial \varphi_0}{\partial x_0}}{\left(1 + ix_0^*\,\dfrac{\partial \varphi_0}{\partial x_0}\right)^2} = \psi\, \frac{-i\,\dfrac{\partial \varphi_0}{\partial x_0}}{1 + ix_0\,\dfrac{\partial \varphi_0}{\partial x_0}} = \psi\psi_1 ,$$

where ψ_1 possesses the same properties as the indexed functions ψ considered above. Note that ψ vanishes identically outside $R_{0.7LL_0^*}$; therefore, when calculating the derivatives of $\psi\tilde{F}_{ju}^*$, we can limit ourselves to the expression of $\tilde{F}_{ju}^*$ inside $R_{0.8LL_0^*}$ (this expression is obtained from the definition of $\tilde{F}_{ju}^*$ with $\psi_1 \equiv 1$). Denote the expressions obtained in this way by F_{ju}^*. Then we have

$$\frac{d(\psi\tilde{F}_{ju}^*)}{dx_0^*} = \frac{d(\psi F_{ju}^*)}{dx_0} = \psi_1\psi F_{ju}^* + \psi\,\frac{dF_{ju}^*}{dx_0^*}\;. \tag{77}$$

Next, we prove that

$$\frac{dF_{ju}^*}{dx_0^*} = \psi\,\frac{dF_{ju}^*}{dx_0} + \sum_{\sigma k} \psi_{j\sigma k}\,\frac{\partial Q_\sigma}{\partial x_{ks}} + \sum_\sigma \psi_{j\sigma}Q_\sigma +$$

$$+ \sum_{\sigma k} \psi_{j\sigma(k)}P_{\sigma(k)} + \sum_{j_1 j_2} \psi_{j j_1 j_2}P_{j_1 j_2}\;. \tag{78}$$

This relation can be established without any difficulty, if the following observations are taken into account: F_j^* has been obtained from a function analytic in all u_ρ, their derivatives, and the variable $x_0 + ix_0^*$, by replacing x_0^* by φ, the functions u_ρ by u_ρ^*, and the derivatives of u_ρ by the respective linear combinations of $p_{\rho m_0 \ldots m_l}$. Therefore, if the functions $p_{\rho m_0 \ldots m_l}$ were the derivatives of some u_ρ^* which is related to a function u_ρ analytic in $x_0 + ix_0^*$, then we would have

$$\frac{dF_j^*}{dx_0^*} = \psi\,\frac{dF_j^*}{dx_0}\;.$$

However, we do not know whether $p_{\rho m_0 \ldots m_l}$ are indeed derivatives of such functions u_ρ^*. Nevertheless, the relations (78) can be arrived at by expressing, in the left-hand side, the derivatives in x_0^* of the functions $p_{\rho m_0 \ldots m_l}$ by means of Q_σ, the other $p_{\ldots}$'s and their derivatives in x_0, and making use of the method M. It can be easily shown that the right-hand side of (78) will be independent of the derivatives of $P_{\sigma(k)}$ and $P_{j_1 j_2}$. Indeed, the terms containing the second derivatives of u^* in x_{ks} and x_0 will be mutually reduced in the first two terms in the right-hand side of the equality under consideration.

Taking into account (77) and (78), we obtain

$$\frac{d(\psi\tilde{F}_{ju}^*)}{dx_0^*} = \psi^2\,\frac{d\tilde{F}_{ju}^*}{dx_0} + \psi\psi_1\tilde{F}_{ju}^* + \sum_{\sigma k} \psi_{j\sigma k}^{(1)}\,\frac{\partial Q_\sigma}{\partial x_{ks}} + \sum_\sigma \psi_{j\sigma}^{(1)}Q_\sigma +$$

$$+ \sum_{\sigma k} \psi_{j\sigma(k)}^{(1)}P_{\sigma(k)} + \sum_{j_1 j_2} \psi_{j j_1 j_2}^{(1)}P_{j_1 j_2}\;. \tag{79}$$

Thus, we have

$$\frac{\partial^2}{\partial x_0^{*2}} \int u_i^* \, e^{-\frac{\pi i}{L}(\alpha,x)} \, dx_0 \dots dx_n = \int \left(\frac{\partial \psi}{\partial x_0^*} \frac{\partial u_i^*}{\partial x_0} + \psi^2 \frac{\partial^2 u_i^*}{\partial x_0^2} + \right.$$

$$+ \sum_j \psi_{ij}^{(7)} u_j^* + \sum_{\substack{m_\sigma \le k_\sigma \\ \sum m_\sigma < n_\rho - 1}} \psi \psi_{\rho k_0 \dots k_l}^{m_0 \dots m_l} \frac{\partial p_{\rho m_0 + 1 \dots m_l}}{\partial x_0} + \sum_j \psi_{ij}^{(6)} Q_j +$$

$$+ \sum_{j_1 j_2} \psi_{ij_1 j_2}^{(6)}(\alpha) P_{j_1 j_2} + \sum_{j_1 j_2} \psi_{ij_1 j_2}(\alpha) \psi \frac{\partial P_{j_1 j_2}}{\partial x_0} + \sum_{jk} \psi_{ij(k)}^{(6)}(\alpha) P_{j(k)} +$$

$$+ \sum_{jk} \psi_{ij(k)}(\alpha) \psi \frac{\partial P_{j(k)}}{\partial x_0} + \sum_j \frac{\alpha_{0s}^{k_0+1} \dots \alpha_{ls}^{k_l}}{\alpha'^{n_\rho}} \, k_{ij}^{(1)}(\alpha) \psi_{ij}^*(\alpha) \psi \tilde{F}_{ju}^* +$$

$$+ \sum_j \frac{\alpha_{0s}^{k_0+1} \dots \alpha_{ls}^{k_l}}{\alpha'^{n_\rho}} \, k_{ij}(\alpha) \psi \, \frac{d(\psi \tilde{F}_{ju}^*)}{dx_0} \right) e^{-\frac{\pi i}{L}(\alpha,x)} \, dx_0 \dots dx_n , \qquad (80)$$

$$i = N' + 1, \dots N_1 .$$

The functions $k_{ij}^{(1)}(\alpha)$ are continuously differentiable in x_k up to the order $n + 3$, the derivatives being bounded functions, $2L$-periodic with respect to all variables.

From (80) we find that

$$\frac{\partial^2}{\partial x_0^{*2}} \int u_i^* \, e^{-\frac{\pi i}{L}(\alpha,x)} \, dx_0 \dots dx_n =$$

$$= \int \left(\frac{\partial \psi}{\partial x_0^*} \frac{\partial u_i^*}{\partial x_0} + \psi^2 \frac{\partial^2 u_i^*}{\partial x_0^2} + \sum_j \psi_{ij}^{(7)} u_j^* + \right.$$

$$+ \sum_{\substack{m_\sigma \le k_\sigma \\ \sum m_\sigma < n_\rho - 1}} \psi \psi_{\rho k_0 \dots k_l}^{m_0 \dots m_l} \frac{\partial p_{\rho m_0 + 1 \dots m_l}}{\partial x_0} \right) e^{-\frac{\pi i}{L}(\alpha,x)} \, dx_0 \dots dx_n +$$

$$+ \sum_{j\nu} \psi_{ij6}^{(\alpha-\nu)}(\alpha) Q_j^{(\nu)} + \sum_{j_1 j_2 \nu} \psi_{ij_1 j_2 6}^{(\alpha-\nu)} P_{j_1 j_2}^{(\nu)} +$$

$$+ \frac{\pi i}{L} \sum_{j_1 j_2 \delta \nu} \psi_{ij_1 j_2}^{(\alpha-\delta-\nu)}(\alpha) \psi^{(\delta)} \nu_0 P_{j_1 j_2}^{(\nu)} +$$

$$+ \sum_{jk\nu} \psi_{ij(k)6}^{(\alpha-\nu)}(\alpha) P_{j(k)}^{(\nu)} + \frac{\pi i}{L} \sum_{jk\nu\delta} \psi_{ij(k)}^{(\alpha-\delta-\nu)}(\alpha) \nu_0 \psi^{(\delta)} P_{j(k)}^{(\nu)} +$$

$$+ \frac{\alpha_{0s}^{k_0+1} \dots \alpha_{ls}^{k_l}}{\alpha'^{n_\rho}} \sum_{j\nu\delta} k_{ij1}^{(\alpha-\delta-\nu)} \Phi_j^{(\nu)} \psi_{ij}^{*(\delta)}(\alpha) +$$

$$+ \frac{\pi i}{L} \frac{\alpha_{0s}^{k_0+1} \cdots \alpha_{ls}^{k_l}}{\alpha'^{n_\rho}} \sum_{j\nu\delta} k_{ij}^{(\alpha-\delta-\nu)}(\alpha) \nu_0 \Phi_j^{(\nu)} \psi^{(\delta)} . \tag{81}$$

In the above formulas we have set

$$f^{(\nu)} = f^{(\nu_0,\ldots,\nu_n)} = \int f e^{-\frac{\pi i}{L}(\nu,x)} dx_0 \ldots dx_n ,$$

$$f^{(\alpha-\nu)} = f^{(\alpha_0-\nu_0,\ldots,\alpha_n-\nu_n)} = \int f e^{-\frac{\pi i}{L}(\alpha-\nu,x)} dx_0 \ldots dx_n ,$$

$$\Phi_j = \psi \tilde{F}_{ju}^* ,$$

and $\sum_\nu$ implies summation over all integer (positive, negative, and 0) values of $\nu_0, \ldots, \nu_n$. To make the notations more readable, the indices of the functions ψ and k are written as subscripts, and not superscripts as before; this convention will apply in similar situations below.

5d. Now we turn to the determination of the Fourier coefficients for $\psi \, \partial^2 u_i^* / \partial x_0 \partial x_0^*$. To this end, we multiply both sides of (64) by $\frac{\pi i}{L}\alpha_0$ and integrate by parts in such a way that this factor disappear. Then we get

$$\int \frac{\partial^2 u_i^*}{\partial x_0 \partial x_0^*} e^{-\frac{\pi i}{L}(\alpha,x)} dx_0 \ldots dx_n =$$

$$= \int \left(\frac{\partial}{\partial x_0} \sum_{\substack{m_\sigma \le k_\sigma \\ \sum m_\sigma < n_\rho - 1}} \psi_{\rho k_0 \ldots k_l}^{m_0 \ldots m_l} \, P_{\rho m_0 + 1 \ldots m_l} \right. +$$

$$+ \frac{\partial \psi}{\partial x_0} \frac{\partial u_i^*}{\partial x_0} + \psi \frac{\partial^2 u_i^*}{\partial x_0^2} \right) e^{-\frac{\pi i}{L}(\alpha,x)} dx_0 \ldots dx_n +$$

$$+ \frac{\alpha_{0s}^{k_0+1} \cdots \alpha_{ls}^{k_l}}{\alpha'^{n_\rho}} \sum_{j\nu} k_{ij0}^{(\alpha-\nu)}(\alpha) \Phi_j^{(\nu)} +$$

$$+ \frac{\pi i}{L} \frac{\alpha_{0s}^{k_0+1} \cdots \alpha_{ls}^{k_l}}{\alpha'^{n_\rho}} \sum_{j\nu\delta} k_{ij}^{(\alpha-\nu)}(\alpha) \nu_0 \Phi_j^{(\nu)} +$$

$$+ \frac{\pi i}{L} \sum_{jk\nu} \psi_{ij(k)}^{(\alpha-\nu)}(\alpha) P_{j(k)}^{(\nu)} \nu_0 + \frac{\pi i}{L} \sum_{j_1 j_2 \nu} \psi_{ij_1 j_2}^{(\alpha-\nu)}(\alpha) \nu_0 P_{j_1 j_2}^{(\nu)} +$$

$$+ \sum_{jk\nu} \psi_{ij(k)7}^{(\alpha-\nu)}(\alpha) P_{j(k)}^{(\nu)} + \sum_{j_1 j_2 \nu} \psi_{ij_1 j_2 7}^{(\alpha-\nu)}(\alpha) P_{j_1 j_2}^{(\nu)} , \tag{82}$$

where the coefficients $k_{ij}^{(\alpha-\nu)}(\alpha)$, $\psi_{ij(k)}^{(\alpha-\nu)}(\alpha)$, and $\psi_{ij_1 j_2}^{(\alpha-\nu)}(\alpha)$ are the same as in right-hand side of (81). We denote by $k_{ij0}^{(\alpha-\nu)}$ the Fourier coefficients for $\partial k_{ij}(\alpha)/\partial x_0$.

From (82) we obtain

$$\int \psi \frac{\partial^2 u_i^*}{\partial x_0 \partial x_0^*}\, e^{-\frac{\pi i}{L}(\alpha,x)}\, dx_0 \ldots dx_n =$$

$$= \int \left(\psi \frac{\partial}{\partial x_0} \sum_{\substack{m_\sigma \le k_\sigma \\ \sum m_\sigma < n_\rho - 1}} \psi_{\rho k_0 \ldots k_l}^{m_0 \ldots m_l}\, P_{\rho m_0 + 1 \ldots m_l} + \right.$$

$$\left. + \psi \frac{\partial \psi}{\partial x_0} \frac{\partial u_i^*}{\partial x_0} + \psi^2 \frac{\partial^2 u_i^*}{\partial x_0^2} \right) e^{-\frac{\pi i}{L}(\alpha,x)}\, dx_0 \ldots dx_n +$$

$$+ \frac{\pi i}{L} \sum_{jv\delta} \frac{(\alpha_{0s} - \delta_{0s})^{k_0+1} \cdots (\alpha_{ls} - \delta_{ls})^{k_l}}{|\alpha' - \delta'|^{n_\rho}}\, k_{ij}^{(\alpha-\delta-\nu)}(\alpha - \delta)\nu_0 \Phi_j^{(\nu)} \psi^{(\delta)} +$$

$$+ \sum_{jv\delta} \frac{(\alpha_{0s} - \delta_{0s})^{k_0+1} \cdots (\alpha_{ls} - \delta_{ls})^{k_l}}{|\alpha' - \delta'|^{n_\rho}}\, k_{ij0}^{(\alpha-\delta-\nu)}(\alpha - \delta)\Phi_j^{(\nu)} \psi^{(\delta)} +$$

$$+ \frac{\pi i}{L} \sum_{jkv\delta} \psi^{(\delta)} \psi_{ij(k)}^{(\alpha-\delta-\nu)}(\alpha - \delta) P_{j(k)}^{(\nu)} \nu_0 +$$

$$+ \frac{\pi i}{L} \sum_{j_1 j_2 v\delta} \psi^{(\delta)} \psi_{ij_1 j_2}^{\alpha-\delta-\nu}(\alpha - \delta)\nu_0 P_{j_1 j_2}^{(\nu)} +$$

$$+ \sum_{jkv\delta} \psi^{(\delta)} \psi_{ij(k)7}^{(\alpha-\delta-\nu)}(\alpha - \delta) P_{j(k)}^{(\nu)} + \sum_{j_1 j_2 v\delta} \psi^{(\delta)} \psi_{ij_1 j_2 7}^{\alpha-\delta-\nu}(\alpha - \delta) P_{j_1 j_2}^{(\nu)}. \quad (83)$$

Here

$$|\alpha' - \delta'| = \sqrt{(\alpha_{0s} - \delta_{0s})^2 + \cdots + (\alpha_{ls} - \delta_{ls})^2}\; ;$$

and it is assumed, as above, that

$$\frac{(\alpha_{0s} - \delta_{0s})^{k_0+1} \cdots (\alpha_{ls} - \delta_{ls})^{k_l}}{|\alpha' - \delta'|^{n_\rho}} = 0 \quad \text{if} \quad |\alpha' - \delta'| = 0\;.$$

We denote by $k(\alpha - \delta)$ and $\psi(\alpha - \delta)$ (the indices are omitted) the values of k and ψ corresponding to $\alpha_{0s} - \delta_{0s}, \ldots, \alpha_{ls} - \delta_{ls}$.

5e. Substituting into (74″) the corresponding expressions given by (81) and (83), we obtain

$$\frac{\partial Q_i^{(\alpha)}}{\partial x_0^*} = \int \left(\sum_{jk} \psi_{ij(k)}^{(8)} P_{j(k)} + \sum_j \psi_{ij}^{(8)} Q_j \right) e^{-\frac{\pi i}{L}(\alpha,x)}\, dx_0 \ldots dx_n +$$

$$+ \sum_{jv} \psi_{ij6}^{(\alpha-\nu)}(\alpha) Q_j^{(\nu)} - \sum_{jkv\delta} \psi^{(\delta)} P_{j(k)}^{(\nu)} \psi_{ij(k)7}^{(\alpha-\delta-\nu)}(\alpha - \delta) -$$

$$- \sum_{j_1 j_2 \nu \delta} \psi^{(\delta)} P^{(\nu)}_{j_1 j_2} \psi^{(\alpha-\delta-\nu)}_{ij_1 j_2 7} (\alpha - \delta) +$$

$$+ \sum_{j_1 j_2 \nu} \psi^{(\alpha-\nu)}_{ij_1 j_2 6} P^{(\nu)}_{j_1 j_2} + \sum_{jk\nu} \psi^{(\alpha-\nu)}_{ij(k)6}(\alpha) P^{(\nu)}_{j(k)} +$$

$$+ \frac{\pi i}{L} \sum_{jk\nu\delta} \nu_0 \psi^{(\delta)} P^{(\nu)}_{j(k)} \left(\psi^{(\alpha-\delta-\nu)}_{ij(k)}(\alpha) - \psi^{(\alpha-\delta-\nu)}_{ij(k)}(\alpha - \delta) \right) +$$

$$+ \frac{\pi i}{L} \sum_{j\nu\delta} \nu_0 \Phi^{(\nu)}_j \psi^{(\delta)} \left(\frac{\alpha_{0s}^{k_0+1} \cdots \alpha_{ls}^{k_l}}{\alpha'^{n_\rho}} k^{(\alpha-\delta-\nu)}_{ij}(\alpha) - \right.$$

$$\left. - \frac{(\alpha_{0s} - \delta_{0s})^{k_0+1} \cdots (\alpha_{ls} - \delta_{ls})^{k_l}}{|\alpha' - \delta'|^{n_\rho}} k^{(\alpha-\delta-\nu)}_{ij}(\alpha - \delta) \right) +$$

$$+ \sum_{j\nu\delta} \frac{\alpha_{0s}^{k_0+1} \cdots \alpha_{ls}^{k_l}}{\alpha'^{n_\rho}} k^{(\alpha-\delta-\nu)}_{ij1}(\alpha) \psi^{*(\delta)}_{ij} \Phi^{(\nu)}_j -$$

$$- \sum_{j\nu\delta} \frac{(\alpha_{0s} - \delta_{0s})^{k_0+1} \cdots (\alpha_{ls} - \delta_{ls})^{k_l}}{|\alpha' - \delta'|^{n_\rho}} k^{(\alpha-\delta-\nu)}_{ij1}(\alpha - \delta) \psi^{(\delta)} \Phi^{(\nu)}_j ,$$

$$i = N' + 1, \ldots, N_1 . \tag{84}$$

Applying the method M, one can easily show that all the terms in (74″), (81), and (83) containing the functions $p_{\rho m_0 \ldots m_l}$ and their derivatives with respect to x_0 and x_0^* yield linear combinations of $P_{j(k)}$ and Q_j.

6. The quantities denoted by $\psi(\alpha)$ and $k(\alpha)$ (the indices are omitted here) depend, for $\alpha' > 0$, not only on the variables x_k but also on the directions of the axes $Ox'_{0s}, \ldots, Ox'_{ls}$. Let us choose the axes $Ox'_{1s}, \ldots, Ox'_{ls}$ such that their direction cosines be continuously differentiable functions of the direction cosines of Ox'_{0s}, i.e., of the variables $\alpha_{0s}/\alpha', \ldots, \alpha_{ls}/\alpha'$. Then, for $\alpha' > 0$, the first order partial derivatives of the functions $\psi(\alpha)$, $k(\alpha)$ with respect to α_{ks}/α' $(k = 0, \ldots, l)$ are continuously differentiable in x_k at least up to the order $n + 3$; all these derivatives being continuous, bounded, $2L$-periodic in all x_k. Denote by $D^{(m)}\psi$ any m-th order partial derivative of ψ with respect to the variables x_k $(k = 0, \ldots, n)$. Let $\rho(\alpha, \beta)$ be the geodesic distance on the sphere

$$\sum_k \frac{\alpha_{ks}^2}{\alpha'^2} = 1 \tag{85}$$

between the points

$$\alpha'_{ks} = \frac{\alpha_{ks}}{\alpha'} \quad \text{and} \quad \beta'_{ks} = \frac{\beta_{ks}}{\beta'}, \qquad k = 0, 1, \ldots, l, \tag{86}$$

of that sphere. Then, for $\alpha' > 0$ and $|\alpha' - \delta| > 0$, we have

$$\left| D^{(m)}\psi(\alpha) - D^{(m)}\psi(\alpha - \delta) \right| = \left| \frac{\partial D^{(m)}\psi(\tilde{\alpha})}{\partial \rho} \right| \rho(\alpha, \alpha - \delta) \leq$$

$$\leq \pi \sum_k \left| \frac{\partial D^{(m)}\psi(\tilde{\alpha})}{\partial \alpha'_{ks}} \frac{\partial \alpha'_{ks}}{\partial \rho} \right| \left[\sum_k \left(\frac{\alpha_{ks}}{\alpha'} - \frac{\alpha_{ks} - \delta_{ks}}{|\alpha' - \delta'|} \right)^2 \right]^{\frac{1}{2}} \leq$$

$$\leq \pi \sum_k \left| \frac{\partial D^{(m)}\psi(\tilde{\alpha})}{\partial \alpha'_{ks}} \right| \sum_k \left| \frac{\alpha_{ks}}{\alpha'} - \frac{\alpha_{ks} - \delta_{ks}}{|\alpha' - \delta'|} \right| \leq$$

$$\leq M(m)\pi(n+1) \sum_k \left| \frac{\alpha_{ks}}{\alpha'} - \frac{\alpha_{ks} - \delta_{ks}}{|\alpha' - \delta'|} \right| . \tag{86'}$$

Here $\tilde{\alpha}$ is a point of the arc on the equator of the sphere (85) joining the two points (86); $M(m)$ is the upper bound for

$$\left| \frac{\partial D^{(m)}\psi(\alpha)}{\partial x'_{ks}} \right| .$$

By virtue of (86') we have

$$\left| \psi^{(\alpha-\delta-\nu)}(\alpha) - \psi^{(\alpha-\delta-\nu)}(\alpha - \delta) \right| \leq \frac{M_1 C_1 \sum_k \left| \frac{\alpha_{ks}}{\alpha'} - \frac{\alpha_{ks} - \delta_{ks}}{|\alpha' - \delta'|} \right|}{(1 + |\alpha - \delta - \nu|)^{n+3}}, \tag{87}$$

where $M_1 = \max\{M(0), \ldots, M(n+3)\}$, and C_1 is a constant depending only on L and n.

On the other hand, the functions ψ defined by (21) possesses continuous $2L$-periodic derivatives up to the order $n+4$ with respect to all x_k; denoting by M_2 the upper bound for the absolute values of these derivatives, we obtain

$$|\psi^{(\delta)}| \leq \frac{M_2 C_2}{(1 + \delta)^{n+4}} . \tag{88}$$

The inequalities (87), (88) yield

$$\left| \psi^{(\delta)} \left(\psi^{(\alpha-\delta-\nu)}(\alpha) - \psi^{(\alpha-\delta-\nu)}(\alpha - \delta) \right) \nu_0 \right| \leq$$

$$\leq \frac{M_1 M_2 C_1 C_2 \sum_k \left| \frac{\alpha_{ks}}{\alpha'} - \frac{\alpha_{ks} - \delta_{ks}}{|\alpha' - \delta'|} \right| |\nu_0|}{(1 + \delta)^{n+4}(1 + |\alpha - \delta - \nu|)^{n+3}} .$$

We further have

$$\frac{\alpha_{ks}}{\alpha'} - \frac{\alpha_{ks} - \delta_{ks}}{|\alpha' - \delta'|} = \frac{\delta_{ks}}{\alpha'} + (\alpha_{ks} - \delta_{ks})\frac{|\alpha' - \delta'| - \alpha'}{\alpha'|\alpha' - \delta'|} =$$

$$= \frac{\delta_{ks}}{\alpha'} + (\alpha_{ks} - \delta_{ks})\frac{\sum_p \delta_{ps}(-2\alpha_{ps} + \delta_{ps})}{\alpha'|\alpha' - \delta'|(|\alpha' - \delta'| + \alpha')},$$

whence

$$\left|\frac{\alpha_{ks}}{\alpha'} - \frac{\alpha_{ks} - \delta_{ks}}{|\alpha' - \delta'|}\right| \leq \frac{|\delta_{ks}| + \sum_p |\delta_{ps}|}{\alpha'}.$$

Therefore, if both α' and $|\alpha' - \delta'|$ are other than zero, we have

$$\left|\psi^{(\delta)}\left(\psi^{(\alpha-\delta-\nu)}(\alpha) - \psi^{(\alpha-\delta-\nu)}(\alpha - \delta)\right)\nu_0\right| \leq$$

$$\leq \frac{M_1 M_2 C_1 C_2 (n+2)\sum_p |\delta_{ps}|(|\alpha_0 - \delta_0 - \nu_0| + |\alpha_0| + |\delta_0|)}{(1+\delta)^{n+4}(1 + |\alpha - \delta - \nu|)^{n+3}\alpha'} \leq$$

$$\leq \frac{3M_1 M_2 C_1 C_2 (n+1)(n+2)}{(1+\delta)^{n+2}(1 + |\alpha - \delta - \nu|)^{n+2}},$$

since $\alpha' \geq 1$ for $\alpha' > 0$.

Now, let us consider in detail the case $\alpha' = 0$. We can write

$$\left|\psi^{(\delta)}\left(\psi^{(\alpha-\delta-\nu)}(\alpha) - \psi^{(\alpha-\delta-\nu)}(\alpha - \delta)\right)\nu_0\right| \leq$$

$$\leq |\psi^{(\delta)}|\left(|\psi^{(\alpha-\delta-\nu)}(\alpha)| + |\psi^{(\alpha-\delta-\nu)}(\alpha - \delta)|\right)|\nu_0| \leq$$

$$\leq \frac{2M_2 C_2 M_3 C_2 (|\alpha_0 - \delta_0 - \nu_0| + |\delta_0|)}{(1+\delta)^{n+4}(1 + |\alpha - \delta - \nu|)^{n+3}} \leq$$

$$\leq \frac{4M_2 M_3 C_2^2}{(1+\delta)^3(1 + |\alpha - \delta - \nu|)^{n+2}},$$

where M_3 is the upper bound for the absolute values of the derivatives of $\psi(\alpha)$ with respect to x_k up to the order $n + 3$. Similar inequalities are obtained in the case of $|\alpha' - \delta'| = 0$.

Thus, the following inequality is always valid:

$$\left|\psi^{(\delta)}\left(\psi^{(\alpha-\delta-\nu)}(\alpha) - \psi^{(\alpha-\delta-\nu)}(\alpha - \delta)\right)\nu_0\right| \leq \frac{M_4}{(1+\delta)^{n+2}(1 + |\alpha - \delta - \nu|)^{n+2}},$$

$$(89)$$

where M_4 is a constant.

In a similar way we establish the inequality

$$\left|\psi^{(\delta)}\left(k^{(\alpha-\delta-\nu)}(\alpha) - k^{(\alpha-\delta-\nu)}(\alpha - \delta)\right)\nu_0\right| \leq \frac{M_5}{(1+\delta)^{n+2}(1 + |\alpha - \delta - \nu|)^{n+2}},$$

$$(90)$$

where M_5 is a constant.

Now, let us estimate the tenth sum in the right-hand side of (84). We start with the case of both α' and $|\alpha' - \delta'|$ being different from zero. Then the first derivatives of the product

$$\frac{\alpha_{0s}^{k_0+1} \alpha_{1s}^{k_1} \cdots \alpha_{ls}^{k_l}}{\alpha'^{n_\rho}} k(\alpha)$$

with respect to the variables α_{ks}/α' ($k = 0, 1, \ldots, l$), possess, in their turn, partial derivatives with respect to the variables x_k up to the order $n+3$, the latter derivatives being continuous functions, $2L$-periodic in all x_k. Let M_6 be the upper bound for the absolute values of all these derivatives. Then, as above, we have for $\nu' \neq 0$:

$$\left| \nu_0 \Phi^{(\nu)} \psi^{(\delta)} \left(\frac{\alpha_{0s}^{k_0+1} \alpha_{1s}^{k_1} \cdots \alpha_{ls}^{k_l}}{\alpha'^{n_\rho}} k^{(\alpha-\delta-\nu)}(\alpha) - \right. \right.$$

$$\left. \left. - \frac{(\alpha_{0s} - \delta_{0s})^{k_0+1} \cdots (\alpha_{ls} - \delta_{ls})^{k_l}}{|\alpha' - \delta'|^{n_\rho}} k^{(\alpha-\delta-\nu)}(\alpha - \delta) \right) \right| \leq$$

$$\leq \left| \frac{\nu_0}{\nu'} \Phi^{(\nu)} \nu' \right| \frac{M_6 C_1 \sum_k \left(\frac{|\delta_{ks}|}{\alpha'} + \sum_p \frac{|\delta_{ps}|}{\alpha'} \right) M_2 C_2}{(1+\delta)^{n+4}(1+|\alpha-\delta-\nu|)^{n+3}} \leq$$

$$\leq \left| \frac{\nu_0}{\nu'} \Phi^{(\nu)} \right| \frac{M_2 M_6 C_1 C_2 (n+2) \sum_k |\delta_{ks}|(|\alpha' - \delta' - \nu'| + \delta' + \alpha')}{\alpha'(1+\delta)^{n+4}(1+|\alpha-\delta-\nu|)^{n+3}} \leq$$

$$\leq 3 \left| \frac{\nu_0}{\nu'} \Phi^{(\nu)} \right| \frac{M_2 M_6 C_1 C_2 (n+1)(n+2)}{(1+\delta)^{n+2}(1+|\alpha-\delta-\nu|)^{n+2}} .$$

When deriving the penultimate estimate, we have used the inequality

$$\nu' \leq |\alpha' - \delta' - \nu'| + \delta' + \alpha' ,$$

which is a consequence of the triangle inequality in the $(\alpha_{0s}, \ldots, \alpha_{ls})$-space. Thus, for $\nu' \neq 0$, $\alpha' \neq 0$, $|\alpha' - \delta'| \neq 0$, we have

$$\left| \nu_0 \Phi^{(\nu)} \psi^{(\delta)} \left(\frac{\alpha_{0s}^{k_0+1} \alpha_{1s}^{k_1} \cdots \alpha_{ls}^{k_l}}{\alpha'^{n_\rho}} k^{(\alpha-\delta-\nu)}(\alpha) - \right. \right.$$

$$\left. \left. - \frac{(\alpha_{0s} - \delta_{0s})^{k_0+1} \cdots (\alpha_{ls} - \delta_{ls})^{k_l}}{|\alpha' - \delta'|^{n_\rho}} k^{(\alpha-\delta-\nu)}(\alpha - \delta) \right) \right| \leq$$

$$\leq \frac{M_7 |A^{(\nu)}|}{(1+\delta)^{n+2}(1+|\alpha-\delta-\nu|)^{n+2}} , \tag{91}$$

where

$$M_7 = M_2 M_6 C_1 C_2 (n+1)(n+2) \quad \text{and} \quad A_j^{(\nu)} = \frac{\nu_0}{\nu'} \Phi_j^{(\nu)}.$$

If $\nu' = 0$, then *a fortiori* $\nu_0 = 0$, and the inequalities (91) become trivial under the convention that $1/\nu' = 1$ for $\nu' = 0$.

Let us consider the case $|\alpha' - \delta'| = 0$. Then we have

$$\left| \nu_0 \Phi^{(\nu)} \psi^{(\delta)} \left(\frac{\alpha_{0s}^{k_0+1} \alpha_{1s}^{k_1} \cdots \alpha_{ls}^{k_l}}{\alpha'^{n_\rho}} k^{(\alpha-\delta-\nu)}(\alpha) - \right. \right.$$

$$\left. \left. - \frac{(\alpha_{0s} - \delta_{0s})^{k_0+1} \cdots (\alpha_{ls} - \delta_{ls})^{k_l}}{|\alpha' - \delta'|^{n_\rho}} k^{(\alpha-\delta-\nu)}(\alpha - \delta) \right) \right| \le$$

$$\le \left| \frac{\nu_0}{\nu'} \Phi^{(\nu)} \right| \frac{M_2 M_8 C_2^2 |\alpha' - \delta' - \nu'|}{(1+\delta)^{n+4}(1+|\alpha-\delta-\nu|)^{n+3}} \le$$

$$\le \frac{M_9 |A^{(\nu)}|}{(1+\delta)^{n+4}(1+|\alpha-\delta-\nu|)^{n+2}}.$$

Similar inequalities are obtained in the case of $\alpha' = 0$. Thus, the inequalities (91) are always valid.

Now, let us evaluate the terms of the last two sums in the right-hand side of (84). The terms of each of these sums can be estimated in a similar way; therefore, we consider only those of the first sum. In what follows, we omit the indices and assume for a start that $\alpha' \ne 0$ and $\nu' \ne 0$. Then we have

$$\Phi^{(\nu)} \psi^{*(\delta)} \frac{\alpha_{0s}^{k_0+1} \cdots \alpha_{ls}^{k_l}}{\alpha'^{n_\rho}} k^{(\alpha-\delta-\nu)}(\alpha) =$$

$$= \Phi^{(\nu)} \psi^{*(\delta)} \frac{\alpha_{0s}^{k_0} \cdots \alpha_{ls}^{k_l}}{\alpha'^{n_\rho-1}} k^{(\alpha-\delta-\nu)}(\alpha) \left\{ \frac{\nu_0}{\nu'} + \left(\frac{\alpha_0}{\alpha'} - \frac{\nu_0}{\nu'} \right) \right\}. \quad (92)$$

On the other hand,

$$\frac{\alpha_0}{\alpha'} - \frac{\nu_0}{\nu'} = \frac{\alpha_0 - \nu_0}{\nu'} - \frac{\alpha_0}{\alpha'} \frac{\sum\limits_p (\alpha_{ps} + \nu_{ps})(\alpha_{ps} - \nu_{ps})}{\nu'(\alpha' + \nu')},$$

and therefore,

$$\left| \frac{\alpha_0}{\alpha'} - \frac{\nu_0}{\nu'} \right| \le \frac{|\alpha_0 - \nu_0|}{\nu'} + \frac{1}{\nu'} \sum_p |\alpha_{ps} - \nu_{ps}| \le \frac{2}{\nu'} \sum_p |\alpha_{ps} - \nu_{ps}| \le$$

$$\le \frac{2}{\nu'} \left(\sum_p |\alpha_{ps} - \delta_{ps} - \nu_{ps}| + \sum |\delta_{ps}| \right). \quad (93)$$

Therefore, by virtue of (92) we obtain

$$\left| \Phi^{(\nu)} \psi^{*(\delta)} \frac{\alpha_{0s}^{k_0+1} \dots \alpha_{ls}^{k_l}}{\alpha'^{n_\rho}} k^{(\alpha-\delta-\nu)}(\alpha) \right| \le$$

$$\le \frac{M_{10}}{(1+\delta)^{n+2}(1+|\alpha-\delta-\nu|)^{n+2}} \left(|A^{(\nu)}| + \left| \frac{\Phi^{(\nu)}}{\nu'} \right| \right) . \qquad (94)$$

If $\nu_0 \ne 0$, then $\left| \dfrac{\Phi^{(\nu)}}{\nu'} \right| \le |A^{(\nu)}|$. Set

$$B_j^{(\nu)} = \begin{cases} A_j^{(\nu)} & \text{if } \nu_0 \ne 0, \\ \dfrac{\Phi_j^{(\nu)}}{\nu'} & \text{if } \nu_0 = 0. \end{cases}$$

Then, we obtain from (94)

$$\left| \Phi^{(\nu)} \psi^{*(\delta)} \frac{\alpha_{0s}^{k_0+1} \dots \alpha_{ls}^{k_l}}{\alpha'^{n_\rho}} k^{(\alpha-\delta-\nu)}(\alpha) \right| \le \frac{M_{11}|B^{(\nu)}|}{(1+\delta)^{n+2}(1+|\alpha-\delta-\nu|)^{n+2}} .$$

$$(95)$$

The case $\alpha' = 0$ is trivial, since the left-hand side of (95) is then equal to 0. The estimates (95) are also obvious for $\nu' = 0$; it should only be kept in mind that we have adopted the convention that $\frac{1}{\nu'} = 1$, and therefore, $B^{(\nu)} = \Phi^{(\nu)}$.

7. On the basis of (84), (89) − (91), and (95), we come to the conclusion that

$$\left| \frac{\partial Q_i^{(\alpha)}}{\partial x_0^*} \right| \le M_{12} \frac{\sum_j |Q_j^{(\nu)}| + \sum_{jk} |P_{j(k)}^{(\nu)}| + \sum_{j_1 j_2} |P_{j_1 j_2}^{(\nu)}| + \sum_j |B_j^{(\nu)}|}{(1+\delta)^{n+2}(1+|\alpha-\delta-\nu|)^{n+2}} .$$

It follows from (65), (67), and (76), that similar estimates hold for

$$\left| \frac{\partial P_{j(k)}^{(\alpha)}}{\partial x_0^*} \right|, \quad \left| \frac{\partial P_{j_1 j_2}^{(\alpha)}}{\partial x_0^*} \right|, \quad \left| \frac{\partial Q_i^{(\alpha)}}{\partial x_0^*} \right|, \qquad i = 1, \dots, N'.$$

Moreover, one can do without δ and B_j in the right-hand sides of these inequalities. However, for the sake of symmetry, we preserve δ and B_j. Let us introduce a unified notation R_i ($i = N+1, \dots, N_4$) for the functions $P_{j(k)}, P_{j_1 j_2}, Q_j$. Then, for every i, we get

$$\left| \frac{\partial R_i^{(\alpha)}}{\partial x_0^*} \right| \le M_{13} \sum_{j\delta\nu} \frac{|R_j^{(\nu)}| + |B_j^{(\nu)}|}{(1+\delta)^{n+2}(1+|\alpha-\delta-\nu|)^{n+2}}, \qquad i = 1, \dots, N_4.$$

From the above inequalities we find that

$$\frac{\partial |R_i^{(\alpha)}|^2}{\partial x_0^*} = \frac{\partial\left(R_i^{(\alpha)}\overline{R_i^{(\alpha)}}\right)}{\partial x_0^*} = \overline{R_i^{(\alpha)}}\frac{\partial R_i^{(\alpha)}}{\partial x_0^*} + R_i^{(\alpha)}\frac{\partial \overline{R_i^{(\alpha)}}}{\partial x_0^*} \leq$$

$$\leq 2M_{14}\sum_{j\delta\nu}\frac{|R_j^{(\nu)}||R_i^{(\alpha)}| + |B_j^{(\nu)}||R_i^{(\alpha)}|}{(1+\delta)^{n+2}(1+|\alpha-\delta-\nu|)^{n+2}} \leq$$

$$\leq M_{15}\sum_{\delta\nu}\frac{|R_i^{(\alpha)}|^2}{(1+\delta)^{n+2}(1+|\alpha-\delta-\nu|)^{n+2}} +$$

$$+ M_{15}\sum_{j\delta\nu}\frac{|R_i^{(\nu)}|^2 + |B_j^{(\nu)}|^2}{(1+\delta)^{n+2}(1+|\alpha-\delta-\nu|)^{n+2}}.$$

Summing the above inequalities over all non-negative integer values of $\alpha_0, \alpha_1\ldots,\alpha_n$, and i, we obtain

$$\frac{\partial}{\partial x_0^*}\sum_{i\alpha}|R_i^{(\alpha)}|^2 \leq M_{15}\sum_{i\delta\nu\alpha}\frac{|R_i^{(\alpha)}|^2}{(1+\delta)^{n+2}(1+|\alpha-\delta-\nu|)^{n+2}} +$$

$$+ M_{15}N_4\sum_{j\delta\nu\alpha}\frac{|R_i^{(\nu)}|^2 + |B_j^{(\nu)}|^2}{(1+\delta)^{n+2}(1+|\alpha-\delta-\nu|)^{n+2}}.$$

In the first sum in the right-hand side, let us perform summation, first, over ν and then over δ; in the second sum, we perform summation with respect to α first, and then, with respect to δ. We thus obtain the following inequality:

$$\frac{\partial}{\partial x_0^*}\sum_{i\alpha}|R_i^{(\alpha)}|^2 \leq M_{15}(1+N_4)C^2\sum_{i\alpha}|R_i^{(\alpha)}|^2 + M_{15}N_4C^2\sum_{i\alpha}|B_i^{(\alpha)}|^2, \qquad (96)$$

where C stands for the sum of the series (57).

8. Let us multiply both sides of (79) by

$$\frac{1}{\alpha'}e^{-\frac{\pi i}{L}(\alpha,x)'}, \qquad (\alpha,x)' = \sum_{k=1}^{n}\alpha_k x_k,$$

and integrate the resulting equation over the cube Q_l, assuming that $1/\alpha' = 1$ for $\alpha' = 0$. Then we integrate by parts the terms in the right-hand side containing derivatives in x_k. We thus find that

$$\frac{\partial}{\partial x_0^*}\frac{1}{\alpha'}\int \psi \tilde{F}_{iu}^* e^{-\frac{\pi i}{L}(\alpha,x)'}\, dx_0\ldots dx_n = \frac{1}{\alpha'}\int\left(-2\frac{\partial\psi}{\partial x_0}\,\psi\tilde{F}_{iu}^* +\right.$$

$$+\psi\psi_1 \tilde{F}_{iu}^* - \sum_{jk} \frac{\partial \psi_{ijk}^{(1)}}{\partial x_{ks}} Q_j + \sum_{jk} \frac{\pi i}{L} \alpha_{ks} \psi_{ijk}^{(1)} Q_j + \sum_j \psi_{ij}^{(1)} Q_j +$$

$$+ \sum_{jk} \psi_{ij(k)}^{(1)} P_{j(k)} + \sum_{j_1 j_2} \psi_{ij_1 j_2}^{(1)} P_{j_1 j_2} \Bigg) e^{-\frac{\pi i}{L}(\alpha,x)'} dx_0 \dots dx_n .$$

Letting $B_{i0}^{(\alpha)}$ denote the values of $B_i^{(\alpha)}$ for $\alpha_0 = 0$, we conclude that

$$\frac{\partial B_{i0}^{(\alpha)}}{\partial x_0^*} = \frac{1}{\alpha'} \sum_\nu \left(-2 \left[\frac{\partial \psi}{\partial x_0} \right]^{(\alpha-\nu)} \Phi_i^{(\nu)} + \psi_1^{(\alpha-\nu)} \Phi_i^{(\nu)} + \right.$$

$$+ \sum_{jk} \left[-\frac{\partial \psi_{ijk}^{(1)}}{\partial x_{ks}} + \frac{\pi i}{L} \alpha_{ks} \psi_{ijk}^{(1)} + \psi_{ij}^{(1)} \right]^{(\alpha-\nu)} Q_j^{(\nu)} +$$

$$\left. + \sum_{jk} \psi_{i(k)1}^{(\alpha-\nu)} P_{j(k)}^{(\nu)} + \sum_{j_1 j_2} \psi_{ij_1 j_2 1}^{(\alpha-\nu)} P_{j_1 j_2}^{(\nu)} \right) , \qquad (97)$$

where $[\dots]^{(\alpha-\nu)}$ denote the Fourier coefficients for the functions in brackets. It is convenient to assume that $1/\alpha' = 1$ (resp., $1/\nu' = 1$) for $\alpha' = 0$ (resp., $\nu' = 0$). Therefore, for all values of α_{ks} and ν_{ks} the following inequality is satisfied:

$$\left| \frac{1}{\alpha'} - \frac{1}{\nu'} \right| \le \frac{|\alpha' - \nu'|}{\alpha'\nu'} \le \frac{\sum_k |\alpha_{ks} - \nu_{ks}| \, |\alpha_{ks} + \nu_{ks}|}{\alpha'\nu'(\alpha' + \nu')} \le \frac{\sum_k |\alpha_{ks} - \nu_{ks}|}{\alpha'\nu'} .$$

Then, by virtue of (97) we obtain

$$\left| \frac{\partial B_{i0}^{(\alpha)}}{\partial x_0^*} \right| \le \sum_\nu \frac{M_{16} |\Phi_j^{(\nu)}|}{(1 + |\alpha - \nu|)^{n+3}} \left(1 + \frac{\sum_k |\alpha_{ks} - \nu_{ks}|}{\alpha'} \right) \frac{1}{\nu'} +$$

$$+ \sum_\nu \frac{M_{16}}{(1 + |\alpha - \nu|)^{n+2}} \left(\sum_j |Q_j^{(\nu)}| + \sum_{jk} |P_{j(k)}^{(\nu)}| + \sum_{j_1 j_2} |P_{j_1 j_2}^{(\nu)}| \right) .$$

Consequently,

$$\frac{\partial |B_{i0}^{(\alpha)}|^2}{\partial x_0^*} = \overline{B_{i0}^{(\alpha)}} \frac{\partial B_{i0}^{(\alpha)}}{\partial x_0^*} + B_{i0}^{(\alpha)} \frac{\partial \overline{B_{i0}^{(\alpha)}}}{\partial x_0^*} \le$$

$$\le \sum_\nu \frac{M_{17} |B_{i0}^{(\alpha)}|}{(1 + |\alpha - \nu|)^{n+2}} \left(\sum_j |B_j^{(\nu)}| + \sum_j |Q_j^{(\nu)}| + \sum_{jk} |P_{j(k)}^{(\nu)}| + \right.$$

$$+\sum_{j_1 j_2} |P_{j_1 j_2}^{(\nu)}|\Big) \le M_{18} \sum_{\nu} \frac{|B_{i0}^{(\alpha)}|^2}{(1+|\alpha - \nu|)^{n+2}} +$$

$$+ M_{18} \sum_{\nu} \frac{\sum_j |B_j^{(\nu)}|^2 + \sum_j |Q_j^{(\nu)}|^2 + \sum_{jk} |P_{j(k)}^{(\nu)}|^2 + \sum_{j_1 j_2} |P_{j_1 j_2}^{(\nu)}|^2}{(1+|\alpha - \nu|)^{n+2}}.$$

Let us carry out summation with respect to $i, \alpha_1, \ldots, \alpha_n$ in the left-hand side of the above inequality and in the first sum of its right-hand side; in the second sum in the right-hand side let us perform summation with respect to $i, \alpha_0, \alpha_1, \ldots, \alpha_n$. Then

$$\frac{\partial}{\partial x_0^*} \sideset{}{'}\sum_{i\alpha} |B_{i0}^{(\alpha)}|^2 \le \sum_{i\nu} \sideset{}{'}\sum_{\alpha} \frac{M_{18} |B_{i0}^{(\alpha)}|^2}{(1+|\alpha - \nu|)^{n+2}} +$$

$$+ M_{18} \sum_{i\alpha\nu} \frac{\sum_j |B_j^{(\nu)}|^2 + \sum_j |Q_j^{(\nu)}|^2 + \sum_{jk} |P_{j(k)}^{(\nu)}|^2 + \sum_{j_1 j_2} |P_{j_1 j_2}^{(\nu)}|^2}{(1+|\alpha - \nu|)^{n+2}},$$

where $\sum'$ implies that the sum is taken over α_k such that $k > 0$. In the first sum in the right-hand side, let us first carry out summation with respect to ν, and the second sum is to be first summed with respect to α. Thus we obtain

$$\frac{\partial}{\partial x_0^*} \sum_{i\alpha} |B_{i0}^{(\alpha)}|^2 \le M_{18} C \sideset{}{'}\sum_{i\alpha} |B_{i0}^{(\alpha)}|^2 + M_{18} C \sum_{\alpha} \Big(\sum_i |B_i^{(\alpha)}|^2 + \sum_i |Q_i^{(\alpha)}|^2 +$$

$$+ \sum_{ik} |P_{i(k)}^{(\alpha)}|^2 + \sum_{j_1 j_2} |P_{j_1 j_2}^{(\alpha)}|^2 \Big) \le M_{19} \sum_{\alpha} \Big(\sum_i |B_i^{(\alpha)}|^2 + \sum_i |Q_i^{(\alpha)}|^2 +$$

$$+ \sum_{ik} |P_{i(k)}^{(\alpha)}|^2 + \sum_{j_1 j_2} |P_{j_1 j_2}^{(\alpha)}|^2,$$

where C is the same constant as in (57). Taking the sum of the inequality just obtained with the inequality (96), we find that

$$\frac{\partial}{\partial x_0^*} \Big(\sum_{i\alpha} |R_i^{(\alpha)}|^2 + \sideset{}{'}\sum_{i\alpha} |B_{i0}^{(\alpha)}|^2 \Big) \le M_{20} \Big(\sum_{i\alpha} |R_i^{(\alpha)}|^2 +$$

$$+ \sideset{}{'}\sum_{i\alpha} |B_{i0}^{(\alpha)}|^2 + \sum_{i\alpha} |A_i^{(\alpha)}|^2 \Big). \tag{98}$$

9. Let us multiply both sides of (73) by

$$C_{n_\rho - 1}^{k_0 \ldots k_l} \frac{\alpha_{0s}^{k_0} \cdots \alpha_{ls}^{k_l}}{\alpha'^{n_\rho - 1}},$$

where $C_{n_p-1}^{k_0\ldots k_l}$ is the coefficient by the term $\alpha_{0s}^{2k_0}\cdots\alpha_{ls}^{2k_l}$ in the polynomial $(\alpha_{0s}^2+\cdots+\alpha_{ls}^2)^{n_p-1}$. Let us sum the equalities thus obtained over all i corresponding to one and the same ρ ($u_i^* = p_{\rho k_0\ldots k_l}$), and over all non-negative integers $k_0,\ldots,k_l$ such that $k_0+\cdots+k_l=n_\rho-1$. In the equalities to be summed, the functions Q_i, $\psi_{ij(k)}$, $\psi_{ij_1j_2}$ may vary, while the coefficients $k_{ij}(\alpha)$ for all values of i under consideration remain unchanged; thus, the summation results in

$$\int\sum_j \frac{\alpha_0}{\alpha'}\psi k_{ij}(\alpha)\tilde{F}_{ju}^* \,e^{-\frac{\pi i}{L}(\alpha,x)}\,dx_0\ldots dx_n = \int\left(\sum_j c_{ij}(\alpha)Q_j+\right.$$

$$\left.+\sum_{jk}\psi_{ij(k)}^{(9)}(\alpha)P_{j(k)} + \sum_{j_1j_2}\psi_{ij_1j_2}^{(9)}(\alpha)P_{j_1j_2}\right)e^{-\frac{\pi i}{L}(\alpha,x)}\,dx_0\ldots dx_n =$$

$$=\int\sum_j k_{ij}^{(2)}(\alpha)R_j\,e^{-\frac{\pi i}{L}(\alpha,x)}\,dx_0\ldots dx_n\ .$$

Here every $c_{ij}(\alpha)$ has the form of a sum whose terms are products of the coefficients $C_{n_p-1}^{k_0\ldots k_l}$ multiplied by products of the corresponding fractions α_{ks}/α'; the functions $\psi_{ij(k)}^{(9)}$ and $\psi_{ij_1j_2}^{(9)}$ are linear combinations of $\psi_{ij(k)}$ and $\psi_{ij_1j_2}$ with constant coefficients.

Denote by $k_{ij}(\alpha,0)$ the values of $k_{ij}(\alpha)$ at the point $(x_0^*,0,\ldots,0)$; and let

$$k_{ij}(\alpha) = k_{ij}(\alpha,0) + \tilde{k}_{ij}(\alpha)\ .$$

Then the preceding relations can be rewritten as

$$\sum_j k_{ij}(\alpha,0)\int\frac{\alpha_0}{\alpha'}\Phi_j\,e^{-\frac{\pi i}{L}(\alpha,x)}\,dx_0\ldots dx_n =$$

$$= -\int\sum_j\frac{\alpha_0}{\alpha'}\tilde{k}_{ij}(\alpha)\Phi_j\,e^{-\frac{\pi i}{L}(\alpha,x)}\,dx_0\ldots dx_n +$$

$$+\int\sum_j k_{ij}^{(2)}(\alpha)R_j\,e^{-\frac{\pi i}{L}(\alpha,x)}\,dx_0\ldots dx_n\ . \tag{99}$$

The square matrix $\|k_{ij}(\alpha)\|$ is split into matrices M_s in the same way as the matrix (3). The determinant of the matrix with elements $k_{ij}(\alpha,0)$, where i,j take the values corresponding to the matrix M_s, is equal to the inverse of the determinant of the matrix

$$\left\{\sum_k \frac{\partial\tilde{F}_{iu}^*}{\partial\left\{\frac{\partial^{n_j}u_j^*}{\partial x_{0s}^{k_0}\cdots\partial x_{ls}^{k_l}}\right\}}\,\alpha_{0s}^{k_0}\cdots\alpha_{ls}^{k_l}\right\},$$

taken at the same point $(x_0^*, 0, \ldots, 0)$, with i, j taking the values corresponding to the matrix M_s. Therefore, the determinant of the matrix formed by
$k_{ij}(\alpha, 0)$ always has its absolute value larger than a positive constant. It
follows that equations (99) can be resolved with respect to the integrals in
the left-hand side, which yields

$$A_i^{(\alpha)} = \int \sum_j \frac{\alpha_0}{\alpha'} \tilde{k}_{ij}^{(1)}(\alpha) \Phi_j \, e^{-\frac{\pi i}{L}(\alpha, x)} \, dx_0 \ldots dx_n +$$

$$+ \int \sum_j k_{ij}^{(3)}(\alpha) R_j \, e^{-\frac{\pi i}{L}(\alpha, x)} \, dx_0 \ldots dx_n \, , \qquad (100)$$

$$i = 1, \ldots, N \, ,$$

where $\tilde{k}_{ij}^{(1)}(\alpha)$ and $k_{ij}^{(3)}(\alpha)$ are linear combinations of $\tilde{k}_{ij}(\alpha)$ and $k_{ij}^{(2)}$ with
coefficients depending only on x_0^* and α_{ks}.

As shown above (see (93)), for $\alpha' \neq 0$ and $\nu' \neq 0$, we have

$$\left| \frac{\alpha_0}{\alpha'} - \frac{\nu_0}{\nu'} \right| \leq \frac{2}{\nu'} \sum_p |\alpha_{ps} - \nu_{ps}| \, .$$

This inequality remains valid for all α' and ν' if we set

$$\frac{1}{\alpha'} = 1 \quad \text{for} \quad \alpha' = 0 \, ,$$

$$\frac{1}{\nu'} = 1 \quad \text{for} \quad \nu' = 0$$

(this convention has already been used above). It follows that

$$\left| \frac{\alpha_0}{\alpha'} \sum_\nu \tilde{k}_{ij1}^{(\alpha-\nu)} \Phi^{(\nu)} \right| \leq \sum_\nu \left| \frac{\nu_0}{\nu'} + \left(\frac{\alpha_0}{\alpha'} - \frac{\nu_0}{\nu'} \right) \right| \left| \tilde{k}_{ij1}^{(\alpha-\nu)} \Phi_j^{(\nu)} \right| \leq$$

$$\leq \sum_\nu \left(\frac{|\nu_0|}{\nu'} + \frac{2}{\nu'} \sum_p |\alpha_{ps} - \nu_{ps}| \right) \left| \tilde{k}_{ij1}^{(\alpha-\nu)} \Phi_j^{(\nu)} \right| \leq$$

$$\leq \sum_\nu \left| \tilde{k}_{ij1}^{(\alpha-\nu)} \right| |A_j^{(\nu)}| + 2 \sum_{\nu p}{}' \left| B_{j0}^{(\nu)} \right| |\alpha_{ps} - \nu_{ps}| \left| \tilde{k}_{ij1}^{(\alpha-\nu)} \right| +$$

$$+ 2 \sum_{\nu p}{}'' |A_j^{(\nu)}| \, |\alpha_{ps} - \nu_{ps}| \left| \tilde{k}_{ij1}^{(\alpha-\nu)} \right| \, , \qquad (101)$$

where $\sum_\nu{}'$ implies summation over all integer values of $\nu_1, \ldots, \nu_n$; ν_0 remains
equal to 0, and $\sum_\nu{}''$ is taken over all $\nu_0, \nu_1, \ldots, \nu_n$ with $\nu_0 \neq 0$.

It follows from (100), (101) that

$$\left| A_i^{(\alpha)} \right| \leq \sum_{j\nu} \left| k_{ij3}^{(\alpha-\nu)} \right| \, |R_j^{(\nu)}| + (2n+1) \sum_{j\nu} (1 + |\alpha - \nu|) |\tilde{k}_{ij1}^{(\alpha-\nu)}| \, |B_j^{(\nu)}| \, . \quad (102)$$

For $k_{ij3}^{(\alpha-\nu)}$ the following inequalities are valid

$$\left|k_{ij3}^{(\alpha-\nu)}\right| \leq \frac{M_{21}}{(2n+1)(1+|\alpha-\nu|)^{n+4}} \; .$$

Later on, it will be shown that

$$\left|\tilde{k}_{ij1}^{(\alpha-\nu)}\right| \leq \frac{\varepsilon}{(2n+1)(1+|\alpha-\nu|)^{n+4}} \; , \tag{103}$$

where ε is arbitrarily small, provided that L and L_0^* are small enough (the latter condition is always assumed). Under the assumption that the inequality (103) has been proved, we obtain from (102)

$$\left|A_i^{(\alpha)}\right| \leq M_{21} \sum_{j\nu} \frac{\left|R_j^{(\nu)}\right|}{(1+|\alpha-\nu|)^{n+2}} + \varepsilon \sum_{j\nu} \frac{\left|B_j^{(\nu)}\right|}{(1+|\alpha-\nu|)^{n+2}} \; ,$$

whence

$$\left|A_i^{(\alpha)}\right|^2 \leq 2M_{21}^2 \left(\sum_{j\nu} \frac{R_j^{(\nu)}}{(1+|\alpha-\nu|)^{n+2}} \right)^2 + 2\varepsilon^2 \left(\sum_{j\nu} \frac{\left|B_j^{(\nu)}\right|}{(1+|\alpha-\nu|)^{n+2}} \right)^2 \; .$$

Summing the above inequalities with respect to α and i, we obtain

$$\sum_{\alpha i} |A_i^{(\alpha)}|^2 \leq 2M_{21}^2 \sum_{ijs\alpha\delta\nu} \frac{|R_j^{(\nu)}|\,|R_s^{(\delta)}|}{(1+|\alpha-\nu|)^{n+2}(1+|\alpha-\delta|)^{n+2}} +$$

$$+2\varepsilon^2 \sum_{ijs\alpha\delta\nu} \frac{|B_j^{(\nu)}|\,|B_s^{(\nu)}|}{(1+|\alpha-\nu|)^{n+2}(1+|\alpha-\delta|)^{n+2}} \leq$$

$$\leq \sum_{ijs\alpha\delta\nu} \frac{M_{21}^2|R_j^{(\nu)}|^2 + \varepsilon^2|B_j^{(\nu)}|^2}{(1+|\alpha-\nu|)^{n+2}(1+|\alpha-\delta|)^{n+2}} +$$

$$+ \sum_{ijs\alpha\delta\nu} \frac{M_{21}^2|R_s^{(\delta)}|^2 + \varepsilon|B_s^{(\delta)}|^2}{(1+|\alpha-\nu|)^{n+2}(1+|\alpha-\delta|)^{n+2}} \; .$$

In the first term in the right-hand side of the last inequality, we perform summation, first, with respect to δ, and then with respect to α; in the second sum, we perform summation with respect to ν, first, and then with respect to α. This yields

$$\sum_{i\alpha} |A_i^{(\alpha)}|^2 \leq 2M_{21}^2 C^2 N_4 N \sum_{j\alpha} |R_j^{(\alpha)}|^2 + 2\varepsilon^2 C^2 N^2 \sum_{j\alpha} |B_j^{(\alpha)}|^2 \; .$$

Let us choose ε so small that

$$2\varepsilon^2 C^2 N^2 < 1 .$$

Then, it follows from the previous inequality that

$$\sum_{i\alpha} |A_i^{(\alpha)}|^2 \leq \frac{2M_{21}^2 C^2 N_4 N \sum_{j\alpha} |R_j^{(\alpha)}|^2 + 2\varepsilon^2 C^2 N^2 \sum_{j\alpha}' |B_{i0}^{(\alpha)}|^2}{1 - 2\varepsilon^2 C^2 N^2} .$$

Let us substitute the estimate obtained for $\sum_{i\alpha} |A_i^{(\alpha)}|^2$ into the right-hand of (98); then we get

$$\frac{\partial}{\partial x_0^*} \left(\sum_{i\alpha} |R_i^{(\alpha)}|^2 + \sum_{i\alpha}' |B_{i0}^{(\alpha)}|^2 \right) \leq M_{22} \left(\sum_{i\alpha} |R_i^{(\alpha)}|^2 + \sum_{i\alpha}' |B_{i0}^{(\alpha)}|^2 \right) .$$

Since

$$\sum_{i\alpha} |R_i^{(\alpha)}|^2 + \sum_{i\alpha}' |B_{i0}^{(\alpha)}|^2 = 0 \quad \text{for} \quad x_0^* = 0 ,$$

it follows that the above sum vanishes over the entire parallelepiped $R_{LL_0^*}$ for sufficiently small L, L_0^*, and $x_0^* \geq 0$.

In a similar manner, this fact can be established for the parallelepiped $R_{LL_0^*}$ and $x_0^* < 0$. By virtue of the Parseval's relation, we can conclude that all Q_i vanish identically in $R_{LL_0^*}$, and therefore, u_i are analytic functions of $x_0 + ix_0^*$.

10. Now it remains to establish the inequalities (103) in $R_{LL_0^*}$ with ε arbitrarily small, provided that L and L_0^* are sufficiently small. To this end, let it be observed, first of all, that the Fourier coefficients for a function $f(x_0, \ldots, x_n)$ defined in cube Q_L are given by

$$f^{(\alpha_0 \ldots \alpha_n)} = \frac{1}{(2L)^{n+1}} \int_{Q_L} f(x_0, \ldots, x_n) e^{-\frac{\pi i}{L}(\alpha, x)} dx_0 \ldots dx_n , \qquad (104)$$

where $\alpha_0, \ldots, \alpha_n$ take integer values (positive, negative, and 0). This definition immediately implies that

$$|f^{(\alpha_0 \ldots, \alpha_n)}| \leq \max |f(x_0, \ldots, x_n)| . \qquad (105)$$

If the function $f(x_0, \ldots, x_n)$ is $2L$-periodic with respect to all its variables x_k (this condition holds in all our considerations), then we can integrate by parts with respect to x_k in the right-hand side of (104); after this integration the boundary terms will be mutually reduced. Let α_M be any one of $\alpha_0, \ldots, \alpha_n$ with the largest absolute value. Assuming $|\alpha_M| \geq 1$, let us integrate by parts, $n+4$ times, the right-hand side of (104) with respect to the variable x_M. Then

$$\left|f^{(\alpha_0,\dots,\alpha_n)}\right| \le \frac{1}{(2L)^{n+1}} \left(\frac{L}{\pi|\alpha_M|}\right)^{n+4} \left|\int_{Q_L} \frac{\partial^{n+4}f}{\partial x_M^{n+4}} e^{-\frac{\pi i}{L}(\alpha,x)}\, dx_0 \dots dx_n\right| \le$$

$$\le \left(\frac{L}{\pi|\alpha_M|}\right)^{n+4} \max\left|\frac{\partial^{n+4}f}{\partial x_M^{n+4}}\right| \le \left(\frac{2L\sqrt{n+1}}{\pi(\alpha+1)}\right)^{n+4} \max\left|\frac{\partial^{n+4}f}{\partial x_M^{n+4}}\right|. \quad (106)$$

Because of (105) and (106), the inequalities of type (103) hold for the Fourier coefficients of the function f, provided that

$$\max|f| \quad \text{and} \quad L^{n+4}\max|D^{(n+4)}f|$$

are sufficiently small; here $D^{(n+4)}f$ denotes any of the $(n+4)$-th order derivatives of f with respect to the variables x_k $(k = 0, 1, \dots, n)$. Therefore, in order to prove the inequalities (103), it suffices to show that

$$\left|\tilde{k}_{ij}^{(1)}\right| < \varepsilon, \quad L^{n+4}\left|D^{(n+4)}\tilde{k}_{ij}^{(1)}\right| < \varepsilon \quad \text{in} \quad R_{LL_0^*}, \quad (107)$$

for all i, j, provided that the constants L, L_0^*, M in relations $(\widetilde{11})$ of §1 are small enough (as usual, we denote by ε an arbitrary small real number). Since $\tilde{k}_{ij}^{(1)}$ continuously depends on x_0^*, it suffices to establish (107) for $x_0^* = 0$. Observe, also, that $\tilde{k}_{ij}^{(1)}$ are linear combinations of $\tilde{k}_{ij}(\alpha)$ with bounded coefficients depending only on $x_0^*, \alpha_0, \dots, \alpha_n$. Therefore, in order to prove (107), it suffices to establish similar inequalities for $\tilde{k}_{ij}(\alpha)$. Moreover, $k_{ij}(\alpha)$ have the form of fractions whose numerators and denominators are polynomials having $\left\{\tilde{a}_{ij}^{*(k_0\dots k_n)}\right\}_u$ as arguments, the coefficients of the polynomials being independent of x_k, x_0^*, but depending on α_k. The subscript u by the right brace is meant to show that inside the braces all derivatives of u_j^* $(j = 1, \dots, N)$ are replaced by the respective u_j^* $(j = N+1, \dots, N_1)$. For all values of α_k and x_k, the fractions representing $k_{ij}(\alpha)$ have denominators whose absolute values are larger than a positive constant. Taking into account that the constant M in the inequalities $(\widetilde{11})$ of §1 can be chosen arbitrarily small, we can assume, while estimating the values of $\tilde{k}_{ij}(\alpha) = k_{ij}(\alpha) - k_{ij}(\alpha, 0)$ and their derivatives in x_k up to the order $n + 4$, that these quantities depend only on the variables x_k explicitly entering into $\left\{\tilde{a}_{ij}^{*(k_0\dots k_n)}\right\}$, and ignore their dependence on x_k through the agency of the functions u_j^* $(j = 1, \dots, N)$; moreover, we can also ignore the dependence on x_k through the agency of the derivatives of the functions φ, since these derivatives are present only as factors by u_j^*. Thus, it remains to estimate $\tilde{k}_{ij}$ and their derivatives, assuming that all functions u_j^* vanish identically, and $x_0^* = 0$.

For $x_0^* = 0$ we have

$$\tilde{a}_{ij}^*(x_0^*, x_0, \dots, x_n; \dots, u_j^*, \dots) \equiv a_{ij}^*(x_0^*, 0, \dots, 0; \dots, u_j^*, \dots) +$$

$$+ \left\{ a_{ij}^*(x_0^*, x_0, \ldots, x_n; \ldots, u_j^*, \ldots) - \right.$$

$$\left. - a_{ij}^*(x_0^*, 0, \ldots, 0; \ldots u_j^*, \ldots) \right\} \varphi_1(x_0, \ldots, x_n) . \qquad (108)$$

If $u_j^* \equiv 0$ and L is sufficiently small, then the absolute value of the expression in curly braces in (108) can be made arbitrarily small over $R_{LL_0^*}$. Thus, the absolute values of all $\tilde{k}_{ij}$ can be made arbitrarily small on Q_L for $x_0^* = 0$, provided that L is chosen sufficiently small and $u_j^* \equiv 0$. The above arguments show that a similar result holds over the whole of the parallelepiped $R_{LL_0^*}$, if L, L_0^*, and M are sufficiently small.

Next, we turn to the evaluation of the absolute values of the products $L^{n+4} \left| D^{(n+4)} \tilde{k}_{ij} \right|$. The derivatives $D^{(n+4)} \tilde{k}_{ij}$, or equivalently, $D^{(n+4)} k_{ij}$, have the form of polynomials (with bounded coefficients) whose arguments are the derivatives in x_k of the functions $\tilde{a}_{ij}^{*(k_0 \ldots k_n)}$. The orders of the derivatives in x_k entering any single term of this polynomial add up to $(n+4)$. Therefore, in order to prove that the product $L^{n+4} \left| D^{(n+4)} k_{ij} \right|$ can be made arbitrarily small, if $L_0^* = M = 0$ and L is chosen small enough, it suffices to verify that for $x_0^* = M = 0$ and small enough L, every product of the form

$$L^p \left| D^{(p)} \tilde{a}_{ij}^{*(k_0 \ldots k_n)} \right| \qquad (1 \le p \le n+4)$$

can be made arbitrarily small in cube Q_L. By virtue of (108), for $x_0^* = 0$ and $M = 0$, we have

$$\tilde{a}_{ij}^{*(k_0 \ldots k_n)} = f(0, \ldots, 0) + [f(x_0, \ldots, x_n) - f(0, \ldots, 0)] \varphi_1(x_0, \ldots, x_n) ,$$

where the functions $f(x_0, \ldots, x_n)$ are independent of L, bounded and continuous, together with their derivatives in x_k up to the order $n + 4$. The function $\varphi_1(x_0, \ldots, x_n)$ depends on L; to indicate this dependence, we write $\varphi_1(L, x_0, \ldots, x_n)$ instead of $\varphi_1(x_0, \ldots, x_n)$. Set

$$\varphi_1(L, x_0, \ldots, x_n) = \varphi_1 \left(1, \frac{x_0}{L}, \ldots, \frac{x_n}{L} \right) .$$

Then, all the conditions imposed on φ_1 in §1 will be satisfied.

Obviously, in order to differentiate $\tilde{a}_{ij}^{*(k_0 \ldots k_n)}$ with respect to x_k ($k = 0, \ldots, n$), it is sufficient to differentiate the product

$$[f(x_0, \ldots, x_n) - f(0, \ldots, 0)] \, \varphi_1 \left(1, \frac{x_0}{L}, \ldots, \frac{x_n}{L} \right) .$$

Any p-th order derivative of this product, with respect to $x_0, \ldots, x_n$, is equal to a sum of terms having the form of a p_1-th order derivative of the function

$$f(x_0, \ldots, x_n) - f(0, \ldots, 0)$$

multiplied by a $(p - p_1)$-th order derivative of $\varphi_1\left(1, \frac{x_0}{L}, \ldots, \frac{x_n}{L}\right)$, where $0 \le p_1 \le p$. Let us multiply each of these terms by L^p, so that the first factor is multiplied by L^{p_1}, and the second one by L^{p-p_1}. Any $(p - p_1)$-th order derivative (with respect to x_k $(k = 0, \ldots, n)$) of $\varphi_1\left(1, \frac{x_0}{L}, \ldots, \frac{x_n}{L}\right)$ multiplied by L^{p-p_1} remains bounded, whereas any p_1-th order derivative of the difference $f(x_0, \ldots, x_n) - f(0, \ldots, 0)$ multiplied by L^{p_1} converges to 0 on Q_L as $L \to 0$. For $p_1 > 0$, the latter statement follows from the fact that $f(x_0, \ldots, x_n)$ is independent of L, and therefore all its derivatives under consideration remain bounded. If $p_1 = 0$, then the oscillation of $f(x_0, \ldots, x_n)$ on Q_L tends to 0 as $L \to 0$, because of the continuity of f.

§6. Two Remarks

1. In the case of linear systems, the above considerations can be substantially simplified. Remaining within the general framework of the above exposition, one can establish a theorem on the analyticity of solutions, with respect to x_0, under weaker assumptions on the solutions and the left-hand sides of the equations.

Theorem. *Consider a system of the form*

$$\sum_j \sum_k a_{ij}^{(k_0\ldots k_n)}(x_0, \ldots, x_n) \frac{\partial^k u_j}{\partial x_0^{k_0} \cdots \partial x_n^{k_n}} + f_i(x_0, \ldots, x_n) = 0 \,,$$

$$i, j = 1, \ldots, N \,.$$

where $\sum_k$ implies summation with respect to all non-negative integers k_σ such that $k_0 + \cdots + k_n \le n_j$. Then any solution u_i $(i = 1, \ldots, N)$ of the above system, which possesses continuous derivatives up to the order $n_i + \left[\frac{n+1}{2}\right] + 2$, is analytic in x_0, provided that the following conditions are satisfied:

1) *all coefficients $a_{ij}^{(k_0\ldots k_n)}$ and functions f_i are analytic in x_0;*

2) *the coefficients $a_{ij}^{(k_0\ldots k_n)}$ such that $\sum k_\sigma = n_j$ possess continuous derivatives in $x_1, \ldots, x_n$ up to the order $\left[\frac{n+2}{2}\right] + n + 6$;*

3) *the rest of $a_{ij}^{(k_0\ldots k_n)}$ and the functions f_i possess continuous derivatives in $x_1, \ldots, x_n$ up to the order $\left[\frac{n+1}{2}\right] + n + 5$;*

4) *the square matrix*

$$\left\| \sum_{(k)} a_{ij}^{(k_0\ldots k_n)} \alpha_0^{k_0} \cdots \alpha_n^{k_n} \right\| \,,$$

where $\sum_{(k)}$ implies summation over all non-negative integers $k_0, \ldots, k_n$
such that $k_0 + \cdots + k_n = n_j \; (n_j > 0)$, has the form

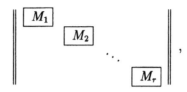

where the elements outside every M_s are identically equal to zero.

Suppose that, for any s, among the coefficients entering into the matrix
M_s the only ones that do not vanish identically are those having all indices
$k_{0s}, \ldots, k_{ls} \; (k_{0s} = k_0)$ different from zero and the rest of k_j equal to zero. It
is then required that the determinant of M_s may never vanish for any real
values of $\alpha_{ks} \; (k = 0, \ldots, l)$ such that $\sum \alpha_{ks}^2 = 1$. Moreover, for the values of
j corresponding to the matrix M_s, it is not only the coefficients $a_{ij}^{(k_0 \ldots k_n)}$ with
$\sum k_\sigma = n_j$ that must vanish identically, but also every coefficient $a_{ij}^{(k_0 \ldots k_n)}$
having a non-zero index from among $k_0, \ldots, k_n$ other than $k_{0s}, k_{1s}, \ldots, k_{ls}$.

2. It follows from the main theorem established in the preceding sections
that any sufficiently smooth solution of an elliptic system of type (1), with
F_i analytic in all arguments, is also an analytic function of any linear combination of its variables with real constant coefficients, and therefore it is analytic in all $x_k \; (k = 0, \ldots, n)$ jointly. Hence, according to the Kowalevskaya
theorem, the Cauchy problem for a normal system of the form

$$\frac{\partial u_i}{\partial t} = \Phi_i(t, x_0, \ldots, x_n; u_1, \ldots, u_N; \ldots), \qquad i = 1, 2, \ldots, N, \quad (109)$$

admits a unique solution analytic in all x_k, if the initial values for the
functions u_i at $t = 0$ are chosen to coincide with a solution of the elliptic
system (1), and Φ_i, F_i are analytic functions of all their arguments. Thus
we see that in the matter of solving the Cauchy problem for normal systems
with analytic right-hand sides, the property of the initial values to satisfy
an analytic elliptic system of type (1) is equivalent to their being analytic
functions, i.e., their satisfying the system of Cauchy–Riemann with respect
to x_k and $x_k^* \; (k = 0, 1, \ldots, n)$.

    In general, elliptic systems possess numerous properties of the Cauchy–
Riemann system, and functions satisfying elliptic systems in many respects
behave like analytic ones. The author intends to give a systematic treatment
of these properties in a separate publication [9], and therefore will limit
himself here to a few remarks:

In §4 we have constructed the system of functions $u_i(x_0^*, x_0, \ldots, x_n)$ $(i = 1, \ldots, N)$ defined in the parallelepiped $R_{0.6LL_0^*}$, on the basis of the functions $v_i(x_0, \ldots, x_n)$ defined in the cube Q_L for $x_0^* = 0$ and satisfying system (1). The functions u_i coincide with $v_i(x_0, \ldots, x_n)$ $(i = 1, \ldots, N)$ in the cube $Q_{0.6L}$ and are analytic in $R_{0.6LL_0^*}$ with respect to $x_0 + x_0^*$. It has been shown that the functions $u_i(x_0^*, x_0, \ldots, x_n)$ $(i = 1, \ldots, N)$ and their derivatives in $x_0, \ldots, x_n$ up to the order $(\delta + n_i - 1)$ have absolute values bounded by a constant M on the entire parallelepiped $R_{0.6LL_0^*}$, provided that $v_i(x_0, \ldots, x_n)$ $(i = 1, \ldots, N)$ and their derivatives up to the order $\delta + \left[\frac{n+2}{2}\right] + n_i$ have absolute values bounded by M_0 in the cube Q_L. The functions $u_i(x_0^*, x_0, \ldots, x_n)$ have been shown to satisfy the Cauchy–Riemann equations in $R_{0.6LL_0^*}$ with respect to x_0 and x_0^*; therefore, all their derivatives

$$\frac{\partial^{k+1} u_i}{\partial x_0^* \partial x_0^{k_0} \cdots \partial x_n^{k_n}}, \qquad k = \sum k_\sigma \leq \delta - 2 + n_i,$$

have absolute values bounded by M, too. Consequently, for any positive ε there is an $\eta > 0$, depending only on ε, M, and system (1), such that the following result is valid. Let

$$v_i^{(1)}(x_0, \ldots, x_n) \quad \text{and} \quad v_i^{(2)}(x_0, \ldots, x_n)$$

be two solutions of system (1) satisfying in Q_L for $x_0^* = 0$ the following inequalities:

$$\left| \frac{\partial^k \left(v_i^{(2)}(x_0, \ldots, x_n) - v_i^{(1)}(x_0, \ldots, x_n) \right)}{\partial x_0^{k_0} \cdots \partial x_n^{k_n}} \right| \leq \eta, \tag{110}$$

$$\left| \frac{\partial^k v_i^{(p)}}{\partial x_0^{k_0} \cdots \partial x_n^{k_n}} \right| \leq M_0, \quad p = 1, 2, \; [8]$$

for any i and any non-negative integers $k_0, \ldots k_n$ such that

$$\sum_{k_\sigma} k_\sigma \leq \delta + \left[\frac{n+1}{2}\right] + n_i.$$

Then the functions

$$u_i^{(1)}(x_0^*, x_0, \ldots, x_n) \quad \text{and} \quad u_i^{(2)}(x_0^*, x_0, \ldots, x_n), \qquad i = 1, \ldots, N,$$

constructed from the given ones, satisfy the inequalities

$$\left| u_i^{(2)}(x_0^*, x_0, \ldots, x_n) - u_i^{(1)}(x_0^*, x_0, \ldots, x_n) \right| \leq \varepsilon \quad \text{in} \quad R_{0.6LL_0^*}. \tag{111}$$

---

[8]This condition implies no loss of generality, as mentioned in the Introduction.

Otherwise, we can construct an infinite sequence of functions

$$u_{is}^{(2)}(x_0^*, x_0, \ldots, x_n) \, , \qquad i = 1, 2, \ldots, N \, , \quad s = 1, 2, \ldots \, , \qquad (112)$$

defined in $R_{0.6LL_0^*}$, uniformly bounded (with respect to s), together with their derivatives of the form

$$\frac{\partial^k u_{is}^{(2)}}{\partial x_0^{k_0} \partial x_1^{k_1} \cdots \partial x_n^{k_n}} \, , \qquad \frac{\partial^k u_{is}^{(2)}}{\partial x_0^* \partial x_0^{k_0-1} \partial x_1^{k_1} \cdots \partial x_n^{k_n}} \, ,$$

$$\sum_\sigma k_\sigma \le \delta + n_i - 1 \, , \quad i = 1, 2, \ldots, N \, , \quad s = 1, 2, \ldots \, ,$$

and satisfying the inequalities

$$\left| u_{is}^{(2)}(0, x_0, \ldots, x_n) - u_i^{(1)}(0, x_0, \ldots, x_n) \right| \le \frac{1}{s} \, , \qquad i = 1, \ldots, N \, ; \quad (113)$$

at the same time, there exists a point of the parallelepiped $R_{0.6LL_0^*}$ such that, for all s, at least one of the differences

$$u_{is}^{(2)}(x_0^*, x_0, \ldots, x_n) - u_i^{(1)}(x_0^*, x_0, \ldots, x_n) \, , \qquad i = 1, \ldots, N \, ,$$

has absolute value larger than a positive constant ε. Then, according to Arzelà's theorem, from the sequence of systems consisting of the functions (112) we can extract a subsequence uniformly convergent on $R_{0.6LL_0^*}$, together with the derivatives $\partial u_{is}^{(2)}/\partial x_0$. Since the functions $u_{is}^{(2)}$ satisfy the Cauchy–Riemann equations with respect to x_0 and x_0^*, the elements of this subsequence have partial derivatives in x_0^* uniformly convergent on $R_{0.6LL_0^*}$. It follows that the limit functions, denoted by $u_i^{(2)}(x_0^*, x_0, \ldots, x_n)$, satisfy the Cauchy–Riemann equations with respect to x_0 and x_0^*. Because of (113), every difference

$$u_i^{(2)}(x_0^*, x_0, \ldots, x_n) - u_i^{(1)}(x_0^*, x_0, \ldots, x_n)$$

must vanish for $x_0^* = 0$. Therefore, according to a well-known result from the theory of analytic functions, every one of these differences must vanish identically. On the other hand, it is impossible that all differences $u_i^{(2)} - u_i^{(1)}$ vanish identically on $R_{0.6LL_0^*}$, since for every s, at least one of the functions $u_{is}^{(2)}$ $(i = 1, \ldots, N)$ differs from $u_i^{(1)}$ by a quantity larger than $\varepsilon > 0$ at some point of $R_{0.6LL_0^*}$.

Thus, the inequalities (111) have been established under the assumption (110). It follows that if two solutions

$$u_i^{(1)}(x_0, \ldots, x_n) \qquad \text{and} \qquad u_i^{(2)}(x_0, \ldots, x_n)$$

of the same elliptic system (1) with analytic left-hand sides, as well as their respective partial derivatives up to the order $\delta + \left[\frac{n+2}{2}\right] + n_i$, differ by a sufficiently small quantity in a real domain G_0 in the $(x_0, \ldots, x_n)$-space, then the difference of the extensions of these solutions to a complex domain in the $(x_0 + ix_0^*, x_1 + ix_1^*, \ldots, x_n + ix_n^*)$-space is also sufficiently small. Then, the usual proof of S. Kowalevskaya's theorem allows us to claim that two solutions of system (109) taking the initial values $u_i^{(1)}(x_0, \ldots, x_n)$ and $u_i^{(2)}(x_0, \ldots, x_n)$ at $t = 0$ differ by a sufficiently small quantity in a fixed neighborhood of the domain G_0. It is easy to see that the latter statement is still valid if the functions $u_i(x_0, \ldots, x_n)$ $(i = 1, \ldots, N)$ satisfy two different analytic elliptic systems of type (1) (and not the same system, as above), provided that the left-hand sides of these systems and their respective partial derivatives of sufficiently high orders, with respect to all their arguments, are sufficiently close to one another.

This result seems to be especially interesting, since, as shown by J. Hadamard (see [5], pp. 39–40), there exist equations and systems of initial functions such that the corresponding Cauchy problem admits two solutions which differ by an arbitrarily large amount in an arbitrarily small neighborhood of the plane $t = 0$, no matter how close the initial functions and their respective derivatives of any given order may be in a real domain G_0, for $t = 0$.

§7. Systems whose Characteristic Equation has Non-Zero Roots

Consider the following system

$$\frac{\partial^{n_i} u_i}{\partial x_0^{n_i}} = \sum_{j,k} a_{ij}^{(k_0 \ldots k_n)} \frac{\partial^k u_j}{\partial x_0^{k_0} \cdots \partial x_n^{k_n}} + f_i(x_0, x_1, \ldots, x_n), \qquad (114)$$

$$i, j = 1, \ldots, N, \quad k = \sum k_\sigma \leq n_j, \quad k_\sigma \geq 0,$$

$$n_j > 0, \quad k_0 < n_j,$$

with constant coefficients $a_{ij}^{(k_0 \ldots k_n)}$ and the functions $f_i(x_0, x_1, \ldots, x_n)$ analytic in all their arguments.

Theorem. *System (114) admits solutions non-analytic[9] in x_0 but possessing continuous derivatives of any order, if the determinant of the square matrix*

---

[9]See Remark 2.

$$\left\| \begin{matrix} \lambda^{n_1} & & & \\ & \lambda^{n_2} & & \\ & & \ddots & \\ & & & \lambda^{n_N} \end{matrix} \right\| - \left\| \sum_{(k)} a_{ij}^{(k_0 \ldots k_n)} \lambda^{k_0} \alpha_1^{k_1} \cdots \alpha_n^{k_n} \right\| \qquad (115)$$

(the sum $\sum_{(k)}$ is taken over all non-negative k_σ such that $k_0 + \cdots + k_n = n_j$)
admits at least one real non-zero root λ for some real values of $\alpha_1, \ldots, \alpha_n$.

Corollary. *If the determinant of the matrix (115) for system (114) has
at least one real root λ, which may be equal to 0 for some real $\alpha_1, \ldots, \alpha_n$,
$\sum \alpha_k^2 \neq 0$, then system (114) admits solutions non-analytic with respect to
all variables x_k $(k = 0, 1, \ldots, n)$ simultaneously.*

Indeed, for a non-zero root λ, this result follows directly from the above
theorem. If $\lambda = 0$, then we can rotate the axis Ox_0 in the $(x_0, x_1, \ldots, x_n)$-
space so as to obtain a new system of type (114) having constant coefficients
and the characteristic matrix whose determinant admits a real non-zero
root; therefore, the above theorem can be applied once again.
    In order to make a system of type (1) resolvable with respect to the
highest order derivatives in some direction, it suffices to assume that the
determinant of the characteristic matrix for that system does not vanish
identically as a function of $\alpha_0, \alpha_1, \ldots, \alpha_n$ [4].

Proof of the Theorem. Let it be observed, first of all, that system (114)
admits solutions analytic with respect to all variables x_k; this fact follows
from Kowalevskaya's theorem. Therefore, to establish the existence of a
solution non-analytic in x_0, but possessing derivatives of arbitrary order,
it suffices to prove the existence of such a solution for the homogeneous
system of type (114); the latter system will be referred to as (114′).
    Suppose that for

$$\alpha_k = a_k, \qquad k = 1, 2, \ldots, n,$$

where all a_k are real, the determinant of the matrix (115) has a real root
$\lambda = q \neq 0$. It is obvious that $a_j \neq 0$ for some j. Let us change the variables,
setting

$$y_1 = \sum_{k=1}^{n} a_k x_k + q x_0, \qquad y_0 = x_0,$$

and taking as

$$y_2, y_3, \ldots, y_n$$

arbitrary linear combinations of $x_1, x_2, \ldots, x_n$ with constant coefficients
such that the corresponding linear transformation is non-degenerate.
    Assume that, after the above transformation, system (114′) takes the
form

$$\frac{\partial^{n_i} u_i}{\partial y_0^{n_i}} = \sum_{j,k} b_{ij}^{(k_0 \ldots k_n)} \frac{\partial^k u_j}{\partial y_0^{k_0} \partial y_1^{k_1} \cdots \partial y_n^{k_n}} , \tag{116}$$

$$i, j = 1, 2, \ldots, N , \quad k_s \geq 0 , \quad k = \sum k_s \leq n_j ,$$

$$n_i > 0 , \quad k_0 < n_j .$$

The characteristic matrix of the above system is

$$\left\| \begin{matrix} \lambda^{n_1} & & & \\ & \lambda^{n_2} & & \\ & & \ddots & \\ & & & \lambda^{n_N} \end{matrix} \right\| - \left\| \sum_{(k)} b_{ij}^{(k_0 \ldots k_n)} \lambda^{k_0} \alpha_1^{k_1} \cdots \alpha_n^{k_n} \right\| .$$

Obviously, for $\alpha_k = 0$ $(k = 2, 3, \ldots, n)$, this matrix has the form

$$\left\| \begin{matrix} (\lambda + \alpha_1 q)^{n_1} & & & \\ & (\lambda + \alpha_1 q)^{n_2} & & \\ & & \ddots & \\ & & & (\lambda + \alpha_1 q)^{n_N} \end{matrix} \right\| -$$

$$- \left\| \sum_{(k)} a_{ij}^{(k_0 \ldots k_n)} (\lambda + \alpha_1 q)^{k_0} a_1^{k_1} \cdots a_n^{k_n} \alpha_1^{n_j - k_0} \right\| .$$

This statement can be easily verified, if we take into account that the characteristic matrix of system (114′) has the following structure: Let us preserve in system (114′) only the terms that involve the highest order derivatives and transfer these terms to the left-hand sides; after replacing the functions u_j by the expressions

$$c_j \exp \left\{ \sum_{i=1}^n \alpha_k x_k + \lambda x_0 \right\} , \quad j = 1, \ldots, N ,$$

and dividing the resulting equations by $\exp \left\{ \sum_{i=1}^n \alpha_k x_k + \lambda x_0 \right\}$, consider the coefficients by c_j; it is these coefficients that form the characteristic matrix for (114′). The characteristic matrix of system (116) for $\alpha_2 = \alpha_3 = \cdots = \alpha_n = 0$ is formed by the coefficients of c_j obtained after we discard all terms but those of the highest order, transfer these terms to the left-hand sides, substitute for u_j the expressions

$$c_j \exp(\lambda y_0 + \alpha_1 y_1) = c_j \exp \left\{ (\lambda + \alpha_1 q) x_0 + \alpha_1 \sum_{k=1}^n a_k x_k \right\} ,$$

$$j = 1, \ldots, N ,$$

and then divide the resulting equations by $\exp(\lambda_0 y_0 + \alpha_1 y_1)$. The determinant of this matrix has the root $\lambda = 0$ for $\alpha_1 = 1$, $\alpha_2 = \alpha_3 = \cdots = \alpha_n = 0$.

Having made these preliminary remarks, we are in a position to construct the solution of system (116) corresponding to the solution of system (114') whose existence we are going to prove. To this end we apply the Fourier method, as it was done in our paper on the linear Cauchy problem (see [6]). Let us seek the solution in the form

$$ u_i(y_0, y_1) = \sum_s c_{is}\, e^{\alpha_1^{(s)} y_1 + \lambda^{(s)} y_0}\,, \qquad i = 1, \ldots, N\,, \tag{117} $$

where s takes positive integer values. In the case of system (114') having real coefficients, both the real part and the imaginary part of the above series are solutions of equations (114'). Let us define c_{is}, $\lambda^{(s)}$, and $\alpha_1^{(s)}$ in such a way that the functions

$$ c_{is}\, e^{\alpha_1^{(s)} y_1 + \lambda^{(s)} y_0}\,, \qquad i = 1, \ldots, N\,, $$

represent a solution of system (116). Then $\lambda = \lambda^{(s)}$, for $\alpha = \alpha^{(s)}$, must be a root of the determinant $D(\lambda, \alpha)$ of the matrix

$$ \left\| \begin{array}{cccc} (\lambda + \alpha q)^{n_1} & & & \\ & (\lambda + \alpha q)^{n_2} & & \\ & & \ddots & \\ & & & (\lambda + \alpha q)^{n_N} \end{array} \right\| $$

$$ - \left\| \sum_k a_{ij}^{(k_0 \ldots k_n)} (\lambda + \alpha q)^{k_0} a_1^{k_1} \cdots a_n^{k_n} \alpha^{n_j - k_0} \right\|\,, \tag{118} $$

where the sum $\sum_k$ is taken over all non-negative integers k_σ (not only over those whose sum is equal to n_j). For the sake of brevity, we write α instead of α_1. The constants c_{is} must satisfy a linear homogeneous system whose coefficients coincide with the elements of the matrix (118); the latter system will be referred to as (119). We obviously have

$$ D(\lambda, \alpha) = d(\lambda, \alpha)\lambda^r + m(\lambda, \alpha)\,, $$

where $d(\lambda, \alpha)$ is a homogeneous polynomial of λ and α, its degree being equal to $\sum_i n_i - r = N^* - r$; $d(0, \alpha) \neq 0$ for $\alpha \neq 0$; $m(\lambda, \alpha)$ is a polynomial of degree less than N^* and arguments λ, α. Set $\lambda = \mu\alpha$ in the equation

$$ D(\lambda, \alpha) = 0\,, \tag{120} $$

and divide the resulting equation by α^{N^*}. Since any polynomial of the argument μ, with the highest order term μ^{N^*}, has all its roots depending continuously on the coefficients, we see that the equation in question admits a root λ such that

$$\lambda(\alpha) = o(\alpha) \quad \text{as} \quad |\alpha| \to \infty .$$

According to the Puiseux theorem, this root can be expanded as

$$\lambda(\alpha) = \alpha^{\frac{k}{p}} \sum_{s=0}^{\infty} c_s \alpha^{-\frac{s}{p}} ,$$

for large values of $|\alpha|$; here p is a positive integer; k and s are non-negative integers; $c_0 \neq 0$, $k < p$.

Let us consider consecutively the following two cases: 1) $k = 0$ and 2) $0 < k < p$, $c_0 \neq 0$.

1) The case $k = 0$. The root $\lambda(\alpha)$ remains bounded as $|\alpha| \to \infty$, since for large $|\alpha|$ we have

$$\lambda(\alpha) = \sum_{s=0}^{\infty} c_s \alpha^{-\frac{s}{p}} . \tag{121}$$

If $\lambda(\alpha)$ and α satisfy equation (120), then

$$\lambda^*(\alpha) = \lambda(\alpha) + \alpha q \tag{122}$$

satisfies the equation

$$D(\lambda - \alpha q, \alpha) = 0 .$$

It follows from (121) and (122) that

$$\alpha = \frac{1}{q} \lambda^* + \sum_{s=0}^{\infty} c_s^* \lambda^{* - \frac{s}{p}} . \tag{123}$$

This result can be proved as follows: Set

$$\alpha^{\frac{1}{p}} = \alpha_1 \quad \text{and} \quad \lambda^{* \frac{1}{p}} = \lambda_1 .$$

Then, from the relation

$$\lambda^* = \alpha q + \sum_{s=0}^{\infty} c_s \alpha^{-\frac{s-k}{p}} , \qquad k < p ,$$

we obtain

$$\lambda_1^p = \alpha_1^p q + \sum_{s=0}^{\infty} c_s \alpha_1^{k-s} = \alpha_1^p \left(q + \sum_{s=0}^{\infty} c_s \alpha_1^{k-s-p} \right) ;$$

therefore,

$$\lambda_1 = \alpha_1 \left(q^* + \sum_{s=0}^{\infty} c_s^* \alpha_1^{k-s-p} \right) ,$$

where q^* and c_s^* are constants, $q^* \neq 0$. Hence, setting

$$\frac{1}{\alpha_1} = \alpha_2 , \quad \frac{1}{\lambda_1} = \lambda_2 ,$$

we find that

$$\frac{1}{\lambda_2} = \frac{1}{\alpha_2} \left(q^* + \sum_{s=0}^{\infty} c_s^* \alpha_2^{s+p-k} \right) ,$$

and therefore

$$\lambda_2 = \frac{\alpha_2}{q^* + \sum_{s=0}^{\infty} c_s^* \alpha_2^{s+p-k}} = \alpha_2 \left(q^{**} + \sum_{s=0}^{\infty} c_s^{**} \alpha_2^{s+p-k} \right) =$$

$$= \alpha_2 q^{**} + \sum_{s=0}^{\infty} c_s^{**} \alpha_2^{s+p+1-k} , \qquad (124)$$

where q^{**} and c_s^{**} are constants, $q^{**} \neq 0$. It follows that α_2 is a holomorphic function of λ_2 in a neighborhood of $\lambda_2 = 0$. Let

$$\alpha_2 = \sum_{s=1}^{\infty} k_s \lambda_2^s$$

(obviously, $k_0 = 0$). Substituting this series for α_2 in the right-hand side of (124) and comparing the coefficients by the same powers of λ_2, we can easily show that

$$k_2 = k_3 = \cdots = k_{p-k} = 0 .$$

Therefore

$$\alpha_2 = k_1 \lambda_2 + \sum_{s=0}^{\infty} k_{s+p-k+1} \lambda_2^{s+p+1-k} = \lambda_2 \left(k_1 + \sum_{s=0}^{\infty} k_{s+p-k+1} \lambda_2^{s+p-k} \right) .$$

Passing from α_2, λ_2 to α_1, λ_1, we obtain

$$\frac{1}{\alpha_1} = \frac{1}{\lambda_1} \left(k_1 + \sum_{s=0}^{\infty} k_{s+p-k+1} \lambda_1^{k-s-p} \right) ,$$

$$\alpha_1 = \frac{\lambda_1}{k_1 + \sum_{s=0}^{\infty} k_{s+p-k+1} \lambda_1^{k-s-p}} = \lambda_1 \left(k^* + \sum_{s=0}^{\infty} k_s^* \lambda_1^{k-s-p} \right) .$$

Taking the p-th power of both sides of the above equality and passing from α_1, λ_1 to α, λ^*, we find that

$$\alpha = \lambda^* \left(k^{**} + \sum_{s=0}^{\infty} k_s^{**} \lambda^{*\frac{(k-s-p)}{p}} \right) = \lambda^* k^{**} + \sum_{s=0}^{\infty} k_s^{**} \lambda^{*\frac{(k-s)}{p}} \,. \tag{125}$$

It is easy to see that $k^{**} = \frac{1}{q}$ and $k_0^{**} \neq 0$ for $c_0 \neq 0$.

Now, let us pass from the variables y_0 and y_1 in (117) to $x_0 = y_0$ and

$$x' = \sum_{1}^{n} a_k x_k = y_1 - q y_0 \,.$$

Then we have

$$u_i = \sum_{s=s_0}^{\infty} c_{is} \, e^{\alpha(s) x' + (\lambda(s) + q\alpha(s)) x_0} = \sum_{s=s_0}^{\infty} c_{is} \, e^{\alpha(s) x' + \lambda^* x_0} \,. \tag{126}$$

Set

$$\lambda^{*(s)} = is \,,$$

and assume that s_0 is large enough. Then, by virtue of (123), the real parts of $\alpha^{(s)}$ will be bounded for $s \to \infty$. Let us determine the coefficients c_{is} from system (119) in such a way that

$$|c_{is}| \leq \frac{1}{s^m}$$

for all i and s, where m is an arbitrary positive integer such that $m \geq n^* + 2$, n^* being the largest among n_i $(i = 1, \dots, N)$; it is assumed that the above inequality for c_{is} holds as equality for at least one value of i. Then, for any i, the series (126), and those obtained by formally differentiating every term of (126) with respect to the variables x_0 and x' up to the order $m - 2$, are uniformly convergent in each strip $|x'| \leq C$ of the real (x_0, x')-plane, and the sum of every series obtained in this way is bounded. Therefore, the above series represent a solution of system (114') which is continuously differentiable up to the order $m - 2$. It is easy to see that at least one function among the u_i's forming this solution is non-analytic in x_0. Indeed, if all the functions $u_i(x_0, x')$ were analytic in x_0 (these functions are periodic in x_0), then their Fourier coefficients (with respect to x_0) would tend to 0 faster than any power of s, as $s \to \infty$. However, it is easy to see that the latter condition is violated at least for one of u_i's.

The functions $u_i(x_0, x')$ just defined take complex values for real x_0, x'. As mentioned above, if the coefficients of system (114') are real, then the real part and the imaginary part of each function, taken separately, also satisfy the equations of system (114'), and consequently, for m sufficiently large, possess partial derivatives of any given order, the derivatives being continuous and bounded in any strip $|x'| \leq C$ of the real (x_0, x')-plane; moreover, one of these functions, at least, is non-analytic in x_0.

2) The case $o < k < p$, $c_o \neq o$ [⁵]. Set

$$\lambda^{*(s)} = is , \qquad s = 1, 2, \ldots ,$$

if $\Re\left\{ k_0^{**}\, e^{\frac{\pi i}{2}\frac{k}{p}} \right\} \neq 0$, and

$$\lambda^{*(s)} = -is , \qquad s = 1, 2, \ldots ,$$

otherwise; here k_0^{**} is the coefficient in the series (125). Then, for $0 < k < p$, there exist values of

$$k_0^{**}\, \lambda^{*(s)\frac{k}{p}} \qquad (k_0^{**} \neq 0) ,$$

with negative real part. Indeed, $\lambda^{*(s)\frac{k}{p}}$ takes the values

$$s^{\frac{k}{p}}\, e^{\left(2m\pm\frac{1}{2}\right)\pi i \frac{k}{p}} \quad \text{for} \quad m = 0, \pm 1, \pm 2, \ldots ,$$

where $s^{\frac{k}{p}}$ stands for the arithmetic value of the root. Since

$$\frac{\pi}{2}\frac{k}{p} < \frac{\pi}{2} ,$$

it follows that either

$$2\pi \frac{k}{p} \qquad \text{or} \qquad 2\pi - 2\pi \frac{k}{p}$$

is $\leq \pi$, and it is always possible to find m such that the real part of either

$$k_0^{**}\, e^{\left(2m+\frac{1}{2}\right)\pi i \frac{k}{p}} = k_0^{**}\, e^{\left(2m+\frac{1}{2}\right)\pi i \frac{k}{p} - 2m\pi i}$$

or

$$k_0^{**}\, e^{\left(2m-\frac{1}{2}\right)\pi i \frac{k}{p}} = k_0^{**}\, e^{\left(2m-\frac{1}{2}\right)\pi i \frac{k}{p} - 2m\pi i}$$

is negative. In what follows, we denote by $\lambda^{*(s)\frac{k}{p}}$ the values for which $k_0^{**}\lambda^{*(s)\frac{k}{p}}$ have negative real parts. Then, by virtue of (125), the real part of $\alpha^{(s)}$ will also be negative for sufficiently large s. For $s \geq s_0$, where s_0 is a positive constant, we have

$$-cs^\rho \leq \Re\{\alpha^{(s)}\} \leq Cs^\rho , \qquad (127)$$

where c and C are positive constants; $\rho = \frac{k}{p}$.

Let us determine the coefficients c_{is} of the series (126) in such a way that

$$|c_{is}| \leq 1 \quad \text{for all} \quad i, s ,$$

and the above inequality turns into equality for at least one value of i. This condition is insufficient to guarantee the uniqueness of c_{is}; for s fixed, we have to define a factor, the same for all i and having absolute value equal to 1. We choose this factor from the condition that one of the expressions

$$c_{is}\, e^{\alpha(s)}$$

be real and non-negative. Denote by $c_{is}^{(m)}$ $(i = 1, \ldots, N)$ the coefficients c_{is} chosen in such a way that

$$c_{ms}^{(m)}\, e^{\alpha(s)}$$

is non-negative.

It will be shown that among N values of m there is at least one value such that the series

$$u_i^{(m)}(x_0, x') = \sum_{s=s_0}^{\infty} c_{is}^{(m)}\, e^{\alpha(s)x' + \lambda^*(s)x_0}\,, \qquad i = 1, 2, \ldots, N\,, \tag{128}$$

represent components of a solution of system $(114')$ for $x' > \varepsilon$, where ε is an arbitrary positive number; this solution is formed by functions possessing continuous bounded derivatives of any order with respect to x_0 and x', and one of these functions is non-analytic in x_0.

In the case of system $(114')$ having real coefficients, the real and the imaginary parts of the constructed functions $u_i^{(m)}$ have the same properties as the $u_i^{(m)}$'s themselves (this can be easily seen from what follows): they are continuous and bounded, together with their derivatives of any order, in that part of the $(x_0, x_1, \ldots, x_n)$-hyperplane where $x' > \varepsilon > 0$; and, for at least one value of m, the real (resp., the imaginary) part of $u_m^{(m)}$ is non-analytic in x_0.

We observe, first of all, that because of (127), for $x' > \varepsilon$ the series (128), as well as those obtained by differentiating each term of (128) up to any order, are uniformly convergent and have bounded sums. Hence, these series define, for $x' > \varepsilon$, a solution of system $(114')$ possessing continuous bounded derivatives of any order.

Let us show that among $u_m^{(m)}$ $(m = 1, 2, \ldots, N)$ there is a function non-analytic in x_0 on the interval $-\eta < x_0 < \eta$, for $x' = 1$. To this end, we show that the derivatives of one of these functions in x_0 do not satisfy the inequalities

$$\left| \frac{\partial^k u}{\partial x_0^k} \right| \le \frac{M k!}{\eta^k} \tag{129}$$

for $x' = 1$ and $x_0 = 0$, where M, η are positive constants. Let it be observed, first, that for $x' = 1$ and $x_0 = 0$ we have

$$\sum_{m=1}^{N} \frac{\partial^k u_m^{(m)}}{\partial x_0^k} = (\pm i)^k \sum_{m=1}^{N} \sum_{s=s_0}^{\infty} c_{ms}^{(m)} s^k\, e^{\alpha(s)}\,.$$

Note that all the quantities $c_{ms}^{(m)} e^{\alpha(s)}$ are real and non-negative; therefore, taking into account the relations (127) and the fact that

$$|c_{ms}^{(m)}| = 1$$

for all s and at least one of the N values of m, we obtain the following inequalities

$$\left| \sum_{m=1}^{N} \frac{\partial^k u_m^{(m)}}{\partial x_0^k} \right| = \sum_{m=1}^{N} \sum_{s=s_0}^{\infty} c_{ms}^{(m)} s^k e^{\alpha(s)} \geq \sum_{s=s_0}^{\infty} s^k e^{\Re\{\alpha(s)\}} \geq$$

$$\geq \sum_{s=s_0}^{\infty} s^k e^{-cs^\rho} .$$

Here, as elsewhere, s^ρ is understood in arithmetic sense. For sufficiently large k such that

$$s_0 \leq \left(\frac{k}{c\rho} \right)^{\frac{1}{\rho}} ,$$

we have

$$\left| \sum_{m=1}^{N} \frac{\partial^k u_m^{(m)}}{\partial x_0^k} \right| \geq \left[\left(\frac{k}{c\rho} \right)^{\frac{1}{\rho}} \right]^k \exp\left\{ -c\left[\left(\frac{k}{c\rho} \right)^{\frac{1}{\rho}} \right]^\rho \right\} \geq$$

$$\geq \left\{ \left(\frac{k}{c\rho} \right)^{\frac{1}{\rho}} - 1 \right\}^k e^{-\frac{k}{\rho}} , \tag{130}$$

where $[x]$, as usual, stands for the entire part of x. Assume that k is chosen so large that

$$\left\{ \left(\frac{k}{c\rho} \right)^{\frac{1}{\rho}} - 1 \right\}^k \geq \frac{1}{2^k} \left(\frac{k}{c\rho} \right)^{\frac{k}{\rho}} ,$$

or, equivalently,

$$\left(\frac{k}{c\rho} \right)^{\frac{1}{\rho}} - 1 \geq \frac{1}{2} \left(\frac{k}{c\rho} \right)^{\frac{1}{\rho}} ,$$

or,

$$\frac{1}{2^\rho} \frac{k}{c\rho} \geq 1 .$$

It then follows from (130) that

$$\left| \sum_{m=1}^{N} \frac{\partial^k u_m^{(m)}}{\partial x_0^k} \right| \geq \frac{1}{c^{\frac{k}{\rho}} 2^k} \left(\frac{k}{\rho} \right)^{\frac{k}{\rho}} e^{-\frac{k}{\rho}} . \tag{131}$$

On the other hand, if (129) were valid for all $u_m^{(m)}$, we should have

$$\left| \sum_{m=1}^{N} \frac{\partial^k u_m^{(m)}}{\partial x_0^k} \right| \le \frac{M k!}{\eta^k} \le \frac{M k^k e^{-k}}{\eta^k} \sqrt{2\pi \, ek} \,, \tag{132}$$

where M and η are positive constants. Therefore, in order to establish the incompatibility of the inequalities (131) and (132), it suffices to show that for $\rho < 1$ and $k \to \infty$ the ratio

$$\frac{\left(\frac{k}{\rho}\right)^{\frac{1}{\rho}} e^{-\frac{1}{\rho}}}{k\, e^{-1}}$$

is larger than any given constant, which is obviously the case.

We have thus shown that, at the point $x_0 = 0$, $x' = 1$, the derivatives in x_0 of at least one of the $u_m^{(m)}$'s do not satisfy some of the inequalities of the form (129), whereas these inequalities are known to hold for derivatives of any analytic function. Note that for $x_0 = 0$, $x' = 1$, and all m, we have

$$\frac{\partial^k u_m^{(m)}}{\partial x_0^k} = \frac{\partial^k \Re\{u_m^{(m)}\}}{\partial x_0^k} \qquad \text{for} \quad k \quad \text{even} \,,$$

and

$$\frac{\partial^k u_m^{(m)}}{\partial x_0^k} = i \frac{\partial^k \Im\{u_m^{(m)}\}}{\partial x_0^k} \qquad \text{for} \quad k \quad \text{odd} \,.$$

Therefore, one of the functions $u_m^{(m)}$ has real (resp., imaginary) part non-analytic in x_0, for $x_0 = 0$, $x' = 1$.

Remark 6. Assume the left-hand sides of a system of type (1) to be holomorphic with respect to all their arguments in a domain G where the arguments take their values. Assume also that in the domain G the system has a characteristic surface

$$\varphi(x_0, x_1, \ldots, x_n) = 0 \tag{133}$$

corresponding to the holomorphic solution

$$u_i = \varphi_i(x_0, x_1, \ldots, x_n) \,, \qquad i = 1, 2, \ldots, N \,,$$

and containing at least one point

$$P(x_k = x_k^0 + i x_k^{*0}) \,, \qquad k = 0, 1, \ldots, n \,,$$

such that for all points of the surface where

$$\Im\{x_k\} = x_k^{*0} = \text{const} \,, \qquad k = 1, \ldots, n \,,$$

the value of $\Im\{x_0\}$ also remains constant and equal to x_0^{*0}. Assume, further, that the surface

$$\varphi(x_0, x_1, \ldots, x_n) = 0 , \qquad \Im\{x_k\} = x_k^{*0} , \qquad k = 0, 1, \ldots, n ,$$

has no tangent vectors parallel to the real axis Ox_0, and the function $\varphi(x_0, x_1, \ldots, x_n)$ is holomorphic in all its arguments. And, finally, let the matrix

$$\left\| \sum_{(k)} \frac{\partial F_i}{\partial \left\{ \dfrac{\partial^{k_0 + \cdots + k_n} u_j}{\partial x_0^{k_0} \cdots \partial x_n^{k_n}} \right\}} \left(\frac{\partial \varphi}{\partial x_0} \right)^{k_0} \cdots \left(\frac{\partial \varphi}{\partial x_n} \right)^{k_n} \right\| \qquad (134)$$

have rank $N^* - m$, where m is the multiplicity of the root $\partial\varphi/\partial x_0 , \ldots$ $\ldots, \partial\varphi/\partial x_n$ of the determinant of this matrix[10]. The symbol $\sum_{(k)}$ has the same meaning as in the expression for the matrix (3), and the values $\partial F_i/\partial\{\cdots\}$ are calculated at point P for $u_i = \varphi_i$ $(i = 1, \ldots, N)$. Moreover, it is assumed that the determinant of the matrix (134), as calculated above, does not vanish identically with respect to $\partial\varphi/\partial x_0 , \ldots, \partial\varphi/\partial x_n$, and P is a non-singular point of the surface (133) (i.e., the derivatives $\partial\varphi/\partial x_0 , \ldots, \partial\varphi/\partial x_n$ cannot be all equal to zero at this point).

Then system (1) is bound to possess solutions non-analytic in x_0. As above, it is implied that if non-holomorphic functions satisfy system (1), the derivatives in (1) are taken with respect to real parts of the variables $x_0, x_1, \ldots, x_n$.

The proof of the above statement is based on the following result. Using the arguments of E. Goursat (see [7], Vol.2, Note II, pp. 337–342), one can show, under the above assumptions about system (1), that this system admits many solutions holomorphic in $x_0, x_1, \ldots, x_n$ and taking, together with their derivatives up to any given finite order, the same values on the surface (133). Defining the functions which, on the hyperplane

$$\Im\{x_k\} = x_k^{*0} = \text{const} \qquad (k = 0, 1, \ldots) ,$$

coincide with the components of one of the above solutions at the points to one side of the surface $\varphi(x_0, x_1, \ldots, x_n) = 0$, and coincide with the components of another solution at the points to the other side of this surface, we obviously obtain a solution on that hyperplane, with components possessing continuous derivatives of any order, but non-analytic in x_0.

---

[10] A root of multiplicity m for a polynomial $P(\alpha_0, \alpha_1, \ldots, \alpha_n)$ is defined as a system of numbers $c_0, c_1, \ldots, c_n$ such that the expansion of P in powers of

$$\alpha_0 - c_0 , \quad \alpha_1 - c_1 , \ldots, \alpha_n - c_n$$

starts with terms of degree m with respect to the above differences.

Let all functions F_i and φ_i, as well as φ, be real-valued for real values of the independent variables (naturally, the latter are assumed to be within the domain G), and let the point P have real coordinates, too; then, under the above assumptions, system (1) admits real non-analytic solutions with continuous derivatives of any order.

Remark 7. It would be erroneous to believe that systems of type (1), whose left-hand sides are holomorphic with respect to all their arguments and admit real characteristic surfaces, can be definitely claimed to possess solutions non-analytic in some of the variables $x_0, x_1, \ldots, x_n$. Consider, for instance, the following system

$$\frac{\partial u_1}{\partial x_0} = \frac{\partial u_2}{\partial x_1},$$

$$\frac{\partial u_2}{\partial x_0} = -\frac{\partial u_1}{\partial x_1},$$

$$\frac{\partial^2 u_i}{\partial x_0^2} + \frac{\partial^2 u_i}{\partial x_1^2} = \sum_{j=1}^{i-1} \sum_{k_0+k_1=0}^{k_0+k_1=n_j} a_{ij}^{(k_0 k_1)} \frac{\partial^k u_j}{\partial x_0^{k_0} \partial x_1^{k_1}},$$

$$n_1, n_2 > 1, \quad k_0, k_1 \geq 0, \quad i = 3, 4, \ldots, N,$$

with constant coefficients $a_{ij}^{(k_0 k_1)}$. If $N > 2$, then any surface is characteristic for this system. However, it can be easily seen that all its solutions are analytic in x_0 and x_1 [6].

§8. Systems whose Characteristic Equation has Zero Root

In some cases, such systems have all their solutions analytic in x_0 (see Example I); and sometimes, they admit non-analytic solutions (see Example II).

Example I. The system

$$\frac{\partial u_1}{\partial x_0} = u_2,$$

$$\frac{\partial u_2}{\partial x_0} = \frac{\partial u_1}{\partial x_1},$$

which is equivalent to the heat equation

$$\frac{\partial^2 u_1}{\partial x_0^2} = \frac{\partial u_1}{\partial x_1},$$

is well known to have all its solutions analytic in x_0, whereas the determinant of its characteristic matrix vanishes for $\alpha_1 = 1$, $\alpha_0 = 0$.

Example II. The system

$$\frac{\partial u_1}{\partial x_0} = u_2 \,,$$

$$\frac{\partial u_2}{\partial x_0} = u_3 \,, \tag{135}$$

$$\frac{\partial u_3}{\partial x_0} = \frac{\partial u_1}{\partial x_1} \,,$$

which is equivalent to the equation

$$\frac{\partial^3 u_1}{\partial x_0^3} = \frac{\partial u_1}{\partial x_1} \,, \tag{136}$$

admits solutions non-analytic in x_0. Such non-analytic solution of equation (136), as well as of system (135), can be obtained by setting

$$u_1 = \sum_{k=1}^{\infty} c_k \, e^{ikx_0 - ik^3 x_1} \,,$$

where $c_k = k^{-m}$ and $m \geq 5$. Choosing m sufficiently large, we can make the function u_1 possess continuous bounded derivatives up to any order with respect to x_0 and x_1 at the points of the real (x_0, x_1)-plane, However, no matter how large m may be, this solution is non-analytic in x_0, since otherwise, the function u_1 being periodic in x_0, the coefficients c_k must tend to 0 faster that any power of k^{-1}, as $k \to \infty$.

References

[1] Petrowsky, I.G. On systems of differential equations which can have only analytic solutions. *Doklady Akad. Nauk SSSR* **17:7** (1937) 339–342.

[2] Bernstein, S.N. Sur la nature analytique des solutions des équations aux dérivées partielles du second ordre. *Math. Ann.* **59** (1929) 20–76.

[3] Lewy, H. Neuer Beweis des analytischen Characters der Lösungen elliptischer Differentialgleichungen. *Math. Ann.* **101** (1929) 605–619.

[4] Petrowsky, I.G. Über das Cauchysche Problem für Systeme von partiellen Differentialgleichungen. *Matem. Sbornik* **2:5** (1937) 815–870. (See this volume, Article 1).

[5] Hadamard, J. *Le problème de Cauchy et les équations aux dérivées partielles linéaires hyperboliques.* Hermann, Paris, 1932.

[6] Petrowsky, I.G. On the Cauchy problem for systems of linear partial differential equations in a class of non-analytic functions. *Bulletin of Moscow State Univ. Math., Mech.* **1**:7 (1938) 1–72. (See this volume, article 2)

[7] Goursat, E. *Leçons sur l'Integration des Équations aux Dérivées Partielles du Second Ordre à Deux Variables Indépendantes.* Vol. 1. Hermann, Paris, 1921, pp. 337–342.

NOTES

(by L.R. Volevich)

[1] (Introduction). See Appendix for a commentary by O.A. Oleinik and L.R. Volevich.

[2] (§4). Convergence of the series (52) is equivalent to $u(x)$ possessing square-summable weak derivatives (in the sense of S.L. Sobolev) up to the order $\left[\frac{n+1}{2}\right] + 1$. Uniform and absolute convergence of the Fourier series for the function $u(x)$ implies its continuity. Therefore, Sobolev's imbedding theorem: $C \subset W_2^{(l)}, l \geq \left[\frac{n+1}{2}\right] + 1$, has actually been proved here in the case of periodic functions.

Actually, the second part of Remark 5 establishes continuity of any function $u(x)$ with square-summable mixed derivatives

$$\frac{\partial^{n_1 + \cdots + n_l} u(x)}{\partial x_{n_1} \cdots \partial x_{n_l}},$$

where $n_1, \ldots, n_l$ take the values $1, \ldots, n$, are different from one another, and $l \geq \left[\frac{n+1}{2}\right] + 1$.

[3] (§6). Unfortunately, this intention of I.G. Petrowsky has remained unrealized.

[4] (§7). Observe that if $(u_1, \ldots, u_N)$ is a smooth solution of the following system of differential equations with constant coefficients:

$$\sum_{j=1}^{N} A_{ij}\left(\frac{\partial}{\partial x_1}, \ldots, \frac{\partial}{\partial x_n}\right) u_j = 0, \qquad i = 1, \ldots, N, \qquad (*)$$

then every function u_j $(1 \leq j \leq N)$ is a solution of the equation

$$D\left(\frac{\partial}{\partial x_1}, \ldots, \frac{\partial}{\partial x_n}\right) v = 0, \qquad (**)$$

where

$$D(\alpha_1, \ldots, \alpha_n) = \det \|A_{ij}(\alpha_1, \ldots, \alpha_n)\|. \qquad (***)$$

Conversely, for any given smooth solution of $(**)$, one can construct a solution of system $(*)$, making use of the differential operators corresponding to the cofactors of the elements of the matrix $\|A_{ij}\|$. This simple observation and Petrowsky's arguments imply that the ellipticity condition for the determinant $D\left(\partial/\partial x_1, \ldots, \partial/\partial x_n\right)$ is necessary for all smooth solutions of system $(*)$ to be analytic functions. Taking into account the main result of the article, we come to the conclusion that every classical solution of a system of partial differential equations with constant coefficients is analytic if and only if the determinant of the system is an elliptic operator.

[5] (§7). In this case, the polynomial $D(\lambda, \alpha)$ is *hypoelliptic* in the sense of Hörmander. The differential equation associated with this polynomial admits no solutions of finite smoothness. I.G. Petrowsky has even constructed a solution belonging to the Gevrey class of order $\rho > 1$ (and not belonging to the Gevrey class of any order $\rho' < \rho$, in particular, not belonging to the class of analytic functions). As pointed out by Hörmander,[1] the algebraic conditions of hypoellipticity have actually been suggested by Petrowsky's proof given below.

[6] (§7). The system considered by Petrowsky is elliptic in a wider sense of Douglis and Nirenberg (see Commentary for details). Indeed, let us define the numbers $s_1, \ldots, s_n, t_1, \ldots, t_n$ as follows:

$$t_1 = t_2 = 0, \quad t_j = 2 - s_j, \quad j = 2, \ldots, N, \quad s_1 = s_2 = 1,$$

the numbers $s_3, \ldots, s_n$ are recursively chosen to satisfy the inequalities

$$s_i + t_j > n_j, \quad j \leq i - 1,$$

or equivalently,

$$s_i > n_j - 2 + s_j, \quad j \leq i - 1.$$

For the set $\{s_j, t_j\}$ defined above, the principal part (in the sense of Douglis and Nirenberg) of the system under consideration is given by the following elliptic system

$$\frac{\partial u_1}{\partial x_0} = \frac{\partial u_2}{\partial x_1},$$

$$\frac{\partial u_2}{\partial x_1} = -\frac{\partial u_1}{\partial x_0},$$

$$\frac{\partial^2 u_i}{\partial x_0^2} + \frac{\partial^2 u_i}{\partial x_1^2} = 0, \quad i = 2, \ldots, N.$$

---

[1] Hörmander, L. *Linear Partial Differential Operators*. Springer-Verlag, Berlin, 1963.

5

On the Diffusion of Waves and the Lacunas for Hyperbolic Systems[*][1]

Consider the wave equation

$$\frac{\partial^2 u}{\partial t^2} = a^2 \left(\frac{\partial^2 u}{\partial x_1^2} + \cdots + \frac{\partial^2 u}{\partial x_p^2} \right) . \tag{1}$$

For any odd $p > 1$, the value of a solution of the Cauchy problem for this equation at a point $(t, x_1, \ldots, x_p)$ is known to depend merely on the initial values taken on the periphery of the base of the characteristic cone with vertex at $(t, x_1, \ldots, x_p)$. For p even, or $p = 1$, the solution $u(t, x_1, \ldots, x_p)$ depends on the initial values on the entire base of that cone.

Let the initial values for u and $\partial u / \partial t$ (we assume the initial values to be given at $t = 0$) be different from 0 only in a small domain G containing a point $(0, x_1^0, \ldots, x_p^0)$. Let us fix $x_1, \ldots, x_p$ and watch the behavior of u at the point $(t, x_1, \ldots, x_p)$ for growing $t > 0$. For any odd $p > 1$, the value of $u(t, x_1, \ldots, x_p)$ can be different from zero only on a small portion of a straight line lying in the $(t, x_1, \ldots, x_p)$-space and parallel to the t-axis; namely, on that portion of the line which contains the vertices of the characteristic cones for equation (1) whose base, at its periphery, has a non-empty intersection with the domain G. If p is even or $p = 1$, then u must be equal to zero only at those points of the line which coincide with vertices of the characteristic cones whose bases are outside G; obviously, these points form a segment on the line with an end-point at $(0, x_1, \ldots, x_p)$; at all other points of this line, the solution $u(t, x_1, \ldots, x_p)$ is, in general, different from 0.

---

[*]Originally published in *Izvestiya Akad. Nauk SSSR. Ser. Mat.* **8** (1944) 101–106. See Appendix for a commentary by A.M. Gabrielov & V.P. Palamodov.

[1]This paper is a review of the results in the field (see [1]). It was presented by I.G. Petrowsky at an annual meeting of the Department of Mathematics and Physics of the Academy of Sciences of the USSR.

Thus, in the case of p odd, $p > 1$, a perturbation at the initial moment in a small neighborhood of the point $(x_1^0, \ldots, x_p^0)$ affects the values of u for $t > 0$ only at the points of the $(x_1, \ldots, x_p)$-space which are close to the sphere of radius at with center at $(x_1^0, \ldots, x_p^0)$. Thus, the initial perturbation near the point $(x_1^0, \ldots, x_p^0)$ produces a spherical wave with center at this point; this wave possesses a forefront and a rear. For p even, or $p = 1$, the initial perturbation near the point $(x_1^0, \ldots, x_p^0)$ affects, in general, the values of u at all points inside the sphere of radius at centered at $(x_1^0, \ldots, x_p^0)$. The wave produced by this perturbation has a prominent forefront and a diffuse rear. In this case, we can speak of *diffusion of waves*; for odd $p > 1$, there is no diffusion of waves.

At the beginning of this century, Hadamard [2] showed that for linear second order hyperbolic equations with variable coefficients, in the case of p even, diffusion of waves can always be observed. As to the diffusion of waves for general linear hyperbolic equations with p odd, the problem remained open for a long time. It was only in 1939 that a paper of Mathisson [3] appeared, where the following result was proved. For $p = 3$, the only class of linear hyperbolic second order equations with no diffusion of waves consists of equations derived from (1) as a result of the following three operations:

1) multiplication of both sides of (1) by a function of $t, x_1, \ldots, x_p$;

2) linear transformation of the unknown function;

3) passing to new independent variables, instead of $t, x_1, \ldots, x_p$.[2]

The methods used by Mathisson to prove this result are applicable, in principle, to second order hyperbolic equations for any odd $p > 1$. However, the case $p > 3$ requires more lengthy calculations, which had not been completed.

The author has been studying similar questions for general hyperbolic systems. Consider a linear hyperbolic system with sufficiently smooth coefficients:

$$\frac{\partial^{n_i} u_i}{\partial t^{n_i}} = \sum_{j,k} a_{ij}^{(k_0,k_1,\ldots,k_p)}(t, x_1, \ldots, x_p) \frac{\partial^{k_0+k_1+\cdots+k_p} u_j}{\partial t^{k_0} \partial x_1^{k_1} \ldots \partial x_p^{k_p}}, \qquad (2)$$

$$k_0 < n_j, \quad k_0 + k_1 + \cdots + k_p \le n_j, \quad i, j = 1, \ldots, N .$$

The lateral surface of the characteristic cone with vertex at $(t^*, x_1^*, \ldots, x_p^*)$ splits its base on the plane $t = t_0$ into several domains, in general. Any one of these domains, say G_{t_0}, is called a *lacuna*, if arbitrary variation of

---

[2]This theorem of Mathisson turned out to be false (see Commentary in Appendix).
– *Ed. note.*

the initial data inside G_{t_0} (provided that we are still within the class of sufficiently smooth initial functions) produces no change in the solution of the Cauchy problem for equation (2) at the point $(t^*, x_1^*, \ldots, x_p^*)$. For definiteness, we assume that $t_0 < t^*$. Consider the following system with constant coefficients

$$\frac{\partial^{n_i} u_i}{\partial t^{n_i}} = \sum_{j,k} a_{ij}^{(k_0, k_1, \ldots, k_p)}(t^*, x_1^*, \ldots, x_p^*) \frac{\partial^{n_j} u_j}{\partial t^{k_0} \partial x_1^{k_1} \ldots \partial x_p^{k_p}}, \qquad (3)$$

$$k_0 < n_j, \quad k_0 + k_1 + \cdots + k_p = n_j, \quad i, j = 1, \ldots, N.$$

If t_0 is close enough to t^*, then any of the domains $G_{t_0}^*$, which the lateral surface of the characteristic cone for system (3) with vertex at $(t^*, x_1^*, \ldots, x_p^*)$ cuts out of its base on the plane $t = 0$, corresponds to a unique domain G_{t_0} (close to $G_{t_0}^*$) constructed for system (2).

Theorem 1. *If for all t_0 close enough to t^* the domain G_{t_0} corresponding to $G_{t_0}^*$ is a lacuna for system (2), then $G_{t_0}^*$ is a lacuna for system (3).*

An obvious corollary of this result is the above mentioned Hadamard's theorem about the diffusion of waves for all linear second order hyperbolic equations with an even number of spatial coordinates, since diffusion of waves is always observed for all linear second order hyperbolic equations with constant coefficients, which can always be reduced to equation (1) by a linear transformation in the $(t, x_1, \ldots, x_p)$-space.

All subsequent considerations will be concerned with the study of lacunas for linear hyperbolic systems of type (3) with constant coefficients. A necessary and sufficient condition for the absence of diffusion for such systems is, obviously, the existence of a lacuna crossed by a straight line passing through the vertex of the characteristic cone and parallel to the t-axis. Lacunas that cannot be destroyed by any sufficiently small variation of the coefficients will be referred to as *stable lacunas*.

Let us regard the operators $\partial/\partial t$, $\partial^k/\partial x^k$ as symbolic multiplication, and equations (3) as a system of linear algebraic equations with respect to $u_1, \ldots, u_N$. Consider the determinant of the matrix formed by the coefficients of the latter system. This determinant can be written in explicit form as

$$\frac{\partial^n}{\partial t^n} + \sum_{(k)} a^{(k_0, k_1, \ldots, k_p)} \frac{\partial^n}{\partial t^{k_0} \partial x_1^{k_1} \ldots \partial x_p^{k_p}},$$

where the sum is taken over all integers $k_1, \ldots, k_p$ such that

$$k_1 + \cdots + k_p = n, \qquad n = n_1 + \cdots + n_N.$$

The equation

$$1 + \sum_{(k)} a^{(k_0.k_1,\ldots,k_p)} z_1^{k_1} \cdots z_p^{k_p} = 0 \tag{4}$$

is called the *characteristic equation* for system (3). It may be regarded as a tangential equation for the surface which forms the intersection of the plane $t_0 = 0$ and the lateral surface of the characteristic cone K with vertex at $(t, 0, \ldots, 0)$ for system (3), if we represent planes in the $(x_1, \ldots, x_p)$-space by equations

$$-t + x_1 z_1 + \cdots + x_p z_p = 0 . \tag{5}$$

By duality, each point $(x_1, \ldots, x_p)$ in this space is associated with a plane in the $(z_1, \ldots, z_p)$-space. In order to decide whether a particular point $(x_1, \ldots, x_p)$ belongs to a stable lacuna at the base of the cone K, let us take a point A on the plane (5) in the $(z_1, \ldots, z_p)$-space. Let us draw all possible straight lines issuing from A, given by equations with real coefficients, and belonging to this plane. Consider points of intersection of these lines with the surface (4). The real (as well as complex) points of intersection form $(p-2)$-dimensional cycles, denoted by C_{real} (*resp.*, C_{imag}). Obviously, the cycle C_{real} does not depend on the choice of A. If we change A, then C_{imag} will be replaced by a cycle homologous to C_{imag} in the complex intersection of (4) and (5).

Under the assumption that the surface (4) has no singular points, we can formulate the following theorems.

Theorem 2. *Let $n_i < p + 1$ for all i. Then system (3) possesses a stable lacuna if and only if the cycles C_{real} for p odd (resp., C_{imag} for p even) corresponding to its points are homologous to zero in the complex intersection of (4) and (5).*

There are some examples of such lacunas.

In order to solve the problem of the existence of lacunas in the case of $n_i \geq p + 1$ for some i, the cycles $C_{\text{real}}(\tau)$ for p odd (*resp.*, $C_{\text{imag}}(\tau)$ for p even) should be constructed for each plane of the form

$$-\tau + x_1 z_1 + \cdots + x_p z_p = 0 \qquad (0 \leq \tau \leq t) . \tag{6}$$

Let S denote the surface formed by these cycles.

Theorem 3. *Let $n_i \geq p + 1$ for some i. Then the following is a necessary condition for the existence of a stable lacuna for equation (3): the cycles $C_{\text{real}}(t)$ for p odd (resp., $C_{\text{imag}}(t)$ for p even), corresponding to its points, are homologous to zero in the complex intersection of (4) and (5).*

Assume the latter condition to be fulfilled. Let us extend films over the cycles $C_{\text{real}}(\tau)$ (*resp.*, $C_{\text{imag}}(\tau)$) in the complex cross-sections of the surface

(4) by the planes (6) for $\tau = t$ and $\tau = 0$. This is always possible because of the hyperbolicity of system (3). An easy proof of this fact can be obtained if we notice that, due to hyperbolicity, the real surface (4) consists of $n/2$ ovals consecutively enclosed in one another, if n is even; in the case of n odd, these ovals should be supplemented with a so-called unpaired piece homologous, in real projective $(x_1, \ldots, x_p)$-space, to a plane. The construction of the said films extended over the cycles $C_{\text{real}}(\tau)$ and $C_{\text{imag}}(\tau)$ for $\tau = 0$ and $\tau = t$ can be performed in various ways.

For p even, all cycles Σ formed by the surfaces S and these films are homologous to each other on the complex surface (4), since, for p even, all $(p-1)$ cycles on a general algebraic surface P of complex dimension $(p-2)$, which is formed by the intersections of the plane (6) for $\tau = 0$ or $\tau = 1$ and the surface (4), are homologous to zero in these cross-sections. In the case of an odd p, different ways of constructing the extended films may lead to the cycles Σ non-homologous to each other on the complex surface (4), since on a general algebraic surface P, for p odd, there exists a $(p-1)$-dimensional cycle non-homologous to zero, namely, an algebraic cycle.

Theorem 4. *Let the conditions of Theorem 3 be satisfied and $n_i \geq p+1$ for some i. Then there exists a stable lacuna for system (3) if and only if the cycles Σ corresponding to its points are homologous to zero on the complex surface (4), for p even (are algebraic, for p odd).*

Note that if this condition is satisfied, then the films over the cycles $C_{\text{real}}(\tau)$ for $\tau = 0$ and $\tau = t$ can be extended in such a way that the cycle Σ is homologous to zero on the complex surface (4).

Apart from the wave equation mentioned above, there are some other equations of primary importance for mathematical physics which admit lacunas, for instance, the system of linear elasticity and the system of crystal optics in three-dimensional space. Either of these systems consists of second order equations, and their real characteristic surfaces are formed by finite ovals consecutively enclosed in one another. Real planes crossing these ovals correspond to the exterior of the characteristic cones.

As indicated earlier, the cycles C_{real} corresponding to these planes are homologous to zero. The planes that do not cross these ovals correspond to lacunas, since the cycles C_{real} are void in this case. In contrast to the above assumption that the characteristic surfaces have no singular points, such points do exist for systems of these two types. Therefore, we can only make a provisional reference to the stability of lacunas for the system of elasticity or the system of crystal optics, if we allow only such variations of these systems that do not violate their hyperbolicity.

References

[1] Petrowsky, I.G. On the dependence of the solution of Cauchy's problem on the initial data. *Doklady Akad. Nauk SSSR* **38:5-6** (1943) 163–165.

[2] Hadamard, J. *Le problème de Cauchy et les équations aux dérivées partielles linéaires hyperboliques.* Hermann, Paris, 1932.

[3] Mathisson, M. Le problème de Hadamard relatif à la diffusion des ondes. *Acta Math.* **71:3-4** (1939) 238–282.

On the Diffusion of Waves and the Lacunas for Hyperbolic Equations[*]

Introduction

Consider the following equation

$$\frac{\partial^2 u}{\partial t^2} = \sum_{k=1}^{p} \frac{\partial^2 u}{\partial x_k^2} \, , \tag{1}$$

and let u be its solution. For $t^* > 0$ and any odd $p > 1$, the value of u at a point $(t^*, x_1^*, \ldots, x_p^*)$ is known[1] to depend only on the values at $t = 0$ (the initial values) of the function u and its derivative with respect to t at the points of the sphere

$$\sum_{k=1}^{p} (x_k - x_k^*)^2 = (t - t^*)^2 \tag{2}$$

belonging to the plane $t = 0$ and obtained as the intersection of that plane with the characteristic cone for equation (1) with vertex at the point $(t^*, x_1^*, \ldots, x_p^*)$. For p even, or $p = 1$, the value of $u(t^*, x_1^*, \ldots, x_p^*)$ depends on the initial data *not only on* the sphere (2) but also within that sphere.

An implication of the above result is that, in the case of an odd $p > 1$, the spherical wave produced by a perturbation in a small neighborhood of a point Q in the $(x_1, \ldots, x_p)$-space has its forefront and its rear both sharp, whereas, in the case of p even or $p = 1$, the wave has only a sharp forefront, its rear being diffuse. In the first case, equation (1) is said to admit no diffusion of waves, whereas in the second case, it is said that there is diffusion of waves for equation (1) (see [3], p. 238).

---

[*]Originally published in *Matem. Sbornik* **17** (1945) 289–370. See Appendix for a commentary by A.M. Gabrielov and V.P. Palamodov, and a supplementary article by A.M. Gabrielov: "Proof of Petrowsky's theorem."
[1]Cf. [1], [2].

It may happen for some partial differential equations or systems that the value of the solution of the Cauchy problem at point $P = (t, x_1, \ldots, x_n)$ does not depend on the values of the initial data over some of the domains into which the base of the characteristic cone with vertex at P is split by its lateral surface. In particular, this is the case for the principal equations of elasticity. We shall then say that such domains form *lacunas* for the systems in question.

In the present paper we consider hyperbolic systems (see [4]) of the form

$$\frac{\partial^{n_i} u_i}{\partial t^{n_i}} = F_i(t, x_1, \ldots, x_p; u_1, \ldots, u_N, \ldots) \qquad (i = 1, 2, \ldots, N) \,, \qquad (3)$$

where partial derivatives of the form

$$\frac{\partial^k u_j}{\partial t^{k_0} \partial x_1^{k_1} \cdots \partial x_p^{k_p}} \,, \qquad k = \sum k_s \le n_j \,, \quad k_0 < n_j \,, \quad j = 1, 2, \ldots, N \,,$$

are not written out explicitly as arguments of the functions F_i in the right-hand sides. The functions F_i are supposed to possess continuous derivatives of sufficiently high orders with respect to all their arguments (see my paper [4] on the Cauchy problem).

Let us give a rigorous definition of lacunas for such systems. We say that in a neighborhood of the initial data

$$\frac{\partial^k u_i}{\partial t^k} = \bar{u}_i^{(k)}(x_1, \ldots, x_p) \qquad (i = 1, \ldots, N \,; \ k = 0, 1, \ldots, n_i - 1)$$

there is a lacuna L for system (3) at point $P = (t^*, x_1^*, \ldots, x_p^*)$, if the functions $u_1, \ldots, u_N$ satisfying this system have their values at point $P = (t^*, x_1^*, \ldots, x_p^*)$ independent of the values of the initial data at $t = t_0 = \text{const}$ prescribed on the domain L belonging to the base of the characteristic cone with vertex at P. Saying that $u_i(t^*, x_1^*, \ldots, x_p^*)$ are independent of the initial data on L, we mean that $u_i(t^*, x_1^*, \ldots, x_p^*)$ do not vary, if the initial functions $\bar{u}_i^{(k)}(x_1, \ldots, x_p)$ are replaced by $\bar{\bar{u}}_i^{(k)}(x_1, \ldots, x_p)$ such that $\bar{u}_i^{(k)}$ and $\bar{\bar{u}}_i^{(k)}$, as well as all their respective derivatives up to a certain order, differ but slightly everywhere, and coincide outside a domain L^* lying, together with its boundary, within L. We shall consider only the domains L whose boundary belongs to the lateral surface of the characteristic cone with vertex at P. The characteristic cone is constructed for the solution determined by the initial functions $\bar{u}_i^{(k)}(x_1, \ldots, x_p)$, and therefore, in the case of a non-linear system, this cone, as well as the lacuna L, depends on the initial data. On the contrary, in the case of linear systems, there is no such dependence.

In Chapter II of the present paper we study lacunas for linear systems in the case $p = 1$.

In Chapter I we establish a relation between the existence of lacunas for general nonlinear hyperbolic systems and the existence of lacunas for the corresponding linear systems with constant coefficients, the latter containing only the highest order derivatives of all the unknown functions. For a single hyperbolic equation of this type, the formulas for the solution of the Cauchy problem, in the case $p > 1$, were given by Herglotz. Since these formulas have to be generalized and modified for our purposes, their detailed derivation, in the form required and without unnecessary restrictions, is given in Chapter III.

In Chapter V, on the basis of the formulas obtained, we give necessary and sufficient conditions for the existence of stable lacunas for a linear hyperbolic equation with constant coefficients, which contains only the highest order derivatives. A lacuna L for such an equation is said to be *stable*, if it is not destroyed by any sufficiently small variation of the coefficients.

At the beginning of Chapter V, we make the assumption that the characteristic equation for each hyperbolic equation under consideration represents an irreducible algebraic surface without singular points. In the case of a reducible surface, the lacuna problem is studied in §19. The existence of lacunas for a system of equations is discussed in §20. Applications of the results obtained to some equations of mathematical physics are described in §21.

While studying the lacunas for hyperbolic equations with constant coefficients, we have substantially relied on the fact that their characteristic equation is algebraic, i.e., its left-hand side is a polynomial. In order to emphasize the importance of the algebraic nature of the equation, we suggest in §22 a generalization of the Cauchy problem in question, which results in a non-algebraic equation; in this case, even the lacuna corresponding to the outside of the base of the characteristic cone may disappear. Then, the value of the solution at the vertex of the characteristic cone becomes dependent not only on the values on the base of the cone, but also outside the base.

In Chapter IV, some theorems about integrals over algebraic surfaces are proved; these results are necessary for Chapter V.

CHAPTER I

Lacunas for General Hyperbolic Systems

§1. Relations between Lacunas for General Hyperbolic Systems and Those for Some Linear Systems with Constant Coefficients

1. Consider a hyperbolic system of type (3) given in a domain G_{tx} of the $(t, x_1, \ldots, x_p)$-space and defined in a neighborhood of a solution $\bar{u}_1, \ldots \bar{u}_N$. The lateral surface of the characteristic cone having its vertex at the point $P = (t^*, x_1^*, \ldots, x_p^*)$ and corresponding to the solution $\bar{u}$ splits its base on the initial data plane $t = t_0$ into several domains G, in general. For definiteness, we assume that $t_0 < t^*$.

In addition to system (3), we consider the following linear hyperbolic system

$$\frac{\partial^{n_i} z_i}{\partial t^{n_i}} = \sum_{j, k_s} a_{ij}^{(k_0, k_1, \ldots, k_p)}(t, x) \frac{\partial^k z_j}{\partial t^{k_0} \partial x_1^{k_1} \cdots \partial x_p^{k_p}} \,, \tag{4}$$

$$i, j = 1, \ldots, N \,, \quad k = \sum k_s \leq n_j \,, \quad k_0 < n_j \,.$$

Here $a_{ij}^{(k_0, k_1, \ldots, k_p)}(t, x)$ are equal to the values of the derivatives

$$\partial \left\{ \frac{\partial F_i}{\dfrac{\partial^k u_j}{\partial t^{k_0} \partial x_1^{k_1} \cdots \partial x_p^{k_p}}} \right\}$$

at the point $(t, x_1, \ldots, x_p)$, provided that every $u_s(t, x_1, \ldots, x_p)$ is replaced by $\bar{u}_s(t, x_1, \ldots, x_p)$. It is evident that the characteristic cone for system (3) with vertex at $P = (t^*, x_1^*, \ldots, x_p^*)$, constructed for the solution $\bar{u}_s$, coincides with the characteristic cone for system (4) with vertex at the same point.

Theorem I. *If a domain G is a lacuna for system* (3) *in a neighborhood of the solution $\bar{u}$, then G is a lacuna for system* (4), *as well.*

It is assumed here that for any initial functions which, together with their derivatives up to a certain order, have values sufficiently close to those of $\bar{u}_i(t, x_1, \ldots, x_p)$ and their derivatives at $t = t_0$, systems (3) and (4) possess solutions satisfying these initial conditions and defined over the whole characteristic cone with vertex at P. According to my paper on the Cauchy problem (see [4]), this condition is always fulfilled whenever t_0 is sufficiently close to t^*; the latter condition on t_0 is assumed in what follows.

Proof. Without loss of generality we may assume that

$$\bar{u}_i(t, x_1, \ldots, x_p) \equiv 0 \quad \text{for all } i \quad \text{and} \quad t^* = x_1^* = \cdots = x_p^* = 0 .$$

Let G be a p-dimensional domain lying on the plane $t = t_0 < 0$ and bounded by a portion of the lateral surface of the characteristic cone with vertex at the origin. Assume that G is not a lacuna for system (4). Then one can construct functions $\varphi_i^{(h)}(x_1, \ldots, x_p)$ with the following properties:

1. All $\varphi_i^{(h)}$ vanish outside a domain $\tilde{G}$ which belongs to G, together with its boundary.

2. All $\varphi_i^{(h)}$ possess sufficiently many derivatives, whose absolute values may be assumed sufficiently small.

3. If the functions z_i satisfy system (4) and the initial conditions

$$\frac{\partial^k z_i}{\partial t^k} = \varphi_i^{(k)} \quad (k = 0, 1, \ldots, n_i - 1 ; \ i = 1, \ldots, N) \tag{5}$$

at $t = t_0$, then for some i we have $z_i \neq 0$ at the origin.

Consider the solution of system (3) satisfying the initial conditions

$$\frac{\partial^k u_i}{\partial t^k} = C\varphi_i^{(k)}(x_1, \ldots, x_p) \quad (i = 1, \ldots, N; \ k = 0, 1, \ldots, n_i - 1) \tag{6}$$

for $t = t_0 < 0$, where C is a constant, $0 \le C \le 1$, and $\varphi_i^{(k)}(x_1, \ldots, x_p)$ are the functions just mentioned. Such a solution exists for any C sufficiently small, whenever t_0 is close enough to 0; this solution continuously depends on C (see my paper [4] on the Cauchy problem). Using Hadamard's formula for representing the difference of functions of several variables (see Hadamard [5], Note 1), one can establish differentiability with respect to C for the solution of system (3) satisfying the initial conditions (6), in exactly the same way as in the case of ordinary differential equations.
    Set

$$\frac{\partial u_i}{\partial C} = v_i ,$$

where u_i satisfy system (3), as well as the initial conditions (6). Obviously, the functions v_i satisfy the following equations

$$\frac{\partial^{n_i} v_i}{\partial t^{n_i}} = \sum_{j,k_s} \frac{\partial F_i}{\partial \left\{ \dfrac{\partial^k u_j}{\partial t^{k_0} \partial x_1^{k_1} \cdots \partial x_p^{k_p}} \right\}} \frac{\partial^k v_j}{\partial t^{k_0} \partial x_1^{k_1} \cdots \partial x_p^{k_p}}$$

$$(i, j = 1, \ldots, N) ,$$

and the initial conditions at $t = t_0$:

$$\frac{\partial^k v_i}{\partial t^k} = \varphi_i^{(k)}(x_1, \ldots, x_p) \qquad (i = 1, \ldots, N; \quad k = 0, 1, \ldots, n_i - 1).$$

Here the derivatives

$$\overline{\partial \left\{ \frac{\partial^k u_j}{\partial t^{k_0} \partial x_1^{k_1} \cdots \partial x_p^{k_p}} \right\}}^{\partial F_i}$$

are calculated under the assumption that

$$u_i(t, x_1, \ldots, x_p) \qquad (i = 1, \ldots, N)$$

is a solution of system (3) satisfying the initial conditions (6) at $t = t_0$.

If the domain G is a lacuna for system (3) in a neighborhood of the identically vanishing solution, then the functions $v_i(t, x_1, \ldots, x_p)$ must vanish at the origin for all sufficiently small C, in particular, for $C = 0$. On the other hand, because of the continuous dependence of v_i on C, every v_i must tend to z_i as $C \to 0$, where the functions z_i satisfy system (4) and the initial conditions (5) for $t = t_0$. But, according to our assumption, the functions z_i do not vanish simultaneously at the origin. We have thus come to a contradiction, which proves Theorem I.

2. Consider the following system of linear hyperbolic equations

$$\frac{\partial^{n_i} u_i}{\partial t^{n_i}} = \sum_{j, k_s} a_{ij}^{(k_0, k_1, \ldots, k_p)}(t, x_1, \ldots, x_p) \frac{\partial^k u_j}{\partial t^{k_0} \partial x_1^{k_1} \cdots \partial x_p^{k_p}} \qquad (7)$$

$$(i, j = 1, 2, \ldots, N; \quad k = k_0 + k_1 + \cdots + k_p \le n_j; \quad k_0 < n_j).$$

The coefficients $a_{ij}^{(k_0, k_1, \ldots, k_p)}(t, x_1, \ldots, x_p)$ are assumed sufficiently smooth. Generally, the lateral surface of the characteristic cone with vertex at $(t^*, x_1^*, \ldots, x_p^*)$ splits its base on the initial plane $t = t_0$ into several domains G. For definiteness, we assume that $t_0 < t^*$.

Let t_0 be sufficiently close to t^*; and let G^* denote any one of the domains cut out of the plane $t = t_0$ by the lateral surface of the characteristic cone with vertex at $(t^*, x_1^*, \ldots, x_p^*)$, for the following system with constant coefficients:

$$\frac{\partial^{n_i} u_i}{\partial t^{n_i}} = \sum_{j, k_s} a_{ij}^{(k_0, k_1, \ldots, k_p)}(t^*, x_1^*, \ldots, x_p^*) \frac{\partial^{n_j} u_j}{\partial t^{k_0} \partial x_1^{k_1} \cdots \partial x_p^{k_p}} \qquad (8)$$

$$(i, j = 1, 2, \ldots, N; \quad k = k_0 + k_1 + \cdots + k_p = n_j; \quad k_0 < n_j);$$

then, for any such domain G^*, there is a unique corresponding domain G constructed for system (7) and close to G^*.

Theorem II. *If for all t_0 close enough to t^* the domain G corresponding to G^* is a lacuna for system (7), then the domain G^* is a lacuna for system (8).*

Proof. Again, let us assume that

$$t^* = x_1^* = \cdots = x_p^* = 0 \,.$$

Let G^* be a p-dimensional domain cut out of the plane $t = t_0 < 0$ by a portion of the lateral surface of the characteristic cone with vertex at the origin; suppose that G^* is not a lacuna for system (8). Then we can construct functions $\varphi_i^{(k)}(x_1, \ldots, x_p)$ with the following properties:

1. These functions identically vanish outside a domain $\tilde{G}^*$ belonging to G^*, together with its boundary.

2. The functions $\varphi_i^{(k)}$ possess sufficiently many derivatives, whose absolute values may be assumed sufficiently small.

3. If the functions v_i satisfy system (8) and the initial conditions

$$\frac{\partial^k v_i}{\partial t^k} = \varphi_i^{(k)} \qquad (k = 0, 1, \ldots, n_i - 1 \,; \ i = 1, \ldots, N)$$

   at $t = t_0$, then for some i we have $v_i \neq 0$ at the origin.

Set

$$t = \mu\tau \,, \quad x_k = \mu y_k \,, \quad u_i = \mu^{n_i} z_i \,,$$

where μ is a constant parameter whose value may vary from 0 to 1. If the functions u_i satisfy system (7), then z_i satisfy the following system

$$\frac{\partial^{n_i} z_i}{\partial \tau^{n_i}} = \sum_{j, k_s} a_{ij}^{(k_0, k_1, \ldots, k_p)}(\mu\tau, \mu y_1, \ldots, \mu y_p) \frac{\partial^k z_j}{\partial \tau^{k_0} \partial y_1^{k_1} \cdots \partial y_p^{k_p}} \, \mu^{n_j - k_0 - k_1 - \cdots - k_p}$$

$$(9)$$

$$(i, j = 1, \ldots, N \,, \quad k = k_0 + k_1 + \cdots + k_p \leq n_j \,; \ k_0 < n_j).$$

Assume that system (7) is hyperbolic in the domain

$$0 < -t < m \,, \quad |x_k| < m \,.$$

Then system (9) is hyperbolic for

$$0 < -\tau < m\mu^{-1} \,, \quad |y_k| < m\mu^{-1} \,.$$

Consider the solution of system (9) satisfying the initial conditions

$$\frac{\partial^k z_i}{\partial t^k} = \varphi_i^{(k)}(y_1, \ldots, y_p) \qquad (i = 1, \ldots, N \; ; \; k = 0, 1, \ldots, n_i - 1) \qquad (10)$$

at $\tau = \tau_0 < 0$, where $\varphi_i^{(k)}$ are the functions constructed above. According to the results of my paper on the Cauchy problem (see [4]), this solution of system (9) exists in a neighborhood of the origin, for all μ from 0 to 1, provided that τ_0 has a sufficiently small absolute value; moreover, this solution continuously depends on μ. Since the domain G is supposed to be a lacuna for system (7), all the functions z_i must vanish at the origin for sufficiently small μ such that $\tilde{G}^* \subset G$.

Let us make the parameter μ tend to 0. Then the functions z_i satisfying system (9) and the initial conditions (10) will uniformly converge to the respective functions v_i constructed above and not vanishing simultaneously at the origin. We thus come to a contradiction, which proves Theorem II.

CHAPTER II

Lacunas for Systems with Two Independent Variables

§2. On the Diffusion of Waves and the Lacunas for Linear Hyperbolic Systems with Two Independent Variables

We first note that it suffices to consider only first order systems, i.e., systems of the form

$$\frac{\partial u_i}{\partial t} = \sum_j a_{ij} \frac{\partial u_j}{\partial x} + \sum_j b_{ij} u_j + f_i \qquad (i = 1, 2, \ldots, N) \; ; \qquad (11)$$

indeed, every hyperbolic system with two independent variables can be easily reduced to the form (11) (see, for instance, [4], pp. 861–863). Because of the hyperbolicity of system (11), there exists a non-singular linear transformation

$$u_i = \sum_j c_{ij} v_j$$

with coefficients having continuous bounded derivatives with respect to t and x, which reduces system (11) to the form

$$\frac{\partial v_i}{\partial t} = \lambda_i \frac{\partial v_i}{\partial x} + \sum_j b_{ij}^* v_j + f_i^* , \tag{12}$$

where λ_i, b_{ij}^*, and f_i^* are continuous bounded functions.

In what follows, it is assumed that the number of equal $\lambda_i(t, x)$ in (12) does not depend on t, x. Then *the characteristics of system* (12), i.e., the integral curves of the equations

$$\frac{dx}{dt} = -\lambda_i , \tag{13}$$

that issue from any one point and do not coincide everywhere have no other point in common. Indeed, let λ_1 and λ_2 correspond to a pair of characteristics which do not coincide everywhere; then λ_1 cannot be identically equal to λ_2, and consequently, according to the above hypothesis, λ_1 and λ_2 do not coincide anywhere. Assume now that the characteristics in question have several points of intersection. Let A and B be any two of such points, with no other points of intersection in between. Then the values of $\lambda_1 - \lambda_2$ at the points A and B are necessarily of opposite sign. Consequently, each of the two characteristics contains a point where λ_1 and λ_2 coincide; but this is in contradiction with our assumption that the number of equal $\lambda_i(t, x)$ is independent of t and x.

In what follows, the integral curve of equation (13) will be referred to as the i-th characteristic. Denote by $P_{i,0}$ the point of intersection of the i-th characteristic issuing from P with the straight line $t = t_0$. Then we obtain the following

Theorem. *The condition*

$$b_{ij}^* \equiv 0 \tag{14}$$

$$\text{for} \quad i \leq k, \ j \geq k+1, \quad \text{or} \quad i > k, \ j < k+1 ,$$

is necessary and sufficient for system (12) *to have the domain G between $P_{k,0}$ and $P_{k+1,0}$ as a lacuna for every point $P = (t^*, x^*)$ with $t_0 < t^*$.*

Under this condition, system (12) is decomposed into two mutually independent systems. One of them consists of the first k equations, the other is formed by the rest. The portions of the straight line $t = t_0$, with the initial data on these portions determining the values of the functions v_i at the point P, do not meet. Therefore, the sufficiency of condition (14) is obvious. Let us proceed now to the proof of its necessity.

The homogeneous system (12) can be rewritten in the form

$$\frac{dw_i}{ds_i} = \sum_j b_{ij}^{**} w_j ,$$ (15)

where d/ds_i denotes the derivative along the integral curves of the ordinary differential equation (13), and

$$b_{ij}^{**} = \frac{b_{ij}^{*}}{\sqrt{1 + \lambda_i^2}} .$$

From (15) we find that

$$w_i(P) = \int_{P_{i,0}}^{P} \sum_j b_{ij}^{**} w_j \, ds_i + w_i(P_{i,0}) ,$$ (16)

where the integration is along the i-th characteristic, i.e., the integral curve of equation (13) passing through the point P.

Assume that the interval $(P_{k,0}, P_{k+1,0})$ is a lacuna, and there is at least one b_{lm}^{**}, with $l \leq k$, $m > k$, which does not vanish identically. Then, let P be a point at which $b_{lm}^{**} \neq 0$. The straight line, where the initial data are prescribed, we choose close enough to P, so that the following conditions are satisfied:

1) the segments of the characteristics starting at P and stretching up to this straight line have their lengths smaller than a positive ε;

2) the values of every coefficient b_{ij}^{**} on these segments differ from its value at P less than by ε.

Then formulas (16) yield

$$|w_i(P) - w_i(P_{i,0})| < NB^{**}\varepsilon W ,$$ (17)

where B^{**} is the upper bound for the absolute values of the coefficients b_{ij}^{**} in the domain under consideration; W is the upper bound for the absolute values of w_i.

In order to determine W, we use the inequality

$$W \leq NB^{**}Ws + W_0 ,$$ (18)

which can be easily derived from (16); here s is the upper bound for the lengths of the characteristics in our domain; W_0 is the upper bound for the absolute values of $w_i(P_{i,0})$.

Assume that the domain under consideration is so small that

$$NB^{**}s < \frac{1}{2} .$$

Then (18) yields

$$W < 2W_0 .$$

Substituting $2W_0$ for W in (17), we get

$$|w_i(P) - w_i(P_{i,0})| < 2NB^{**}\varepsilon W_0 . \tag{19}$$

Let us now prescribe the initial data on the line $t = t_0$ as follows: for $t = t_0$, all w_i vanish identically, except for w_m; the latter function is equal to 0 everywhere outside an interval L belonging, together with its end-points, to the interval $(P_{k,0}, P_{k+1,0})$; moreover, $w_m = 1$ on the major portion of L, while on subintervals L_1 and L_2 (whose length may be assumed less than η) adjacent to the end-points of L, the function w_m drops down to zero in a monotone and smooth fashion.

Then the inequalities (19) yield for the entire domain:

$$|w_i(Q)| < 2NB^{**}\varepsilon \quad \text{if} \quad i \neq m .$$

For $i = m$ we obtain

$$|w_m(Q) - 1| < 2N^{**}\varepsilon \quad \text{if} \quad Q_{m,0} \subset L - L_1 - L_2 ,$$

$$|w_m(Q)| < 2NB^{**}\varepsilon \quad \text{if} \quad Q_{m,0} \not\subset L ,$$

$$|w_m(Q)| < 1 + 2NB^{**}\varepsilon \quad \text{elsewhere} .$$

Here the points Q and $Q_{i,0}$ are related in the same way as P and $P_{i,0}$. Using these estimates for w_j at intermediate points of the l-th characteristic on the segment $PP_{l,0}$, and applying (16) with $i = l$, we obtain

$$|w_l(P)| > (|B_{lm}(P)| - \varepsilon)[1 - 2NB^{**}\varepsilon](s_L - 2\eta)c - 2\eta[1 + 2NB^{**}\varepsilon]B^{**}C -$$
$$- 2NB^{**2}s\varepsilon - (N-1)B^{**}2NB^{**}s\varepsilon , \tag{20}$$

where s_L is the length of the interval L.

The first three terms in the right-hand side of the above inequality estimate the integral

$$\int_{P_{l,0}}^{P} b_{lm}^{**} w_m \, ds_l . \tag{21}$$

The first term estimates the value of this integral over the portion of the characteristic $PP_{l,0}$ crossed by the characteristics with index m passing through the points of $L - L_1 - L_2$; here c (resp., C) is the positive lower (resp., upper) bound for the ratio of the lengths of the segments cut out from the characteristic $PP_{l,0}$ and from the line $t = t_0$ by the two characteristics with index m. The second term estimates the value of the integral (21) over that part of the characteristic $PP_{l,0}$ which is crossed by the characteristics with index m passing through the points of L_1 and L_2. The third term corresponds to the remaining part of the integral. The last term in the right-hand side of (20) estimates the integrals of the other summands of the integrand in (16).

For sufficiently small ε and η, the right-hand side of (20) is positive, and therefore

$$w_l(P) \neq 0 ,$$

which inequality is incompatible with our assumption of the interval $(P_{k,0}, P_{k+1,0})$ being a lacuna.

Thus, in the case $l \leq k$, $m \geq k + 1$, the assumption that there is a coefficient b_{lm}^{**} (and therefore, b_{lm}^{*}) different from 0 at some point has lead us to a contradiction. The fact that all b_{ij}^{*} vanish identically, in the case $i > k$, $j < k + 1$, can be established in the same way.

CHAPTER III

Derivation of the Formulas for the Solution of the Cauchy Problem for a Single Hyperbolic Equation with Constant Coefficients

In what follows, we shall widely use a slightly modified version of the formulas obtained by Herglotz (see [6], [7]) for the solution of the Cauchy problem. Since our objectives require that Herglotz's formulas be modified and generalized, and since the original, extremely important, works of Herglotz have been published in a hardly available journal, it seems appropriate to give here a detailed derivation of these formulas in a form suitable for our purposes.

§3. Setting of the Problem

1. Consider a homogeneous linear hyperbolic equation of order[2] n, involving only the n-th order derivatives of a single unknown function u. Let us write this equation in the form

---

[2] Herglotz considers only the case of n being even; in our case, n is an arbitrary positive integer. For p even, Herglotz's formulas are different from ours.

$$\Delta u = 0 , \tag{22}$$

where the operator Δ can be symbolically represented as a homogeneous polynomial of degree n with respect to

$$\frac{\partial}{\partial t} , \frac{\partial}{\partial x_1} , \ldots , \frac{\partial}{\partial x_p} .$$

The corresponding characteristic equation

$$\Delta(\lambda, \xi_1, \ldots, \xi_p) = 0 \tag{23}$$

has, in its left hand side, the homogeneous polynomial of degree n obtained from the operator Δ by replacing

$$\frac{\partial}{\partial t} , \frac{\partial}{\partial x_1} , \ldots , \frac{\partial}{\partial x_p} \quad \text{by} \quad \lambda, \xi_1, \ldots, \xi_p ,$$

respectively. For the sake of definiteness, we assume the coefficient by λ^n to be equal to 1.

2. The Cauchy problem can be formulated as follows: *for $t \geq 0$, find a solution of equation* (22) *satisfying the conditions*

$$u = \frac{\partial u}{\partial t} = \cdots = \frac{\partial^{n-2} u}{\partial t^{n-2}} = 0 , \quad \frac{\partial^{n-1} u}{\partial t^{n-1}} = f(x_1, \ldots, x_p) \tag{24}$$

for $t = 0$.

If this solution be denoted by $u_{n-1}(f)$, then the solution $u_k(f)$ of the same equation supplemented by the following conditions at $t = 0$:

$$\frac{\partial^k u}{\partial t^k} = f(x_1, \ldots, x_p) ,$$

$$\frac{\partial^m u}{\partial t^m} = 0 \quad \text{for} \quad m \neq k, \ 0 \leq m \leq n - 1 ,$$

is given by

$$\frac{\partial u_{n-1}(f)}{\partial t} - u_{n-1} \left(\frac{\partial^n u_{n-1}(f)}{\partial t^n} \right) \quad \text{for} \quad k = n - 2 , \tag{25}$$

and

$$\frac{\partial^2 u_{n-1}(f)}{\partial t^2} - u_{n-2} \left(\frac{\partial^n u_{n-1}(f)}{\partial t^n} \right) -$$

$$- u_{n-1} \left(\frac{\partial^{n+1} u_{n-1}(f)}{\partial t^{n+1}} - \frac{\partial^{n-1}}{\partial t^{n-1}} \left(u_{n-2} \left(\frac{\partial^n u_{n-1}(f)}{\partial t^n} \right) \right) \right) , \tag{25'}$$

for $k = n - 3$, etc.

Therefore, without loss of generality, we may restrict ourselves to the initial conditions of the form (24).

By assumption, equation (22) is hyperbolic; therefore, equation (23) specifies mutually non-intersecting ovals in the $(\xi_1/\lambda, \ldots, \xi_p/\lambda)$-space; the number of these ovals is $[n/2]$, and they are consecutively enclosed in one another so that every one of them contains the origin; in addition to these ovals, if n is odd, (23) specifies a single unpaired piece, because every real straight line passing through the origin must have n separate real points of intersection with the surface given by (23) (see [8], Ch. 3, §2). It is clear that these ovals cannot have real singular points; otherwise, any such point would at least be a double one, and therefore, being joined with the origin, would produce a straight line having more than n points of intersection with the surface (23) of order n.

Some statements and proofs suggested by Herglotz pertain only to the case of the ovals having no infinite points; however, this condition is non-essential for our analysis.

Throughout the next few sections, we assume that

$$n \geq p + 2 .$$

It is shown in §10 and §11 how this restriction can be dropped.

§4. Application of the Fourier Integrals

1. Let the initial function $f(x_1, \ldots, x_p)$ be represented by the Fourier integral

$$f(x_1, \ldots, x_p) = \int g(\xi_1, \ldots, \xi_p) \, e^{i(x,\xi)} \, d\xi_1 \ldots d\xi_p ,$$

where $(x, \xi) = \sum_{k=1}^{p} x_k \xi_k$,

$$g(\xi_1, \ldots, \xi_p) = \frac{1}{(2\pi)^p} \int f(y, \ldots, y_p) \, e^{-i(y,\xi)} \, dy_1 \ldots dy_p . \qquad (26)$$

Set

$$\Gamma(t, x, \xi) = \frac{1}{2\pi i^n} \oint \frac{e^{i(x,\xi) + it\tau}}{\Delta(\tau, \xi)} \, d\tau ,$$

where the integral is taken over a closed curve belonging to the complex τ-plane and encircling all the roots τ of the polynomial

$$\Delta(\tau, \xi) = 0 . \qquad (27)$$

It is easy to see that

$$\Delta\Gamma(t, x, \xi) = 0 ;$$

and for $t = 0$ we have

$$\Gamma = \frac{\partial \Gamma}{\partial t} = \cdots = \frac{\partial^{n-2}\Gamma}{\partial t^{n-2}} = 0 \; ; \qquad \frac{\partial^{n-1}\Gamma}{\partial t^{n-1}} = e^{i(x,\xi)} \; .$$

Formally, let us construct the following integral

$$u(x,t) = \int g(\xi)\Gamma(t,x,\xi)\,d\xi \; ,$$

which clearly satisfies the initial conditions, as well as the given differential equation, provided that the differentiation under the sign of the integral is admissible. Let us replace $g(\xi)$ by its expression (26). Changing the order of integration, we obtain

$$u(x,t) \;=\; \int K(t,x-y)f(y)\,dy \; , \tag{28}$$

$$K(t,x) \;=\; \frac{1}{(2\pi)^{p+1}i^n}\int e^{i(x,\xi)}\oint \frac{e^{it\tau}}{\Delta(\tau,\xi)}\,d\tau\,d\xi \; . \tag{29}$$

2. The transformations in the preceding section are purely formal; no attention has been given to their justification. In order to justify them, we have to show that the right-hand side of (29) makes sense, and the right-hand side of (28) satisfies the initial conditions (24), as well as the differential equation (22).

Let us first prove that the right-hand side of (29) makes sense. To this end, we note that

$$\oint \frac{e^{it\tau}}{\Delta(\tau,\xi)}\,d\tau = 2\pi i\sum_\tau \frac{e^{it\tau}}{\Delta'_\tau(\tau,\xi)} = 2\pi i\sum_\tau \frac{e^{it\tau} - g_\nu(it\tau)}{\Delta'_\tau(\tau,\xi)} \; , \tag{30}$$

$$0 \le \nu \le n-2 \; ,$$

where the summation is extended over the roots τ of the polynomial $\Delta(\tau,\xi)$, and

$$g_\nu(x) = 1 + \frac{x}{1} + \frac{x^2}{1\cdot 2} + \cdots + \frac{x^\nu}{\nu!} \; .$$

The last transformation in (30) is based on Euler's formula which yields

$$\sum_\tau \frac{M(\tau)}{P'(\tau)} = 0 \tag{31}$$

for any polynomial $M(\tau)$ of degree ν with respect to τ, provided that M has no multiple roots and $\nu < n-1$; the summation is extended over all roots of the polynomial $P(\tau)$ of degree n. Set

$$\frac{\xi_\alpha}{\sqrt{\xi_1^2 + \cdots + \xi_p^2}} = \eta_\alpha , \qquad \frac{\tau}{\sqrt{\xi_1^2 + \cdots + \xi_p^2}} = T .$$

Then η_α and T satisfy the equation

$$\Delta(T, \eta_1, \ldots, \eta_p) = 0 , \tag{32}$$

where

$$\eta_1^2 + \cdots + \eta_p^2 = 1 . \tag{33}$$

Because of the assumptions about the polynomial $\Delta(\tau, \xi)$ (see §3), T is bounded on the entire sphere (33), and therefore

$$|\tau| \le C\sqrt{\xi_1^2 + \cdots + \xi_p^2} , \tag{34}$$

where C is an absolute constant and τ is an arbitrary root of equation (27). On the other hand, all the roots T of equation (32) on the entire real sphere (33) being different from one another, we have

$$\Delta_T'(T, \eta_1, \ldots, \eta_p) \ne 0$$

on this sphere.

Furthermore, since Δ_T' is continuous on this sphere, we have

$$C_1 \ge |\Delta_T'(T, \eta_1, \ldots, \eta_p)| \ge C_0 > 0$$

therein, where C_0 and C_1 are absolute constants. Thus, for any real $\xi_1, \ldots, \xi_p$, we have

$$C_1(\xi_1^2 + \cdots + \xi_p^2)^{\frac{n-1}{2}} \ge |\Delta_\tau'(\tau, \xi_1, \ldots, \xi_p)| \ge C_0(\xi_1^2 + \cdots + \xi_p^2)^{\frac{n-1}{2}} .$$

Hence, using the inequality (34), we obtain for the function

$$\varphi(t, \xi) = \sum_\tau \frac{e^{it\tau}}{\Delta_\tau'(\tau, \xi)} = \sum_\tau \frac{e^{it\tau} - g_\nu(it\tau)}{\Delta_\tau'(\tau, \xi)}$$

the following estimates:

$$|\varphi(t, \xi)| = \left| \sum_\tau \frac{e^{it\tau} - g_{n-2}(it\tau)}{\Delta_\tau'(\tau, \xi)} \right| \le C_2$$

if $\xi_1^2 + \cdots + \xi_p^2 < 1$, and

$$|\varphi(t, \xi)| = \left| \sum_\tau \frac{e^{it\tau}}{\Delta_\tau'(\tau, \xi)} \right| \le C_3(\xi_1^2 + \cdots + \xi_p^2)^{\frac{n-1}{2}}$$

if $\xi_1^2 + \cdots + \xi_p^2 \ge 1$. It follows that the integral

$$K(t, x) = \frac{1}{(2\pi)^p i^{n-1}} \int e^{i(x,\xi)} \varphi(t, \xi) \, d\xi$$

exists for all real t and x, provided that $n - p \geq 2$.

It is easy to see that the derivatives of $\varphi(t, \xi)$, of any order with respect to ξ_α, exist and are absolutely integrable (cf. (30)); here τ is regarded as a function of ξ_α determined by equation (27).

Consequently, the Fourier inversion formula is valid for $\varphi(t, \xi)$, and therefore, in the case of

$$f(y) = e^{i(\xi, y)}$$

the formula

$$u(t, x) = \int K(t, x - y) e^{i(\xi, y)} \, dy$$

yields

$$u(t, x) = \frac{e^{i(x, \xi)}}{2\pi i^n} \oint \frac{e^{it\tau}}{\Delta(\tau, \xi)} \, d\tau \,,$$

which function is the solution of the Cauchy problem.

In a neighborhood G_0 of any given point $P = (t, x)$, the solution of the Cauchy problem for the hyperbolic equation depends on the values of initial functions only at the points of a certain bounded domain G_1 on the plane $t = 0$, and therefore, while studying this solution in G_0, we can arbitrarily define the initial data outside G_1. Keeping this in mind, suppose that the initial function $f(x)$ vanishes outside a bounded domain. Since, in the present paper, we are not concerned with solving the Cauchy problem for the widest possible class of initial functions, we assume $f(x)$ to have sufficiently many continuous derivatives. Such functions can be expanded into Fourier series in a sufficiently large rectangular parallelepiped, the rate of convergence of the Fourier coefficients to zero being fast enough. We may then solve the Cauchy problem for every single term of the Fourier series (as the initial function), and after that take the sum of all the solutions obtained. It is not difficult to show that the resulting series is uniformly convergent, together with the series obtained from it after term-by-term differentiation, and therefore represents the solution of the Cauchy problem for the given initial function (cf. a similar argument in [8]). It follows that (28) also yields the solution of the Cauchy problem with the initial conditions (24).

§5. Transformation of $K(t, x)$

For our further purposes we transform the expression

$$K(x, t) = \frac{1}{(2\pi n)^p i^{n-1}} \int e^{i(x, \xi)} \sum_\tau \frac{e^{it\tau} - g_m(it\tau)}{\Delta'_\tau(\tau, \xi)} \, d\xi \,,$$

where $m = n - p - 1$, as follows. Instead of $\xi_1, \ldots, \xi_p$ we introduce new independent variables $p_1, \ldots, p_{p-1}, \tau$ related to the former ones by

$$\xi_1 = p_1\tau, \quad \ldots, \quad \xi_{p-1} = p_{p-1}\tau, \quad \xi_p = p_p\tau,$$

where τ is any one of the n functions of $\xi_1, \ldots, \xi_p$ determined by the equation

$$\Delta(\tau, \xi_1, \ldots, \xi_p) \equiv \tau^n H(p_1, \ldots, p_p) = 0.$$

We thus obtain

$$d\xi_1 \ldots d\xi_p = \frac{\tau^{p-1}}{H_p}[H_1 p_1 + \cdots + H_p p_p]\, dp_1 \ldots dp_{p-1}\, d\tau, \qquad (35)$$

where

$$H_k = \frac{\partial H}{\partial p_k}.$$

Applying the Euler formula to the homogeneous polynomial

$$\Delta(\tau, \xi_1, \ldots, \xi_p)$$

of degree n and arguments $\tau, \xi_1, \ldots, \xi_p$, we get

$$\sum \frac{\partial \Delta}{\partial \xi_k}\xi_k + \frac{\partial \Delta}{\partial \tau}\tau \equiv n\Delta(\tau, \xi) = 0,$$

or

$$\sum \tau^n H_k p_k + \tau \Delta_\tau'(\tau, \xi) = 0. \qquad (36)$$

Finding $\sum H_k p_k$ from (36) and substituting the obtained expression into (35), we get

$$d\xi_1 \ldots d\xi_p = -\frac{\tau^{p-n}}{H_p}\Delta_\tau'(\tau, \xi)\, dp_1 \ldots dp_{p-1}\, d\tau. \qquad (37)$$

The differentials $d\xi_1, \ldots, d\xi_p$ are always assumed to be positive; the same condition will also be imposed on $dp_1 \ldots, dp_{p-1}$ and $d\tau$. Then (37) can be rewritten in the form

$$d\xi_1 \ldots d\xi_p = \left|\frac{\tau^{p-n}\Delta_\tau'(\tau, \xi)}{H_p}\right| dp_1 \ldots dp_{p-1}\, d\tau.$$

We notice that (36) implies

$$\Delta_\tau'(\tau, \xi) = -\tau^{n-1}\sum H_k p_k =$$

$$= -\frac{\tau^{n-1}}{2}\left(\operatorname{grad} H(p_1, \ldots, p_p) \cdot \operatorname{grad}\left(p_1^2 + \cdots + p_p^2\right)\right).$$

Hence

$$d\xi_1 \cdots d\xi_p = -\frac{\Delta'_\tau(\tau, \xi)\tau^{p-n}}{|H_p|} \operatorname{sgn} \tau^{p-1} \times$$

$$\times \operatorname{sgn} \left[\operatorname{grad} H(p) \cdot \operatorname{grad} (p_1^2 + \cdots p_p^2) \right] dp_1 \ldots dp_{p-1} \, d\tau \; .$$

Consequently, we have

$$K(t, x) = \tag{38}$$

$$= -\frac{1}{(2\pi)^p i^{n-1}} \int d\omega \int_{-\infty}^{+\infty} \frac{e^{i t \tau} - g_m(i t \tau)}{\tau^{m+1}} \exp\left(i \sum x_k p_k \tau\right) \operatorname{sgn} \tau^{p-1} \, d\tau \; ,$$

where the integration is over the entire real hypersurface

$$H(p_1, \ldots, p_p) = 0 \; ; \tag{39}$$

$$d\omega = \frac{dp_1 \ldots dp_{p-1}}{|H_p|} \operatorname{sgn} \left[\operatorname{grad} H(p) \cdot \operatorname{grad} \left(p_1^2 + \cdots + p_p^2 \right) \right] \; , \tag{40}$$

the sign of $d\omega$ being chosen positive or negative, according as the function H increases or decreases in the vicinity of the corresponding point on the hypersurface (39), while passing along the straight line from the origin to that point. Consecutive integration, first with respect to τ, and then with respect to $p_1, \ldots, p_{p-1}$, is possible because of the absolute convergence of the integral over the entire $(p_1, \ldots, p_{p-1}, \tau)$-space.

§6. Calculation of the Integral $\int\limits_{-\infty}^{+\infty} e^{i a \tau} \frac{e^{i b \tau} - g_m(i b \tau)}{\tau^{m+1}} \operatorname{sgn} \tau^{p-1} \, d\tau$

By the formula for the remainder term in the Taylor series, we have

$$\frac{e^x - g_m(x)}{x^{m+1}} = \frac{1}{m!} \int_0^1 e^{\theta x} (1 - \theta)^m \, d\theta \; .$$

Therefore

$$e^{i a \tau} \frac{e^{i b \tau} - g_m(i b \tau)}{(i b \tau)^{m+1}} = \frac{1}{m!} \int_0^1 e^{(a + b\theta)\tau i} (1 - \theta)^m \, d\theta$$

and

$$\begin{aligned}
I(M) &= \int_0^M \frac{e^{i b \tau} - g_m(i b \tau)}{\tau^{m+1}} e^{i a \tau} \, d\tau = \\
&= \frac{(ib)^{m+1}}{m!} \int_0^1 \frac{e^{(a + b\theta)Mi} - 1}{(a + b\theta)i} (1 - \theta)^m \, d\theta = \\
&= \frac{i^m}{m!} \int_0^b \frac{e^{(a + \theta)Mi} - 1}{a + \theta} (b - \theta)^m \, d\theta \; .
\end{aligned}$$

We assume a and b to be real numbers such that $a \neq 0$ and $a + b \neq 0$. Since the integrand is holomorphic in any finite domain of the values of θ, the integral does not depend on the path of integration. We choose the path of integration in the upper halfplane θ if $M \to +\infty$, and in the lower one if $M \to -\infty$. Then we get

$$I(\pm\infty) = -\frac{i^m}{m!} \int_0^b \frac{(b - \theta)^m}{a + \theta} \, d\theta ,$$

where the path of integration lies in the upper halfplane for $I(+\infty)$, and in the lower one for $I(-\infty)$. The last integral can be easily written in terms of following two integrals:

$$\int_0^b \frac{(a + b)^m - (b - \theta)^m}{a + \theta} \, d\theta = \sum_{\nu=1}^m \frac{(a + b)^{m-\nu} b^\nu}{\nu} ,$$

$$\int_0^b \frac{d\theta}{a + \theta} = \ln\left|\frac{a + b}{a}\right| \mp \frac{\pi i}{2}[\operatorname{sgn}(a + b) - \operatorname{sgn} a] ,$$

the upper sign being taken for $I(+\infty)$ and the lower one for $I(-\infty)$. Set

$$\Phi_m(b, a) = (a + b)^m \ln\left|\frac{a + b}{a}\right| - \sum_{\nu=1}^m \frac{(a + b)^{m-\nu} b^\nu}{\nu} ,$$

$$\tag{41}$$

$$\Psi_m(b, a) = \frac{\pi}{2}(a + b)^m[\operatorname{sgn}(a + b) - \operatorname{sgn} a] .$$

Then

$$I(\pm\infty) = \frac{i^m}{m!}[-\Phi_m(b, a) \pm i\Psi(b, a)] .$$

Hence

$$\int_{-\infty}^{+\infty} \frac{e^{ib\tau} - g_m(ib\tau)}{\tau^{m+1}} e^{ia\tau} \operatorname{sgn} \tau^{p-1} \, d\tau = 2\frac{i^{m+1}}{m!}\Psi_m(b, a) \tag{42}$$

for p odd, and

$$\int_{-\infty}^{+\infty} \frac{e^{ib\tau} - g_m(ib\tau)}{\tau^{m+1}} e^{ia\tau} \operatorname{sgn} \tau^{p-1} \, d\tau = 2\frac{i^{m+1}}{m!}\Phi_m(b, a) \tag{43}$$

for p even.

§7. Transformation of $K(t, x)$ (continued)

1. The integrand in the right-hand side of (38) contains integrals of the type just considered, where both cases $a = 0$ and $a + b = 0$ are possible. However, the set of points $(p_1, \ldots, p_{p-1})$ where this may happen is of zero measure, and therefore can be disregarded.

Using (42) and (43), we obtain, on account of (38), the following relations

$$K(t,x) = \frac{-2i}{(2\pi i)^p m!} \int \Psi_m \left(t, \sum x_k p_k \right) d\omega \tag{44'}$$

for p odd, and

$$K(t,x) = \frac{2}{(2\pi i)^p m!} \int \Phi_m \left(t, \sum x_k p_k \right) d\omega \tag{44''}$$

for p even.

By virtue of (41), the first of these equalities can be rewritten as

$$K(t,x) = \frac{-2\pi i}{(2\pi i)^p m!} \int_{(0,t)} \left(\sum x_k p_k + t \right)^m d\omega , \tag{44}$$

where the symbol $\int_{(0,t)} d\omega$ implies integration over the part of the real hypersurface (39) lying between the planes

$$\sum x_k p_k + t = 0 \quad \text{and} \quad \sum x_k p_k = 0 .$$

2. Remark. Formula (44) is valid for the solution of the *generalized Cauchy problem*: find a function $u(t, x_1, \ldots, x_p)$ which satisfies the conditions (24) for $t = 0$ and can be represented by the integral

$$u(t,x) = \int g(\xi) v(t,x) \, e^{(x,\xi)} \, d\xi_1 \ldots d\xi_p ,$$

where $g(\xi)$ is the Fourier transform of the initial function $f(x_1, \ldots, x_p)$; $v(t, \xi_1, \ldots, \xi_p)$ satisfies the differential equation

$$H \left(\frac{d}{dt}, \xi_1, \ldots, \xi_p \right) v = 0 .$$

Here $H(\lambda, \xi_1, \ldots, \xi_p)$ is assumed to be a sufficiently smooth homogeneous function of degree n and arguments $\lambda, \xi_1, \ldots, \xi_p$; moreover, H is a polynomial of degree n with respect to λ, and admits only real roots that are finite and different from one another for all real $\xi_1, \ldots, \xi_p$ such that $|\xi| \neq 0$; $H(\lambda, 0, \ldots, 0) \equiv \lambda^n$. Under these conditions we obviously have

$$v(t,\xi) = \frac{1}{2\pi i^n} \oint \frac{e^{it\tau}}{\Delta(\tau, \xi)} \, d\tau .$$

For the usual Cauchy problem, formulas (44) can be derived in a much simpler way.

3. In what follows, it would be convenient, especially in the case of p even, to have an expression for $K(t,x)$ somewhat different from that obtained in Sect. 2. To this end, while integrating over the hypersurface (39), we introduce new independent variables $q_1,\ldots,q_{p-1}$ instead of $p_1,\ldots,p_{p-1}$, setting

$$p_1 = qq_1,\ldots,p_{p-1} = qq_{p-1},p_p = q, \tag{45}$$

where $q = p_p$ is regarded as a function of $p_1,\ldots,p_{p-1}$ uniquely determined by its equation (39) in a neighborhood of each point of our hypersurface. Then we have

$$|d\omega| = \left|\frac{dp_1 \ldots dp_{p-1}}{H_p}\right| = \left|\frac{q^{p-1} dq_1 \ldots dq_{p-1}}{\frac{dH}{dq}}\right|,$$

where

$$\frac{dH}{dq} = \frac{d}{dq} H(qq_1,\ldots,qq_{p-1},q).$$

While performing the last differentiation, we must take into account the dependence of H on q as many times as q enters H and assume $q_1,\ldots,q_{p-1}$ to be constant. Recalling our convention about the sign of $d\omega$, we can finally write

$$d\omega = \frac{|q|^{p-1}\operatorname{sgn} q\, dq_1 \ldots dq_{p-1}}{\frac{dH}{dq}}. \tag{46}$$

Hence, applying Euler's formula (31) for the roots q of the polynomial

$$H(qq_1,\ldots,qq_{p-1},q)$$

of degree n and argument q, we easily find that

$$\int h(q)\,d\omega = \int \sum_q \frac{h(q)|q|^{p-1}\operatorname{sgn} q}{\frac{dH}{dq}}\,dq_1 \ldots dq_{p-1} = 0,$$

provided that $h(q)$ is a polynomial in q of a degree less than $n - p$ for p even, or

$$h(q) = M(q)\operatorname{sgn} q,$$

where $M(q)$ is a polynomial in q of a degree less than $n - p$, for p odd. Hence we obtain some important corollaries.

a) In the case of p odd we have

$$K(t,x) = -\frac{1}{2(2\pi i)^{p-1}m!}\int \left(\sum x_k p_k + t\right)^m \operatorname{sgn}\left(\sum x_k p_k + t\right)\,d\omega, \tag{44*}$$

since

$$\int \left(\sum x_k p_k + t\right)^m \text{ sgn} \left(\sum x_k p_k\right) d\omega = 0 \ .$$

For the same reason, $K(0, x) = 0$ and $K(t, x) = 0$, if the hyperplane $\sum x_k p_k + t = 0$ crosses every oval of the hypersurface (39). The latter case can be reduced to the case $t = 0$ by a parallel translation of the coordinate axes which shifts the origin to the inside of the smallest oval.

b) In the case of p even, we have[3]

$$K(x, t) = \frac{2}{(2\pi i)^p m!} \int \left(\sum x_k p_k + t\right)^m \ln \left|\frac{\sum x_k p_k + t}{\sum x_k p_k}\right| d\omega \ . \qquad (47)$$

In particular, according to the above remark, we have

$$K(0, x) = 0 \ .$$

It can be easily seen that if the plane $\sum x_k p_k + t = 0$ crosses every oval of the hypersurface (39), then we also have $K(t, x) = 0$, since this case can be reduced to the case $t = 0$ by a parallel translation of the coordinate axes which places the origin inside the smallest oval.

§8. Further Transformation of $K(t, x)$

The expressions (47) are still not quite convenient for our further analysis. In order to have more suitable formulas, let us consider the function

$$\varphi(s, x) = \int \frac{d\omega}{\sum x_k p_k + s} \ , \qquad (48)$$

where s is a complex number, and the integral is taken over the whole real hypersurface (39). Since we have assumed that $n \geq p + 2$, the integral (48) exists and

$$\int_{i\varepsilon}^{t+i\varepsilon} \varphi(s, x) \, ds = \int_{i\varepsilon}^{t+i\varepsilon} ds \int \frac{d\omega}{\sum x_k p_k + s} \ ,$$

where ε is a positive real number, and the integral is along the rectilinear segment joining the points $i\varepsilon$ and $t + i\varepsilon$. While calculating the last integral, we are justified to change the order of integration with respect to s and ω. We get

$$\int_{i\varepsilon}^{t+i\varepsilon} \frac{ds}{X + s} = \ln \left|\frac{X + t + i\varepsilon}{X + i\varepsilon}\right| + i[\,\text{ampl}\,(X + t + i\varepsilon) - \text{ampl}\,(X + i\varepsilon)] \ ,$$

$$\int_{i\varepsilon}^{t+i\varepsilon} \varphi(s, x) \, ds = \int \ln \left|\frac{X + t + i\varepsilon}{X + i\varepsilon}\right| d\omega +$$

$$+ i \int [\,\text{ampl}\,(X + t + i\varepsilon) - \text{ampl}\,(X + i\varepsilon)] \, d\omega \ ,$$

---

[3] Formulas (44*), (47) coincide with those suggested by Herglotz.

where

$$X = \sum x_k p_k .$$

Hence, passing to the limit as $\varepsilon \to 0$, we obtain

$$\int_0^t \varphi(s, x) \, ds = \int \ln \left| \frac{E}{X} \right| d\omega - \pi i \int_{(0,t)} d\omega ,$$

where $E = \sum x_k p_k + t$.

Integrating both sides of this equality k more times with respect to t from 0 to t, we get

$$\int_0^t (t - s)^k \varphi(s, x) \, ds = \int E^k \ln \left| \frac{E}{X} \right| d\omega - \pi i \int_{(0,t)} E^k \, d\omega =$$

$$= \Phi_k^*(t, x) + i \Psi_k^*(t, x) ,$$

where $\Phi_k^*(t, x)$ and $\Psi_k^*(t, x)$ are real valued functions. Comparing these formulas with (44') and (44''), we obtain

$$K(t, x) = \frac{2i}{(2\pi i)^p m!} \Psi_m^*(t, x) = \frac{2i}{(2\pi i)^p m!} \int_0^t (t - s)^m v(s, x) \, ds \qquad (49)$$

for p odd, and

$$K(t, x) = \frac{2}{(2\pi i)^p m!} \Phi_m^*(t, x) = \frac{2}{(2\pi i)^p m!} \int_0^t (t - s)^m u(s, x) \, ds \qquad (50)$$

for p even, provided that $\varphi(s + i0, x) = u(s, x) + iv(s, x)$.

§9. The Functions $\varphi(t, x), \Phi_m^*(t, x), \Psi_m^*(t, x)$

1. For our further purposes it is necessary to transform the expression for $\varphi(t, x)$. This transformation turns out to be different for p odd and p even. First, consider the case of p odd.[4] To simplify calculations, we assume that $x_p \neq 0$ and the rest of x_k are equal to 0 (the case of all x_k being equal to zero is not considered here). Then

$$E \equiv x_p^* p_p^* + t .$$

This case is obtained from the general one by a rotation of the coordinate axes in the $(p_1, \ldots, p_p)$-space, so that the plane

$$x_1 p_1 + \cdots + x_p p_p = 0$$

---

[4]In this case no substantially new results are obtained here, and it is merely for the sake of completeness that we consider odd p.

becomes

$$x_1 p_1 + \cdots + x_p p_p \equiv x_p^* p_p^* = 0 . \tag{51}$$

It is evident that this rotation does not change the surface element do and $\sqrt{H_1^2 + \cdots + H_p^2}$; therefore, their ratio $d\omega$ also remains invariant. Moreover, $x_p^* = \sqrt{x_1^2 + \cdots + x_p^2}$, and p_p^* is equal to the distance from the origin to the plane $E = 0$. Keeping in mind that, to within infinitesimals of higher order, the product of an infinitesimal η by the integral of $d\omega$ over a part of the real surface (39) is equal to thé volume enclosed between the said part of the surface $H = 0$ and the corresponding part of the surface $H = \eta$, we can easily see that

$$|d\omega| = \left| \frac{do}{\sqrt{H_1^2 + \cdots + H_p^2}} \right| = \left| \frac{do_q \, dp_p^*}{\sqrt{H_1^{*2} + \cdots + H_{p-1}^{*2}}} \right| ,$$

where do_q is an element of the $(p-2)$-dimensional surface cut off from the hypersurface (39) by the hyperplane $p_p^* = q = $ const,

$$H_k^* = \frac{\partial}{\partial p_k^*} H^*(p_1^*, \ldots, p_p^*) , \qquad H^*(p_1^*, \ldots, p_p^*) \equiv H(p_1, \ldots, p_p) .$$

Set

$$d\omega_q = \frac{|do_q|}{\sqrt{H_1^{*2} + \cdots + H_{p-1}^{*2}}} \, \mathrm{sgn} \left(\mathrm{grad}\, H(p) \cdot \mathrm{grad}\, \sqrt{p_1^2 + \cdots + p_p^2} \right)$$

if

$$p_p^* = q , \tag{51'} ;$$

and let

$$\chi(x_1, \ldots, x_p, q) = \int d\omega_q ,$$

where the integral is taken over the entire real intersection of the hypersurface (39) and the hyperplane (51'). Then we can write

$$\varphi(t + i\varepsilon, x_1, \ldots, x_p) = \int_{-\infty}^{+\infty} \frac{\chi(x_1, \ldots, x_p, q) \, dq}{x_p^* q + t + i\varepsilon} .$$

Set $\varphi = u + iv$, where u and v take real values. Applying the well-known arguments of the potential theory, we obtain

$$v(t, x_1, \ldots, x_p) = \frac{1}{2i}[\varphi(t + 0i, x) - \varphi(t - 0i, x)] =$$

$$= -\frac{\pi}{x_p^*} \chi \left(x_1, \ldots, x_p, -\frac{t}{x_p^*} \right) .$$

2. Remark. If the complex surface

$$H^*(p_1^*, \ldots, p_{p-1}^*, p_p^*) = 0 , \qquad p_p^* = q = \text{const} , \qquad (52)$$

has no singular points, then the differential

$$\frac{do_q}{\sqrt{H_1^{*2} + \cdots + H_{p-1}^{*2}}} = \frac{dp_1^* \ldots dp_{p-2}^*}{H_{p-1}^*} = \frac{dp_1^* \ldots dp_{k-1}^* dp_{k+1}^* \ldots dp_{p-1}^*}{H_k^*}$$

on this surface is of the 1st kind, and therefore, any one of its integrals is finite. But the requirement that the complex surface (52) should have no singular points is quite unnecessary for justifying the arguments of the present section and thereby attributing a certain sense to the integrals obtained. It suffices that the real cycles, along which the corresponding integrals are taken, do not pass through singular points. Under this condition, our considerations become applicable, in particular, to the case of a reducible surface (39). As to the formulas (44) and (47), they are always valid if the surface (39) has no real singular points.

3. Consider now the case of p even. Let us introduce new independent variables $q_1, \ldots, q_{p-1}$ related to $p_1, \ldots, p_{p-1}$ by (45). Then, by virtue of (46), we get

$$\varphi(t + i\varepsilon, x_1, \ldots, x_p) = \int \sum_q \frac{q^{p-1}}{(q \sum x_k q_k + t + i\varepsilon) \frac{dH}{dq}} dq_1 \ldots dq_{p-1} ,$$

where $dq_1 \ldots dq_{p-1}$ is assumed positive; $q_p = 1$; the integral is taken over the entire real hyperplane $(q_1, \ldots, q_{p-1})$, and the sum is taken over all roots q of the equation

$$H(qq_1, \ldots, qq_{p-1}, q) = 0$$

corresponding to the point $(q_1, \ldots, q_{p-1})$ under consideration.

Let us apply the Euler formula to the following function of q:

$$\left(q \sum x_k q_k + t + i\varepsilon \right) H(qq_1, \ldots, qq_{p-1}, q) \equiv (Qq + t + i\varepsilon)H(q) ,$$

where $Q = \sum_{k=1}^{p-1} x_k q_k + x_p$. Then

$$\sum_q \frac{q^{p-1}}{\frac{dH}{dq}(q \sum x_k q_k + t + i\varepsilon)} = \frac{(t + i\varepsilon)^{p-1}}{H\left(-\frac{t+i\varepsilon}{Q}\right) Q^p} =$$

$$= \frac{(t + i\varepsilon)^{p-1} Q^{n-p}}{G(t + i\varepsilon, q_1, \ldots, q_{p-1})} ,$$

where

$$G(t + i\varepsilon, q_1, \ldots, q_{p-1}) = Q^n H\left(-\frac{t + i\varepsilon}{Q}\right)$$

is a polynomial in q_k of degree n. Consequently,

$$\varphi(t + i\varepsilon, x_1, \ldots, x_p) = (t + i\varepsilon)^{p-1} \int \frac{\left(\sum\limits_{k=1}^{p-1} x_k q_k + x_p\right)^{n-p}}{G(t + i\varepsilon, q_1, \ldots, q_{p-1})} \, dq_1 \ldots dq_{p-1}.$$

Let us rotate the coordinate axes in the $(p_1, \ldots, p_p)$-space, as described in the preceding section, and mark the corresponding variables in new coordinates by the asterisk. Then we find

$$\varphi(t + i\varepsilon, x_1, \ldots, x_p) = \varphi(t + i\varepsilon, x_p^*) =$$

$$= (t + i\varepsilon)^{p-1} x_p^{*n-p} \int \frac{dq_1^* \ldots q_{p-1}^*}{G^*(t + i\varepsilon, q_1^*, \ldots, q_{p-1}^*)}. \qquad (53)$$

The integral is taken over the whole real plane $(q_1^*, \ldots, q_{p-1}^*)$.

First, let us perform integration along the real q_{p-1}^*-axis from $-\infty$ to $+\infty$. Applying the theorem about residues of an analytic function and noticing that the sum of the residues corresponding to the roots above the real axis is equal to the minus sum of the residues with respect to the roots below the real axis, we obtain

$$\varphi(t + i\varepsilon, x_1, \ldots, x_p) =$$

$$= (t + i\varepsilon)^{p-1} x_p^{*n-p} \pi i \int \sum \frac{\operatorname{sgn} \operatorname{Re}\left(\frac{q_{p-1}^*}{i}\right)}{G_{p-1}^*} \, dq_1^* \ldots dq_{p-2}^*. \qquad (53')$$

Here G_{p-1}^* stands for the derivative of G^* with respect to q_{p-1}^*; the sum under the integral is taken over all, inevitably complex, roots q_{p-1}^* of the polynomial G^* which correspond to the real point $(q_1^*, \ldots, q_{p-2}^*)$. It can be easily shown that all these roots are complex for $\varepsilon \neq 0$. Indeed, if there exist real $q_1^*, \ldots, q_{p-1}^*$ annihilating the expression

$$G^*(t + i\varepsilon, q_1^*, \ldots, q_{p-1}^*) =$$

$$= x_p^{*n} H^*\left(-\frac{q_1^*(t + i\varepsilon)}{x_p^*}, \ldots, -\frac{q_{p-1}^*(t + i\varepsilon)}{x_p^*}, \frac{t + i\varepsilon}{x_p^*}\right) =$$

$$= (-t - i\varepsilon)^n \Delta^*\left(-\frac{x_p^*}{t + i\varepsilon}, q_1^*, \ldots, q_{p-1}^*, 1\right),$$

then the equation

$$\Delta^*(\lambda, q_1^*, \ldots, q_{p-1}^*, 1) = 0$$

must have complex roots λ, which is impossible because of the hyperbolicity of the equation in question.

As in the preceding subsection, we set

$$\varphi(t + i0, x) = u(t, x) + iv(t, x) .$$

Then

$$u(t, x_1, \ldots, x_p) = \pi i t^{p-1} x_p^{*n-p} \int \sum \frac{\operatorname{sgn} \operatorname{Re} \left(\frac{q_{p-1}^*}{i} \right)}{G_{p-1}^*(t, q_1^*, \ldots, q_{p-1}^*)} dq_1^* \ldots dq_{p-2}^* =$$

$$= -\frac{\pi i}{x_p^*} \int \sum \frac{\operatorname{sgn} \operatorname{Re} \frac{p_{p-1}^*}{i}}{H_{p-1}^* \left(p_1^*, \ldots, p_{p-1}^*, -\frac{t}{x_p^*} \right)} dp_1^* \ldots dp_{p-2}^* , \qquad (54)$$

where the sum under the first (resp., second) integral is taken over all complex (non-real) roots q_{p-1}^* (resp., p_{p-1}^*) of the polynomial G^* (resp., H) corresponding to the real point $(q_1^*, \ldots, q_{p-2}^*)$ (resp., $(p_1^*, \ldots, p_{p-2}^*)$). It is only near the points of the hyperplane $(q_1^*, \ldots, q_{p-2}^*)$ where $G_{p-1}^* = 0$, or in a neighborhood of infinite points of that hyperplane, that the admissibility of passing to the limit under the sign of the integral (53'), as $\varepsilon \to 0$, may be subject to doubt.

As to the integrals near the points where $G_{p-1}^* = 0$, we can proceed as follows. Let us eliminate q_{p-1}^* from the equations

$$G^*(t, q_1^*, \ldots, q_{p-1}^*) = 0 \qquad (55)$$

and

$$G_{p-1}^*(t, q_1^*, \ldots, q_{p-1}^*) = 0 .$$

Denote the equation obtained in this way by

$$R(t, q_1^*, \ldots, q_{p-2}^*) = 0 . \qquad (56)$$

Let us divide the real surface (56), outside a neighborhood of its infinite points, into a finite number of non-overlapping regions $G_k^{(s)}$ such that on every one of them the same derivative of G^* with respect to some q_k^* has absolute value larger than a positive constant. The part of the integral (53') over a small domain $B_k^{(s)}$ on the real $(q_1^*, \ldots, q_{p-2}^*)$-plane such that $G_k^{(s)}$ belongs to its interior, or equivalently, over a $(p-2)$-dimensional portion $C_k^{(s)}$ belonging to the surface

$$G(t + i\varepsilon, q_1^*, \ldots, q_{p-1}^*) = 0$$

whose projection is contained in $B_k^{(s)}$, can be represented as the integral

$$\int \frac{dq_1^* \dots dq_{k-1}^* \, dq_{k+1}^* \dots dq_{p-1}^*}{G_k^*(t + i\varepsilon, q_1^*, \dots, q_{p-1}^*)} \operatorname{Re} \left(\frac{q_{p-1}^*}{i} \right).$$

Obviously, for all ε sufficiently small, the integrand remains uniformly bounded, and therefore, it is possible to pass to the limit under the sign of the integral as $\varepsilon \to 0$, since the domain of integration is bounded.

Consider now the neighborhoods of infinite points of the hyperplane $(q_1^*, \dots, q_{p-2}^*)$. By a suitable projective transformation, each neighborhood can be transformed into a finite domain, according to the usual procedure for establishing finiteness of an integral of the 1st kind over an infinite domain. Thereupon, it becomes evident that the integrands over each neighborhood containing no points of the surface (56) are uniformly bounded as $\varepsilon \to 0$; thus we get a justification for passing to the limit under the sign of the integral over such neighborhoods. In the case of a neighborhood containing points of the surface (56), the arguments of the preceding paragraph should be applied.

A statement similar to the remark of Sect. 2 is also valid in the present situation.

4. Let us formulate the results obtained in Sects. 1 and 2. In order to construct $v(t, x_1, \dots, x_p)$ or $u(t, x_1, \dots, x_p)$, according to p being odd or even, respectively, we can proceed as follows. On the hyperplane $\sum x_k p_k + t = 0$, consider all straight lines parallel to the axis Op_{p-1}^*, and consider the following two groups of points: the real points of intersection of these lines with the hypersurface (39), and, separately, the complex points, together with the real ones that are limit points for the set of the complex points. Each group of points forms a $(p-2)$-dimensional cycle, denoted by $C_{\text{real}}(t, x)$ and $C_{\text{imag}}(t, x)$, respectively, and lying on the algebraic surface which is the intersection of the hypersurface (39) and the hyperplane $\sum x_k p_k + t = 0$, the complex dimension of the algebraic surface being $p - 2$. It is obvious that the set $C_{\text{real}}(t, x)$ forms a cycle.

In order to show that $C_{\text{imag}}(t, x)$ is also a cycle, we note that the pieces of this manifold can be projected to the real $(p_1^*, \dots, p_{p-2}^*)$-hyperplane as one-sheeted pieces joining each other either along the plane at infinity or along the real surface

$$R\left(p_1^*, \dots, p_{p-2}^*, -\frac{t}{x_p^*} \right) = 0 \tag{56'}$$

obtained by eliminating p_{p-1}^* from the equations

$$H^*\left(p_1^*, \dots, p_{p-2}^*, p_{p-1}^*, -\frac{t}{x_p^*} \right) = 0 \tag{55'}$$

and

$$H_{p-1}^*\left(p_1^*,\ldots,p_{p-2}^*,p_{p-1}^*,-\frac{t}{x_p^*}\right)=0\,. \tag{55*}$$

To every $(p-3)$-dimensional piece of the surface (56′) there corresponds, in the real $(p_1^*,\ldots,p_{p-2}^*)$-plane, either a double real root of the equation (55′), or a pair of double complex conjugate roots of this equation. The roots p_{p-1}^* of equation (55′) can have multiplicity larger than 2 only on a set of points of the real surface (56′) of dimension less than $p-3$, since at these points we have

$$\frac{\partial^2}{\partial p_{p-1}^{*2}}H^*\left(p_1^*,\ldots,p_{p-2}^*,p_{p-1}^*,-\frac{t}{x_p^*}\right)=0\,,$$

and the last equation can be prevented from being a consequence of equations (55′) and (55*) by an arbitrarily small rotation of the coordinate axes. Over the pieces of the first kind, two complex conjugate pieces of $C_{\text{imag}}(t,x)$, being projected into the same piece of the real $(p_1^*,\ldots,p_{p-2}^*)$-plane, join one another. These pieces, as well as their projections, have opposite orientation, since their respective signs of $\operatorname{Im}p_{p-1}^*$ are opposite. Hence, the boundaries of these pieces along (56′) are mutually cancelled. Over the pieces of the second kind, two pieces of $C_{\text{imag}}(t,x)$, for which $\operatorname{Im}p_{p-1}^*>0$, are joined, as well as two pieces with $\operatorname{Im}p_{p-1}^*<0$. Contrary to the situation in the first case, the projections of all these pieces do not end at (56′), but extend, with the same orientations, beyond the piece of the real surface (56′) under consideration. Thereby, the boundaries of all these pieces along (56′) are mutually cancelled. The pieces joining along the plane at infinity also have opposite orientation, since p_{p-1}^* changes sign while passing through infinity. Therefore, if p is even, the boundaries of the pieces of $C_{\text{imag}}(t,x)$ joining at infinity are also mutually cancelled.

In order to construct $v(t,x_1,\ldots,x_p)$, we have to consider the real cycles $C_{\text{real}}(t,x)$. They coincide with the entire intersection of the real hypersurface (39) with he real hyperplane

$$\sum x_k p_k + t = 0\,, \tag{57}$$

and we have

$$v(t,x_1,\ldots,x_p)= \tag{58}$$

$$=-\frac{\pi}{x_p^*}\int\frac{\operatorname{sgn}\left(\operatorname{grad}H(p)\cdot\operatorname{grad}(p_1^2+\cdots+p_p^2)\right)\cdot\operatorname{sgn}H_{p-1}^*}{H_{p-1}^*\left(p_1^*,\ldots,p_{p-1}^*,-\frac{t}{x_p^*}\right)}\,dp_1^*\ldots dp_{p-2}^*\,,$$

where the integral is taken along these real cycles $C_{\text{real}}(t,x)$. Their orientation is determined by the condition that the projections of the simplexes, from which these cycles can be constructed, into the $(p_1^*, \ldots, p_{p-2}^*)$-plane have the signs

$$\text{sgn} \, [\text{grad} \, H(p) \cdot \text{grad} \, (p_1^2 + \cdots + p_p^2)] \cdot \text{sgn} \, H_{p-1}^*.$$

Hence, it is easy to see that if, for instance, the hyperplane (57) crosses all the ovals of the hypersurface (39), then the intersection consists of $[n/2]$ real ovals enclosing each other and having the same orientation. If the hyperplane (57) crosses one of the real ovals of the surface (39), which has no points at infinity, and the intersection consists of several real ovals, i.e., of several surfaces homeomorphic to the $(p-2)$-dimensional sphere, then the outer oval and those consecutively enclosed in an even number of ovals have the same (say, positive) orientation; while the ovals consecutively enclosed in an odd number of ovals have the opposite (negative) orientation. The statement "the oval A is enclosed in the oval B" is understood in the following sense. Let us regard the projective $(p_1, \ldots, p_{p-1})$-hyperplane as a $(p-1)$-dimensional sphere with end-points of each diameter identified. Then, every oval in this space corresponds to a cone with vertex at the center of the sphere. If an oval A is enclosed in an oval B, then the cone corresponding to the oval A is enclosed in the cone corresponding to B.

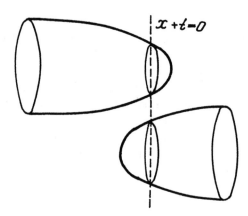

$x + t = 0$

Fig. 1.

Consider another example. Assume that the surface (39) contains an oval crossing the plane at infinity and having two real ovals as its intersection with the hyperplane (57). For the sake of simplicity, we assume these two ovals to be convex, as shown in Fig. 1. Then, these two ovals are of different orientation, since their respective values of the scalar product

$$\operatorname{grad} H(p) \cdot \operatorname{grad} (p_1^2 + \cdots p_p^2)$$

have opposite signs, while H_{p-1}^* are of the same sign on the upper (resp. lower) halves of the ovals.

5. Note that the integral representing $u(t,x_1,\ldots,x_p)$ for p even, as well as the integral representing $v(t,x_1,\ldots,x_p)$ for p odd, is of the 1st kind, provided that the surface

$$H^*(p_1^*,\ldots,p_{p-1}^*,p_p^*) = 0 , \qquad p_p^* = -\frac{t}{x_p^*} , \qquad (59)$$

has no singular points.

In the case of $p = 2$, the cycles along which the integrals representing $u(t,x_1,x_2)$ are taken reduce to the complex points of intersection of the straight line (57) and the curve (39), and the respective integrals are replaced by finite sums consisting of terms which correspond to these points.

Sometimes, in order to obtain $u(t,x_1,\ldots,x_p)$, it is more convenient to perform integration in the right-hand side of (54) along a cycle other than C_{imag} constructed in Sect. 3, namely, along a cycle C_{imag}^Q homologous to C_{imag} in the complex surface (59). Since the integral in the right-hand side of (54) is of the 1st kind, its value remains unchanged after the replacement of the cycle, because of the assumed absence of singular points of the surface (59). In order to construct C_{imag}^Q, take a real point Q on the plane (57) and consider all straight lines that belong to this plane, pass through Q, and can be defined by equations with real coefficients. Consider the complex points of intersection of these straight lines with the hypersurface (39), as well as the real limit points for the complex ones; all these points form the cycle C_{imag}^Q. In order to show that the cycle C_{imag}^Q is homologous to C_{imag}, we note that if the point Q be continuously moved to the point at infinity of the p_{p-1}^*-axis, the cycle C_{imag}^Q would be continuously deformed and tend to coincide with C_{imag}. In exactly the same way, while constructing the cycle C_{real} for p odd, we could have taken any real point Q on the hyperplane (57) and all straight lines belonging to this hyperplane, passing through Q, and given by equations with real coefficients; then we could have considered the real points of intersection of these straight lines with the hypersurface (39), instead of drawing the straight lines parallel to the Op_{p-1}^*-axis and considering the real points of their intersection with the surface (39). However, in this case the resulting cycle would coincide with C_{real}.

The admissibility of integration along the cycle $C_{\mathrm{imag}}^Q(t,x)$ instead of $C_{\mathrm{imag}}(t,x)$, while constructing $u(t,x_1,\ldots,x_p)$, can also be demonstrated as follows. Let us subject the $(p_1^*,\ldots,p_{p-1}^*)$-space to a projective transformation P with real coefficients, so as to move the point Q to infinity along the Op^*_{p-1}-axis, in its negative direction. Then we apply to the transformed integral (53) the arguments that have led us to (54), and return to the original coordinates $(p^*_1, \ldots, p^*_{p-1})$ by the transformation inverse to P.

Hence, as an immediate consequence, we find that $u(t, x_1, \ldots, x_p) = 0$, if the intersection of the surface (39) with the plane (57) consists of $[n/2]$ real ovals consecutively enclosed into each other and (in the case of n odd) of an unpaired piece. Indeed, taking the point Q inside the smallest of these ovals, we see that every real straight line passing through Q in the hyperplane (57) crosses the hypersurface (39) at real points only, and therefore, the complex cycle C^Q_{imag} which, being integrated along, yields u is absent. This remark will be often used in the sequel.

6. Let us show that in the case of p odd we have

$$v(t, x_1, \ldots, x_p) = 0 \ ,$$

if the hyperplane (57) crosses all the ovals of the hypersurface (39). In this situation, the intersection in question consists of $[n/2]$ ovals consecutively enclosed into each other, and (for n odd) of an unpaired piece. Take the origin of the coordinates $(p^*_1, \ldots, p^*_{p-1})$ inside the smallest of these ovals. Then (58) can be rewritten in the form

$$v(t, x_1, \ldots, x_p) =$$

$$= -\frac{\pi}{x^*_p} \int \frac{\text{sgn}\left(\text{grad } H^*\left(p^*_1, \ldots, p^*_{p-1}, -\frac{t}{x^*_p}\right) \cdot \text{grad}\left(p^{*2}_1 + \cdots + p^{*2}_{p-1}\right)\right)}{\left|H^*_{p-1}\left(p^*_1, \ldots, p^*_{p-1}, -\frac{t}{x^*_p}\right)\right|} \times$$

$$\times dp^*_1 \ldots dp^*_{p-2} = -\frac{\pi}{x^*_p} \int d\omega_{-\frac{t}{x^*_p}} \ . \tag{60}$$

Here grad $H^*\left(p^*_1, \ldots, p^*_{p-1}, -t/x^*_p\right)$ is taken in the hyperplane $(p^*_1, \ldots, p^*_{p-1})$; $d\omega_{-t/x^*_p}$ has the same form as $d\omega$ in (40), and therefore, we can apply to (60) the arguments of §7.3, where p was larger by 1. It is assumed here that the intersection of the hyperplane (57) with the Op^*_p-axis lies within the smallest of the ovals.

Comparing (49) and (50) with (60) and (54), we come to the following conclusion. In order to obtain $K(t, x)$ in the case of $n > p+1$, it is necessary to construct the cycles $C_{\text{real}}(\tau, x)$, if p is odd (resp., $C_{\text{imag}}(\tau, x)$, if p is even), for all τ from 0 to t. Let $S(t, x)$ be the $(p-1)$-dimensional surface formed by these cycles. Then $K(t, x)$ can be expressed in terms of the following integrals of the 1st kind over this surface:

$$K(t, x) = C_{pn} \int_{S(t,x)} \frac{(t + x^*_p p^*_p)^m \, dp^*_1 \ldots dp^*_{p-2} \, dp^*_p}{H^*_{p-1}(p^*_1, \ldots, p^*_{p-1}, p^*_p)} =$$

$$= C_{pn} \int\limits_{S(t,x)} \frac{(t + \sum x_k p_k)^m \, dp_1 \ldots dp_{p-1}}{H_p(p_1, \ldots, p_p)}, \qquad (61)$$

where C_{pn} is a constant depending on p and n; the sign of $dp_1^* \ldots dp_{p-2}^*$ is chosen according to the above rules.

§10. Formulas for $K(t, x)$ in the Case of $n = p + 1$

For the derivation of these formulas consider the following $(n + 2)$-th order equation

$$\left(\frac{\partial^2}{\partial t^2} - \sum a_{ik} \frac{\partial^2}{\partial x_i \partial x_k} \right) \Delta \left(\frac{\partial}{\partial t}, \frac{\partial}{\partial x_1}, \ldots, \frac{\partial}{\partial x_p} \right) w = 0$$

instead of the n-th order equation (22). Instead of the initial conditions (24) for $t = 0$, consider the following ones

$$w = \frac{\partial w}{\partial t} = \cdots = \frac{\partial^n w}{\partial t^n} = 0, \quad \frac{\partial^{n+1} w}{\partial t^{n+1}} = f(x_1, \ldots, x_p), \quad t = 0.$$

We choose real coefficients a_{ik} such that the form

$$\sum a_{ik} p_i p_k$$

is positive definite. Then, obviously,

$$u = \left(\frac{\partial^2}{\partial t^2} - \sum a_{ik} \frac{\partial^2}{\partial x_i \partial x_k} \right) w .$$

Now, let us assume the coefficients a_{ik} to be so large in absolute value that the real ellipsoid

$$\widetilde{H}(p_1, \ldots, p_p) \equiv - \sum a_{ik} p_i p_k + 1 = 0$$

lies within the smallest of the ovals of the hypersurface (39). Then Herglotz's formulas (44), (47) are valid for the kernel $K_w(t, x)$ corresponding to $w(t, z)$. Our aim is to derive the expression for the kernel $K_u(t, x)$ corresponding to $u(t, x)$.

For $n \leq p + 1$ (actually, one can show that the following can be claimed only for $n \leq p$) the expression for the function $u(t, x_1, \ldots, x_p)$ involves, in general, not only the integrals of the products of the initial functions $f(y_1, \ldots, y_p)$ by the kernel $K(t, x_1 - y_1, \ldots, x_p - y_p)$ over the whole plane of the initial data, but also integrals over some lower dimensional manifolds forming the intersection of the initial plane with the lateral surface of the characteristic cone with vertex at $(t, x_1, \ldots, x_p)$, as well as the derivatives of these integrals with respect to t. However, for the analysis of lacunas it suffices to consider sufficiently smooth initial functions which vanish everywhere outside domains belonging, together with their boundaries, to the interior of the domains cut out from the base of the characteristic cone by its lateral surface. For such initial data, the function $u(t, x_1, \ldots, x_p)$ can always be represented by integrals of the said products over the entire initial plane. It can be easily seen that in the latter case we can differentiate the expression for $u(t, x_1, \ldots, x_p)$ under the sign of the integral, and the kernel $K_u(t, x)$ corresponding to u can be obtained from $K_w(t, x)$ by the formula

$$K_u(t, x_1, \ldots, x_p) = \left(\frac{\partial^2}{\partial t^2} - \sum a_{ik} \frac{\partial^2}{\partial x_i \partial x_k} \right) K_w(t, x_1, \ldots, x_p) .$$

Let us consider in detail the case of p being odd. The case of p even can be treated in a similar fashion. For the kernel $K_w(t, x)$ formula (44) holds, and therefore

$$K_w(t, x) = C_{p,n+2} \int_{S_{H\widetilde{H}}(t,x)} \frac{\pm (\sum x_k p_k + t)^2}{(H\widetilde{H})'_p} \, dp_1 \ldots dp_{p-1} .$$

Hence

$$K_u(t, x) = \left(\frac{\partial^2}{\partial t^2} - \sum a_{ik} \frac{\partial^2}{\partial x_i \partial x_k} \right) K_w(t, x) =$$

$$= 2C_{p,n+2} \int_{S_{H\widetilde{H}}(t,x)} \frac{\pm \widetilde{H}}{H_p \widetilde{H} + \widetilde{H}_p H} \, dp_1 \ldots dp_{p-1} =$$

$$= C_{pn} \int_{S_{\widetilde{H}}(t,x)} \frac{\pm \widetilde{H}}{\widetilde{H}_p H} \, dp_1 \ldots dp_{p-1} + C_{pn} \int_{S_H(t,x)} \frac{\pm dp_1 \ldots dp_{p-1}}{H_p} .$$

The first of the above integrals is equal to 0, since $\widetilde{H} = 0$ on the surface $\widetilde{H}$; the second integral yields $K_u(t, x)$.

Having the expression (47) for $K_u(t, x)$, we can pass to (50) and (61), as described above.

§11. The Case $p \le n$

1. Instead of the Cauchy problem (22), (24), consider the following one: find a function $w(t, x)$ satisfying the initial conditions at $t = 0$:

$$w(0, x) = \frac{\partial w}{\partial t} = \cdots = \frac{\partial^{n+\tilde{n}-2} w}{\partial t^{n+\tilde{n}-2}} = 0 , \quad \frac{\partial^{n+\tilde{n}-1} w}{\partial t^{n+\tilde{n}-1}} = f(x_1, \ldots, x_p) ,$$

and the differential equation

$$\tilde{\Delta}\left(\frac{\partial}{\partial t}, \frac{\partial}{\partial x_1}, \ldots, \frac{\partial}{\partial x_p}\right) \Delta\left(\frac{\partial}{\partial t}, \frac{\partial}{\partial x_1}, \ldots, \frac{\partial}{\partial x_p}\right) w = 0 .$$

Here $\tilde{\Delta}(\lambda, \xi_1, \ldots, \xi_p)$ is a form of an even degree $\tilde{n}$ such that $\tilde{n} \geq p - n + 1$. The equation

$$\widetilde{H}(p_1, \ldots, p_p) \equiv \tilde{\Delta}(1, p_1, \ldots, p_p) = 0$$

defines, in real domain, $\tilde{n}/2$ ovals consecutively enclosed into each other, all of them contained by the smallest real oval of the hypersurface (39). Obviously,

$$u(t, x_1, \ldots, x_p) = \tilde{\Delta}\left(\frac{\partial}{\partial t}, \frac{\partial}{\partial x_1}, \ldots, \frac{\partial}{\partial x_p}\right) w .$$

Formulas (44) and (47) are clearly valid for the function w. As mentioned in the preceding section, the initial functions considered in our study of lacunas allow us to represent the function $u(t, x_1, \ldots, x_p)$ by an integral of the form

$$\int K_u(t, x_1 - y_1, \ldots, x_p - y_p) f(y_1, \ldots, y_p) \, dy_1 \ldots dy_p$$

over the whole plane of initial data. It is evident that

$$K_u(t, x_1, \ldots, x_p) = \tilde{\Delta}\left(\frac{\partial}{\partial t}, \frac{\partial}{\partial x_1}, \ldots, \frac{\partial}{\partial x_p}\right) K_w(t, x_1, \ldots, x_p) .$$

If p is odd, no other conditions, apart from those stated above, need be imposed on the coefficients of $\tilde{\Delta}$. For p even, we take, on the hyperplane (57), a real point Q outside the hypersurface (39) and contract the entire hypersurface (62) in all directions perpendicular to the line OQ to such an extent that the complex cycles formed by the complex points of intersection of the hypersurface (62) with various straight lines given by equations with real coefficients, passing through Q, and lying on (57), do not meet similar cycles constructed for the hypersurface (39).

Passing to the coordinates x_k^*, p_k^*, we can rewrite (44), (47) in the form

$$K_w(t, x) \equiv K_w^*(t, x^*) =$$

$$= -\frac{2\pi i}{(n + \tilde{n} - p)!(2\pi i)^p} \frac{\partial}{\partial t} \int_{(0,t)} \left(\sum x_k^* p_k^* + t\right)^{n + \tilde{n} - p} d\omega_w , \qquad (\widetilde{44}')$$

$$K_w(t, x) \equiv K_w^*(t, x^*) =$$

$$= \frac{2}{(n + \tilde{n} - p)!(2\pi i)^p} \frac{\partial}{\partial t} \int \left(\sum x_k^* p_k^* + t\right)^{n + \tilde{n} - p} \ln\left|\frac{\sum x_k^* p_k^* + t}{\sum x_k^* p_k^*}\right| d\omega_w . \qquad (\widetilde{47}')$$

These integrals exist, even if the real hypersurface (39) is infinite, since $\sum x_k^* p_k^* + t$ in the first integral remains bounded, while in the second integral we have

$$\left| \ln \left| \frac{\sum x_k^* p_k^* + t}{\sum x_k^* p_k^*} \right| \right| < \frac{t}{|\sum x_k^* p_k^* + t|}$$

for sufficiently large $\sum x_k^* p_k^*$.

The integration in $(\widetilde{47}')$ can first be performed over that portion of the domain where $-a \le p_p^* \le +a$ (a is a positive constant); on this set $\sum x_k^* p_k^* + t$ is bounded for all $x_1^*, \ldots, x_{p-1}^*$ equal to zero. In order to calculate the integral over the remaining portion, let us change the variables as follows:

$$p_k^* = \frac{p_k^{**}}{p_p^{**}} \quad (k = 1, \ldots, p-1) ; \qquad p_p^* = \frac{1}{p_p^{**}} , \qquad (63)$$

and thereby transform the integral in question into an integral over the portion of the domain considered first.

It suffices to assume here that the hyperplane (57) does not touch the real surface $H\widetilde{H} = 0$.

It is possible to show that $K(t, x)$ is an analytic function in a neighborhood of those values of t and x_k which correspond to a non-singular intersection of the surface (39) with the hyperplane (57).

2. Differentiating $(\widetilde{44}')$ (resp., $(\widetilde{47}')$) $n + \tilde{n} - p$ times with respect to t and x_k^*, and using the results of §7.3, we obtain

$$\frac{\partial^{n+\tilde{n}-p} K_w^*(t, x^*)}{\partial t^{k_0} \partial x_1^{*k_1} \ldots \partial x_p^{*k_p}} = -\frac{2\pi i}{(2\pi i)^p} \frac{\partial}{\partial t} \int_{(0,t)} p_1^{*k_1} \ldots p_p^{*k_p} \, d\omega_w , \qquad (\widetilde{44}'')$$

$$\frac{\partial^{n+\tilde{n}-p} K_w^*(t, x^*)}{\partial t^{k_0} \partial x_1^{*k_1} \ldots \partial x_p^{*k_p}} = \frac{2}{(2\pi i)^p} \frac{\partial}{\partial t} \int p_1^{*k_1} \ldots p_p^{*k_p} \ln \left| \frac{\sum x_k^* p_k^* + t}{\sum x_k^* p_k^*} \right| \, d\omega_w . \quad (\widetilde{47}'')$$

In order to justify differentiation under the sign of the integral, it suffices to establish uniform (with respect to t, x_k) convergence of the integrals obtained, if the domain of integration is infinite. To this end we perform integration in $(\widetilde{47}'')$ in the same order as in $(\widetilde{47}')$. To prove uniform convergence of the integral in $(\widetilde{44}'')$ and of the part of the integral in $(\widetilde{47}'')$ over the domain $-a \le p_p^* \le +a$ (a is sufficiently large), we assume, first, that no plane of the family $x_p^* = x_p^{*0} = \text{const}$ (x_p^* is finite) touches the real surface $H\widetilde{H} = 0$ at infinity. This condition excludes a surface of dimension $\le p - 1$ in the real $(x_1, \ldots, x_p)$-plane. An expression for $K(t, x)$ at the points of this surface will be obtained later by passing to the limit. Recall that the plane $\sum x_k^* p_k^* + t = 0$ and the plane $\sum x_k^* p_k^* = 0$ do not touch the real surface

$H\widetilde{H} = 0$. Therefore, the domain of integration can be divided into the following three parts, the integral being bounded and uniformly convergent over each part.

The first part is defined by the inequalities $p_p^{(1)} \leq p_p^* \leq p_p^{(2)}$, where $p_p^{(1)}$ and $p_p^{(2)}$ are chosen in such a way that the value $-t/x_p^*$ lies between $p_p^{(1)}$ and $p_p^{(2)}$ and no plane of the form $p_p^* = $ const $(p_p^{(1)} \leq p_p^* \leq p_p^{(2)})$ touches the real surface $H\widetilde{H} = 0$. The first part of the domain also includes the points with $|p_p^*| \leq \varepsilon$, where $\varepsilon > 0$ is chosen such that no plane of the form $p_p^* = $ const, $|p_p^*| \leq \varepsilon$, touches the surface $H\widetilde{H} = 0$. Integrating over this part, first with respect to $p_1^*, \ldots, p_{p-2}^*$ for constant p_p^*, and moving the factor $\ln |(\sum x_k^* p_k^* + t) / \sum x_k^* p_k^*|$ to the left of the integral sign, we obtain uniformly convergent integrals of the 1st kind. Therefore, the total integrals $(\widetilde{44''})$ and $(\widetilde{47''})$ over this part of the domain are uniformly convergent.

Now, let us define the second portion of the domain $|p_p^*| \leq a$ (after removing the points of the first portion) by the inequality

$$p_1^{*2} + \cdots + p_{p-2}^{*2} \leq R^2 ,$$

the constant R being so large that all the points where the plane $p_p^* = $ const is tangential to the surface $H\widetilde{H} = 0$ (all these points are finite) belong to this portion of the domain. Since the integrand has no singular points within this portion, the corresponding integral is uniformly convergent.

Consider the integral over the third (remaining) part of the domain $|p_p^*| \leq a$. Integrating, first, with respect to the variables $p_1^*, \ldots p_{p-2}^*$, we obtain integrals uniformly convergent in $t, x_1^*, \ldots, x_p^*, p_p^*$. Since the segment of integration with respect to p_p^* is finite, therefore the integrals $(\widetilde{44''})$ and $(\widetilde{47''})$, in total, are uniformly convergent over the third portion of the domain.

The part of the integral over the remaining domain $|p_p^*| > a$ can be dealt with in a similar manner after the transformation (63).

The integrals in $(\widetilde{44''})$ and $(\widetilde{47''})$ for the values of t and x under consideration can be calculated in the same way as $v(t, x)$ and $u(t, x)$ in §§9.1, 9.3. Thus, we obtain

$$\frac{\partial^{n+\bar{n}-p} K_w^*}{\partial t^{k_0} \partial x_1^{*k_1} \ldots \partial x_p^{*k_p}} = -\frac{2i}{(2\pi i)^p} V^*(t, x^*) = -\frac{2i}{(2\pi i)^p} V(t, x) , \quad (64)$$

$$\frac{\partial^{n+\bar{n}-p} K_w^*}{\partial t^{k_0} \partial x_1^{*k_1} \ldots \partial x_p^{*k_p}} = \frac{2}{(2\pi i)^p} U^*(t, x^*) = \frac{2}{(2\pi i)^p} U(t, x) , \quad (65)$$

where

$$V^*(t, x^*) = -\frac{\pi}{x_p^*} \int_{C(t,x)} p_1^{*k_1} \dots p_p^{*k_p} \times$$

$$\times \frac{\text{sgn} \left[\text{grad } H(p) \cdot \text{grad } (p_1^2 + \dots + p_p^2) \right] dp_1^* \dots dp_{p-2}^*}{\left| \left[H^* \left(p_1^*, \dots, p_{p-1}^*, -\frac{t}{x_p^*} \right) \widetilde{H}^* \left(p_1, \dots, p_{p-1}, -\frac{t}{x_p^*} \right) \right]_{p-1}' \right|},$$

$$U^*(t, x^*) =$$

$$= \frac{\pi}{x_p^*} \int_{C(t,x)} \frac{p_1^{*k_1} \dots p_p^{*k_p} \text{ sgn Re} \left(\frac{p_{p-1}^*}{i} \right) dp_1^* \dots dp_{p-2}^*}{\left| \left[H^* \left(p_1^*, \dots, p_{p-1}^*, -\frac{t}{x_p^*} \right) \widetilde{H}^* \left(p_1, \dots, p_{p-1}, -\frac{t}{x_p^*} \right) \right]_{p-1}' \right|}.$$

Here

$$C(t, x) = C_{1\,\text{real}}(t, x) + C_{2\,\text{real}}(t, x) \qquad \text{if } p \text{ is odd,}$$

$$C(t, x) = C_{1\,\text{imag}}(t, x) + C_{2\,\text{imag}}(t, x) \qquad \text{if } p \text{ is even.}$$

The equalities (64) and (65) can be written in shorter form as

$$\frac{\partial^{n+\tilde{n}-p} K_w^*}{\partial t^{k_0} \partial x_1^{*k_1} \dots \partial x_p^{*k_p}} =$$

$$= -\frac{C_{pn}}{x_p^*} \int_{C(t,x)} \pm \frac{p_1^{*k_1} \dots p_p^{*k_p}}{(H^* \widetilde{H}^*)_s'} dp_1^* \dots dp_{s-1}^* \dots dp_{s+1}^* \dots p_{p-1}^* \cdot (66)$$

The cycles $C_{1\,\text{real}}(t, x)$ or $C_{1\,\text{imag}}(t, x)$ are taken on the surface $H = 0$, and the cycles $C_{2\,\text{real}}(t, x)$ or $C_{2\,\text{imag}}(t, x)$ on the surface $\widetilde{H} = 0$. The cycles $C_1(t, x)$ and $C_2(t, x)$ have no intersection.

The cycles C_1 may happen to be infinite. In this case, as well as in some other cases, it is convenient to replace integration along these cycles by integration along some other finite cycles. To this end we note that, according to a theorem to be proved later on (see §16), every $(p - 2)$-dimensional cycle C_1 on a general algebraic surface of complex dimension $(p - 2)$ with no singular points is homologous to a finite cycle C_1^* and also, possibly (for p even), to an algebraic surface of complex dimension $(p-2)/2$ imbedded into H; it should be kept in mind that any integral over the latter surface is equal to zero (see [9], pp. 135–136).

3. Moreover, it is easy to show that our cycle C_1 can only be homologous to a finite one. Indeed, for p odd, it cannot contain any algebraic cycle, since the dimension of any algebraic cycle is even. For an even p, let us take as Q the point at infinity of the axis Op^*_{p-1} and choose the axis Op^*_{p-1} in such a way that it does not cross the surface $H = 0$ at infinity. Then it is possible to find a real c so large that the surface

$$p^{*2}_{p-1} = c^2 \left(1 + \sum_{k=1}^{p-2} p^{*2}_k + p^{*2}_p \right) , \qquad p^*_p = -\frac{t}{x^*_p} , \qquad (67)$$

has no real points in common with the surface $H^* = 0$ at infinity. But it also has no imaginary points in common with $C^Q_{1\,\text{imag}}$, since the coordinates $p^*_1, \ldots, p^*_{p-2}, p^*_p$ of every point of the constructed cycle $C^Q_{1\,\text{imag}}$ are all real, and real coordinates $p^*_1, \ldots, p^*_{p-2}, p^*_p$ on the surface (67) correspond to real points. Therefore, the cycle $C^Q_{1\,\text{imag}}$ has no intersection with the algebraic surface (67). Consequently, it does not meet the surface S of complex dimension $p/2$ which lies in the same plane $p^*_p = -t/x^*_p$ as the cycle $C^Q_{1\,\text{imag}}$ and which coincides with the intersection of the surface $H = 0$ and the planes

$$p^*_k = p^0_k = \text{const} , \qquad k = p, p - 2, p - 3, \ldots, \frac{p}{2} + 1 . \qquad (68)$$

Let us choose the constants p^0_k so as to prevent the plane (68) from crossing the finite cycle C^*_1. Then the index of intersection of S and $C^Q_{1\,\text{imag}}$, equal to zero, will coincide with the index of intersection of S and the sum consisting of C^*_1 and the supposed algebraic cycle A of complex dimension $(p-2)/2$, the sum being homologous to $C^Q_{1\,\text{imag}}$. Consequently, the index of intersection of A and S is equal to zero, which is impossible, since the algebraic surfaces S and A lie in the same $(p-1)$-dimensional plane, have complex dimensions $(p-2)/2$ and $p/2$, respectively, and therefore, their intersection index must be equal to the positive number of their common points. It should be noticed that the result just proved for $C^Q_{1\,\text{imag}}$ in the case of a particular Q is also valid for other Q, since all cycles $C^Q_{1\,\text{imag}}$ constructed for various Q are homologous to each other.

4. For the differential under the sign of the integral (66), the only singular points on the surface (39), (57) are those common to this surface and the surface (62), all the other points, even those at infinity, being non-singular. Consider a $(p-1)$-dimensional film Π which can be extended over the cycle $C_1(t, x)$ and the finite cycle C^*_1 homologous to $C_1(t, x)$ in the complex surface (39), (57). In Π, let us surround the points of intersection of the surface $\widetilde{H}$ with the surface (39), (57) by thin tubes T. Such tubes T can be obtained, for instance, by considering in Π small neighborhoods of the cycles formed by the intersection of the surface $\widetilde{H}$ and the surface (39), (57). Denote by C^{**}_1 the $(p-2)$-dimensional surfaces of the tubes T within the film Π. Then the integral (66) over the cycle C_1 is obviously equal to the integral of the same differential over $C_1^* + C_1^{**}$. Therefore, in what follows, we may assume that

$$C = C_1^* + C_1^{**} + C_2 .$$

5. In the sequel, we have to compare $U(t, x)$ and $V(t, x)$ for close values of t and x_k. Note that in the expressions for $U(t, x)$ and $V(t, x)$ we can replace integration of the above differentials of the 1st kind over the cycle $C(t, x)$ by integration over a cycle close to $C(t, x)$. To simplify notations, we use the same symbol for these cycles. The cycle

$$C(t + \Delta t, x_1 + \Delta x_1, \ldots, x_p + \Delta x_p)$$

is obtained from $C(t, x)$ as follows. Let K_Δ be the intersection of the surface $H\widetilde{H} = 0$ with the plane

$$\sum (x_k + \Delta x_k)p_k + t + \Delta t = 0 . \tag{57_Δ}$$

Consider the projection of the set K_Δ along straight lines parallel to the Op_p^*-axis into the plane $x_p^* = \text{const}$ containing the cycle $C(t, x)$. Starting at each point of the cycle $C(t, x)$, let us draw straight lines normal to the intersection of the surface $H\widetilde{H} = 0$ with (57) until they meet K_Δ. Denote by $\widetilde{C}(t, x)$ the cycle thus obtained. Now, let $C(t+\Delta t, x+\Delta x)$ be the projection of the cycle $\widetilde{C}(t, x)$ into the plane (57_Δ) along straight lines parallel to the Op_p^*-axis.

Having established this correspondence between the points of the cycles $C(t, x)$ and $C(t + \Delta t, x + \Delta x)$, we can differentiate, with respect to t, x_k and p_p^*, an integral of the form (66) over a portion of the cycle $C(t, x)$, or an element of such an integral. The integral

$$I_B(t, x_1^*, \ldots, x_p^*) = -\frac{1}{x_p^*} \int_{B_{p-2}} \frac{p_1^{*k_1} \ldots p_p^{*k_p}}{(H^*\widetilde{H}^*)_s'} \, dp_1^* \ldots dp_{s-1}^* \, dp_{s+1}^* \ldots dp_{p-1}^*$$

taken over a finite part B_{p-2} of the cycle $C(t, x_1, \ldots, x_p)$ can be represented by the integral

$$\frac{1}{\nu!} \frac{\partial^{\nu+1}}{\partial t^{\nu+1}} \int_{B_{p-1}} \left(\sum x_k^* p_k^* + t \right)^\nu \frac{p_1^{*k_1} \ldots p_p^{*k_p}}{(H^*\widetilde{H}^*)_s'} \, dp_1^* \ldots dp_{s-1}^* \, dp_{s+1}^* \ldots dp_{p-1}^* \, dp_p^*$$

over a $(p - 1)$-dimensional domain B_{p-1} formed by the points of the cycles $C(\tau, x_1, \ldots, x_p)$, $t_0 \le \tau \le t$, related to the points of B_{p-2} by the above construction. Here t_0 is independent of t and is slightly less than t; $(H^*\widetilde{H}^*)_s' \ne 0$ on B_{p-1}. It follows that

$$\frac{\partial^\nu I_B(t, x_1^*, \ldots, x_p^*)}{\partial t^{\nu_0} \partial x_1^{*\nu_1} \ldots \partial x_p^{*\nu_p}} =$$

$$= \frac{\partial^{\nu+1}}{\partial t^{\nu+1}} \int_{B_{p-1}} \frac{p_1^{*k_1+\nu_1} \ldots p_p^{*k_p+\nu_p}}{(H^* \widetilde{H}^*)'_s} \, dp_1^* \ldots dp_{s-1}^* \, dp_{s+1}^* \ldots dp_p^* =$$

$$= -\frac{1}{x_p^*} \frac{\partial^\nu}{\partial t^\nu} \int_{B_{p-2}} \frac{p_1^{*k_1+\nu_1} \ldots p_p^{*k_p+\nu_p}}{(H^* \widetilde{H}^*)'_s} \, dp_1^* \ldots dp_{s-1}^* \, dp_{s+1}^* \ldots dp_{p-1}^* =$$

$$= -\frac{1}{x_p^*} \int_{B_{p-2}} \frac{\partial^\nu}{\partial t^\nu} \left[\frac{p_1^{*k_1+\nu_1} \ldots p_p^{*k_p+\nu_p}}{(H^* \widetilde{H}^*)'_s} \, dp_1^* \ldots dp_{s-1}^* \, dp_{s+1}^* \ldots dp_{p-1}^* \right].$$

Therefore, because of (66), we obtain for every $k \geq n + \tilde{n} - p$

$$\frac{\partial^k K_w(t, x)}{\partial t^{k_0} \partial x_1^{*k_1} \ldots \partial x_p^{*k_p}} = \tag{69}$$

$$= -\frac{C_{pn}}{x_p^*} \int_{C(t, x^*)} \frac{\partial^{k-n-\tilde{n}+p}}{\partial t^{k-n-\tilde{n}+p}} \left[\frac{p_1^{*k_1} \ldots p_p^{*k_p}}{(H^* \widetilde{H}^*)'_s} \, dp_1^* \ldots dp_{s-1}^* \, dp_{s+1}^* \ldots dp_{p-1}^* \right].$$

In order to verify the existence of the last integral and justify the differentiation with respect to t, x_k, in the case allowing, in particular, the cycle $C(t, x)$, i.e., $C^{**}(t, x)$, to have points at infinity, we proceed as follows. Let us represent $C(t, x)$ as a union of finitely many domains B_{p-2}^k such that every B_{p-2}^k can be transformed by a projective mapping into a finite domain which, in its turn, can be projected, without "folds" or "wrinkles", from a point A into a plane. Let us choose new coordinates p^{**} such that A becomes a point at infinity on the Op_{p-1}^{**}-axis.

We assume that the complex surface of intersection of $H\widetilde{H} = 0$ with (57) does not touch the plane at infinity in the $(p_1^*, \ldots, p_p^*)$-space. After the corresponding change of the variables in the integral over the domain B_{p-2}^k, we obtain an integral over a finite domain in which the integrand has no singularities (the denominator $(H^* \widetilde{H}^*)'_{p-1}$ will be replaced by an expression of the form $(H^{**} \widetilde{H}^{**})'_{p-1}$, which does not vanish at any point). Moreover, in the new integral, the integrand and the domain of integration depend smoothly on $t, x_1^*, \ldots, x_p^*$. The elements of the latter integral corresponding to those of the original one will also depend smoothly on these parameters and on the point of the domain. The elements corresponding to the points at infinity of the original integral will have no singularities.

Some difficulties arise, however, if the intersection of the surfaces $H\widetilde{H} = 0$ and (57) touches the plane at infinity in the $(p_1^*, \ldots, p_p^*)$-space, at a point of the cycle $C^{**}(t, x)$. In this situation, the projective transformation of a neighborhood of such point into a finite domain may happen to map a normal to the intersection of $H\widetilde{H} = 0$ and (57) into a tangent. This case, however, will not be considered separately, since the corresponding formulas can be obtained by passing to the limit. It should also be mentioned that, instead of the normals, we can consider any other family of straight lines which do not touch the surface $H\widetilde{H} = 0$ and whose direction smoothly depends on the point of this surface.

We may also project the whole complex projective $(p_1, \ldots, p_p)$-space, together with the surface $H\widetilde{H} = 0$, into a finite sphere S_p, consider the normals to the image of the surface $H\widetilde{H} = 0$, $p_p^* = -t/x_p^*$, and execute all further constructions in S_p. The finite sphere S_p is defined as follows. Let us regard the complex projective $(p_1, \ldots, p_p)$-space as imbedded in the complex projective $(p_1, \ldots, p_p, p_{p+1})$-space. Then S_p is defined by the equation

$$|p_1|^2 + \cdots + |p_p|^2 + |p_{p+1} - 1|^2 = 1 .$$

The projective $(p_1, \ldots, p_p)$-space is projected into S_p along straight lines passing through the point $P = (0, \ldots, 0, 1)$; the coordinate p_{p+1} is assumed real.

From the above formulas we obtain

$$\widetilde{\Delta}^* \left(\frac{\partial}{\partial t}, \frac{\partial}{\partial x_1^*}, \ldots, \frac{\partial}{\partial x_p^*} \right) K_w^*(t, x^*) = K_u^*(t, x^*) = K_u(t, x) =$$

$$= -\frac{C_{pn}}{x_p^*} \int_{C(t,x)} \pm \frac{\partial^{p-n}}{\partial t^{p-n}} \left(\frac{\widetilde{H}^*}{(H^*\widetilde{H}^*)_s'} dp_1^* \ldots dp_{s-1}^* \, dp_{s+1}^* \ldots dp_{p-1}^* \right) .$$

Hence, observing that $\widetilde{H} = 0$ on C_2, and $H = 0$ on C_1^* and C_1^{**}, we get

$$K_u(t, x) = -\frac{C_{pn}}{x_p^*} \int_{C_1^*(t,x)} \pm \frac{\partial^{p-n}}{\partial t^{p-n}} \left(\frac{1}{H_s^{*'}} dp_1^* \ldots dp_{s-1}^* \, dp_{s+1}^* \ldots dp_{p-1}^* \right) -$$

$$- \frac{C_{pn}}{x_p^*} \int_{C_1^{**}(t,x)} \pm \frac{\partial^{p-n}}{\partial t^{p-n}} \left(\frac{1}{H_s^{*'}} dp_1^* \ldots dp_{s-1}^* \, dp_{s+1}^* \ldots dp_{p-1}^* \right) . \qquad (70^*)$$

The last integral is equal to zero; indeed, by the above transformation it can be reduced to an integral of a function without singularities taken over the cycle $C^{**}(t, x)$ which can be contracted into a $(p - 3)$-dimensional intersection of the $(p - 1)$-dimensional film Π with the surface $\widetilde{H} = 0$.

Since, for $n < p$, the differential

$$\frac{dp_1^* \ldots dp_{s-1}^* \, dp_{s+1}^* \ldots dp_{p-1}^*}{H_s^{*'}}$$

has a singularity at infinity, we replace it by the differential

$$\frac{dp_1^* \ldots dp_{s-1}^* \, dp_{s+1}^* \ldots dp_{p-1}^*}{H_s^{*'} H_k^{**}}$$

in every domain B_{p-2}^k containing points at infinity. Here H_k^{**} is a polynomial with respect to $p_1^*, \ldots, p_p^*$ which does not vanish in B_{p-2}^k; its degree n^{**} is taken large enough to guarantee that the new differential has no singularities at infinity. Set

$$H_k^{**}(p_1^*, \ldots, p_p^*) = \Delta_k^{**}(1, p_1^*, \ldots, p_p^*) \,,$$

where $\Delta_k^{**}(\lambda, p_1^*, \ldots, p_p^*)$ is a homogeneous polynomial with respect to all its arguments. Applying the operator

$$\Delta_k^{**}\left(\frac{\partial}{\partial t} \,, \frac{\partial}{\partial x_1^*} \,, \ldots, \frac{\partial}{\partial x_p^*} \right)$$

to the new differential, we arrive at the integral in question. Therefore, the integrand in (70^*) has no singularities.

Thus, we finally obtain

$$K_u(t,x) = -\frac{C_{pn}}{x_p^*} \frac{\partial^{p-n}}{\partial t^{p-n}} \int_{C_1^*(t,x)} \pm \frac{dp_1^* \ldots dp_{p-2}^*}{H_{p-1}^{*'}} =$$

$$= \frac{C_{pn}}{(-x_p^*)^{p-n+1}} \frac{\partial^{p-n}}{\partial p_p^{* \, p-n}} \int_{C_1^*(t,x)} \pm \frac{dp_1^* \ldots dp_{p-2}^*}{H_{p-1}^{*'}} \,. \tag{70}$$

Here the signs of $dp_1^* \ldots dp_{p-2}^*$ are chosen according to the same rule as for the integrals (54), (58).[5]

---

[5]Similar arguments can be used to show that the integral

$$-\frac{C_{pn}}{x_p^*} \int_{C(t,x)} \frac{\partial^{p-n}}{\partial t^{p-n}} \left(\frac{1}{H_{p-1}^{*'}} \right) dp_1^* \ldots dp_{p-2}^*$$

exists and is equal to $K_u(t,x)$, even in the case of $C_1(t,x)$ having points at infinity.

CHAPTER IV

Cycles in Algebraic Surfaces; Integrals over Such Cycles

§12. Theorems on Abelian Integrals

1. Consider the following integral of the 1st kind:

$$I = \int \frac{M_m(x_1, \ldots, x_p)}{H'_p(x_1, \ldots, x_p)} \, dx_1 \ldots dx_{p-1} \, ,$$

where $M_m(x)$ is a polynomial of degree $m \leq n-p-1$, and $H(x_1, \ldots, x_p) = 0$ is an algebraic surface of order n without singular points. The integral is taken over a $(p-1)$-dimensional cycle C belonging to this hypersurface. It is well known that, for all cycles homologous to one another in the hypersurface $H = 0$, the value of an integral of this type is the same. The coefficients of H can be continuously varied, so that the cycle C is continuously deformed, as, for instance, in §11.5. In this situation, the following results can be stated.

Theorem III. *The integral I is an analytic function of the coefficients of H in a neighborhood of any of their values corresponding to a surface without singularities.*

Theorem IV. *The integral I is identically equal to 0 in a domain formed by the values of the coefficients of H, if and only if the cycle C is homologous to zero, or (in the case of p odd) to a lower-dimensional algebraic surface in $H = 0$.*

2. Remark. A continuous variation of the coefficients of the polynomial H results in a continuous deformation of the corresponding complex manifold, of the cycles in that manifold and the films spread over these cycles. Therefore, if a cycle C_0 is homologous to zero in a hypersurface $H_0(x) = 0$, then, in a hypersurface $H(x) = 0$ obtained from $H_0 = 0$ by a continuous deformation, the cycle obtained from C_0 by this continuous deformation is also homologous to zero.

3. The sufficiency of latter the conditions in Theorem IV is obvious. Before proving their necessity, we formulate another theorem, whose proof is similar to that of Theorem IV. Consider the integral

$$I(\lambda) = \int_C \frac{M_m(x_1,\ldots,x_p)}{H'_p(x_1,\ldots,x_p,\lambda)}\, dx_1 \ldots dx_{p-1}\,, \tag{71}$$

where $H(x_1,\ldots,x_p,\lambda)$ is a polynomial in x_k and λ of degree n, and $M_m(x_1,\ldots,x_p)$ is a polynomial in x_k of degree m. Let us regard λ as a parameter and assume that for the values of λ under consideration the surface

$$H(x_1,\ldots,x_p,\lambda) = 0 \tag{72}$$

has no singular points. The integral is taken over a $(p-1)$-dimensional normal cycle in the complex surface (72), the cycle having no points at infinity. The arguments, used in §11 to prove that the integral of

$$\frac{\partial^{p-n}}{\partial t^{p-n}}\left(\frac{dp_1^* \ldots dp_{p-1}^*}{H_{p-1}^*}\right)$$

over C_1^{**} is equal to 0, show that the residue of $\partial^k I/\partial \lambda^k$, $k > p - n + m$, over any cycle in the plane at infinity is equal to 0. Therefore, the value of $\partial^k I/\partial \lambda^k$ does not depend on the finite cycle C over which the integral is taken, provided that C is chosen from a class of cycles homologous to one another.

Theorem III′. *For any $k > p-n+m$, the derivative $\partial^k I/\partial \lambda^k$ is an analytic function of the coefficients of equation (72) and λ in a neighborhood of any of their values which correspond to a surface (72) without singular points.*

Theorem IV′. *The equations $\partial^k I/\partial \lambda^k = 0$, $(k > p - n + m)$ hold in a domain formed by the values of the coefficients of equation (72) and λ (and therefore $\partial^k I/\partial \lambda^k$ are identically equal to 0), if and only if the cycle C is homologous to zero in the algebraic hypersurface (72), or (in the case of p odd) to a lower-dimensional algebraic surface contained in (72). The latter possibility is never realized, since the cycle C is finite by assumption.*

Again, the sufficiency of the latter conditions in Theorem IV′ is obvious. Their necessity and the necessity of the conditions in Theorem IV will be established simultaneously. In both theorems we have to deal with functions holomorphic with respect to the coefficients of equation (72); these functions are assumed to vanish in a domain formed by the values of the coefficients, which means that they are identically equal to zero. Therefore, in order to establish the necessity of the above conditions, it suffices to show that for any cycle non-homologous to zero, there exists a surface (72) without singular points and such that the corresponding function is different from zero.

§13. Normal Representation for a $(p-1)$-Dimensional Cycle; Proof of Theorem III for Arbitrary p

1. Assume that $p \geq 2$ and $n > 2$. Let $H(x_1, \ldots, x_p) = 0$ be an algebraic surface of order n. It is assumed that this surface has no singular points and does not touch the plane at infinity. Consider the family of intersections of this surface and the planes $x_p = $ const. Each intersection is an algebraic surface of order n whose complex dimension is equal to the dimension of the surface $H(x_1, \ldots, x_p) = 0$ minus 1. These surfaces of complex dimension $p - 2$ have singular points only when x_p takes certain, generally, distinct values, whose number is $\mu = n(n-1)^{p-1}$. These values of x_p are determined by the following system of equations:

$$H(x_1, \ldots, x_p) = 0, \quad H'_1(x_1, \ldots, x_p) = 0, \ldots, H'_{p-1}(x_1, \ldots, x_p) = 0 . \quad (73)$$

Let us denote these values by $x_p^{(1)}, x_p^{(2)}, \ldots, x_p^{(\mu)}$. For every $x_p^{(k)}$, the $(p-2)$-dimensional surface

$$H(x_1, \ldots, x_p) = 0 , \quad (74)$$

where x_p is regarded as a parameter, contains a cycle $C_{p-2}^{(k)}(x_p)$ of real dimension $p - 2$, contracting into a point as x_p tends to $x_p^{(k)}$ (a so-called *vanishing cycle*).

The coordinate axes in the $(x_1, \ldots, x_p)$-space can always be rotated (if necessary) in such a way that all the points $(x_1, \ldots, x_p)$ satisfying system (73) become finite[6]; this condition will be assumed in what follows.

In the complex x_p-plane, consider a finite point a_p different from every $x_p^{(k)}$. Let us join a_p with all $x_p^{(k)}$ by mutually disjoint lines l_k having no points at infinity. Denote by $T_{p-1}^{(k)}$ the trace of the cycle $C_{p-2}^{(k)}(x_p)$ as x_p moves along the path l_k. Then for any $(p-1)$-dimensional cycle C_{p-1} on the surface $H = 0$, there exists (see [10], p.93) a cycle, called *normal representation of* C_{p-1}, which is homologous to C_{p-1} on that surface and has the form

$$\sum_k c_k T_{p-1}^{(k)} + (C_a)_{p-1} , \quad (75)$$

where c_k are integers, and $(C_a)_{p-1}$ is a $(p-1)$-dimensional piece of the surface

---

[6]Indeed, in the dual space, it is a straight line L passing through the origin that corresponds to the family of planes $x_p = $ const. We may assume that: 1) the line L crosses the surface S dual to $H(x_1, \ldots, x_p) = 0$ only at regular points P_k; 2) L never touches S; 3) the tangent planes to S at the points P_k do not contain the origin. The points of intersection of S and L, μ in number, correspond to μ planes, each plane touching the surface $H = 0$ at a unique finite point.

In what follows, it is convenient to choose the family of planes $x_p = $ const such that their intersections with the surface $H(x_1, \ldots, x_p) = 0$ do not touch their intersection with the hyperplane at infinity. It is easy to see that such planes can indeed be constructed.

$$H(x_1, \ldots, x_{p-1}, a_p) = 0$$

extended over the cycle

$$\sum_k c_k C_{p-2}^{(k)}(a_p)$$

homologous to zero on that surface.

On the other hand, for $p \geq 3$, every vanishing cycle $C_{p-2}^{(k)}(x_p)$, used for the construction of the tube $T_{p-1}^{(k)}$, can be represented as a difference

$$T'_{p-2} - T''_{p-2} \tag{76}$$

of two tubes belonging to the intersection of the surface $H = 0$ with the plane $x_p = $ const (see [10], p.91). These two tubes on the surface (74), where x_p is regarded as a fixed parameter, correspond, on the x_{p-1}-plane, to two paths l' and l'' issuing from the same point a_{p-1}. The cycles $C'_{p-3}(x_{p-1})$ and $C''_{p-3}(x_{p-1})$ forming these tubes coincide at $x_{p-1} = a_{p-1}$; the paths l' and l'' can be assumed to coincide in the vicinity of the point a_{p-1}. Obviously, the portions of the tubes T'_{p-2} and T''_{p-2} corresponding to the common portion of these two paths are mutually reduced and, therefore, can be neglected while constructing the cycle $C_{p-2}^{(k)}(x_p)$. It is for this reason that the cycle $C_{p-2}^{(k)}(x_p)$ contracts into a point as the end-points of the paths l' and l'' become close.

The paths $l'(x_p)$ and $l''(x_p)$ may be assumed finite; this follows from the argument used for showing the paths l_k to be finite. Considering a similar construction for the cycles

$$C_{p-3}(x_p, x_{p-1}), \quad C_{p-4}(x_p, x_{p-1}, x_{p-2}), \ldots,$$

and taking into account that the cycles $l_k, l(x_p), l(x_p, x_{p-1}), \ldots$ are finite, we see that all tubes $T_{p-1}^{(k)}$ in (75) are also finite. A similar construction applied to the sections of the surface $H(x_1, \ldots, x_p) = 0$ by the planes

$$x_p = x_p^0 = \text{const}, \ldots, \quad x_{k+1} = x_{k+1}^0 = \text{const},$$

allows us to change the coordinate axes $Ox_1, \ldots, Ox_k$ as $x_{k+1}^0, \ldots, x_p^0$ vary.

Thus, for any p, all infinite points of the cycle (75) can be assumed to belong to $(C_a)_{p-1}$ only. Cycles of this type will be called *quasi-finite*.

2. *Proof of Theorem III.* The above statements concerning the normal representation show that every integral of the 1st kind

$$\int \frac{M(x_1, \ldots, x_p)}{H'_1(x_1, \ldots, x_p)} \, dx_2 \ldots dx_p \tag{77}$$

over a $(p-1)$-dimensional cycle C_{p-1} belonging to the surface $H = 0$ can be represented in the form

$$\sum_{k=1}^{\mu} c_k \int_{T_{p-1}^{(k)}} \frac{M(x_1,\ldots,x_p)}{H_1'(x_1,\ldots,x_p)} \, dx_2 \ldots dx_p =$$

$$= \sum_{k=1}^{\mu} c_k \int_{l_k} dx_p \int_{C_{p-2}^{(k)}(x_p)} \frac{M(x_1,\ldots,x_p)}{H_1'(x_1,\ldots,x_p)} \, dx_2 \ldots dx_{p-1} \, .$$

Let us assume Theorem III to be proved for algebraic surfaces of complex dimension $p-2$.

Obviously, Theorem III holds for surfaces of dimension 0, since in this case the integral (77) is reduced to a sum of the integrand's values at finitely many points.

Let us prove Theorem III for algebraic surfaces of complex dimension $p-1$. For this purpose, consider the integral

$$I(\varepsilon) = \sum_{k=1}^{\mu} c_k \int_{l_k-\varepsilon} dx_p \int_{C_{p-2}^{(k)}(x_p)} \frac{M(x_1,\ldots,x_p)}{H_1'(x_1,\ldots,x_p)} \, dx_2 \ldots dx_{p-1} \, ,$$

where $l_k - \varepsilon$ denotes the path l_k with an interval of length ε adjacent to the point $x_p^{(k)}$ removed. For any sufficiently small $\varepsilon > 0$, the integral $I(\varepsilon)$ is a holomorphic function of the coefficients of H, since we integrate, with respect to x_p, a holomorphic function of these coefficients, when the latter are sufficiently close to the coefficients of the surface under consideration. On the other hand, the values of $I(\varepsilon)$, as $\varepsilon \to 0$, remain bounded if the coefficients of H are within this domain, since $I(\varepsilon)$ converges to an integral of the 1st kind over a cycle in this surface. On the basis of the arguments usually employed to prove that an integral of the 1st kind over a surface without singularities is finite, one can also show that the above integral is uniformly bounded, provided that the coefficients of this surface are sufficiently close to those of a certain surface without singularities. Therefore $I(0)$, i.e., the integral (77), is a holomorphic function of the coefficients of H in any domain formed by the values of these coefficients such that the surface $H = 0$ has no singularities.

3. *Proof of Theorem III'.* The above proof of analytic dependence of an integral of the 1st kind on the coefficients of the algebraic surface $H = 0$ is fully applicable to the analytic dependence on the coefficients of H in the case of an arbitrary integral of the form (77) taken over a $(p-1)$-dimensional finite cycle of type (75), provided that the algebraic surface $H = 0$ has no singular points.

Independently (of the theorem under discussion), it will be shown later on (see §15, Theorem VI) that a multiple of any $(p-1)$-dimensional cycle on the surface $H = 0$ is homologous to a finite cycle of type (75), or (in the case of p odd), possibly, to the sum of a finite cycle of type (75) and an algebraic surface of complex dimension $(p-1)/2$ belonging to $H = 0$. It follows that every finite cycle is homologous to a finite cycle of type (75) only, since otherwise we would have an algebraic cycle homologous to a finite one, and therefore, its index of intersection with every algebraic surface of complex dimension $(p-1)/2$ lying in $H = 0$ and in the plane at infinity would be equal to 0, which is impossible (see [9]). Thereby Theorem III$'$ is proved for any finite cycle C.

§14. Further Study of Normal Representation

1. In order to find out how many tubes $T_{p-1}^{(k)}$ correspond to any cycle $C_{p-2}^{(k)}(a_p)$, and with a view to clarifying the formation of vanishing cycles from the tubes $T_{p-1}^{(k)}$, consider first the following special case:

$$H^0(x_1,\ldots,x_p) \equiv f_1(x_1) + f_2(x_2) + \cdots + f_p(x_p) = 0 , \qquad (78)$$

where $f_k(x)$ is a polynomial in x of degree n, all its roots being finite and different from zero and one another.

2. Before studying the sections of this surface by the planes $x_p = \text{const}$, let us consider the equation

$$f_1(x_1) = -\xi \qquad (79)$$

depending on the parameter ξ. By assumption, for $\xi = a_1$, this equation has n roots different from one another, where a_1 is any number different from every ξ_k. Denote by

$$x_1^{(1)}, x_1^{(2)}, \ldots, x_1^{(n-1)}$$

the roots of the equation

$$f_1'(x_1) = 0$$

and set

$$f_1(x_1^{(k)}) = -\xi_k . \qquad (80)$$

Let us choose the function f_1 such that all ξ_k be mutually distinct. Let $l_1,\ldots,l_{n-1}$ be the lines on the ξ-plane that join the point $\xi = a_1$ to $\xi_1,\ldots,\xi_{n-1}$, respectively, and have no points in common but the initial one. If we move the point ξ along l_1 from $\xi = a_1$ to $\xi = \xi_1$, we see that among the roots of the equation

$$f_1(x_1) = -a_1 \qquad (80')$$

two roots, say $x_{10}^{(1)}$ and $x_{10}^{(2)}$, become equal. Among the paths l_k, there exists at least one, say l_2, such that by moving along l_2 we can make the root $x_{10}^{(2)}$

merge with another root of equation (80), say $x_{10}^{(3)}$; otherwise, the Riemann surface of equation (79), and consequently, equation (79) itself would split into two, which is impossible.

Attaching suitable indices to the roots of equation (79), we obtain a path l_1 along which the roots $x_{10}^{(1)}$ and $x_{10}^{(2)}$ merge into each other, and the path l_2 along which $x_{10}^{(2)}$ and $x_{10}^{(3)}$ approach one another.

Consider a table where, for each path l_k, the superscripts associated with the roots x_{10} tending to coincide when we move along l_k are listed below the corresponding l_k:

$$
\begin{array}{ccccccc}
l_1 & \cdots & l_{k-1} & l_k & \cdots & l_{n_1} \\
(1,2) & \cdots & (k-1,k) & (k,k+1) & \cdots & (n_1, n_1+1)
\end{array}
\qquad (81)
$$

Among the rest of the paths l, there is always a path l_{n_1+1} such that by moving along l_{n_1+1} we can make the root $x_{10}^{(k_1)}$, $k_1 \le n_1 + 1$, merge with $x_{10}^{(k_2)}$, where $k_2 > n_1 + 1$; otherwise, (79) would split into several equations. If $k_1 = n_1 + 1$, then appending l_{n_1+1} to the paths in (81) and changing, if necessary, the numeration of $x_{10}^{(k)}$ for $k > n_1 + 1$, we obtain a system of $n_1 + 1$ paths and the following table

$$
\begin{array}{cccccc}
l_1 & \cdots & l_{k-1} & l_k & \cdots & l_{n_1} & l_{n_1+1} \\
(1,2) & \cdots & (k-1,k) & (k,k+1) & \cdots & (n_1, n_1+1) & (n_1+1, n_1+2)
\end{array}
\quad .
$$

If $k_1 < n_1 + 1$, then we deform the path l_{n_1+1} in such a way that its beginning and its end remain fixed, while in its motion the paths in (81) are left behind, in consecutive order, one after another as they issue from the point $\xi = a_1$. When l_{n_1+1} passes across the paths in (81) such that their respective x_{10} tending to coincide along these paths have superscripts other than k_1 and k_2, the superscripts of x_{10} corresponding to the deformed path l_{n_1+1}, as well as the superscripts of x_{10} corresponding to the path crossed, do not change. If the path l_{n_1+1} crosses in its movement l_{k_1} (resp., l_{k_1-1}) then the table

$$
\begin{array}{cccccc}
l_1 & \cdots & l_{k_1-1} & l_{k_1} & \cdots & l_{n_1} & , \; l_{n_1+1} \\
(1,2) & \cdots & (k_1-1,k_1) & (k_1,k_1+1) & \cdots & (n_1,n_1+1) & (k_1,k_2)
\end{array}
$$

is replaced by

$$
\begin{array}{cccccc}
l_1 & \cdots & l_{k_1-1} & l_{k_1} & \cdots & l_{n_1} & l_{n_1+1} \\
(1,2) & \cdots & (k_1-1,k_1) & (k_1+1,k_2) & \cdots & (n_1,n_1+1) & (k_1,k_2)
\end{array}
\quad ,
$$

or, respectively, by

$$
\begin{array}{cccccc}
l_1 & \cdots & l_{k_1-1} & l_{k_1} & \cdots & l_{n_1} & l_{n_1+1} \\
(1,2) & \cdots & (k_1-1,k_2) & (k_1,k_1+1) & \cdots & (n_1,n_1+1) & (k_1,k_2)
\end{array}
\quad ;
$$

(cf. [11], p. 206). In either case, enumerating, if necessary, the points x_{10} in a different way, we can easily come to table (81).

Thus we obtain $n - 1$ paths

$$l_1, l_2 \ldots, l_{n-1}$$

having no other point in common but the initial one, and joining the point $\xi = a_1$ to $\xi_1, \ldots, \xi_{n-1}$, respectively; by moving along l_k from $\xi = a_1$ to ξ_k, the roots $x_{10}^{(k)}$ and $x_{10}^{(k+1)}$ of equation (79) are made to merge.

3. Now we turn to the construction of the paths, to be used in the sequel, on the Riemannian surface for the equation

$$f_1(x_1) + f_2(x_2) = -\eta , \tag{82}$$

where η is a parameter. Denote by $x_2^{(1)}, \ldots, x_2^{(n-1)}$ the roots of $f_2'(x_2)$ and set

$$f_1(x_1^{(k)}) + f_2(x_2^{(p)}) = -\eta_{kp} .$$

In order simplify our further exposition (in Sect. 5), we assume that all η_{kp} are different from zero and one another.

Set

$$\eta + f_2(x_2) = \xi . \tag{83}$$

Assume first that $\eta \neq \eta_{kp}$ for all k, p, and that neither of the paths l_k can pass through any of the points $\eta + f_2(x_2^{(s)})$.

Then we can establish a correspondence between each path l_k on the ξ-plane and n non-intersecting paths on the x_2-plane ending at the points $x_2^{(kp)}(\eta)$ and satisfying the equation

$$f_1(x_1^{(k)}) + f_2(x_2^{(kp)}(\eta)) = -\eta . \tag{84}$$

The set of $n - 1$ paths (considered above) issuing from the same point on the ξ-plane corresponds to n groups, each consisting of $n - 1$ paths on the x_2-plane, all the paths belonging to the same group issue from the same point and, except at this point, are mutually disjoint. Let us consider one of these groups. Denote by $\tilde{l}_k$ the path from this group corresponding to the path l_k on the ξ-plane.

Before constructing the other paths on the x_2-plane, we make the following observation. By the arguments similar to those of Sect. 2, we can define $n - 1$ paths l_{kp}^* issuing from the point $\eta = 0$, ending at η_{kp} ($k, p = 1, 2, \ldots, n - 1$), and having the following property: while η moves along the path l_{kp}^*, the roots $x_2^{(kp)}(\eta)$ and $x_2^{(k,p+1)}(\eta)$ of equation (84) tend to merge. As shown in Sect. 2, these paths can be chosen to have no other point in common but the initial one.

Let us return to the construction of the other paths on the x_2-plane. Let $x_2^{(k1)}(\eta)$ be the terminal point of the path $\tilde{l}_k$, for which we use the corresponding notation $l_{k1}(\eta)$. Let us make η move along the path l_{k1}^*, so as to make the points $x_2^{(k1)}(\eta)$ and $x_2^{(k2)}(\eta)$ coincident. Next, we subject the path l_{k1}, constructed above, to a continuous deformation, so that:

1) its initial point remains at $x_2 = a_2$;

2) its end-point remains at $x_2^{(k1)}(\eta)$;

3) during the deformation, this path does not meet any of the other points $x_2^{(kp)}(\eta)$.

When η reaches η_{k_1}, i.e., $\eta = \eta_{k_1}$, the points $x_2^{(k1)}(\eta)$ and $x_2^{(k2)}(\eta)$ coincide. Then we follow the path l_{k1}^* in the opposite direction. But now we place the end-point of $l_{k1}(\eta)$ at $x_2^{(k2)}(\eta)$, while retaining the above three conditions imposed on the path's deformation. Denote by $l_{k2}(\eta)$ the path obtained in this way. Thus we return to the point $\eta = 0$. Next, we follow the path l_{k2}^* on the η-plane, from $\eta = 0$ to $\eta = \eta_{k2}$. Let the path $l_{k2}(\eta)$ be deformed under the above three conditions. The points $x_2^{(k2)}(\eta)$ and $x_2^{(k3)}(\eta)$ will coincide at $\eta = \eta_{k2}$. Then we follow $l_{k2}^*(\eta)$ in the opposite direction, placing the end-point of $l_{k2}(\eta)$ at the point $x_2^{(k3)}(\eta)$, again retaining the above three conditions. Denote the new path obtained in this manner by $l_{k3}(\eta)$. Once again we return to the point $\eta = 0$. Then we follow the path l_{k3}^*, etc., and proceed in this way until all the paths $l_{k1}^*, l_{k2}^*, \ldots, l_{k,n-1}^*$ are exhausted.

The following observation regarding the paths $l_{kp}(\eta)$ constructed above will be helpful in the sequel. The paths

$$l_{k1}(\eta), l_{k2}(\eta), \ldots, l_{kn}(\eta)$$

can always be deformed in such a way that they will have no other point in common but 0. The images of all these paths (corresponding to the same k) on the ξ-plane, obtained by the formula (83), will have common initial points, common terminal points, and no point ξ_k between them.

4. Assume that the Riemannian surface has somehow been constructed for the curve (82). The term *loop* $l_{kp}(\eta)$, or *tube* $l_{kp}(\eta)$, will be applied to a curve on that surface whose projection to the x_2-plane coincides with the line $l_{kp}(\eta)$ which has its initial as well as its terminal point at $x_2 = a_2$, and which, being passed along, brings us from the value $x_{10}^{(k)}$ to $x_{10}^{(k+1)}$. The loop obtained from $l_{kp}(\eta)$ by passing in the opposite direction will be denoted by $l_{kp}^{-1}(\eta)$.

Every cycle on the Riemannian surface R is homologous to a sequence of loops. The loops $l_{kp}(\eta)$ and $l_{kp}^{-1}(\eta)$ in this sequence are mutually cancelled. When η, moving along the line l_{kp}^*, approaches one of the points η_{kp}, the branch points $x_2^{(kp)}(\eta)$ and $x_2^{(k,p+1)}(\eta)$ (these are the terminal points of the paths $l_{kp}(\eta)$ and $l_{k,p+1}(\eta)$) become merged. A cycle on R whose projection on the x_2-plane encloses these points, and which is homologous to $l_{kp}(\eta) + l_{k,p+1}^{-1}(\eta)$, can be contracted into a single point as $\eta \to \eta_{kp}$. A cycle of this type, as well as any other cycle homologous to it, is said to be *vanishing*.

Theorem. *Every cycle on the curve* (82) *is homologous to a linear combination of vanishing cycles.*

Proof. As mentioned above, every cycle C on R is homologous to a sequence of loops. Let the cycle C contain a loop l_{k_1}. Adding to, or subtracting from C a vanishing cycle homologous to

$$l_{k1}(\eta) + l_{k2}^{-1}(\eta) \,,$$

as many times as necessary, we obtain a cycle that does not contain the loop $l_{k1}(\eta)$. If the resulting cycle contains the loop $l_{k2}(\eta)$, we can arrive at a cycle containing neither the loop $l_{k1}(\eta)$ nor $l_{k2}(\eta)$ by adding or subtracting, as many times as necessary, a vanishing cycle homologous to

$$l_{k2}(\eta) + l_{k3}^{-1}(\eta) \,.$$

Proceeding in this way, we come to a cycle C^0 that does not contain any of the loops

$$l_{k1}(\eta), l_{k2}(\eta), \ldots, l_{k,n-1}(\eta) \quad \text{and} \quad l_{k1}^{-1}(\eta), l_{k2}^{-1}(\eta), \ldots, l_{k,n-1}^{-1}(\eta) \,.$$

Let

$$C^0 = \sum_s l_{k_s n}^{c_s}(\eta) \,,$$

where $c_s = \pm 1$, $k_s = 1, 2, \ldots, n-1$; $l_{k_s n}^{+1}(\eta) = l_{k_s n}(\eta)$. Since the cycle $l_{kp}(\eta) + l_{kp}^{-1}(\eta)$ is always homologous to zero, it may be assumed that for every s we have

$$l_{k_{s+1} n}^{c_{s+1}}(\eta) \neq l_{k_s n}^{-c_s}(\eta) \,. \tag{84'}$$

Let us show that C^0 cannot consist of the loops $l_{k_s n}^{c_s}(\eta)$ only. Assume the contrary. We can take $c_1 = +1$, since, otherwise, the sign of C^0 could be reversed.

Consider the point $x_2 = a_2$ with the value $x_1 = x_{10}^{(k_1)}$. Starting from this point and moving along the loop $l_{k_1 n}(\eta)$, we return to the same point $x_2 = a_2$ with the value $x_1 = x_{10}^{(k_1+1)}$. Because of (84'), this value can be changed only if we move along those of the subsequent loops whose first subscript is $k_1 + 1$. Therefore, $k_2 = k_1 + 1$ and $c_2 = +1$. Next, moving along $l_{k_2 n}(\eta)$, we come to the point $x_2 = a_2$ with the value $x_1 = x_{10}^{(k_2+1)}$. Continuing this process, we return again and again to the point $x_2 = a_2$ with values $x_{10}^{(k)}$ having increasing k; the initial value of x_1 will never occur again. Therefore, C^0 is not a cycle, which contradicts our assumption. Thereby, the proof is complete.

5. Now, in order to facilitate our further analysis, let us turn to the construction of the normal representation of two-dimensional cycles on surfaces of the form

$$f_1(x_1) + f_2(x_2) + f_3(x_3) = -\zeta \; ; \tag{85}$$

here ζ is a parameter. Denote by $x_3^{(1)}, \ldots, x_3^{(n-1)}$ the roots of $f_3'(x_3)$ and set

$$f_1(x_1^{(k)}) + f_2(x_2^{(p)}) + f_3(x_3^{(s)}) = -\zeta_{kps} \; .$$

For the sake of simplicity, we assume all ζ_{kps} to be different from zero and one another. Set

$$\zeta + f_3(x_3) = \eta \; . \tag{85'}$$

Then equation (85) is reduced to (82). Assume, first, that

$$\zeta \neq \zeta_{kps} \quad \text{for all} \quad k, p, s \; ,$$

and that neither of the paths l_{kp}^* on the η-plane contains any of the points $\zeta + f_3(x_3^{(s)})$. Then every path l_{kp}^* on the η-plane corresponds, by virtue of (85'), to n mutually non-intersecting paths on the x_3-plane ending at the points $x_s^{(kps)}$ which satisfy the equation

$$f_1(x_1^{(k)}) + f_2(x_2^{(p)}) + f_3(x_3^{(kps)}) = -\zeta \; . \tag{86}$$

The entire set of $(n-1)^2$ paths (considered above) issuing from a single point on the η-plane corresponds to n groups, each having $(n-1)^2$ paths on the x_3-plane; all paths from the same group issue from one point and, except at this point, are mutually disjoint. Consider one of these groups. Let $l_{kp}(\zeta)$ denote the path from this group corresponding to the path l_{kp}^* on the x_3-plane.

Arguments similar to those of Sect. 2 show that for any pair of values $x_1^{(k)}$, $x_2^{(p)}$, there exist $(n-1)$ paths l_{kps}^* on the ζ-plane, each starting at $\zeta = 0$ and ending at ζ_{kps}, respectively; moreover, these paths have the following property: as ζ moves along l_{kps}^*, the root $x_3^{(kps)}(\zeta)$ of equation (86) approaches the root $x_3^{(kp,s+1)}(\zeta)$, $k, p, s = 1, \ldots, n-1$. These paths can be chosen to have no point in common but the initial one.

Using the paths l_{kps}^* on the ζ-plane, we construct the paths $l_{kps}(\zeta)$, $k, p = 1, \ldots, n-1$, $s = 1, \ldots, n$, on the x_3-plane in exactly the same manner as the paths $l_{kp}(\eta)$ on the x_2-plane were constructed from the paths l_{kp}^* on the η-plane (see Sect. 3). Every such path $l_{kps}(\zeta)$, $k, p = 1, \ldots, n-1$, $s = 1, \ldots, n$, corresponds to a vanishing cycle on the curve (85), where x_3 is regarded as a parameter. All paths $l_{kps}(\zeta)$ differing only by the values of s, at $\zeta = 0$, correspond to the same cycle on (85). Every path $l_{kps}(\zeta)$ corresponds to a two-dimensional tube $T_{kps}(\zeta)$ on the surface (85). When ζ moves along any path l_{kps}^* from $\zeta = 0$ to $\zeta = \zeta_{kps}$, the tubes $T_{kps}(\zeta)$ and $T_{kp,s+1}(\zeta)$ will approach one another and merge at $\zeta = \zeta_{kps}$. The parts of these tubes corresponding to the values of x_3 close to a_3 can be made to coincide by a small deformation. Therefore, the cycle composed of these tubes, when x_3 moves along $l_{kps}(\zeta)$ from $x_3 = x_3^{(kps)}$ to $x_3 = a_3$, and then along $l_{kp,s+1}(\zeta)$ from $x_3 = a_3$ to $x_3 = x_3^{(kp,s+1)}$, is homologous to a cycle which contracts to a single point as $\zeta \to \zeta_{kps}$.

Using these vanishing cycles, we can similarly construct tubes $T_{kpst}(\tau)$ for the surface

$$f_1(x_1) + f_2(x_2) + f_3(x_3) + f_4(x_4) = -\tau \, .$$

Continuing this process, we can construct tubes T for any surface (78).

6. The tubes T for $p > 2$ and the loops l for $p = 2$, constructed in the special case (78), induce tubes or loops for any algebraic surfaces or algebraic curves of order n which have no singularities and do not touch the plane at infinity. Indeed, continuously changing the coefficients of equation (78), we can transform the surface, or, in the case of $p = 2$, the curve, defined by this equation into an arbitrary algebraic surface $\widetilde{H}(x_1, \ldots, x_p) = 0$ of order n without singularities or touching points with the plane at infinity; moreover, this deformation can be performed in such a way that in its process we never obtain a surface with singularities, or a surface tangential to the plane at infinity, or a surface $H(x_1, \ldots, x_p) = 0$ such that the system

$$H(x_1, \ldots, x_p) = 0 \, , \quad H_1(x_1, \ldots, x_p) = 0 \; ; \ldots , H_{p-1}(x_1, \ldots, x_p) = 0 \quad (87)$$

admits solutions with coinciding x_p, or solutions $(x_1, \ldots, x_p)$ that are infinite. It is assumed here that the surface $\widetilde{H} = 0$ itself satisfies this condition; in order that the coordinates x_p of the solutions of system (87) for $H = \widetilde{H}$ be mutually distinct, and the solutions $(x_1, \ldots, x_p)$ be finite, we can always rotate, if necessary, the coordinate axes $x_1, \ldots, x_p$, provided that the surface $\widetilde{H} = 0$ itself has no singular points (cf., §13, Sect. 1). In fact, the condition that the coordinates x_p of some solutions of system (87) coincide, or be infinite, or that the surface $H = 0$ have singular points is expressed by a single algebraic equation with respect to the complex coefficients of the equation $H = 0$.

The solution of system (87) varies continuously, while, under the above conditions, the surface $H^0 = 0$ is being deformed into the surface $\widetilde{H} = 0$. The paths l on the plane $x_p = $ const and their respective vanishing cycles in the intersections of the surface $H(x_1, \ldots, x_p) = 0$ and the planes $x_p = $ const, as well as the tubes T corresponding to these paths, can also be assumed to vary continuously during the above deformation. Therefore, on the surface $\widetilde{H} = 0$, as well as on all intermediate surfaces, the paths l are split into groups, each having n lines with the initial point $x_p = a_p$ corresponding to the same cycle on the surface

$$H(x_1, \ldots, x_{p-1}, x_p) = 0 , \qquad x_p = a_p .$$

§15. Proof of Theorems IV and IV' for Arbitrary p

1. In the $(x_1, \ldots, x_p)$-space, consider the surface

$$f_1(x_1) + \cdots + f_{p-1}(x_{p-1}) + f_p(x_p) + f_{p+1}(\lambda) \equiv H^*(x_1, \ldots, x_{p-1}) +$$

$$+ f_p(x_p) + f_{p+1}(\lambda) \equiv H(x_1, \ldots, x_p, \lambda) \equiv H(\lambda) = 0 , \qquad (88)$$

where

$$H(x_1, \ldots, x_p, 0) = 0$$

is an algebraic surface of order n in the $(x_1, \ldots, x_p)$-space, having no singular points and satisfying the conditions of §14.6; $f_k(x_k)$ and $f_{p+1}(\lambda)$, just as in §14, are polynomials of degree n whose derivatives have no multiple roots; λ is regarded as a parameter. The surface (88), as well as any other surface of order n with coefficients sufficiently close to those of (88), has no singular points for the values of λ belonging to the domain under consideration.

Consider the integral

$$I^\nu(\lambda) = \frac{\partial^\nu}{\partial \lambda^\nu} \int \frac{M(x_1, \ldots, x_p)}{H'_1(x_1, \ldots, x_p, \lambda)} \, dx_2 \ldots dx_{p-1} \, dx_p \qquad (89^\nu)$$

over a cycle C_{p-1} on the surface $H(\lambda) = 0$ (the cycle C_{p-1} must be finite, unless the integral $I(\lambda)$ is of the 1st kind), where $M(x_1, \ldots, x_p)$ is a polynomial of degree $\nu > p - n + m$.

Since C_{p-1} is homologous to a quasi-finite cycle of the form (75) on the surface $H(\lambda) = 0$, it follows that $I^\nu(\lambda)$ can be represented as the ν-th derivative of the integral

$$I(\lambda) = \sum_{k=1}^{\mu} c_k \int_a^{a_k(\lambda)} \Omega_k(x_p, \lambda) \, dx_p , \qquad (89)$$

where

$$\Omega_k(x_p, \lambda) = \int \frac{M(x_1, \ldots, x_p)\, dx_2 \ldots dx_{p-1}}{H_1^{*'}} . \tag{89*}$$

Here the integral is taken over the cycle $C_{p-2}^{(k)}(x_p, \lambda)$ mentioned in §13.1.

It is easy to see that

$$\Omega_k(x_p, \lambda) = \sum_{\nu=0}^{m} x_p^{\nu} \Omega_k^{\nu}(x_p, \lambda) ,$$

where $\Omega_k^{\nu}(x_p, \lambda)$ are analytic functions of

$$\eta = f_p(x_p) + f_{p+1}(\lambda) \tag{90}$$

in any domain of the values of η containing no points η_e defined next. In (89), the limits of integration $a_k(\lambda)$ are equal to the values of x_p such that the equations

$$\eta_e = f_p(x_p) + f_{p+1}(\lambda)$$

are satisfied, where η_e is any value of η for which the system

$$H^*(x_1, \ldots, x_{p-1}) + \eta = 0 ,$$

$$H_1^*(x_1, \ldots, x_{p-1}) = 0 ,$$

$$\ldots\ldots\ldots\ldots\ldots\ldots\ldots$$

$$H_{p-1}^*(x_1, \ldots, x_{p-1}) = 0$$

can be solved with respect to $x_1, \ldots, x_{p-1}$.

By a small variation of H^* we can always guarantee that for different solutions of the above system the values of η are different, i.e., we may assume all η_e to be different from one another. Indeed, the condition for two of η_e to be equal is expressed in terms of a polynomial of the coefficients of H^* being equal to zero, which cannot hold identically; here H^* may be assumed to have the form (78).

Then (89) can be rewritten as

$$I(\lambda) = \sum_{k=1}^{\mu} c_k \int_{f_p(a)+f_{p+1}(\lambda)}^{f_p(a_k(\lambda))+f_{p+1}(\lambda)} \frac{\Omega_k(x_p, \eta)}{f_p'(x_p)}\, d\eta .$$

Instead of a single subscript k, let us assign to $a(\lambda)$ two subscripts: i and j. We assign the same subscript i to all $a(\lambda)$ yielding the same value of f_p. Obviously, j varies from 1 to n. Accordingly, in other cases two subscripts i, j will also be used instead of k. Then we get

$$I(\lambda) = \sum_i \sum_j c_{ij} \int_{\eta_0}^{\eta_i} \frac{\Omega_{ij}(x_p^j, \eta)\, d\eta}{f_p'(x_p^j(\eta, \lambda))} \tag{92}$$

where we have set

$$f_p(a_{ij}) + f_{p+1}(\lambda) = \eta_i , \quad f_p(a) + f_{p+1}(\lambda) = \eta_0 ,$$

and $x_p^j(\eta, \lambda)$ is regarded as one of the functions of η and λ determined by equation (90). The paths l_{ij} on the x_p-plane starting at the point a and ending at a_{ij}, $i, j = 1, \ldots, n$, are transformed by (90) into the paths l_{ij} on the η-plane, l_{ij} having the same starting point η_0 and the same terminal point η_i independent of λ. Since a can be chosen arbitrary, the point η_0 can also be assumed independent of λ.

Because of the special choice of the paths l on the x_p-plane (see §14), the paths of integration in η from η_0 to η_i corresponding to the same i can be continuously deformed into one another without crossing any of the points η_s; all cycles $C_{p-2}^{ij}(x_p)$ corresponding to the tubes T_{p-1}^{ij} with the same i can be assumed homologous in the intersection of the surface (88) with the plane $x_p =$ const. Consequently, all functions $\Omega_{ij}(x_p^j, \eta)$ with equal second subscripts coincide, and therefore (92) can be rewritten in the form

$$I(\lambda) = \sum_{ij} c_{ij} \int_{\eta_0}^{\eta_i} \frac{\Omega_i(x_p^j, \eta)}{f_p'(x_p^j(\eta, \lambda))} \, d\eta . \tag{93}$$

The second subscript by Ω has been omitted here as being of no importance.

A value of λ is said to be *critical*, if equation (91) admits a multiple root x_p for this λ. It is obvious that for a critical value of λ the surface (88) will have a singular point. To simplify our further exposition, in similarity to what has been done in §14, the second subscript assigned to the tube T_{p-1}^{ij} will be chosen such that the tube T_{p-1}^{ij} would approach the tube $T_{p-1}^{i,j+1}$ as λ approaches a critical value λ_0.

2. Let us evaluate $\Omega_i(x_p, \eta)$. Assume that for $k < p$ we have $x_k = 0$ at the end-point of the tube T_{p-1}^{ij}, and $\eta_i = 0$. Then (88) yields

$$H \equiv H^*(x_1, \ldots, x_{p-1}) + \eta \equiv f_1(x_1) + \cdots + f_{p-1}(x_{p-1}) + \eta \equiv$$

$$\equiv \eta + \sum_{j=1}^{p-1} a_j x_j^2 + \cdots ,$$

where the terms of the higher order (with respect to $x_1, \ldots, x_{p-1}$) are not written out explicitly. Since the derivatives of the polynomials $f_k(x_k)$ admit no multiple roots, all a_j are different from 0. We can limit ourselves to the case $a_j = -1$, $j = 1, \ldots, p - 1$, by changing the variables, if necessary. Assume also that η takes only real negative values in the vicinity of the end-point of the tube T_{p-1}^{ij}. For sufficiently small $|\eta|$, $\eta < 0$, the cycle $C_{p-1}^i(\eta)$ can be obtained by drawing all possible straight lines with real coefficients in the

$(x_1, \ldots, x_{p-1})$-plane, starting at the point A, where $x_j = 0$, $j = 1, \ldots, p-1$, and going up to their points of intersection (nearest to A) with the surface $H^*(x_1, \ldots, x_{p-1}) + \eta = 0$. Let us change the variables in the integral

$$\Omega_i(x_p, \eta) = \int_{C^i_{p-1}(\eta)} \frac{M(x_1, \ldots, x_p)\, dx_1 \ldots dx_{p-2}}{H^{*'}_{p-1}} \, ,$$

setting $x_j = \sqrt{\eta}\, r y_j$, $j = 1, \ldots, p-1$, where $\sum y_j^2 = 1$. Then we obtain

$$H^* + \eta \equiv \eta[1 - r^2] + \sqrt{\eta}\, M_1(y, \sqrt{\eta}, r) = 0 \, ,$$

$$M(x_1, \ldots, x_p) = M(0, \ldots, 0, x_p) + \sqrt{\eta}\, M_2(y, \sqrt{\eta}, r, x_p) \, ,$$

where the functions M_ν are polynomials with respect to all their arguments.

Assume that $M(0, \ldots, 0, x_p) = C \neq 0$. Let us pass in the above integral to the surface element $d\sigma$ on the real $(p-2)$-dimensional sphere Σ: $y_1^2 + \cdots + y_{p-1}^2 = 1$. Then, for sufficiently small $|\eta|$, we have $r \approx 1$ and

$$\Omega_i(x_p, \eta) = \frac{\eta^{\frac{p-3}{2}}}{2} \int \frac{M(0, \ldots, 0, x_p) + \sqrt{\eta}\, M_1(y, \sqrt{\eta}, r, x_p)}{-1 + \sqrt{\eta} M_3(y, \sqrt{\eta}, r)}\, d\sigma \approx C C_1 \eta^{\frac{p-3}{2}} \, ,$$

$$(*)$$

where $C_1 \neq 0$, M_3 is a polynomial with respect to y_j, $\sqrt{\eta}$, r.

For $p = 2$ we have

$$\Omega_i(x_2, \xi) = \frac{M(x_1^{(i)}, x_2)}{f_1'(x_1^{(i)})} - \frac{M(x_1^{(i+1)}, x_2)}{f_1'(x_1^{(i+1)})} \, ,$$

where $x_1^{(i)}$ and $x_1^{(i+1)}$ are the values of x_1 at the end-points of the path on the ξ-plane (cf. §14.2). For sufficiently small $|\xi|$ we have

$$\Omega_i(x_2, \xi) \approx \pm \frac{M(0, x_2)}{\sqrt{|\xi|}} \quad \text{if} \quad \xi_i = 0 \, , \ M(0, x_2) \neq 0 \, .$$

3. Assume that $I(\lambda) = 0$ for all λ. Then we must have $I'(\lambda) \equiv 0$, and (93) implies that

$$\frac{dI}{d\lambda} = -\sum_{ij} c_{ij} \int_{\eta_0}^{\eta_i} \frac{\Omega_i'(x_p^j, \eta)}{\left\{ f_p'(x_p^j(\eta, \lambda)) \right\}^2} \frac{dx_p^j}{d\lambda}\, d\eta \, ,$$

where

$$\Omega_i'(x_p, \eta) \equiv \Omega_i(x_p, \eta) f_p''(x_p) - f_p'(x_p) \sum_{\nu=1}^{m} \nu(x_p)^{\nu-1} \Omega_i^\nu(\eta) \equiv$$

$$\equiv \sum_{\nu=0}^{m+n-2} x_p^\nu \Omega_i'^\nu(\eta) = - \int_{C^i_{p-2}(\eta)} \frac{M(x) f_p''(x_p) + M_1(x) f_p'(x_p)}{H_1^{*'}}\, dx_2 \ldots dx_{p-1} \, ,$$

$M_1(x)$ is a polynomial in $x_1, \ldots, x_p$. It follows from (90) that

$$\frac{\mathrm{d}x_p}{\mathrm{d}\lambda} = -\frac{f'_{p+1}(\lambda)}{f'_p(x_p)} .$$

Therefore

$$I'(\lambda) \equiv f'_{p+1}(\lambda) \sum_{ij} c_{ij} \int_{\eta_0}^{\eta_i} \frac{\Omega'_i(x_p^j, \eta)}{\left\{ f'_p(x_p^j(\eta, \lambda)) \right\}^3} \, \mathrm{d}\eta \equiv 0 . \tag{94}$$

Since $f'_{p+1}(\lambda) \not\equiv 0$, it follows from the identity $I'(\lambda) \equiv 0$ that the coefficient by $f'_{p+1}(\lambda)$ in the right-hand side of (94) is identically equal to 0. Denoting this coefficient by $I_1(\lambda)$ and applying transformations similar to those just used, we find that

$$I'_1(\lambda) \equiv f'_{p+1}(\lambda) \sum_{ij} c_{ij} \int_{\eta_0}^{\eta_i} \frac{\Omega''_i(x_p^j, \eta)}{\left\{ f'_p(x_p^j(\eta, \lambda)) \right\}^5} \, \mathrm{d}\eta \equiv 0 ,$$

where

$$\Omega''_i(x_p, \eta) = 3\Omega'_i(x_p, \eta)f''_p(x_p) - f'_p(x_p) \sum_{\nu=1}^{m+n-2} \nu x_p^{\nu-1} \Omega'^{\nu}_i(\eta) =$$

$$= \sum_{\nu=0}^{2n+m-3} x_p^{\nu} \Omega''^{\nu}_i(\eta) = \int_{C^i_{p-2}(\eta)} \frac{3M(x)[f''_p(x_p)]^2 + M_2(x)f'_p(x_p)}{H_1^{*'}} \, \mathrm{d}x_2 \ldots \mathrm{d}x_{p-1} ;$$

here $M_2(x)$ is a polynomial in $x_1, \ldots, x_p$.

Since $f'_{p+1}(\lambda) \not\equiv 0$, its coefficient in the right-hand side of the last equality must vanish identically. Proceeding in like manner, we come to the conclusion that, for m arbitrarily large, the sum

$$\sum_{ij} c_{ij} \int_{\eta_0}^{\eta_i} \frac{\Omega_i^{(m)}(x_p^j, \eta)}{\left\{ f'_p(x_p^j(\eta, \lambda)) \right\}^{2m+1}} \, \mathrm{d}\eta = \sum_{ij} c_{ij} \int_a^{a_{ij}} \frac{\Omega_i^{(m)}(x_p^j, \eta)}{\left\{ f'_p(x_p^j(\eta, \lambda)) \right\}^{2m}} \, \mathrm{d}x_p \tag{95}$$

must identically vanish with respect to λ.

We have assumed $f'_p(x_p)$ to admit no multiple roots, and therefore each critical value of λ has been linked to a unique pair of coinciding roots among all x_p satisfying equations (91). For the same reason, the polynomial $f''_p(x_p)$ does not vanish as these roots merge.

On the other hand, since the polynomial $H(x_1, \ldots, x_p, \lambda)$ has been constructed independently of $M(x_1, \ldots, x_p)$, we may assume that $M(x_1, \ldots, x_p) \neq 0$ at the ends of the two tubes T_{p-1} in the case of these ends being merged at a critical value $\lambda = \lambda_0$ (otherwise, we can replace $H(x_1, \ldots, x_p, \lambda)$ by

$H(x_1 - x_1^0, \ldots, x_p - x_p^0, \lambda)$). Consequently, in the vicinity of the ends of these tubes, for λ close to λ_0, we have

$$\left| \Omega_i^{(m)}(x_p, \lambda) \right| \geq c \left| \Omega_i(x_p, \lambda) \right| , \tag{**}$$

where c is a positive constant.

Assume that for $i = i^0$, $j = j^0$ we have

$$c_{i^0 j^0} \neq 0 , \qquad c_{i^0, j^0+1} = 0 . \tag{96}$$

Then there exists a critical value $\lambda = \lambda_0$ such that there is exactly one fraction in (95) whose denominator tends to zero as $\lambda \to \lambda_0$, $\eta \to \eta_{i^0}$. Denote by x_p^0 the double root of the corresponding equation (91). Then the difference $x_p - x_p^0$ tends to 0 at the same rate as $\sqrt{\eta - \eta_{i^0}}$, provided that x_p and η are related by the equation

$$\eta = f_p(x_p) + f_{p+1}(\lambda_0) .$$

Thus, for $\lambda = \lambda_0$ and $\eta \to \eta_{i^0}$, the denominator of the said fraction tends to 0 at the same rate as $(\eta - \eta_{i^0})^m$. Then, it follows from (*), (**) that, for m being large enough, the integral of exactly one summand in (95) is divergent, if the condition (96) holds. Therefore, if (96) holds, the sum in (95) becomes very large for λ close to λ_0 and sufficiently large m. It follows that the integral $I(\lambda)$ and its derivative with respect to λ cannot be identically equal to 0.

4. In order to establish Theorems IV and IV', it remains to prove the following two theorems.

Theorem V. *For $p \geq 2$, every $(p-1)$-dimensional cycle on the surface (88), or (according to §14.6) on any algebraic surface of complex dimension $p-1$ without singular points, is homologous to a cycle of the form*

$$\sum_{ij} c_{ij} T_{p-1}^{ij} + cA + (C_a)_{p-1} \tag{97}$$

described in §14.6, where $c_{in} = 0$ for all i; A is zero cycle for p even; A is an algebraic cycle for p odd.

Theorem VI. *Every $(p-1)$-dimensional cycle C_{p-1} on a surface of type (88) is homologous to a linear combination of finite vanishing cycles (and an algebraic cycle, if C_{p-1} is infinite and p odd). Therefore, the term $(C_a)_{p-1}$ in (97) can be dropped, provided that the index j takes the values $1, \ldots, n$.*

For $p = 2$, Theorem VI has already been proved in §14.4. Theorem VI for $p > 2$ can be deduced from Theorem V.

Proof. According to Lefschetz's theorem (see [10], p.93, Th. IV), for $p >$ 2, every $(p-1)$-dimensional cycle is homologous to a linear combination of invariant and vanishing cycles. Let us show that a cycle of the form (97) cannot be invariant if it satisfies the condition (96) at least once. To this end, according to another Lefschetz's theorem (see [10], p.93, Th.I), it suffices to find, for each cycle of this type, a vanishing cycle whose index of intersection with the cycle (97) is different from 0. This vanishing cycle can be constructed as follows.

Take a point a' on the x_p-plane other than a. Let us join a' with a_{ij} by lines l'_{ij} which do not cross the previously constructed l_{ij} joining a with a_{ij}. Next, we construct the tubes T'^{ij}_{p-1} for l'_{ij} in the same way as T^{ij}_{p-1} have been constructed for l_{ij}. Then the index of intersection of the vanishing cycle

$$T'^{i_0 j_0}_{p-1} - T'^{i_0, j_0+1}_{p-1}$$

with the cycle (97) is equal to $c_{i_0 j_0}$ (see [10], p.96, Th. XIV), and therefore, is non-zero.

It follows that for an invariant cycle (97), according to Theorem V, all c_{ij} are equal to 0 and we get either a zero cycle or an algebraic one for p odd, and a zero cycle for p even. Therefore, Theorem VI is a consequence of Theorem V.

5. *The proof of Theorem V and of the Fundamental Theorems IV and IV'* will be performed by induction. Assume Theorem VI to be proved for algebraic surfaces (88) of order n and complex dimension $p-2$. For $p-2=1$ the theorem has already been proved in §14.4.

Consider all linear combinations of the form

$$\sum_{ij} c_{ij} T^{ij}_{p-1} , \qquad (98)$$

where T^{ij}_{p-1} are $(p-1)$-dimensional tubes; c_{ij} take integer values; $i = 1, 2, \ldots, (n-1)^{p-1}$, $j = 1, \ldots, n-1$. According to the result proved in Sect. 2, neither any Abelian integral (89) over (98), nor its derivatives with respect to λ, can vanish, in general, unless all c_{ij} vanish simultaneously. The linear system (98) with arbitrary integer coefficients c_{ij} possesses $(n-1)^p$ linearly independent generators. As pointed out in §13.1, a system of this type can form a cycle only if there exists a linear combination of cycles $C^{ij}_{p-2}(a_p)$ homologous to zero in the intersection of the surface (88) with the plane $x_p = a_p$ (here the notations of §13 are used).

Let us show that the last condition can be formulated in terms of a system of as many as $B_{p-2} - \beta_{p-2}$ linearly independent equations with respect to the coefficients c_{ij}, and therefore, there exists as many as A_{p-1} linearly independent generators for the system (98) of $(p-1)$-dimensional cycles such that the integrals (89) over these cycles and their derivatives with respect to λ are, in general, different from 0, where

$$A_{p-1} \geq (n-1)^p - B_{p-2} + \beta_{p-2} \, .$$

Here $\beta_{p-2} = 1$ for p even, and $\beta_{p-2} = 0$ for p odd; B_{p-2} is the $(p-2)$-dimensional Betti number for the section of the hypersurface (88) by the hyperplane $x_p = a_p$, or equivalently, the $(p-2)$-dimensional Betti number for a general algebraic surface of order n and complex dimension $p-2$ without singularities.

Proof. Keeping in mind the method used in §14.5, §14.6 for the construction of vanishing cycles from the tubes T, we can easily verify that the sum (98) corresponds to

$$\sum_{i=1}^{(n-1)^{p-1}} \sum_{j=1}^{n-1} c_{ij} C_{p-2}^{ij}(a_p) \tag{99}$$

on the cross-section $x_p = a_p$. Since Theorem VI is assumed to hold for algebraic surfaces of complex dimension $p-2$, it follows that from among the vanishing cycles $C_{p-2}^{ij}(a_p)$ in the above sum we can choose

$$B_{p-2}^* = B_{p-2} - \beta_{p-2}$$

linearly independent ones, and thus each of the remaining cycles $C_{p-2}^{ij}(a_p)$ is a linear combination of the chosen ones. Let

$$C_{p-2}^i(a_p) = \sum_{s=1}^{B_{p-2}^*} k_{is} C_{p-2}^s(a_p) \, ; \qquad i = B_{p-2}^* + 1, \ldots, (n-1)^p - \beta_{p-2} = \nu \, .$$

The second superscript in C_{p-2}^{ij} has been dropped here because of its unimportance. Substituting the above expression into (99), we get

$$\sum_{i=1}^{B_{p-2}^*} \sum_{j=1}^{n-1} c_{ij} C_{p-2}^i(a_p) + \sum_{i=B_{p-2}^*+1}^{\nu} \sum_{j=1}^{n-1} c_{ij} \sum_{s=1}^{B_{p-2}^*} k_{is} C_{p-2}^s(a_p) =$$

$$= \sum_{i=1}^{B_{p-2}^*} C_{p-2}^i(a_p) \sum_{j=1}^{n-1} \left[c_{ij} + \sum_{s=B_{p-2}^*+1}^{\nu} c_{sj} k_{si} \right] \, .$$

In order that this expression be homologous to zero, it is necessary and sufficient that the coefficients by every C_{p-2}^i be equal to 0, i.e.,

$$\sum_{j=1}^{n-1} \left[c_{ij} + \sum_{s=B_{p-2}^*+1}^{\nu} c_{sj} k_{si} \right] = 0 \quad \text{for all} \quad i = 1, 2, \ldots, B_{p-2}^* \, .$$

In order to establish linear independence of these equations, it suffices to find a non-degenerate square matrix of size B_{p-2}^{*} whose elements are the coefficients by c_{ij}. For this purpose we can take, in particular, the unit matrix whose elements are the coefficients by c_{i1}, $i = 1, \ldots, B_{p-2}^{*}$.

Thus we have found as many as

$$A_{p-1} = (n-1)^p - B_{p-2} + \beta_{p-2} \tag{100}$$

linearly independent $(p-1)$-dimensional cycles of the form (99) described in Theorem V, the integrals of type (89^{ν}) over these cycles, in general, being different from 0.

On the other hand, let us calculate the number B_{p-1} of linearly independent $(p-1)$-dimensional cycles on a general algebraic surface of complex dimension $p-1$. According to what has been said in §14.6, all non-singular algebraic surfaces of order n are mutually homologous. Therefore, in order to find B_{p-1}, it suffices to determine the $(p-1)$-dimensional Betti number for any particular algebraic surface of order n, complex dimension $p-1$, and no singular points; for instance, we can take a surface of the form (78).

For $p = 2$, it is well known that $B_1 = (n-1)(n-2)$, which proves the formula

$$A_1 = (n-1)^2 - (n-1) = B_1 .$$

For any algebraic surface of complex dimension d we have

$$R_d - R_{d-2} = n(n-1)^d - 2(r_{d-1} - R_{d-1}) - (r_{d-2} - R_{d-2})$$

(see [10], p. 94). Here R_k is the k-dimensional Betti number for the surface under consideration; r_{d-1} (resp., r_{d-2}) is the $(d-1)$-dimensional (resp., $(d-2)$-dimensional) Betti number for the generic section of the surface under consideration

$$H(x_1, \ldots, x_{d+1}) = 0 \tag{101}$$

by the planes $x_{d+1} = $ const (resp., $x_d = $ const, $x_{d+1} = $ const). Denote by B_k the k-dimensional Betti number for a general algebraic surface of complex dimension k. Then the preceding formula yields for $k = p - 1$:

$$B_{p-1} - B_{p-3} = n(n-1)^{p-1} - 2(B_{p-2} - R_{p-2}) - (B_{p-3} - R_{p-3}) .$$

It is a known fact (see [10], p. 89) that any $(p-2)$-dimensional cycle on an algebraic surface (101) of complex dimension $p-1$ is homologous, on this surface, to a $(p-2)$-dimensional cycle on the generic intersection of the surface and the plane $x_p = $ const. This latter cycle C_{p-2}, according to our assumption that Theorem VI holds for the number of coordinates equal to $p-1$, is homologous on this intersection to a linear combination of vanishing cycles and an algebraic one, for p even. Each of these vanishing cycles is homologous to zero on the entire surface H, since a $(p-1)$-dimensional film can be extended over it, the film being the trace of a vanishing cycle as x_p approaches a critical value. Consequently, on a general complex surface (101), the cycle C_{p-2} is homologous to zero, or, possibly, to an algebraic cycle, for p even. Thus, for a general algebraic surface of complex dimension $p - 1$, we have

$$R_{p-2} = \beta_{p-2} .\tag{102}$$

The constant R_{p-3} is equal to the $(p-3)$-dimensional Betti number for a general algebraic surface of complex dimension $p - 2$ (see [10], p. 89). We have just proved this number to be equal to β_{p-3}.

Thus, we find that

$$B_{p-1} - \beta_{p-3} = n(n-1)^{p-1} - 2(B_{p-2} - \beta_{p-2}) - (B_{p-3} - \beta_{p-3}) ,$$

or

$$B_{p-1} + B_{p-2} - \beta_{p-2} - \beta_{p-3} = n(n-1)^{p-1} - (B_{p-2} + B_{p-3} - \beta_{p-2} - \beta_{p-3}) .$$

This relation can be written in the form

$$B_{p-1} + B_{p-2} - 1 = n(n-1)^{p-1} - (B_{p-2} + B_{p-3} - 1) ,\tag{103}$$

since $\beta_d + \beta_{d-1} = 1$ for all d. It has also been shown elsewhere (see, for instance, [9], p. 113) that

$$R_2 = n(n-1)^2 - 4p + 2R_1 - n + 2 ,\tag{104}$$

where R_2 is the two-dimensional Betti number for an algebraic surface of order n and complex dimension 2; R_1 is its one-dimensional Betti number; p is the genus of a generic plane cross-section of this surface. For the general algebraic surface of complex dimension 1 without singularities, the following formula is also known:

$$2p = (n-1)(n-2) .$$

Because of (102) we have
$$R_1 = 0 .$$

Thus, for an algebraic surface of order n and complex dimension 2, it follows from (104) that

$$R_2 = n(n-1)^2 - 2(n-1)(n-2) - (n-2) = (n-1)^3 - B_1 + 1 ,$$

where B_1 is the one-dimensional Betti number for an algebraic curve of order n without singular points. Therefore, if $p = 3$, we have

$$B_{p-1} + B_{p-2} - 1 = (n-1)^3 .$$

It follows that for any p formula (103) yields

$$B_{p-1} + B_{p-2} - 1 = (n-1)^p .$$

Comparing this relation with (100), we obtain

$$A_{p-1} = B_{p-1} - 1 + \beta_{p-2} = B_{p-1} - \beta_{p-1} .$$

Hence we see that, for p even, the Betti number for all linearly independent $(p-1)$-dimensional cycles on a general algebraic surface (88) of complex dimension $p-1$ and no singular points coincides with A_{p-1}, the number of linearly independent $(p-1)$-dimensional cycles (98) on that surface, such that for any non-trivial linear combination of these cycles the integrals of the form (89$^\nu$), as well as their derivatives mentioned in Theorem IV', are different from 0, in general. For p odd, the Betti number B_{p-1} is equal to $A_{p-1} + 1$, where 1 is due to an algebraic cycle. Thus, Theorems IV, IV', V are proved for general algebraic surfaces of complex dimension $p-1$ with no singular points; therefore, these theorems are valid for surfaces of any complex dimension.

CHAPTER V

Basic Theorems about Lacunas; Examples and Applications

§16. Proof of the Basic Theorems about Lacunas

1. We say that a lacuna is *stable* if it is not destroyed by any sufficiently small variation of the coefficients of H. Assume first that $n \leq p$. Let us place the vertex of the characteristic cone at $(t, 0, \ldots, 0)$, and consider one of the domains G cut out of the plane $t = 0$ by its lateral surface. It follows from (70) and Theorem IV', proved in the preceding chapter, that G is a stable lacuna if and only if the cycles $C_{\text{imag}}(t, x)$, for p even (resp., $C_{\text{real}}(t, x)$, for p odd) are homologous to zero on the complex intersection of the surface (39) and the plane (57) corresponding to the point $(-x_1, \ldots, -x_p)$ considered in the domain G, or these cycles are algebraic. The latter case is obviously excluded because of the results of §11.

Let us make the following observation. If the cycle $C(t,x)$ constructed for a particular point of the domain G is homologous to zero, then the same can be claimed regarding the cycle C constructed for any other point of this domain, since the latter can be obtained from the former by continuous extension (cf. §12.2).

The above arguments prove the following

First Basic Theorem. *The domain G is a stable lacuna for equation (22) with $n \leq p$, if and only if the cycle $C_{real}(t,x)$, for p odd (the cycle $C_{imag}(t,x)$, for p even), corresponding to any one of its interior points $(-x_1, \ldots, -x_p)$ is homologous to zero in the complex intersection of the plane (57) and the surface (39).*

2. For $n \geq p+1$, as we have seen in §9, the functions $K(t,x)$ can be represented as integrals of the 1st kind over the surfaces $S(t,x)$ formed by the cycles $C_{real}(\tau, x)$, for p odd (cycles $C_{imag}(\tau,x)$, for p even), where τ varies from 0 to t. Clearly, if for a fixed t the function $K(t,x)$ vanishes in a domain G of the $(x_1, \ldots, x_p)$-space, and G forms a lacuna on the plane $t = 0$, then all the derivatives of K with respect to x, and therefore, with respect to t, vanish in that domain. It follows that the functions $u(t,x)$, for p even (the functions $v(t,x)$, for p odd, respectively), which can be represented in terms of the integrals of the 1st kind over the cycles $C_{imag}(t,x)$ (resp., $C_{real}(t,x)$), must also vanish for the given algebraic surface, as well as for any other sufficiently close algebraic surface, provided that the lacuna is stable. The arguments which have lead us to the preceding theorem can be used now to prove the

Second Basic Theorem. *The following is a necessary condition for the domain G to be a stable lacuna for equation (22) with $n > p$: the cycles $C_{real}(t,x)$, for p odd (the cycles C_{imag}, for p even) are homologous to zero in the complex intersection of the surface (39) and the plane (57).*

Assume that the latter condition is satisfied. Let us extend films over the cycles $C_{real}(t,x)$ (resp., $C_{imag}(t,x)$) in the complex intersection of the surface (39) and the planes (57) for $\tau = t$ and $\tau = 0$. This is always possible, since these cycles are homologous to zero. This fact is obvious for p even, since the cycle $C_{imag}^Q(0,x)$ is void if Q coincides with the origin. For p odd, this fact is due to Theorem IV and §9.6. The said films having been constructed, the surface $S(t,x)$ can be extended to form a cycle Σ, and $K(t,x)$ can be represented as an integral of the 1st kind over Σ.

Hence, using Theorem IV, we obtain the

Third Basic Theorem. *The domain G is a stable lacuna for equation (22) with $n > p$, if and only if the cycle Σ corresponding to any one of its*

interior points P is homologous to zero on the complex surface (39), *for p even* (*is algebraic on that surface, for p odd*).

Here we can limit ourselves to the cycle Σ corresponding to any interior point P, since the above condition, being fulfilled for a particular point, turns out to be satisfied for all other points of the same domain G at the base of the characteristic cone (cf. §12.2).

For p even, the cycle Σ cannot be algebraic because of its odd dimension. For p odd, there exists a $(p-1)$-dimensional algebraic cycle A lying in the intersection of the general complex surface (39) and the complex hyperplane (57). Any other $(p-1)$-dimensional algebraic cycle on the complex surface (39) is homologous to a multiple of A on that surface. Adding a multiple of A to the film extended over $C_{\text{real}}(t, x)$ on the plane (57), we can always make the cycle Σ homologous to zero on (39), provided that before this operation Σ were algebraic.

3. Remark. We can express the preceding statement in more precise terms. It follows from (25) that, for sufficiently smooth functions $\varphi_s(x_1, \ldots, x_p)$ supported in any one of the domains G cut out of the base of the characteristic cone with vertex at $(t, x_1, \ldots, x_p)$ by its lateral surface, the solution u of equation (22) satisfying the initial conditions for $t = 0$:

$$\frac{\partial^s u}{\partial t^s} = \varphi(x_1, \ldots, x_p), \qquad s = 0, \ldots, n-1,$$

can be represented in the form

$$u(t, x_1, \ldots, x_p) =$$

$$= \int \sum K_s(t, x_1 - y_1, \ldots, x_p - y_p) \varphi_s(y_1, \ldots, y_p) \, dy_1 \ldots dy_p \, . \, (105)$$

The expression for $K_{n-1}(t, x)$ has been obtained above. Let us construct $K_{n-2}(t, x)$ using (25). Denote by $u_s(t, x_1, \ldots, x_p, \varphi)$ the solution of equation (22) whose derivatives in t of orders $0, 1, 2, \ldots, s-1, s+1, \ldots, n-1$ vanish identically at $t = 0$, and its derivative of order s is equal to $\varphi(x_1, \ldots, x_p)$. Then we have

$$u_{n-2}(t, x_1, \ldots, x_p, \varphi_{n-2}) =$$

$$= \frac{\partial}{\partial t} \int K_{n-1}(t, x_1 - y_1, \ldots, x_p - y_p) \varphi_{n-2}(y_1, \ldots, y_p) \, dy_1 \ldots dy_p -$$

$$- \int K_{n-1}(t, x_1 - y_1, \ldots, x_p - y_p) \frac{\partial^n u_{n-1}(0, y, \varphi_{n-2})}{\partial t^n} \, dy_1 \ldots dy_p \, . \, (106)$$

The function u_{n-1} satisfies equation (22) at $t = 0$, and all its derivatives in t of order less than $n-1$ vanish at $t = 0$, while the derivative of order $n-1$ is equal to φ_{n-2}; therefore, at $t = 0$ we have

$$\frac{\partial^n u_{n-1}(0, y, \varphi_{n-2})}{\partial t^n} = \sum_{k=1}^{p} a_k \frac{\partial \varphi_{n-2}}{\partial y_k} \; ,$$

where a_k are constant. Substituting this expression into (106) and integrating by parts, we obtain

$$u_{n-2}(t, x, \varphi_{n-2}) = \int \varphi_{n-2}(y) \left[\frac{\partial}{\partial t} - \sum_{k=1}^{p} a_k \frac{\partial}{\partial x_k} \right] K_{n-1}(t, x - y) \, dy \; .$$

Using similar arguments in the case of arbitrary s, we find $K_s(t, x)$ to be a linear combination (with constant coefficients) of the derivatives of $K_{n-1}(t, x)$ of order $n - 1 - s$ with respect to t and x_k. Hence, on the basis of what has been said in §10 and §11 about differentiating $K(t, x)$, it follows that $K_s(t, x)$, with $s < p$, can be represented by an integral of the 1st kind over the cycles $C_{\text{imag}}(t, x)$, or $C_{\text{real}}(t, x)$, according to p being even or odd; and $K_s(t, x)$, with $s \geq p$, can be represented by an integral of the 1st kind over the surface $S(t, x)$. Therefore, the value of the solution u at the point (t, x), in the case of equation (22) or any other equation close enough to (22), is independent of the values of all φ_s in a domain G, if and only if the cycle C_{imag} (resp., C_{real}) constructed for any point P of this domain is homologous to zero on the cross-section (corresponding to P) of the surface (39) by the plane (57), provided that $s < p$. For $s \geq p$, it is necessary that the cycles $C_{\text{imag}}(t, x)$ (resp., $C_{\text{real}}(t, x)$) be homologous to zero on the corresponding complex cross-section. The latter condition being fulfilled, the solution $u(t, x)$ is independent of the values of φ_s in the domain G, if and only if the cycles Σ generated by $C(t, x)$ (as described in Sect. 2) are homologous to zero on the complex surface (39) (cf. §20.7).

§17. The Case $p = 2$

For $p = 2$, the cycle $C_{\text{imag}}(t, x)$ reduces to the set S formed by complex (non-real) points of intersection of the curve

$$H(p_1, p_2) = 0 \tag{107}$$

and the straight line

$$x_1 p_1 + x_2 p_2 + t = 0 \; . \tag{108}$$

Performing, if necessary, a suitable affine transformation of the (p_1, p_2)-plane, we can always assume that $x_1 = 0$ and $x_2 > 0$. Then the integral over the cycle $C_{\text{imag}}(t, x)$ corresponds to the sum

$$\frac{\pi i}{x_2} \sum \frac{\operatorname{sgn} \operatorname{Re} \frac{p_1}{i}}{H_1 \left(p_1, -\frac{t}{x_2} \right)} \tag{109}$$

over all complex roots p_1 of the polynomial $H\left(p_1, -t/x_2\right)$.

Let us show that *the above sum vanishes in a domain formed by the values of the coefficients of H, if and only if the set S is empty, i.e., all points of intersection of* (108) *and* (109) *are real.*

The sufficiency of the latter condition is obvious; let us establish its necessity. First we note that the structure of the sum in question shows it to be an analytic function of the coefficients of H. Therefore, in order to prove that this sum does not vanish identically, it suffices to find a curve H for which the sum is not equal to zero. Such a curve can be constructed as follows. Let us make any two complex-conjugate points $p_1^{(1)}, p_1^{(2)}$ belonging to the intersection of (107) and (108) approach one another in such a way that they do not go to infinity, while all the other intersection points $p_1^{(k)}$ approach different finite real points. Then the two summands in (109), corresponding to $p_1^{(1)}$ and $p_1^{(2)}$, viz.,

$$\frac{1}{\left(p_1^{(1)} - p_1^{(2)}\right)\left(p_1^{(1)} - p_1^{(3)}\right)\cdots\left(p_1^{(1)} - p_1^{(n)}\right)} -$$

$$-\frac{1}{\left(p_1^{(2)} - p_1^{(1)}\right)\left(p_1^{(2)} - p_1^{(3)}\right)\cdots\left(p_1^{(2)} - p_1^{(n)}\right)} =$$

$$= \frac{1}{p_1^{(1)} - p_1^{(2)}}\left[\frac{1}{\left(p_1^{(1)} - p_1^{(3)}\right)\cdots\left(p_1^{(1)} - p_1^{(n)}\right)} + \right.$$

$$\left. + \frac{1}{\left(p_1^{(2)} - p_1^{(3)}\right)\cdots\left(p_1^{(2)} - p_1^{(n)}\right)}\right]$$

will grow to infinity, while all the other terms remain bounded. Therefore, the imaginary part of the total sum cannot vanish identically.

§18. Examples

1. An algebraic surface with the cycle C_{imag} homologous to zero.

Consider an ellipse which crosses each of the positive semi-axes at two different points, as shown in Fig. 2. Let

$$H(p, p_p) = 0$$

be its equation. Consider a surface of the 4th order

$$H\left(p_1^2 + \cdots + p_{p-1}^2, p_p^2\right) = 0 \tag{110}$$

which consists of two ovals, one inside another. The smaller oval corresponds to the arc FK, and the larger one to the ark $ABCD$. The intersection of

the surface (110) and the hyperplane Π obtained by rotating the straight
line MN around the Op_p-axis (see Fig. 2), as well as any other hyperplane
close enough to Π, consists of two ovals, one inside another. If the point Q,
used for the construction of the cycle C^Q_{imag}, be taken inside the smaller oval,
then any real straight line belonging to Π and passing through Q crosses
the surface (110) at real points only. Therefore, the cycles C^Q_{imag} for the
plane Π, as well as for any other plane close enough to Π, are homologous
to zero.

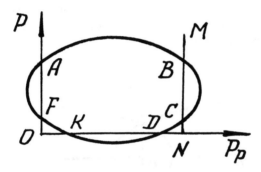

Fig. 2.

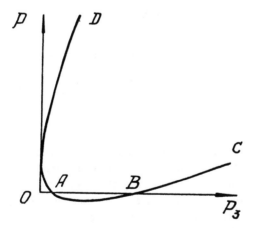

Fig. 3.

2. An algebraic surface with the cycle C_{real} homologous to zero.

Consider a real hyperbola

$$H(p, p_3) = 0 \tag{111}$$

whose position with respect to the coordinate axes is shown in Fig. 3. This hyperbola crosses the positive semi-axes Op_3 at two different real points and touches the positive semi-axis Op. Let $H(0,0) < 0$. The surface of the 4th order

$$H\left(p_1^2 + p_2^2, p_3^2\right) = 0$$

is obtained by revolving the curve outlined in Fig. 4 around the axis Op_3. This surface consists of two ovals, one inside another. The interior oval corresponds to the arc BC, and the exterior one to the arc AD. The latter oval contains a singular circle.

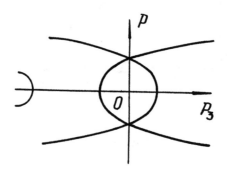

Fig. 4.

For real ε, c, C_1 sufficiently small in absolute value, the surface

$$F\left(p_1^2, p_2^2, p_3\right) \equiv H\left(p_1^2 + p_2^2, p_3^2\right) - \varepsilon p_3 \left(p_1^2 - p_2^2\right) +$$

$$+ c\varepsilon^2 \left(p_1^2 - p_2^2\right)^2 - c c_1 \varepsilon^3 = 0 \tag{112}$$

also consists of two ovals, one inside another and without singular points. Then the intersection of the surface (112) and the plane $p_3 = 0$ will consist of four ovals: two of them are produced by the intersection of this plane with the right-hand portion of the larger oval, the other two – by the intersection with its left-hand portion. The relative position of these ovals is shown in Fig. 5; the first two of them (2 and 4) are indicated by the thicker line.

Let us shift the origin in the (p_1, p_2, p_3)-space to a point within the smaller of the two ovals forming the surface (112). Then $\widetilde{H} = 0$ can be considered as the characteristic equation for some hyperbolic equation. According to what has been said in §9.4, the ovals drawn in thicker line, and

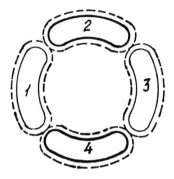

Fig. 5.

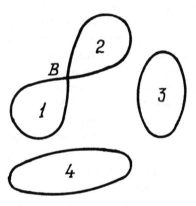

Fig. 6.

those in thinner one (see Fig. 5), which form the cycle C_{real} cut out of the surface (112) by the former plane $p_3 = 0$, should be regarded as having opposite orientations on that plane. Let us show that, being oriented in this way, they form a cycle homologous to zero on the complex curve

$$H_0\left(p_1, p_2\right) \equiv H\left(p_1^2 + p_2^2, 0\right) + c\left(p_1^2 - p_2^2\right)^2 \varepsilon^2 - cc_1\varepsilon^3 = 0 . \qquad (113)$$

First we note that, according to a well-known theorem (see [11], pp. 230–235), four real ovals of a 4th order curve without singularities always split the Riemannian surface of that curve into two parts. Thus, our problem is to prove that the four ovals, defined above and having the above specified orientation, are homologous to zero. To this end, set $c_1 = 0$. Then the curve of the 4th order $H_0(p_1, p_2) = 0$ defined above would consist of four ovals having points of contact as shown by the dash-lines in Fig. 5. If we take any of the points of contact as the new origin and write the equation of the resulting curve as

$$H^*(p_1, p_2) = 0 \, ,$$

then the equation

$$H^{**}(p_1, p_2) \equiv H^*(p_1, p_2) + b \left(p_1^2 + p_2^2 \right) = 0$$

would represent the curve outlined in Fig. 6, provided that b has a sufficiently small absolute value and a suitably chosen sign. If we first add $\varepsilon \left(p_1^2 + p_2^2 \right)$, with ε varying between 0 and b, to the left-hand side of equation (113), and then, in the resulting equation, continuously transform H_0 into H^*, we thereby convert the equation $H_0(p_1, p_2) = 0$ (continuously changing its coefficients) into the equation $H^{**}(p_1, p_2) = 0$ corresponding to the curve shown in Fig. 6, all intermediate curves possessing no singular points. The genus of the 4th order curve $H^{**}(p_1, p_2) = 0$ with one singular point will be equal to 2. Consequently, its three real pieces (see Fig. 6) divide the corresponding Riemannian surface into two parts. Here the portion consisting of the ovals 1 & 2 is regarded as one piece. Expressing p_2 as a function of p_1 near the contact point B of the ovals 1 and 2, we can easily verify that the upper (resp., lower) branch of the oval 2 near B is a continuation of the lower (resp., upper) branch of the oval 1. Obviously, the partition of the Riemannian surface for the curve $H_0(p_1, p_2) = 0$ by its four ovals is continuously transformed (by the continuous change of the coefficients described above) into the partition of the Riemannian curve for $H^{**}(p_1, p_2) = 0$ by its real pieces shown in Fig. 6; therefore, the ovals 1 & 2 of the former partition must have opposite orientation. In exactly the same way, the ovals 2 & 3, as well as 4 & 1, can be shown to have opposite orientation.

The surface $F \left(p_1^2 + p_4^2 + \cdots + p_p^2, p_2^2, p_3 \right) = 0$ furnishes a similar example for $p > 3$, since every real intersection of this surface and the plane $c_1 p_1 = c_4 p_4 = \cdots = c_p p_p, \ p_3 = 0$, is homologous to zero in their complex intersection, where $c_1, c_4, \ldots, c_p$ are arbitrary real constants such that $c_1^2 + c_4^2 + \cdots + c_p^2 > 0$.

3. Algebraic surfaces with the cycle Σ homologous to zero. Consider a real algebraic surface of order n

$$\widetilde{H}(p_1, \ldots, p_p) = 0$$

consisting of $[n/2]$ finite ovals consecutively enclosed in one another, as well as an unpaired piece, if n is odd. Let us perform, in the $(p_1, \ldots, p_p)$-space, a projective transformation with real coefficients such that a fixed hyperplane crossing all the ovals moves to infinity. Let us place the origin inside the least of the ovals. Assume that after this transformation the equation $\widetilde{H} = 0$ becomes

$$H(p_1, \ldots, p_p) = 0 \, . \tag{114}$$

For the sake of brevity, we preserve the previous notation for the coordinates. The equation thus obtained can be considered as the characteristic equation for some hyperbolic equation. The interior of the smallest oval of the surface $H = 0$ is divided into two parts by the plane at infinity. The part containing the origin will be referred to as the first, and the other one as the second. As shown above, the cycles $C_{\text{real}}(t, x)$ for p odd (resp., the cycles $C_{\text{imag}}(t, x)$ for p even) that correspond to the planes (57) crossing the smallest oval are homologous to zero in the corresponding complex cross-sections of the surface $H = 0$. Therefore, the surface $S(t, x)$ that corresponds to the planes crossing only the second part of the smallest oval can be extended to form the cycle $\varSigma$. The fact that the latter cycle is homologous to zero on the complex surface (114) admits the following simple verification. The kernel $K(t, x)$, which corresponds to these planes and can be represented by Herglotz's formulas (44*) or (47), has been shown (see §7) to vanish for the surface (114) under consideration, as well as for any other surface sufficiently close to it. On the other hand, this kernel can be represented by an integral of the first kind over the cycle $\varSigma$. Therefore, by Theorem IV, this cycle must be homologous to zero on the complex surface (114).

The above examples (for p even, as well as odd) are trivial in the sense that they pertain to the case of $K(t, x)$ vanishing in the exterior of the base of the characteristic cone. It is of great interest to find out whether there exist examples of another nature.

§19. Remarks about the Case of a Reducible Surface $H = 0$

1. The surface $H = 0$ has been assumed so far to be irreducible and without singular points. Now let

$$\varDelta(\lambda, \xi_1, \ldots, \xi_p) \equiv \prod_s \varDelta_s(\lambda, \xi_1, \ldots, \xi_p) \, , \tag{115}$$

where each surface

$$H_s(p_1, \ldots, p_p) \equiv \varDelta_s(1, p_1, \ldots, p_p) = 0 \tag{116}$$

has no singular points and is irreducible. If equation (115) is hyperbolic, then it is obvious that each surface (116) consists of ovals consecutively enclosed in one another and, possibly, of an unpaired piece, if the order n_s of $\varDelta_s$ is odd; the number of these ovals is equal to $[n_s/2]$. Therefore, each equation

$$\Delta_s\left(\frac{\partial}{\partial t}\,,\,\frac{\partial}{\partial x_1}\,,\,\dots,\,\frac{\partial}{\partial x_p}\right)u = 0 \tag{117}$$

is hyperbolic.

Theorem VII. *Let G be a domain belonging to the base of the characteristic cone for the equation*

$$\Delta\left(\frac{\partial}{\partial t}\,,\,\frac{\partial}{\partial x_1}\,,\,\dots,\,\frac{\partial}{\partial x_p}\right)u = 0\,. \tag{118}$$

Assume that G is a lacuna for this equation. Then G forms a part of a lacuna for every differential equation (117), as well as for any equation of the form

$$\prod_s{}'\Delta_s\left(\frac{\partial}{\partial t}\,,\,\frac{\partial}{\partial x_1}\,,\,\dots,\,\frac{\partial}{\partial x_p}\right)u = 0\,, \tag{119}$$

where the product $\prod_s'$ is composed of some of the factors in $\prod_s$.

Proof. Let us take in the initial conditions for u at $t = 0$ sufficiently smooth functions that vanish identically outside a domain G_1 belonging, together with its boundary, to the interior of G. Let us also impose the condition that the function

$$u'' = \prod_s{}''\Delta_s\left(\frac{\partial}{\partial t}\,,\,\frac{\partial}{\partial x_1}\,,\,\dots,\,\frac{\partial}{\partial x_p}\right)u\,, \tag{120}$$

itself, or any of its derivatives in t of order $\leq n'-1$, be equal to an arbitrary sufficiently smooth function. Here

$$\prod_s{}''\Delta_s \cdot \prod_s{}'\Delta_s \equiv \prod_s \Delta_s\,;$$

n' and n'' are the orders of $\prod_s'\Delta_s$ and $\prod_s''\Delta_s$, respectively.[7] Then the function u satisfying equation (118) must vanish in a neighborhood of the vertex of the characteristic cone, since the domain G is assumed to be a lacuna for this equation. Therefore, the function u'' must also vanish in a neighborhood of vertex of the characteristic cone. As mentioned above, u'' satisfies equation (119) of hyperbolic type. The value of u'' at the vertex of the characteristic cone is equal to 0 for any initial data of u'', provided that the initial functions are sufficiently smooth and vanish identically outside the domain $\tilde{G}$. Therefore, $\tilde{G}$ is part of a lacuna for equation (119), q.e.d.

---

[7] As the initial values for u' and its derivatives in t of order $\leq n'-1$ we can choose sufficiently smooth arbitrary functions, since every derivative $\partial^k u''/\partial t^k$ determined by (120) involves a derivative of u with respect to t which does not enter the preceding derivatives of u'' with respect to t.

Theorem VII'. *Let the domain $\tilde{G}$ be a lacuna for equation* (117), *as well as for all equations obtained from* (117) *by sufficiently small variations of the coefficients of the polynomials H_s. Then $\tilde{G}$ forms a part of stable lacunas for each equation* (119).

2. Assume that

$$\Delta(\lambda, \xi_1, \dots, \xi_p) \equiv \Delta_1(\lambda, \xi_1, \dots, \xi_p)\Delta_2(\lambda, \xi_1, \dots, \xi_p) , \qquad (121)$$

where $\Delta_s(\lambda, \xi_1, \dots, \xi_p) = 0$ are irreducible algebraic surfaces without singular points $(s = 1, 2)$. Assume further that the domain G is part of a stable lacuna for each equation

$$\Delta_s\left(\frac{\partial}{\partial t}, \frac{\partial}{\partial x_1}, \dots, \frac{\partial}{\partial x_p}\right) u = 0 , \qquad s = 1, 2 .$$

For the sake of simplicity, we assume that $n_1 + n_2 \geq p$.

Then, according to the First and the Third Basic Theorems, the cycles $C_s(t, x)$, $s = 1, 2$, as well as the cycles Σ_s such that $n_s > p$, for any point of the domain G must be homologous to zero on the corresponding complex algebraic surfaces. Let us assume these conditions to be fulfilled.

We consider in detail only the following case: for any point $(x_1, \dots, x_p)$ of the domain G the cycles $C_1(t, x)$ and $C_2(t, x)$ are mutually disjoint, and the cross-sections of the complex surfaces $H_1 = 0$ and $H_2 = 0$ by the corresponding plane $p_p^* = $ const do not touch one another. Then the above formulas for $K(t, x)$ remain valid, as indicated in §9.2. Therefore, $K(t, x)$ (or one of its derivatives with respect to t) can be represented in the form

$$cI = c \int_{C_1 + C_2} \frac{dp_1^* \dots dp_{p-2}^*}{H_{p-1}^*\left(p_1^*, \dots, p_p^*\right)} .$$

Taking into account the relation (121), we obtain

$$I = \int_{C_1} \frac{dp_1^* \dots dp_{p-2}^*}{H_{1,p-1}^*\left(p_1^*, \dots, p_{p-1}^*, p_p^*\right) H_2^*\left(p_1^*, \dots, p_p^*\right)} +$$

$$+ \int_{C_2} \frac{dp_1^* \dots dp_{p-2}^*}{H_1^*\left(p_1^*, \dots, p_{p-1}^*, p_p^*\right) H_{2,p-1}^*\left(p_1^*, \dots, p_p^*\right)} = I_1 + I_2 .$$

Since $K(t, x)$ vanishes in a neighborhood of (t, x), we must have $I = 0$. Over each cycle $C_s(t, x)$ homologous to zero on the corresponding complex surface (116), let us spread a film lying on that surface. Consider the first of these films S_1 lying on the surface

$$H_1^* \left(p_1^*, \ldots, p_{p-1}^*, p_p^* \right) = 0 , \quad p_p^* = \text{const} . \tag{122}$$

Since

$$\frac{dp_1^* \ldots dp_{p-2}^*}{H_{1,p-1}^*} = \frac{dp_1^* \ldots dp_{k-1}^* \, dp_{k+1}^* \ldots dp_{p-1}^*}{H_{1k}^*}$$

for any k, and the surface (122) has no singular points by assumption, it follows that the integrand in I turns into ∞ only at the points of the intersection R_1 of the film S_1 and the surface

$$H_2^* \left(p_1^*, \ldots, p_p^* \right) = 0 , \quad p_p^* = \text{const} . \tag{123}$$

Enclosing R_1 in a thin tube in S_1, integrating with respect to p_{p-2}^* first, and applying the theorem about residues, we obtain

$$I_1 = 2\pi i \int_{R_1} \frac{dp_1^* \ldots dp_{p-3}^*}{H_{1,p-1}^* \frac{dH_2^*}{dp_{p-2}^*}} .$$

Here dH_2^*/dp_{p-2}^* stands for the derivative of H_2^* with respect to p_{p-2}^*, where we have taken into account not only the explicit dependence of H_2^* on p_{p-2}^* but also its dependence on p_{p-2}^* through the agency of p_{p-1}^* related to p_{p-2}^* by (122), while $p_1^*, \ldots, p_{p-3}^*, p_p^*$ are kept constant.

Performing the necessary calculations, we get

$$I_1 = 2\pi i \int_{R_1} \frac{dp_1^* \ldots dp_{p-3}^*}{H_{1,p-1}^* H_{2,p-2}^* - H_{1,p-2}^* H_{2,p-1}^*} .$$

In a similar way we obtain

$$I_2 = 2\pi i \int_{R_2} \frac{dp_1^* \ldots dp_{p-3}^*}{H_{1,p-1}^* H_{2,p-2}^* - H_{1,p-2}^* H_{2,p-1}^*} .$$

Hence

$$I = 2\pi i \int_R \frac{dp_1^* \ldots dp_{p-3}^*}{H_{1,p-1}^* H_{2,p-2}^* - H_{1,p-2}^* H_{2,p-1}^*} ,$$

where $R_1 + R_2 = R$. We see that I is an integral of the first kind over a $(p-3)$-dimensional cycle R lying in the intersection of the surfaces (122) and (123). If R is homologous to zero in the complex intersection of these surfaces, then $I = 0$. If $n = p$, then, according to the formulas of §11 and the assumptions on the cycles $C_s(t, x)$, this condition is sufficient to guarantee that $K(t, x)$ vanishes in a neighborhood of this point.

§20. Systems of Equations

1. Here we limit ourselves to hyperbolic linear systems with constant coefficients and derivatives of the highest order only, i.e., systems of the form

$$\frac{\partial^{n_i} u_i}{\partial t^{n_i}} = \sum_{j,k} a_{ij}^{(k_0,k_1,\dots,k_p)} \frac{\partial^{n_j} u_j}{\partial t^{k_0} \partial x_1^{k_1} \dots \partial x_p^{k_p}} , \qquad (124)$$

where $i, j = 1, \dots, N;$, $k_0 + k_1 + \dots + k_p = n_j$; $k_0 < n_j$. Moving all the terms to the left-hand side, let us rewrite this system as

$$\sum_{j,k} A_{ij}^{(k_0,k_1,\dots,k_p)} \frac{\partial^{n_j} u_j}{\partial t^{k_0} \partial x_1^{k_1} \dots \partial x_p^{k_p}} = 0 . \qquad (124')$$

Consider the matrix

$$\left\| \sum_{j,k} A_{ij}^{(k_0,k_1,\dots,k_p)} \frac{\partial^{n_j}}{\partial t^{k_0} \partial x_1^{k_1} \dots \partial x_p^{k_p}} \right\| , \qquad (125)$$

and denote by A_{ij} the cofactor of the element

$$\sum_{j,k} A_{ij}^{(k_0,k_1,\dots,k_p)} \frac{\partial^{n_j}}{\partial t^{k_0} \partial x_1^{k_1} \dots \partial x_p^{k_p}} .$$

Let us multiply the i-th equation in $(124')$ by A_{ij} and take the sum of the resulting equations with respect to i. We then find that for any j the function u_j must satisfy the same hyperbolic equation

$$\Delta \left(\frac{\partial}{\partial t}, \frac{\partial}{\partial x_1}, \dots, \frac{\partial}{\partial x_p} \right) u = 0 , \qquad (126)$$

where $\Delta(\partial/\partial t, \partial/\partial x_1, \dots, \partial/\partial x_p)$ is the determinant of the matrix (125).

If the initial data at $t = 0$ for the functions u_i satisfying equations (124) have been given in the form

$$\frac{\partial^k u_i}{\partial t^k} = \varphi_i^{(k)} , \qquad i = 1, \dots, N; \quad k = 0, 1, \dots, n_i - 1 ,$$

then the function u_i can be determined as a solution of equation (126) satisfying the following initial conditions

$$\frac{\partial^k u_i}{\partial t^k} = \varphi_i^{(k)} , k = 0, 1, \dots, n_i - 1 ,$$

$$\frac{\partial^{n_i} u_i}{\partial t^{n_i}} = \sum_{j,k} a_{ij}^{(k_0,k_1,\dots,k_p)} \frac{\partial^{n_j - k_0} \varphi_j^{(k_0)}}{\partial x_1^{k_1} \dots \partial x_p^{k_p}} .$$

$$(127)$$

We can find the initial data at $t = 0$ for the derivatives of u_i of order $n_i + 1$ with respect to t, if we differentiate the i-th equation of system (124) with respect to t once, and then substitute in the right hand side the values of

$$\frac{\partial^{k_0} u_j}{\partial t^{k_0}} , \quad \frac{\partial^{k_0+1} u_j}{\partial t^{k_0+1}}$$

given by (127), etc.

In general, the initial data at $t = 0$ for the k-th order derivatives of u_i, $k \geq n_i$, can be expressed by linear combinations (with constant coefficients) of the derivatives of $\varphi_j^{(s)}(x_1, \ldots, x_p)$ of order $\geq k - n_i - 1$.

Hence we immediately deduce that *every lacuna for equation* (126) *is at the same time a lacuna for system* (124). However, system (124) may have other lacunas we well.

2. Assume that the surface

$$H(p_1, \ldots, p_p) \equiv \Delta(1, p_1, \ldots, p_p) = 0 \tag{128}$$

related to equation (126) has no singular points, and that to every point $(0, x_1, \ldots, x_p)$ within a domain G at the base of the characteristic cone with vertex at $(t, 0, \ldots, 0)$ there corresponds a cycle $C_{\text{real}}(t, x)$, for p odd (resp., $C_{\text{imag}}(t, x)$ for p even), homologous to zero in the intersection of the complex surface (128) and the complex plane

$$t + x_1 p_1 + \cdots + x_p p_p = 0 . \tag{129}$$

Then *the domain* G *is a lacuna for system* (124), *if*

$$n_i \leq p \quad \text{for all} \quad i .$$

Indeed, according to (105) we have

$$u_i(t, x) = \sum_{s=0}^{n-1} \int K_s(t, x - y) \varphi_i^{(s)}(y) \, dy , \tag{130}$$

where $\varphi_i^{(s)}$ stand for the initial values of $\partial^s u_i / \partial t^s$ at $t = 0$, and $n = \sum n_i$; the integral is taken over the entire $(y_1, \ldots, y_p)$-plane. It is assumed here that all functions $\varphi_i^{(s)}$ are sufficiently smooth and differ from 0 only inside one of the domains G cut out of the plane $t = 0$ by the lateral surface of the characteristic cone with vertex at $(t, x_1, \ldots, x_p)$. According to §16.3, the function $K_s(t, x - y)$ is a linear combination of the derivatives of $K(t, x - y)$ of order $\geq n - 1 - s$ with respect t, x. On the other hand, as we have just noted, every $\varphi_i^{(s)}$, $s \geq n_i$, is a linear combination (with constant coefficients) of the derivatives of $\varphi_j^{(k)}$ of order $\geq s - n_i + 1$ with respect to $x_1, \ldots, x_n$.

In every term of the sum (130) with $s \geq n_i$, let us perform integration by parts with respect to x_k, so that only the products of $\varphi_j^{(k)}$ by expressions of the form

$$\frac{\partial^k K(t, x - y)}{\partial t^{k_0} \partial y_1^{k_1} \dots \partial y_p^{k_p}} = \pm \frac{\partial^k K(t, x - y)}{\partial t^{k_0} \partial x_1^{k_1} \dots \partial x_p^{k_p}}$$

remain in (130). It should be noticed that we always have $k \geq n - n_i$. Therefore, according to what has been said in §17.3, we find that for $n_i \leq p$ every function $\varphi_j^{(k)}$ in the integrand is multiplied by an integral of the 1st kind over the cycle $C_{\mathrm{imag}}^*(t, x)$, or $C_{\mathrm{real}}^*(t, x)$, homologous to zero by assumption. Consequently, all these integrals are equal to 0, q.e.d.

3. It should be noticed that in order that a domain G be a stable lacuna for system (124), it is necessary that the cycles C_{real}, for p odd (resp., C_{imag}, for p even), corresponding to the points of G be homologous to zero in the complex intersection of the surface (128) and the plane (129), and the cycle Σ be homologous to zero on the complex surface (128), if $n_i > p$ for some i. Indeed, the functions $\varphi_i^{(s)}$, $s = 0, 1, \dots, n_i - 1$, in the integrand in (130) represent the initial data at $t = 0$ for the function u_i and its first $n_i - 1$ derivatives with respect to t. Therefore, if the domain G is a lacuna for system (124), then the functions $K_s(t, x - y)$, $s = 0, 1, \dots, n_i - 1$, must vanish at every point of G. In order that this condition could hold *in a stable manner*, it is necessary (as will be shown later on) that the corresponding cycles $C_{\mathrm{real}}(t, x)$, for p odd (resp., $C_{\mathrm{imag}}(t, x)$, for p even), be homologous to zero in the complex intersection of (128) and (129), and that the cycle Σ be homologous to zero on the complex surface (128), if $n_i > p$ for some i.

This statement would follow from §16.3, were it possible to choose the coefficients of the system in such a way that (128) would represent an arbitrary surface of order n. In the general case, however, this cannot be done. Thus, in the case of $p \geq 2N$, the surface (128) inevitably contains singular points; for instance, singular are the points at which two rows of the matrix (125) vanish simultaneously. To make Theorems IV and IV′ applicable to the cycle Σ, it suffices to show that there is a surface, among those given by (128), such that the integral $I(\lambda)$ over any quasi-finite cycle of the form (75), which is non-homologous to zero, and all derivatives of $I(\lambda)$ with respect to λ are different from 0 on that surface. Here the integral $I(\lambda)$ has the form (71) (M is a polynomial in $p_1, \dots, p_p, \lambda$)) and is taken over the section $p_{p+1} = \lambda = \mathrm{const}$ of the surface (128) constructed for the $(p+1)$-th variable. The following sections of §20 deal with the construction of this surface. We assume that $p > 1$.

Similar arguments allow us to use statements analogous to Theorems IV and IV′ for the proof of the necessity of the condition that the cycles

C_{real} and C_{imag} be homologous to zero in the complex intersection of (128) and (129). For this purpose, it suffices to reduce p by 1.

4. We first note that if the surface (128) of complex dimension $p - 1$ has no singular points, then the surface of type (128) whose complex dimension is p can have a finite number of singular points and no singular curves, surfaces, etc. Indeed, let L be a singular curve on the complex surface (128) of complex dimension p (the curve L must be algebraic); then every section of this surface by the plane $p_{p+1} = \text{const}$ must contain singular points, in particular, the points of intersection of L and the plane $p_{p+1} = \text{const}$. For the sake of simplicity, we also assume that the p-dimensional surface (128) has no singular points.

5. Consider a given regular surface (128) of complex dimension $p-1$, which corresponds to the value $\lambda = \lambda_1$ and does not touch the plane at infinity. Let us construct for this surface the finite normal representation of the $(p-1)$-dimensional cycle (to within algebraic cycles, if p is odd), as described in §14.6. In this representation, the coordinates $p_1, \ldots, p_p$ of the end-points of the tubes T_{p-1} are determined by the equations[8]

$$H(p_1, \ldots, p_p, \lambda) = 0 , \quad H_1(p_1, \ldots, p_p, \lambda) = 0 , \ldots,$$

$$H_{p-1}(p_1, \ldots, p_p, \lambda) = 0 . \tag{131}$$

Let us eliminate $p_1, \ldots, p_{p-1}$ from (131), and denote the resulting equation by $F(p_p, \lambda) = 0$. Assume that the latter equation has finite and mutually distinct roots p_p for $\lambda = \lambda_1$. This equation is irreducible; therefore, for each pair of its roots $p'_p(\lambda_1)$ and $p''_p(\lambda_1)$ we can choose a path L on the λ-plane such that these roots coincide at the end-point of L (for $\lambda = \lambda_0$), while all the other roots p_p remain finite and mutually distinct. Moreover, at the points of L other than the end-point λ_0, all roots $p_p(\lambda)$ are finite and mutually distinct. In particular, let us take the points $p'_p(\lambda_1)$ and $p_p''(\lambda_1)$

---

[8] In order that system (131) admit infinite solutions $(p_1, \ldots, p_p)$ for any finite value of λ, it is necessary that there exist non-trivial solutions of the following system of homogeneous equations

$$H(p_1, \ldots, p_p) = 0 , \quad H_1(p_1, \ldots, p_p) = 0, \ldots, H_{p-1}(p_1, \ldots, p_p) = 0 ,$$

where $H(p_1, \ldots, p_p)$ is obtained from $H(p_1, \ldots, p_p, \lambda)$ by deleting the terms of degree $< n$ with respect to $p_1, \ldots, p_p$. Then the surface $H(p_1, \ldots, p_p, \lambda) = 0$, with λ regarded as a parameter, would touch the hyperplane at infinity in the projective $(p_1, \ldots, p_p)$-space, for every λ. Subjecting, if necessary, the variables $p_1, \ldots, p_p, \lambda$ to a projective transformation, we can come to the case such that this surface does not touch the hyperplane at infinity for any finite value of λ. Let us assume the latter condition to be initially satisfied for the surface $H(p_1, \ldots, p_p, \lambda) = 0$ under consideration.

to be the projections of the end-points of the tubes T_{p-1}^{ij} and $T_{p-1}^{i,j+1}$ to the p_p-plane. Then the finite end-points of the two and only two tubes T_{p-1}^{ij} and $T_{p-1}^{i,j+1}$ tend to coincide at a finite point C as $\lambda \to \lambda_0$, while those of the other tubes remain finite and mutually distinct. To simplify the notations we set $\lambda_0 = 0$.

The surface $H(p_1, \ldots, p_p, \lambda_0) = 0$ is the intersection of p-dimensional surface (128) and the plane $\lambda = \lambda_0 = \text{const}$, which touches the surface (128) at one and only one point.

6. Let us show that the integral over the tube T_{p-1}^{ij} and its derivatives with respect to λ of a sufficiently high order become infinitely large as $\lambda \to \lambda_0$. Then it can be claimed that every integral, as well as its derivatives, mentioned in §20.3, over any finite or quasi-finite cycle non-homologous to zero does not vanish identically.

In similarity to what has been stated in §15, we can represent this integral over the tube T_{p-1}^{ij} as

$$\int_a^{a_{ij}(\lambda)} \Omega_{ij}(p_p, \lambda)\, dp_p \ ,$$

where $\Omega_{ij}(p_p, \lambda)$ stands for the integral

$$\int \frac{M(p_1, \ldots, p_p, \lambda)\, dp_1 \ldots dp_{p-2}}{H_{p-1}'(p_1, \ldots, p_p, \lambda)} \tag{132}$$

over a cycle $C_{p-2}^{ij}(p_p, \lambda)$ contracting to a single point as p_p approaches the critical value $a_{ij}(\lambda)$ corresponding to the end-point of the tube T_{p-1}^{ij}.

7. According to §20.2, $M(p_1, \ldots, p_p, \lambda)$ is a polynomial with respect to the variables $p_1, \ldots, p_p, \lambda$. In order to establish the necessity of the condition that the $(p-2)$-dimensional cycles C_{real} and C_{imag} should be homologous to zero, we have to consider the integral over these cycles which yields $K_0(t, x - y)$ in (105). The polynomial $M = M_{n-1}$ for this integral has degree $n - 1$ and depends on $p_1, \ldots, p_{p-1}, p_p = \lambda$ only. In this case, H also depends only on $p_1, \ldots, p_p$. To keep the symmetry of notation we shall sometimes write p_{p+1} instead of p_p. In order to prove that the cycle Σ must be homologous to zero, we have to consider, in the case of $n_i > p$ for some i, the integrals over Σ which yield $K_p(t, x - y)$ in (105). For this integral, $M_{n-p-1}(p_1, \ldots, p_p, \lambda)$ has degree $n - p - 1$ with respect to $p_1, \ldots, p_p$. It is not very difficult to show, by virtue of (25), (25′), that the polynomial M_{n-1} can be obtained from the polynomial $H(p_1, \ldots, p_p)$ in the left-hand side of (128) by excluding the terms whose degree is higher than $n - 1$ with respect to $p_1, \ldots, p_p$. In order to construct the polynomial $M_{n-p-1}(p_1, \ldots, p_p, \lambda)$, we have to proceed as follows:

1) to write out a system similar to (124) for the independent variables $t, x_1, \ldots, x_p, x_{p+1}$;

2) to replace the operator $\partial/\partial x_{p+1}$ by $p_{p+1}\,\partial/\partial t$;

3) to write out for the system thus obtained equation (128') similar to (128);

4) to delete the terms of degree $> n - p - 1$ with respect to $p_1, \ldots, p_p$ in the polynomial $H(p_1, \ldots, p_p, p_{p+1})$ in the left-hand side of the equation obtained $(p_{p+1} = \lambda)$.

Let us show that M_{n-1} (resp., M_{n-p-1}) can be assumed different from 0 at the point C mentioned in §20.5. For this purpose we can always find a suitable linear transformation of the variables. Let $p_1^0, 0, \ldots, 0$ be the coordinates of the point C, where p_1^0 is finite. Let us replace p_1^0 by $p_1^0 + \varepsilon$ in equation (128), where ε is a small real number. In this way we obtain the equation $H(p_1 + \varepsilon, p_2, \ldots, p_p, \lambda) = 0$. For this equation, the polynomials M_{n-1} (resp., M_{n-p-1}) constructed according to the above procedure take the form

$$M_{n-1}^\varepsilon(p_1^0 + \varepsilon, p_2, \ldots, p_p, \lambda) \quad \left(\text{resp.,} \quad M_{n-p-1}^\varepsilon(p_1^0 + \varepsilon, p_2, \ldots, p_p, \lambda)\right) \ .$$

Let us show that $M_k^\varepsilon(p_1^0 + \varepsilon, 0, \ldots, 0, \lambda_0)$, $k = n - 1$, $n - p - 1$, undergo variation if ε varies, for every value of $p_1^0 + \varepsilon$; this fact is sufficient to prove that M_{n-1}^ε and M_{n-p-1}^ε do not vanish identically at the point $C\,(p_1^0 + \varepsilon, 0, \ldots, 0, \lambda_0)$. Set

$$M_k^\varepsilon(p, 0, \ldots, 0, \lambda_0) = f_k^\varepsilon(p) \ , \qquad H(p, 0, \ldots, 0, \lambda_0) = f(p) \ .$$

Obviously, $f_k^\varepsilon(p + \varepsilon)$ can be obtained from $f(p + \varepsilon)$ by excluding the terms of degree $> k$ with respect to p. Our aim is to show that $f_k^\varepsilon(p+\varepsilon)$ undergoes change if ε changes, for every fixed finite value of $p + \varepsilon$. Set

$$\varphi_k(p) = f(p) - f_k^0(p) \ .$$

Then $f_k^\varepsilon(p + \varepsilon) = f_k^0(p + \varepsilon) + \varphi_k^\varepsilon(p + \varepsilon)$, where $\varphi_k^\varepsilon(p + \varepsilon)$ is obtained from $\varphi(p + \varepsilon)$ in the same way as $f_k^\varepsilon(p + \varepsilon)$ from $f(p + \varepsilon)$. Let

$$\varphi_k(p) \equiv a_n p^n + a_{n-1} p^{n-1} + \cdots + a_{k+1} p^{k+1} \ , \qquad a_n \neq 0 \ .$$

The assumption $a_n \neq 0$ can be made without loss of generality, since, even if $a_n = 0$, this inequality can be obtained by a small rotation of the coordinate axes in the $(p_1, \ldots, p_p, p_{p+1})$-space. Since

$$\varphi_k^\varepsilon(p + \varepsilon) = \varphi(\varepsilon) + p\varphi'(\varepsilon) + \cdots + \frac{p^k}{k!}\varphi^{(k)}(\varepsilon) \ ,$$

we see that $\varphi_k^\varepsilon(p + \varepsilon)$ is a polynomial, both with respect to $p + \varepsilon$ and ε, which for no value of $p + \varepsilon$ can vanish identically with respect to ε, since its highest order (with respect to ε) coefficient is equal to

$$a_n \left[1 - n + \frac{n(n-1)}{2} - \cdots \pm \frac{n(n-1)\cdots(n-k+1)}{k!} \right] =$$

$$= \pm \frac{(n-1)(n-2)\cdots(n-k)}{k!} a_n \neq 0 \ .$$

8. Choosing as the new origin in the $(p_1, \ldots, p_p)$-space the point that is being approached by the end-point of T_{p-1}^{ij} as $\lambda \to 0$, and reducing the sum of all second order terms to canonical form, we can write the equation of the surface under consideration as

$$H(p_1, \ldots, p_p, \lambda) \equiv k\lambda + \sum_{j=1}^{p} a_j p_j^2 + b\lambda^2 + \cdots , \qquad (133)$$

where the terms of order > 2 with respect to p and λ are not written out explicitly. Every coefficient a_j, as well as k, can be assumed non-vanishing. Indeed, consider a similar expansion for a general p-dimensional surface (128) in a neighborhood of any non-singular point and assume one of the coefficients a_j to be equal to zero; then the determinant of the matrix

$$\left| \frac{\partial^2 H}{\partial p_i \partial p_j} \right| \qquad (i, j, = 1, \ldots, p)$$

must vanish at every point of this surface. However, since the equation $H = 0$ of the form (128) has its highest order terms involving only one variable p_j with an arbitrary coefficient, we see that the diagonal elements of the above determinant contain terms which do not enter any other element. It follows that the determinant cannot vanish identically. The coefficient k is different from 0, since the normal derivative of the function $H(p_1, \ldots, p_p, \lambda)$ at a regular point of the surface $H = 0$ does not vanish.

Thus, the coefficients k and a_j, $j = 1, \ldots, p$, are all different from 0. Passing to the new variables, if necessary, we can limit ourselves to the case

$$a_j = 1, \quad j = 1, \ldots, p-1 ; \quad a_p = -1 , \quad k = 1 \ .$$

Then (133) becomes

$$H(p_1, \ldots, p_p, \lambda) \equiv \lambda - p_p^2 + \sum_{j=1}^{p-1} p_j^2 + b\lambda^2 + \cdots , \qquad (134)$$

where the higher order terms (with respect to p and λ) are not written out explicitly. We can always get rid of the terms having the form

$$p_j \sum a^*_{jk_0k_p}\lambda^{k_0}p_p^{k_p}, \quad j=1,\ldots,p-1,$$

by replacing, if necessary, p_j by $p_j + B_j(p_p,\lambda)$, where the functions $B_j(p_p,\lambda)$ satisfy the equations

$$H_i(B_1,\ldots,B_{p-1},p_p,\lambda) \equiv 2B_i + \cdots = 0, \quad i=1,\ldots,p-1.$$

Here, the higher order terms are also not written out explicitly. Solving this equation by the method of successive approximations, we find that, for $|p_p|$ and $|\lambda|$ sufficiently small, the functions $B_j(p_p,\lambda)$ are analytic with respect to p and λ, and their expansions in powers of p_p and λ start with the terms whose order is ≥ 2. Substituting $p_i + B_i$ for p_i, $i=1,\ldots,p-1$, in $H(p_1,\ldots,p_p,\lambda)$, we see that the new $p_1,\ldots,p_p$ satisfy the equation

$$H^*(p_1,\ldots,p_p,\lambda) \equiv \lambda - p_p^2 + \sum_{j=1}^{p-1}p_j^2 + b\lambda^2 + \cdots = 0, \qquad (135)$$

where H^* is a polynomial of degree n with respect to $p_1,\ldots,p_{p-1}$, and at the same time, H^* is a holomorphic function of λ, p_p for sufficiently small $|\lambda|, |p_p|$. The lower order terms in the expansion of H^* in powers of λ and p_p coincide with the corresponding terms in equation (134). Assume that $p_k = 0$ for $k < p$ and

$$p_p - p_p^0 = -\sqrt{\lambda} + o\left(\sqrt{\lambda}\right)$$

at the end-point of the tube T^{ij}_{p-1}. Let λ be positive. Assume also that p_p takes only real values on T^{ij}_{p-1}, and $|p_p| > |p_p^0|$ in a neighborhood of the end-point of this tube.

Let us evaluate $\Omega(p_p,\lambda)$ (henceforth, the indices are dropped). For sufficiently small real λ and p_p, the cycle $C_{p-1}(p_p,\lambda)$ can be obtained by considering on the plane $p_p = $ const all possible straight lines (with real coefficients) issuing from the point $A(0,\ldots,0,p_p,\lambda)$ up to the point nearest to A where they meet the surface $H^*(p_1,\ldots,p_p,\lambda) = 0$. Let us change the variables in the integral

$$\Omega(p_p,\lambda) = \int\limits_{C_{p-1}(p_p,\lambda)} \frac{M^*\,dp_1\ldots dp_{p-1}}{H^{*\prime}_{p-1}},$$

setting

$$p_j = \rho r y_j, \quad j=1,\ldots,p-1,$$

where $\sum y_j^2 = 1$ and

$$\rho^2 = -H^*(0,\ldots,0,p_p,\lambda) \equiv -\lambda + p_p^2 - b\lambda^2 + \cdots \qquad (136)$$

We then obtain

$$H^*(p_1, \ldots, p_p, \lambda) \equiv H^{**}(y_1, \ldots, y_{p-1}, p_p, \lambda, \rho, r) =$$
$$= \rho^2 \left[-1 + r^2 + r^2 M_0(y, p_p, \lambda) + \rho r^3 M_1(y, p_p, \lambda, \rho, r) \right] = 0 , \quad (137)$$

$$M^*(p_1, \ldots, p_p, \lambda) \equiv C + M_2(y, p_p, \lambda, \rho, r) , \quad C \neq 0 ,$$

where M_ν are analytic functions of all their variables, and

$$M_0(y, 0, 0) = 0 , \quad C = \text{const} , \quad M_2(y, 0, 0, r) = 0 .$$

Let $d\sigma$ be the area element on the surface of the real $(p - 2)$-dimensional sphere $\Sigma : y_1^2 + \cdots + y_{p-1}^2 = 1$. Then

$$\Omega(p_p, \lambda) = \frac{\rho^{p-3}}{2} \int_\Sigma \frac{r^{p-3}[C + M_2(y, p_p, \lambda, \rho, r)] \, d\sigma}{1 + M_0(y, p_p, \lambda) + \rho M_3(y, p_p, \lambda, \rho, r)} =$$

$$= \rho^{p-3} \Phi(p_p, \lambda, \rho) ,$$

where $2M_3 = \partial \left[r^3 M_1 \right] / \partial r$.

For the values of λ close to 0, we have

$$r \approx 1$$

on the tube T_{p-1}, near its end-point, i.e., for small p_p. Let us regard p_p as a function of λ and ρ. From (136), (137), we find that

$$\frac{\partial p_p}{\partial \lambda} = -\frac{-1 + \cdots}{2p_p + \cdots} \approx \frac{1}{2p_p} , \quad (138)$$

$$\frac{\partial r}{\partial \lambda} = -\frac{H_{p_p}^{***'} \frac{\partial p_p}{\partial \lambda} + H_\lambda^{***'}}{H_r^{***'}} = \frac{O(1)O\left(\frac{1}{p_p}\right) + O(1)}{2r + 2r M_0 + \rho M_4} = O\left(\frac{1}{p_p}\right), \quad (138')$$

where $H^{***} \equiv H^{**}/\rho^2$. Hence it is easy to see that

$$\Omega(p_p, \lambda) \equiv \Omega^*(\rho, \lambda) = C\rho^{p-3} , \quad C > 0 ,$$

$$\frac{\partial \phi}{\partial \lambda} = O\left(\frac{1}{p_p}\right) , \quad \Omega_\lambda^{*'}(\rho, \lambda) = \rho^{p-3} O\left(\frac{1}{p_p}\right) .$$

In general, for any $k \geq 1$ we have

$$\frac{\partial^k \Omega^*(\rho, \lambda)}{\partial \lambda^k} = \rho^{p-3} O\left(\frac{1}{p_p^{2k-1}}\right) . \quad (139)$$

Next, we turn to the evaluation of the integral

$$I^*(\lambda) = \int^{p_p^0(\lambda)} \Omega(p_p, \lambda)\, dp_p$$

over the tube $T(\lambda)$; the lower limit of integration being dropped, since it can be chosen arbitrary (see §15.1). Let us take ρ as the variable of integration, instead of p_p. We then obtain

$$I^*(\lambda) = \int_{\rho_0}^{0} \Omega^*(\rho, \lambda)\, \frac{\partial p_p}{\partial \rho}\, d\rho$$

(ρ_0 is supposed to be independent of λ),

$$\frac{\partial p_p}{\partial \rho} = \frac{2\rho}{H_{p_p}^*(0,\ldots,0,p_p,\lambda)} = \frac{2\rho}{2p_p + \cdots} \approx \frac{\rho}{p_p}\,. \tag{140}$$

Taking into account relations (138), (138′), (139), (140), we see that the integrand in

$$\frac{\partial^k I^*(\lambda)}{\partial \lambda^k} = \int_{\rho_0}^{0} \frac{\partial^k}{\partial \lambda^k}\left(\Omega^* \frac{\partial p_p}{\partial \rho}\right) d\rho$$

is of the same order as

$$\Omega^* \frac{\partial^k}{\partial \lambda^k}\left(\frac{\partial p_p}{\partial \rho}\right)$$

for small λ and ρ, namely, of order ρ^{p-2}/p_p^{2k-1}. If $p_p \to \rho$, $\lambda \to 0$, therefore, $\partial^k I^*/\partial \lambda^k$ becomes infinitely large for sufficiently large k, q.e.d.

The above results yield

Theorem VIII. *Assume that the surface (129) has no singular points and $n_i \le p$ for all i. Then the necessary and sufficient condition for the domain G to·be a stable lacuna for system (124) is as follows: the cycles $C_{\text{real}}(t, x)$, for p odd (the cycles $C_{\text{imag}}(t, x)$, for p even, respectively), corresponding to the points P of the domain G, are homologous to zero in the complex inter- section of the surface (128) and the planes (129) corresponding to the points P. In the case of $n_i > p$ for some i, the necessary and sufficient condition for the domain G to form a stable lacuna is the above condition supple- mented with the following one: the cycle Σ from the Third Basic Theorem is homologous to zero on the complex surface (128).*

This theorem can be expressed in more precise terms, in the same way as it has been done for the Third Basic Theorem in §16.3.

While proving the necessity of the condition that the cycle Σ be homolo- gous to zero, we have assumed the surface $H(p_1, \ldots, p_p, \lambda) = 0$ of dimension p to have no singular points.

§21. Examples

1. Consider the wave equation

$$\frac{\partial^2 u}{\partial t^2} = \frac{\partial^2 u}{\partial x_1^2} + \cdots + \frac{\partial^2 u}{\partial x_p^2} \, . \tag{141}$$

Its characteristic surface is the sphere

$$1 = \sum_{k=1}^{p} p_k^2 \, . \tag{142}$$

In the case of p odd, the cycles C_{real} are always homologous to zero (see [12]) in the intersection of the complex surface (142) and the complex plane (129). If this plane does not meet the real surface (142), then the cycle $C_{\text{real}}(t, x)$ is simply void. Consequently, the entire domain on the plane $t = 0$ coinciding with the interior of the base of the characteristic cone with vertex at $(t, x_1, \ldots, x_p)$ forms a lacuna; this is also true for the domain coinciding with the exterior of the base.

In the case of p even, the cycle $C_{\text{imag}}^Q(t, x)$ is void, if the plane (128) has a non-empty real intersection with the surface (142) and the point Q (the starting point of the straight lines used for the construction of the cycle $C_{\text{imag}}^Q(t, x)$) lies within this real intersection. If the plane (129) has no real intersection with the surface (142), then the cycle $C_{\text{imag}}(t, x)$ cannot be homologous to zero in the corresponding cross-section of the complex surface (142). Let us prove this statement for $p = 4$; similar arguments can also be used for other even values of $p > 4$. As to the case $p = 2$, according to §17, we have to consider, instead of the cycle $C_{\text{imag}}(t, x)$, two complex points of intersection of the straight line (129) and the algebraic curve (142).

Now, consider the intersection of the surface

$$p_1^2 + p_2^2 + p_3^2 + p_4^2 = 1 \tag{143}$$

and the plane

$$p_1 x_1 + p_2 x_2 + p_3 x_3 + p_4 x_4 + t = 0 \, .$$

By a rotation of the coordinate axes in the (x_1, x_2, x_3, x_4)-space, we can transform this plane into

$$p_4 x_4 + t = 0 \, ,$$

whence

$$p_4 = -\frac{t}{x_4} \, . \tag{144}$$

If there is no real intersection of the plane (144) and the surface (143), then

$$\left| \frac{t}{x_4} \right| > 1 \, .$$

Substituting p_4 given by (144) into (143), we get

$$p_1^2 + p_2^2 + p_3^2 = 1 - \frac{t}{x_4^2} = -\rho^2 , \tag{145}$$

where ρ is a positive constant. In order to construct the cycle $C_{\text{imag}}(t, x)$, let us consider the family of straight lines parallel to the axis Op_3. The points of their intersection with the surface (145) will have real coordinates p_1, p_2 and a purely imaginary one p_3.

In order to show that the cycle $C_{\text{imag}}(t, x)$, thus obtained, is non-homologous to zero on the complex surface (145), we find a two-dimensional cycle on this surface whose intersection index with $C_{\text{imag}}(t, x)$ is equal to 1. We can choose this cycle to be a rectilinear generatrix of the surface (145)

$$p_1 + ip_2 = a(p_3 + i\rho) , \qquad p_1 - ip_2 = -a^{-1}(p_3 - i\rho) , \tag{146}$$

where p_1, p_2, p_3, a are, in general, complex numbers; a is a constant. It is easy to see that the cycle $C_{\text{imag}}(t, x)$ has a unique point in common with the cycle (146), namely,

$$p_1 = \frac{2a_2\rho}{|a|^2 - 1} , \qquad p_2 = \frac{-2a_1\rho}{|a|^2 - 1} , \qquad p_3 = i\rho \,\frac{1 + |a|^2}{1 - |a|^2} ,$$

where $a_1 = \operatorname{Re} a$, $a_2 = \operatorname{Im} a$. This point cannot be a point of contact of these cycles.

It is known that, for p even, the domain corresponding to the points (t, x) such that $C_{\text{imag}}(t, x)$ is non-homologous to zero does not form a lacuna. Thus, our basic theorems can be extended to the case $n = 2$.

2. Consider the system of elasticity

$$\frac{\partial^2 u}{\partial t^2} = \mu \nabla^2 u + (\nu + \mu) \frac{\partial \Delta}{\partial x_1} ,$$

$$\frac{\partial^2 v}{\partial t^2} = \mu \nabla^2 v + (\nu + \mu) \frac{\partial \Delta}{\partial x_2} , \tag{147}$$

$$\frac{\partial^2 w}{\partial t^2} = \mu \nabla^2 w + (\nu + \mu) \frac{\partial \Delta}{\partial x_3} ,$$

where

$$\nabla^2 = \frac{\partial^2}{\partial x_1^2} + \frac{\partial^2}{\partial x_2^2} + \frac{\partial^2}{\partial x_3^2} ,$$

$$\Delta = \frac{\partial u}{\partial x_1} + \frac{\partial v}{\partial x_2} + \frac{\partial w}{\partial x_3} .$$

Using the notation of §20, we find that

$$\Delta(\lambda, \xi_1, \xi_2, \xi_3) = \left(\lambda^2 - \mu\rho^2\right)^2 \left[\lambda^2 - (2\mu + \nu)\rho^2\right]$$

to within a constant factor, where

$$\rho^2 = \xi_1^2 + \xi_2^2 + \xi_3^2 \ .$$

The equation

$$\Delta\left(\frac{\partial}{\partial t}, \frac{\partial}{\partial x_1}, \frac{\partial}{\partial x_2}, \frac{\partial}{\partial x_3}\right) u = 0$$

is equivalent to the following system

$$\frac{\partial^2 u}{\partial t^2} = \mu\left(\frac{\partial^2 u}{\partial x_1^2} + \frac{\partial^2 u}{\partial x_2^2} + \frac{\partial^2 u}{\partial x_3^2}\right) + u_1\ ,$$

$$\frac{\partial^2 u_1}{\partial t^2} = \mu\left(\frac{\partial^2 u_1}{\partial x_1^2} + \frac{\partial^2 u_1}{\partial x_2^2} + \frac{\partial^2 u_1}{\partial x_3^2}\right) + u_2\ , \qquad (148)$$

$$\frac{\partial^2 u_2}{\partial t^2} = (2\mu + \nu)\left(\frac{\partial^2 u_2}{\partial x_1^2} + \frac{\partial^2 u_2}{\partial x_2^2} + \frac{\partial^2 u_2}{\partial x_3^2}\right)\ .$$

Let us introduce a new system

$$\frac{\partial^2 u}{\partial t^2} = \mu\left(\frac{\partial^2 u}{\partial x_1^2} + \frac{\partial^2 u}{\partial x_2^2} + \frac{\partial^2 u}{\partial x_3^2}\right) + u_1\ ,$$

$$\frac{\partial^2 u_1}{\partial t^2} = (\mu + \varepsilon)\left(\frac{\partial^2 u_1}{\partial x_1^2} + \frac{\partial^2 u_1}{\partial x_2^2} + \frac{\partial^2 u_1}{\partial x_3^2}\right) + u_2\ , \qquad (148_\varepsilon)$$

$$\frac{\partial^2 u_2}{\partial t^2} = (2\mu + \nu)\left(\frac{\partial^2 u_2}{\partial x_1^2} + \frac{\partial^2 u_2}{\partial x_2^2} + \frac{\partial^2 u_2}{\partial x_3^2}\right)\ .$$

According to §9, Part 1, of my paper [4], the solution of the Cauchy problem for the latter system continuously depends on $\varepsilon \to 0$, and therefore, converges to a solution of the Cauchy problem for system (148) with the same initial data. The characteristic surface for system (148_ε) has no singular points, if $\varepsilon \neq 0$. In order to find the function u, let us use formula (130). We replace $\varphi^{(s)}$ in (130) by the expression obtained from system (147) with the initial data for the functions $u, v,$ and w. On the basis of the arguments similar to those which have lead us to Theorem VIII, we find that the planes of the (p_1, p_2, p_2)-space that do not meet the surface

$$\Delta_\varepsilon(1, p_1, p_2, p_3) = 0$$

correspond to the points of a lacuna for u, i.e., the value of u at the vertex of the characteristic cone remains invariable when the values of u, v, w and their first derivatives are changed at these points. Since the value of u at the vertex of the characteristic cone continuously depends on ε (as indicated above), the planes in the (p_1, p_2, p_3)-space that do not meet the surface

$$\Delta(1, p_1, p_2, p_3) = 0$$

correspond to the points of a lacuna for (147). Applying this argument to v and w, we see that these points constitute a lacuna for the whole system (147).

In order to verify the absence of lacunas for the system of plane elasticity, one can use Theorem VII of §19.

3. Consider the system of crystal optics

$$\sigma_1 \frac{\partial^2 u}{\partial t^2} = \nabla^2 u - \frac{\partial \Delta}{\partial x_1} ,$$

$$\sigma_2 \frac{\partial^2 v}{\partial t^2} = \nabla^2 v - \frac{\partial \Delta}{\partial x_2} , \tag{149}$$

$$\sigma_3 \frac{\partial^2 w}{\partial t^2} = \nabla^2 w - \frac{\partial \Delta}{\partial x_3} .$$

Here we use the notations of the preceding section. For this system, we have (see, for instance, [13])

$$\Delta(\lambda, \xi_1, \xi_2, \xi_3) \equiv \lambda^2 \Delta^*(\lambda, \xi_1, \xi_2, \xi_3) \equiv$$

$$\equiv \lambda^6 - \lambda^4 \Psi(\xi_1, \xi_2, \xi_3) + \lambda^2 \rho^2 \Phi(\xi_1, \xi_2, \xi_3) ,$$

to within a constant factor, where

$$\Psi(\xi_1, \xi_2, \xi_3) \equiv \frac{\xi_2^2 + \xi_3^2}{\sigma_1} + \frac{\xi_3^2 + \xi_1^2}{\sigma_2} + \frac{\xi_1^2 + \xi_2^2}{\sigma_3} ,$$

$$\Phi(\xi_1, \xi_2, \xi_3) \equiv \frac{\xi_1^2}{\sigma_2 \sigma_2} + \frac{\xi_2^2}{\sigma_3 \sigma_1} + \frac{\xi_3^2}{\sigma_1 \sigma_2} ,$$

$$\rho^2 = \xi_1^2 + \xi_2^2 + \xi_3^2 .$$

The surface

$$\Delta^*(1, p_1, p_2, p_3) = 0 \tag{150}$$

consists (see [13]) of two real ovals enclosed in one another and having 4 points in common. Let

$$D_k(1, p_1, p_2, p_3) = 0$$

be the equation of the k-th oval, where D_k is chosen such that

$$D_k(\lambda, p_1, p_2, p_3)$$

is a homogeneous function of the second order with respect to λ, p_1, p_2, p_3, and has the form of a polynomial in λ with the coefficient 1 by λ^2.

The Cauchy problem for the equation

$$\Delta^* \left(\frac{\partial}{\partial t}, \frac{\partial}{\partial x_1}, \frac{\partial}{\partial x_2}, \frac{\partial}{\partial x_3} \right) u = 0$$

consists in finding a function

$$u(t, x_1, x_2, x_3) = \int e^{i(x_1 \xi_1 + x_2 \xi_2 + x_3 \xi_3)} \varphi(t, \xi_1, \xi_2, \xi_3) \, d\xi_1 \, d\xi_2 \, d\xi_3$$

satisfying the given initial conditions, where $\varphi(t, \xi_1, \xi_2, \xi_3)$ satisfies the system

$$D_1 \left(\frac{d}{dt}, \xi_1, \xi_2, \xi_3 \right) \varphi = \varphi_1 \,,$$

$$D_2 \left(\frac{d}{dt}, \xi_1, \xi_2, \xi_3 \right) \varphi_1 = 0 \,,$$

and the initial data for φ_1 are determined by those for φ.

Consider a function $u_\varepsilon(t, x_1, x_2, x_3)$ satisfying the equation

$$\Delta_\varepsilon^* \left(\frac{\partial}{\partial t}, \frac{\partial}{\partial x_1}, \frac{\partial}{\partial x_2}, \frac{\partial}{\partial x_3} \right) u = 0 \,, \tag{151}$$

and the same initial conditions as the function u considered above. Here Δ_ε^* is a polynomial of degree 4 with respect to $\partial/\partial t$, $\partial/\partial x_1$, $\partial/\partial x_2$, $\partial/\partial x_3$, defined the equation

$$\Delta_\varepsilon^*(\lambda, \xi_1, \xi_2, \xi_3) \equiv \Delta^*(\lambda, \xi_1, \xi_2, \xi_3) + \varepsilon \lambda^4 \,,$$

where ε is a small real parameter whose sign is chosen in such a way that the real surface

$$\Delta^*(1, p_1, p_2, p_3) + \varepsilon = 0 \tag{152}$$

consists of two ovals enclosed in one another and having no points in common. Let

$$D_k(1, p_1, p_2, p_3, \varepsilon) = 0 \tag{153}$$

be the equation of the k-th oval, where $D_k(\lambda, \xi_1, \xi_2, \xi_3, \varepsilon)$ is a homogeneous function of the second order with respect to $\lambda, \xi_1, \xi_2, \xi_3$, and has the form of a polynomial with respect to λ with 1 as the coefficient by λ^2. We represent the function u_ε in the form

$$u_\varepsilon(t, x_1, x_2, x_3) = \int e^{i(x_1\xi_1 + x_2\xi_2 + x_3\xi_3)} \varphi_\varepsilon(t, \xi_1, \xi_2, \xi_3)\, d\xi_1\, d\xi_2\, d\xi_3\ ,$$

where $\varphi_\varepsilon(t, \xi_1, \xi_2, \xi_3)$ satisfies the system of ordinary differential equations

$$D_1\left(\frac{d}{dt}, \xi_1, \xi_2, \xi_3, \varepsilon\right)\varphi_\varepsilon = \varphi_{1\varepsilon}\ , \tag{154$_1$}$$

$$D_2\left(\frac{d}{dt}, \xi_1, \xi_2, \xi_3, \varepsilon\right)\varphi_{1\varepsilon} = 0\ , \tag{154$_2$}$$

and the same initial conditions at $t = 0$ as the function $\varphi(t, \xi_1, \xi_2, \xi_3)$; the initial values of $\varphi_{1\varepsilon}$ being determined by those of φ_ε. According to §2, Ch. III, of my paper [8], we have

$$u_\varepsilon(t, x_1, x_2, x_3) \to u(t, x_1, x_2, x_3) \quad \text{as} \quad \varepsilon \to 0\ . \tag{155}$$

Since the characteristic surface (152) has no singular points for the values of $\varepsilon \neq 0$ under consideration, we can use (61), (130) in relation to the solution of the Cauchy problem for the corresponding differential equation. Therefore the planes in the (p_1, p_2, p_3)-space that do not meet the surface (152) correspond to a lacuna for equation (151) with the initial data taken from (149). Consequently, the planes in the (p_1, p_2, p_3)-space that do not meet the real surface (150) also correspond to the points of a lacuna for system (149).

§22. Two Remarks

1. Following the terminology of Hadamard [3], we say that *there is no diffusion of waves* for a given hyperbolic system, if the following condition holds: let the initial data be different from 0 only within a small domain G; then the value of the solution at any later moment t may be different from 0 only within a certain spherical layer bounded by two concentric spheres containing the domain G and expanding with the growth of t.

　　There is no diffusion of waves for a hyperbolic system, if and only if there exists a lacuna for this system crossed by a straight line parallel to the t-axis and passing through the vertex of the characteristic cone in the $(t, x_1, \ldots, x_p)$-space.

2. As a generalization of the Cauchy problem it is natural to consider the following one: *find a function*

$$u_{1\epsilon}(t, x_1, x_2, x_3) = \int e^{i(x_1\xi_1 + x_2\xi_2 + x_3\xi_3)} \varphi_{1\epsilon}(t, \xi_1, \xi_2, \xi_3)\, d\xi_1\, d\xi_2\, d\xi_3\,,$$

given the values of $u_{1\epsilon}$ *and the values of* $\partial u_{1\epsilon}/\partial t$ *at* $t = 0$, *provided that the function* $\varphi_{1\epsilon}(t, \xi_1, \xi_2, \xi_3)$ *satisfies equation* (154$_2$). If the left-hand side of (154$_2$) is a polynomial in ξ_1, ξ_2, ξ_3, then this problem reduces to the usual Cauchy problem.

Similar generalizations can be considered in the case of $D_2(\lambda, \xi_1, \xi_2, \xi_3)$ in the left-hand side of equation (154$_2$) being a homogeneous function of any order n with respect to $\lambda, \xi_1, \ldots, \xi_p$ and having the form of a polynomial of degree n with respect to λ. This problem (we call it *the generalized Cauchy problem*) is well-posed[9] if and only if, for n even, the surface

$$D_2(1, p_1, \ldots, p_p) = 0 \tag{156}$$

consists of $n/2$ ovals consecutively enclosed in one another, having no points of contact, and with the origin belonging to the least of the ovals; for n odd, these ovals must be supplemented by an *unpaired piece* bearing similarity to the real plane (a proof of this result can be found in [8]). If all these surfaces are sufficiently smooth and mutually disjoint, then for p odd and $p < n$ the solution of the generalized Cauchy problem is given in terms of (44). It should be noticed that in this generalized case we cannot claim $K(t, x)$ to vanish when the plane (129) crosses all the ovals of the surface (156), and therefore, the solution of the generalized Cauchy problem cannot be claimed independent of the initial data outside the base of the characteristic cone. To construct a counter example in the case of an odd n, it should be kept in mind that the ovals of the surface (156) can be chosen arbitrary.

References

[1] Petrowsky, I.G. On the dependence of the solution of Cauchy's problem on the initial data. *Doklady Akad. Nauk SSSR* **38:5-6** (1943) 163–165.
[2] Petrowsky, I.G. On the diffusion of waves and the lacunas for hyperbolic systems. *Izvestiya Akad. Nauk SSSR, Ser. Mat.* **8** (1944) 101–106. (Article 5 of this volume).
[3] Hadamard, J. *Le Problème de Cauchy et les Équations aux Dérivées Partielles Linéaires Hyperboliques.* Paris: Hermann, 1932.,
[4] Petrowsky, I.G. Über das Cauchysche Problem für Systeme von partiellen Differentialgleichungen. *Matem. Sbornik* **2:5** (1937) 815–870. (Article 1 of this volume).
[5] Hadamard, J. *Leçons sur la Propagation des Ondes et les Équations de l'Hydrodynamique.* Hermann, Paris, 1903.

---

[9]For the definition of a well-posed problem see, for instance, [4].

[6] Herglotz, G. Über die Integration linearer partieller Differentialgleichungen mit Konstanten Koeffizienten, I, II. *Berichte Sächs. Akad. d. Wiss.* **78** (1926) 93–126, 287–318.

[6] Herglotz, G. Über die Integration linearer partieller Differentialgleichungen mit Konstanten Koeffizienten, III. *Berichte Sächs. Akad. d. Wiss.* **80** (1928) 69–114.

[8] Petrowsky, I.G. On the Cauchy problem in classes of non-analytic funcions. *Bulletin of Moscow Univ. Math., Mech.* **1:7** (1938) 1–72. (Article 2 of this volume).

[9] Zarisky, O. *Algebraic Surfaces.* Springer-Verlag, Berlin, 1935.

[10] Lefschetz, S. *Analysis Situs et Géométrie Algebrique.* Gauthier-Villars, Paris, 1924).

[11] Severi, F. *Vorlesungen über algebraische Geometrie.* Teubner, Leipzig, 1921.

[12] Van der Waerden, B. Zur algebraische Geometrie, IV. *Math. Ann.* **109** (1934) 8–11.

[13] Hilbert, D; Courant, R. *Methoden der Mathematischen Physik.* Berlin, 1937, Bd. 2.

PETROWSKY'S ARTICLES
ON ALGEBRAIC TOPOLOGY

7

On the Topology of Real Plane
Algebraic Curves*

Introduction

As early as 1876, Harnack [1] showed that the maximal number of components (maximal connected subsets) of a real algebraic curve of order n in the projective plane is precisely $\frac{1}{2}(n-1)(n-2)+1$. At the same time, Harnack suggested a process for the construction of curves with this maximal number of components. In the sequel, such curves will be referred to as *M-curves*. Harnack showed that these *M*-curves have no singular points.

Let us take a sphere in the three-dimensional space [*], wherein the projective plane containing our algebraic curve is situated, and join the center of this sphere to every point of the projective plane by a straight line. We thus project the plane on the sphere. A component of an algebraic curve is called an *oval* (or an *even* component), if its projection on the sphere consists of two ordinary closed curves. If this projection on the sphere S consists of a single closed curve, the corresponding component is called *odd*. Algebraic curves having no real[1] singular points possess at most one odd component. Hence, every algebraic curve (having no real singular points) of even order consists of ovals only, while a curve of odd order has (besides ovals) exactly one odd component.

In 1891, Hilbert [2] proposed a new method for the construction of *M*-curves. In the same work, Hilbert announced without proof that an *M*-curve of order 6 cannot have all its ovals lying outside each other; at least one of these ovals must lie within another oval. Here the words "an oval lies within another oval" mean that the cone projecting the first oval on the sphere S lies within the projecting cone of the second oval. Hilbert considers this a

---

*Originally published in *Ann. Math.* **39** (1938) 189–209. Notes [1-16] by V.M. Kharlamov. See Appendix for a supplementary article by V.M. Kharlamov.

[1]In the sequel, whenever we mention singularities of the curve, we mean only *real* singular points.

remarkable fact, since it proves that M-curves cannot have a too simple topological structure. In his report [3] to the International Mathematical Congress in 1900 on modern problems of mathematics, Hilbert considers the investigation of the topology of M-curves and of the corresponding algebraic surfaces as most timely. After a series of attempts, the above mentioned theorem stated by Hilbert was at last proved in 1911 by Rohn [4]. In the same work, Rohn proved that an algebraic curve of the sixth order cannot possess an oval with ten other ovals of the same curve interior to it [^{2}].

The object of the present paper is to give a general method which would enable us to obtain some of the above mentioned results of Rohn and to extend them to plane curves of arbitrary order. We obtain here the following results: *For n even, an algebraic curve of order n consists of at most $\frac{1}{8}(3n^2 - 6n) + 1$ ovals exterior to each other. Curves having this number of ovals exterior to each other do exist.* They can be constructed by the same method as that used by Harnack for the construction of M-curves.

Hilbert's construction of M-curves leads us to curves having among their components $\left[\frac{1}{2}n - 1\right]$ (where $[k]$ stands for the greatest integer $\leq k$) consecutive ovals lying within each other (i.e., every next oval within the preceding one). In order to extend our theorem to such curves, let us call an oval O of a curve $F(x, y) = 0$ of even order *positive* (resp., *negative*), if the value of the function $F(x, y)$ decreases (resp., increases) when we cross the oval in the outward direction. In the case of an even n, we can always suppose (changing the sign of $F(x, y)$, if necessary) that the ovals not lying within other ovals or lying inside an even number of consecutive ovals are positive, while the ovals lying within an odd number of ovals are negative. Then we can prove the following general theorem, which is an extension of the one formulated above.

Denote by p the number of positive ovals and by m the number of negative ovals of an algebraic curve of an even order n. Then

$$|p - m| \leq \frac{3n^2 - 6n}{8} + 1 ,$$

and there exist curves on which this bound is reached.

While proving this result, we obtain a somewhat more accurate bound for $|p - m|$ [^{3}] We shall see, in particular, that the number of ovals lying outside each other and all of them within the same oval cannot exceed $\frac{1}{8}(3n^2 - 6n) + 1$, if the algebraic curve does not contain other ovals. We can prove by examples that this bound cannot exceed the precise one by more than 3 in the case of $n = 4k$ (k is an integer), and by more than 1 for $n = 4k + 2$. Rohn's result (that there are no curves of order 6 with ten ovals lying inside the eleventh one) shows that this bound is not precise [^{4}].

If the order n is odd, Harnack's process already yields the M-curves consisting (in addition to an odd branch) of ovals, none of which lies inside another oval. This case may be dealt with by the same method that we have used in the case of n even, with the following modifications.

Just as in the case of n even, any finite [5] oval of the curve $F(x,y) = 0$ of odd order is called *positive*, if we pass from the values of $F(x,y) > 0$ to the values of $F(x,y) < 0$ when crossing the oval from the inside; the oval is called *negative* in the opposite case. This definition evidently does not apply to ovals of a curve of odd order which cross the line at infinity. Ovals of the latter type will be called *zero-ovals*. Now we can prove that:

For an odd n the following inequality holds:

$$\left| m - p - \Delta + \frac{k+1}{2} \right| \leq \frac{3n^2 - 4n + 1}{8} ,$$

where m is the number of negative ovals; p is the number of positive ovals; k is the number of real points of intersection of the curve $F(x,y) = 0$ with the line at infinity; Δ is a certain positive integer $\leq k$ depending on the character of intersection of $F(x,y) = 0$ with the line at infinity. A more accurate definition of this number is given in §2 (see Lemma 3 and the proof of the Second Fundamental Theorem).[2]

Below we give examples showing that *this bound for $|p - m|$ is precise in the sense that it is attained on some curves.*

As an immediate corollary of this theorem we obtain the following result. The odd branch of an algebraic curve of an odd order n, together with the line at infinity, divides the projective plane into several regions. If the algebraic curve $F(x,y) = 0$ contains only ovals exterior to each other and lying in one and the same of these regions, then their number is less or equal to

$$\frac{3n^2 - 4n + 1}{8} + \left| \Delta - \frac{k+1}{2} \right| \leq \frac{3n^2 - 4n + 1}{8} + \frac{k-1}{2} .$$

Since the line at infinity does not differ (in the sense of projective geometry) from any other straight line on the plane, we can substitute in this theorem any other straight line for the line at infinity.

The proofs of all these theorems are based on a Jacobi-Euler's formula pertaining to solutions of systems of algebraic equations (see [6], [7], and also [8]) and involve consideration of the deformation of lines $F(x,y) = C$ when C crosses the critical values of $F(x,y)$. This type of analysis is similar to that used by Morse [9] in relation to the critical points of a function.

---

[2]I failed to take into account the number Δ in my note [5].

The method that we apply to the investigation of plane algebraic curves admits a natural generalization to algebraic hypersurfaces

$$F(x_1, \ldots, x_d) = 0$$

in a projective space of arbitrary dimension d. We intend to give a thorough treatment to his question in another paper.

The results of the present paper could be obtained as a corollary from the theorems concerning hypersurfaces in d-dimensional space (to be considered in a subsequent paper). Nevertheless, we prefer to treat them separately, because of their complete geometrical clarity and their capability to elucidate the general method of our analysis.

It should be kept in mind that all algebraic curves considered in what follows have no (real) singular points.

§1. Some Preliminary Lemmas

Lemma 1. *Let*

$$F(x, y) = 0 \tag{1}$$

be the equation of a real[3] algebraic curve. If we continuously vary its coefficients, the topological structure of the curve changes only when the coefficients pass through the values for which the curve has a singularity, i.e., through the values taken at the points which satisfy (1), *as well as the following equations*

$$F'_x(x, y) = 0 , \qquad F'_y(x, y) = 0 . \tag{2}$$

Proof. Evidently, the topological structure of the curve (1) can change only when two different points of this curve unite, or when an isolated point appears or disappears, i.e., when the curve has singular points. It may happen, of course, that these singularities lie at infinity. Then we should pass to homogeneous coordinates, in order to take into account these points.

Lemma 2. *Let*

$$F(x, y) = 0 \tag{3}$$

be the equation of a (plane) algebraic curve of order n without singularities. Then, without changing the order of the curve or its topological structure, we can change its equation in such a way that system (2) *will have* $(n-1)^2$ *different finite solutions, real or imaginary, and for any two different real solutions* (x_1, y_1) *and* (x_2, y_2) *of* (2) *we have*

$$F(x_1, y_1) \neq F(x_2, y_2) .$$

---

[3]In what follows, we limit ourselves to equations of algebraic curves with real coefficients.

Proof. By Lemma 1, if we vary the coefficients of equation (3) to such a small extent that no singularities arise, then the curve does not change its topological structure (if, as we have assumed, the curve (3) has no singular points). On the other hand, the conditions which guarantee system (2) to have infinite solutions, or multiple solutions, or infinitely many solutions, are expressed in terms of certain polynomials of the coefficients of (3) being equal to zero. Therefore, unless these polynomials vanish identically, we can always slightly vary the coefficients in such a way that the modified equation satisfies the conditions of Lemma 2. The fact that the said polynomials do not vanish identically can be easily established by constructing particular functions satisfying the conditions of Lemma 2.

Lemma 3. *Let (x_0, y_0) be a real finite critical point of the function $F(x, y)$, i.e., a point satisfying system (2). If*

$$D(x_0, y_0) = \begin{vmatrix} \dfrac{\partial^2 F}{\partial x^2} & \dfrac{\partial^2 F}{\partial x \partial y} \\ \dfrac{\partial^2 F}{\partial x \partial y} & \dfrac{\partial^2 F}{\partial y^2} \end{vmatrix} > 0$$

at this point (we call such critical points plus-points), then, as C varies from $F(x_0, y_0) + \varepsilon$ to $F(x_0, y_0) - \varepsilon$ ($\varepsilon > 0$), the difference $p - m$ increases by 1, where p is the number of positive ovals of the curve $F(x, y) = C$, and m is the number of negative[4] ovals of this curve.

If $D(x_0, y_0) < 0$ (we call such critical points minus-points), and the polynomial $F(x, y)$ is of even order, then, as C varies from $F(x_0, y_0) + \varepsilon$ to $F(x_0, y_0) - \varepsilon$, the difference $p - m$ always decreases by 1, except once, when it decreases by 2.

Finally, if $D(x_0, y_0) < 0$ and $F(x, y)$ is of odd order, then the difference $p - m$ either decreases by 1 or remains fixed.

It is supposed here that all critical values of $F(x, y)$ (i.e., its values at critical points) are different and ε is so small that the interval

$$(F(x_0, y_0) - \varepsilon, \ F(x_0, y_0) + \varepsilon)$$

contains no critical values of $F(x, y)$ other than $F(x_0, y_0)$.

Proof. If $D(x_0, y_0) > 0$, then at the critical point (x_0, y_0) the function $F(x, y)$ has a maximum or a minimum. If $F(x_0, y_0)$ is a maximum, then *a new positive oval arises* as C, while decreasing, passes through the critical value. If $F(x_0, y_0)$ is a minimum of $F(x, y)$, then *a negative oval disappears* as C passes through the value $F(x_0, y_0)$.

---

[4]Cf. Introduction.

If $D(x_0, y_0) < 0$, then (x_0, y_0) is a saddle point for the function $F(x, y)$. We must distinguish here between the case of n even and the case of n odd.

If the order n of $F(x, y)$ is even, then there are three possibilities:

1. A positive oval touches another oval (positive or negative). In this case *a positive oval disappears.*

2. An oval (positive or negative) touches itself. Here we must distinguish between two subcases:

2a. An "odd branch" [6] can be traced in the region where $F(x, y) > F(x_0, y_0) + \varepsilon$ or in the region where $F(x, y) < F(x_0, y_0) - \varepsilon$ ($\varepsilon > 0$). Then, as C varies from $F(x_0, y_0) + \varepsilon$ to $F(x_0, y_0) - \varepsilon$, *a new negative oval appears.*

2b. Although we can trace odd branches in the regions $F(x, y) > F(x_0, y_0) - \varepsilon$ and $F(x, y) < F(x_0, y_0) + \varepsilon$, no odd branches can be traced either in the region $F(x, y) < F(x_0, y_0) - \varepsilon$ or in the region $F(x, y) > F(x_0, y_0) + \varepsilon$. It is evident that this case may present itself *at most once* as C varies from $-\infty$ to $+\infty$. When this happens, *a positive oval turns into a negative one* as C passes from $F(x_0, y_0) + \varepsilon$ to $F(x_0, y_0) - \varepsilon$.

The case 2a (resp., the case 2b) can also be defined as describing the situation when the sign of the outer ovals (i.e., those not contained in the other ovals of the curve) does not change (resp., changes from $+$ to $-$).

Two negative ovals cannot come into contact when C decreases, since they move away from each other.

If n is odd (and $D(x_0, y_0) < 0$), we can distinguish the following cases:

1. A positive oval touches another oval or the odd branch. Then this *positive oval disappears.*

2. A (positive or negative) oval touches itself. Then *a new negative oval arises.*

3. A zero-oval or the odd branch touches itself or another zero-oval. In order to describe, in the simplest manner, all possibilities which can arise in this case, we consider the projection of our projective plane upon a hemisphere S bounded by a great circle L representing the line at infinity. Consider the set of all points of this hemisphere which correspond to the points of the plane where $F(x, y) > 0$. This (open) set contains a number of regions $G_1, G_2, \ldots, G_{k_1}$, whose boundaries contain segments $l_1, l_2, \ldots, l_{k_1}$ of L. When C decreases, these regions expand, while the segments $l_1, l_2, \ldots, l_{k_1}$ remain intact. Now, if the boundary of one of these regions G_i touches itself, *a new negative oval arises*; if it touches the boundary of another such region, then *the regions coalesce and a zero-oval disappears, or a new zero-oval arises* (the later is a zero-oval since it crosses the line at infinity). In this case the number of the regions G_i decreases by 1. Thus, in all three cases Lemma 3 is proved.

Lemma 4. *Suppose that the curve $F(x, y) = C$ meets the line at infinity at k different points (obviously, k does not depend on C), and all critical points[5] of $F(x, y)$ are finite and different; then the difference D between the number of minus-points and the number of plus-points of this function is equal to $k - 1$.*

Proof. Denote by C_M the maximal critical value of $F(x, y)$ and by C_m its minimal critical value. Consider again the projection of the plane (x, y) upon the hemisphere S bounded by the great circle corresponding to the line at infinity. Let M_c be the set of all points of the hemisphere which correspond to the points (x, y) of the plane such that $F(x, y) > C$.

If $k \geq 1$ then, for $C > C_M$, the set M_c consists of k regions $G_1, \ldots, G_k$, k being the number of (real) points at which the curve meets the line at infinity; and, for $C < C_m$, the set M_c consists of a single simply connected region. Therefore, when C varies from $C > C_M$ to $C < C_m$, all k regions G_i must coalesce. This would give us exactly $k - 1$ minus-points, since two of the regions G_i can coalesce only when C passes through a minus-point. Moreover, new ovals may arise and subsequently disappear while C varies from $C_M + \varepsilon$ to $C_m - \varepsilon$. A new positive oval can arise only at a plus-point and disappear at a minus-point, while a negative oval arises at a minus-point and disappears at a plus-point; at every plus-point a positive oval arises or a negative one disappears; at a minus point either a negative oval arises or a positive one disappears or two of the initial domains G_i merge; all these phenomena do not affect the difference D, which remains equal to $k - 1$.

If $k = 0$ then, for $C > C_M$, the curve consists of a single negative oval; and, for $C < C_m$, the curve is imaginary, containing no real points of the plane; or conversely, for $C > C_M$, the curve is imaginary; and, for $C < C_m$, it consists of a single positive oval. In both cases there must be a plus-point at which this oval arises or disappears. All other ovals, arising and subsequently disappearing as C varies from $C_M + \varepsilon$ to $C_m - \varepsilon$, give rise to an equal number of plus and minus points. Therefore, the difference D is equal to -1 and the statement of Lemma 4 holds in this case, too.

Lemma 5. [7] *Let $F_1(x, y)$ and $F_2(x, y)$ be two polynomials of degree n in x, y. Assume that F_1 and F_2 vanish simultaneously at exactly n^2 different (finite) points; and let $f(x, y)$ be a polynomial of degree $l < n$ in x, y which does not vanish identically. Then $f(x, y)$ cannot vanish at more than nl points at which both $F_1(x, y)$ and $F_2(x, y)$ vanish.*

Proof. The polynomials F_1 and F_2 have no common divisor, since they vanish simultaneously on a discrete set of points. Denote by $M_1(x, y)$ the greatest common divisor of F_1 and f; and by $M_2(x, y)$ the greatest common

---

[5]The number of these points is finite.

divisor of F_2 and f. Let $M_1(x,y)$ and $M_2(x,y)$ be of degree n_1 and n_2, respectively (n_1 and n_2 may, of course, be equal to 0). Since F_1 and F_2 have no common divisor, we can write

$$F_1(x,y) = M_1(x,y)\widetilde{M}_1(x,y) ,$$

$$F_2(x,y) = M_2(x,y)\widetilde{M}_2(x,y) ,$$

$$f(x,y) = M_1(x,y)M_2(x,y)M(x,y) ,$$

where $\widetilde{M}_1(x,y)$, $\widetilde{M}_2(x,y)$ and $M(x,y)$ are polynomials in x,y.

The functions F_1, F_2 and f can vanish simultaneously only at those (finite) points of the plane (x,y) which satisfy at least one of the following three systems

1. $M_1(x,y) = 0 ,\quad F_2(x,y) = 0 ,$

2. $M_2(x,y) = 0 ,\quad F_1(x,y) = 0 ,$

3. $M(x,y) = 0 ,\quad F_1(x,y) = 0 \quad$ (or $F_2(x,y) = 0$).

The left-hand sides of each of these systems are relatively prime. Therefore, the first system has at most $n_1 n$ (finite) solutions, the second one at most $n_2 n$, and the third one at most $(l - n_1 - n_2)n$ (cf. [10], p.263). The sum of these numbers is equal to $ln < n^2$, since $l < n$. Thereby Lemma 5 is proved.

Lemma 6. *Let A be a fixed imaginary number different from 0; and let $f(x,y)$ be a polynomial in x,y with real coefficients. Then the condition that the real part of the product $A[f(x,y)]^2$ is equal to 0 at a given point (x_0, y_0) (real or imaginary) can be expressed in the form of a linear homogeneous equation with respect to the coefficients of the polynomial $f(x,y)$, the coefficients of this equation being real [8].*

Proof. Let $A = a + bi$ and $f(x_0, y_0) = c + di$, where a, b, c, d are real numbers. Then

$$A[f(x_0, y_0)]^2 = (a + bi)(c + di)^2$$

and

$$\Re\left\{A[f(x_0, y_0)]^2\right\} = ac^2 - ad^2 - 2bcd = c^2\left(a - 2bx - ax^2\right) ,$$

where $x = d/c$ and $\Re\{k\}$ denotes the real part of k. The equation

$$ax^2 + 2bx - a = 0$$

has (for any real a and b) at least one real finite root. Denote this root by k. Then $d = kc$ implies that $\Re\left\{A[f(x,y)]^2\right\} = 0$, and the relation $d = kc$ is a linear homogeneous equation with respect to the coefficients of $f(x,y)$; the coefficients of this equation are real, q.e.d.

§2. Two Fundamental Theorems

First Fundamental Theorem. *Let p be the number of positive ovals of a curve $F(x, y) = 0$ of an even order n, and let m be the number of its negative ovals. Then*

$$-\frac{3n^2 - 6n}{8} - \delta \leq p - m \leq \frac{3n^2 - 6n}{8} + 1 - \delta ,$$

where $\delta = 0$ if the outer ovals are positive, and $\delta = 1$ if the outer ovals are negative. This formula implies, of course, that

$$|p - m| \leq \frac{3n^2 - 6n}{8} + 1 .$$

Proof. The critical points of the function $F(x, y)$ are given by (2). On the basis of Lemma 2, we may always assume that this system possesses $(n-1)^2$ different finite solutions, and for different solutions (x_1, y_1) and (x_2, y_2) we have $F(x_1, y_1) \neq F(x_2, y_2)$. In this situation the following Euler-Jacobi's theorem holds:

$$\sum_{i=1}^{(n-1)^2} \frac{P(x_i, y_i)}{J(x_i, y_i)} = 0 , \tag{4}$$

where $J(x, y)$ is the Jacobian of (2), and $P(x, y)$ is an arbitrary polynomial in x, y of degree lower than $2n - 4$. The sum in the left-hand side of (4) is taken over all solutions of system (2) (see [7], [8]). Since system (2) admits no multiple solutions, the denominator $J(x_i, y_i)$ in every term of (4) is different from 0.

In particular, (4) holds if we take

$$P(x, y) = F_1(x, y)[f(x, y)]^2 , \tag{5}$$

where $f(x, y)$ is an arbitrary polynomial of degree $\frac{1}{2}(n - 4)$ and

$$F_1(x, y) = nF(x, y) - xF_x'(x, y) - yF_y'(x, y) .$$

The function $F_1(x, y)$ thus defined is of degree $n - 1$ at the most (this follows immediately from Euler's theorem on homogeneous functions); at any critical point of F, the function F_1 has the same sign as F. The function $f(x, y)$ can be chosen such that its coefficients are real and it vanishes at

$$\frac{1}{2}\left[\frac{n-4}{2} + 1\right]\left[\frac{n-4}{2} + 2\right] - 1$$

arbitrarily chosen critical points of $F(x, y)$. Lemma 5 shows that $f(x, y)$ cannot vanish at every critical point of $F(x, y)$.

After these preliminary remarks, we turn to the proof of the First Fundamental Theorem. Denote by k the number of points at which the curve $F(x,y) = C$ meets the line at infinity. If $k > 0$ and $C > C_M$, then this curve consists of $k/2$ positive ovals. When C decreases from $C_M + \varepsilon$ to 0, the curve acquires

$$p + \alpha - \frac{k}{2} \quad \text{new positive ovals,}$$

$$m + \beta \quad \text{new negative ovals,}$$

and loses

$$\alpha \quad \text{positive ovals,}$$

$$\beta \quad \text{negative ovals.}$$

Accordingly, C passes through the critical values of $F(x,y)$ at

$$p + \alpha + \beta - \frac{k}{2} \quad \text{plus-points (}A\text{-points) and}$$

$$m + \alpha + \beta - \delta \quad \text{minus-points (}B\text{-points)}$$

(see Lemma 3). Here $\delta = 0$ if the outer ovals [^{9}] are positive, and $\delta = 1$ if they are negative. The latter formulas giving the number of A- and B-points, proved under the assumption that $k > 0$, are also valid for $k = 0$. This fact can be proved by considering directly the two possible cases when we have at the outset $(C > C_M)$: 1) an imaginary curve $(\delta = 0)$, or 2) a negative oval $(\delta = 1)$. We have thus proved that $F(x,y) > 0$ at

$$p + \alpha + \beta - \frac{k}{2} \quad \text{plus-points and}$$

$$m + \alpha + \beta - \delta \quad \text{minus-points of } F.$$

But, according to Lemma 4, in addition to these, there exist

$$\frac{(n-1)^2 - 2\gamma + 1}{2} - p - \alpha - \beta \quad \text{plus-points (}B'\text{-points) and}$$

$$\frac{(n-1)^2 - 2\gamma - 1}{2} - m - \alpha - \beta + \frac{k}{2} + \delta \quad \text{minus-points (}A'\text{-points)}$$

at which $F(x,y) < 0$. Here 2γ denotes the number of imaginary critical points of $F(x,y)$.

At A- and A'-points we have $F/J > 0$. There are all told

$$\frac{(n-1)^2 - 1}{2} - \gamma + p - m + \delta$$

such points. At B- and B'-points we have $F/J < 0$. The number of these points is

$$\frac{(n-1)^2 + 1}{2} - \gamma - p + m - \delta .$$

On the other hand, it is easy to see that the sum of the number of A-points and the number of A'-points is greater than $\frac{1}{8}n(n-2) - \gamma - 1$. Indeed, assuming the contrary, let us choose $\frac{1}{8}n(n-2)$ coefficients of the polynomial $f(x, y)$ entering (4) and (5) in such a way that it vanishes at all A- and A'-points and, moreover, the real parts of the terms $P(x_i, y_i)/J(x_i, y_i)$ in (4) corresponding to imaginary solutions of system (2) vanish, too (we suppose that $f(x, y)$ does not vanish identically). The last condition can be fulfilled in consequence of Lemma 6. There is at least one real critical point of $F(x, y)$ where $f(x, y)$ differs from 0. Indeed, if

$$\frac{n(n-2)}{8} - \gamma - 1 \geq 0$$

(evidently, only this case needs consideration), then the number of real critical points of $F(x, y)$, which is equal to $(n-1)^2 - 2\gamma$, is greater than

$$(n-1)^2 - \frac{n(n-2)}{4} + 2 - 1 = \frac{3}{4}n^2 - \frac{3}{2}n + 2 ,$$

while the number of critical points of $F(x, y)$ at which $f(x, y)$ vanishes cannot exceed (by Lemma 5)

$$\frac{(n-1)(n-4)}{2} .$$

Then the left-hand side of (4) reduces to a sum of non-positive numbers which cannot vanish simultaneously, while the right-hand side is equal to 0; but this is impossible. In the same way we can prove that the number of B- and B'-points is greater than $\frac{1}{8}n(n-2) - \gamma - 1$.

A combination of the latter results with those obtained above [10] yields

$$-\frac{3n^2 - 6n}{8} \leq p - m + \delta \leq \frac{3n^2 - 6n}{8} + 1 , \qquad q.e.d.$$

In particular, for $n = 6$ we obtain

$$-9 - \delta \leq p - m \leq 10 .$$

For $m = 0$ we get $p \leq 10$, i.e., an algebraic curve of the sixth order cannot consist of more than 10 positive ovals; in other words, of more than 10 ovals lying exterior to each other (Hilbert-Rohn's Theorem).

Second Fundamental Theorem. *Again, let p be the number of positive ovals and m the number of negative ovals of a curve of order n without singularities; assume, this time, the order n to be odd. Then*

$$\left| m - p - \Delta + \frac{k+1}{2} \right| \le \frac{3n^2 - 4n + 1}{8} ,$$

where k is the number of real points of our curve on the line at infinity; Δ is the number of the regions G_i corresponding to the curve $F(x, y) = 0$ and defined in Lemma 3 $(1 \le \Delta \le k)$. We can also write this formula in the following, more symmetrical, form:

$$\left| m - p + \frac{\Delta^- - \Delta^+}{2} \right| \le \frac{3n^2 - 4n + 1}{8} ,$$

where $\Delta^+ = \Delta$, and $\Delta^- = k + 1 - \Delta$ is the number of the regions G_i corresponding to the curve $-F(x, y) = 0$.

Proof. Our starting point is again the identity (4) with

$$P(x, y) = F_1(x, y)[f(x, y)]^2 , \tag{6}$$

this time, $f(x, y)$ being a polynomial of degree $\frac{1}{2}(n - 5)$. Suppose that the curve

$$F(x, y) = C \tag{7}$$

meets the line at infinity at k points. Then the number k is odd and independent of C. For $C > C_M$, the curve (7) evidently consists only of its odd branch. If the curve $F(x, y) = 0$ consists (in addition to the odd branch) of p positive and m negative ovals, and Δ is the number of the regions G_i (defined in Lemma 3), then for C varying from $C_M + \varepsilon$ to 0

$$p + \alpha \quad \text{new positive ovals,}$$

$$m + \beta \quad \text{new negative ovals}$$

appear (where α and β are non-negative integers) and

$$\alpha \quad \text{positive ovals,}$$

$$\beta \quad \text{negative ovals,}$$

are lost, while the number of the regions G_i is decreased by $k - \Delta$. Consequently, there exist

$$p + \alpha + \beta \quad \text{plus-points (A-points),}$$

$$m + \alpha + \beta + k - \Delta \quad \text{minus-points (B-points),}$$

at which $F(x, y) > 0$.

From Lemma 4, it follows that in addition to these points there exist

$$\frac{(n-1)^2 - k + 1}{2} - p - \alpha - \beta - \gamma \quad \text{plus-points } (B'\text{-points}),$$

$$\frac{(n-1)^2 + k - 1}{2} - m - \alpha - \beta - \gamma - k + \Delta \quad \text{minus-points } (A'\text{-points}),$$

at which $F(x, y) < 0$; 2γ again denotes the number of imaginary critical points of $F(x, y)$.

At

$$\frac{(n-1)^2 + k - 1}{2} + p - m - \gamma - k + \Delta$$

A- and A'-points we have $F/J > 0$, while at

$$\frac{(n-1)^2 + k + 1}{2} - p + m - \gamma + k - \Delta$$

B- and B'-points we have $F/J < 0$. Hence it is easy to see that both numbers

$$\frac{(n-1)^2 + k - 1}{2} + p - m - k + \Delta,$$

$$\frac{(n-1)^2 - k + 1}{2} - p + m + k - \Delta$$

are greater than

$$\frac{1}{2}\left(\frac{n-5}{2} + 1\right)\left(\frac{n-5}{2} + 2\right) - 1 = \frac{(n-1)(n-3)}{8} - 1.$$

Indeed, assuming the contrary, we can choose the $\frac{1}{8}(n-1)(n-3)$ coefficients of the function $f(x, y)$ in such a way that f does not vanish identically, but vanishes at all A- and A'-points (resp., at all B- and B'-points) and, moreover, the real parts of the complex terms $P(x_i, y_i)/J(x_i, y_i)$ in (4) corresponding to the imaginary critical points of $F(x, y)$ vanish, too. Lemma 6 shows that such a choice is possible. Using Lemma 5, we can prove (in the same way as we proved a similar assertion above; see the proof of the First Fundamental Theorem) that the function $f(x, y)$ differs from 0 no less than at one real critical point of $F(x, y)$. Then the left-hand side of (4) is a sum of real terms having the same sign, one of the terms, at least, being different from 0. Therefore, the sum cannot be equal to 0. Hence

$$\left| m - p - \Delta + \frac{k+1}{2} \right| \leq \frac{3n^2 - 4n + 1}{8}, \tag{8}$$

which completes the proof of our theorem.

§3. Discussion of the Results; Examples

First of all, we should prove that *the bounds obtained for* $|p - m|$, *viz.*,

$$|p - m| \leq \frac{3n^2 - 6n}{8} + 1 \quad \text{for} \quad n \quad \text{even}$$

and

$$\left| m - p - \Delta + \frac{k+1}{2} \right| \leq \frac{3n^2 - 4n + 1}{8} \quad \text{for} \quad n \quad \text{odd}$$

are the best possible.

To this end, for every even n, we construct an algebraic curve of order n consisting of $\frac{1}{8}(3n^2 - 6n) + 1$ ovals lying exterior to each other; and, for every odd n, we construct a curve of order n consisting of $\frac{1}{8}(3n^2 - 4n + 1) - \frac{1}{2}(n - 1)$ positive ovals lying exterior to each other, all of them belonging to one and the same region bounded by the odd branch and the line at infinity; moreover, the latter curve shall be such that the number of regions G_i defined in Lemma 3 is equal to n (odd). The process we use for the construction of these curves is a slight modification of that used by Harnack for M-curves. The existence of such curves will be proved by induction.

First of all, such curves exist for $n = 1$ and $n = 2$.

Assume now that for a certain even n there exists a curve $a^{(n)} = 0$ of order n consisting of $\frac{1}{8}(3n^2 - 6n) + 1$ ovals lying exterior to each other. Assume, further, that one of these ovals crosses a straight line $a^{(1)} = 0$ at two real points $A_n^{(1)}$ and $A_n^{(n)}$ and touches the same line at $\frac{1}{2}(n - 2)$ points $A_n^{(2i)} \equiv A_n^{(2i+1)}$ $\left(i = 1, 2, \ldots, \frac{1}{2}(n - 2) \right)$; we assume that the order of the points $A_n^{(1)}$, $A_n^{(2i)}$, $A_n^{(n)}$ on $a^{(1)} = 0$ coincides with the order of their indices. In the initial case, $n = 2$, we can take for $a^{(2)} = 0$ an ellipse meeting the line $a^{(1)} = 0$ at two different real points.

Consider the curve

$$a^{(n+1)} \equiv a^{(1)} a^{(n)} + \lambda q^{(n+1)} = 0$$

(here λ is a real constant and $q^{(n+1)} = 0$ is the equation, with real coefficients, of an algebraic curve of order $n + 1$ without real singular points) which meets the line $a^{(1)} = 0$ at $n+1$ different points $A_{n+1}^{(1)}$, $A_{n+1}^{(2)}, \ldots, A_{n+1}^{(n+1)}$, all of them lying outside the segment $A_n^{(1)} A_n^{(2)} A_n^{(n)}$.[6]

---

[6]In order to construct such a curve, take any real algebraic curve of order $n + 1$ meeting the line $a^{(1)} = 0$ at $n + 1$ different real points none of which lie on the segment $A_n^{(1)} A_n^{(2)} A_n^{(n)}$. If this curve has singularities, we can slightly vary its equation, so that the new curve becomes free from singularities; the displacement of the points of intersection of the curve and the line $a^{(1)} = 0$ will be arbitrarily small, so that these points will remain real, distinct, and exterior to the segment $A_n^{(1)} A_n^{(2)} A_n^{(n)}$.

Then, taking λ sufficiently small, we can choose its sign in such a way that the curve $a^{(n+1)} = 0$ consists of

1) an odd branch meeting the line $a^{(1)} = 0$ at $n+1$ points $A_{n+1}^{(1)}, A_{n+1}^{(2)}, \ldots,$ $A_{n+1}^{(n+1)}$ and

2) ovals lying in one and the same region bounded on the projective plane by the odd branch of the curve $a^{(n+1)} = 0$ and the line $a^{(1)} = 0$, the number of these ovals being equal to

$$\frac{3n^2 - 6n}{8} + \frac{n}{2} = \frac{3n^2 - 2n}{8} = \frac{3(n+1)^2 - 4(n+1) + 1}{8} - \frac{(n+1) - 1}{2}.$$

Moreover, if all these ovals are positive, then $n + 1$ is the number of the regions G_i (defined as in Lemma 3) bounded by the line $a^{(1)} = 0$ (which plays here the part of the line at infinity) and the odd branch of the curve $a^{(n+1)} = 0$. All these constructions are represented schematically in Fig. 1 where the curve $a^{(n+1)} = 0$ is indicated by the thicker line and the curves $a^{(1)} = 0$, $a^{(n)} = 0$, and $q^{(n+1)} = 0$, by thinner lines.

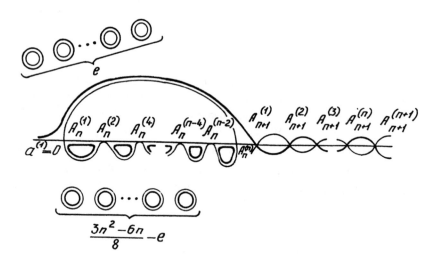

Fig. 1.

Remark. We give here a detailed proof of the fact that for sufficiently small $\lambda \neq 0$ the curve $a^{(n+1)} = 0$ has no singularities and contains no ovals except those shown in Fig. 1.

We begin with the first of these statements, namely, that the curve $a^{(n+1)} = 0$ has no singular points. We say that λ is a *critical* value if the curve $a^{(n+1)} \equiv a^{(1)}a^{(n)} + \lambda q^{(n+1)} = 0$ has critical points. Evidently, our statement will be proved if we show that there is only a finite number

of critical values of λ. A value of λ is critical if it satisfies the following
equations (with respect to x, y, λ)

$$a^{(n+1)} = 0 , \qquad \frac{\partial a^{(n+1)}}{\partial x} = 0 , \qquad \frac{\partial a^{(n+1)}}{\partial y} = 0 . \qquad (9)$$

Eliminating x and y, we obtain an algebraic equation for λ. Therefore, in
order to prove our statement (that there is only a finite number of critical
values of λ), it suffices to demonstrate that this equation is not satisfied
identically for all values of λ. This is obvious, since the said equation is not
satisfied for $\lambda = \infty$, the curve $q^{(n+1)} = 0$ having no singularities. It follows
that for λ_0 sufficiently small there are, in the interval $(-\lambda_0, +\lambda_0)$, no critical
values of λ with the exception of $\lambda = 0$; therefore, all curves $a^{(n+1)} = 0$,
$0 < \lambda < \lambda_0$, have the same topological structure, and this is also the case
for all curves $a^{(n+1)} = 0, -\lambda_0 < \lambda < 0$.

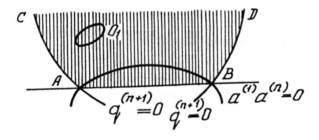

Fig. 2.

On the other hand, it is evident that for small values of $|\lambda|$ the curve
$a^{(n+1)} = 0$ passes "near" the curve $a^{(1)}a^{(n)} = 0$. Here the word "near" is
understood in the following sense: if we project the projective plane on a
hemisphere, then the image of the curve $a^{(n+1)} = 0$ will lie in a neighborhood
(in the ordinary sense) of the image of the curve $a^{(1)}a^{(n)} = 0$. Evidently, the
curve $a^{(n+1)} = 0$ lies on that side of every portion of the curve $a^{(1)}a^{(n)} = 0$
where $a^{(1)}a^{(n)}$ and $\lambda q^{(n+1)}$ have opposite signs. Let us consider one of these
portions and continue it until it crosses the curve $q^{(n+1)} = 0$. While passing
through a point $A_n^{(i)}$, we keep the continued curve near the thick line of
Fig. 1. The segment AB of the curve $a^{(1)}a^{(n)} = 0$ is schematically represented in Fig. 2; its end-points A and B belong to the curve $q^{(n+1)} = 0$; the
shaded portion represents the part of the plane where $a^{(1)}a^{(n)}$ and $\lambda q^{(n+1)}$

have opposite signs. This region of the plane we denote by G. The thick line represents the curve $a^{(n+1)} = 0$. If this curve (for small $|\lambda|$) contains any other oval O_1, in addition to those shown in Fig. 1, then this oval must lie entirely within the region G (Fig. 1) or another such region. Suppose now that $\lambda \to 0$. Then, only the following two cases are possible (since $a^{(1)}a^{(n)}$ and $q^{(n+1)}$ preserve their sign in G):

First case. The oval O_1 contracts. Then this oval cannot disappear without passing through a critical point. Therefore (since there are no critical values of λ within the interval under consideration) it must approach either an oval or a point, which has to be singular for the curve $a^{(1)}a^{(n)} = 0$. Both these situations are impossible.

Second case. The oval O_1 expands. Then it cannot approach the segment AB of the curve $a^{(1)}a^{(n)} = 0$ so as to merge into that segment at $\lambda = 0$.

To conclude our remark, we point out that the ovals of the curve $a^{(n+1)} = 0$, which in Fig. 1 are lying within the corresponding ovals of the curve $a^{(n)} = 0$, may happen to have the opposite mutual position, i.e., the oval of the curve $a^{(n+1)} = 0$ may contain the oval of the curve $a^{(n)} = 0$ in its interior. These ovals can also have a non-empty intersection; this would happen if they are crossed by the curve $q^{(n+1)} = 0$ (then the intersection points of the two ovals would be those at which the curve $q^{(n+1)} = 0$ crosses one of these ovals). Similar observations apply to all other figures in this paper. ☐

Suppose now that for a certain *odd* n there exists a curve $a^{(n)} = 0$ of order n consisting of:
1) an odd branch crossing the line $a^{(1)} = 0$ at n different real points

$$A_n^{(1)}, A_n^{(2)}, \ldots, A_n^{(n)},$$

2) $\frac{1}{8}(3n^2 - 4n + 1) - \frac{1}{2}(n-1)$ ovals exterior to each other and lying in the same region bounded by the odd branch and the line $a^{(1)} = 0$.
Consider the curve

$$a^{(n+1)} \equiv a^{(1)}a^{(n)} + \lambda q_{n+1}^{(1)} q_{n+1}^{(2)} \cdots q_{n+1}^{(n+1)} = 0,$$

where λ is a constant; $q_{n+1}^{(i)} = 0$ ($i = 1,\ldots,n+1$) are the equations of the straight lines crossing the line $a^{(1)} = 0$ at the respective points $A_{n+1}^{(1)}, A_{n+1}^{(2)}, \ldots, A_{n+1}^{(n+1)}$ outside the segment $A_n^{(1)} A_n^{(2)} A_n^{(n)}$. We also assume that $q_{n+1}^{(2i)} \equiv q_{n+1}^{(2i+1)}$ for $i = 1,\ldots,\frac{1}{2}(n-1)$, while the points $A_{n+1}^{(1)}, A_{n+1}^{(2i)}, A_{n+1}^{(n+1)}$ are mutually different.
Then we can choose the sign of λ in such a way that for small enough $|\lambda|$ the curve $a^{(n+1)} = 0$ will consist of

$$\frac{3n^2 - 4n + 1}{8} - \frac{n-1}{2} + n = \frac{3(n+1)^2 - 6(n+1)}{8} + 1$$

ovals lying exterior to each other; the line $a^{(1)} = 0$ crosses one of these ovals at two points $A_{n+1}^{(1)}$ and $A_{n+1}^{(n+1)}$ and touches this oval at $\frac{1}{2}(n-1)$ more points $A_{n+1}^{(2i)}$, $i = 1, \ldots, \frac{1}{2}(n-1)$, while it does not meet the other ovals of the curve.

This construction is outlined schematically in Fig. 3. Here again (as in Fig. 1) the thicker line indicates the curve $a^{(n+1)} = 0$, while the thinner one the curve $a^{(n)} = 0$ and the lines $a^{(1)} = 0$, $q_{n+1}^{(i)} = 0$.[7]

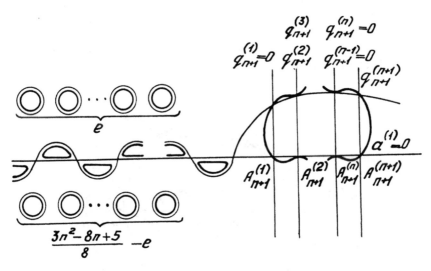

Fig. 3.

Next, we apply the inequality obtained for n even (see the First Fundamental Theorem) *to the case of an algebraic curve (of order n) formed by a number of ovals exterior to each other and an outer oval containing in its interior all the other ovals.*

Changing the sign of $F(x, y)$, if necessary, we can make the outer oval positive; all the other (inner) ovals will then be negative. Then we must take $\delta = 0$ in the First Fundamental Theorem. Denoting by S the number of inner (negative) ovals, we obtain (the number of positive ovals being equal to 1) the following inequality

---

[7]We can prove, just as in the case of n even (see Remark), that for sufficiently small $|\lambda|$ the curve $a^{(n+1)} = 0$ will have no singularities and contain no ovals other than those shown in thicker line in Fig 3. In this case we see that it is not for every λ that system (9) has a solution, because of the Bertini theorem [11].

$$S \leq \frac{3n^2 - 6n}{8} + 1 . \tag{10}$$

On the other hand, for any $n = 4k + 2$ (here k is a non-negative integer) there exists an algebraic curve of order n consisting of an outer oval and $S = \frac{1}{8}(3n^2 - 6n)$ ovals lying interior to it and exterior to each other [12].

To prove this statement we apply Hilbert's process with some modifications. We proceed by induction. Suppose that for a certain $n = 4k + 2$ there exists a curve $a^{(n)} = 0$ of order n which is formed by an outer oval containing in its interior $\frac{1}{8}(3n^2 - 6n)$ other ovals lying outside each other. Moreover, we assume that the outer oval crosses an ellipse $b^{(2)} = 0$ at two real points $A_n^{(1)}$ and $A_n^{(2n)}$ and touches it internally at $n-1$ points $A_n^{(2)}, A_n^{(4)}, \ldots, A_n^{(2n-2)}$ which coincide with $A_n^{(3)}, A_n^{(5)}, \ldots, A_n^{(2n-1)}$ (see Fig. 4, thinner line), respectively, and all belong to one of the two arcs of the ellipse $b^{(2)} = 0$ having their ends at the points $A_n^{(1)}$ and $A_n^{(2n)}$ [13].

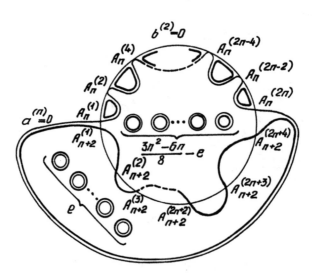

Fig. 4.

In the initial case, $n = 2$, we must take as $a^{(2)} = 0$ an ellipse touching another one $b^{(2)} = 0$ internally (at the point $A_n^{(2)}$) and crossing the latter ellipse at two other points $A_2^{(1)}$ and $A_2^{(4)}$.

On the arc $A_n^{(1)}A_n^{(2n)}$ of the ellipse $b^{(2)} = 0$ which does not contain the points $A_n^{(i)}$, let us take $2n + 4$ different points $A_{n+2}^{(1)}, \ldots, A_{n+2}^{(2n+4)}$ and consider a curve $q^{(n+2)} = 0$ of order $n + 2$ passing through these points and having no real singular points. This curve can be constructed, for instance, if we slightly vary the coefficients of the equation

$$q_{n+1}^{(1)} q_{n+2}^{(2)} \cdots q_{n+2}^{(n+2)} = 0 \,,$$

where $q_{n+2}^{(i)} = 0$ $(i = 1, \ldots, n + 2)$ is the equation of the straight line passing through the neighboring points $A_{n+2}^{(i)}$. Then, for λ sufficiently small in absolute value and having a suitable sign, the curve given by the equation

$$a^{(n+2)} \equiv b^{(2)} a^{(n)} + \lambda q^{(n+2)} = 0$$

has the form resembling that of the curve drawn in thicker line in Fig. 4 and in thinner line in Fig. 5.

This curve consists of:

1) $\frac{1}{8}(3n^2 - 6n) - l + n$ ovals lying exterior to each other and within the ellipse $b^{(2)} = 0$. Here l denotes the number of ovals of the curve $a^{(n)} = 0$ lying outside the ellipse $b^{(2)} = 0$;

2) an oval crossing the ellipse $b^{(2)} = 0$ at $2b + 4$ different real points $A_{n+2}^{(i)}$;

3) l ovals lying within the preceding oval and outside the ellipse.

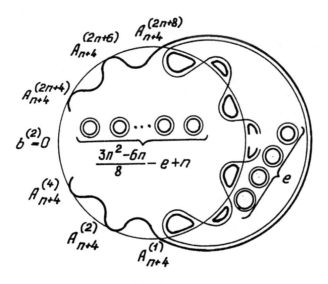

Fig. 5.

For sufficiently small $|\lambda|$, the curve $a^{(n+2)} = 0$ will have neither singular points nor ovals other than those shown in Fig. 5. This fact can be proved in the same way as a similar statement established in the above Remark.

On the arc $A_{n+2}^{(1)} A_{n+2}^{(2n+4)}$ of the ellipse $b^{(2)} = 0$, containing no other points $A_{n+2}^{(i)}$, we take $2n + 8$ points

$$A_{n+4}^{(1)}, A_{n+4}^{(2)}, \ldots, A_{n+4}^{(2n+8)}$$

such that the points $A_{n+4}^{(2i)}$ and $A_{n+4}^{(2i+1)}$ $(i = 1, 2, \ldots, n+3)$ coincide, while all the points $A_{n+4}^{(1)}, A_{n+4}^{(2)}, A_{n+4}^{(4)}, \ldots, A_{n+4}^{(2n+6)}, A_{n+4}^{(2n+8)}$ are different (see Fig. 5). Let us join the points $A_{n+4}^{(1)}$ and $A_{n+4}^{(2)}$; $A_{n+4}^{(3)}$ and $A_{n+4}^{(4)}$; $\ldots$; $A_{n+4}^{(2n+7)}$ and $A_{n+4}^{(2n+8)}$ by the straight lines

$$q_{n+4}^{(1)} = 0 \; ; \quad q_{n+4}^{(2)} = 0 \; ; \quad \ldots ; \quad q_{n+4}^{(n+4)} = 0 \; .$$

Then, for λ with a sufficiently small absolute value and suitably chosen sign, the curve (Fig. 5, thicker curve)

$$a^{(n+4)} \equiv b^{(2)} a^{(n+2)} + \lambda q_{n+4}^{(1)} q_{n+4}^{(2)} \cdots q_{n+4}^{(n+4)} = 0$$

consists of

$$\frac{3n^2 - 6n}{8} + n + 2n + 3 = \frac{3(n+4)^2 - 6(n+4)}{8}$$

ovals exterior to each other and interior to another oval which crosses the ellipse $b^{(2)} = 0$ at two points $A_{n+4}^{(1)}$ and $A_{n+4}^{(2n+8)}$ and touches this ellipse internally at $(n+3)$ other points $A_{n+4}^{(2i)}$ $(i = 1, \ldots, n+3)$, all of which lie on the same arc $A_{n+4}^{(1)} A_{n+4}^{(2n+8)}$ of the ellipse. Consequently, our curve $a^{(n+4)} = 0$ of order $n+4$ possesses all the properties required, q.e.d.[8]

In the above mentioned work on the curves of order 6, Rohn [4] proves that such a curve cannot consist of an outer oval and ten other ovals interior to the outer one and exterior to each other. This shows that the upper bound for the number of such ovals, which we have found to be

$$\frac{3n^2 - 6n}{8} + 1 \; ,$$

cannot be attained in the case of $n = 6$, nor in the case of $n = 2$ [14].

For $n = 4k$ (k is integer), there exist algebraic curves of order n consisting of an outer oval and

$$S = \frac{3n^2 - 6n}{8} - 2$$

more ovals exterior to each other and lying within the outer oval.

To start the induction, consider the case $k = 1$ $(n = 4)$. Let $a^{(2)} = 0$ and $b^{(2)} = 0$ be the equations of two ellipses, of which the second one lies in the interior of the first. Take 5 different points $A_4^{(1)}, A_4^{(2)} \equiv A_4^{(3)}, A_4^{(4)} \equiv$

---

[8]The fact that, for sufficiently small $|\lambda|$, the curve $a^{(n+4)} = 0$ contains no ovals other than those indicated in Fig. 5 and has no singular points can be established in the same way as similar statements proved above (see Remark and footnote 7).

$A_4^{(5)}$, $A_4^{(6)} \equiv A_4^{(7)}$, and $A_4^{(8)}$ on the ellipse $b^{(2)} = 0$ and draw four lines through the points $A_4^{(1)}$ and $A_4^{(2)}$ $\left(q_4^{(1)} = 0\right)$; $A_4^{(3)}$ and $A_4^{(4)}$ $\left(q_4^{(2)} = 0\right)$; $A_4^{(5)}$ and $A_4^{(6)}$ $\left(q_4^{(3)} = 0\right)$; $A_4^{(7)}$ and $A_4^{(8)}$ $\left(q_4^{(4)} = 0\right)$. Then the curve

$$a^{(4)} \equiv a^{(2)}b^{(2)} + \lambda q_1 q_2 q_3 q_4 = 0,$$

for λ with a sufficiently small absolute value and suitably chosen sign, consists of two ovals (one within another) close to the ellipses $a^{(2)} = 0$ and $b^{(2)} = 0$. The inner oval crosses the ellipse $b^{(2)} = 0$ at two points $A_4^{(1)}$ and $A_4^{(8)}$ and touches it from the outside at three points $A_4^{(2)}$, $A_4^{(4)}$, $A_4^{(6)}$ belonging to the same arc $A_4^{(1)}A_4^{(8)}$ of the ellipse.

Suppose now that for a certain $n = 4k$ we have constructed an algebraic curve $a^{(n)} = 0$ of order n consisting of an outer oval and $\frac{1}{8}(3n^2 - 6n) - 2$ other ovals lying within the outer one and outside each other; let one of these ovals be such that its crosses the ellipse $b^{(2)} = 0$ at two points $A_n^{(1)}$ and $A_n^{(2n)}$ and touches it from the outside at $n - 1$ points $A_n^{(2)}, A_n^{(4)}, \ldots, A_n^{(2n-2)}$ coinciding with $A_n^{(3)}, A_n^{(5)}, \ldots, A_n^{(2n-1)}$, respectively, and belonging to the same arc $A_n^{(1)}A_n^{(2n)}$ of the ellipse (see Fig. 6, thinner curve).

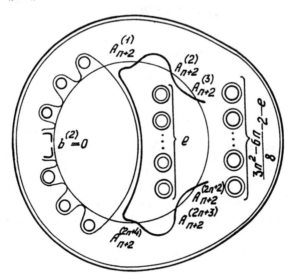

Fig. 6.

Let us take on the other arc $A_n^{(1)}A_n^{(2n)}$ (i.e., on the arc $A_n^{(1)}A_n^{(2n)}$ that does not contain $A_n^{(2)}, A_n^{(4)}, \ldots, A_n^{(2n-1)}$) $2n + 4$ different points $A_{n+2}^{(1)}, A_{n+2}^{(2)}, \ldots,$ $A_{n+2}^{(2n+4)}$. Through these points we draw an algebraic curve $q^{(n+2)} = 0$ of order $n + 2$ without singularities. Then, for λ with a sufficiently small absolute value and a suitable sign, the curve

$$a^{(n+2)} \equiv a^{(n)}b^{(2)} + \lambda q^{(n+2)} = 0$$

will have a form similar to that outlined in Fig. 6 (thicker curve) and Fig. 7 (thinner curve).

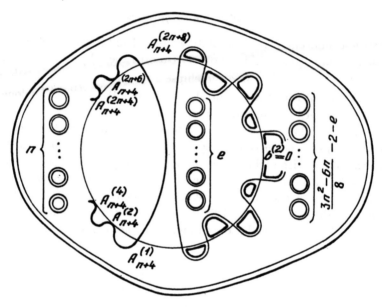

Fig. 7.

Next, on the arc $A^{(1)}_{n+2}A^{(2n+4)}_{n+2}$ forming part of the ellipse $b^{(2)} = 0$ and containing no other points $A^{(i)}_{n+2}$, we take $n+5$ different points $A^{(1)}_{n+4}, A^{(2)}_{n+4} \equiv A^{(3)}_{n+4}, A^{(4)}_{n+4} \equiv A^{(5)}_{n+4}, \ldots, A^{(2n+6)}_{n+4} \equiv A^{(2n+7)}_{n+4}, A^{(2n+8)}_{n+4}$ and draw straight lines $q^{(1)}_{n+4} = 0, q^{(2)}_{n+4} = 0, \ldots, q^{(n+4)}_{n+4} = 0$ through $A^{(1)}_{n+4}$ and $A^{(2)}_{n+4}, A^{(3)}_{n+4}$ and $A^{(4)}_{n+4}, \ldots, A^{(2n+7)}_{n+4}$ and $A^{(2n+8)}_{n+4}$, respectively. Then, for λ with a sufficiently small absolute value and a suitable sign, the curve of order $n + 4$

$$a^{(n+4)} \equiv a^{(n+2)}b^{(2)} + \lambda q^{(1)}_{n+4}q^{(2)}_{n+4} \cdots q^{(n+4)}_{n+4} = 0$$

(see Fig. 7, thicker curve) consists of an outer oval and

$$\frac{3n^2 - 6n}{8} - 2 + n + 2n + 4 - 1 = \frac{3(n + 4)^2 - 6(n + 4)}{8} - 2$$

inner ovals (all of them within the outer oval and outside each other), one of which crosses the ellipse $b^{(2)} = 0$ at two points $A^{(1)}_{n+4}$ and $A^{(2n+8)}_{n+4}$ and touches it from the outside at $n + 3$ other points $A^{(2)}_{n+4}, A^{(4)}_{n+4}, \ldots, A^{(2n+6)}_{n+4}$

coinciding with $A_{n+4}^{(3)}, A_{n+4}^{(5)}, \ldots, A_{n+4}^{(2n+7)}$, respectively, and belonging to the same arc $A_{n+4}^{(1)} A_{n+4}^{(2n+8)}$.[9]

Similar constructions, as well as inequalities of type (10), can be obtained, of course, in the case of n odd.

I close this paper with an observation that in all instances known to the author not only do we have

$$|p - m| \leq \frac{3n^2 - 6n}{8} + 1 \quad \text{for } n \text{ even,}$$

and

$$|p - m| \leq \frac{3n^2 - 4n + 1}{8} + \left| \frac{k+1}{2} - \Delta \right| \quad \text{for } n \text{ odd,}$$

but the same inequalities are valid for p and m separately. We could have proven this fact $[^{15}]$, had there been a method allowing to eliminate an arbitrary oval of a curve of order n, without affecting either its order or the topology of the other ovals. All curves constructed by the processes of Harnack and Hilbert admit such elimination of ovals. In fact, the above examples amount to such Harnack's or Hilbert's curves with some of their ovals wiped out. But whether it is possible to eliminate ovals (without affecting the topology of other ovals) for an arbitrary algebraic curve remains unknown to the author.

In the same manner, we could prove the inequality $S \leq \frac{1}{8}(3n^2 - 6n))$, if it were possible to make the outer oval, being expanded, touch itself without affecting the topology of the other ovals. Then, either a new negative oval would appear or the outer positive oval would become negative. In both cases, the inequality $S \leq \frac{1}{8}(3n^2 - 6n)$ would be established.

References

[1] Harnack. *Math. Ann.* **10** (1876) 189–199.
[2] Hilbert, D. *Math. Ann.* **38** (1891) 115–138.
[3] Hilbert, D. *Nachr. Kgl. Ges. Wiss. Göttingen*, 1900, 253–297.
[4] Rohn, K. Die ebene Kurven 6 Ordnung mit elf Ovalen. *Ber. Verhandl. Leipzig.* **63** (1911) 540–555.
[5] Petrowsky, I.G. Sur la topologie des courbes planes réelles et algébriques. *C.R. Acad Sci. Paris* **197** (1933) 1270–1272.
[6] Euler, L. *Inst. Calc. Integr. Petrop.* **2** (1768–70) 1169.
[7] Jacobi, C.G.J. *Gesammelte Werke.* Bd. 3. Berlin, 1885.

---

[9]In the same way as it has been done for $n = 4k + 2$, we can prove that the curve $a^{(n+4)} = 0$ does not have a more complicated topological structure, i.e., it does not have any ovals other than those represented in Fig. 7, nor does it have singular points.

[8] Kronecker, L. Über einige Interpolationsformeln für ganze Funktionen mehrer Variabeln. *Werke.* Bd. 1 (1895) 133–141.

[9] Morse, M. Critical points. *Trans. Amer. Math. Soc.* **27** (1925) 345–346.

[10] Kronecker'sche Methoden. In: *Encyklopädie der Mathematischen Wissenschaften.* Bd. 1. 1899.

NOTES

(by V.M. Kharlamov)

[1] The projective space is meant here; it is assumed that the center of the sphere is outside the plane of the curve.

[2] The proof of both theorems suggested by Rohn is incomplete (see Gudkov, D.A. *Uspekhi Mat, Nauk* **29:4** (1974)). A complete proof of the first theorem (stated by Hilbert) has been given by Petrowsky in the present article. A complete proof of the second theorem is due to Gudkov (*Ucheniye Zapiski Gork. Univ.* **87** (1968)).

[3] Apparently, the author means $p - m$ instead of $|p - m|$.

[4] For $n = 6$, the estimate given by Petrovski takes the form $m \leq 10$, while Rohn's result (see note 2) yields $m \leq 9$.

[5] It is assumed that a straight line is fixed on the projective plane, the so-called *line at infinity*. Ovals that do not meet this line are called *finite*.

[6] The author does not mean branches of the considered algebraic curves, but rather refers to arbitrary topological circles belonging to the projective plane in a unilateral fashion and crossing the line at infinity at finitely many points. The domain is thought of as containing such a circle, if it contains all its points but those at infinity.

[7] Lemma 5 deals with a complex domain.

[8] Lemma 6 gives a sufficient condition (it is by no means necessary) for $\Re\{A[f(x_0, y_0)]^2\}$ to be equal to 0.

[9] Outer ovals of the curve $F(x, y) = 0$ are meant here.

[10] The author means the formulas for the number of the points of type A, A' and the number of the points of type B, B'.

[11] See, for instance: Griffiths & Harris: *Principles of Algebraic Geometry*, New York, 1978.

[¹²] By now it has been proved that $S \leq \frac{1}{8}(3n^2 - 6n)$ for $n = 4k + 2$ (see Appendix for a supplementary article to Petrowsky's works on the topology of real algebraic manifolds, Sect. 3.1)

[¹³] These two arcs are not interchangeable, and it is important for what follows that the arc indicated in Fig. 4 is chosen.

[¹⁴] See note [¹²] above.

[¹⁵] The following conjecture is meant here: for any (non-singular) curve of order n, the numbers p and m are $\leq \frac{1}{8}(3n^2 - 6n) + 1$ for n even; and $\leq \frac{1}{8}(3n^2 - 4n + 1) + \left|\frac{k+1}{2} - \varDelta\right|$ for n odd. So far, this statement has neither been proved true nor false (see Appendix for a commentary on Petrowsky's works on algebraic manifolds, Introduction).

[¹⁶] See note [¹²] above.

8

On the Topology of Real Algebraic Surfaces[*]

(In collaboration with O.A. Oleinik)

Estimates for the Euler characteristic are obtained here for real algebraic surfaces

$$F(x_1, x_2, \ldots, x_m) = 0$$

in projective space; the surfaces are assumed to have no real singular points. We also obtain estimates for the Euler characteristic of the closure, in projective space, of the set on which the polynomial $F(x_1, x_2, \ldots, x_m)$ is non-negative.

In what follows, $F(x_1, x_2, \ldots, x_m)$ stands for a polynomial of degree n with respect to the variables $x_1, x_2, \ldots, x_m$, the coefficients of F are real. Assume that the system of equations

$$F = 0, \quad F_1 = 0, \ldots, \quad F_m = 0, \tag{1}$$

where F_k is the derivative of F with respect to x_k, admits neither finite nor infinite real solutions. Let Γ_0 be the set formed by the points $(x_1, x_2, \ldots, x_m, x_{m+1})$ belonging to the m-dimensional real projective space $\mathcal{P}_m$ and such that

$$x_{m+1}^n F\left(\frac{x_1}{x_{m+1}}, \ldots, \frac{x_m}{x_{m+1}}\right) = 0.$$

Then Γ_0 is a closed manifold of dimension $m-1$ (i.e., Γ_0 is an algebraic surface without real singular points). Denote by $E(\Gamma_0)$ the Euler characteristic of Γ_0, i.e.,

$$E(\Gamma_0) = \sum (-1)^r p^r,$$

---

[*]Originally published in *Izvestiya Akad. Nauk SSSR, Ser. Mat.* **13** (1949) 389–402. See Appendix for a commentary by V.M. Kharlamov.

where p^r is the r-dimensional Betti number modulo 2. The following inequality will be proved here for any odd m:

$$|E(\Gamma_0)| \leq (n-1)^m - 2S(m,n) + 1 . \qquad (2)$$

The number $S(m,n)$ is defined in Lemma 3. For m even, $E(\Gamma_0)$ is known to be always equal to zero.

Denote by M_0 the closure, in the projective space $\mathcal{P}_m$, of the set formed by all finite points such that

$$x_{m+1}^n F\left(\frac{x_1}{x_{m+1}}, \ldots, \frac{x_m}{x_{m+1}}\right) \geq 0 \quad \text{for} \quad x_{m+1} = 1$$

(the set $x_{m+1} = 0$ is assumed to be the plane at infinity). Then, for n even, the following estimate

$$|E(M_0)| \leq \frac{1}{2}(n-1)^m - S(m,n) + \frac{1}{2} \qquad (3)$$

holds for any m. If n is odd, then

$$|E(M_0)| \leq \frac{1}{2}(n-1)^m - S(m,n) + 1 \qquad (4)$$

for m odd; and

$$|E(M_0)| \leq \frac{1}{2}(n-1)^m - S(m,n) + \frac{1}{2}(n-1)^{m-1} - S(m-1,n) + 1 \qquad (5)$$

for m even.

For $m = 2$, the inequalities (3) and (5) have been obtained earlier by I.G. Petrowsky [1]. In this case I.G. Petrowsky constructed polynomials F which turn the inequalities (3), (5) into equalities. For $m = 3$ and $n = 4$, it is also possible to find a polynomial F such that (2) and (3) turn into equalities (see [2]). The surface Γ_0 corresponding to this F consists of 10 ovals.

Lemma 1. *The topology of the sets Γ_0 and M_0 does not change, if we vary the coefficients of the polynomial $F(x_1, x_2, \ldots, x_m)$ within a sufficiently small neighborhood of their original values.*

Proof. As usual, let us map the projective space $\mathcal{P}_m$ onto the sphere S_m in such a way that each point of $\mathcal{P}_m$ would correspond to a pair of opposite end-points of a diameter of S_m. Let Γ_0' and M_0' correspond to a polynomial $F'(x_1, x_2, \ldots, x_m)$ whose coefficients are close to those of the original polynomial $F(x_1, x_2, \ldots, x_m)$. We can easily establish a continuous one-to-one correspondence between the images of the manifolds Γ_0 and Γ_0' on the sphere

S_m, by issuing a normal $\tilde{n}$ on S_m from each point of the image of Γ_0 up to its intersection with the image of Γ_0', where diametrically opposite points correspond to diametrically opposite points. In order to show that M_0 and M_0' are homeomorphic, let us choose $\varepsilon > 0$ so small that the normals $\tilde{n}$ of length ε to the image of Γ_0 on the sphere S_m would be mutually disjoint, no matter what side of Γ_0 these normals may be directed to. Let L be a layer formed by these normal segments near the image of Γ_0. Assume that the coefficients of F' are so close to the respective coefficients of F that the image of Γ_0' belongs to the interior of L. Then it is easy to see that there exists a continuous one-to-one correspondence between the intersection of L with the image of M_0 on S_m and the intersection of L with the image of M_0' on the same sphere, such that the end-points of the normals $\tilde{n}$ remain fixed if they (the end-points) belong to the interior of the image of M_0. This, again, can be accomplished in such a manner that diametrically opposite points would correspond to diametrically opposite points. This homeomorphism can be extended to the entire M_0 and the entire M_0', if the points of M_0 outside L are left intact.

Lemma 2. *There exists an arbitrarily small variation of the coefficients of $F(x_1, x_2, \ldots, x_m)$ which yields a polynomial $F'(x_1, x_2, \ldots, x_m)$ such that the system*

$$F_1' = 0, \quad F_2' = 0, \ldots, \quad F_m' = 0 \qquad (6)$$

has $(n-1)^m$ mutually distinct finite solutions $(x_1^\alpha, x_2^\alpha, \ldots, x_m^\alpha)$. These solutions may be real or complex (they are called critical points P_α *of $F'(x_1, x_2, \ldots, x_m)$). The function F' takes different non-zero values at different critical points.*

Proof. Infinite solutions of system (6) can be obtained by equating to zero the highest order terms of each equation in (6). A system consisting of m homogeneous equations with m unknown quantities admits a non-trivial solution only if its resultant is equal to zero (see [3]). The resultant has the form of a polynomial of the coefficients of these equations, and therefore, of the coefficients of F; and it does not vanish identically. It follows that, changing the coefficients of F, if necessary, we can obtain a polynomial for which system (6) admits no infinite solutions.

We can also make system (6) to have only a finite number of solutions. Indeed, this may be the case if and only if there exist, among the resultants R_i obtained by eliminating $x_1, \ldots, x_{i-1}, x_{i+1}, \ldots, x_m$ from equations (6), functions which do not vanish identically with respect to x_i (see [3]). These resultants have the form of polynomials whose arguments are the coefficients of F and x_i, and they do not vanish identically with respect to these arguments: it is possible to construct polynomials F for which these resultants do not vanish identically in x_i. Therefore, changing, if necessary, the coefficients of F by an arbitrarily small amount, we can come to a system (6) with finitely many solutions, all of them being finite.

Now subjecting, if necessary, the coefficients of F to an additional small change, we can obtain a system (6) admitting mutually different finite solutions only. A necessary and sufficient condition for this is known to be

$$I(P_\alpha) = \frac{D(F_1, \ldots, F_m)}{D(x_1, \ldots, x_m)} \neq 0$$

at every critical point, where $I(P_\alpha)$ is the Jacobian. Therefore, it is necessary and sufficient that the product of these determinants over all critical points is different from zero. But this product, being a symmetrical function of the solutions of system (6), can be expressed as a rational function of the coefficients of (6) (see [4], p. 274), and therefore, of the coefficients of F. Examples show that this function does not vanish identically. It follows that by an arbitrarily small variation of the coefficients of F we can come to the situation when system (6) has only different solutions; their number is known to be equal to $(n-1)^m$.

Let $(x'_1, x'_2, \ldots, x'_m)$ and $(x''_1, x''_2, \ldots, x''_m)$ be two different critical points of the polynomial F, and let $I(P) \neq 0$ at these points. Assuming that the identity

$$F(x'_1, \ldots, x'_m) \equiv F(x''_1, \ldots, x''_m)$$

holds when the coefficients of F vary in a neighborhood of their fixed values $(a_1, a_2, \ldots, a_s)$ and differentiating this identity with respect to a_k, we get

$$\frac{\partial F}{\partial a_k} + \sum_j F_j(x') \frac{\partial x'_j}{\partial a_k} \equiv \frac{\partial F}{\partial a_k} + \sum_j F_j(x'') \frac{\partial x''_j}{\partial a_k} . \tag{7}$$

It follows from (7) and (6) that

$$\frac{\partial F(x')}{\partial a_k} \equiv \frac{\partial F(x'')}{\partial a_k} \quad \text{for all } k ,$$

and therefore, $x'_1 = x''_1, \ldots, x'_m = x''_m$, which contradicts our initial assumption.

Adding, if necessary, an arbitrarily small constant to $F(x_1, \ldots, x_m)$, we can always obtain a polynomial F' which, apart from satisfying all the conditions specified above, is different from 0 at every critical point.

Lemma 3. *Changing, if necessary, the coefficients of F by an arbitrarily small amount, we can obtain a polynomial $F(x_1, \ldots, x_m)$ for which there exists another polynomial $f(x_1, \ldots, x_m)$ of degree*

$$l = \left[\frac{mn - 2m - n}{2} \right]$$

([N] *stands for the entire part of* N), *with real coefficients and the following properties:*

1. $f(x_1, \ldots, x_m)$ *vanishes at each of* $S(m,n) - \gamma - 1$ *arbitrarily chosen real critical points of* $F(x_1, \ldots, x_m)$, *provided that* $S(m,n) - \gamma - 1 > 0$. *Here* 2γ *is the number of complex critical points of the function* F, *and* $S(m,n)$ *is the number of the terms of the polynomial*

$$\prod_{i=1}^{m} \frac{x_i^{n-1} - 1}{x_i - 1}, \tag{8}$$

such that their degree does not exceed l.

2. *At every complex critical point* P_α, *the expression* $A_\alpha f^2(P_\alpha)$ *has a non-negative real part. Here every* A_α *is a constant independent of* f, $A_\alpha \neq 0$; *the numbers* A_α *are determined only by the function* F, *so that complex-conjugate critical points* P_α *correspond to complex-conjugate numbers* A_α.

3. *There exists either a real critical point* P_α *at which* f *is different from* 0, *or a complex critical point of* F *at which* $A_\alpha f^2(P_\alpha)$ *is different from* 0.

Proof. First of all, let us slightly change the coefficients of F, so as to make F have all the properties listed in Lemma 2. The resulting polynomial is still denoted by F. Let $P_\alpha(\xi_1^\alpha, \ldots, \xi_m^\alpha)$ be any critical point of the function F. Then every F_i can be represented in the form

$$F_i \equiv \sum_{j=1}^{m} \left(x_j - \xi_j^\alpha\right) F_{ij}^\alpha, \tag{9}$$

where F_{ij}^α is a polynomial in $x_1, \ldots, x_m$, $\xi_1^\alpha, \ldots, \xi_m^\alpha$.

The determinant $\varphi_\alpha = \left|F_{ij}^\alpha\right|$, regarded as a function of $x_1, \ldots, x_m$, vanishes at every critical point P_β other than P_α. Indeed, let us substitute for $x_1, \ldots, x_m$ in the right-hand sides of (9) the values $\xi_1^\beta, \ldots, \xi_m^\beta$ corresponding to the critical point P_β. Since the resulting right-hand sides must be equal to 0, we thus obtain a system of m linear homogeneous equations with respect to $\left(\xi_j^\beta - \xi_j^\alpha\right)$, which admits a non-trivial solution; therefore, the said determinant is equal to 0. On the other hand, at P_α the determinant $\left|F_{ij}^\alpha\right|$ coincides with the value of the Jacobian $I(P_\alpha)$, and is therefore different from 0 (see [5]).

There are many ways to represent F_i in the form (9). Our choice of F_{ij}^α is as follows. Set

$$F_i \equiv \Phi_i^{n-1} + \Phi_i^{n-2} + \cdots + \Phi_i^0,$$

where Φ_i^k is a homogeneous polynomial of degree k with respect to $x_1, \ldots, x_m$. Let us take as F_{ij}^α the following polynomial

$$\Phi_{ij\alpha}^{n-2} + \Phi_{ij\alpha}^{n-3} + \cdots + \Phi_{ij\alpha}^{0} \, ,$$

where

$$\Phi_{ij\alpha}^{n-2} = \frac{1}{n-1} \frac{\partial \Phi_i^{n-1}}{\partial x_j} \, ,$$

$$\Phi_{ij\alpha}^{n-3} = \frac{1}{n-2} \frac{\partial}{\partial x_j} \left(\Phi_i^{n-2} + \frac{1}{n-1} \sum_{s=1}^{m} \xi_s^{\alpha} \frac{\partial \Phi_i^{n-1}}{\partial x_s} \right) \, ,$$

$$\Phi_{ij\alpha}^{n-4} = \frac{1}{n-3} \frac{\partial}{\partial x_j} \left[\Phi_i^{n-3} + \right.$$

$$\left. + \frac{1}{n-2} \sum_{k=1}^{m} \xi_k^{\alpha} \frac{\partial}{\partial x_k} \left(\Phi_i^{n-2} + \frac{1}{n-1} \sum_{s=1}^{m} \xi_s^{\alpha} \frac{\partial \Phi_i^{n-1}}{\partial x_s} \right) \right] \, ,$$

etc. Obviously, $\Phi_{ij\alpha}^{k}$ is a homogeneous polynomial of degree k with respect to $x_1, \ldots, x_m$; however, with respect to the arguments $\xi_1^{\alpha}, \ldots, \xi_m^{\alpha}$, the function $\Phi_{ij\alpha}^{k}$ is, in general, a non-homogeneous polynomial of degree $n - 2 - k$. If, in the expression for the determinant $\left| F_{ij}^{\alpha} \right|$, we bring together all the terms of degree $> l$ in $x_1, \ldots, x_m$, we obtain a polynomial of degree $\leq m(n-2)-l-1$ in $\xi_1^{\alpha}, \ldots, \xi_m^{\alpha}$. Denote the latter polynomial by $\psi^{\alpha}(x, \xi)$. It is easy to see that the coefficients of ψ^{α} by the powers of $\xi_1^{\alpha}, \ldots, \xi_m^{\alpha}$ are independent of α. Using the fact that every $P_{\alpha}(\xi_1^{\alpha}, \ldots, \xi_m^{\alpha})$ is a solution of system (6), we can express all powers of the form $(\xi_1^{\alpha})^{k_1} \cdots (\xi_m^{\alpha})^{k_m}$, where

$$k_1 + \cdots + k_m \leq m(n - 2) - l - 1$$

and $k_i \geq n - 1$ for some i, in terms of linear combinations of the powers

$$(\xi_1^{\alpha})^{k_1} \cdots (\xi_m^{\alpha})^{k_m} \tag{10}$$

with $k_i < n - 1$ for all $i = 1, \ldots, m$, and

$$k_1 + k_2 + \cdots + k_m \leq m(n - 2) - l - 1 \, ,$$

the coefficients of these linear combinations being independent of α. Indeed, in the special case of F having the form

$$F(x_1, \ldots, x_m) \equiv f_1(x_1) + f_2(x_2) + \cdots + f_m(x_m) \, ,$$

where $f_k(x_k)$ is a polynomial of degree n with respect to x_k, this expression in terms of linear combinations can be implemented as follows. Let us multiply each equation of the system

$$F_1(\xi_1, \ldots, \xi_m) = 0, \quad \ldots, \quad F_m(\xi_1, \ldots, \xi_m) = 0 \tag{11}$$

by a product $(\xi_1)^{k_1} \cdots (\xi_m)^{k_m}$ for all possible values of k_j such that

$$k_1 + k_2 + \cdots + k_m = R , \qquad R \le m(n-2) - l - n .$$

We thus obtain a system of equations, from which we can choose a sub-system D_R with the following properties: every equation of this subsystem expresses a product of the form

$$(\xi_1)^{k_1} \cdots (\xi_m)^{k_m} \tag{12}$$

with

$$\max\{k_1, k_2, \ldots, k_m\} = \mu > n - 2 , \quad k_1 + k_2 + \cdots + k_m = R - n - 1 ,$$

through similar products with

$$\max\{k_1, k_2, \ldots, k_m\} \le n - 2 , \quad k_1 + k_2 + \cdots + k_m \le R + n + 1 ,$$

or with

$$\max\{k_1, k_2, \ldots, k_m\} > n - 2 , \quad k_1 + k_2 + \cdots + k_m < R + n + 1 .$$

The system D_R has as many equations as there are different expressions of the form (12). Using the subsystems $D_0, D_1, \ldots, D_{m(n-2)-l-n}$, it is easy to perform the required transformation of $\psi^\alpha(x, \xi)$ in the special case of F under consideration.

In order to pass from this special case to the general one, let us make the following observation. For the general polynomial F in question, let us construct a system

$$D'_R \qquad (R = 0, 1, \ldots, m(n-2) - l - n) ,$$

multiplying each equation in (11) by the same powers $(\xi_1)^{k_1}, \cdots (\xi_m)^{k_m}$ as those used in the construction of system D_R. The determinant for this system, whose matrix consists of the coefficients by the powers of the form (12), is a polynomial with respect to the coefficients of F. Clearly, in the special case considered above, this polynomial is different from 0. Therefore, we can change, if necessary, the coefficients of the original polynomial F by an arbitrarily small amount and thus come to the situation where all systems D'_R $(R = 0, 1, \ldots, m(n-2) - l - n)$ have their determinants different from zero. Thereby, the terms of the form (12) can be expressed through linear combinations of terms of the form (10). In what follows, the coefficients of F are assumed to have been changed in this manner, if ever this change is necessary.

If we express, in the polynomial $\psi^\alpha(x, \xi)$, all powers of the form (12), such that $k_i > n - 2$ for at least one i, through the powers (10) with all

$k_i < n - 1$, then $\psi^\alpha(x, \xi)$ will contain no more than $\tilde{S}(m, n)$ terms with respect to $\xi_1^\alpha, \ldots, \xi_m^\alpha$. The coefficients of $\psi^\alpha(x, \xi)$ by the same powers of $\xi_1^\alpha, \ldots, \xi_m^\alpha$ will be independent of α. Here $\tilde{S}(m, n)$ is the number of terms of the polynomial (8), such that their degree is $\leq m(n - 2) - l - 1$.

First we consider the case of all critical points P_α being real. For any choice of $\tilde{S}(m, n) + 1$ points from among these, it is always possible to construct a polynomial

$$f(x_1, \ldots, x_m) = \sum_\alpha C_\alpha \varphi_\alpha \tag{13}$$

(here the sum is over these points) possessing the following properties: its degree with respect to $x_1, \ldots, x_m$ is $\leq l$; it vanishes at the remaining $(n - 1)^m - \tilde{S}(m, n) - 1$ critical points; among the points over which the sum (13) is taken, there exists one, at least, where this polynomial is different from 0. To construct this polynomial, it suffices to choose the constants C_α in (13) such that the terms of degree $> l$ with respect to $x_1, \ldots, x_m$ are eliminated in (13), and $C_\alpha \neq 0$ for at least one α. Accordingly, we have to solve a system consisting of $\tilde{S}(m, n)$ linear homogeneous equations for the unknown constants C_α, $\tilde{S}(m, n) + 1$ in number. This system is bound to possess a non-trivial solution. Obviously,

$$(n - 1)^m - \tilde{S}(m, n) = S(m, n) ,$$

since $(n - 1)^m$ is the number of terms of the polynomial (8), $S(m, n)$ is the number of its terms of degree $\leq l$, and $\tilde{S}(m, n)$ is the number of its terms of degree $> l$.

Next, we turn to the case when the set of critical points contains 2γ complex ones. Let φ_α and $\varphi_{\tilde{\alpha}}$ be the determinants corresponding to the complex-conjugate critical points P_α and $P_{\tilde{\alpha}}$. We assume that these determinants have been transformed in the above fashion, so that their respective $\psi^\alpha(x, \xi)$ and $\psi^{\tilde{\alpha}}(x, \xi)$ involve no powers of ξ_i^α greater than $n - 2$. Then

$$U_\alpha = \frac{\varphi_\alpha + \varphi_{\tilde{\alpha}}}{2} \quad \text{and} \quad V_\alpha = \frac{\varphi_\alpha - \varphi_{\tilde{\alpha}}}{2i}$$

are polynomials in $x_1, \ldots, x_m$ with real coefficients. The functions U_α and V_α vanish at all critical points P_β other than P_α and $P_{\tilde{\alpha}}$. Let

$$A_\alpha = \tilde{A} + i\tilde{B} , \quad A_{\tilde{\alpha}} = \tilde{A} - i\tilde{B} ,$$

and let

$$dU_\alpha + eV_\alpha = a + ib \quad \text{at} \quad P_\alpha ,$$

where $\tilde{A}, \tilde{B}, a, b, d, e$ are real numbers; $\tilde{A}$ and $\tilde{B}$ have given values such that $\tilde{A}^2 + \tilde{B}^2 > 0$, and a, b, d, e are the parameters to be defined later on.

Obviously the value of

$$\mathrm{Re}\left\{(\tilde{A}+i\tilde{B})(dU_\alpha + eV_\alpha)^2\right\} \quad \text{at the point} \quad P_\alpha$$

is equal to the value of

$$\mathrm{Re}\left\{(\tilde{A}-i\tilde{B})(dU_\alpha + eV_\alpha)^2\right\} \quad \text{at the point} \quad P_{\bar{\alpha}} \, ;$$

and both these values are equal to

$$R_\alpha = \tilde{A}a^2 - \tilde{A}b^2 - 2\tilde{B}ab \, .$$

Let us choose d and e such that $R_\alpha > 0$. This choice is always possible. Indeed, we have

$$dU_\alpha + eV_\alpha = \frac{1}{2}(d - ie)\alpha_\alpha \quad \text{at } P_\alpha.$$

Assume that $\varphi_\alpha = q_1 \times iq_2$ at this point; then, according to what has been said above, φ_α is different from 0. Therefore

$$a = \frac{1}{2}(dq_1 + eq_2) \, , \qquad b = \frac{1}{2}(dq_2 - eq_1) \, .$$

For any a and b, this system can be solved uniquely with respect to d and e, since its determinant is equal to

$$-\left(q_1^2 + q_2^2\right) \neq 0 \, .$$

In order that R_α be positive, let us take $b = 0$, $a = 1$ if $\tilde{A} > 0$; and $a = 0$, $b = 1$ if $\tilde{A} < 0$; finally, if $\tilde{A} = 0$, let us choose a and b such that $\tilde{B}ab < 0$. Denote the polynomial $dU_\alpha + eV_\alpha$ obtained in this way by ω_α.

Taking the functions φ_α, $\tilde{S}(m,n) + 1 - \gamma$ in number, corresponding to the $\tilde{S}(m,n) + 1 - \gamma$ arbitrarily chosen real points P_α (provided that this number is positive), and taking as many as γ functions ω_α corresponding to all complex critical points, we can construct a polynomial (with real coefficients) having the form

$$f(x_1,\ldots,x_m) = \sum_{\alpha=1}^{\gamma} C_\alpha \omega_\alpha + \sum_{\alpha=\gamma+1}^{\tilde{S}+1} C'_\alpha \varphi_\alpha \, ,$$

where C_α and C'_α are real numbers chosen in such a way that the degree of this polynomial with respect to $x_1,\ldots,x_m$ is $\leq l$. It is obvious that f vanishes at

$$(n-1)^m - 2\gamma - [\tilde{S}(m,n) + 1 - \gamma] = S(m,n) - \gamma - 1$$

arbitrarily chosen real critical points, and $A_\alpha f^2$ has a non-negative real part at all complex critical points P_α. Since, among the C_α's and C'_α's, there is at least one different from 0, it is easy to see that condition 3 is satisfied. This completes the proof of Lemma 3.

In exactly the same manner we can construct a polynomial $f(x_1, \ldots, x_m)$ having all the properties listed in Lemma 3, with the following modification: at every complex critical point P_α, the real part of $A_\alpha f^2$ must be ≤ 0.

Let us introduce the following *notations*:

M_C is the set of points in the projective space $\mathcal{P}_m$ such that

$$x_{m+1}^n F\left(\frac{x_1}{x_{m+1}}, \ldots, \frac{x_m}{x_{m+1}}\right) \geq C \quad \text{for} \quad x_{m+1} = 1 \; ;$$

$E(M_C)$ is the Euler characteristic of the closure of M_C;

M^* is the boundary of the set M;

$\overline{M}$ is the closure of the set M.

Lemma 4. *Let $F(x_1, \ldots, x_m)$ be any polynomial such that the system $F_1 = 0, \ldots, F_m = 0$ has the properties stated in Lemma 3. Then*

$$\left| E(M_0) - \frac{E(M_{C_1}) + E(M_{C_2})}{2} \right| \leq \frac{1}{2}(n-1)^m - S(m,n) + 1 , \qquad (14)$$

where $S(m,n)$ is defined as in Lemma 3; C_1 and C_2 are real constants such that

$$C_2 < F(P_\alpha) < C_1$$

for all real critical points P_α.

Proof. Let us make the constant C decrease from C_1 to 0, assuming that $C_1 > 0$. Then $E(M_C)$ may undergo change only when C passes through its critical values, i.e., the values of $F(x_1, \ldots, x_m)$ at real critical points. Let F be represented as

$$F(x_1, \ldots, x_m) = C^\alpha + \sum_{i=1}^k x_i'^2 - \sum_{i=k+1}^m x_i'^2 + \varepsilon \qquad (15)$$

in a neighborhood of the critical point P_α, where C^α is the value of F at the point $P_\alpha(\xi_1^\alpha, \ldots, \xi_m^\alpha)$; x_i' are linear combinations of the differences $(x_j - \xi_j^\alpha)$ with coefficients depending on P_α; ε stands for small terms of higher order as compared to $\sum_{i=1}^m x_i'^2$.

The number of squared quantities in (15) is equal to m, since the Jacobian for system (11) has been assumed different from 0 at this point. Let p^r be the r-dimensional Betti number for the set $\overline{M}_C$. Here and in what follows, the Betti numbers are counted modulo 2. When C, being decreased, passes through the said critical value C^α, then either the number p^k becomes increased by 1, or p^{k-1} is decreased by 1. This statement can be proved by the arguments used in [7] in relation to functions defined on manifolds, after introducing the metric of the sphere in the projective space $\mathcal{P}_m$ (cf. [6]).

Assume that for the values of C between 0 and C_1 there are A critical points with k even and B critical points with k odd; and for the values of C between 0 and C_2 there are A' critical points with k even and B' critical points with k odd. Then

$$(n-1)^m - 2\gamma = A + B + A' + B' \, ,$$

$$E(M_0) - E(M_{C_1}) = A - B \, ,$$

$$E(M_0) - E(M_{C_2}) = B' - A' \, .$$

It follows that

$$A + B' = \frac{1}{2}(n-1)^m - \gamma + E(M_0) - \frac{E(M_{C_1}) + E(M_{C_2})}{2} \, ,$$

$$A' + B = \frac{1}{2}(n-1)^m - \gamma - E(M_0) + \frac{E(M_{C_1}) + E(M_{C_2})}{2} \, .$$

Moreover, the sign of the function E/I, where

$$I = \frac{D(F_1, \ldots, F_m)}{D(x_1, \ldots, x_m)} \, ,$$

is $(-1)^m$ at $A + B'$ real critical points, and $(-1)^{m+1}$ at $A' + B$ real critical points.

Let us show that

$$A + B' > S(m, n) - \gamma - 1 \, ,$$
$$A' + B > S(m, n) - \gamma - 1 \, .$$
\hfill (16)

Indeed, since the polynomial $F(x_1, \ldots, x_m)$ is assumed to possess all the properties stated in Lemma 2, therefore, for any polynomial $f_1(x_1, \ldots, x_m)$ of degree $\leq [m(n-2) - 1]$ the Euler-Jacobi formula yields

$$\sum_\alpha \frac{f_1(P_\alpha)}{I(P_\alpha)} = 0 \, ,$$
\hfill (17)

where the sum is taken over all critical points of F (see [5]). Set $f_1 = F_0 f^2$, where

$$F_0 = nF - \sum_{i=1}^{m} x_i F_i ,$$

and f is the polynomial of degree l constructed in Lemma 3 and vanishing at $A + B'$ real critical points at which $(-1)^m F/I > 0$, provided that

$$A + B' \leq S(m, n) - \gamma - 1 .$$

To construct this polynomial, we take for A_α the value of F/I at the complex critical point P_α and choose the sign of R_α to be $(-1)^{m+1}$. Then the left-hand side of (17) will be different from 0, since all non-vanishing real parts of the terms in (17) corresponding to complex critical points, as well as all non-vanishing terms in (17) corresponding to real critical values, have the same sign. Moreover, there is at least one term in (17) different from 0, because of the condition 3 of Lemma 3. This contradiction shows that

$$A + B' > S(m, n) - \gamma - 1 .$$

The second inequality in (16) can be proved in exactly the same way. Hence we obtain (14).

Now we can turn to the proof of relations (2)–(5), under the assumption that system (1) has no real solutions, i.e., the surface Γ_0 contains no real singular points. Without loss of generality, we can assume that the polynomial $F(x_1, \ldots, x_m)$ possesses all the properties stated in Lemma 2 and Lemma 3. Indeed, if this is not the case, we can subject the coefficients of F to an arbitrarily small variation and thus, by Lemma 2 and Lemma 3, obtain a polynomial F' with the required properties. But Lemma 1 shows that $E(M_0)$ for the polynomial F is the same as $E(M_0)$ for F'.

Theorem 1. *For n even, we have*

$$|E(M_0)| \leq \frac{1}{2}(n - 1)^m - S(m, n) + \frac{1}{2} ,$$

where $S(m, n)$ is defined in Lemma 3.

Proof. Let us show that for n even we have

$$E(M_{C_1}) + E(M_{C_2}) = 1 ,$$

where C_1 and C_2 have been defined in Lemma 4. Then Theorem 1 will follow from Lemma 4.

We shall need the following well-known results:

1. Any real algebraic surface without real singular points is a polyhedron, to within one-to-one continuous mappings. (see [8], p. 333).

2. Let K_1 and K_2 be two polyhedrons; let $K_1 + K_2$ and $K_1 K_2$ be their union and their intersection, respectively. Then (see [9], p. 287)

$$E(K_1 + K_2) = E(K_1) + E(K_2) - E(K_1 K_2) \,. \qquad (18)$$

As usual, let us map $\mathcal{P}_m$ onto the sphere S_m so that each point of $\mathcal{P}_m$ would correspond to a pair of diametrically opposite points of S_m. Denote by K^s the image in S_m of the polyhedron K from $\mathcal{P}_m$ by this mapping. It is easy to see that

$$E(K^s) = 2E(K) \,,$$

since the number of simplexes of every dimension is doubled by this mapping. Let us show that

$$E\left(M_{C_1}^s\right) + E\left(M_{C_2}^s\right) = 2 \,.$$

Denote by Θ the intersection of $\overline{M}_{C_1}$ with the hyperplane at infinity. We obviously have

$$E(\Theta^s) = E\left(M_{C_1}^s\right) \,,$$

since $\overline{M}_{C_1}^s$ can be continuously deformed into Θ^s. It easily follows from (18) that

$$E(\Theta^s) + E(S_{m-1} - \Theta^s) = E(S_{m-1}) + E(\Theta^{s*}) \,, \qquad (19)$$

where S_{m-1} is the image on the sphere S_m of the hyperplane at infinity; Θ^{s*} is the boundary of Θ^s.

Further, it is easy to see that

$$E\left(M_{C_2}^s\right) + E(S_m - \overline{M}_{C_2}^s) = E(S_m) + E(\overline{M}_{C_2}^{s*}) \,, \qquad (20)$$

$$E(S_m - \overline{M}_{C_2}^s) = E(S_{m-1} - \Theta^s) \,. \qquad (21)$$

Let us show that

$$E(\overline{M}_{C_2}^{s*}) = 2E(S_{m-1} - \Theta^s) - E(\Theta^{s*}) \,. \qquad (22)$$

Indeed, $\overline{M}_{C_2}^{s*}$ can be represented as a sum of two complexes glued along their common boundary Θ^{s*} and homeomorphic to $\overline{S_{m-1} - \Theta^s}$.

Taking the sum of (20) and (19), we find, by virtue of (21), (22), that

$$E(\Theta^s) + E(M_{C_2}^s) = E(S_{m-1}) + E(S_m) \,.$$

Since

$$E(S_{m-1}) + E(S_m) = 2 \quad \text{and} \quad E(\Theta^s) = E\left(M_{C_1}^s\right) \,,$$

we have

$$E\left(M_{C_1}^s\right) + E\left(M_{C_1}^s\right) = 2 \,,$$

and thus

$$E\left(M_{C_1}\right) + E\left(M_{C_2}\right) = 1.$$

Theorem 2. *For n even and m odd, we have*

$$|E(\Gamma_0)| \le (n-1)^m - 2S(m,n) + 1 \tag{23}$$

(As usual, it is assumed here that the surface has no real singular points).

Proof. By virtue of (18), it is easy to show that

$$E(M_0) + E(\overline{\mathcal{P}_m - M_0}) = E(\mathcal{P}_m) + E(\Gamma_0) \,, \tag{24}$$

where $\overline{\mathcal{P}_m - M_0}$ is the closure of the set consisting of points such that

$$-x_{m+1}^n F\left(\frac{x_1}{x_{m+1}}, \ldots, \frac{x_m}{x_{m+1}}\right) \ge 0 \quad \text{for} \quad x_{m+1} = 1 \,.$$

Therefore, $E(\overline{\mathcal{P}_m - M_0})$ satisfies the inequality (3). Moreover, $E(\mathcal{P}_m) = 0$ for m odd. Thus, (23) follows from (24) and (3).

Theorem 3. *If n and m are both odd, then*

$$|E(M_0)| \le \frac{1}{2}(n-1)^m - S(m,n) + 1 \,, \tag{25}$$

$$|E(\Gamma_0)| \le (n-1)^m - 2S(m,n) + 1 \,. \tag{26}$$

If n is odd and m even, then

$$|E(M_0)| \le \frac{1}{2}(n-1)^m - S(m,n) + \frac{1}{2}(n-1)^{m-1} - S(m-1,n) + 1 \,, \tag{27}$$

$$|E(\Gamma_0)| = 0 \,.$$

Proof. For n odd, the boundary of M_0 consists of Γ_0 and the plane at infinity $x_{m+1} = 0$. Moreover,

$$E\left(M_{C_1}\right) = E(\mathcal{P}_{m-1}) \,, \tag{28}$$

since $\overline{M}_{C_1}$ can be continuously deformed into the hyperplane at infinity, and therefore, their Betti numbers are the same.

Using (18) we find

$$E\left(M_{C_2}\right) + E(\overline{\mathcal{P}_m - M_{C_2}}) = E(\mathcal{P}_m) + E\left(M_{C_2}^*\right) \,. \tag{29}$$

Let ω be the intersection of Γ_0 and the hyperplane $x_{m+1} = 0$; then

$$E\left(M_{C_2}^*\right) = 2E(\mathcal{P}_{m-1}) = E(\omega) . \tag{30}$$

This fact can be easily verified, if one takes into account that $M_{C_2}^*$ consists of the hyperplane at infinity glued along ω to the surface

$$x_{m+1}^n F\left(\frac{x_1}{x_{m+1}}, \ldots, \frac{x_m}{x_{m+1}}\right) - C_2 x_{m+1}^n = 0$$

homeomorphic to the hyperplane at infinity. We also have

$$E(\overline{\mathcal{P}_m - M_{C_2}}) = E(\mathcal{P}_{m-1}) , \tag{31}$$

since the set $\overline{\mathcal{P}_m - M_{C_2}}$ can be continuously deformed into the hyperplane $x_{m+1} = 0$.

From (30), (31) and (29), we get

$$E(M_{C_2}) = E(\mathcal{P}_m) + E(\mathcal{P}_{m-1}) - E(\omega) . \tag{32}$$

Since ω is an $(m-2)$-dimensional manifold and, thus, $E(\omega) = 0$ for m odd, it follows that

$$E\left(M_{C_2}\right) = 1 . \tag{33}$$

Relations (28) and (33) imply (25), because of Lemma 4.

Using (18), we can establish the following relation for n odd

$$E(M_0) + E(\overline{\mathcal{P}_m - M_0}) = E(\mathcal{P}_m) + E(\mathcal{P}_{m-1}) + E(\Gamma_0) - E(\omega) . \tag{34}$$

For m odd we have $E(\omega) = 0$, and it follows from (34) that

$$E(\Gamma_0) = E(M_0) - 1 + E(\overline{\mathcal{P}_m - M_0}) .$$

For $E(\overline{\mathcal{P}_m - M_0})$ the estimate (25) holds, since $\overline{\mathcal{P}_m - M_0}$ is the closure of the set consisting of the points such that

$$-x_{m+1}^n F\left(\frac{x_1}{x_{m+1}}, \ldots, \frac{x_m}{x_{m+1}}\right) \geq 0 \quad \text{for} \quad x_{m+1} = 1 .$$

It follows from (14), (28) and (33) that

$$|E(M_0) - 1| \leq \frac{1}{2}(n-1)^m - S(m,n) .$$

Therefore, the estimate (26) holds for m odd. On the basis of (26) and Lemma 4, i' is easy to get (27). Indeed, for m even we have

$$E\left(M_{C_1}\right) = E(\mathcal{P}_{m-1}) = 0 .$$

It follows from (32) that $E\left(M_{C_2}\right) = 1 - E(\omega)$. Applying the estimate (26) to $|E(\omega)|$, we obtain

$$|E\left(M_{C_2}\right)| \le (n-1)^{m-1} - 2S(m-1,n) + 2 \ .$$

Hence, by virtue of Lemma 4, we find

$$|E(M_0)| \le \frac{1}{2}(n-1)^m - S(m,n) + \frac{1}{2}|E\left(M_{C_2}\right)| \le$$

$$\le \frac{1}{2}(n-1)^m - S(m,n) + \frac{1}{2}(n-1)^{m-1} - S(m-1,n) + 1 \ .$$

Remark 1. It is easy to calculate that

$$S(2,n) = \frac{1}{2}\left[\frac{n}{2}\right]\left(\left[\frac{n}{2}\right]-1\right) \ , \qquad S(3,n) = \frac{1}{6}n(n-1)(n-2) \ .$$

Remark 2. If the surface

$$x_{m+1}^n F\left(\frac{x_1}{x_{m+1}}, \dots, \frac{x_m}{x_{m+1}}\right) = 0$$

has finitely many isolated real singular points, then $E(M_0)$ can be easily shown to coincide with the Euler characteristic of the set M_0 constructed for the polynomial

$$F'(x_1, \dots, x_m) = F(x_1, \dots, x_m) + \varepsilon \ ,$$

where ε is a small positive constant. Therefore, in this case $E(M_0)$ also satisfies the estimates (3) – (5).

References

[1] Petrowsky, I.G. On the topology of real plane algebraic curves. *Ann. Math.* **39**:1 (1938) 189–209. (Article 7 of this volume).
[2] Hilbert, D. Über die Gestalt einer Fläche vierter Ordnung. *Gesammelte Abhandlungen*. Bd. 2, Springer-Verlag, Berlin, 1933, pp. 449–453.
[3] Van der Waerden, B.L. *Algebra*. Bd. 2. Springer-Verlag, 1967.
[4] Diskriminante eines Gleichungssystems. In: *Encyklopädie der mathematischen Wissenschaften*. Bd. 1. Leipzig, 1899.
[5] Kronecker, L. Über einige Interpolationsformeln für ganze Funktionen mehrer Variabeln. *Werke*. Bd. 1 (1895) 133–141.
[6] Morse, M. Critical points. *Trans. Amer. Math. Soc.* **27** (1925) 345–346.
[7] Seifert, H.; Threlfall, W. *Variationsrechnung im Grossen*. Berlin, 1938.
[8] Lefschetz, S. *Algebraic Topology*. New York, 1942.
[9] Alexandroff, P. Hopf, H. *Topologie*. Berlin, 1935.

APPENDIX

COMMENTARIES

SUPPLEMENTARY ARTICLES

Hyperbolic Equations*

by V.Ya. Ivrii and L.R. Volevich

The remarkable memoirs of I.G. Petrowsky on systems of partial differential equations have played a crucial role in the creation of modern general theory of partial differential equations and, in many respects, have determined the main features of this theory. Starting from the classical equations of mathematical physics – *the equation of string vibrations, the heat equation, and the Laplace equation,* – I.G. Petrowsky singled out three wide classes of systems (nonlinear, in general), which have become known as *hyperbolic in the sense of Petrowsky (or strictly hyperbolic), parabolic, and elliptic in the sense of Petrowsky.* During the next three decades, deep and far reaching theories were developed for these three types of systems.

In this review, we consider the questions of well-posedness for the Cauchy problem and the mixed initial boundary value problem for higher order hyperbolic equations and systems. The support of a solution (the existence of lacunas) and the singular support of a solution (its singularities), as well as their structure and behavior in the case of hyperbolic systems, are described in commentaries on Articles 5 and 6 of this volume (see Appendix). Beyond the scope of the present review are such topics as the existence of global solutions for nonlinear hyperbolic equations and systems, and the application of rapidly oscillating integrals for the construction of the parametrix for the Cauchy problem.

1. Well-Posedness of the Cauchy Problem for Strictly Hyperbolic Systems

1.1. Setting of the Problem

Here we use the following notations, which have become quite common by now: $x = (x_0, x') = (x_0, x_1, \ldots, x_n)$ is the generic point of the Euclidean

---

*A commentary on Articles 1 and 3 of this volume.

space $\mathbb{R}^{n+1}$; the dual variables are denoted by $\xi = (\xi_0, \xi') = (\xi_0, \xi_1, \ldots, \xi_n)$; and

$$D = (D_0, D') = (D_0, \ldots, D_n), \quad D_j = -i\frac{\partial}{\partial x_j} .$$

For a multi-index $\alpha = (\alpha_0, \alpha_1, \ldots, \alpha_n)$ with α_j taking non-negative integer values, we write

$$\xi^\alpha = \xi_0^{\alpha_0} \xi_1^{\alpha_1} \cdots \xi_n^{\alpha_n}, \quad |\alpha| = \alpha_0 + \alpha_1 + \cdots + \alpha_n .$$

Let

$$P(x, D) = \sum_{|\alpha| \le m} a_\alpha(x) D^\alpha \tag{1.1}$$

be a differential operator with matrix valued coefficients; the coefficients $a_\alpha(x)$ are $N \times N$-matrices whose elements are infinitely differentiable functions. The matrix $P(x, \xi) = \sum_{|\alpha| \le m} a_\alpha(x) \xi^\alpha$ is called the *symbol* of the operator (1.1), and the matrix

$$p(x, \xi) = \sum_{|\alpha| = m} a_\alpha(x) \xi^\alpha \tag{1.2}$$

is called the *principal symbol* of this operator.

Let Ω be a layer

$$\Omega = \left\{ (x_0, x') \in \mathbb{R}^{n+1}, \ 0 \le x_0 \le T, \ x' \in \mathbb{R}^n \right\} .$$

In Ω we consider the following Cauchy problem

$$P(x, D)u(x) = f(x), \quad x \in \Omega, \tag{1.3}$$

$$D_0^j u(x)\Big|_{x_0=0} = g_j(x'), \quad j = 0, \ldots, m - 1 . \tag{1.4}$$

For the sake of simplicity, we limit ourselves to the case of a layer and assume the matrices $a_\alpha(x)$ to be independent of x for $|x'| > R$.

Assume that (1.3), (1.4) is a non-characteristic problem in the sense that

$$\det p(x, 1, 0, \ldots, 0) \ne 0 . \tag{1.5}$$

Remark. Because of (1.5), system (1.3) can be written as the following system of Kowalevskaya type:

$$D_0^m u = \sum_{j=1}^m P_j(x, D') D_0^{m-j} u + F \quad (\text{ord } P_j \le j), \tag{1.3'}$$

and it is in this form that Petrowsky writes his systems. The operator (1.1) is said to be *strictly hyperbolic* with respect to x_0, if the condition (1.5) is satisfied, and all roots ξ_0 of the characteristic equation

$$\det p(x, \xi_0, \xi') = 0 \tag{1.6}$$

are real and different for all $(x, \xi') \in \Omega \times (\mathbb{R}^n \setminus 0)$.   □

Definition. *The Cauchy problem* $(1.3), (1.4)$ *is* L_2*-well-posed, if for any*

$$f \in C^\infty(\Omega), \quad g_j \in C^\infty(\Omega)$$

it admits a solution $u \in C^\infty(\Omega)$ *which is unique and satisfies the following energy inequality:*

$$\sum_{|\alpha| \le m-1+q} \|D^\alpha u; t\| \le C \sum_{|\alpha| \le m-1+q} \|D^\alpha u; 0\| + C' \int_0^t \sum_{|\alpha| \le q} \|D^\alpha Pu; s\| \, ds,$$

$$\tag{1.7}$$

where $q = 0, 1, \ldots$; $t \in (0, T]$; *and*

$$\|v; t\|^2 = \int_{\mathbb{R}^n} |v(t, x')|^2 \, dx' .$$

(It is assumed that the expression in right-hand side makes sense.)

In fact, the crucial part of Petrowsky's article[1] is the proof of L_2-well-posedness for problem $(1.3), (1.4)$.

1.2. The Main Ideas of Petrowsky's Method

The principal point in the proof of L_2-well-posedness for problem $(1.3), (1.4)$ is the derivation of an *a priori* estimate of type (1.7) for an arbitrary smooth function $u(x)$. After the usual replacement of the variables, the case $m > 1$ in (1.1) is reduced to the case $m = 1$, so that we have to deal with the following first order system

$$Pu = \frac{\partial u}{\partial x_0} - i \sum A_j(x) D_j u - B(x) u = f . \tag{1.8}$$

In order to establish the inequality (1.7) in the latter case, Petrowsky constructs a quadratic form $Q(u, u)$ with the following properties:

(i) there exist positive constants C_1, C_2 such that for $t \in (0, T)$ we have

$$C_1 \|u; t\| \le Q(u(t, \cdot), u(t, \cdot)) \le C_2 \|u; t\| ;$$

(ii) there exist positive constants C_3, C_4 such that

$$\frac{dQ(u(t, \cdot), u(t, \cdot))}{dt} \le C_3 Q(u(t, \cdot), u(t, \cdot)) + C_4 \|(Pu)(t, \cdot)\|^2 .$$

---

[1]The operators considered by Petrowsky are of a more general type than (1.1). We shall return to this subject in Sect. 1.4.

The conditions (i) and (ii) obviously imply the estimate (1.7) for $q = 0$. Differentiating (1.8) and applying the estimate (1.7) with $q = 0$ to the corresponding derivatives, we obtain (1.7) for $q = 1$, etc.

Especially transparent is the construction of the quadratic form Q for an operator with constant coefficients

$$P(D)u = \frac{\partial u}{\partial x_0} - iA(D')u = f , \qquad A(D') = \sum_{j=1}^{n} A_j D_j . \qquad (1.8')$$

Strong hyperbolicity of the operator $P(D)$ implies that $A(\xi')$ can be reduced to diagonal form, in the sense that there exists a matrix $C(\xi')$ with the following properties:

(a) the elements of $C(\xi')$ are piecewise continuous functions, homogeneous of order 0;

(b) $|\det C(\xi')| > \delta > 0$, $\forall \xi' \in \mathbb{R}^n \setminus 0$;

(c) $C(\xi')A(\xi') = K(\xi')C(\xi')$, where $K(\xi')$ is a diagonal matrix with real elements.

Now we set

$$Q(u(t, \cdot), u(t, \cdot)) = \int \langle C(\xi')\hat{u}(t, \xi'), C(\xi')\hat{u}(t, \xi') \rangle \, d\xi' ,$$

where $\hat{u}(x_0, \xi')$ is the Fourier transform of $u(x)$ with respect to x'; $\langle \cdot, \cdot \rangle$ is the Hermitian scalar product in $\mathbb{C}^N$. Differentiating the last identity with respect to t, we find

$$\frac{dQ(u, u)}{dt} = 2\mathrm{Re} \int \left\langle C(\xi') \frac{d\hat{u}(t, \xi)}{dt}, C(\xi')\hat{u}(t, \xi') \right\rangle \, d\xi' . \qquad (1.9)$$

Applying the Fourier transformation to (1.8'), we obtain

$$\frac{d\hat{u}(t, \xi')}{dt} = iA(\xi')\hat{u}(t, \xi') + \hat{f}(t, \xi') .$$

This equation allows us to rewrite (1.9) in the form

$$\frac{dQ(u, u)}{dt} = 2\mathrm{Re}\, i \int \langle K(\xi')C(\xi')\hat{u}, C(\xi')\hat{u} \rangle \, d\xi' +$$

$$+ 2\mathrm{Re} \int \langle C(\xi')\hat{f}, C(\xi')\hat{u} \rangle \, d\xi' .$$

Since $K(\xi')$ is a real valued matrix, it follows that the first term in the right-hand side vanishes; estimating the second term by the Schwartz inequality, we obtain (ii) with $C_3 = C_4 = 1$.

In the case of variable coefficients, strong hyperbolicity of system (1.8) implies the existence of a matrix $C(x, \xi')$ with properties similar to (a), (b), and (c):

(A) The elements of $C(x, \xi')$ are homogeneous piecewise continuous functions of ξ'; they depend smoothly on x and satisfy the inequalities

$$|D_x^\alpha C(x, \xi')| < C_\alpha \varepsilon , \quad |\alpha| = 1, 2, \dots ; \qquad (1.10)$$

(B) $|\det C(x, \xi')| > \delta > 0 , \quad \forall (x, \xi) \in \Omega \times (\mathbb{R}^n \setminus 0) ;$

(C) $C(x, \xi')A(x, \xi') = K(x, \xi')A(x, \xi'), \quad A(x, \xi') = \sum_{j=1}^{n} \xi_j A_j(x)$, where $K(x, \xi')$ is a diagonal matrix with real elements satisfying the condition (A).

As noticed by Petrowsky, it suffices to prove the estimate (1.7) for functions with a compact support which is small enough with respect to x'. Assuming T to be sufficiently small, we can make the constant ε in (1.10) sufficiently small, and thereby, make the coefficients of system (1.8) close to constant ones.

Proceeding in a bold manner, Petrowsky considered the quadratic form

$$Q(u, u) = \int \langle a(t, \xi'), a(t, \xi') \rangle \, d\xi' ,$$

where

$$a(y_0, \xi') = (2\pi)^{-n} \int e^{-(y', \xi')} C(y, \xi')u(y) \, dy'.\,^2 \qquad (1.11)$$

Let us clarify the meaning of the quadratic form Q. By Parseval's relation we have

$$Q(u, u) = \int \langle Cu, Cu \rangle \, dx' , \qquad (1.12)$$

where

$$(Cu)(x) = \int e^{i(x', \xi')} a(x_0, \xi') \, d\xi' =$$
$$= \iint e^{i(x'-y', \xi')} C(x_0, y', \xi')u(x_0, y') \, dy' \, d\xi' ,$$

i.e., C is a *pseudo-differential operator* (a P.D.O., according to modern terminology) associated with the symbol $C(x, \xi)$.

The verification of conditions (i) and (ii) occupies, in the original Petrowsky's article, 17 pages filled with lengthy chain-like inequalities and almost no comments. Implicit in these formulas are all the major facts from the theory of pseudo-differential operators, namely:

---

[2] In order to emphasize the relation of Petrowsky's construction to the theory of pseudo-differential operators, we have slightly modified the former by replacing the periodic functions by compactly supported ones, and the Fourier series by the integrals.

1) estimates for L_2-norms of operators with bounded symbols;

2) lower bounds for elliptic operators whose coefficients are close to constant ones;

3) commutation relations for the product of two operators;

4) the estimate for the difference of an operator with a real valued symbol and the corresponding conjugate operator.

All this proved a strong hindrance to the study of Petrowsky's article. At the end of the '50s, Calderon, Mizohata, and Yamaguti (see [1]–[4]) obtained the energy inequalities for hyperbolic systems, using essentially the same method as that of Petrowsky; however, unlike the latter, they could rely on the technique, already available at that time, of singular integro-differential operators (P.D.O.'s, in more recent terminology) developed by Mikhlin, Zygmund, and Calderon (see [5], [6]). This technique allowed them to simplify the derivation of the energy estimate and to make the method of Petrowsky fully comprehensible and clear.

1.3. New Proof of Petrowsky's Results

In 1953, J.Leray published his lectures [7], which have their origin in a systematic analysis of Petrowsky's results on the Cauchy problem for hyperbolic equations. The first part of these lectures is aimed at systems with constant coefficients and is in close relation to Petrowsky's paper about lacunas (see Articles 5 and 6 of this volume). The second part is a study of systems with variable coefficients.

Leray considered the general problem of constructing quadratic forms Q satisfying conditions (ii); however, in contrast to Petrowsky, he limited himself to the forms associated with differential operators, and therefore obtained estimates less accurate than (1.7).

In the case of a single scalar equation of higher order, Leray suggested a new method for the derivation of *a priori* estimates. His method is based on a simple result about polynomials of a single complex variable: *let $p(\lambda)$ and $q(\lambda)$ be two polynomials of degree m and $m-1$, respectively, with mutually different real roots such that the roots of $p(\lambda)$ are separated from one another by the roots of $q(\lambda)$. Then*

$$\frac{p(\lambda)\overline{q(\lambda)}}{\operatorname{Im}\lambda} > 0, \qquad \forall \lambda \in \mathbb{C}^1 \setminus 0 .$$

Starting from this fact, Leray showed that the form

$$-\operatorname{Im} \int_{\mathbf{R}^{n+1}} e^{-2\gamma x_0} P(x,D)u(x)\overline{Q(x,D)u(x)}\, dx , \qquad \gamma = \text{const} , \qquad (1.13)$$

has definite sign, where P and Q are strictly hyperbolic operators of order m and $m-1$, respectively, with principal symbols $p(x,\xi)$ and $q(x,\xi)$ such that the roots (with respect to ξ_0) of $p(x,\xi)$ are separated from one another by the roots of $q(x,\xi)$. Estimates from below for the quadratic form (1.13) allowed to obtain the following estimate for the operator P in the entire space:

$$C_\gamma \sum_{|\alpha| \leq m-1+q} \left\| e^{-\gamma x_0} D^\alpha u \right\| \leq \sum_{|\alpha| \leq q} \left\| e^{-\gamma x_0} D^\alpha P u \right\|, \qquad \gamma \geq \gamma_0. \qquad (1.14)$$

A counterpart of the estimate (1.14) for negative q is, apparently, also due to Leray.

The method suggested by Leray (the so-called *method of separating operator*) was further developed in the works of Gårding (see [8]), who applied this method to establish Petrowsky's inequality (1.7) and similar inequalities in dual norms for the formally conjugate operator. By virtue of the Hahn-Banach theorem, the dual inequality obtained by Gårding allowed him to avoid the Cauchy-Kowalevskaya theorem (used by Petrowsky and Leray, who, in their turn, followed Schauder) and make substantially more simple and accurate the existence theorem for the solutions of the Cauchy problem.

Volevich and Gindikin (see [9]–[12]) have extended Leray's method of separating operator to a wide class of problems well-posed in the sense of Petrowsky; this class includes, as a special case, the Cauchy problem for hyperbolic equations.

Ladyzhenskaya [13] has proved the existence of solutions of the Cauchy problem for hyperbolic equations, using the method of finite differences. By Galiorkin approximations, Vishik [14] established solvability for the Cauchy problem in the case of a single higher order hyperbolic operator (see also [10]).

1.4. General Hyperbolic Systems

In his article, I.G. Petrowsky considers hyperbolic systems with each unknown function having its own highest order of differentiation. An even more general class of systems has been introduced by Leray in this lectures [7]. Instead of (1.1), he considers a differential operator with matrix valued coefficients, namely, the operator

$$P(x, D) = \| P_{ij}(x, D) \|_{i,j=1,\ldots,N}, \qquad (1.15)$$

where

$$P_{ij}(x, D) = \sum_{|\alpha| \leq \pi_{ij}} a_{ij\alpha}(x) D^\alpha. \qquad (1.16)$$

Assume that there exist integer $m_1, \ldots, m_N, n_1, \ldots, n_N$ such that

$$\pi_{ij} \le n_j - m_i , \tag{1.17}$$

and $P_{ij} \equiv 0$ if $n_j - m_i < 0$. Given these $n_1, \ldots, m_N$, we define the principal symbol for the operator (1.15), setting

$$p(x, \xi) = \|p_{ij}(x, \xi)\| , \tag{1.18}$$

$$p_{ij}(x, \xi) = \begin{cases} \displaystyle\sum_{|\alpha| = n_j - m_i} a_{ij\alpha}(x)\xi^\alpha , \\[2mm] 0 , \quad \pi_{ij} < n_j - m_i . \end{cases}$$

Let the matrix p be formed by separate blocks (in particular, it may consist of a single block)

$$p(x, \xi) = \left\| \begin{array}{ccc} H_1(x, \xi) & & 0 \\ & \ddots & \\ 0 & & H_J(x, \xi) \end{array} \right\| . \tag{1.18'}$$

The operator (1.15) is said to be *strictly hyperbolic in the direction* x_0, if all (homogeneous) symbols

$$h_j(x, \xi) = \det H_j(x, \xi) \tag{1.18''}$$

are strictly hyperbolic in the direction ξ_0 (i.e., $h_j(x, 1, 0, \ldots, 0) \ne 0$ and the roots of every equation $h_j(x, \xi) = 0$ with respect to ξ_0 are mutually different).

Remark. In the above definition, different h_j are allowed to have equal roots. □

In general, operators (1.15), hyperbolic in the sense of the above definition, do not represent systems of Kowalevskaya type. Therefore, it is the *generalized Cauchy problem* which is usually considered for such operators; this problem can be formulated as follows:

Since $m_1, \ldots, m_N$ satisfying the inequalities (1.17) are defined to within a constant, we can normalize them by the condition $\min m_i = 0$; then we must have $n_j \ge 0$ $(j = 1, \ldots, N)$.

The generalized Cauchy problem for the system

$$P(x, D)u(x) = f(x) \tag{1.19}$$

consists in finding a (smooth) vector valued function $u = (u_1, \ldots, u_N)$ which satisfies equations (1.19) supplemented by the initial conditions

$$u_j(x) - v_j(x) \text{ has zero of order } n_j \text{ for } x_0 = 0, \qquad (1.20)$$
$$j = 1, \ldots, N ,$$

where $v = (v_1, \ldots, v_N)$ is a given (smooth) vector valued function.

A version of L_2-well-posedness can also be introduced for problem (1.19), (1.20); however, we shall not dwell on this subject but limit ourselves to the observation that for the existence of a smooth solution of problem (1.19), (1.20) the following *compatibility conditions* are necessary:

$$f_i - \sum_{j=1}^{N} P_{ij} v_j \text{ has zero of order } m_i \text{ for } x_0 = 0, \qquad (1.21)$$
$$i = 1, \ldots, N .$$

Thus, for $m_1 = \cdots = m_N = 0$, we get the definition of a hyperbolic system suggested by Petrowsky. In this case, the conditions (1.21) are dropped, and (1.20) becomes equivalent to the usual initial conditions for the Cauchy problem, viz.:

$$D^k u_j(x)\Big|_{x_0=0} = g_{k_j}(x') , \qquad k = 0, \ldots, n_j - 1 , \quad j = 1, \ldots, N .$$

The definition of hyperbolicity in the sense of Leray (in contrast to that in the sense of Petrowsky) seems not quite effective, since it is an open question whether the integers m_i, n_j exist. In this connection the following observation should be made. Each set of $m_1, \ldots, n_N$ satisfying (1.17) is associated to the principal part (1.18), where $\det p(x, \xi)$ is either identically equal to 0 or coincides with the principal part of $\det P(x, \xi)$. The latter condition holds if and only if

$$\deg \det P(x, \xi) = R , \qquad (1.22)$$

$$\sum n_j - \sum m_j = R , \qquad (1.23)$$

where
$$R = \max_{\gamma} \left(\pi_{1\gamma(1)} + \cdots + \pi_{N_\gamma(N)} \right) ,$$

and γ takes its values among all permutations of $1, \ldots, N$. The methods of linear programming (see [15]) yield a proof for the existence of integers satisfying relations (1.17), (1.23). Therefore, hyperbolicity in the sense of Leray means that condition (1.22) is fulfilled and the principal part can be written in the form (1.18'), where (1.18'') are strictly hyperbolic polynomials.

2. Well-Posedness of the Cauchy Problem for Non-Strictly Hyperbolic Equations

As mentioned in Sect. 1, the condition of strict hyperbolicity of the operator (1.1) guarantees that the Cauchy problem for this operator is L_2-well-posed. In the case of a scalar operator (i.e. $N = 1$), the energy inequality (1.7) for some q implies that the operator is strictly x_0-hyperbolic (see Ivrii & Petkov [16]). For systems, however, this result does not hold; moreover, there is no simple algebraic criterion of L_2-well-posedness in this case. It has only been proved in this connection that x_0-hyperbolicity is a necessary condition for L_2-well-posedness. In the case of first order systems (see [16]), the necessity of the so-called uniform diagonalizability has been established; the latter condition amounts to the existence of a matrix $C(x, \xi')$ with the following properties: $C(x, \xi')$ and its inverse have bounded norms, and $C^{-1}(x, \xi')A(x, \xi')C(x, \xi')$ is a diagonal matrix (cf. Sect. 1.2). It has actually been proved by Petrowsky (see Sect. 1) that the Cauchy problem for the system is L_2-well-posed if the matrix $C(x, \xi')$ smoothly depends on x, ξ'. A sufficient condition for L_2-well-posedness can be formulated as follows: there exists a smooth positive definite symmetric matrix $S(x, \xi')$ such that $S(x, \xi')A(x, \xi')$ is also symmetric (see Friedrichs & Lax [18], Mizohata [19]). On the other hand, it has been shown by Strang [17] that the estimate (1.17) may be absent in the case of a non-smooth diagonalizing matrix $C(x, \xi')$.

L_2-well-posedness is by no means the only type of well-posedness known for the Cauchy problem. Among the other types, the most important and thoroughly studied is C^∞-well-posedness. Its definition is as follows.

The Cauchy problem (1.3), (1.4) *is C^∞-well-posed, if it admits a solution*

$$u \in C^\infty(\Omega), \quad \forall f \in C^\infty(\Omega), \quad \forall g_j \in C^\infty(\Omega_0),$$

where $\Omega_t = \{x \in \Omega, \ x_0 = t\}$, and the following implication holds

$$\forall t \in (0, T] \ \{f = 0 \text{ for } x_0 < t, \ g_j = 0\} \Longrightarrow \{u = 0 \text{ for } x_0 < t\}.$$

The fact that x_0-hyperbolicity is a necessary condition for the non-characteristic Cauchy problem for a system of type (1.1) to be C^∞-well-posed has been established, under additional assumptions, by Lax [20], and by Mizohata [19] in the general case. A relatively simple proof of this result can be found in [16].

Hyperbolic equations (systems) that do not satisfy the condition of strict hyperbolicity will be referred to as *non-strictly hyperbolic*, or *weakly hyperbolic*. In this section we give a description of the most important classes of weakly hyperbolic operators, for which necessary and sufficient conditions of well-posedness have been studied in detail.

2.1. Operators with Constant Coefficients

Exhaustive results about operators with constant coefficients are due to Lax [21] and Gårding [22] (see also [23]). It has been shown that the Cauchy problem is C^∞-well-posed if and only if the operator $P(D)$ is hyperbolic in the sense of Gårding, i.e., if there exists a constant C such that

$$\det P(\xi_0, \xi') \neq 0 \,, \quad \xi' \in \mathbb{R}^n, \quad |\mathrm{Im}\, \xi_0| > C \,. \tag{2.1}$$

It follows that C^∞-well posedness of the Cauchy problem for scalar weakly hyperbolic equations with constant coefficients can be guaranteed only if additional conditions are imposed on the lower order terms.

For first order systems, the restrictions on the lower order terms can be avoided if and only if

$$\left\| p^{-1}(\xi) \right\| \leq \frac{C}{\mathrm{Im}\, \xi_0} \,, \quad \mathrm{Im}\, \xi_0 \neq 0 \,, \quad \xi' \in \mathbb{R}^n \,. \tag{2.2}$$

This condition is equivalent to uniform diagonalizability of $p(\xi)$, which in its turn is equivalent to an L_2-estimate of type (1.7). Suitable conditions for higher order systems can be found in [24].

2.2. Scalar Operators whose Characteristics have (Locally) Constant Multiplicity

This is another fully studied class of operators. Consider a scalar operator of type (1.1) and assume that each root $\lambda_j(x, \xi')$ of its characteristic equation (1.6) has constant multiplicity r_j $(j = 1, \ldots, s)$.[3]

Let us associate to each root the following eikonal equation

$$\frac{\partial \varphi_j}{\partial x_0} - \lambda_j(x, \mathrm{grad}_{x'} \varphi_j) = 0 \,. \tag{2.3}$$

We say that Condition (L) (the generalized Levi condition) holds for the scalar operator P, if

$$P(f \, e^{i\rho\varphi}) = O(\rho^{m-r_j}) \quad \text{as} \quad \rho \to +\infty \,, \quad j = 1, \ldots, s \,, \tag{2.4}$$

---

[3]This condition can be equivalently expressed by

$$p(x, \xi) = g_1^{r_1}(x, \xi) \cdots g_s^{r_s}(x, \xi) \,,$$

where g_j are strictly hyperbolic polynomials such that

$$\{g_j(x, \xi) = g_k(x, \xi) = 0 \,, \ \xi \neq 0\} \Longrightarrow j = k \,.$$

for any $\hat{x} \in \Omega$, any $f \in C_0^\infty$ with support in a sufficiently small neighborhood of $\hat{x}$, and any real valued function $\varphi_j(x)$ satisfying (2.3) in a neighborhood of $\hat{x}$ and such that $\mathrm{grad}_{x'}\varphi_j \neq 0$.

Strang and Flashka [25] established the necessity of Condition (L) for the Cauchy problem to be C^∞-well-posed; Chazarain [26] proved its sufficiency and, assuming Condition (L) to be satisfied, constructed the parametrix for problem (1.3),(1.4) in the form of an integral Fourier operator, and also studied the propagation of singularities.

The results of Strang and Flashka imply that to ensure C^∞-well-posedness of the Cauchy problem for a scalar weakly hyperbolic equation with characteristics of constant multiplicity, it is always necessary to impose restrictions on the terms of lower order.

2.3. First Order System with Characteristics of Constant Multiplicity

In this situation, the structure of the principal symbol is not completely determined by the multiplicities of the characteristic roots of its determinant, but depends on the Jordanian normal representation.

Petkov [27] established necessary conditions for well-posedness of the Cauchy problem in the case of systems with invariable structure of the Jordanian representation of the principal symbol; sufficient conditions in this case were obtained by Petkov [28] and Demay [29]. Here we set forth only the sufficient conditions, since the necessary ones are nearly the same, but the method used for their derivation requires an additional assumption, namely, that the multiplicity of every characteristic root be ≤ 3.

Assume that every characteristic root $\lambda_j(x, \xi')$ has constant multiplicity r_j and

$$\mathrm{rank}\, p\,(x, \lambda_j(x,\xi'),\xi') = N - r_j + \kappa_j \,, \quad \kappa_j = \mathrm{const} \,,$$

$$\{\kappa_j > 0\} \Longrightarrow \{\kappa_j = r_j - 1 \ \text{or} \ \kappa = 1\} \,,$$

which means that there cannot be more than one Jordan box associated with each eigenvalue λ_j, and the order of this box is either equal to 2 ($\kappa_j = 1$), or to r_j ($\kappa_j = r_j - 1$).

Denote by $R_{k,j}(x, \xi')\,(\neq 0)$, $k = 1, \ldots, \kappa_j$, the generalized eigenvectors corresponding to the eigenvalue $\lambda_j(x, \xi')$:

$$R_{k,j}(x,\xi') = p\,(x, \lambda_j(x,\xi'),\xi')\,R_{k+1,j}(x,\xi')\,, \qquad k = 1, \ldots, \kappa_j - 1 \,.$$

In the case of first order systems, an analogue of the generalized Levi condition can be formulated as follows:

Let $\varphi_j(x)$ be a real valued solution of the eikonal equation (2.3) in a neighborhood of $\hat{x} \in \Omega$, and let $\operatorname{grad}_{x'}\varphi_j(\hat{x}) = \hat{\xi} \neq 0$. Assume that for any $f \in C_0^\infty$ in a neighborhood of $\hat{x}$ there exist vector valued functions $V_{k,j}(x)$, $k = 1, \ldots, \kappa_j$, and scalar functions $g_{k,j}$, $k = 2, \ldots, r_j - 1$, $g_{k,j}\big|_{x_0=0} = 0$, such that

$$
e^{-i\rho\varphi_j} P\left(e^{i\rho\varphi_j} \left[f(x)R_{1,j}(x, \operatorname{grad}_{x'}\varphi_j) + \sum_{k=2}^{r_j-1} g_{k,j}(x)R_{k,j}(x, \operatorname{grad}_{x'}\varphi_j) + \right.\right.
$$

$$
\left.\left. + \sum_{k=1}^{\kappa_j} V_{k,j}(x)\rho^{-k} \right] \right) = O(\rho^{-\kappa_j}) \quad \text{as} \quad \rho \to +\infty .
$$

If the above condition holds for any $\hat{x} \in \Omega$, then the Cauchy problem is C^∞-well-posed.

2.4. Necessary Conditions for C^∞-Well-Posedness for Operators with Characteristics of Variable Multiplicity

In order to formulate the most general and essential conditions for the Cauchy problem to be well-posed in this case, the notion of *fundamental matrix* and that of *subprincipal symbol* are needed.

Let $(\hat{x}, \hat{\xi})$ be a critical point of the principal symbol of the scalar operator (1.1), i.e.,

$$
p(\hat{x}, \hat{\xi}) = 0 , \quad \operatorname{grad}_{x,\xi} p(\hat{x}, \hat{\xi}) = 0 .
$$

Then the *fundamental matrix* for $p(x, \xi)$ at the point $(\hat{x}, \hat{\xi})$ is, by definition, the $2(n+1) \times 2(n+1)$-matrix

$$
F_p(\hat{x}, \hat{\xi}) = \begin{pmatrix} p_{\xi x} & p_{\xi\xi} \\ -p_{xx} & -{}^t p_{\xi x} \end{pmatrix} ,
$$

and

$$
p^s(\hat{x}, \hat{\xi}) = p_{m-1}(x, \xi) + \frac{i}{2} \sum_{j=0}^{n} \frac{\partial^2}{\partial x_j \partial \xi_j} p(x, \xi)
$$

is the *subprincipal symbol* of P; here $p_{m-1}(x, \xi)$ stands for the sum of the terms of order $m - 1$ entering the symbol $P(x, \xi)$.

If $(\hat{x}, \hat{\xi})$ is a critical point of the hyperbolic symbol $p(x, \xi)$, then either the matrix $F_p(\hat{x}, \hat{\xi})$ has a couple of real simple non-zero eigenvalues μ and $-\mu$, and the remaining eigenvalues are purely imaginary, or every eigenvalue of $F_p(\hat{x}, \hat{\xi})$ is purely imaginary.[4] It has been proved by Ivrii & Petkov [16]

---

[4] The symbol $p(x, \xi)$ is assumed to be real valued; this assumption does not lead to a loss of generality because of hyperbolicity.

and Hörmander [30] that if $(\hat{x}, \hat{\xi})$ is a critical point of $p(x, \xi)$ and there are no real non-zero eigenvalues of $F_p(\hat{x}, \hat{\xi})$, then the following condition is necessary for the Cauchy problem to be C^∞-well-posed:

$$p^s(\hat{x}, \hat{\xi}) \in \mathbb{R}^1, \qquad \left| p^s(\hat{x}, \hat{\xi}) \right| \leq \frac{1}{2} \operatorname{Tr}^+ F_p(\hat{x}, \hat{\xi}),$$

where $\operatorname{Tr}^+ = \frac{1}{2} \sum_j \omega_j$, and $\pm i \omega_j$, $\omega_j \geq 0$, form the set of all eigenvalues of the matrix $F_p(\hat{x}, \hat{\xi})$.

According to [16], the operator P is called *regularly* (or *effectively*) *hyperbolic*, if the Cauchy problem is C^∞-well-posed for any operator $P + Q$, where Q is an arbitrary operator of lower order. The above statements imply that for regular hyperbolicity it is necessary that the matrix $F_p(x, \xi)$ has a pair of real non-zero eigenvalues at each critical point of the principal symbol $p(x, \xi)$.

According to [16], a pair of non-negative integers (r, q) is called the *index of well-posedness* for the Cauchy problem $(1.3), (1.4)$, if the Cauchy problem has a unique solution $u \in C^r(\Omega)$ for any $f \in C^q(\Omega)$ and any $g_j \in C^{q+m-j}(\Omega)$.

It has been shown in [16] that if $(\hat{x}, \hat{\xi})$ is a critical point of $p(x, \xi)$ and $\pm \mu(\hat{x}, \hat{\xi})$ is a pair of real non-zero eigenvalues of $F_p(\hat{x}, \hat{\xi})$, then a pair (r, q) can be the index of well-posedness for the Cauchy problem for P only if

$$\left| \operatorname{Im} p^s(\hat{x}, \hat{\xi}) \right| \leq cn(q + m - r) |\mu(\hat{x}, \hat{\xi})|,$$

where c is an absolute constant.

A notion of *completely regular hyperbolicity* has been introduced in [16] for operators P such that the index of well-posedness for the Cauchy problem for $P + Q$ does not depend on Q, where Q is any operator whose order is less than the order of P. The above statements show that for the completely regular hyperbolicity of P it is necessary that the principal symbol $p(x, \xi)$ has no critical points.

These results imply that for a completely regular hyperbolic operator P the symbol $p(x, \xi)$ has simple characteristic roots inside the layer Ω, and for $x_0 = 0$ or $x_0 = T$, the characteristic roots may have double multiplicity. For a regularly hyperbolic operator P, the characteristic roots inside Ω may be either single of double, and for $x_0 = 0$ or $x_0 = T$ their multiplicity may reach 3.

Petkov & Kutev [31] found necessary conditions for regular hyperbolicity in the case of systems with characteristics of variable multiplicity. These authors considered a system in two independent variables

$$P = D_0 E + A(x) D_1 + B(x),$$

where $A(\hat{x})$ has an eigenvalue $\hat{\lambda}$ of multiplicity $r \geq 2$, and $\operatorname{rank}(A(\hat{x}) - \hat{\lambda}) = N - 1$. It has been shown in [31] that for a regularly hyperbolic operator P

the matrix $F_p(x, \xi)$ has a pair of real non-zero eigenvalues. A necessary condition of well-posedness has also been obtained in the case of F_p having no real eigenvalues.

2.5. Necessary Conditions for C^∞-Well-Posedness in the Case of Characteristics of Variable Multiplicity; Conditions on Lower Order Terms

In the case of variable multiplicity, one can state a condition which is an analogue of the generalized condition of Levi. Assume that in a neighborhood of $(\hat{x}, \hat{\xi})$ the characteristic equation has smooth roots $\lambda_j(x, \xi)$, $j = 1, \ldots, r$, which coincide at the point $(\hat{x}, \hat{\xi})$; and let the remaining roots at this point be different from $\lambda_1(\hat{x}, \hat{\xi})$. As shown in [16], if problem $(1.3), (1.4)$ is well-posed and possesses a finite cone of dependence, then for any φ satisfying the eikonal equation with λ_1 and any $f \in C_0^\infty$ with a sufficiently small support we have

$$P\left(f\, e^{i\varphi\rho}\right)(\hat{x}) = O\left(\rho^{m-[(r+1)/2]}\right) \quad \text{as} \quad \rho \to +\infty .$$

This condition, as well as its counterpart in the case of roots of invariable multiplicity, is a consequence of the following general result obtained in [16].

Let the coordinates x be divided into two groups:

$$x = (x^{(1)}, x^{(2)}) = (x_0, \ldots, x_k, x_{k+1}, \ldots, x_n) ,$$

and let the dual variables and multi-indices be split accordingly. Consider $\hat{x} \in \Omega$ and assume that for some given rational numbers $p \geq q \geq 0$ we have

$$p^{(r,0,\ldots,0)}(\hat{x}, 0, \xi^{(2)}) \neq 0 ,$$

$$p_{(\beta)}^{(\alpha)}(\hat{x}, 0, \xi^{(2)}) = 0 \quad \text{if} \quad |\alpha| + p|\beta^{(1)}| + q|\beta^{(2)}| < r .$$

Assume also that the Cauchy problem for P is well-posed and has a finite cone of dependence. Then

$$p_{s(\beta)}^{(\alpha)}(\hat{x}, 0, \xi^{(2)}) = 0 , \quad \forall \xi^{(2)} \in \mathbb{R}^{n-k} ,$$

provided that

$$|\alpha| + p|\beta^{(1)}| + q|\beta^{(2)}| + (m - s)(1 + p) < r .$$

The above results can be effectively applied to wide classes of equations whose characteristic roots "coalesce" to within a finite order. In some cases of infinite order "coalescence", the results of Nersesjan [38] are applicable.

2.6. Sufficient Conditions of Well-Posedness for Scalar Weakly Hyperbolic Equations

As shown by Ivrii [32], the absence of critical points of the principal symbol $p(x, \xi)$ is not only a necessary but also a sufficient condition for completely regular hyperbolicity. Moreover, such operators admit *a priori* estimates resembling the estimate obtained by Petrowsky and differing from the latter by weighted norms with weight functions having zeros or singularities of order $1/2$ at $t = 0$ or $t = T$. An example of such an estimate is provided by

$$\sum_{|\alpha| \leq q+m-1} \|D^\alpha u; t\| \leq C \sum_{|\alpha| \leq q+m} \|D^\alpha u; 0\| + C' \sum_{|\alpha| \leq q} \left(\int_0^t \|D^\alpha P u; \tau\|^\kappa \, \tau^{\frac{\kappa}{2}-1} \, d\tau \right)^{\frac{1}{\kappa}},$$

where $u \in C^\infty$, $1 \leq \kappa \leq 2$, $0 \leq t \leq T' < T$.

Some sufficient conditions for regular hyperbolicity have also been obtained in [32]. Namely, assume that the following factorization

$$p(x, \xi) = p_1(x, \xi) p_2(x, \xi), \quad p_i(\hat{x}, \hat{\xi}) = 0, \quad i = 1, 2, \qquad (2.5)$$

holds in a neighborhood of every critical point $(\hat{x}, \hat{\xi})$. Then $F_p(\hat{x}, \hat{\xi})$ has eigenvalues $\pm\{p_1, p_2\}(\hat{x}, \hat{\xi})$, and the remaining eigenvalues are all equal to 0. Thus, in the case of factorization (2.5),

$$\{p_1, p_2\}(\hat{x}, \hat{\xi}) \neq 0$$

is a necessary condition for regular hyperbolicity. This condition also happens to be sufficient, and we can take as the index of well-posedness any pair $(m-1+q, q+N)$, where $q = 1, \ldots$, and N is determined by the values of the subprincipal symbol and the values of the fundamental matrix at critical points of the principal symbol p.

A similar result holds even in the case of $n = 1$, if we drop the factorization condition but assume, in addition, that in a neighborhood of its critical point $(\hat{x}, \hat{\xi})$ the principal symbol $p(x, \xi)$ can be represented in the form

$$p(x, \xi) = q_0^2(x, \xi) - \sum_{j=1}^{\hat{s}} q_j^2(x, \xi),$$

where $q_j \in C^\infty$ are real valued symbols, $q_j(\hat{x}, \hat{\xi}) = 0$, and the vectors $\operatorname{grad}_{\xi^\bullet - x} q_j(\hat{x}, \hat{\xi})$, $j = 0, \ldots, \hat{s}$, are linearly independent. It follows from the results obtained by Oleinik [37] for degenerate equations that a similar statement is valid for $m = 2$ and $n = 1$ without any additional assumptions.

Ivrii [33] and Hörmander [30] have examined sufficient conditions for C^∞-well-posedness in the case of operators whose fundamental matrices have no real non-zero eigenvalues.

Assume that P is an x_0-hyperbolic operator, and

$$\Sigma = \left\{ (x, \xi) \in \Omega \times \mathbb{R}^{n+1}, \; p(x, \xi) = 0, \; \operatorname{grad} p(x, \xi) = 0 \right\}$$

is a C^∞-manifold such that for every point $\rho \in \Sigma$ the tangent space to Σ coincides with the kernel of the fundamental matrix: $T_\rho \Sigma = \operatorname{Ker} F_p(\rho)$, and $\operatorname{Ker} F_p^2(\rho) \cap \operatorname{Ran} F_p^2(\rho) = 0$.

Assume, further, that there is a vector field

$$z(\rho) \in \operatorname{Ker} F_p(\rho) \cap \operatorname{Ran} F_p(\rho)$$

such that $z(\rho)$ smoothly depends on $\rho \in \Sigma$ and

$$\sigma(z(\rho), \zeta) = 0 \Longrightarrow \varepsilon \sigma(F_p(\rho)\zeta, \zeta) \geq 0 ,$$

where the right-hand side turns into equality only for $\zeta \in \operatorname{Ker} F_p(\rho)$. Here $\varepsilon = \pm 1$ and does not depend on ζ; σ is a simplectic bilinear form.

Assume that for any $\rho \in \Sigma$ the strong Ivrii-Petkov-Hörmander condition is satisfied:

$$\operatorname{Im} p^s(\rho) = 0 , \quad |\operatorname{Re} p^s(\rho)| < \frac{1}{2} \operatorname{Tr}^+ F_p(\rho) , \qquad \rho \in \Sigma .$$

Then the Cauchy problem for P is C^∞-well-posed. Moreover, an energy estimate (different from that of Petrowsky) is valid in this case.

Hörmander [30] has also proved the following result: if Σ is a C^∞-manifold and

$$T_\rho \Sigma = \operatorname{Ker} F_p(\rho) , \quad F_p^2(\rho) = 0 ,^5 \quad p^s(\rho) = 0, \quad \forall \rho \in \Sigma ,$$

then the Cauchy problem for P is C^∞-well-posed. All these results have been obtained by the method of separating operator.

Some generalizations of the results described in this section to the case of operators with characteristics of higher multiplicity can be found in [34].

2.7. Degenerate Hyperbolic Equations

Numerous studies are dedicated to equations which are strictly hyperbolic only outside certain smooth manifolds; such equations are usually referred to as *degenerate hyperbolic equations*. Most of these studies are concerned with second order equations with two independent variables; a review of these results can be found in the monographs of Smirnov [35] and Bitsadze [36].

---

[5] Under these assumptions, all eigenvalues of $F_p(\rho)$ are equal to 0.

General second order equations

$$Pu = u_{x_0 x_0} - \sum_{i,j=1}^{n} a_{ij}(x)u_{x_i x_j} + b_0(x)u_{x_0} + \sum_{j=1}^{n} b_j(x)u_{x_j} + c(x)u = f(x) \quad (2.6)$$

were studied by Oleinik [37], who established the following sufficient condition for the Cauchy problem to be well-posed:

$$\alpha x_0 \left(\sum b_j(x)\xi_j\right)^2 \le A \sum a_{ij}(x)\xi_i\xi_j + \sum \frac{\partial a_{ij}(x)}{\partial x_0}\xi_i\xi_j , \qquad \forall \xi \in \mathbb{R}^n ,$$

where α, A are positive constants. Under this condition, an energy estimate of type (1.7) was proved. This estimate implies that we can take as the index of well-posedness any pair $(k, k + p + 2)$ with integer $k \ge 2$ and $p \ge -1$ such that $\alpha > (2p + 6)^{-1}$ (see also [94], [95]).

Degenerate hyperbolic equations of higher order, as well as systems, have been studied by Nersesjan [38], Menikoff [39], and others. The construction of a parametrix for such equations has been considered by Yoshikawa [40] and Alinhac [41].

Sakamoto [42] considered differential operators whose symbols can be expanded in asymptotic (with respect to x_0) series of the form

$$P(x, \xi) = \sum_{\mu \in N} x_0^\mu P^{(\mu)}(x', \xi) , \qquad (2.7)$$

where $N = \{\mu_0, \mu_1, \ldots, \mu_k, \ldots\}$, $\mu_0 < \mu_1 < \cdots \to +\infty$; and it is assumed that the principal symbol is strictly hyperbolic for $x_0 > 0$. Sakamoto imposes some additional restrictions on the behavior of the symbol for $x_0 \to +0$. These conditions yield a representation of the principal symbol in the form

$$P(x, \xi) = \prod_{i=1}^{l} \prod_{j=1}^{m_i} (\xi_0 - x_0^{\tau_i} \lambda_{ij}(x, \xi')) ,$$

where $\lambda_{ij}(x, \xi')$ are real and mutually different.

For operators (2.7) the following version of the Cauchy problem is proposed. Let $H(\mathbb{R}^{n-1}) = \bigcap_s H^{(s)}(\mathbb{R}^{n-1})$ and

$$C^\infty \left((0, T]; H(\mathbb{R}^{n-1})\right) =$$

$$= \left\{ f(x_0; x') , \sup_{T_0 < t < T} \int_{\mathbb{R}^{n-1}} \left|D_0^i D_x^\alpha f(x)\right|^2 dx < \infty \quad \forall i, \alpha, \forall T_0 > 0 \right\}.$$

Consider a function $f(x) \in C^\infty \left((0, T]; H(\mathbb{R}^{n-1})\right)$ and assume that in this space $f(x)$ admits the following asymptotic expansion

$$f(x) \sim x_0^{-m} \sum_{\nu \in N} x_0^{\nu} f_{\nu}(x') , \quad f_{\nu}(x') \in H(\mathbb{R}^{n-1}) .$$

Let $u_{\nu}(x) \in H(\mathbb{R}^{n-1})$, $\nu \in N$, be a given sequence related to $f(x)$ by the corresponding compatibility conditions, which can be easily written out. Then the Cauchy problem is understood in the sense that one needs to find a function $u(x)$ such that

$$u(x) \in C^{\infty}\left((0,T]; H(\mathbb{R}^{n-1})\right) ; \quad u(x) \sim \sum_{\nu \in N} x_0^{\nu} u_{\nu}(x') ,$$

$$P(x,D)u(x) = f(x) \quad x \in \Omega = (0,T] \times \mathbb{R}^{n-1} . \tag{2.8}$$

Sakamoto proved existence and uniqueness theorems for problem (2.8), and also obtained *a priori* estimates for weighted L_2-norms of its solutions, taking as weight functions various powers of x_0.

In [43], Sakamoto extended the results of [42] to operators with zeros of constant multiplicity 2 for $x_0 > 0$.

In close relation to the studies on degenerate equations are the works on the characteristic Cauchy problem [45], [46]. These papers deal with strictly hyperbolic equations (systems) with the Cauchy data prescribed on a smooth surface $\{s(x) = 0\}$ which may have characteristic direction at some points, i.e.,

$$P(x^0, \text{grad } s(x^0)) = 0 \quad \text{for some} \quad x^0 . \tag{2.9}$$

Leray, Gårding & Kotake [44] considered the Cauchy problem for a hyperbolic system with analytic coefficients. For an analytic initial manifold, assuming this manifold to be characteristic (in the sense of (2.9)) at the points of its analytic submanifold, these authors examined the asymptotic behavior of the solution near the characteristic submanifold and showed that this solution is an algebroid function. Kondratiev [45] considered questions of existence, uniqueness and asymptotic behavior of solutions in the case when the coefficients, the initial surface and the characteristic submanifold are C^{∞}-smooth. Weinberg & Mazja [46] examined the following case: P is a strictly hyperbolic operator; the characteristic points of the initial surface fill up a compact set V, whose nature is left unspecified, and V is even allowed to have positive measure. In this case, the initial conditions (1.4) are replaced by

$$D_0^j u(x)\Big|_{x \in S} = g_j , \quad j = 0, \ldots, m-2 ,$$

$$D_0^{m-1} u(x)\Big|_{x \in S \setminus V} = g_{m-1} , \tag{1.4'}$$

where m is assumed even. In [46], energy estimates in some special weighted norms are obtained for solutions of problem (1.3), (1.4'), and existence and uniqueness theorems are established.

2.8. Well-Posedness of the Cauchy Problem in Gevrey Classes

As we have seen above, the Cauchy problem for weakly hyperbolic equations of general type is C^∞-well-posed only if additional restrictions are imposed on the lower order terms. In order to have a well-posed Cauchy problem for any weakly hyperbolic operator with analytic coefficients, the notion of C^∞-well-posedness should be abandoned and C^∞ should be replaced by a more narrow class of functions, for instance, the Gevrey class $G^{(\kappa)}$. First, we give some necessary definitions.

Let $1 \le \kappa < \infty$. The *Gevrey class* $G^{(\kappa)}$ is the set formed by all $f \in C^\infty$ such that

$$\exists C > 0: \quad \forall \alpha \quad |D^\alpha f| \le C^{1+|\alpha|}(|\alpha|!)^\kappa .$$

Together with the sets $G^{(\kappa)}$, which should be called *inductive Gevrey classes*, it is natural to consider more narrow classes of functions, the so-called *projective Gevrey classes* $G^{(\kappa)}$ formed by all $f \in C^\infty$ such that

$$\forall \varepsilon \quad \exists C = C(\varepsilon) > 0: \quad |D^\alpha f| < C\varepsilon^{|\alpha|}(|\alpha|!)^\kappa \quad \forall \alpha .$$

The Cauchy problem $(1.3),(1.4)$ is said to be $G^{(\kappa)}$-*well-posed*, if the following conditions are satisfied: for a given layer $0 \le x_0 \le T$ and any $f, g_j \in G^{(\kappa)}$ there is a solution $u \in G^{(\kappa)}$; moreover, the relations

$$Pu = 0 \quad \text{for} \quad x_0 < t , \qquad D_0^j u\big|_{x_0=0} = 0 , \quad j = 0,\ldots, m-1 ,$$

imply that $u(x) = 0$ for $x_0 \le t$.

If for $f, g_j \in G^{(\kappa)}$ a solution exists only in a thin layer $0 \le x_0 \le T_0$, with T_0 depending on f, g_j, and the above uniqueness condition is satisfied, then problem $(1.3),(1.4)$ is said to be *locally* $G^{(\kappa)}$-*well-posed*.

$G^{(\kappa)}$-*well-posedness* can be defined in a similar way.

The questions of $G^{(\kappa)}$-well-posedness have been considered by Leray & Ohya [47] – [49], Hörmander (see [23], Ch. 5), Beals [50], Nersesjan [51]. All these authors impose on the operator P, in addition to its hyperbolicity, some other condition such as the possibility of representing the principal part of the operator as a product of strictly hyperbolic differential operators (Leray, Ohya), or the operator having constant coefficients (Hörmander), etc. Without any additional assumptions, $G^{(\kappa)}$-well-posedness of the Cauchy problem for hyperbolic operators has been established by Ivrii [52] and Bronstein [53]. Next we formulate the result of [53] in more precise terms.

Let P be a hyperbolic operator, and let r be the largest multiplicity of its characteristic roots, $\kappa^* = r/(r-1)$. Assume that the coefficients of P belong to $G^{(\kappa)}$. If $\kappa < \kappa^*$, then the Cauchy problem $(1.3),(1.4)$ is $G^{(\kappa)}$-well-posed; if $\kappa = \kappa^*$, then this problem is locally $G^{(\kappa)}$-well-posed. If the coefficients of P belong to the class $G^{(\kappa)}$ and $\kappa < \kappa^*$, then the Cauchy problem is $G^{(\kappa)}$-well-posed. All these results are extended in [53] to systems hyperbolic in the sense of Sect. 1.4.

As shown by examples in [48], [23], [53], it is impossible to go beyond the Gevrey classes $G^{(\kappa)}$ with $\kappa \leq r/(r-1)$, if there is no additional information about the hyperbolic operator.

As proved by Ivrii [54] for an operator with analytic coefficients, its hyperbolicity is a necessary condition of $G^{(\kappa)}$-well-posedness, or $G^{(\kappa)}$-well-posedness.

Ivrii [54] and, independently, Komatsu [55] have found necessary conditions of well-posedness in Gevrey classes for scalar operators with characteristics of constant multiplicity (the sufficiency of these conditions follows from the results of [47]). Let the phase φ be a solution of the eikonal equation (2.3) corresponding to a characteristic root of multiplicity r. Then a necessary condition for $G^{(\kappa)}$-well-posedness can be expressed by

$$P\left(e^{i\varphi\rho+i\psi\rho^{\frac{1}{\kappa}}h}\right) = \rho^{m-r\left(1-\frac{1}{\kappa}\right)} e^{i\varphi\rho+i\psi\rho^{\frac{1}{\kappa}}} \left\{ \sum_{|\alpha|\leq r} \frac{\psi_x^\alpha}{\alpha!} P_m^{(\alpha)}(x,\varphi_x)h + O(1) \right\},$$

as $\rho \to +\infty$, for any real valued $\psi \in C^\infty$ and any $h \in C_0^\infty$. A necessary condition for local $G^{(\kappa)}$-well-posedness can be written in the form

$$P\left(e^{i\varphi\rho+i\psi\rho^{\frac{1}{\kappa}}h}\right) = O\left(\rho^{m-r\left(1-\frac{1}{\kappa}\right)}\right) \quad \text{as} \quad \rho \to +\infty,$$

for any real valued $\psi \in C^\infty$ and any $h \in C_0^\infty$. It should be observed that the latter condition turns into the generalized Levi condition (2.4) as $\kappa \to +\infty$.

In [56], Ivrii obtained necessary conditions for $G^{(\kappa)}$-well-posedness in the case of scalar operators with characteristics of variable multiplicity. These conditions are a generalization of the conditions for C^∞-well-posedness given in Sect. 2.5.

Nersesjan [51] and Ivrii [57] also studied sufficient conditions of well-posedness in Gevrey classes for degenerate equations; as a rule, the necessary and sufficient conditions obtained in [57] are the same. The examples analyzed in [57] allow us to make the following remarks of general character. The above three types of well-posedness (namely, $G^{(\kappa)}$-, local $G^{(\kappa)}$- and $G^{(\kappa)}$-well-posedness) differ from one another, in general. For operators with analytic coefficients, it may be observed that there exists a constant κ^*, $1 < \kappa^* \leq \infty$, such that for $\kappa < \kappa^*$ all three types of well-posedness take place, while for $\kappa = \kappa^*$ any of these three types can either hold or be violated, and the following implications are valid:

$$\left\{G^{(\kappa)}\text{-well-posedness}\right\} \Longrightarrow \left\{\text{local } G^{(\kappa)}\text{-well-posedness}\right\} \Longrightarrow$$

$$\Longrightarrow \left\{G^{(\kappa)}\text{-well-posedness}\right\};$$

however, the inverse implications do not hold, in general.

The case $\kappa = 1$ corresponds to analytic functions, and local $G^{(1)}$-well-posedness (for equations with analytic coefficients) follows from the Cauchy-Kowelevskaya theorem. Bony & Schapira [58] have shown that for an operator whose principal symbol is hyperbolic the domain of existence of an analytic solution depends on the operator only. For hyperbolic equations, Bony & Schapira have also established well-posedness of the Cauchy problem in classes on hyperfunctions. Komatsu [55] proved the converse statement: let Ω be a neighborhood of the origin independent of the radius of analyticity of the Cauchy data and independent of the right-hand side; assume that in Ω there always exists a classical solution (or even a solution in the class of hyperfunctions) of the Cauchy problem; then the operator P is x_0-hyperbolic.

3. The Mixed Problem for Hyperbolic Equations

3.1. Solvability of the mixed problem for second order hyperbolic equations with classical boundary conditions (of Dirichlet or Neumann type) was proved for the first time in the mid-'30s by Krzyzansky & Schauder [59] (see also [60]). For more than three decades, no adequate theory could have been developed for the mixed problem (in contrast to the Cauchy problem) in the cases other than that of second order equations. The results existing at that time were of fragmentary nature and pertained either to the case of two independent variables or some special classes of operators (a fairly complete bibliography can be found in reviews [61], [62]). Of relatively general nature were the results on the solvability of positive local boundary value problems for symmetric first order hyperbolic systems (Friedrichs [63], [64], Dezin [65], [66]) and for symmetrizable systems [18]. More references can be found in [61].

The first important result of general character is due to Agmon [67], who considered the mixed problem in half-space for a scalar higher order operator with principal part having constant coefficients. Agmon introduced the so-called *uniform Lopatinsky condition* and obtained an energy estimate, assuming this condition to be fulfilled in addition to another one, namely, that every root of the principal symbol with respect to the variable dual to the normal has multiplicity ≤ 2 (Agmon's condition). At the end of the '60s, a complete theory of mixed problems satisfying the uniform Lopatinsky condition appeared simultaneously in two versions: for first order systems, the theory was developed by Kreiss [68] (see also the works of Ralston [69], Rauch [70] and Agranovich [71], [72], which contain important supplementary results to those obtained by Kreiss), and for a single higher order equation, the theory is due to Sakamoto [73] (see also the article of Balaban [74], whose results rely on the Agmon condition, and the review

by Volevich & Gindikin [75]). Below we describe in detail the results of
Sakamoto and Kreiss. At the end of this section we briefly mention some
works where the mixed problem is studied under weaker assumptions than
those implied by the uniform Lopatinsky condition.

3.2. Let Ω be a cylinder, $\Omega = [0, \infty] \times \omega$, where ω is a domain in $\mathbb{R}^n$,
and let $\partial \omega$ be the boundary of ω. The mixed problem consists in finding a
solution $u(x)$, $x \in \Omega$, of the following equations

$$Pu = f \quad \text{in} \quad \Omega , \tag{3.1}$$

$$B_j u = \varphi_j \quad \text{on} \quad [0, \infty] \times \partial \omega , \quad j = 1, \ldots, \mu , \tag{3.2}$$

$$D^j u = g_j \quad \text{on} \quad \Omega_0 = \{x \in \Omega, \ x_0 = 0\} , \quad j = 0, \ldots, m-1 . \tag{3.3}$$

Here P is a differential operator of order m with principal symbol $p(x, \xi)$; B_j
are differential operators of orders $m_j < m$ with principal symbols $b_j(x, \xi)$,
respectively. For the sake of simplicity, we assume that ω is the half-space

$$\omega = \{x' \in \mathbb{R}^n, \ x_n \geq 0\} ,$$

and the Cauchy data g_j are identically equal to 0. In this case, it can be
assumed that equation (3.1) holds for all $x \in \mathbb{R}^1 \times \omega$ and

$$u = f = g_j = 0 \quad \text{for} \quad x_0 < 0 . \tag{3.3'}$$

It is also assumed that the following conditions are satisfied:

(A) The operator P is strictly hyperbolic with respect to x_0.

(B) The boundary $(0, \infty] \times \partial \omega$ is non-characteristic with respect to P (i.e.,
the coefficient by D_n^m does not vanish for $x_n = 0$).

By virtue of (A) and (B), the equation (with respect to λ)

$$p(x_0, \ldots, x_{n-1}, 0, \xi_0 - i\gamma, \xi_1, \ldots, \xi_{n-1}, \lambda) = 0 \tag{3.4}$$

has m complex roots $\lambda_j^{\pm}(\theta)$, which continuously depend on the parameter

$$\theta = (x_0, \ldots, x_{n-1}, \xi_0, \gamma, \xi_1, \ldots, \xi_{n-1})$$

and satisfy the inequalities $\pm \operatorname{Im} \lambda_j^{\pm} > 0$; moreover, the respective numbers
m^+ and m^- of such roots are independent of θ.

(C) The number of conditions in (3.2) is equal to the number of the roots
of (3.4) belonging to the upper half-plane, i.e., $\mu = m^+$.

(D) Consider the factorization of the symbol $p(\theta, \lambda)$

$$p(\theta, \lambda) = p_+(\theta, \lambda)p_-(\theta, \lambda) = \prod_{j=1}^{m_+}\left(\lambda - \lambda_j^+(\theta)\right) \prod_{j=1}^{m_-}\left(\lambda - \lambda_j^-(\theta)\right).$$

The polynomials $\{b_j(\theta, \lambda)\}_{j=1,\ldots,\mu}$ have the property of uniform linear independence mod $p_+(\theta, \lambda)$ for $(x_0, \ldots, x_{n-1}) \in \mathbb{R}^n$ and

$$\gamma^2 + \xi_0^2 + \cdots + \xi_{n-1}^2 = 1, \qquad \gamma \geq 0. \tag{3.5}$$

Condition (D) is known as the *uniform Lopatinsky condition*.

In order to formulate the existence theorem for problem $(3.1)-(3.3)$, we need some well-known functional spaces. Let $H_\gamma^{(s)}$ be the space of functions $\varphi(x_0, \ldots, x_{n-1})$ such that $e^{-\gamma x_0}\varphi \in H^{(s)}(\mathbb{R}^n)$, and let $\langle\cdot\rangle_{(s),\gamma}$ be the norm in this space. Denote by $H_\gamma^{(m,s)}$ the space of all $u(x)$, $x \in \mathbb{R}_+^n$, having a finite norm

$$\|u\|_{(m,s),\gamma}^2 = \sum_{|\alpha|\leq m}\int_0^\infty \langle D^\alpha u(\cdot, x_n)\rangle_{(s),\gamma}^2\, dx_n.$$

By $\overset{\circ}{H}_\gamma^{(m,s)}$, $\overset{\circ}{H}_\gamma^{(s)}$ we denote the respective subspaces consisting of functions that vanish for $x_0 < 0$.

Problem $(3.1)-(3.3')$ is said to be L_2-*well-posed*, if $\forall s \in \mathbb{R}^1$, $\exists \gamma(s) > 0$ such that $\forall \gamma > \gamma(s)$, $\forall f \in \overset{\circ}{H}_\gamma^{(0,s)}$, $\forall \varphi_j \in \overset{\circ}{H}^{(m+s-1-m_j)}$, $j = 1, \ldots, \mu$, problem $(3.1) - (3.3')$ admits a unique solution $u \in \overset{\circ}{H}_\gamma^{(m,s-1)}$, and this solution satisfies the inequality

$$\gamma\|u\|_{(m,s-1),\gamma}^2 + \sum_{j=0}^{m-1}\left\langle(D_n^j u)(x_0, \ldots, x_{n-1}, 0)\right\rangle_{(m-1+s-j),\gamma} \leq$$

$$\leq C\left(\frac{1}{\gamma}\|Pu\|_{(0,s),\gamma} + \sum_{j=1}^{\mu}\langle(B_j u)(x_0, \ldots, x_{n-1}, 0)\rangle_{(m-1+s-m_j,\gamma)}\right). \tag{3.6}$$

Theorem. *The mixed problem $(3.1) - (3.3')$ is L_2-well-posed if and only if the conditions* (A), (B), (C), (D) *are satisfied.*

The crucial point in the proof of this theorem is the derivation of the estimate (3.6) from the conditions (A),(B),(C),(D). This estimate is due to Sakamoto, who obtained it by micro-localization of (3.6) and constructed for this purpose analogues of the separating Leray operators (cf. Sect. 1.3) for micro-local canonical forms P. Solvability of the mixed problem was proved by Sakamoto under an additional assumption on the operators B_j (the so-called condition of normality in the sense of Schechter). Agranovich [71] proved the estimate (3.6) and the existence of solutions for the mixed problem by reducing this problem to a system of first order pseudo-differential operators; to the latter system this author applied the method of Kreiss [68]. A complete proof of the above theorem can be found in the article by Volevich & Gindikin [75].

3.3. Now we discuss the setting of the mixed problem for first order hyperbolic systems. The mixed problem with zero Cauchy data amounts to finding a vector valued function $u(x) = (u_1(x), \ldots, u_N(x))$ satisfying the system

$$Pu = D_0 u + \sum_{j=1}^{n} A_j(x) D_j u + A'(x) u = f(x) , \quad x \in \omega \times \mathbb{R}^1 , \qquad (3.7)$$

the boundary conditions

$$B(x)u\Big|_{x_n=0} = \varphi , \qquad (3.8)$$

and the initial conditions

$$u(x) = f(x) = 0 , \quad \varphi(x) = 0 , \quad x_0 < 0 . \qquad (3.9)$$

Here $A_1, \ldots, A_n, A'$ are square $N \times N$-matrices; $B(x)$ is a matrix with N columns and μ rows.

Let us formulate for problem $(3.7)-(3.9)$ the counterparts of conditions (A)–(D) of the preceding section.

($\tilde{A}$) System (3.7) is strictly hyperbolic.[6]

($\tilde{B}$) The boundary $\mathbb{R}^1 \times \partial\omega$ is non-characteristic with respect to P, i.e., $\exists A_n^{-1}(x)$.

Condition ($\tilde{B}$) allows us to consider the matrix

$$M(x, \xi_0 - i\gamma, \xi_1, \ldots, \xi_n) = -A_n^{-1}(x) \left((\xi_0 - i\gamma)E + \sum_{j=1}^{n} \xi_j A_j(x) \right) . \qquad (3.10)$$

By virtue of ($\tilde{A}$) and ($\tilde{B}$) the following factorization is valid:

$$\det \| \lambda E - M(x_0, \ldots, x_{n-1}, 0, \xi_0 - i\gamma, \xi_1, \ldots, \xi_{n-1}, 0) \| = \Delta_+(\theta, \lambda) \Delta_-(\theta, \lambda) ,$$

where $\Delta_\pm(\theta, \lambda)$ are polynomials (with respect to λ) of degree $m^\pm$ independent of $\theta = (x_0, \ldots, x_{n-1}, \xi_0, \gamma, \xi_1, \ldots, \xi_{n-1})$; moreover, for $\gamma > 0$, all roots of $\Delta_\pm$ belong to the upper (lower) half of the complex plane.

---

[6]Kreiss [68] thought of hyperbolicity in a narrow sense, assuming the roots of equation (1.6) (with respect to ξ_0) to be different from one another, while Agranovich [72] replaced this assumption by the condition of uniform diagonalizability of the matrix $\xi_0 E + \sum \xi_j A_j$.

(C̃) The matrix $B(x)$ has m^+ rows, i.e., $\mu = m^+$.

(D̃) (Uniform Lopatinsky condition.) The rows of the matrix

$$B(x_0, \ldots, x_{n-1}, 0)L(\theta, \lambda) ,$$

where L is the cofactor matrix of (3.10), are linearly independent mod $\Delta_+(\theta, \lambda)$ at the points of the half-sphere (3.5) for $(x_0, \ldots, x_{n-1}) \in \mathbb{R}^n$.

Problem $(3.7) - (3.9)$ is said to be L_2-well-posed, if for some $\gamma_0 > 0$ and any f, φ such that $e^{-\gamma x_0} f \in L_2(\mathbb{R}^1 \times \omega)$, $e^{-\gamma x_0} \varphi \in L_2(\mathbb{R}^1 \times \partial\omega)$, there exists a vector valued function $u(x)$ with the following properties:

1) $e^{-\gamma x_0} u(x) \in L_2(\mathbb{R}^1 \times \omega)$;

2) $u(x)$ satisfies (3.1) in the sense of distributions;

3) $u(x)$ admits a restriction $u(x_0, \ldots, x_{n-1}, 0)$ to the hyperplane $x_n = 0$ such that
$$e^{-\gamma x_0} u(x_0, \ldots, x_{n-1}, 0) \in L_2(\mathbb{R}^1 \times \partial\omega) ;$$

4) if $f = 0$ and $\varphi = 0$ for $x_0 < 0$, then $u = 0$ for $x_0 < 0$;

5) the following estimate holds:

$$\gamma \left\| e^{-\gamma x_0} u \right\| + \left\| e^{-\gamma x_0} u(\cdot, 0) \right\| \leq$$
$$\leq C \left(\gamma^{-1} \left\| e^{-\gamma x_0} Pu \right\| + \left\| e^{-\gamma x_0} (Bu)(\cdot, 0) \right\| \right), \quad (3.11)$$

where C is a constant.

Theorem. *Problem* $(3.7) - (3.9)$ *is* L_2-*well-posed, if and only if the conditions* (Ã), (B̃), (C̃), (D̃) *are satisfied.*

Actually, this theorem is due to Kreiss [68] (see also [69], [71], [72]). Its proof is essentially based on Kreiss' construction of a symmetrizing operator given by a matrix $R(\theta)$ which smoothly depends on θ, satisfies the inequality

$$\left| D_x^\alpha D_0^{\beta_0} \ldots D_{n-1}^{\beta_{n-1}} \left(\frac{\partial}{\partial\gamma} \right)^{\beta_n} R(\theta) \right| < \text{const} \left(|\xi_0| + \cdots + |\xi_{n-1}| + |\gamma| \right)^{-|\beta|} ,$$

and possesses the following properties:

1) $R = R^*$;

2) $\exists C > 0$ such that $\operatorname{Im} R \cdot M > C\gamma E$;

3) $\exists c > 0$, $\exists C \in \mathbb{R}^1$ such that $\forall w \in \mathbb{C}^N$ we have

$$-(Rw, w) > c|w|^2 - C|Bw|^2 .$$

Using the symmetrizing operator R, the Gårding inequality and the standard technique of pseudo-differential operators, one can easily establish the *a priori* estimate (3.11)

The problem of constructing a symmetrizing operator is purely algebraic; its solution is based on fairly delicate results about perturbations of a matrix that smoothly depends on a multi-dimensional parameter in a neighborhood of a point at which the matrix has non-trivial Jordan boxes.

3.4. The uniform Lopatinsky condition, whose versions have been considered in Sects. 3.2 and 3.3, substantially restricts the class of problems under discussion; in particular, this condition is violated in the case of the mixed problem for the wave equation with the Neumann boundary conditions. On the other hand, in the case of zero Neumann conditions, the hyperbolic equation admits the classical energy estimate, and existence and uniqueness theorems are valid. In this connection, Agemi & Shirota [76] introduced the notion of weak L_2-well-posedness for problem (3.1) $-$ (3.3), if the estimate (3.6) holds in the case of zero right-hand sides in (3.2), (3.3).[7] Some necessary and sufficient conditions of weak L^2-well-posedness for higher order equations and systems have been obtained in [77]–[80].

A large number of works by Japanese authors are dedicated to weak L_2-well-posedness in the case of mixed problems for second order equations. Thus, Miyatake (see [81]–[83]) found necessary and sufficient conditions in the case of first order boundary operators with real or complex coefficients; it turned out that weak L_2-well-posedness for a problem with variable coefficients is equivalent to the same property for the problem with "frozen" coefficients.

Shirota [84] has shown that, in the case of constant coefficients, the speed of propagation of disturbances for weakly L_2-well-posed problems in half-space is finite and does not exceed that for the Cauchy problem (the speed of sound). As noticed by Gårding [85], for second order equations this condition is equivalent to the Miyatake conditions. The condition of weak L_2-well-posedness is equivalent to an estimate of type (3.6), where "1/2 of

---

[7]Agemi, Shirota, and other Japanese mathematicians refer to such problems as L_2-well-posed; we, however, use the term "weak L_2-well-posedness", in order to reserve the term "L_2-well-posedness" for the problems considered in Sects. 3.2 and 3.3.

the order of the derivative is lost" in the norms along the boundary. It is also possible to consider problems with a greater loss of smoothness, both in the norms along the boundary and the norms over the domain itself; problems of this type are usually associated with C^∞-well-posedness.

C^∞-well-posedness for mixed problems with constant coefficients has been studied by Hersh [87], Chazarain & Piriou [88], Sakamoto [89], Godunov & Gordienko [90]. C^∞-well-posedness for the oblique derivative problem for the wave equation is the subject of a series of papers by Ikawa (see [91] and references therein). Precise results on well-posedness (and ill-posedness) of general boundary value problems for second order equations are due to Eskin [92], [93].

References

[1] Calderon, A. *Integrales Singulares y sus Aplicaciones a Ecuaciones Diferenciales Hiperbolicas.* Buenos Aires: Univ. Press, 1960.

[2] Mizohata, S. Systémes hyperboliques. *J. Math. Soc. Jap.* **11** (1959) 205–233.

[3] Mizohata, S. Note sur le traitement par les opérateurs d'intégrale singuliere du problème de Cauchy. *J. Math. Soc. Jap.* **11** (1959) 234–240.

[4] Yamaguti, M. Sur l'inegalité d'énergie pour les systémes hyperboliques. *Proc. Jap. Acad. Sci.* **35** (1959) 37–41.

[5] Mikhlin, S.G. *Multi-Dimensional Singular Integrals and Integral Equations.* Moscow, Fizmatgiz, 1962.

[6] Calderon, A.; Zygmund, A. Singular integral operators and differential equations. *Amer. J. Math.* **79:4** (1957) 901–921.

[7] Leray, J. *Hyperbolic Differential Equations.* Princeton, N.J., 1953.

[8] Gårding, L. *Cauchy's Problem for Hyperbolic Equations.* Chicago Univ., 1957.

[9] Volevich, L.R.; Gindikin, S.G. The Cauchy problem for differential operators with dominating principal part. *Funk. Anal. Prilozh.* **2:3** (1968) 22–40.

[10] Volevich, L.R.; Gindikin, S.G. The Cauchy problem for pluri-parabolic differential equations. II. *Mat. Sbornik* **78:2** (1969) 214–236.

[11] Volevich, L.R. The Energy method for the Cauchy problem for differential operators well-posed in the sense of Petrowsky. *Trudy Mosc. Mat. Ob.-va* **31** (1974) 147–187.

[12] Gindikin, S.G. Energy estimates related to Newton's polyhedron. *Trudy Mosc. Mat. Ob.-va* **31** (1974) 189–236.

[13] Ladyzhenskaya, O.A. On solving the Cauchy problem for hyperbolic systems by the method of finite differences. *Uchen. Zapiski Leningrad. Univ. Ser. Mat.* **23** (1952) 192–246.

438   Appendix. Comentary on Articles 1 & 3

[14] Vishik, M.I. On boundary value problems for quasilinear parabolic systems and the Cauchy problem for hyperbolic equations. *Dokl. Akad. Nauk SSSR* **140:6** (1961) 98–101.

[15] Volevich, L.R. On a problem of linear programming arising in the theory of differential equations. *Uspekhi Mat. Nauk* **18:3** (1963) 155–162.

[16] Ivrii, V.Ya.; Petkov, V.M. Necessary conditions for well-posedness of the Cauchy problem for weakly hyperbolic equations. *Uspekhi Mat. Nauk* **29:5** (1974) 3–70.

[17] Strang, G. Necessary and sufficient conditions for well-posedness of the Cauchy problem. *J. Diff. Eqs.* **2:1** (1966) 107–114.

[18] Friedrichs, K.O.; Lax, P.D. On symmetrizable differential operators. In: *Singular Integrals: Proc. Symp. Pure Math. 1967*, Providence, Amer. Math. Soc. 1967, Vol. 9, pp. 128–137.

[19] Mizohata, S. Some remarks on the Cauchy problem. *J. Math. Kyoto Univ.* **1:1** (1961) 109–127.

[20] Lax, P.D. Asymptotic solutions of oscillatory initial value problems. *Duke Math. J.* **24:4** (1957) 627–646.

[21] Lax, A. On Cauchy's problem for partial differential equations with multiple characteristics. *Comm. Pure Appl. Math.* **9:2** (1956) 135–139.

[22] Gårding, L. Linear hyperbolic partial differential equations with constant coefficients. *Acta Math.* **85:1** (1950) 1–62.

[23] Hörmander, L. *Linear Partial Differential Operators.* Berlin, Springer-Verlag, 1963.

[24] Swensson, L. Necessary and sufficient conditions for hyperbolicity of polynomials with hyperbolic principal part. *Ark. Mat.* **8** (1969) 145–162.

[25] Flashka, G.; Strang, G. The correctness of the Cauchy problem. *Advances in Math.* **6:3** (1971).

[26] Chazarain, J. Opérateurs hyperboliques à caractéristiques de multiplicité constante. *Ann. Inst. Fourier* **24:1** (1974) 173–202.

[27] Petkov, V.M. Necessary conditions for well-posedness of the Cauchy problem for non-symmetrizable hyperbolic systems. *Trudy Seminara imeni Petrovsk.* **1** (1975) 211–236.

[28] Petkov, V.M. Parametrix for the Cauchy problem for non-symmetrizable hyperbolic systems with characteristics of constant multiplicity. *Trudy Mosc. Mat. Ob.-va* **37** (1978) 3–47.

[29] Demay, Y. Paramètrix pour des systémes hyperboliques du premier ordre a multiplicité constante. *J. Math. Pures et Appl.* **56:4** (1977) 393–422.

[30] Hörmander, L. The Cauchy problem for differential equations with double characteristics. *J. Anal. Math.* **34** (1977) 110–196.

[31] Petkov, V.M., Kutev, N.D. On regularly hyperbolic first order systems. *Godishn. Sofia Univ.* **67** (1976) 375–389. '

[32] Ivrii, V.Ya. Sufficient conditions for regular and completely regular hyperbolicity. *Trudy Mosc. Mat. Ob.-va* **33** (1976) 3–65.

[33] Ivrii, V.Ya. Well-posedness of the Cauchy problem for weakly hyperbolic operators. Energy integral. *Trudy Mosc. Mat. Ob.-va* **34** (1977) 151–170.

[34] Zaitseva, O.V.; Ivrii, V.Ya. On well-posedness of the Cauchy problem for a class of hyperbolic operators with characteristics of large multiplicity. *Uspekhi Mat. Nauk* **37:3** (1982) 187–188.

[35] Smirnov, M.M. *Degenerate Equations of Elliptic and Hyperbolic Type.* Moscow, Nauka, 1966.

[36] Bitsadze, A.V. *Equations of Mixed Type.* Moscow, Izd.-vo Akad. Nauk SSSR, 1959.

[37] Oleinik, O.A. On the Cauchy problem for weakly hyperbolic equations. *Comm. Pure Appl. Math.* **23** (1970) 569–589.

[38] Nersesjan, A.B. The Cauchy problem for a symmetric system degenerating on the initial hyperplane. *Dokl. Akad. Nauk SSSR* **196:2** (1971) 289–292.

[39] Menikoff, A. The Cauchy problem for weakly hyperbolic equations. *Amer. J. Math.* **97** (1975) 548–558.

[40] Yoshikawa, A. Construction of a parametrix for the Cauchy problem for some weakly hyperbolic equations. *Hokkaido Math. J.* **7:1** (1978) 1–26, 127–141.

[41] Alinhac, S. Paramétrixes et propagation des singularites pour un problème de Cauchy a multiplicité variable. *Astérisqe, Colloq. Intern. CNRS* **34&35** 3–26.

[42] Sakamoto, R. Cauchy's problem for degenerate hyperbolic equations. *Comm. Pure Appl. Math.* **33** (1980) 785–816.

[43] Sakamoto, R. Some remarks on degenerate hyperbolic Cauchy problems. *Comm. Pure Appl. Math.* **33** (1980) 817–830.

[44] Gårding, L.; Leray, J.; Kotake, T. Problème de Cauchy, I–IV. *Bull. Soc. Math. Fr.* **92** (1964) 263–361.

[45] Kondratiev, V.A. The Cauchy problem with characteristic points on the initial surface. *Vestnik Mosk. Univ. Ser. Mat. Mekh.* **1** (1974) 84–92.

[46] Weinberg, B.R.; Mazja, V.G. Characteristic Cauchy problem for a hyperbolic equation. *Trudy Seminara Petrovsk.* **7** (1981) 101–117.

[47] Leray, J. Équation hyperboliques non-strict, contre-exemples du type De Giorgi aux théorèmes d'existence et unicité. *Math. Ann.* **162:2** (1966) 3–11.

[48] Ohya, J.Yu. Systémes linéaires, hyperboliques non-stricts. *Deuxiéme Colloq. Anal. Funct. Louvain, CBRM*, 1964, pp. 145–190.

[49] Leray, J.; Ohya, J.Yu. Équations et systémes non-linéaires, hyperboliques non-stricts. *Math. Ann.* **170:3** (1967) 167–205.

[50] Beals, R. Hyperbolic equations and systems with multiple characteristics. *Arch. Rat. Anal. Appl.* **48:2** (1972) 123–152.

[51] Nersesjan, A.B. On infinitely smooth solutions of the Cauchy problem for degenerate hyperbolic equations. *Izvest. Akad. Nauk. SSSR, Ser. Mat.* **4:3** 182–191.

[52] Ivrii, V.Ya. Well-posedness in Gevrey classes for the Cauchy problem for weakly hyperbolic operators. *Mat. Sbornik* **96:3** (1975) 390–413.

[53] Bronstein, M.D. The Cauchy problem for hyperbolic operators with characteristics of variable multiplicity. *Trudy Mosc. Mat. Ob.-va* **41** (1980) 83–99.

[54] Ivrii, V.Ya. Conditions of well-posedness in Gevrey classes for the Cauchy problem for weakly hyperbolic operators. *Sib. Mat. Zh.* **17:3** (1980) 547–563.

[55] Komatsu, N. Irregularity of characteristic elements and hyperbolicity. Tokyo Univ Press, 1978.

[56] Ivrii, V.Ya. Conditions of well-posedness in Gevrey classes for the Cauchy problem for hyperbolic operators with characteristics of variable multiplicity. *Sib. Mat. Zh.* **17:6** (1976) 1256–1270.

[57] Ivrii, V.Ya. Well-posedness in Gevrey classes for the Cauchy problem for weakly hyperbolic operators. *Izvestiya Vuzov, Mat.* **2** (1978) 26-35.

[58] Bony, J.M.; Shapira, P. Solutions hyperfonctions du problème de Cauchy. Preprint.

[59] Krzyzanski, M.; Schauder, J. Quasilineare Differentialgleichungen zweiter Ordnung vom hyperbolischen Typus. Gemischte Bandwertaufgaben. *Stud. Math.* **6** (1933) 162–189.

[60] Ladyzhenskaya, O.A. *The Mixed Problem for Hyperbolic Equations.* Moscow, Gostekhizdat, 1953.

[61] Agranovich, M.S. Boundary value problems for systems of pseudo-differential first order operators. *Uspekhi Mat. Nauk* **24:1** (1969) 61–125.

[62] Duff, G.F.D. Hyperbolic differential equations and waves. In: *Boundary Value Problems for Linear Evolution Partial Differential Equations.* Ed. by H.G.Garnir, Dordrecht, Holland; Boston, Reidel Publ. Co., 1977, pp.27–155.

[63] Friedrichs, K.O. Symmetric hyperbolic linear differential equations. *Comm. Pure Appl Math.* **7** (1954) 345–392.

[64] Friedrichs, K.O. Symmetric positive linear differential equations. *Comm. Pure Appl. Math.* **11** (1958) 333–418.

[65] Dezin, A.A. Boundary value problems for some linear symmetric first order systems. *Mat. Sbornik* **49:3** (1959) 459–484.

[66] Dezin, A.A. Existence and uniqueness theorems for solutions of boundary value problems for partial differential equations in function spaces. *Uspekhi Mat. Nauk* **14:3** (1959) 21–74.

[67] Agmon, S. Problème mixte pour les équations hyperboliques d'ordre supérieur. *Colloq. Intern. CNRS,* Paris, 1962, pp. 1–6.

[68] Kreiss, H.O. Initial boundary value problems for hyperbolic systems. *Comm Pure Appl. Math.* **23** (1970) 277–298.

[69] Ralston, F.V. Note on a paper of Kreiss. *Comm. Pure Appl. Math.* **24** (1971) 759–762.

[70] Rauch, J. L_2 is a continuable initial condition for Kreiss' mixed problems. *Comm. Pure Appl. Math.* **25** (1972) 265–285.

[71] Agranovich, M.S. Boundary value problems for systems with a parameter. *Mat. Sbornik* **84:1** (1971) 27–65.

[72] Agranovich, M.S. A problem for a matrix depending on a parameter and its application to hyperbolic systems. *Funkt. Anal. Prilozh.* **6:2** (1972) 1–11.

[73] Sakamoto, R. Mixed problems for hyperbolic equations. I,II. *J. Math. Kyoto Univ.* **10:3** (1970) 349–373, 403–417.

[74] Balaban, T. On the mixed problem for a hyperbolic equation. *Mem. Amer. Math. Soc.* 1971.

[75] Volevich, L.R.; Gindikin, S.G. The method of energy estimates for the mixed problem. *Uspekhi Mat. Nauk* **35:5** (1980) 53–120.

[76] Agemi, R.; Shirota, T. On necessary and sufficient conditions for L^2-well-posedness of mixed problems for hyperbolic equations. I, II. *J. Fac. Sci. Hokkaido Univ. Ser. I* (1970) **21:2** 133–151, **21:3** 137–149.

[77] Sakamoto, R. L^2-posedness for hyperbolic mixed problems. *Publ. RIMS Kyoto Univ.* **8** (1973) 265–293.

[78] Sakamoto, R. On a class of hyperbolic mixed problems. *J. Math. Kyoto Univ.* **16:2** (1976) 429–474.

[79] Ohkubo, T.; Shirota, T. On structures of certain L^2-well-posed mixed problems for hyperbolic systems of first order. *Hokkaido Math. J.* **4:1** (1975) 82–158.

[80] Agemi, R. Iterated mixed problems for d'Alambertians. I,II. *Hokkaido Math. J.* (1974) **3:1** 104–128, **4:2** 281–294.

[81] Miyatake, S. Mixed problem for a hyperbolic equation of second order. *J. Math. Kyoto Univ.* **13:3** (1973) 435–487.

[82] Miyatake, S. Mixed problems for hyperbolic equations of second order with first order complex boundary operators. *Jap. Math. New Ser.* **1:1** (1975) 111–158.

[83] Miyatake, S. A sharp form of existence theorem for hyperbolic mixed problems of second order. *J. Math. Kyoto Univ.* **17:2** (1977) 199–223.

[84] Shirota, T. On the propagation speed of hyperbolic operator with mixed boundary conditions. *J. Fac. Sci. Hokkaido Univ. Ser. I* **22:1** 25–31.

[85] Gårding, L. Le problème de la dérivée oblique pour l'équation des ondes. Rectification à la Note. *C.R. Acad. Sci. Paris* **285** (1977) 773–775.

[86] Kubota, K. Remarks on boundary value problems for hyperbolic equations. *Hokkaido Math. J.* **2:2** (1973) 202–213.

[87] Hersh, R. Mixed problems in several variables. *J. Math. Mech.* **12** (1963) 317–334.

[88] Chazarain, J.; Piriou, A. Caractŕisation des problèmes mixtes hyperboliques bien posés. *Ann. Inst. Fourier* **22:4** (1972) 193–237.

[89] Sakamoto, R. E-well-posedness for hyperbolic mixed problems with constant coefficients. *J. Math. Kyoto Univ.* **14:1** (1974) 93–118.

[90] Godunov, S.K.; Gordienko, V.M. The mixed problem for the wave equation. *Trudy Seminara Soboleva, Novosibirsk* **2** (1978) 93–118.

[91] Ikawa, M. On the mixed problem for the wave equation in an interior domain. *Comm. Part. Diff. Eqs.* **3** (1978) 249–295.

[92] Eskin, G. Well-posedness and propagation of singularities for initial boundary value problem for second order hyperbolic equation with general

boundary condition. *Sém. Goulaouic-Schwartz.* 1979–1980. Paris. Ecole
Polytecnique, 1980.

[93] Eskin, G. General initial-boundary problems for second order hyperbolic equations. In: *Singularities in Boundary Value Problems.* Edited by
H.G. Garnir. Dordrecht; Boston, Reidel Publ. Co., 1981.

[94] Oleinik, O.A. The Cauchy problem and boundary value problems for
second order hyperbolic equations degenerating inside the domain or on
its boundary. *Dokl. Akad. Nauk SSSR* **169:3** (1966) 525–528.

[95] Oleinik, O.A. On second order hyperbolic equations degenerating inside
the domain or on its boundary. *Uspekhi Mat. Nauk* **24:2** (1969) 229–230.

Theory of Systems of Partial Differential Equations*

by I.M. Gelfand, I.G. Petrowsky, G.E. Shilov

After the appearance of the well-known Petrowsky's papers in the '30s, a necessity has been felt for passing from the consideration of some special partial differential equations and systems to the study of general systems of partial differential equations.

The theory of systems, in its present form, is by no means a theory of general systems, and it is not quite clear whether one should strive to develop a theory embracing all possible systems of partial differential equations.

The modern theory of systems of partial differential equations is mainly concerned with the description of two types of processes: those evolving in time and stationary ones.

Processes of evolution are described by systems which naturally allow to formulate the Cauchy problem (in the entire space) or the mixed problem (in a domain with a boundary). Such systems will be referred to as *evolutionary*.

Stationary processes can be naturally associated with various types of boundary value problems. It should be emphasized that problems modeling stationary processes are closely connected with evolutionary problems. Thus, it is not by accident that a solution of a boundary value problem for a stationary process can often be more easily found by considering an evolutionary problem for the corresponding process of stabilization; in this situation, the solution of the stationary boundary value problem is obtained as the limit, as $t \to \infty$, of the solution for the corresponding evolutionary problem. For instance, the solution of the Dirichlet problem for the Laplace equation is well-known to be the limit of the solution of the heat equation

---

*A supplement to Article 2 of this volume. Originally published in the *Proceedings of the 3d All-Union Mathematical Congress, Moscow, 1956*. Vol. 3. Izd.-vo Akad. Nauk SSSR, Moscow, 1958, pp. 65–72.

with fixed boundary values, as $t \to \infty$. A similar result has been established by M.I. Vishik for strongly elliptic systems.

Another example of such limiting behavior is provided by the problem of stationary flow of a viscous fluid described by the nonlinear Navier-Stokes system. A solution of this problem can be obtained as the limit of a solution of the problem for non-stationary flow as $t \to \infty$.

The interest in regarding a stationary process as a limit of an evolutionary one is also due to the fact that solutions of evolutionary problems can be used for the numerical calculation of solutions of boundary value problems. In numerical analysis, boundary value problems are usually solved by the net method; this involves finding solutions of large algebraic systems of equations (linear or nonlinear). Numerical solution of an evolutionary problem by the same net method is much simpler, since the values of the unknown functions at any subsequent moment are directly expressed through their values at the preceding moment. Of course, passing to the limit as $t \to \infty$ may require calculations for a large number of steps, and it would be interesting to know how an "optimal" (i.e., with the fastest rate of convergence for $t \to \infty$) evolutionary system can be constructed for a given stationary one.

In view of all this, our chief object in this note will be evolutionary systems; and we limit ourselves only to a few remarks about stationary systems.

As mentioned above, for stationary systems it is natural to consider boundary value problems; for a long time such problems have been associated with elliptic systems introduced by I.G. Petrowsky. On the other hand, there is an example (constructed by A.V. Bitsadze) of non-uniqueness for the Dirichlet problem for an elliptic system. For this reason it seemed proper to narrow the class of elliptic systems: M.I. Vishik, in his impressive works, introduced the class of strongly elliptic systems, for which the Dirichlet problem has a unique solution. Nevertheless, in spite of the great importance of strongly elliptic systems, it seems to us inconsistent to limit the studies to this class only, for, apart from the absence of an invariant definition of such systems (for instance, the definition given by M.I Vishik is not invariant when the equations are replaced by their linear combinations), the requirement of the uniqueness of a solution for the Dirichlet problem is, in our opinion, unnatural. The Dirichlet problem for systems is not particularly different from other boundary value problems, which are known to admit several solutions; and therefore, the requirement of uniqueness should be replaced by the condition on the corresponding operator to be of Fredholm type, i.e., the problem should not have more than a finite number of solutions, and an analogue of the Fredholm alternative should hold. It seems that the operators associated with more or less arbitrary elliptic systems are of Fredholm type.

We also call attention to the fact that it is still unknown whether any elliptic system can be viewed as a stationary system for some non-stationary one.

The subject of our second remark will be made clear if we consider, for instance, a system of second order equations with constant coefficients and n independent variables. Every system of this type can be associated with a point in N-dimensional space determined by the coefficients of the system.

Of great interest is the problem of finding how many connected components form the set of all these points, and how these components can be characterized. Thus, for systems with two independent variables, M.I. Vishik has shown that there are only two connected components, and for systems associated with the points of one of these components the Dirichlet problem has only one solution, while for systems corresponding to the points of another component a solution can never be unique.

As mentioned above, in this note we intend to discuss in detail evolutionary systems. The simplest model which allows us to clarify all characteristic features of partial differential equations is provided by linear systems with constant coefficients; and, for evolutionary systems, the simplest problem is the Cauchy problem in the entire space.

Here we shall not dwell on mixed problems, since they have been discussed elsewhere, and moreover, their consideration must be preceded by a study of well-posedness for the corresponding Cauchy problem without boundary conditions.

Thus, we focus our attention on the Cauchy problem for an evolutionary system whose coefficients are independent of the spatial variables; in many cases, such systems can be written in the form

$$\frac{\partial u_j(x,t)}{\partial t} = \sum_{l=1}^{m} P_{jl}\left(t, i\frac{\partial}{\partial x}\right) u_l(x,t), \qquad j = 1, 2, \ldots, m, \qquad (1)$$

where $P_{jl}(t,s)$ are polynomials in $s = (s_1, \ldots, s_k)$ of maximal degree p (called the order of the system), with coefficients continuous in t. Some of the results mentioned here are also valid for systems with coefficients depending on the spatial variables.

System (1) is supplemented by the initial conditions

$$u_j(x,0) = U_j(x), \qquad j = 1, 2, \ldots, m. \qquad (2)$$

The Cauchy problem consists in finding functions $u_j(x,t)$ which satisfy system (1) for $t \geq 0$ and the initial conditions (2) for $t = 0$.

A set of functions $K = \{f(x)\}$ is said to form a *class of well-posedness* for system (1), if, for any (sufficiently regular) initial functions $U_j(x)$, problem (1), (2) has a solution belonging to the class K for each value of $t \geq 0$, this solution is unique in K and continuously depends on the initial functions $U_j(x)$; the meaning of the latter condition must be specified together with the definition of the class K. Thus, as shown by Tikhonov [1] in 1935, for the heat equation

$$\frac{\partial u}{\partial t} = \frac{\partial^2 u}{\partial x^2}$$

a class of well-posedness K can be defined as the set of all functions $f(x)$ satisfying the inequality $|f(x)| \leq C_1 \exp(Cx^2)$ (in this case, a sequence $f_n(x)$ is said to be convergent to zero in K, if these functions uniformly converge to zero on every finite interval and satisfy the estimate $|f_n(x)| \leq C_1 \exp(Cx^2)$ with constants C, C_1 independent of n). If we consider a wider class, replacing in the aforesaid estimate the exponent 2 by $2 + \varepsilon$, we obtain a set which cannot be called a class of well-posedness for the heat equation, since the corresponding Cauchy problem is known to have more than one solution within this class.

In this connection, one can consider a simpler problem of specifying classes of uniqueness for solutions of the Cauchy problem $(1), (2)$. A set of functions $L = \{f(x)\}$ is said for be a *class of uniqueness* for system (1), if within this class the Cauchy problem can have only one solution with given initial functions $U_j(x)$. Obviously, every class of well-posedness is also a class of uniqueness; however, the converse statement is not true.

The statement of the problem about the classes of well-posedness and the classes of uniqueness for general systems of type (1) (and even more general systems), and the first fundamental results in this direction are due I.G. Petrowsky [2], who indicated a necessary and sufficient condition for the set of all functions $f(x)$ bounded in the entire space, together with a certain number of their derivatives, to form a class of well-posedness for system (1). This condition (Condition A) is as follows: *Consider the corresponding formally dual system of ordinary differential equations*

$$\frac{dv_j(s, t)}{dt} = \sum_{l=1}^{m} P_{jl}(t, s) v_l(s, t), \qquad j = 1, 2, \ldots, m. \tag{3}$$

Then all solutions of system (3), considered as functions of real parameter s, should not grow faster than a certain power of $|s|$ as $|s| \to \infty$. For systems with constant coefficients this condition can be easily expressed in terms of characteristic roots of the matrix $P(s) = \|P_{jl}(s)\|$; namely, *the roots must have their real parts (as functions of s) bounded, or growing not faster than $C \ln|s|$*; for $k = 1$, the latter case is reduced to the former one, and the same can be expected for any k.

I.G. Petrowsky singled out two types of systems which unfailingly satisfy Condition A, namely, parabolic and hyperbolic systems; the corresponding definitions (their modern versions) will be given later on. The definition of hyperbolicity, in particular, was generalized and expressed in its natural final form by L. Gårding [11], whose main principle in defining hyperbolic systems was a finite speed of propagation of the initial disturbances for such systems.

The article of I.G. Petrowsky opens a whole series of studies. Thus, it was noticed that Condition A ensures uniqueness of solutions for the Cauchy problem not only in the class of bounded functions but also among all functions of polynomial growth (Lyantse [3], Schwartz [4]). The class of uniqueness (and well-posedness) was further widened by O.A. Ladyzhen-skaya [5] for parabolic equations and then by S.D. Eidelman [6] for parabolic systems. It turned out that for such systems the class of well-posedness is characterized by the inequality

$$|f(x)| \leq C_1 \exp\left\{C|x|^{\frac{p}{p-1}}\right\} ,$$

and the exponent $p/(p-1)$ cannot be increased. This result implies, in particular, the aforementioned Tikhonov's theorem for the heat equation (in which case $p = 2$).

Application of a new general method – the method of the Fourier transform of rapidly growing functions – made it possible for I.M. Gelfand & G.E. Shilov [7] to find a general solution for the problem of the uniqueness classes. The result obtained by Gelfand and Shilov can be formulated as follows. Consider the matrix $Q_{t_0}^t(s)$ formed by the normal fundamental system of solutions for the system of ordinary differential equations (3). In contrast to what has been previously done by I.G. Petrowsky, now the matrix is considered not only for real but also for all complex values of the parameter $s = \sigma + i\tau$. The elements of this matrix are entire analytic functions of s having a certain maximal order $p_0 \leq p$, where p is the order of system (1). If $p_0 > 1$, then the uniqueness class K for the Cauchy problem (1), (2) is determined by the inequality

$$|f(x)| \leq C_1 \exp\left\{C|x|^{p_0'-\varepsilon}\right\} , \qquad \frac{1}{p_0} + \frac{1}{p_0'} = 1 ,$$

for any fixed $\varepsilon > 0$.

Subsequently, some improvements were introduced into this construction by K.I. Babenko, B.L. Gurevich, S.D. Eidelman, independently; this allowed them to drop the parameter ε in the exponent. On the other hand, the replacement of p_0' by $p_0' + \varepsilon$ in the exponent is always responsible for the loss of uniqueness, as shown by G.N. Zolotariov. For $p_0 \leq 1$, the uniqueness is guaranteed in the class of all functions, without any restrictions on their growth.

Thereby the problem of characterizing classes of uniqueness, as formulated by I.G. Petrowsky in [2], was essentially exhausted.

Next, we briefly describe a modern approach to this problem, as suggested by I.M. Gelfand and G.E. Shilov in 1955 (see [8]). Consider the basic (linear topological) spaces S_α^β formed by infinitely differentiable functions $\varphi(x)$ such that

$$\left|x^{(k)}\varphi^{(q)}(x)\right| \le CA^{|k|}B^{|q|}|k|^{|k|\alpha}|q|^{|q|\beta} , \qquad k_j, q_m = 0,1,2,\dots . \quad (4)$$

The inequalities (4) impose restrictions on the decay of basic functions at infinity ($\sim C_1 \exp\{-C|x|^{1/\alpha}\}$), and on the growth of their derivatives for $q \to \infty$ (if $\beta < 1$, we have entire functions of growth rate $(1-\beta)^{-1}$).

In general, the above conditions are so strong that for some α, β the space S_α^β may happen to be trivial (i.e., consist of zero element only). The necessary and sufficient condition for this space to be non-trivial is the inequality $\alpha + \beta \ge 1$, with the exception of the cases $\alpha = 0, \beta = 1$, and $\alpha = 1$, $\beta = 0$. In these spaces, differential operators of the form $F(i\partial/\partial x)$ are defined (and continuous), provided that the function $F(s)$ is of order $< 1/\beta$. In particular, the operator $e^{tP(i\partial/\partial x)}$, which yields the solution of the Cauchy problem for system (1) (in the case of constant coefficients; if the coefficients depend on t, we should consider the corresponding multiplicative integral), is defined in the space S_α^β, provided that the entire function $e^{tP(i\partial/\partial x)}$ has its growth order p_0 less than $1/\beta$, or, in the case of p_0 given and β to be determined, if $\beta < 1/p_0$. Now, in order to have a non-trivial space S_α^β, we should set $\alpha \ge 1 - \beta \ge 1 - 1/p_0 = 1/p_0'$, or equivalently, $1/\alpha < 1/p_0'$; therefore, at infinity, the basic functions from the space S_α^β have exponential decay rate of order $\ge p_0'$. Thus we have constructed a space S_α^β in which the Cauchy problem is solvable. Hence, using the classical construction of Holmgren, we can come to the conclusion that the Cauchy problem has a unique solution even in the space of continuous linear functionals on S_α^β; in particular, a solution is always unique in the class of functions with exponential growth at infinity of order less than p_0', and this is exactly what we stated.

Now we describe the construction of classes of well-posedness (see [10]). Consider the following properties of the matrix $Q_{t_0}^t(s)$:

a) the order of growth p_0 in the complex plane;

b) the order of growth (or decay) h along the real axis:

$$\left\|Q_0^t(\sigma)\right\| \le C_t e^{tC|\sigma|^h} ; \quad (5)$$

for $h = 0$, this inequality is replaced by

$$\left\|Q_0^t(\sigma)\right\| \le C_t(1 + |\sigma|)^q ;$$

c) the genus ν of the system; this number determines, on the s-plane, a domain of the form

$$|\tau| \leq C(1 + |\sigma|)^{\nu} \, ,$$

wherein the inequality (5) is preserved, so that

$$\left\| Q_0^t(s) \right\| \leq C_t' e^{Ct|\sigma|^h} \, .$$

For some ν, which may be positive or negative, domains of this type exist, and we always have $\nu \geq 1 - (p_0 - h)$, as shown by B.Ya. Levin.

In the case of constant coefficients, it is easier to specify classes of well-posedness in terms of the characteristic roots $\lambda_j(s)$ of the matrix $P(s)$, or more precisely, in terms of the function

$$\Lambda(s) = \max_j \operatorname{Re} \lambda_j(s) \, .$$

Namely, we have the following inequality

$$e^{t\Lambda(s)} \leq \left\| Q_0^t(s) \right\| \leq C(1 + |s|)^{mp} e^{t\Lambda(s)} \, ,$$

which allows us to express all the conditions imposed on the form of $Q_0^t(s)$ as restrictions on the function $\Lambda(s)$.

The results of this investigation are given in Table 1. To obtain these results, the following method has been used. The solution of the Cauchy problem $(1), (2)$ in the sense of distributions has the form of a convolution

$$u(x, t) = G(x, t) * U_0(x) \, , \tag{6}$$

where the Green matrix $G(x, t)$ is the Fourier transform of the resolving matrix $Q(\sigma, t)$. The information about the numbers p_0, h, ν allows us to calculate to which of the spaces S_α^β the matrix $Q_0^t(\sigma)$ belongs (or in which of these spaces this matrix is a multiplicator). The Fourier transform of S_α^β coincides with S_β^α, therefore, we can find which of these spaces contains the Green matrix $G(x, t)$ (or in which of these spaces this matrix is a convolutor). Finally, this allows us to indicate the cases when the convolution (6) represents an ordinary function, and this is exactly what we need.

As we have done regarding the uniqueness classes, it is natural to ask whether the growth order of the functions forming a class of well-posedness can be increased without the loss of well-posedness. Obviously, this increase is impossible for systems parabolic in the sense of Petrowsky, as well as for regular systems. However, there are no results of this kind in other cases. For systems of type III_2 (well-posed in the sense of Petrowsky and of genus $\nu < 0$), A.G. Kostyuchenko and G.E. Shilov proved, for any given $l > 0$, the existence of $q_1 = q_1(l)$ such that in the class

$$K_{l,q_1} = \left\{ \left| f^{(q)}(x) \right| \leq C(1 + |x|)^l \, , q \leq q_1 \right\}$$

the Cauchy problem is ill-posed. Thus, for the Schrödinger equation

$$\frac{\partial u}{\partial t} = i\frac{\partial^2 u}{\partial x^2}$$

and given l, the set $K_{l,l+4}$ is a class of well-posedness, while $K_{l,l-3}$ is not. Therefore, one cannot substantially improve the result given in Table 1 for systems of type III$_2$, i.e., well-posedness in the class of sufficiently smooth functions of polynomial growth.

Another series of questions concerns the determination of a system's type by the class of its well-posedness; the answers to these questions are desirable in order to justify the classification of systems given in the table.

It has been proved that if the Cauchy problem for a system is well-posed in the class of all sufficiently smooth functions (without any restrictions on their growth at infinity), then the system is hyperbolic (see Gårding [11], Schwartz [4], Borok [12]). It is a very interesting problem to establish a similar result for parabolic systems.

Parabolic systems have another important property: even for a discontinuous initial function $U_0(x)$, the solution of the corresponding Cauchy problem (from the class of well-posedness) is always infinitely differentiable in x (and even analytic, if $h > 1$; see [13]). This property is characteristic of parabolic systems; as shown by V.M. Borok, system (1) must be parabolic, if for any bounded initial functions it has a solution $u(x,t)$ differentiable with respect to x for $t > 0$.

Next, we describe some generalizations of the above results to the case of systems with variable coefficients. First results in this direction are due to S.Z. Bruck [14], and pertain to systems parabolic in the sense of Petrowsky, with coefficients depending on x (a system of this type is said to be parabolic if the conditions of parabolicity hold for any formally fixed values of x in the coefficients). Using Petrowsky's estimates and the method of variation of constants (this method allows one to reduce the problem to an integral equation of Volterra's type; see Levi [15]), S.Z. Bruck incorporated parabolic systems with bounded coefficients into the scheme of Petrowsky. Further, having obtained more accurate estimates for the resolving matrix $Q_0^t(s)$ and using results about the basic spaces described above (see [7]), S.D. Eidelman [16] established, under similar assumptions, that the class of well-posedness for the corresponding systems is determined by the same inequality as for systems of this type with constant coefficients.

Some other facts are also known about the uniqueness classes for systems with variable coefficients. Using the results of S.D. Eidelman [16], A.G. Kostyuchenko [17] constructed analogues of the spaces S_α^β in the case of the operator d/dx in the definition of these spaces being replaced by an arbitrary elliptic operator. This construction made it possible to extend

Well-Posedness Classes for Systems of the Form $\dfrac{\partial u_j(x,t)}{\partial t} = \sum\limits_{k=1}^{m} P_{jk}\left(t, i\dfrac{\partial}{\partial x}\right) u_k(x,t)$

No.	System's type	Definition in terms of matrix	Definition in terms of characteristic roots	The class of well-posedness
I_1	Parabolic $\nu > 0$, including parabolic in the sense of Petrowsky ($\nu = 1$);	$|Q_0^t(\sigma)| \le C_t \exp\{-Ct|\sigma|^h\}$ $h = p_0$	$\Lambda(\sigma) \le -C|\sigma|^h + C_1$ $h = p_0$	$|f(x)| \le C_1 \exp\{C|x|^{p_0/(p_0-\nu)}\}$ $|f(x)| \le C_1 \exp\{C|x|^{p_0/(p_0-1)}\}$
II_2	$\nu \le 0$			$|f(x)| \le C_\varepsilon \exp\{\varepsilon|x|^{h/(h-\nu)}\}$
II	Hyperbolic	$|Q_0^t(s)| \le C_t \exp\{Ct|s|\}$ $|Q_0^t(\sigma)| \le C_t(1+|\sigma|)^q$	$\Lambda(s) \le C|s| + C_1$ $\Lambda(\sigma) \le C$	Functions of finite smoothness ($\sim q$) with no restrictions on growth
III_1	Well-posed in the sense of Petrowsky $\nu > 0$, including regular ($\nu = 1$)	$|Q_0^t(\sigma)| \le C_t(1+|\sigma|)^q$	$\Lambda(\sigma) \le C$	$|f^{(r)}(x)| \le C_1 \exp\{C|x|^{p_0/(p_0-\nu)}\}$ $|f^{(r)}(x)| \le C_1 \exp\{C|x|^{p_0/(p_0-1)}\}$ ($r \le q$)
III_2	$\nu \le 0$			$|f^{(r)}(x)| \le C_1(1+|x|)^l$ ($r \le r_0(l)$)
IV_1	Provisionally well-posed, $\nu > 0$	$|Q_0^t(\sigma)| \le C_t \exp\{Ct|\sigma|^h\}$, $h < 1$	$\Lambda(\sigma) \le C|\sigma|^h + C_1$ $h < 1$	Infinitely smooth functions $|f^{(r)}(x)| \le C_1 A^r r^{-r/h} \exp\{C|x|^{p_0/(p_0-\nu)}\}$ for $\nu \ge p_0$ with no growth restrictions
IV_2	$\nu \le 0$			$|f^{(r)}(x)| \le C_3 A^r r^{-r/h} \exp\{C|x|^{h/(h-\nu)}\}$
V	Analytically well-posed	$|Q_0^t(s)| \le C_1 \exp\{Ct|s|^{p_0}\}$	$\Lambda(s) \le C|s|^{p_0} + C_1$	Entire functions with order of growth $\le p_0'$

Table 1.

the above mentioned results of I.M. Gelfand and G.E. Shilov to systems involving analytic operators instead of d/dx.

A.G. Kostyuchenko and G.E. Shilov have shown that any class of functions having a finite exponential growth along the real axis is a class of uniqueness for systems of Kowalevskaya type

$$\frac{\partial u_j(x,t)}{\partial t} = \sum_k \frac{\partial}{\partial x} [a_{jk}(x,t)u_k(x,t)] ,$$

where the coefficients $a_{jk}(x,t)$ are analytic functions with respect to $z = x+iy$ in a strip $|y| \le y_0$ and can be represented by Fourier-Stiltjes integrals with an exponentially decaying measure. This condition on the coefficients is stronger than their boundedness in the said strip, and weaker than their integrability in x.

The first proof of well-posedness of the Cauchy problem for hyperbolic equations and systems with variable coefficients is due to I.G. Petrowsky [18]. Alternative proofs and some special cases can be found in: Leray [19], Friedrichs [20].

In conclusion, we mention some studies on the Cauchy problem for systems of a more general form compared to (1). In his two notes [21], [22], S.L. Sobolev indicated some cases of well-posed Cauchy problems for systems containing, in their left-hand sides, differential operators with respect to x, in addition to the differentiation with respect to t. In general form, namely,

$$\frac{\partial}{\partial t} P_1 \left(\frac{\partial}{\partial x} \right) u(x,t) = P_2 \left(\frac{\partial}{\partial x} \right) u(x,t) ,$$

such systems were considered by S.A. Galpern [23], who established the necessity of Condition A for the corresponding Cauchy problems to be well-posed in the class of bounded functions; this condition is also sufficient for well-posedness, if some additional assumptions are made, for instance, if $\det P(s)$ is assumed to have no real roots. Moreover, using the methods of L. Schwartz [4], one can prove the uniqueness in the class of polynomially growing functions, provided that Condition A is satisfied. As shown by G. Gusachenko, even in the absence of Condition A, such equations admit a uniqueness class consisting of functions with exponential decay as $|x| \to \infty$.

References

[1] Tikhonov, A.N. Uniqueness theorems for the heat equation. *Mat. Sbornik* **42:2** (1935) 199–216.

[2] Petrowsky, I.G. On the Cauchy problem for linear systems of partial differential equations in classes of non-analytic functions. *Bull. Mosk. Univ. Mat. Mekh.* **1:7** (1938) 1–72.

[3] Lyantse, V.E. On the Cauchy problem in the class of functions of real variable. *Ukr. Mat. Zh.* **1:4** (1949) 42–63.

[4] Schwartz, L. Les équations d'évolution liées au product de composition. *Ann. Inst. Fourier* **2** (1950) 19–49.

[5] Ladyzhenskaya, O.A. On the uniqueness of solution for the Cauchy problem for a linear parabolic equation. *Mat. Sbornik* **27** (1950) 175–184.

[6] Eidelman, S.D. On the Cauchy problem for parabolic systems. *Dokl. Akad. Nauk SSSR* **96:6** (1954) 913–915.

[7] Gelfand, I.M.; Shilov, G.E. Fourier transform of rapidly growing functions and some questions of uniqueness for the Cauchy problem. *Uspekhi Mat. Nauk* **8:6** (1953) 3–54.

[8] Gelfand, I.M.; Shilov, G.E. On a new approach to the uniqueness theorems for the Cauchy problem. *Dokl. Akad. Nauk SSSR* **102:6** (1955) 1065–1068.

[9] Shilov, G.E. On a problem of quasi-analyticity. *Dokl. Akad. Nauk SSSR* **102:5** (1955) 893–895.

[10] Shilov, G.E. On the conditions of well-posedness for the Cauchy problem for systems of partial differential equations with constant coefficients. *Uspekhi Mat. Nauk* **10:4** (1955) 89–100.

[11] Gårding, L. Linear hyperbolic partial differential equations with constant coefficients. *Acta Math.* **85** (1950) 1–62.

[12] Borok, V.M. On solving the Cauchy problem for some types of linear partial differential equations. *Mat. Sbornik* **36(78):2** (1955) 281–298.

[13] Eidelman, S.D. On the analyticity of solutions of parabolic systems. *Dokl. Akad. Nauk SSSR* **103:1** (1955) 27–30.

[14] Bruck, C.Z. The Cauchy problem for a parabolic system of differential equations. *Izvest. Akad. Nauk SSSR, Ser. Mat.* **10:2** (1946) 105–120.

[15] Levi E.E. On linear elliptic partial differential equations. *Uspekhi Mat. Nauk* **8** (1941) 249–292.

[16] Eidelman, S.D. On fundamental solutions of parabolic systems. *Mat. Sbornik* **38:1** (1956) 51–92.

[17] Kostyuchenko, A.G. Uniqueness theorems for solutions of the Cauchy problem and the mixed problem for some linear systems of partial differential equations. *Dokl. Akad. Nauk SSSR* **103:1** (1965) 13–16.

[18] Petrowsky, I.G. On the Cauchy problem for a system of partial differential equations. *Mat. Sbornik* **2:5** (1937) 815–870.

[19] Leray, J. *Hyperbolic Differential Equations*. Princeton, N.J. Inst Adv. Study, 1953.

[20] Friedrichs, K.O. Symmetric hyperbolic linear differential equations. *Comm. Pure Appl. Math.* **7:2** (1954) 345–392.

[21] Sobolev, S.L. The Cauchy problem for a special system that does not belong to the Kowalevskaya type. *Dokl. Akad. Nauk SSSR* **82:2** (1952) 205–208.

[22] Sobolev, S.L. On a new problem for systems of partial differential equations. *Dokl. Akad. Nauk SSSR* **81:6** (1951) 1007–1009.

[23] Galpern, S.A. The Cauchy problem for equations of Sobolev's type. *Uspekhi Mat. Nauk* **8:5** (1953) 191–193.

Systems of Linear Partial Differential Equations[*]

by O.A. Oleinik and V.P. Palamodov

I.G. Petrowsky's work on the Cauchy problem, published in this volume as Article 2, served as a starting point of numerous investigations. In particular, it gave birth to three vast branches in the theory of partial differential equations: the study of uniqueness classes and classes of well-posedness for general evolutionary systems of partial differential equations; the theory of the Cauchy problem and the mixed problem for higher order parabolic equations and parabolic systems; the theory of hyperbolic equations and systems.

This fact can be explained by the important place these topics have in modern analysis, as well as by the clarity and relative simplicity of Petrowsky's article. The necessity to develop these theories had been stressed by the author in his publications.

A review of the results obtained before 1956 was presented by I.M. Gelfand, I.G. Petrowsky, G.E. Shilov at the Third All-Union Mathematical Congress [1] (see the preceding article of this Appendix), and also in the monograph of Gelfand & Shilov [2]. Therefore, we shall only discuss here some of the works pertaining to the aforementioned branches of the theory that were published after the appearance of [1], [2]. Here we do not touch upon hyperbolic equations and systems, for a separate commentary on Articles 1 and 3 deals with this subject (see the first article of this Appendix).

1. Classes of Uniqueness

The main result, as established in [2] and improved by Babenko in [4], about the classes of uniqueness for the Cauchy problem for systems of linear partial differential equations with constant coefficients can be stated as follows: an

---

[*]A commentary on Article 2 of this volume.

estimate of the type

$$|u(t,x)| \leq C \exp\left\{A|x|^{p'}\right\}, \quad A, C = \text{const},$$

for a solution $u(t,x)$ of the Cauchy problem ensures its uniqueness, if $p' = p/(p-1)$, where p is the so-called modified order of the system calculated from its symbol. Zolotariyov [5]–[7] found a necessary and sufficient condition on the function $\Phi(x)$ for the estimate

$$|u(t,x)| \leq C \exp\{\Phi(x)\}, \quad C = \text{const} > 0$$

to determine a class of uniqueness for such a system. These results generalize a theorem obtained by Täcklind [8] for the heat equation to the case of evolutionary systems with constant coefficients. In particular, Zolotariyov's results imply that for no constant $\varepsilon > 0$ the exponent p' in the inequality $|u(t,x)| \leq C \exp\{|x|^{p'}\}$ of Gelfand-Shilov's theorem can be replaced by $p' + \varepsilon$. Uniqueness classes were also studied by Chaus [9]. In [10]–[14] uniqueness theorems are obtained for solutions of the Cauchy problem for equations and systems with constant coefficients, in the case of equations that cannot be resolved with respect to the highest order derivatives in t.

Palamodov [15] established a uniqueness theorem for general (in particular, overdetermined) systems with constant coefficients. This theorem is an extension of the result proved by Gelfand and Shilov; it states that any solution of such a system, whose growth at infinity (in general, with respect to the variables t and x) satisfies a suitable exponential estimate, is uniquely determined by the values of all its derivatives on a subspace E, provided that the system is weakly hypoelliptic in the direction of the subspace conormal to E.

The question of uniqueness for the characteristic Cauchy problem becomes much more complicated if the coefficients of the equation or system depend on both the time variable and the spatial coordinates. It is for parabolic systems that the most comprehensive results have been obtained in this case. These results are described in Section 4.

Uniqueness theorems in the case of general evolutionary systems with variable coefficients have been obtained only for some fairly narrow classes of systems. Thus, Yamanaka [16], [17] assumes the coefficients of the system to be polynomials of some special degree, while Zhytomirsky [18] imposes restrictions on the coefficients in the form of estimates. Several papers by Zhytomirsky contain detailed analysis of equations with a single spatial variable. In [19], a bound on the growth of the coefficients (as $|x| \to \infty$) is found: for coefficients with a slower growth at infinity, the uniqueness class depends only on the order of the equation and coincides with the uniqueness class in the case of constant coefficients. In [20], uniqueness classes are specified for equations with rapidly growing coefficients. Papers [21], [22] are closely linked to [19], [20]. Uniqueness theorems for some special classes of equations with variable coefficients have also been studied in [23]–[25]. The classes of uniqueness considered in these papers consist of functions having different asymptotic behavior in different directions.

2. Classes of Well-Posedness for Equations and Systems with Constant Coefficients

The condition found by I.G. Petrowsky is necessary and sufficient for well-posedness of the Cauchy problem for linear evolutionary systems with coefficients depending on the time variable, provided that the solutions are considered in a class of sufficiently smooth bounded functions. Systems that satisfy Condition A are presently referred to as well-posed in the sense of Petrowsky. This class covers all systems hyperbolic in the sense of Petrowsky and those hyperbolic in the sense of Gårding. L. Schwartz [26] gave an exposition of Petrowsky's results from the standpoint of distributions and showed this condition to be necessary and sufficient for well-posedness of the Cauchy problem in the class of distributions belonging to S' with respect to the spatial coordinates and depending smoothly on the time variable. Moreover, L. Schwartz noticed that in Petrowsky's results differential operators with respect to spatial variables can be replaced by operators of convolution. A systematic study of well-posedness of the Cauchy problem for such convolution equations has been performed by Volevich & Gindikin [27]. These authors found analogues of Condition A for such equations; their conditions essentially depend on the kind of the Cauchy problem under consideration: a homogeneous problem (with zero initial data), a non-homogeneous problem in a class of functions (distributions) that grow or decay at infinity as a degree of a polynomial, a problem on a finite or semi-infinite time interval.

G.E. Shilov studied questions of well-posedness for the Cauchy problem in various classes of distributions. In [28], as well as in a joint paper with Kostyuchenko [29], expanding sequences of classes of systems with constant coefficients are indicated, together with the corresponding well-posedness classes for the Cauchy problem. In such a sequence, parabolic systems are taken as the narrowest class, then go the systems well-posed in the sense of Petrowsky, which also differ according to their genus. In particular, for systems well-posed in the sense of Petrowsky and having non-positive genus, as for instance, is the case of the Schrödinger equation, the following result has been proved: there are infinitely many well-posedness classes not contained in each other. The widest class of systems is formed by the so-called provisionally well-posed systems. The class of well-posedness for such systems consists of infinitely differentiable functions whose growth is subject to estimates depending on $|x|$ and the order of the estimated derivative.

If a system is not provisionally well-posed, then its well-posedness classes of the type just described contain only analytic functions. A detailed review of these results is given in [30].

Another method to single out well-posedness classes has been suggested by Dikopolov & Shilov [31], [32], Shilov [33], and Palamodov [34]. In these papers, the Cauchy problem is considered for non-stationary equations and systems with constant coefficients, and the solution is sought in a space of distributions whose growth cannot be faster than polynomial. The class of well-posedness is specified by conditions on the Fourier transform of the initial data. These conditions require that the vector of the initial functions must have the value of its Fourier transform, at any point of the real space where this transform is defined, belonging to the subspace spanned by the eigenvectors and root vectors of the characteristic matrix which correspond to the eigenvalues with non-positive real parts. Lavrov [35] described similar classes of well-posedness which may contain distributions of arbitrary growth at infinity.

A differential operator with symbol $P(\tau,\xi)$ is called *strongly well-posed* if all its "shifts" of the form $P(\tau,\xi+i\omega) = P_\omega(\tau,\xi)$ correspond to differential operators well-posed in the sense of Petrowsky [36]. Each strongly well-posed operator can be associated with a maximal tube-shaped complex domain in which its symbol does not vanish. It is shown in [37], [38] that, knowing this tube-shaped domain, one can give a complete description of exponential classes of well-posedness for the corresponding Cauchy problem; in this case, there exist multi-dimensional essentially non-isotropic classes of well-posedness.

Among the operators that satisfy Petrowsky's condition of well-posedness, there is also a group of operators whose class of well-posedness allows functions of exponential growth in some directions and polynomial growth in others. Such operators (the simplest example is given by the equation $u_t = au_{xxx}$) have been studied by Fedoryuk [39] and Gindikin [40].

As a generalization of the Cauchy problem one can consider a boundary value problem in a layer $\mathbb{R}^n \times [0,T]$, with boundary values prescribed at $t = 0$ and $t = T$. Well-posedness and the questions of existence for such problems in various classes of finitely differentiable functions have been studied by Borok [4], [42].

If a given differential operator P with constant coefficients satisfies Petrowsky's Condition A, then there exists a fundamental solution supported in the half-space $t \geq 0$, and therefore, as shown by Malgrange, the Cauchy problem has a solution in the space $\mathcal{D}_F$ (formed by distributions of finite order), and no restrictions on the growth at infinity are required. However, Petrowsky's Condition A is *not* a necessary condition of solvability in this class of distributions. A delicate analysis performed by Hörmander [43], [44] allowed him to find a necessary and sufficient condition for the Cauchy problem to have a solution in $\mathcal{D}'_F$. The form of Hörmander's condition resembles that of Petrowsky's Condition A.

3. Well-Posedness of the Cauchy Problem for Equations and Systems with Variable Coefficients

In contrast to the case of equations and systems with constant coefficients, the problem of well-posedness in the case of variable coefficients is very far from being exhausted; here, beyond the theory of hyperbolic and parabolic equations, a relatively small number of results are known.

The case of a single higher order equation and that of a system with a single spatial variable have been studied by Zhytomirsky [45], [46], and Borok & Zhytomirsky [47], [48]. It is assumed in these papers that the coefficients of the equation tend to constants as $|x| \to \infty$, and for the limit equation Petrowsky's Condition A is fulfilled. These authors indicate conditions on the rate of convergence of the coefficients to their limits at infinity which guarantee that the classes of well-posedness for the limit equation with constant coefficients are the same as the classes of well-posedness for the given equation with variable coefficients.

Differential operators with variable coefficients and many independent variables are considered in [36]; the symbols of these operators satisfy the condition of strong well-posedness for any fixed (t, x), and the values of these symbols at different points (t, x) are related by the condition of equipotentiality in the sense of Hörmander. In this case, the parametrix can be constructed, and on its basis well-posedness of the Cauchy problem in spaces H^s is proved. Pseudo-differential operators with holomorphic symbols are studied in [27]. The methods of the theory of such operators make it possible to establish well-posedness of the Cauchy problem in classes of functions which may have exponential growth or decay at infinity (see [27], [49], [50]). Leray's method of separating operator (see [51]) is applied in [52]–[56] to study strongly well-posed problems for differential operators with variable coefficients; here no conditions of equipotentiality are imposed on the operators. This method is based on energy estimates and allows one to examine exponential classes of well-posedness (see [54], [56]).

4. Parabolic Systems

There are numerous studies dedicated to the investigation of parabolic systems introduced in the paper of I.G. Petrowsky under discussion; nowadays these systems are called parabolic in the sense of Petrowsky. At present, the theory of parabolic equations and systems has become a large and intensively growing branch of the theory of partial differential equations. The uniqueness theorem for the Cauchy problem for parabolic systems in the class of bounded functions has been proved by I.G. Petrowsky. A generalization of Tikhonov's theorem for parabolic equations is due to Ladyzhenskaya [57], and for parabolic systems – to Eidelman (see [58]). An analogue of Täcklind's theorem for such systems is established in [5].

O.A. Oleinik [59]–[63] suggested a new approach to the proof of uniqueness theorems for the Cauchy problem and for evolutionary boundary value problems in infinite domains. This method is based on the examination of a certain auxiliary system with an additional independent variable, and on the application of estimates of its solutions for complex values of this additional variable (see [59]–[63]). A modification of this procedure is the so-called method of the introduction of a parameter suggested in [64]; applications of the latter are given in [65], which also contains the description of an abstract procedure to derive uniqueness theorems for evolutionary equations by the introduction of a parameter. A prerequisite for the application of this method is a local *a priori* estimate of special type. The method of the introduction of a parameter yields uniqueness theorems which generalize Täcklind's result to the case of general parabolic systems considered by Solonnikov in [66]. In some sense, these systems are a counterpart of hyperbolic systems considered by Leray [51], and of elliptic systems examined by Douglis & Nirenberg [67]. The uniqueness theorems have been established for the Cauchy problem and for mixed boundary value problems in infinite domains. In [65], one can also find a review of results about uniqueness of the Cauchy problem and mixed problems in unbounded domains, as well as asymptotic properties of solutions of parabolic equations and systems. A detailed bibliography on the subject is also given in [65].

Eidelman [58] has constructed the Green matrices for systems $2b$-parabolic in the sense of Petrowsky in the case of variable coefficients. These Green matrices were applied to establish well-posedness of the Cauchy problem in classes of bounded as well as rapidly growing functions.

In the early '60s, Zagorsky [68], Agranovich & Vishik [69], and Eidelman [58] introduced an analogue of Lopatinsky's condition for parabolic systems and proved solvability and well-posedness of the mixed problem in cylindrical domains for systems $2b$-parabolic in the sense of Petrowsky. Solonnikov [66] has studied the mixed problem for parabolic systems of a more general type corresponding to systems elliptic in the sense of Douglis-Nirenberg. For such systems, Solonnikov constructed the Green matrices, established *a priori* estimates, and proved the existence of solutions for the mixed problem and the Cauchy problem in various functional spaces.

A generalization of the class of systems parabolic in the sense of Petrowsky is introduced in [27]. The corresponding definition relies on the notion of the principal part of the operator with respect to Newton's polygon. An operator parabolic in the sense of [27] is a composition of $2b$-parabolic operators with different b. The Cauchy problem for such operators in classes of exponentially growing functions is studied in [49]. In [65] the method of the introduction of a parameter is applied to a class of such equations in order to establish uniqueness theorems for the Cauchy problem and the mixed boundary value problem in an infinite domain. More general parabolic systems than those introduced in [27] have been examined in [70].

Asymptotic behavior of fundamental solutions for parabolic equations with constant coefficients is considered in: Fedoryuk [39], Efgrafov & Postnikov [71], Gindikin & Fedoryuk [72]–[74]. Asymptotic properties of fundamental solutions and Green matrices for general parabolic systems have been studied by Oleinik & Radkevich [65] by the method of the introduction of a parameter.

Among the results obtained so far for parabolic systems, the following should also be mentioned (see [58], [75]): *a priori* estimates for various norms of the solutions; hypoellipticity in the case of coefficients analytic in x; analyticity of solutions with respect to the spatial variables; theorems of Liouville and Phragmen-Lindelöf's type; asymptotic behavior of solutions for $t \to \infty$.

References

[1] Gelfand, I.M.; Petrowsky, I.G.; Shilov, G.E. Theory of systems of partial differential equations. In: *Proceedings of the Third All-Union Mathematical Congress, 1956.* Moscow, Izd.-vo Akad. Nauk SSSR, 1958, pp. 65–72.

[2] Gelfand, I.M.; Shilov, G.E. *Some Questions of the Theory of Differential Equations.* Moscow, Fizmatgiz, 1958.

[3] Petrowsky, I.G. Über das Cauchysche Problem für Systeme von partiellen Differentialgleichungen. *Matem. Sbornik* 2:5 (1937) 815–870. (See this volume, Article 1).

[4] Babenko, K.I. On a new problem of quasi-analyticity and the Fourier transform of entire functions. *Trudy Mosk. Mat. Ob.-va* 5 (1956) 523–542.

[5] Zolotaryiov, G.N. On the uniqueness of solutions of the Cauchy problem for systems parabolic in the sense of Petrowsky. *Izvest. Vuzov. Mat.* 2 (1958) 118–135.

[6] Zolotaryiov, G.N. Estimates from above for the uniqueness classes of solutions of the Cauchy problem for systems of partial differential equations. *Nauchnye Doklady Visshei Shkoli* **2** (1958) 37–40.

[7] Zolotaryiov, G.G. Maximal classes of uniqueness for the Cauchy problem for partial differential equations. *Ucheniye Zapiski, Ivanov. Pedag. Inst.* **34** (1963) 34–45.

[8] Täcklind, S. Sur le class quasianalytique des solutions des équations aux dérivées partielles du type parabolique. *Nova Acta Regiae Soc. Sci. Upsal. Ser. 4*, **10:3** (1936) 3–55.

[9] Chaus, N.N. On the uniqueness of solutions of the Cauchy problem for systems of partial differential equations. *Ukr. Mat. Zh.* **17:1** (1965) 126–130.

[10] Galpern, S.A. The Cauchy problem for general systems of linear partial differential equations. *Trudy Mosk. Mat. Ob.-va* **9** (1960) 401–423.

[11] Kostyuchenko, A.G.; Eskin, G.I. On Sobolev-Galpern equations. *Uspekhi Mat. Nauk* **15:2** (1960) 211–212.

[12] Eskin, G.I. On the uniqueness of solutions of the Cauchy problem for non-Kowalevskian systems. *Trudy Mosk. Mat. Ob.-va* **10** (1961) 285–295.

[13] Borok, V.M. On the Cauchy problem for general linear equations. *Doklady Akad. Nauk SSSR* **177:4** (1967) 759–762.

[14] Borok, V.M.; Iokhvidovich, N.Yu. On the Cauchy problem for general linear systems. In: *Theory of Functions, Functional Analysis, and Their Applications.* **27**, Kharkov, 1977, pp. 26–33.

[15] Palamodov, V.P. *Linear Differential Operators with Constant Coefficients.* Moscow, Nauka, 1967.

[16] Yamanaka, T. A note on the Cauchy problem for equations with polynomial coefficients. *Funkcialaj Ekvacioj* **10** (1967) 35–41.

[17] Yamanaka, T. On the uniqueness of solutions of the Cauchy problem for systems of linear partial differential equations. *Funkcialaj Ekvacioj* **9** (1966) 219–249.

[18] Zhitomirsky, Ya.I. On differential operators of infinite order in S-type spaces. *Mat. Sbornik* **80:8** (1969) 405–410.

[19] Zhitomirsky, Ya.I. Uniqueness calsses for solutions of the Cauchy problem for linear equations with growing coefficients. *Izvest. Akad. Nauk SSSR, Ser. Mat.* **31:4** (1967) 763–782.

[20] Zhitomirsky, Ya.I. Uniqueness calsses for solutions of the Cauchy problem for linear equations with rapidly growing coefficients. *Izvest. Akad. Nauk SSSR, Ser. Mat.* **31:5** (1967) 1159–1178.

[21] Palyutkin, V.G. On calsses of uniqueness for solutions of the Cauchy problem for equations with rapidly growing coefficients. *Ukr. Mat. Zh.* **27:2** (1975) 262–265.

[22] Palyutkin, V.G. Uniqueness classes for solutions of the Cauchy problem for equations with variable coefficients. *Matem. Zametki* **26:6** (1979) 835–849.

[23] Chaus, N.N. On the uniqueness of solutions of the Cauchy problem for some systems with variable coefficients. *Ukr. Mat. Zh.* **23:1** (1971) 122–125.

[24] Chaus, N.N. Conditions of uniqueness for solutions of the Cauchy problem for some special systems with variable coefficients. *Ukr. Mat. Zh.* **24:3** (1972) 421–426.

[25] Chaus, N.N. On triviality classes for solutions of some systems with variable coefficients. *Ukr. Mat. Zh.* **33** (1981) 133–136.

[26] Schwartz, L. Les équations d'évolution liées au product de composition. *Ann. Inst. Fourier* **2** (1950) 19–49.

[27] Volevich, L.R.; Gindikin, S.G. The Cauchy problem and related problems for convolution equations. *Uspekhi Mat. Nauk* **27:4** (1972) 65–143.

[28] Shilov, G.E. On the conditions of well-posedness of the Cauchy problem for systems of partial differential equations with constant coefficients. *Uspekhi Mat. Nauk* **10:4** (1955) 89–100.

[29] Kostyuchenko, A.G.; Shilov, G.E. On the solution of the Cauchy problem for regular systems of partial differential equations. *Uspekhi Mat. Nauk* **9:3** (1954) 141–148.

[30] Shilov, G.E. Correctly posed problems for linear constant coefficients partial differential equations in half-space. In: *Proc. Intern. Congr. Math. 1962, Stockholm.* 1963, pp.403–404.

[31] Dikopolov, G.V.; Shilov, G.E. On well-posed problems in half-space for partial differential equations with a non-zero right-hand side. *Sib. Mat. Zh.* **1:1** (1960) 141–148.

[32] Dikopolov, G.V.; Shilov, G.E. On well-posed boundary value problems for partial differential equations in half-space. *Izvest. Akad. Nauk SSSR, Ser. Mat.* **24:3** (1960) 69–100.

[33] Shilov, G.E. On well-posed problems in half-space for linear partial differential equations with constant coefficients. *Uspekhi Mat. Nauk* **19:3** (1964) 3–52.

[34] Palamodov, V.P. Conditions for correct solvability in the large for a class of equations with constant coefficients. *Sib. Mat. Zh.* **4:5** (1963) 1137–1146.

[35] Lavrov, A.M. Ill-posed Cauchy problems and a uniqueness theorem for partial differential equations. *Doklady Akad. Nauk SSSR* **258:2** (1981) 299–303.

[36] Volevich, L.R.; Gindikin, S.G. Pseudo-differential operators and the Cauchy problem for differential equations with variable coefficients. *Funkt. Anal. Prilozh.* **1:4** (1967) 8–25.

[37] Gindikin, S.G. Convolutors in spaces of generalized functions with exponential asymptotics. *Funkt. Anal. Prilozh.* **6:2** (1972) 83–85.

[38] Volevich, L.R.; Gindikin, S.G. Convolutors in spaces of distributions and related problems for convolution equations. *Sel. Math. Sov.* **2:1** (1982) 9–30.

[39] Fedoryuk, M.V. Asymptotics of Green functions for equations well-posed in the sense of Petrowsky with constant coefficients and classes of well-

posedness for the solutions of the Cauchy problem. *Mat. Sbornik* **62:4** (1963) 397–463.

[40] Gindikin, S.G. Differential opeartors with real characteristics of large multiplicity. *Doklady Akad. Nauk SSSR* **204:5** (1972) 1037–1040.

[41] Borok, V.M. Correctly solvable boundary value problems for systems of linear partial differential equations in an infinite layer. *Izvest. Akad. Nauk SSSR. Ser. Mat.* **35:1** (1971) 185–201.

[42] Borok, V.M. On correct solvability of a boundary value problem for linear equations with constant coefficients in an infinite layer. *Izvest. Akad. Nauk SSSR. Ser. Mat.* **35:4** (1971) 922–939.

[43] Hörmander, L. On the characteristic Cauchy problem. *Ann. Math.* **88:2** (1968) 341–370.

[44] Hörmander, L. *The Analysis of Linear Partial Differential Operators. II.* Berlin, Springer-Verlag, 1983.

[45] Zhytomirski, Ya.I. A condition for correct solvability of the Cauchy problem for systems of linear partial differential equations with variable coefficients. *Izvest. Vuzov. Mat.* **4** (1960) 79–88.

[46] Zhytomirsky, Ya.I. The Cauchy problem for some systems of partial differential equations parabolic in the sense of Shilov with variable coefficients. *Izvest. Akad. Nauk SSSR. Ser. Mat.* **23:6** (1959) 925–932.

[47] Borok, V.M.; Zhytomirsky, Ya.I. On the fundamental solution for equations well-posed in the sense of Petrowsky. *Izvest. Akad. Nauk SSSR. Ser. Mat.* **30:4** (1966) 951–968.

[48] Borok, V.M.; Zhytomirsky, Ya.I. The Cauchy problem for odd order equations that is well-posed in the sense of Petrowsky. *Izvest. Akad. Nauk SSSR. Ser. Mat.* **30:5** (1966) 1133–1146.

[49] Volevich, L.R. The Cauchy problem for hypoelliptic differential operators in classes of rapidly growing functions. *Uspekhi Mat. Nauk* **25:1** (1970) 191–192.

[50] Gindikin, S.G. Classes of well-posedness for the Cauchy problem for differential equations with variable coefficients. *Funkt. Anal. Prilozh.* **4:2** (1970) 83–84.

[51] Leray, J. *Hyperbolic Differential Equations.* Princeton, N.J. Inst. Adv. Study, 1953.

[52] Volevich, L.R.; Gindikin, S.G. The Cauchy problem for pluri-parabolic differential equations. II. *Mat. Sbornik* **78:2** (1969) 214–236.

[53] Volevich, L.R.; Gindikin, S.G. The Cauchy problem for differential operators with dominating principal part. *Funkt. Anal. Prilozh.* **2:3** (1968) 22–40.

[54] Volevich, L.R. Energy method in the Cauchy problem for differential operators well-posed in the sense of Petrowsky. *Trudy Mosk. Mat. Ob.-va* **31** (1974) 147–187.

[55] Gindikin, S.G. Energy estimates related to the Newton's polyhedron. *Trudy Mosk. Mat. Ob.-va* **31** (1974) 189–236.

[56] Volevich, L.R.; Gindikin, S.G. The method of energy estimates for the Cauchy problem for differential operators with variable coefficients. In:

Proceedings of the All-Union Congress on PDE's dedicated to the 75th anniversary of I.G. Petrowsky. Moscow, Moscow Univ. Press, 1973, pp. 75–77.

[57] Ladyzhenskaya, O.A. On the uniqueness of solutions of the Cauchy problem for a linear parabolic equation. *Mat. Sbornik* **27** (1950) 175–184.

[58] Eidelman, S.D. *Parabolic Systems.* Moscow, Nauka, 1964.

[59] Oleinik, O.A. On the uniqueness of solutions of the Cauchy problem for general parabolic systems in classes of rapidly growing functions. *Uspekhi Mat. Nauk* **29:5** (1974) 229–230.

[60] Oleinik, O.A. On the uniqueness of solutions of boundary value problems and the Cauchy problem for general parabolic systems. *Doklady Akad. Nauk SSSR* **220:6** (1975) 34–37.

[61] Oleinik, O.A. On the behavior of solutions of linear parabolic systems of differential equations in unbounded domains. *Uspekhi Mat. Nauk* **30:2** (1975) 219–220.

[62] Oleinik, O.A. On the behavior of solutions of the Cauchy problem and the boundary value problem for parabolic systems of partial differential equations in unbounded domains. *Rend. Mat. Ser 6,* **8:2** (1975) 545.

[63] Oleinik, O.A. The analyticity of solutions of partial differential equations and its applications. In: *Trends in Application of Pure Mathematics to Mechanics: Proc. Symp. Lecce (Italy).* London, Pitman, 1975, pp. 281–298.

[64] Oleinik, O.A. The method of introducing a parameter for the study of evolutionary equations. *Uspekhi Mat. Nauk* **33:3** (1978) 126.

[65] Oleinik, O.A.; Radkevich, E.V. The method of introducing a parameter for the study of evolutionary equations. *Uspekhi Mat. Nauk* **33:5** (1978) 7–76.

[66] Solonnikov, V.A. On boundary value problems for linear parabolic differential equations of general type. *Trudy MIAN SSSR* **83** (1965) 3–162.

[67] Douglis, A.; Nirenberg, L. Interior estimates for elliptic systems of partial differential equations. *Comm. Pure Appl. Math.* **8:4** (1955) 503–538.

[68] Zagorsky, T.Ya. *The Mixed Problem for Systems of Partial Differential Equations of Parabolic Type.* Lvov, Lvov Univ. Press, 1981.

[69] Agranovich, M.S.; Vishik, M.I. Elliptic problems with a parameter and parabolic problems of general type. *Uspekhi Mat. Nauk* **19:3** (1964) 53–161.

[70] Gindikin, S.G. General parabolic systems. *Uspekhi Mat. Nauk* **26:3** (1971) 219–220.

[71] Evgrafov, M.A.; Postnikov, M.M. Asymptotics of Green's functions for parabolic and elliptic equations with constant coefficients. *Mat. Sbornik* **82:1** (1970) 3–29.

[72] Gindikin, S.G.; Fedoryuk, M.V. Asymptotics of the fundamental solution for a differential equation parabolic in the sense of Petrowsky with constant coefficients. *Mat. Sbornik* **91:4** (1973) 500–522.

[73] Gindikin, S.G.; Fedoryuk, M.V. Saddle-points for parabolic polynomials. *Mat. Sbornik* **94:3** (1974) 385–406.

[74] Gindikin, S.G.; Fedoryuk, M.V. Asymptotics of Green's function for hypoelliptic differential operators well-posed in the sense of Petrowsky with constant coefficients. In: *Problems in Mechanics and Mathematical Physics*. Moscow, Nauka, 1976, pp.98–116.

[75] Oleinik, O.A.; Radkevich, E.V. Analyticity and theorems of Liouville and Phragmen-Lindelöf's type for general parabolic systems of differential equations. *Funkt. Anal. Prilozh.* **8:4** (1974) 59–70.

On the Analyticity of Solutions of Systems of Partial Differential Equations*

by O.A. Oleinik and L.V. Volevich

1. The famous 19th problem of Hilbert is closely related to the description of the classes of (nonlinear) differential equations whose sufficiently smooth solutions can only be analytic functions (see [1]).

In his classical works, Bernstein (see [2]) gave the first proof of the fact that a solution of a nonlinear second order elliptic equation with two independent variables is analytic, provided that the solution is three times continuously differentiable and the equation is given by an analytic function. In subsequent publications this result of Bernstein has been elaborated, generalized, and repeatedly proved by other methods. In the case of a single nonlinear second order elliptic equation with many independent variables, analyticity of smooth solutions has been established by Hopf. A review of the results on the problem of analyticity, obtained before 1940, is given in an article by Bernstein & Petrowsky [3], and the results up to 1969 have been described by Oleinik [4].

Substantial progress in the study of analyticity is due to Petrowsky's paper [5], where a wide class of nonlinear systems (now called elliptic in the sense of Petrowsky) is singled out, and analyticity of all sufficiently smooth solutions of such systems is established. (Regarding this paper of Petrowsky see also [6]).

I.G. Petrowsky was the first to prove the corresponding converse theorems. Thus, in the case of systems with constant coefficients, he showed ellipticity to be a necessary condition for all (sufficiently smooth) solutions to be analytic. While proving this result, Petrowsky thoroughly examined how the smoothness of solutions depends on the behavior, in the complex region, of the roots of the determinant of the system's characteristic matrix (nowadays called "symbol"). Analyzing the proof of Petrowsky's converse theorem, Hörmander [7] singled out a class of differential equations and sys-

---

*A commentary on Article 4 of this volume.

tems with constant coefficients which can have only infinitely differentiable solutions; thereby Hörmander solved a problem posed by L. Schwartz. The technique of distributions, developed by the early '50s, allowed one to consider the problem of smoothness from a more general standpoint: thus, in the case of linear equations, the *a priori* assumption of sufficient smoothness of a solution has been replaced by the condition that the corresponding generalized function should satisfy the equation in the sense of distributions. In modern terminology, differential operators corresponding to equations with all their distribution solutions being infinitely differentiable are called hypoelliptic. Hörmander showed that all solutions of hypoelliptic equations actually belong to some Gevrey classes and are analytic along certain directions. A detailed exposition of these topics can be found in the monograph of Hörmander [7] (see also [8]).

2. In the concluding part of his article [5], Petrowsky gives a simple example of a system which is not elliptic in the sense of his definition but has all its sufficiently smooth solutions analytic. This fact gave rise to the necessity of generalizing the definition of elliptic systems.

J. Leray, in his lectures on hyperbolic systems [9], generalized Petrowsky's definition of hyperbolic systems. A similar generalization for elliptic systems is due to Douglis & Nirenberg [10]. The definition of Douglis and Nirenberg can be formulated as follows.

Consider a linear system of partial differential equations

$$\sum_{j=1}^{N} P_{lj}(x,D)u_j(x) = f_l(x), \qquad l = 1,\dots,N, \qquad (1)$$

where

$$x = (x_1,\dots,x_n), \quad D = (D_1,\dots,D_n),$$

$$D_j = -i\frac{\partial}{\partial x_j}, \quad \alpha = (\alpha_1,\dots,\alpha_n), \quad |\alpha| = \alpha_1 + \cdots + \alpha_n,$$

$$D^\alpha = D_1^{\alpha_1}\cdots D_n^{\alpha_n}, \quad P_{lj}(x,D) = \sum_{|\alpha|\le\alpha_{lj}} a_{lj\alpha}(x)D^\alpha.$$

Each set of integers $n_1,\dots,n_N, m_1,\dots,m_N$ satisfying the inequalities

$$\alpha_{lj} \le n_j - m_l, \quad l,j = 1,\dots,N, \qquad (2)$$

can be related to the principal symbol of system (1)

$$P^0(x,\xi) = \left\| P_{lj}^0(x,\xi) \right\|, \quad l,j = 1,\dots,N,$$

where

$$P_{lj}^0(x,\xi) = \begin{cases} \sum_{|\alpha|=n_j-m_l} a_{lj\alpha}(x)\xi^\alpha \,, \\ 0 \quad \text{for} \quad \alpha_{lj} < n_j - m_l \,. \end{cases}$$

System (1) is said to be *elliptic* if the integers $m_1,\ldots,m_N, n_1,\ldots,n_N$ satisfying (2) can be chosen in such a way that

$$\det P^0(x,\xi) \neq 0 \,, \quad |\xi| \neq 0 \,.$$

If it is possible to take $n_j = \max_l \alpha_{lj}$ and $m_j = 0$, then system (1) is elliptic in the sense of Petrowsky. The definition of ellipticity suggested by Douglis-Nirenberg is not fully effective, since it is uncertain how the integers satisfying conditions (2) can be constructed. A verifiable definition of ellipticity in the sense of Douglis-Nirenberg has been suggested by Volevich in [11]: Set

$$R = \max_\gamma \left(\alpha_{1\gamma(1)} + \cdots + \alpha_{N\gamma(N)} \right) \,,$$

where γ takes its values among all permutations of $1,\ldots,N$. Denote by r the degree of the polynomial $\det \|P_{ij}(x,\xi)\|$ with respect to ξ. System (1) is said to be elliptic, if $R = r$ and

$$P_0(x,\xi) \neq 0 \,, \quad |\xi| \neq 0 \,,$$

where $P_0(x,\xi)$ is the principal homogeneous part of $\det \|P_{lj}(x,\xi)\|$. The equivalence of the latter definition to that of Douglis-Nirenberg is proved in [12].

It should be mentioned that the class of systems elliptic in the sense of Douglis-Nirenberg (in contrast to systems elliptic in the sense of Petrowsky) is closed with respect to passing to the conjugate operator (this fact is important for the construction of fundamental solutions; see [13]).

Analyticity of all (sufficiently smooth) solutions of linear elliptic (in the above sense) systems with analytic coefficients has been proved by Morrey & Nirenberg [14] (see also [15]). Their method is much simpler than that originally used by Petrowsky and is based on Schauder type *a priori* estimates for the norms of solutions in Sobolev spaces.

A similar result for nonlinear systems has been obtained by Morrey [16]. The proof of analyticity of solutions of linear elliptic systems is the subject of [62].

Analyticity of solutions of elliptic pseudo-differential equations and systems is established in [17], [18].

The classes of elliptic systems defined by Petrowsky and those specified by Douglis and Nirenberg have one essential disadvantage: these systems lack invariance with respect to non-degenerate linear transformations of the unknown functions, and they are not closed with respect to a composition of the operators of the same class. Both these operations may turn an elliptic system into another one, which, in general, may not belong to the class of elliptic systems in the sense of either definition. Therefore, it would be natural to search for a correct definition of elliptic systems without these deficiencies. However, it is quite possible that such a definition can only be found for overdetermined systems.

In this connection, it should be observed that in the proof of Petrowsky the original system of nonlinear equations, after differentiation and passing to new unknown functions, is reduced to a quasilinear overdetermined system (unfortunately, without an explicit description in algebraic terms), and then analyticity of solutions of this overdetermined system is established in some chosen direction.

In John's monograph [13], analyticity of solutions of elliptic systems is also proved by reduction to overdetermined systems.

3. As mentioned above, ellipticity is a necessary condition of analyticity of all solutions (smooth, as well as in the class of distributions) of a linear equation with constant coefficients and analytic right-hand side. This statement does not hold in the case of equations with variable (analytic) coefficients. As shown by Mizohata [19] (see also Suzuki [20], Oleinik [21], Oleinik & Radkevich [22], [23]), for any even positive s, the equation

$$\frac{\partial u}{\partial x_0} + i x_0^s \frac{\partial u}{\partial x_1} = 0$$

admits only analytic solutions, whereas for odd s, this equation has non-analytic (even non-smooth) solutions. This fact gave rise to the interesting problem of characterizing classes of non-elliptic equations (or degenerate elliptic equations) which admit only analytic solutions.

Before we describe the results in this direction, it is necessary to recall some definitions.

An operator P is called *hypoelliptic* in a domain Ω, if

$$\{u \in \mathcal{D}'(\Omega),\ Pu \in C^\infty(\Omega)\} \implies \{u \in C^\infty(\Omega)\} . \tag{3}$$

An operator P is called *analytic-hypoelliptic* in Ω, if

$$\{u \in \mathcal{D}'(\Omega),\ Pu \in \mathcal{A}(\Omega)\} \implies \{u \in \mathcal{A}(\Omega)\} , \tag{4}$$

where $\mathcal{A}(\Omega)$ is the space of real analytic functions in Ω.

Along with (4), one considers a more general definition of analytic hypoellipticity: there exists an integer $q \geq 0$ such that

$$\{u \in C^q(\Omega),\ Pu \in \mathcal{A}(\Omega)\} \implies \{u \in \mathcal{A}(\Omega)\} . \tag{5}$$

Definition (5) can be extended further, if we assume that $q = \infty$, i.e.,

$$\{u \in C^{\infty}(\Omega), \ Pu \in \mathcal{A}(\Omega)\} \Longrightarrow \{u \in \mathcal{A}(\Omega)\} \ . \tag{5'}$$

Note that in the case of a hypoelliptic operator, definitions (4), (5) and (5') coincide, and therefore, definitions (5) and (5') become interesting only if applied to non-hypoelliptic operators.

The definitions of analytic hypoellipticity can be formulated in terms of the so-called micro-localization; then one can speak of analytic hypoellipticity in a neighborhood of a point of the cotangent bundle $T^{*}(\Omega)$; these questions will be discussed in more detail later on.

4. The simplest generalization of the class of elliptic operators is the class of operators with simple characteristics, or the so-called *operators of principal type* introduced by Hörmander (see [7]).

Let $P(x, D)$ be a differential operator (or a classical pseudo-differential operator) with symbol

$$P(x, \xi) \sim P_m(x, \xi) + P_{m-1}(x, \xi) + \cdots \ . \tag{6}$$

The function $P_m(x, \xi)$ homogeneous in ξ is called the *principal symbol* of P. Set

$$\text{char}\,(P) = \{(x, \xi) \in \mathbb{R}^n \times \mathbb{R}^n, \ P_m(x, \xi) = 0\} \ .$$

The operator $P(x, D)$ is said to be of *principal type* if

$$\{(x, \xi) \in \text{char}\,(P)\} \Longrightarrow \{\text{grad}\,_\xi\, P_m(x, \xi) \neq 0\} \ .$$

For differential operators of principal type with analytic coefficients, it has been proved by Treves [24] that hypoellipticity (3) implies analytic hypoellipticity (5'). On the other hand, as shown by Treves in [25], [26], for operators of principal type, hypoellipticity is equivalent to subellipticity (in the sense of Hörmander and Egorov), and the latter is equivalent to the following condition on the principal symbol P_m: let (x^0, ξ^0) be an arbitrary point of char(P) such that

$$\text{grad}\,_\xi\, \text{Re}\, P_m(x^0, \xi^0) \neq 0$$

(the operator P can always be multiplied by $z \in \mathbb{C}^1$ such that

$$\text{grad}\,_\xi\, \text{Re}\,(z P_m(x^0, \xi^0)) \neq 0 \ ;$$

and it is assumed in what follows that this multiplication has already been performed). Then, for the hypoelliptic operator $P(x, D)$ of principal type, the restriction of the function $\text{Im}\, P_m(x, \xi)$ to the bicharacteristic corresponding to the symbol $\text{Re}\, P_m(x, \xi)$ and passing through (x^0, ξ^0) must have zeros of even order only. Thus, in the case of differential operators of principal type with analytic coefficients, there is a sufficient condition of analytic hypoellipticity (4) expressed in terms of the principal symbol of the original operator.

5. In the case of operators with multiple characteristics, the conditions of analytic hypoellipticity become much more difficult to express. For such operators, hypoellipticity does not imply analytic hypoellipticity, in general. As shown by Baouendi & Goulaouic [27] (see also [21]–[23]), the operator

$$D_t^2 + D_x^2 + x^2 D_y^2 \tag{7}$$

is hypoelliptic but not analytic-hypoelliptic.

Oleinik and Radkevich (see [21]-[23]) have established the following result of general character. In a domain $\Omega \subset \mathbb{R}^n$, consider a linear homogeneous system of partial differential equations $Pu = 0$ with analytic coefficients. For $G \subset \Omega$, let $Q_\delta^j(G)$ be the cylinder

$$Q_\delta^j(G) = \{x, y_j : x \in G, \ x_j + iy_j \in \mathbb{C}^1, \ |y_j| < \delta\} \,.$$

Let $\mathcal{B}(\Omega)$ be a Banach space formed by generalized solutions $u \in \mathcal{D}'(\Omega)$ of system $Pu = 0$ in Ω; let $\|u\|_{\mathcal{B}(\Omega)}$ be the norm in this space, and assume that any sequence converging in the norm of $\mathcal{B}(\Omega)$ is also convergent in $\mathcal{D}'(\Omega)$. Assume also that system $Pu = f$ has the following property of analyticity in x_j: for any f analytic with respect to x_j in Ω, any generalized solution $u \in \mathcal{D}'(\Omega)$ of system $Pu = f$ is analytic with respect to x_j in Ω. Then, for any subdomain $G \subset \Omega$ such that $\overline{G} \subset \Omega$, there exist δ and C (depending on G) such that

$$\sup_{Q_\delta^j(G)} |U| \le C \|u\|_{\mathcal{B}(\Omega)} \tag{8}$$

for any solution of the system $Pu = 0$ from the space $\mathcal{B}(\Omega)$, where U is an analytic extension of $u(x)$ to $Q_\delta^j(G)$. The estimate (8) for $j = 1, \ldots, n$ holds in the case of analytic hypoellipticity of the operator P in Ω.

This result allows one to reduce the proof of the lack of analytic hypoellipticity for an operator P to the construction of some special families of solutions for $Pu = 0$ which violate the estimate (8). The conditions, which these families must satisfy, are thoroughly analyzed in [23]. In particular, it is shown in [23], with the help of (8), that the operator

$$P_1(D_t) + |t|^{2s} P_2(D_y) + |t|^{2d} P_3(D_z)$$

(here P_1, P_2, P_3 are homogeneous elliptic operators of the variables $t = (t_1, \ldots, t_{k_1})$, $y = (y_1, \ldots, y_{k_2})$, $z = (z_1, \ldots, z_{k_3})$, respectively; the coefficients of P_1, P_2, P_3 are constant, and the order of these operators is 2ν with respect to the corresponding variables; s and d are non-negative integers) is analytic-hypoelliptic in a neighborhood of the origin, if and only if $s = d$. For $s = 0$, $d = 1$, $\nu = 1$, $k_1 = k_2 = k_3 = 1$, we obtain the operator (7). Moreover, [23] contains a description of some classes of first order systems which do not have the property of analytic hypoellipticity. These results are a generalization of [19]. One of the results of [28] is the following theorem:

Let P be a linear operator with analytic complex valued coefficients in a domain Ω:

$$P = \sum_{\substack{|\alpha|=m \\ \alpha_n=0}} a_\alpha(x)D^\alpha + \sum_{|\alpha|\leq m-1} b_\alpha(x)D^\alpha = P_m + P_{m-1} \ .$$

Assume that for some $\hat{x} \in \Omega$ and $\hat{\xi} \in \mathbb{R}^n$, $\hat{\xi} \neq 0$, we have

$$P_m(\hat{x},\hat{\xi}) \neq 0 \ .$$

Then the equation $Pu = 0$ admits a solution that is non-analytic with respect to x_n in any sufficiently small neighborhood of $\hat{x}$.

Some general conditions ensuring analyticity of all solutions are established in [29] for second order equations with two independent variables.

6. A wide class of analytic-hypoelliptic operators has been singled out by Grushin [30]. Let the independent variables x be divided into two groups: $x = (x',y)$, and let P be an operator of the form

$$P(x,D) = \sum_{\mathcal{M}} a_{\alpha\beta\gamma}(x)y^\gamma D_{x'}^\beta D_y^\alpha \ , \tag{9}$$

where $\mathcal{M}$ is a finite set consisting of all multi-indices α,β,γ such that

$$|\alpha| + |\beta| \leq m \ , \quad |\alpha| + (1+\delta)|\beta| - m \leq |\gamma| \leq m\delta \ ,$$

for some rational $\delta > 0$. Let us associate with the operator (9) an auxiliary operator L having polynomial coefficients and depending on the parameter ξ:

$$L(y,\xi,D_y) = \sum_{\mathcal{M}^0} a_{\alpha\beta\gamma}(0)y^\gamma\xi^\beta D_y^\alpha \ , \tag{10}$$

where $\mathcal{M}^0$ consists of $(\alpha,\beta,\gamma) \in \mathcal{M}$ such that

$$|\gamma| = |\alpha| + (1+\delta)|\beta| - m \ .$$

Set

$$L^0(y,\xi,D_y) = \sum_{\substack{(\alpha,\beta,\gamma)\in\mathcal{M}^0 \\ |\alpha|+|\beta|=m}} a_{\alpha\beta\gamma}(0)y^\gamma\xi^\beta D_y^\alpha \ . \tag{11}$$

Grushin assumes that the following conditions are satisfied:

Condition I. The operator $L(y,D_{x'},D)$ is elliptic for $y \neq 0$ (this condition guarantees that the equation

$$L^0(y,\xi,D_y)v(y) = 0 \tag{12}$$

has a finite dimensional kernel in S).

Condition II. For $\xi \neq 0$, equation (12) does not have nontrivial solutions $v \in S$ (here S is the space formed by all C^∞-functions which decay at infinity, together with all their derivatives, faster than $(1+|x|)^{-N}$, $\forall N > 0$).

It has been shown in [30] that if Conditions I and II are fulfilled, then the operator (9) with analytic coefficients $a_{\alpha\beta\gamma}(x)$ is analytic-hypoelliptic (in the sense of (4)).

If the operator P does not contain lower order terms (i.e., if $\mathcal{M} = \mathcal{M}^0$), then Condition I is necessary and sufficient for hypoellipticity, as well as for analytic hypoellipticity.

Thus, for the operators considered by Grushin, the question of analytic hypoellipticity is reduced to the examination of the spectrum of the auxiliary operator (11) involving a smaller number of variables and depending on a parameter. In some cases the spectrum of the auxiliary operator admits a simple description. For instance, such a description allowed Tosques [31] to show that an operator of the form

$$P = \left(\frac{\partial}{\partial t} - t^k a A \right) \left(\frac{\partial}{\partial t} - t^k b A \right) + t^{k-1} c A \tag{13}$$

(here $a, b, c \in \mathbb{C}^1$, $\operatorname{Re} a > 0$, $\operatorname{Re} b < 0$, k is an odd number, A is a positive first order pseudo-differential operator) is analytic-hypoelliptic if and only if

$$\frac{c - a + b}{(k+1)(a-b)}, \qquad \frac{c}{(k+1)(a-b)}$$

do not belong to the set of non-negative integers.

As regards second order partial differential equations with two independent variables, Oleinik & Radkevich [29], using other methods, obtained sufficient conditions of analytic hypoellipticity for operators of the form

$$D_x^2 + x^{2k} D_y^2 + a_1 D_x + a_2 D_y + c , \tag{14}$$

for any positive integer k. These conditions are formulated in terms of inequalities for the coefficients a_1, a_2, c, which are always satisfied if $a_1 \equiv a_2 \equiv c \equiv 0$. A wide class of second order elliptic equations degenerating on a non-characteristic curve can be reduced to equations of type (14). It has also been proved in [29] that any second order equation with a non-negative characteristic form and any number of independent variables is not analytic-hypoelliptic in a neighborhood of the origin, if it possesses a real characteristic passing through the origin.

7. Treves [32] applied Fourier integral operators to reduce the operators arising in the theory of Neumann's $\bar\partial$-problem to operators studied by Grushin. Leaving aside some interesting and deep relations with complex analysis, let us formulate Treves' main result as applied to analytic hypoellipticity of pseudo-differential operators. Treves [32] considered a matrix pseudo-differential operator P satisfying the following conditions:

I. The operator P can be represented in the form

$$P = I_d A + B , \tag{15}$$

where I_d is the identity $d \times d$-matrix ($d \geq 1$); A is a scalar operator; B is a matrix pseudo-differential operator, its order being smaller than that of A by 1 or more.

II. $P_0(x, \xi) \geq 0$, where $P_0(x, \xi)$ is the principal symbol of A.

III. On the characteristic manifold char (P), $P_0(x, \xi)$ has zero of multiplicity 2 precisely. Moreover, it is assumed that at any characteristic point the spectrum of the matrix $B_0(x, \xi)$ (the principal symbol of the matrix operator B) does not belong to a certain set of negative real numbers (this set can be easily described in terms of the operator A); in other words, P satisfies the necessary and sufficient condition of hypoellipticity "with the loss of one derivative", i.e., $Pu \in H^s_{\text{loc}}(\Omega)$ implies that $u \in H^{s+m-1}_{\text{loc}}(\Omega)$, where Ω is an arbitrary domain, $m = \text{ord}\, P$ (see Hörmander [33]).

Next, an additional assumption is made about the characteristic set:

IV. The set char (P) is a simplectic analytic submanifold of the cotangent bundle $T^*(\Omega)$, i.e., the restriction of the canonical simplectic form $\sum_{i=1}^n d\xi_i \wedge dx_i$ of the bundle $T^*(\Omega)$ to char (P) is non-degenerate at any point.

Under these conditions, the operator P is analytic-hypoelliptic.

Treves constructs an analytic parametrix preserving the analytic wave front (see below), so that for the operator P there is micro-local analytic hypoellipticity.

In order to prove local analyticity of solutions of Neumann's $\bar{\partial}$-problem, Treves suggested a method which is extremely complicated, since the reduction of the problem to the operators of type (15) on the boundary requires the technique of hyperfunctions (introduced by Sato [34]), and dozens of pages are needed for the construction and the study of the parametrix for (15).

Another, more natural and direct, approach to the proof of local analyticity for the $\bar{\partial}$-problem has been suggested by Tartakoff [35]. The main result of [35] can be formulated as follows.

On a real analytic manifold M of dimension $2n - 1$, consider $2n - 1$ linearly independent real analytic vector fields $Z_1, Z_2, \ldots, Z_{2n-2}, T$ such that the so-called Levi matrix $c_{jk}(x)$, $x = (x_1, \ldots, x_{2n-1})$, determined by

$$[Z_j, Z_k] = c_{jk}(x)T + \sum_{l=1}^{2n-2} d_{jkl}(x)Z_l , \qquad (16)$$

is non-degenerate at $x^0 \in M$, i.e., $\| \det c_{jk}(x^0) \| \neq 0$.

Consider the following system of differential operators

$$P = \sum_{j,k=1}^{2n-2} a_{jk}(x)Z_j Z_k + \sum_{j=1}^{2n-2} a_j(x)Z_j + a(x) , \qquad (17)$$

where Z_j are the corresponding first order differential operators, and a_{jk} are matrices with analytic elements. It is assumed that the following a priori estimate holds:

$$\sum_{j,k=1}^{2n-2} \| Z_j Z_k v \| \le C(\|Pv\| + \|v\|) , \quad v \in C_0^\infty(\Omega) . \qquad (18)$$

Analytic hypoellipticity (in the sense of (4)) for the operator (17) is proved in [35] under the above assumptions.

Note that for Neumann's $\bar{\partial}$-problem in a pseudo-convex domain the inequality (18) is a direct consequence of the well-known Kohn-Morrey estimate. The condition of non-degeneracy of the Levi matrix coincides with the condition of the characteristic manifold to be simplectic (see condition IV). It should also be observed that the characteristics of operators (17) (in contrast to Treves' operators (15)) may have any multiplicity.

A remarkable generalization of the results of Treves and Tartakoff is due to Metivier [36] (see also [37]). Metivier established the following general theorem.

Let $P(x, D)$ be an analytic pseudo-differential operator whose symbol admits an expansion of the form (6). Assume that char (P) is an analytic simplectic manifold on which $P_m(x, \xi)$ has zero of order k precisely, and the lower order symbol P_{m-j} has zero of an order $\ge k - 2j$, $j \le k/2$. Let P be a hypoelliptic operator with the loss of $k/2$ derivatives. Then P is an analytic-hypoelliptic operator.

Pseudo-differential operators possessing the property of hypoellipticity with the loss of $k/2$ derivatives have been thoroughly studied. For operators P with symbols vanishing on char (P) in the aforesaid manner, this condition is equivalent to the following one: the family of auxiliary operators

$$\sigma_{x^0,\xi^0}^k(y, D_y) = \sum_{|\alpha|+|\beta|+2j=k} \frac{1}{\alpha!}\frac{1}{\beta!} \left(\left(\frac{\partial}{\partial x}\right)^\alpha \left(\frac{\partial}{\partial \xi}\right)^\beta P_{m-j}(x^0, \xi^0) \right) y^\alpha D_y^\beta$$

has no kernel in $S(\mathbb{R}^n)$ for any $(x^0, \xi^0) \in$ char (P) (see [38]).

8. Treves' condition of the characteristic manifold being simplectic is non-essential for hypoellipticity and is very important for analytic hypoellipticity. The first example indicating this importance is due Baouendi and Goulaouic who considered the operator (7); there are some other, more general, examples given by Oleinik & Radkevich [23] and Metivier [39].

The article of Metivier [40] is dedicated to the question whether the condition of the characteristic manifold being simplectic is necessary for analytic hypoellipticity of second order differential operators with a real non-negative principal symbol.

It is assumed in [30] that there is a submanifold V of the characteristic manifold such that the canonical form identically vanishes on V. Metivier proves the absence of analytic hypoellipticity, assuming that for the operator P^* (formally conjugate to P) the inequality

$$\|\varphi\|_\varepsilon \leq C \left(\|P^*\varphi\|_0 + \|\varphi\|_0 \right) , \qquad \varphi \in \mathcal{D}(\Omega) ,$$

holds with some $\varepsilon > 0$, and V satisfies some additional restrictions. These restrictions are quite strong; in particular, they leave out of consideration the operators of the form

$$D_x^2 + (x^2 + y^2)D_y^2 . \tag{19}$$

In [41], using another method, Metivier proved that the operator (19) is not analytic-hypoelliptic. Moreover, it has been shown in [41] that a second order differential operator with real analytic coefficients and a non-negative characteristic form cannot be analytic-hypoelliptic, if its characteristic manifold has the form of a ray

$$\mathrm{char}\,(P) = \{(x^0, \lambda\xi^0),\ \lambda \in \mathbb{R}^1 \setminus 0\}$$

and the principal symbol has zero of order 2, exactly, on that ray.

9. So far we have discussed the description of the classes of operators for which hypoellipticity implies analytic hypoellipticity. All these classes have one common property: the canonical form is non-degenerate on the characteristic manifold. However, there are several examples of non-hypoelliptic operators with the property of analytic hypoellipticity in a wider sense (5), or (5′). The first example of this kind is due to Baouendi, Goulaouic & Lipkin [42] who showed that the operator

$$L_{\lambda,\mu} = r^2\Delta + \mu r \frac{\partial}{\partial r} + \lambda , \quad r = |x| , \quad \Delta = \sum_{j=1}^{n} \frac{\partial^2}{\partial x_j^2} ,$$

is hypoelliptic for no values of $\lambda, \mu \in \mathbb{C}^1$, and at the same time satisfies (5′).

Baouendi & Sjostrand [43] considered operators with polynomial coefficients, namely, the operators of the form

$$P_0(x, D) = \sum_{|\alpha| \leq m} a_\alpha(x) D^\alpha , \qquad (20)$$

where $a_\alpha(x)$ are homogeneous polynomials of degree $|\alpha|$, and the operator P_0 is elliptic outside the origin. Denote by $q_0(\theta, \eta, \xi)$ the principal symbol of the operator (20) in polar coordinates; here θ are the angle coordinates, η are their respective dual variables, and ξ is the dual variable of the radius. On the complex plane, consider two sets Γ and Γ_+ defined by

$$\Gamma = \left\{ z \in \mathbb{C}^1 : \exists (\theta, \eta) \in T^*(S_{n-1}) \setminus \{0\}, \quad q_0(\theta, \eta, -iz) = 0 \right\} ,$$

$$\Gamma_+ = \{ z \in \Gamma, \ \operatorname{Re} z > 0 \} .$$

As another assumption, take **Condition H$_1$**: the set Γ_+ lies within finitely many angles, each of these angles having magnitude $< \pi/(n-1)$.

Under this condition, the operator (20) has been proved to lack hypoellipticity ($\exists v \notin C^\infty$ such that $P_0 v = 0$), and at the same time shown to be analytic-hypoelliptic in the sense of (5).

Assume that Condition H$_1$ is replaced by **Condition H$_2$**: the convex hull of Γ_+ lies within an angle of magnitude $< \pi/n$.

Then, in a sufficiently small neighborhood of the origin, the property (5) holds for the operator

$$P(x, D) = P_0(x, D) + \sum_{|\alpha| \leq m} a'_\alpha(x) D^\alpha ,$$

where a'_α are analytic functions defined near the origin and having zero of order $|\alpha| + 1$ at the origin.

It should be observed that the conditions H$_1$ and H$_2$ are always satisfied if m is even and

$$\sum_{|\alpha| = m} a_\alpha(x) D^\alpha \equiv q(x) \Delta^{m/2} .$$

Close results have been stated in a note by Radkevich [44], who considers operators of the form

$$P(x, D) = \sum_{|\alpha| \leq m, \, |\gamma| = |\alpha|} a_{\alpha\gamma}(x) x^\gamma D^\alpha ,$$

where $a_{\alpha\gamma}(x)$ are real analytic functions defined in a neighborhood of the origin.

A class of non-hypoelliptic operators with the property (5) for finite q has been singled out by Bronstein [45], [46]. Let $P(x, \xi)$ be the symbol of operator P, and let

$$P(x,\xi) = P_m(x,\xi) + P'(x,\xi) , \qquad \deg P' < m .$$

Assume that

$$C^{-1} < \frac{P_m(x,\xi)}{\sum\limits_{j=1}^{n} |x_j|^{q_j} |\xi_j|^m} < C , \qquad q_j < m ,$$

$$|P'(x,\xi)| \le C_1 \sum_{j=1}^{n} \left(|x_j|^{q_j} |\xi_j|^m + |\xi_j|^{m-q_j} \right) ,$$

where C, C_1 are positive constants. Then the operator P is analytic-hypoelliptic in the sense of (5), and the smallest *a priori* order of smoothness q (from (5)) is determined by the lower order terms (i.e., by the operator P'). The inequality $q_j < m$ is a necessary condition for (5). If for some j we have $q_j \ge m$, then there is a point $x^0 \in \mathbb{R}^n$ such that the equation $P = 0$ admits a C^∞-solution in a neighborhood of x^0 and has zero of infinite order at x^0. A close result has been proved by Oleinik & Radkevich in [29] for $m = 2$.

As an illustration of the results obtained in [45], [46], consider the operator

$$P = x^2 D_x^4 + D_y^4 - \gamma D_x^2 .$$

This operator satisfies the condition (5) for $q > 6(\gamma + 10)$. However, P does not belong to the class of hypoelliptic operators, since the equation $Pu = 0$ admits a non-smooth solution $u(x,y) = x^\beta$, where $\beta = 0.25 + \sqrt{0.25 + \gamma}$.

Baouendi & Treves [47] examined local properties of solutions of overdetermined first order systems with one unknown function:

$$L_j(u) = f_j , \quad j = 1, \ldots, N .$$

The system of operators $L_1, \ldots, L_N$ has the property of analytic hypoellipticity in Ω if

$$\{ u \in \mathcal{D}'(\Omega), \, L_j(u) \in \mathcal{A}(\Omega), \, j = 1, \ldots, N \} \Longrightarrow \{ u \in \mathcal{A}(\Omega) \} .$$

Replacing $\mathcal{A}(\Omega)$ by $C^\infty(\Omega)$ we obtain hypoellipticity. We shall not state here all intricately formulated results of [47], but limit ourselves to the example, given in [47], of a system in $\mathbb{R}^4$:

$$L_1 = \frac{\partial}{\partial t_1} + \sqrt{-1} \left[3 \frac{\partial}{\partial x_1} - (4t_1 t_2 + 3)t_1^2 \frac{\partial}{\partial x_2} \right] ,$$

$$L_2 = \frac{\partial}{\partial t_2} - \sqrt{-1} \frac{\partial}{\partial x_2} .$$

In [47], this system is shown to be analytic-hypoelliptic; on the other hand, this system is non-hypoelliptic, as shown in [48].

10. The questions of local analyticity of solutions of differential equations find their natural and fruitful generalization in the study of analytic wave fronts of solutions. Analytic wave fronts, which belong to the cotangent bundle, provide more elaborate and precise information about the structure of analytic singularities of the solutions.

The notion of wave front $WF_a(u)$ for a distribution $u \in \mathcal{D}'(\Omega)$ (Ω is a domain in $\mathbb{R}^n$) was initially introduced by Hörmander [49]. According to [49], a point (x^0, ξ^0) of the cotangent bundle $T^*(\Omega)$ does not belong to $WF_a(u)$, if there is a sequence of compactly supported functions $\{\varphi_n(x)\}$ equal to 1 in a sufficiently small (but independent of n) neighborhood of x^0 and such that the Fourier transforms $\widehat{\varphi_n u}$ satisfy the estimate

$$|\widehat{\varphi_n u}(\xi)| \le C^{n+1} n! (1 + |\xi|)^{-n}$$

for all ξ belonging to a sufficiently small conic neighborhood of ξ^0.

Previously, in the theory of hyperfunctions developed by Sato [34], another notion had been introduced, namely, that of the singular spectrum of a hyperfunction. As shown by Bony [50], the definitions of Sato and Hörmander are equivalent in the case of distributions. There are several other equivalent definitions of an analytic wave front. Here we give the most recent one, which is due to Sjostrand [51] and is a modification of the definition suggested by Bros & Jagolnitzer [52].

Let $u \in \mathcal{E}'(\Omega)$, $(x^0, \xi^0) \in T^*(\Omega) \setminus \{0\}$. We say that $(x^0, \xi^0) \in WF_a(u)$ if and only if the quantity

$$\left| \int_\Omega \exp(-i \langle x, \xi \rangle - |\xi||x - \beta|^2) u(x) \, dx \right|$$

exponentially decays with respect to ξ in a sufficiently small complex neighborhood of ξ^0 and $\beta \in \mathbb{R}^n$ close enough to x^0.

The above definition of the analytic wave front of a distribution implies that it is a closed conic (with respect to ξ) subset of $T^*(\mathbb{R}^n) \setminus \{0\}$. If we denote by P the natural projection

$$P : T^*(\mathbb{R}^n) \to \mathbb{R}^n \qquad ((x, \xi) \mapsto x) \,,$$

then

$$PWF_a(u) = \operatorname{sing\,supp}_a u \,, \tag{21}$$

where $\operatorname{sing\,supp}_a u$ is the complement of the set consisting of all points where the (generalized) function u is analytic. Thus, a function $u(x)$ is analytic in a domain Ω if the projection of its analytic wave front to Ω is empty.

Differential operators with analytic coefficients (as well as analytic pseudo-differential operators) do not enlarge the wave front of a function:

$$WF_a(Pu) \subset WF_a(u) \, .$$

An operator P is said to be *micro-locally analytic-hypoelliptic* if it preserves the analytic wave front of any distribution:

$$WF_a(Pu) = WF_a(u) \, . \tag{22}$$

It follows from (21) and definition (4) that any micro-locally analytic-hypoelliptic operator is analytic-hypoelliptic. The converse statement, in general, does not hold: analytic hypoellipticity does not imply micro-local analytic hypoellipticity. Propagation of analytic wave fronts for operators with multiple characteristics is examined in [53], [54].

At present, the problems of micro-local analytic hypoellipticity are intensively studied; recently, some deep relations of this subject with complex analysis have been discovered (see Trepreau [55]–[57]). A review of the latest results in this field and further references can be found in [57] (see also [61]).

In this commentary we do not touch upon analyticity of solutions of boundary value problems. In recent years, great attention has been given to the propagation of analytic wave fronts on the boundary. Shapira [58], Sjostrand [51], Rauch & Sjostrand [59] obtained deep results about the structure of analytic singularities on the boundary for the Dirichlet problem and some other, more general, problems. Many results pertaining to the behavior and propagation of analytic wave fronts on the boundary can be found in [60] (in particular, see [60] for the results of Melrose, Shapira, Sjostrand and Kataoka, and references in the papers of these authors).

The authors express their gratitude to M.D. Bronstein for his assistance during the work on this commentary.

References

[1] *The Problems of Hilbert*. Moscow, Nauka, 1969.
[2] Bernstein, S.N. Sur la nature analytique des solutions des équations aux dérivées partielles du second ordre. *Math. Ann.* **59** (1904) 20–76.
[3] Bernstein, S.N.; Petrowsky, I.G. On the first boundary value problem (the Dirichlet problem) for elliptic equations and the properties of functions which satisfy these equations. *Uspekhi Mat. Nauk* **8** (1941) 8–26. (See also: Bernstein, S.N. *Selected Works*. Moscow, Izd.-vo Akad. Nauk SSSR, 1960, Vol. 3, pp. 394–419).
[4] Oleinik, O.A. On the 19th problem of Hilbert. In: *The Problems of Hilbert*. Mowcow, Nauka, 1969, pp. 206–208.
[5] Petrowsky I.G. Sur l'analyticité des solutions des systèmes d'équations différentielles. *Mat. Sbornik* **5(47):1** (1939) 3–70.

[6] Oleinik, O.A.; Radkevich, E.V. On the analyticity of solutions of partial differential equations (further development of Petrowsky's theory). *Uspekhi Mat. Nauk* **28:5** (1973) 257–259.

[7] Hörmander, L. *Linear Partial Differential Operators*. Berlin, Springer-Verlag, 1963.

[8] Hörmander, L. *The Analysis of Linear Partial Differential Operators*. Vols. I,II. Berlin, Springer-Verlag, 1983.

[9] Leray, J. *Hyperbolic Differential Equations*. Princeton, N.J., Inst. Adv. Study, 1953.

[10] Douglis, A.; Nirenberg, L. Interior estimates for elliptic systems of partial differential equations. *Comm. Pure Appl. Math.* **8** (1955) 503–538.

[11] Volevich, L.R. On general systems of differential equations. *Doklady Akad. Nauk SSSR* **132:1** (1960) 20-23.

[12] Volevich, L.R. On a problem of linear programming arising in the theory of differential equations. *Uspekhi Mat. Nauk* **18:3** (1963) 155–162.

[13] John, F. *Plane Waves and Spherical Means Applied to Partial Differential Equations*. New York, 1955.

[14] Morrey, C.B.; Nirenberg, L. On the analyticity of solutions of linear elliptic systems of partial differential equations. *Comm. Pure Appl. Math.* **10** (1957) 271–290.

[15] Bers, L.; John, F.; Schechter, M. *Partial Differential Equations*. Interscience Publishers, New York, 1964.

[16] Morrey, C.B. On the analyticity of the solutions of analytic nonlinear elliptic systems of partial differential equations. *Amer. J. Math.* **80:1** (1958) 198–277.

[17] Krée, B.; Boutet de Monvel, L. Pseudo-differential operators and Gevrey classes. *Ann. Inst. Fourier* **17** (1967) 295–324.

[18] Volevich, L.R. Pseudo-differential operators with holomorphic symbols and Gevrey classes. *Trudy Mosk. Mat. Ob.-va* **24** (1971) 43–68.

[19] Mizohata, S. Solutions nulles et solutions non analytiques. *J. Math. Kyoto Univ.* **2:1** (1962) 271–302.

[20] Suzuki, H. Analytic-hypoelliptic differential operators of first order in two independent variables. *J. Math. Soc. Jap.* **16:4** (1964) 367–374.

[21] Oleinik, O.A. On the analyticity of solutions of partial differential equations and systems. *Astérisque, Colloq. Intern. CNRS* **2-3** (1973) 272–285.

[22] Oleinik, O.A.; Radkevich, E.V. On the analyticity of solutions of linear differential equations and systems. *Doklady Akad. Nauk SSSR* **207:4** (1972) 785–788.

[23] Oleinik, O.A.; Radkevich, E.V. On the analyticity of solutions of linear partial differential equations. *Mat. Sbornik* **90:4** (1973) 592–606.

[24] Treves, F. Analytic hypoelliptic partial differential equations of principal type. *Comm. Pure Appl. Math.* **24** (1971) 537–570.

[25] Treves, F. Hypoelliptic equations of principal type with analytic coefficients. *Comm. Pure Appl. Math.* **23** (1970) 637–651.

[26] Treves, F. A new method of proof of subelliptic estimates. *Comm. Pure Appl. Math.* **24** (1971) 71–115.

[27] Baouendi, M.S.; Goulaouic, C. Non-analytic hypoellipticity for some degenerate elliptic operators. *Bull. Amer. Math. Soc.* **78:3** (1972) 483–486.

[28] Oleinik, O.A.; Radkevich, E.V. On the conditions of existence of non-analytic solutions for linear partial differential equations of arbitrary order. *Trudy Mosk. Mat. Ob.-va* **31** (1974) 17–33.

[29] Oleinik, O.A.; Radkevich, E.V. On the analyticity of solutions of linear second order partial differential equations. *Trudy Seminara Petrowsk.* **1** (1975) 163–173.

[30] Grushin, V.V. On a class of elliptic pseudo-differential operators degenerating on a submanifold. *Mat. Sbornik* **84(126)** (1971) 163–195.

[31] Tosques, M. Analytic-hypoellipticity for a class of second order evolution operators with double characteristics. *Bull. Unione Mat. Ital.* **14:1** (1977) 160–171.

[32] Treves, F. Analytic-hypoellipticity of a class of pseudo-differential operators with double characteristic and applications to the $\bar\partial$-Neumann problem. *Comm. Part. Diff. Eqs.* **3:7** (1978) 475–642.

[33] Hörmander, L. A class of hypoelliptic pseudo-differential operators with double characteristic. *Math. Ann.* **217** (1975) 165–188.

[34] Sato, M.; Kawai, T.; Kashiwara, M. Microfunctions and pseudo-differential equations. *Lect. Notes Math.* **287** (1973) 264–524.

[35] Tartakoff, D. The local real analyticity of solution to $\Box_b$ and $\bar\partial$-Neumann problem. *Acta Math.* **145:34** (1980) 177–204.

[36] Metivier, G. Analytic hypoellipticity for operators with multiple characteristics. *Comm. Part. Diff. Eqs.* **6:1** (1981) 1–90.

[37] Tartakoff, D. Hypoellipticité analytique des opérateurs à caracteristiques multiples, démonstration élémentaire. *Sém. Gaulaouic-Meyer-Schwartz, 1980–1981.* Paris, École Polyt. 1981.

[38] Boutet de Monvel, L.; Grigis, A.; Helffer, B. Paramètrixes d'opérateurs pseudodifferentielles à caracteristiques multiples. *Astérisque, Colloq. Intern. CNRS* **34-35** (1976) 585–639.

[39] Metivier, G. Propriété des iterés et ellipticité. *Comm. Part. Diff. Eqs.* **3** (1978) 827–876.

[40] Metivier, G. Une classe d'opérateurs non hypoelliptiques analytiques. *Indiana Univ. Math. J.* **29:6** (1980) 823–860.

[41] Metivier, G. Non hypoellipticité analytique pour $D_x^2 + (x^2 + y^2)D_y^2$. *C.R. Acad. Sci. Paris* **292** (1981) 401–404.

[42] Baouendi, M.S.; Goulaouic, C.; Lipkin, L.F. On the operator $\Delta r^2 + \mu(\partial/\partial r) + \lambda$. *J. Diff. Eqs.* **15** (1974) 499–509.

[43] Baouendi, M.S.; Sjostrand, J. Regularité analytique pour des opérateurs elliptiques singuliers en un point. *Ark. Mat.* **14:1** (1976) 9–33.

[44] Radkevich, E.V. Complex wave front in the problem of analyticity of solutions of linear partial differential equations. *Doklady Akad. Nauk SSSR* **233:4** (1977) 559–562.

[45] Bronstein, M.D. On the analyticity of solutions of equations degenerating on the characteristic manifold. *Uspekhi Mat. Nauk* **31:3** (1976) 205–206.

[46] Bronstein, M.D. On local smoothness of solutions of differential equations. In: *Proceedings of the All-Union Conf. on PDE's*. Moscow, Moscow Univ. Press, 1978, pp. 278–279.

[47] Baouendi, M.S.; Treves, F. A property of the functions and distributions annihilated by a locally integrable system of complex vector fields. *Ann. Math.* **113:2** (1981) 387–422.

[48] Maire, H.M. Hypoelliptic overdetermined systems of partial differential equations. *Comm. Part. Diff. Eqs.* **5** (1980) 331–380.

[49] Hörmander, L. Uniqueness theorems and wave front sets for solutions of linear differential equations with analytic coefficients. *Comm. Pure Appl. Math.* **24** (1971) 671–704.

[50] Bony, J.M. Équivalence des diverses notions de spectre singulier analytique. *Sém. Goulaouic-Schwartz, 1976–1977*. Paris, École Polyt., 1977, Exp. no. 3.

[51] Sjostrand, J. Propagation of analytic singularities for second order Dirichlet problems. I, II, III. *Comm. Part. Diff. Eqs.* **5:1** (1980) 41–94, **5:2** (1980) 128–207, **6:5** (1980) 499–567.

[52] Bros, J.; Jagolnitzer, D. Support essentiel et structure analytique des distributions. *Sém. Goulaouic-Lions-Scwartz, 1975*. Paris, École Polyt., 1975, Exp. no. 18.

[53] Hanges, N. Propagation of analyticity along real bicharacteristics. *Duke Math. J.* **48:1** (1981) 269–277.

[54] Grigis, A.; Schapira, P.; Sjostrand, J. Propagation de singularites analytiques pour des opérateurs à caractéristiques multiples. *C.R. Acad. Sci. Paris* **292** (1981) 397–400.

[55] Trepreau, J.M. Sur la regularité analytique des opérateurs de type principal. *C.R. Acad. Sci. Paris* **293** (1981) 561–563.

[56] Trepreau, J.M. Sur l'hypoellipticité analytique microlocale des opérateurs de type principal. Preprint. Dep. Math. Fac. Reims.

[57] Trepreau, J.M. Systémes différentielles à caractéristiques simples et structures réelles-complexes. *Sém. Bourbaki* **595** (1981-82).

[58] Schapira, P. Propagation at the boundary and reflection of analytic singularities of solutions of linear partial differential equations. *Publ. RIMS Kyoto Univ.* 1977, Vol. 12, Suppl., pp. 441–453.

[59] Rauch, J.; Sjostrand, J. Propagation of analytic singularities along diffracted rays. *Indiana Univ. Math. J.* **30:3** (1981) 389–401.

[60] *Singularities in Boundary Value Problems*. Proc. Nato Adv. Study Inst. at Maratea (Italy). Ed. by H.G. Garnir. Dordrecht, Boston, Reidel Publ. Co., 1981.

[61] Grigis, A.; Rothschild, L. A criterion of analytic hypoellipticity of a class of differential operators with polynomial coefficients. *Ann. Math.* **118:3** (1983) 443–460.

[62] Palamodov, V.P. Expansion in holomorphic waves. *Vestnik Mosk. Univ. Mat. Mekh.* **1** (1974) 135–142.

Huyghens' Principle and Its Generalizations*

by A.M. Gabrielov and V.P. Palamodov

1. The notion of a lacuna is closely related to the Huyghens Principle formulated in his famous treatise on the theory of light [1]. The term "Huyghens
Principle" has been given a precise meaning by Hadamard in his monograph on the Cauchy problem [2]. According to Hadamard, this principle
(in a narrow sense) should be interpreted as follows: if a perturbation of
light arises at the moment $t = 0$, and this perturbation is localized in close
vicinity of a point O, then at the moment $t = t'$ its influence is observed only
near the surface of the sphere of radius $\omega t'$ with center at O, i.e., in a very
thin spherical layer centered at O. Here ω is the speed of light propagation.
An equivalent modern formulation of this statement reads: the fundamental
solution of the wave equation with three spatial variables is supported on
the surface $|x| = \omega t$ (cone of rays), i.e., the fundamental solution vanishes
on its complement. This complement consists of two components: one inside this surface and another outside it; therefore, both these components
produce lacunas for the wave equation (the definition of a lacuna is given
below). The exterior being a lacuna means only that the speed of light
is finite, which fact had been known before Huyghens (O. Römer), and
in modern terminology, means hyperbolicity of the wave operator. On the
other hand, the existence of an interior lacuna is a very rare phenomenon
and is characteristic of the particular equation; in the present case, it can be
explained by the dimension $n = 4$, the fact that the coefficients of the equation are constant, and also by the large size of the symmetry group for this
equation. This phenomenon can be physically interpreted to mean that the
light signal leaves no trace after passing an observer, i.e., an instantaneous
flash of light at $t = 0$ remains instantaneous for all its observers. Obviously
this property affords the inhabitants of our world far greater possibilities
to perceive information carried by light or acoustic waves than, say, the hy-

---

*A commentary on Articles 5 and 6 of this volume.

pothetical inhabitants of a two-dimensional medium, in which Huyghens' Principle (in a narrow sense) does not hold, and every subsequent signal is superimposed by the preceding ones.

Further investigations showed that beyond this remarkable physical principle there is a deep and fascinating mathematical theory (one of its names is *the theory of lacunas*). By the efforts of Poisson, Kirchhoff, Volterra, Tedonet and other mathematicians who considered the wave equation, its solutions, in the case of constant coefficients and arbitrarily many spatial variables, were thoroughly studied and the conclusion was reached that the narrow Huyghens' Principle holds only if the overall number of variables is even. And it was Hadamard himself who constructed the fundamental solution for an arbitrary second order hyperbolic equation with analytic coefficients. Having found the asymptotics of this solution near the conoid of rays, Hadamard proved that the interior of this conoid cannot be a lacuna if the number of independent variables is odd (he says that there is "diffusion of waves" in this case).

2. The creation of the lacuna theory for hyperbolic equations of arbitrary order is to be credited to I.G. Petrowsky [3]. Hyperbolic equations describe any type of disturbance propagation processes occurring with a finite speed in physical systems, for instance, propagation of electromagnetic, elastic and other waves. Propagation of small perturbations is described by linear hyperbolic equations

$$p\left(x, -i\frac{\partial}{\partial x}\right) u = 0 , \qquad x = (x_1, \ldots, x_n) ,$$

where x_n is the time variable. Generally speaking, p is a matrix operator, and u is a vector valued function; however, for the sake of simplicity, we assume here p and u to be of scalar form. The motion of a perturbation caused by an impulse concentrated at a point $x = y$ is described by the fundamental solution $E(x, y)$ of this operator, namely, by a function (a distribution, in general) that vanishes for $x_n < y_n$ and satisfies the equation

$$p\left(x, -i\frac{\partial}{\partial x}\right) E(x, y) = \delta(x - y) .$$

The fact that the propagation speed of this disturbance is finite means that $E(\cdot, y)$ has its support in a certain cone-shaped set with vertex at y, and the intersection of this set with any hyperplane $x_n = $ const is compact. The fundamental solution E is smooth (to the extent permitted by the coefficients of the operator) outside the conoid formed by the rays issuing from the point y (these rays are called *bicharacteristics* of the operator p). The conoid of rays splits the physical space into finitely many connected components. Petrowsky says that a particular component forms a *lacuna* for the fundamental solution, if $E \equiv 0$ on this component. It is well-known that the solution of any Cauchy problem for a hyperbolic equation can be obtained by applying to the initial functions integral operators whose kernels are equal to the fundamental solution or its derivatives in x_n. Therefore, a non-trivial lacuna inside the conoid of rays gives rise to a phenomenon similar to the Huyghens' Principle: if the initial impulse is concentrated in a sufficiently small region, then at any fairly distant point in space a temporary or perpetual equilibrium will be restored after the impulse has travelled through that point. This phenomenon is observed if it is the interior component that forms a lacuna, i.e., the component of the complement to the conoid of rays containing the time axis.

Note that for an operator with coefficients independent of x we have $E(x,y) = E(x - y, 0)$, and therefore, we are dealing with lacunas for the function $E(x,0)$ of the variables x only. The cone of rays containing the singularities of the function E represents the dual cone (the cone of normals) to the cone $P(\xi) = 0$, where $P(\xi)$ is the principal symbol of the operator p. Thus, for the wave equation with constant coefficients, the Huyghens Principle in a space of even dimension means that the function $E(x,0)$ has a lacuna inside the light cone. Previous to Petrowsky's works, only a few examples of non-trivial lacunas for other equations had been known: for the system of elasticity and the system of crystal optics, the interior component forms a lacuna.

The main achievement of Petrowsky's paper is the criterion for the existence of a lacuna for the fundamental solution of a homogeneous strictly hyperbolic equation with constant coefficients. Petrowsky considers a complex projective manifold A_x which is the intersection of the hyperplane $(x, \xi) = 0$ and the surface $P(\xi) = 0$. According to Petrowsky's criterion, the component containing the point x is a *stable lacuna* (i.e., a lacuna for all hyperbolic operators close to the given one), if and only if a certain, specially constructed cycle γ on the manifold A_x is homologous to zero. In modern literature this cycle is referred to as *Petrowsky's cycle*. The proof of this result consists of two parts of about equal length and complexity. In the first part Petrowsky finds an expression for the fundamental solution outside the cone of rays in terms of an integral of a certain rational form over the said cycle. This integral representation generalizes a previous result obtained by Herglotz [4] and is now called the Herglotz-Petrowsky formula. The generalization made by Petrowsky is essential, for, as we now know, there can be no non-trivial lacunas in the case of Herglotz. The Herglotz-Petrowsky formula immediately implies that Petrowsky's condition is sufficient for the existence of a lacuna. It also clearly implies that the fundamental solution is analytic outside the cone of rays.

The proof of the necessity of Petrowsky's condition constitutes the second part of the paper and is performed by the methods of algebraic topology. Petrowsky argues as follows: according to the formula described above, the existence of a stable lacuna means that, for a fairly large set of regular forms on the manifold A_x, the integrals of these forms over the cycle γ must vanish. Using an explicit geometrical construction, Petrowsky finds a basis of relative cycles having the middle dimension of the pair (A_x, H), where H is a hyperplane cross-section of A_x. The cycle γ is represented as a sum of the relative cycle γ', whose border is homologous to zero in H, and a film in H extended over this border. In the integral representation, the cycle γ can be replaced by γ', since the restrictions of all the forms to H are equal to zero. The cycle γ' is represented as a linear combination of the basic cycles supplemented, possibly, by an algebraic cycle. Next, the coefficients of this linear combination are shown to vanish. Assuming a non-zero coefficient by one of the basic cycles, Petrowsky indicates a deformation of the manifold A_x which produces a singular point on this manifold (a point that attracts the given basic cycle). Then the integral of one of the forms over this basic cycle acquires a singularity, if regarded as a function of the parameter characterizing the deformation (asymptotic behavior of this integral is calculated explicitly), whereas the integrals over the rest of the summands in the representation of γ' remain regular. It follows that the integral of one of the forms over γ cannot vanish identically, which is inconsistent with the lacuna being stable. Therefore, the cycle γ is homologous to an algebraic cycle. Calculating the index of intersection of γ with a subspace of complementary dimension, Petrowsky shows this algebraic cycle to be trivial, and thereby completes his chain of arguments.

The proof described above relies on the latest achievements in algebraic geometry of that period and contains new ideas and methods of the topology of complex manifolds characteristic of modern theory of singularities.

3. Petrowsky's studies of the lacunas and fundamental solutions for general hyperbolic equations with constant coefficients have been carried on by his student A.M. Davidova and by V.A. Borovikov. In [5], a remarkable relation is established between the existence of a lacuna and the geometry of the cone of rays in a neighborhood of its non-singular point x_0. Assume that near the point x_0 the intersection of the cone of rays with the hyperplane $x_n = 1$ (x_n is the time variable) is given by the equation $x_1 = f(x_2, \ldots, x_{n-1})$, and moreover, $f(x_0) = 0$, $\mathrm{d}f(x_0) = 0$, and the quadratic form corresponding to $\left\{ \partial^2 f(x_0)/\partial x_i \partial x_j \right\}$ has k positive eigenvalues. If k is odd, then the domain $\{x_1 > f(x_2, \ldots, x_{n-1})\}$ is not a lacuna. It has been shown by Borovikov [6] that in this case the fundamental solution and any of its derivatives can be expanded – on the normal to the cone of rays – into a (non-zero) power series with respect to half-integer powers of x_1. This fact is responsible for the absence of lacunas. In the case of k even, this expansion contains only positive integer powers of x_1. Moreover, using the method suggested by Borovikov, one can easily show that the fundamental solution admits an infinitely differentiable extension from the domain $\{x_1 > f(x_2, \ldots, x_{n-1})\}$ to a neighborhood of x_0. Later, this property of a distribution became known as "sharpness". In [7], Borovikov established a number of sufficient conditions for the absence of lacunas. In particular, he showed that, for an operator that cannot be represented as a composition of operators of positive order, a component of the complement to the cone of rays cannot be a lacuna, if it is bounded by the parts of this cone corresponding to different components of the real part of the cone $P(\xi) = 0$. The interior component, i.e., the portion bounded by the slowest part of the cone of rays, is a lacuna only if $m < n$ and n is even.

Galpern & Kondrashev [8] studied lacunas for operators reducible to a product of wave factors. For n even and the velocities being different from one another, there can be no non-trivial lacunas apart from an interior one for $m < n$. However, the areas between the components of the cone of rays are weak lacunas in the sense that the fundamental solution has the form of a polynomial in each area. Galpern [9] has extended some of Petrowsky's results to a class of equations with constant coefficients which cannot be resolved with respect to the highest order derivative in t. He calls an operator of this type *quasi-hyperbolic* if the highest and the lowest order coefficients in its expansion in powers of $\partial/\partial t$ are both elliptic operators in $\mathbb{R}^{n-1}$, and all roots of the characteristic equation are real and mutually distinct. For such operators an analogue of the Herglotz-Petrowsky formula has been obtained, together with a criterion for an interior component to be a lacuna. Lokshin [10] extended Borovikov's results to the case of quasi-hyperbolic equations.

4. Another approach to the Huyghens Principle has been found by Riesz in his remarkable treatise [11], where fundamental solutions are constructed for arbitrary powers $\square^\lambda$, $\lambda \in \mathbb{C}^1$, of the wave operator $\square$. Riesz introduces a family of distributions

$$Z_h = \frac{s^{\lambda-n}}{\pi^{\frac{n-2}{2}} 2^{\lambda-1} \Gamma\left(\frac{\lambda}{2}\right) \Gamma\left(\frac{\lambda+2-n}{2}\right)}, \qquad \lambda \in \mathbb{C}^1,$$

where $s = \sqrt{x_n^2 - x_1^2 - \cdots - x_{n-1}^2}$ for $x_n > 0$ such that $x_n^2 > x_1^2 + \cdots + x_{n-1}^2$, and $s = 0$ otherwise. This family forms a group with respect to convolution, i.e., $Z_\lambda * Z_\mu = Z_{\lambda+\mu}$ and $Z_{-2k} = \square^k \delta$ for any integer $k \geq 0$. It follows that Z_{2k} is a fundamental solution for the operator $\square^k$. On the other hand, if n is even and $0 < 2k < n$, then the distribution $s^{\lambda-n}$ has a simple pole at $\lambda = 2k$, and this pole is compensated by the pole of the denominator $\Gamma\left(\frac{\lambda+2-n}{2}\right)$. Hence, at the point $\lambda = 2k$ the function Z_λ is equal to the residue of the function $s^{\lambda-n}$, to within a factor independent of x. But s^λ is an entire function of λ for $s \neq 0$. It follows that the fundamental solution vanishes for $s \neq 0$, and therefore is supported in the cone $x_n = \sqrt{x_1^2 + \cdots + x_{n-1}^2}$. Consequently, the Huyghens Principle holds for comparatively low powers ($k < n/2$) of the wave operator in the space of even dimension.

Gårding [12] elaborated the idea of Riesz and considered, in particular, the operator $p = \det\{\partial/\partial x_{ij}\}$ in n^2-dimensional space. He interprets the points of this space as Hermitian quadratic forms of order n. For powers of p, the following stronger version of the Huyghens Principle is valid: the fundamental solution for p^k has its support in the set of non-negative forms of rank $\leq k$. This property is due to the extensive group of symmetries of this space, $G = GL(n, \mathbb{R}^1)$, preserving the operator p (which coincides, to within a factor, with the Cayley operator for this representation). Weinberg & Gindikin [13] established the Huyghens Principle (its stronger version) for some other operators associated with representations of the linear group.

5. In 1970–1973, Atiyah, Bott, and Gårding published a fundamental treatise on the theory of lacunas [14], 15], which they dedicated to I.G. Petrowsky. As stated in the introduction to these publications, the authors meant to clarify and generalize Petrowsky's theory. In fact, they gave another, more algebraically oriented, exposition of the main result of that theory and also obtained a generalization of this result in the case of weakly hyperbolic equations with constant coefficients. In particular, these authors make use of a cycle other than that introduced by Petrowsky; in the case of strictly hyperbolic equations, these cycles are related through the operation of taking a tube. Let $A \subset \mathbb{C}Y^{n-1}$ be the projective manifold formed by all roots of the principal symbol of a given hyperbolic operator; let X be the hyperplane associated with a given point in physical space. Using a construction very similar to that of Leray [16], Atiyah, Bott and Gårding introduce a relative cycle α in the space $(\mathbb{C}P^{n-1} \setminus A, X)$. In brief, this is a deformation of the chain $\mathbb{R}P^{n-1}$ in the direction of an imaginary vector field whose value at any point x belongs to the local cone of hyperbolicity for the principal symbol. Here the orientation of the chain $\mathbb{R}P^{n-1}$ is determined by the sign of a linear form vanishing on the hyperplane X, and the field must be tangential to this hyperplane. The border $\partial\alpha$ determines an element of the group $H_{n-2}(X \setminus A, Z)$. If this element is equal to zero, then the Herglotz-Petrowsky-Leray formulas imply that x belongs to a weak lacuna, i.e., in the component containing x, the fundamental solution has the form of a polynomial of degree $m - n$ (m is the order of the operator). For $m < n$ this lacuna is genuine. These properties are stable with respect to small variations of the operator's coefficients.

The converse result can be stated as follows: if the domain containing the point x is a weak lacuna for a sufficiently large power of the given operator, then the element $\partial \alpha$ is of zero class. It is possible to consider the first power only, provided that the set A can have as singularities only normal intersections, in particular, if A is non-singular[1]. The results obtained by Atiyah, Bott and Gårding are based on serious preliminary work. For weakly hyperbolic equations, even the notion of a cone of rays (wave front) is non-trivial; it is based on the so-called localization in the sense of Hörmander, and the wave front happens to depend continuously on the coefficients of the operator.

If a particular domain bounded by the cone of rays is not a lacuna, then it is natural to inquire what type of singularity the fundamental solution may have at any of its boundary points x_0. The simplest qualitative characteristic is sharpness (see above). The authors of [15] find a local version of Petrowsky's criterion, which allows one to make conclusions about the sharpness of the fundamental solution. The results obtained by Davidova and Borovikov are incorporated into this theory.

Vasiliev [17] performed detailed analysis of the fundamental solution's sharpness near singular points of the cone of rays. By definition, each point of the cone of rays is associated with a hyperplane touching the projective characteristic cone. In a neighborhood of the point of touch, the characteristic cone may be regarded as the graph of a smooth function defined on this hyperplane. The type of the characteristic cone's singularity is determined by the type of the critical point of this function. Vasiliev has proved the necessity of the local Petrowsky's criterion for the cone of rays to have singular points corresponding to critical points of finite multiplicity, and explicitly indicated all cases of sharpness of the fundamental solution near singular points corresponding to critical points of type A, D and E. Sharpness near singularities of a more complicated nature (in particular, X_9) has been examined with the help of a computer.

Using the technique of Fourier integrals, Gårding [18] adapted some of Petrowsky's ideas to the theory of strictly hyperbolic equations with smooth coefficients. According to the method of Lax-Duistermaat-Hörmander [19], the parametrix for such an equation can be written as a sum of as many Fourier integrals as is the order of the equation. Gårding noticed that each separate summand cannot have a sharp front; however, sharp fronts may appear if the summands are joined into couples in accordance with the involution $\xi \mapsto -\xi$ on the characteristic manifold. This coupling corresponds to passing from the sphere to the projective space in the proof of the Herglotz-Petrowsky formulas. This expression of the parametrix almost immediately yields information about its sharpness in non-singular points of the conoid of rays. In order to examine sharpness near singular points, Gårding con-

---

[1]A simplified proof of these results in the case of non-singular A can be found in the supplementary article "Proof of Petrowsky's Theorem" by A.M. Gabrielov (see the next article of this Appendix).

structs a local analogue of Petrowsky's cycle and writes the singular part of the parametrix in the form of an integral over that cycle. Then, the following definition is given: let x_0 be a point on the conoid of rays, and let l be a smooth path issuing from x_0 and belonging to the complement of this conoid; the local Petrowsky criterion is said to hold for x_0, l if there exists a cycle β such that for any point $x \in l$ close enough to x_0 the local Petrowsky's cycle constructed for x is homologous to β. The following statement is regarded by Gårding as having partially hypothetical character: if the local Petrowsky's criterion holds, then the parametrix (or more generally, the paired Fourier integral) is sharp at the point x_0 from the direction of the connected component of the complement to the cone of rays containing l. It is beyond doubt that in many cases the ideas laid down by Gårding in this paper can be turned into a complete proof (for instance, under the assumption that the corresponding Lagrange manifold over the physical space has generic position).

By the method of Gårding, Tvorogov [20] established a sharpness criterion for singularities of the parametrix for an arbitrary quasi-hyperbolic operator with smooth coefficients. This criterion generalizes the aforementioned result of Borovikov-Davidova and amounts to the coincidence of certain two integers modulo 2. One of these integers depends only on the structure of the germ of the Lagrange manifold corresponding to the given point on the conoid of rays. The other integer is the flattening index of the ray reaching this point, i.e., the number of its intersection points in generic position with the instant patterns of the front which have zero Gaussian curvature on this ray.

6. The problem of lacunas for equations with variable coefficients is studied far less, in spite of considerable efforts in this direction. The results in this field are connected with the problem stated by Hadamard [2]: to find second order hyperbolic equations for which the Huyghens principle is valid in narrow sense. As indicated above, this question pertains only to equations in an even-dimensional space and admits a positive answer in the case of the wave equation with constant coefficients, as well as in some other cases which can be easily reduced to this one by changing the independent variables or the unknown function, and by the multiplication of the equation itself by some known function. That it is only in these cases that the Huyghens Principle holds in the narrow sense has become known as the hypothesis of Hadamard, who, however, never made public this statement as it is given here, although this conjecture may have seemed to him quite probable[2]. Several mathematicians (among them Courant, Mathisson and Asgeirsson) tried to verify this conjecture; and Mathisson [21] claimed

---

[2]This opinion belongs to V.M. Babich.

it to be true. Later, however, the work of Mathisson has been found erroneous. In 1953, Stellmacher [22] constructed non-trivial equations of the form $\Box u + c(x)u = 0$ in 6-dimensional space, for which the narrow Huyghens Principle is valid. Lagnese & Stellmacher [23] suggested a method for generating the so-called Huyghens' equations: for a given equation of some special class, a sequence of Huyghens' equations of the same type can be constructed, every preceding equation being defined in the space whose dimension is that of the next equation minus 2.

The desire to see our physical space as a unique object supported the hope that in the four-dimensional case "Hadamard's hypothesis" is nevertheless true. This hope was destroyed by an example discovered by Günter [24], who constructed a second order equation which cannot be reduced to the case of constant coefficients but agrees with the Huyghens Principle. Actually, Günter indicated a whole family of equations depending on an arbitrary function of a single variable (see also [25]). According to Hadamard, the principal part of a second order hyperbolic equation is associated with a Riemannian space of signature $(1,3)$. Ibragimov [26] examined the relation between symmetries of this space and the Huyghens Principle. For a space with a fairly ample conformal group, the Huyghens Principle holds if and only if the equation itself is conformally invariant. In this case the sum of the lower order terms is reduced to the unknown function multiplied by $\frac{1}{6}$th of the scalar curvature. Since all four-dimensional spaces with an ample conformal group are known, Ibragimov's theorem yields a family of Huyghens' equations very similar to the family found by Günter, whose premises had been entirely different. This coincidence holds out a hope that the class of Günter-Ibragimov's equations, even if not exhaustive, covers a considerable portion of all Huyghens' equations with four independent variables.

During the last decade, these topics have been considered by analysts belonging to the school of Günther (see [27]–[29]), and also by McLenaghan [30], [31], Lax & Phillips [32].

References

[1] Huyghens, Ch. *Traité de la Lumière*. Pierre van der Aa. The Haague, 1860.

[2] Hadamard, J. *Lectures on Cauchy's Problem in Linear Partial Differential Equations*. New Heaven, Yale Univ. Press, 1923.

[3] Petrowsky, I.G. On the diffusion of waves and the lacunas for hyperbolic equations. *Mat. Sbornik* **17(59):3** (1945) 289–370.

[4] Herglotz, G. Über die Integration linearer partieller Differentialgleichungen mit konstanten Koeffizienten. *Ber. Verhandl. Sächs. Akad. Wiss. Leipzig* **78** (1926) 93–126, 287–318; **81** (1928) 60–144.

[5] Davidova, A.M. A sufficient condition for the absense of lacunas for a hyperbolic partial differential equation. Thesis, Moscow, Moscow Univ., 1945.

[6] Borovikov, V.A. Fundamental solutions of linear partial differential equations with constant coefficients. *Trudy Mosk. Mat. Ob.-va* **8** (1959) 199–257.

[7] Borovikov, V.A. Some sufficient conditions for the absense of lacunas. *Mat. Sbornik* **55:3** (1961) 237–254.

[8] Galpern, S.A.; Kondrashev, V.E. The Cauchy problem for differential operators that can be split into wave factors. *Trudy Mosk. Mat. Ob.-va* **16** (1967) 109–136.

[9] Galpern, S.A. Lacunas for non-hyperbolic equations. *Doklady Akad. Nauk SSSR* **132:5** (1960) 990–993.

[10] Lokshin, A.A. Fundamental solutions of quasil-hyperbolic equations and polynomials in many variables. *Trudy Seminara Petrowsk.* **3** (1978) 99–116.

[11] Riesz, M. L'intégrale de Riemann-Liouville et le problème de Cauchy. *Acta Math.* **81** (1949) 1–223.

[12] Gårding, L. The solution of Cauchy's problem for two totally hyperbolic differential equations by means of Riesz integrals. *Ann. Math.* **48:4** (1947) 785–826.

[13] Weinberg, B.R., Gindikin, S.G. On a stronger version of Huyghens' principle for a class of differential operators with constant coefficients. *Trudy Mosk. Mat. Ob.-va* **16** (1967) 151–180.

[14] Atiyah, M.F.; Bott, R.; Gårding, L. Lacunas for hyperbolic differential operators with constant coefficients. I. *Acta Math.* **124** (1970) 109–189.

[15] Atiyah, M.F.; Bott, R.; Gårding, L. Lacunas for hyperbolic differential operators with constant coefficients. II. *Acta Math.* **131** (1973) 145–206.

[16] Leray, J. Un prolongement de la transformation de Laplace qui transforme la solution unitaire d'un opérateur hyperbolique en sa solution élémentaire. *Bull. Soc. Math. France* **90** (1962) 39–156.

[17] Vasiliev, V.A. Sharpness and the local Petrowsky condition for strictly hyperbolic equations with constant coefficients. *Izvest. Akad. Nauk SSSR. Ser. Mat.* **50:2** (1986) 243–284.

[18] Gårding, L. Sharp fronts of paired oscillatory integrals. *Publ. RIMS, Kyoto Univ.* 1977, Vol. 12, Suppl., pp 53–68. Corrections: *Publ. RIMS Kyoto Univ.*, 1977, Vol. 13:3, p.821.

[19] Duistermaat, J.J.; Hörmander, L. Fourier integral operators. II. *Acta Math.* **128** (1972) 183–269.

[20] Tvorogov, V.B. Sharp front and singularities of solutions for a class of non-hyperbolic equations. *Doklady Akad. Nauk SSSR* **244:6** (1979) 1327–1331.

[21] Mathisson, M. Le problème de Hadamard relatif à la diffusion des ondes. *Acta Math.* **71:3-4** (1939) 249–282.

[22] Stellmacher, K.L. Ein Beispiel einer Huyghensschen Differentialgleichung. *Nachr. Acad. Wiss. Göttingen. Math.-phys. Kl. IIa* **10** (1953) 133–138.

[23] Lagnese, J.E.; Stellmacher, K.L. A method of generating classes of Huyghens' operators. *J. Math. Mech.* **17:5** (1967) 461–471.

[24] Günter, P. Ein Beispiel einer nichttrivialen Huyghenschen Differentialgleichung mit vier unabhängigen Variablen. *Arch. Rat. Meck. Anal.* **18:2** (1965) 103–106.

[25] Ibragimov, N.Kh.; Mamontov, E.V. Sur le problème de J. Hadamard relatif à la diffusion des ondes. *C.R. Acad. Sci. Paris* **270** (1970) 456–458.

[26] Ibragimov, N.Kh. On the group classification of second order differential equations. *Doklady Akad. Nauk SSSR* **183:2** (1968) 274–277.

[27] Günter, P.; Wünsch, V. Maxwellsche Gleichungen und Huyghenschen Prinzip. I. *Math. Nachr.* **63** (1974) 274–277.

[28] Wünsch, V. Maxwellsche Gleichungen und Huyghenschen Prinzip. II. *Math. Nachr.* **73** (1976) 19–36.

[29] Schimming, R. Das Huyghenschen Prinzip bei linearer hyperbolischer Differentialgleichungen zweiter Ordnung für allgemeine Felder. *Beitr. Analysis* **11** (1978) 45–90.

[30] McLenaghan, R.G. An explicit determination of the empty space-times on which the wave equation satisfies Huyghens' principle. *Proc. Cambridge Phil. Soc. 65* (1969) 139–155.

[31] McLenaghan, R.G. On the validity of Huyghens' principle for second order partial differential equation with four independent variables. *Ann. Inst. Poincaré* **20:2** (1974) 153–188.

[32] Lax, P.D.; Phillips, R.S. An example of Huyghens' principle. *Comm. Pure Appl. Math.* **31** (1978) 415–421.

Proof of Petrowsky's Theorem*

by A.M. Gabrielov

This supplementary article contains a proof of Petrowsky's criterion for the existence of lacunas for strictly hyperbolic operators (see [1]). Essentially, our exposition follows the ideas of Atiyah, Bott & Gårding [2], [3], but does not rely on the intricate technique developed by these authors to make their investigation more general and comprehensive. Moreover, our proof, in contrast to [2] and [3], makes no reference to the Grothendieck theorem.

1. Hyperbolic Polynomials

Let $P(\xi_1, \ldots, \xi_n)$ be a homogeneous polynomial of degree m with real coefficients, and let $\eta \in \mathbb{R}^n \setminus 0$.

Definition 1. The polynomial P is said to be *strictly hyperbolic* with respect to η, if for any $\xi \in \mathbb{R}^n$ linearly independent of η the equation $P(\xi + t\eta) = 0$ admits m different real roots with respect to t.

Proposition 1. *Let $A = \{\xi \in \mathbb{R}^n, \ P(\xi) = 0\}$, and let Γ be a connected component of the set $\mathbb{R}^n \setminus A$ containing η. Then Γ is a convex cone, and for each $\zeta \in \Gamma$ the polynomial P is strictly hyperbolic with respect to ζ.*

Proof. Let $t_1(\xi) < \cdots < t_m(\xi)$ be the roots of the equation $P(\xi + t\eta) = 0$ numbered in increasing order. The set A can be represented as a union of cones $A_j = \{\xi + t_j(\xi)\eta, \ \xi \in \mathbb{R}^n\}$ with common vertex at the origin.

Then the set $\mathbb{R}^n \setminus A$ is split into $m + 1$ domains, so that A_j is the border between the domains indexed by j and $j + 1$. In particular, $-\Gamma$ and Γ coincide with the domains indexed by 1 and $m + 1$, respectively. Let $\zeta \in \Gamma$. Then every straight line in $\mathbb{R}^n \setminus 0$ parallel to ζ crosses all the sets A_j, since one of its points belongs to $-\Gamma$ and another to Γ. It follows that the operator P is strictly hyperbolic with respect to ζ. Further, assume

---

*A supplement to Articles 5 and 6 of this volume.

that Γ is a non-convex cone. Then there is a straight line l in $\mathbb{R}^n$ whose intersection with Γ is unconnected. Let ζ_1 and ζ_2 be two points belonging to different connected components of $l \cap \Gamma$. It is easy to see that the intersection of Γ with the straight line parallel to ζ_1 and passing through ζ_2 is also unconnected. Therefore, this line crosses A_m at more than one point. Since it also crosses every A_j, it must have more than m points of intersection with the set A, which is incompatible with the degree of the polynomial P being equal to m.

Definition 2. *The propagation cone K for the polynomial P is the cone dual to Γ:*

$$K = \{x \in \mathbb{R}^n, \ (x,\zeta) \geq 0, \ \forall \zeta \in \Gamma\} \ .$$

Obviously, K is a closed convex cone.

Proposition 2. *The distribution*

$$u(x) = \frac{1}{(2\pi)^n} \int_{\mathbb{R}^n} \frac{e^{i(x,\xi-i\eta)}}{P(\xi-i\eta)} \, d\xi \tag{1.1}$$

is a fundamental solution for the operator $P(D)$, $D = -i\dfrac{\partial}{\partial x}$; $u(x)$ is supported in K. Here $d\xi = d\xi_1 \wedge \cdots \wedge d\xi_n$, and the orientation of $\mathbb{R}^n$ is such that $d\xi > 0$.

Remark. In general, the integral (1.1) is divergent. □

The value of the distribution $u(x)$ applied to a finite test function $\varphi = \varphi(x)$ is equal to

$$(u,\varphi) = \int_{\mathbb{R}^n - i\eta} \frac{\mathcal{F}^{-1}\varphi(\xi)}{P(\xi)} \, d\xi \ ,$$

where $\mathcal{F}^{-1}$ is the inverse Fourier transformation, $\mathbb{R}^n - i\eta = \{\xi - i\eta, \ \xi \in \mathbb{R}^n\}$.

Proof. Differentiating (1.1), we immediately get

$$P(D)u(x) = \frac{1}{(2\pi)^n} \int_{\mathbb{R}^n} e^{i(x,\xi-i\eta)} \, d\xi = \delta(x) \ .$$

By the Cauchy theorem, the value of u on any test function φ is independent of η for $\eta \in \Gamma$; therefore, $P(\xi)$ does not vanish on $\mathbb{R}^n - i\eta$. On the other hand, we have

$$(u,\varphi) = \int_{\mathbb{R}^n} \frac{\mathcal{F}^{-1}\left(e^{(x,\eta)}\varphi(x)\right)(\xi)}{P(\xi - i\eta)} \, d\xi \ ,$$

and $(u, \varphi) \to 0$ for φ supported in the set $(x, \eta_0) < 0$ and $\eta = t\eta_0$, $t \to +\infty$; therefore $(u, \varphi) = 0$. It follows that $u(x) = 0$ on the set $(x, \eta) < 0$. The same argument applied to each $\zeta \in \Gamma$ yields $u(x) \equiv 0$ outside K.   □

Proposition 3. *For an operator $P(D)$ strictly hyperbolic with respect to η, a fundamental solution supported in the cone K is unique.*

Proof. Let $\operatorname{supp} u \subset K$; then we can extend $\tilde{u}(\xi)$ as an analytic function to the domain $\{\xi - i\eta, \ \eta \in \Gamma\}$, using the formula $\tilde{u}(\xi - i\eta) = \tilde{v}(\xi)$, where $v(x) = u(x)\, e^{-(x,\eta)}$. In this case,

$$\tilde{u}(\xi) = \lim_{\eta \to 0} \tilde{u}(\xi - i\eta)$$

in the sense of distributions. Let u_1 and u_2 be two fundamental solutions for $P(D)$ supported in K. Then $P(\xi)\,(\tilde{u}_1(\xi) - \tilde{u}_2(\xi)) \equiv 0$. Because of the uniqueness of analytic extension, we have

$$P(\xi - i\eta)\,(\tilde{u}_1(\xi - i\eta) - \tilde{u}_2(\xi - i\eta)) \equiv 0 \quad \text{for} \quad \eta \in \Gamma \,,$$

and therefore, $\tilde{u}_1(\xi - i\eta) = \tilde{u}_2(\xi - i\eta)$. Passing to the limit as $\eta \to 0$, we find that $\tilde{u}_1(\xi) \equiv \tilde{u}_2(\xi)$, which means that $u_1 \equiv u_2$.   □

Definition 3. The cone of rays for the operator $P(D)$ is the set W consisting of the vectors $x = \lambda\, dP(\xi)$, where $\xi \in A$, $(x, \eta) \geq 0$.

Obviously, $\partial K \subset W$, and we can easily deduce from the condition of hyperbolicity that $W \subset K$.

Proposition 4. *The function $u(x)$, from Proposition 2 is analytic outside the set W.*

Proof. If $x \notin K$, then $u(x) = 0$; if $x \in K \setminus W$, then the constant vector field η in (1.1) can be deformed into another vector field $\eta(\xi)$ such that $(x, \eta(\xi)) < -\varepsilon|\xi|$ for large $|\xi|$, and in the process of deformation the values of $(x, \eta_s(\xi))$ remain bounded from above, while $|P(\xi - i\eta_s(\xi))|$ remains bounded from below by a positive constant. Here $\eta_s(\xi)$ is the family of vector fields realizing the deformation; $0 \leq s \leq 1$, $\eta_0(\xi) = \eta$, $\eta_1(\xi) = \eta(\xi)$. In this situation we can apply the Cauchy theorem, which yields

$$u(x) = \frac{1}{(2\pi)^n} \int_\beta \frac{e^{i(x,\zeta)}}{P(\zeta)}\, d\zeta \,, \tag{1.2}$$

where $\beta = \{\zeta = \xi - i\eta(\xi), \ \xi \in \mathbb{R}^n\}$; the orientation of β is induced from $\mathbb{R}^n$. This integral is absolutely convergent and analytically depends on x.

In order to construct such a deformation, let us consider on the intersection of the set $A = \{\xi \in \mathbb{R}^n, \ P(\xi) = 0\}$ with the sphere $|\xi| = 1$ a vector field $\theta(\xi) \in \mathbb{R}^n$ such that θ is tangential to A and $(x, \theta(\xi)) = 2 + (x, \eta)$ (this field exists, since $x \notin W$). Let us take a continuous extension of the field $\theta(\xi)$ to the entire sphere $|\xi| = 1$, preserving the relation $(x, \theta(\xi)) = 2 + (x, \eta)$.

Next, we set

$$\eta(\xi) = \varepsilon \left(\eta + |\xi| \left(\eta - \theta \left(|\xi|^{-1} \xi \right) \right) \right) \tag{1.3}$$

and define the deformation $\eta_s(\xi)$ by joining with straight line segments, first, η and $\varepsilon\eta$, and then $\varepsilon\eta$ and $\eta(\xi)$.

It is easy to verify that for sufficiently small positive ε the deformation η_s fulfills all our requirements.

2. Herglotz-Petrowsky's Formulas

Our aim is to pass in (1.2) to polar coordinates, then integrate along the radius and express $u(x)$ as an integral of a rational form over a cycle in projective space (the so-called Petrowsky cycle).

In order to define the Petrowsky cycle, let us introduce orientation in $\mathbb{R}^n$ such that $d\xi_1 \wedge \cdots \wedge d\xi_n > 0$. Consider the corresponding induced orientation on the sphere $S^{n-1} = \partial B^n \subset \mathbb{R}^n$ (where B^n is the unit ball) and the opposite orientation on the hemisphere $\{\xi \in S^{n-1}, (x, \xi) < 0\}$. Denote the resulting chain by $S^\vee$. Let $\eta(\xi)$ be a vector field in $\mathbb{R}^n$ with the following properties:

$$(1) \quad (x, \eta(\xi)) \equiv 0 \,,$$

$$(2) \quad \frac{(\,dP(\xi), \eta(\xi))}{(\,dP(\xi), \eta)} > 0 \quad \text{for} \quad \xi \neq 0, \ P(\xi) = 0 \,.$$

For $x \notin W$, a field with these properties exists, and all such fields are mutually homotopic. Denote by α a relative cycle in $\mathbb{C}^n \setminus A$ obtained from $S^\vee$ by a small translation along the field $-i\eta(\xi)$ and having its border in $X \setminus A$, where

$$A = \{\zeta \in \mathbb{C}^n, \ P(\zeta) = 0\} \,, \qquad X = \{\zeta \in \mathbb{C}^n, \ (x, \zeta) = 0\} \,.$$

Let α^* be the image of α by the natural projection $\mathbb{C}^n \setminus \{0\} \to \mathbb{C}P^{n-1}$. The cycle α^* is a relative cycle of the pair $(\mathbb{C}P^{n-1} \setminus A^*, X^* \setminus A^*)$, where A^* and X^* are the images of A and X by the projection into $\mathbb{C}P^{n-1}$. The cycle α^* is called the *relative Petrowsky cycle*, and $\beta^* = \frac{1}{2} t_X \cdot \partial \alpha^* \subset \mathbb{C}P^{n-1} \setminus (A^* \cup X^*)$ is the *absolute Petrowsky cycle*. Here t_{X^*} is the operation of extending a tube over X^*; under this operation, a correspondence is established between each point $\zeta \in X^*$ and the border of a small neighborhood of this point in the two-dimensional real plane belonging to $\mathbb{C}P^{n-1}$ and orthogonal to X^* at the point ζ. An alternative definition of the cycle β^* can be obtained as follows.

Let $\eta^+(\xi)$ and $\eta^-(\xi)$ be two vector fields in $\mathbb{R}^n$ satisfying the conditions

(1) $(x, \eta^+(\xi)) < 0$ for $\xi \neq 0$,

(2) $(x, \eta^-(\xi)) > 0$ for $\xi \neq 0$,

(3) $\dfrac{(\,\mathrm{d}P(\xi), \eta^{\pm}(\xi))}{(\,\mathrm{d}P(\xi), \eta)} > 0$ for $\xi \neq 0$, $P(\xi) = 0$.

Denote by $\beta^{\pm}$ the cycle in $\mathbb{C}^n \setminus (A \cup X)$ obtained from S^{n-1} by a small translation along the field $-i\eta^{\pm}(\xi)$. Set $\beta' = \beta^+ - \beta^-$, and let $\beta^* \subset \mathbb{C}P^{n-1} \setminus (A^* \cup X^*)$ be the image of the cycle β' by the projective mapping $\mathbb{C}^n \setminus \{0\} \to \mathbb{C}P^{n-1}$.

It is easy to show that both these definitions of the cycle β^* yield cycles homologous to one another in $\mathbb{C}P^{n-1} \setminus (A^* \cup X^*)$.

Recall that α^* and β^* depend on the point $x \notin W$. To indicate this dependence, we denote these cycles by α_x^* and β_x^*, respectively.

Theorem 1 (Herglotz-Petrowsky's Formulas in Leray's Form). *The fundamental solution $u(x)$ and its derivatives are given by the following integrals of rational functions:*

$$\frac{\partial^{|\nu|} u(x)}{\partial x^{\nu}} = \frac{-\pi i^{m-n+1}}{(2\pi)^n q!} \int_{\alpha_x^*} \frac{\zeta^{\nu} \cdot (x, \zeta)^q}{P(\zeta)} \omega, \qquad q \geq 0, \qquad (2.1)$$

$$\frac{\partial^{|\nu|} u(x)}{\partial x^{\nu}} = \frac{(-q-1)!(-1)^{\nu} i^{n-m}}{(2\pi)^n} \int_{\beta_x^*} \frac{\zeta^{\nu} \cdot (x, \zeta)^q}{P(\zeta)} \omega, \qquad q < 0, \quad (2.2)$$

where $\nu = (\nu_1, \ldots, \nu_n)$, $|\nu| = \nu_1 + \cdots + \nu_n$, $q = m - n - |\nu|$,

$$\omega = \sum_{j=1}^{n} (-1)^{j-1} \zeta_j \, \mathrm{d}\zeta_1 \wedge \cdots \wedge \widehat{\mathrm{d}\zeta_j} \wedge \cdots \wedge \mathrm{d}\zeta_n.$$

Proof. To simplify notations, we derive these formulas only for the function $u(x)$ itself. The expressions for its derivatives can be obtained in exactly the same way. For the calculations which follow it is convenient to incorporate $u(x)$ into the family of functions

$$u_{\lambda}(x) = \frac{1}{(2\pi)^n} \int_{\beta} P^{\lambda}(\zeta) \, e^{\,i(x,\zeta)} \, \mathrm{d}\zeta \qquad (2.3)$$

analytically depending on the parameter $\lambda \in \mathbb{C}^1$. Here β is the cycle from (1.2), $P^{\lambda} = e^{\lambda \ln P}$, $\ln P(\zeta)$ is a continuous branch of the logarithm of P on β. For $\lambda = -1$ we have $u(x) = u_{-1}(x)$.

For $\mathrm{Re}\,\lambda \geq 0$, the integrand in (2.3) has no singularities at the origin. Therefore, if we drop the first summand in (1.3), we can replace the vector field $\eta(\xi)$, which determines the cycle β, by another one, positively homogeneous of degree 1 with respect to ξ.

The resulting vector field belongs to the set of mutually homotopic fields which satisfy the conditions:

(1^+) $(x, \eta(\xi)) < 0$ for $\xi \neq 0$,

(2) $\dfrac{(\,\mathrm{d}P(\xi), \eta(\xi))}{(\,\mathrm{d}P(\xi), \eta)} > 0$ for $\xi \neq 0$, $P(\xi) = 0$,

(3) $\eta(t\xi) = t\eta(\xi)$ for $t > 0$.

Using the fact that the vector field $\eta(\xi)$ and the polynomial P are homogeneous, let us pass in (2.3) to polar coordinates:

$$u_\lambda(x) = \frac{1}{(2\pi)^n} \int_{\beta^+} P^\lambda(\zeta) \left[\int_0^\infty r^{q-1}\, e^{\,ir(x,\zeta)}\, \mathrm{d}r \right] \omega \,,$$

where $q = m\lambda + n$, $\beta^+ = \beta \cap \{|\zeta| = 1\}$,

$$\omega = \sum_{j=1}^n (-1)^{j-1} \zeta_j\, \mathrm{d}\zeta_1 \wedge \cdots \wedge \widehat{\mathrm{d}\zeta_j} \wedge \cdots \wedge \mathrm{d}\zeta_n \,;$$

the orientation of β^+ is determined by that of β according to the formula $\beta^+ = \partial(\beta \cap \{|\zeta| \leq 1\})$. The integral with respect to r is equal to

$$\Gamma(q)\, e^{\frac{\pi i q}{2}} (x, \zeta)^{-q} \,, \quad \text{where} \quad (x, \zeta)^{-q} = e^{-q \ln(x,\zeta)} \,,$$

and the branch of the logarithm on the set $\mathrm{Im}\,(x, \zeta) > 0$ is chosen such that $\ln i = \pi i / 2$. Consequently,

$$u_\lambda(x) = \frac{\Gamma(q)\, e^{\frac{\pi i q}{2}}}{(2\pi)^n} \int_{\beta^+} P^\lambda(\zeta)(x, \zeta)^{-q} \omega \,.$$

For $m < n$, by analytic extension in λ we obtain

$$u(x) = \frac{(n - m - 1)!\, i^{n-m}}{(2\pi)^n} \int_{\beta^+} \frac{(x, \zeta)^{m-n}}{P(\zeta)} \omega \,. \tag{2.4}$$

If $m \geq n$, then $\Gamma(q)$ has a pole at the point $n - m$. Since $u_\lambda(x)$ is an analytic function of λ, it follows that the value of $u(x)$ is equal to the integral of the regular part of the integrand at the point -1, i.e., the integral of the coefficient by $(\lambda + 1)^0$ in the Laurent expansion for the function

$$\chi_\lambda(\zeta) = \frac{\Gamma(q)\, e^{\frac{\pi i q}{2}}}{(2\pi)^n} P^\lambda(\zeta)(x, \zeta)^{-q} \,.$$

This coefficient has the form

$$\frac{(-1)^{m-n}i^{n-m}(x,\zeta)^{m-n}}{(2\pi)^n(m-n)!P(\zeta)}\left[\Gamma'(1)+\sum_{k=1}^{m-n}\frac{1}{k}+\frac{\pi i}{2}+\frac{1}{m}\ln P(\zeta)-\ln(x,\zeta)\right].$$

Thus, for $m \geq n$, we have

$$u(x)=\frac{i^{m-n}}{(2\pi)^n(m-n)!}\int_{\beta+}\frac{(x,\zeta)^{m-n}}{P(\zeta)}\left[\Gamma'(1)+\sum_{k=1}^{m-n}\frac{1}{k}+\frac{\pi i}{2}+\right.$$

$$\left.+\frac{1}{m}P(\zeta)-\ln(x,\zeta)\right]\omega\ . \qquad (2.5)$$

In order to transform the expressions (2.4) and (2.5), we assume that $(x,\eta) > 0$ and make use of the fact that $u(-x) = 0$.

If we replace x by $-x$, then the integral over the cycle β^+ in (2.5) will be replaced by the integral over the cycle β^- obtained after replacing the condition (1^+) on the vector field $\eta(\xi)$ by the condition

$$(1^-)\qquad\qquad (-x,\eta(\xi)) < 0\quad\text{for}\quad \xi \neq 0\ .$$

The only difference between the cycles β^- and β^+ is that the set $X = \{(x,\zeta) = 0\}$ is passed around in the complex region from the opposite sides. If $m \geq n$, then the integrand in (2.5) has no poles on this set. Therefore, by subjecting β^+ and β^- to a suitable deformation, these cycles can be made coincident and belonging to the set $\text{Im}\,(x,\zeta) = 0$, so that

$$\ln(-x,\zeta)=\begin{cases}\ln(x,\zeta)+\pi i & \text{for } (x,\zeta) > 0\ ,\\ \ln(x,\zeta)-\pi i & \text{for } (x,\zeta) < 0\ .\end{cases}$$

Thus, for $m \geq n$, we get

$$u(x)=u(x)-(-1)^{m-n}u(-x)=\frac{-\pi i^{m-n+1}}{(2\pi)^n(m-n)!}\int_{\alpha'}\frac{(x,\zeta)^{m-n}}{P(\zeta)}\omega\ , \qquad (2.6)$$

where α' is the chain obtained from β^+ (or β^-) by changing the orientation of the set $\text{Re}\,(x,\zeta) < 0$. Obviously, $\partial\alpha' \subset X \setminus A$.

If $m < n$, then the cycles β^+ and β^- can be deformed so as to make them coincident everywhere outside a neighborhood of the set X containing a pole of the integrand in (2.4). Then we get

$$u(x)=u(x)-(-1)^{m-n}u(-x)=\frac{(n-m-1)!\,i^{n-m}}{(2\pi)^n}\int_{\beta'}\frac{(x,\zeta)^{m-n}}{P(\zeta)}\omega\ , \qquad (2.7)$$

where $\beta' = \beta^+ - \beta^-$.

The integrands in (2.6) and (2.7) have the form $\Phi(\zeta)\omega$, where $\Phi(\zeta)$ is a homogeneous function of degree $-n$ with respect to ζ. Direct calculation shows that this differential form is invariant with respect to transformations of the form $\zeta_j = \varphi(\nu)\nu_j$, and therefore, is the inverse image of a differential form on $\mathbb{C}P^{n-1}$ by the natural projection $\rho : \mathbb{C}^n \setminus \{0\} \to \mathbb{C}P^{n-1}$. Denoting by $\zeta = (\zeta_1, \ldots, \zeta_n)$ the homogeneous coordinates in $\mathbb{C}P^{n-1}$, we obtain the Herglotz-Petrowsky formulas (in Leray's form):

$$u(x) = \frac{\pi i^{m-n+1}}{(2\pi)^n (m-n)!} \int_{\alpha^*} \frac{(x, \zeta)^{m-n}}{P(\zeta)} \omega , \qquad m \geq n ; \qquad (2.8)$$

$$u(x) = \frac{(n-m-1)! \, i^{n-m}}{(2\pi)^n} \int_{\beta^*} \frac{(x, \zeta)^{m-n}}{P(\zeta)} \omega , \qquad m < n , \qquad (2.9)$$

where α^* and β^* are the respective images of α' and β' by the projection $\mathbb{C}^n \setminus \{0\} \to \mathbb{C}P^{n-1}$,

$$\omega = \sum_{j=1}^{n} (-1)^{j-1} \zeta_j \, d\xi_1 \wedge \cdots \wedge \widehat{d\zeta_j} \wedge \cdots \wedge d\zeta_n .$$

One can easily verify that the cycles α^* and β^*, constructed in the proof, are homologous to the Petrowsky cycles defined at the beginning of this section. The theorem is proved.

3. Lacunas and Topology

As before, we assume $P(\xi)$ to be a homogeneous strictly hyperbolic polynomial with respect to η, and denote by W the cone of rays for the operator $P(D)$, by K the propagation cone, and by $u(x)$ the fundamental solution supported in K. According to Proposition 4, the function $u(x)$ is analytic outside W.

Definition 4. A connected component C of the complement to W in $\mathbb{R}^n$ forms a *lacuna* for $u(x)$, if $u(x) \equiv 0$ in C. The component C is a *weak lacuna*, if $u(x)$ is a polynomial in C. A lacuna is said to be *stable*, if it is preserved under small variations of the coefficients of P.

As above, let A^* denote the set of zeros of $P(\xi)$ in $\mathbb{C}P^{n-1}$, and let X^* be the set of zeros of (x, ξ) in $\mathbb{C}P^{n-1}$.

Theorem 2 (Petrowsky). *A connected component C of the complement to W in K forms a stable weak lacuna, if and only if the homology class*

$$[\beta_x^*] \in H_{n-1}(\mathbb{C}P^{n-1} \setminus (A^* \cup X^*), \mathbb{C})$$

of the cycle β_x^* constructed for some point $x \in \mathbb{C}^1$ is equal to zero. For $m < n$, every weak lacuna is a lacuna.

Supplement to Theorem 2 (Atiyah-Bott-Gårding).

a) *If $m \geq n$, then the only stable lacuna is the trivial lacuna $\mathbb{R}^n \setminus K$, i.e., for any weak lacuna C, the function $u(x)$ is different from zero in C or becomes different from zero after a small perturbation of the coefficients of P.*

b) *If A^* is a non-singular set, then all weak lacunas (for $m < n$, all lacunas) are stable.*

Theorem 2 and its supplement will be proved in the following propositions.

Proposition 5. *Let $[\beta_x^*] = 0$ for $x \in C$; then C is a stable weak lacuna; and in the case of $m < n$, C is a stable lacuna.*

Proof. The integrand in (2.2) is a rational differential $(n-1)$-form on $\mathbb{C}P^{n-1} \setminus (X^* \cup A^*)$. Since any rational form of a higher dimension is closed, its integral depends only on the homology class

$$[\beta_x^*] \in H_{n-1}(\mathbb{C}P^{n-1} \setminus (X^* \cup A^*), \mathbb{C})$$

of the cycle β_x^*. In particular, if $[\beta_x^*] = 0$, then the integral vanishes, which means that

$$\frac{\partial^{|\nu|} u(x)}{\partial x^\nu} = 0 \quad \text{for} \quad |\nu| > m - n .$$

Since $u(x)$ is analytic in C, this implies that $u(x)$ is a polynomial in C; if $m - n \geq 0$, then C is a weak lacuna; and, in the case of $m < n$, C is a lacuna. Further, the restrictions (1)–(3) on $\eta(\xi)$ imply that the cycle β_x^*, constructed for P, can be taken as β_x^* for any homogeneous polynomial Q close enough to P; moreover, if $[\beta_x^*] = 0$ in $H_{n-1}(\mathbb{C}P^{n-1} \setminus (A^* \cup X^*), \mathbb{C})$, then the cycle β_x^* is also of zero class with respect to the corresponding homologies for the polynomial Q (this is a consequence of the fact that the set A formed by all zeros of the polynomial $P(\xi)$ continuously depends on its coefficients). Hence we obtain the stability of the lacuna C.

Proposition 6. *If A^* is a non-singular set and C is a weak lacuna, then $[\beta_x^*] = 0$ for any $x \in C$.*

Remark. This immediately implies that for a stable weak lacuna C we have $[\beta_x^*] = 0$, since P can be replaced by a close polynomial whose complex roots form a non-singular set.

Proof of Proposition 6. Since on the set C the function $u(x)$ is a polynomial, it follows from (2.2) that the integrals of all the forms

$$\left(\frac{\zeta^\nu \cdot (x,\zeta)^q}{P(\zeta)}\right)\omega\,, \qquad |\nu| = m - n - q\,, \tag{3.1}$$

over the cycle β_x^* are equal to zero. As shown below, the forms of this type yield $(n-1)$-dimensional cohomologies of the space $\mathbb{C}P^{n-1} \setminus (X^* \cup A^*)$. Hence, $[\beta_x^*] = 0$. $\qquad\qquad\qquad\qquad\qquad\qquad\qquad\qquad\qquad\qquad\qquad\square$

Proposition 7. *If A^* is a non-singular set, then the classes of differential forms of type (3.1) produce the space*

$$H^{n-1}(\mathbb{C}P^{n-1} \setminus (X^* \cup A^*), \mathbb{C})\,.$$

Proof. Let Ω be a bundle of differential $(n-1)$-forms over $\mathbb{C}P^{n-1} \setminus X^*$ holomorphic on $\mathbb{C}P^{n-1} \setminus (X^* \cup A^*)$ and having a pole of order ≤ 1 on $A^* \setminus X^*$. The sections of Ω are forms of type $(f(\xi)/P(\xi))\omega$, where $f(\xi)$ is a homogeneous function of order $m-n$, analytic in $\mathbb{C}P^{n-1} \setminus X^* \cong \mathbb{C}^{n-1}$, i.e., f can be represented as a convergent series

$$f(\xi) = \sum_\nu c_\nu \xi^\nu (x,\xi)^{m-n-|\nu|}\,.$$

Therefore, each form in $\Gamma(\mathbb{C}P^{n-1}\setminus X^*, \Omega)$ can be expanded into a convergent series whose terms are forms of type (3.1); and thus, it suffices to verify that the sections of the bundle Ω yield $H^{n-1}(\mathbb{C}P^{n-1} \setminus (A^* \cup X^*), \mathbb{C})$.

Since the set $A^* = \{P = 0\}$ is non-singular, any form $\psi \in \Omega$ can be written as

$$\psi = \theta_1 \wedge \frac{dP}{P} + \theta_2$$

in a neighborhood of any point $\zeta \in A^* \setminus X^*$; here θ_1, θ_2 are regular forms on $A^* \setminus X^*$. Set

$$\operatorname{res}_{A^*}\psi = 2\pi i \theta_1\big|_{A^* \setminus X^*}.$$

Thereby, we have a well-defined homomorphism of bundles:

$$\operatorname{res}_{A^*}: \ \Omega \to \Omega^{n-2}(A^* \setminus X^*)\,,$$

where $\Omega^{n-2}(A^* \setminus X^*)$ is the bundle of holomorphic differential $(n-2)$-forms on $A^*\setminus X^*$. Obviously, this homomorphism is epimorphic. Since $\mathbb{C}P^{n-1}\setminus X^*$ is a Stein manifold (see, for instance, [4]), it follows that the associated homomorphism of sections

$$\operatorname{res}_{A^*}: \ \Gamma(\mathbb{C}P^{n-1} \setminus X^*, \Omega) \to \Gamma(A^* \setminus X^*, \Omega^{n-2}(A^* \setminus X^*))$$

is also epimorphic. Since $A^* \setminus X^*$ is a Stein manifold, the sections of the bundle $\Omega^{n-2}(A^*\setminus X^*)$ yield $H^{n-2}(A^*\setminus X^*, \mathbb{C})$. Therefore, the homomorphism

$$\operatorname{res}_{A^*}: \ \Gamma(\Omega) \to H^{n-2}(A^* \setminus X^*, \mathbb{C})$$

is epimorphic. This homomorphism can be incorporated into the following commutative diagram

$$
\begin{array}{ccc}
\Gamma(\Omega) & \xrightarrow{\ j\ } & H^{n-1}(\mathbb{C}P^{n-1} \setminus (A^* \cup X^*), \mathbb{C}) \\
\Big\downarrow {\scriptstyle \text{res}\,_{A^*}} & & \Big\downarrow {\scriptstyle \text{Res}\,_{A^*}} \\
& & H^{n-2}(A^* \setminus X^*, \mathbb{C})
\end{array}
\tag{3.2}
$$

where j is the natural representation of cohomologies $\mathbb{C}P^{n-1} \setminus (A^* \cup X^*)$ by regular forms on $\mathbb{C}P^{n-1} \setminus (A^* \cup X^*)$; $\text{Res}\,_{A^*}$ is the dual mapping to the tube operation

$$
t_{A^*} : H_{n-2}(A^* \setminus X^*, \mathbb{C}) \to H_{n-1}(\mathbb{C}P^{n-1} \setminus (A^* \cup X^*), \mathbb{C})
$$

which associates each point $\zeta \in A^*$ to the border of a small neighborhood of this point on the two-dimensional real plane in $\mathbb{C}P^{n-1} \setminus A^*$ orthogonal to A^* at ζ. For any cycle $Z \subset \mathbb{C}P^{n-1} \setminus (A^* \cup X^*)$ there exists a chain C in $\mathbb{C}P^{n-1} \setminus X^* \cong \mathbb{C}^{n-1}$ such that $Z = \partial C$. The cycles Z and $t_{A^*}(C \cap A)$ are obviously homologous. It follows that t_{A^*} is epimorphism, and therefore, $\text{Res}\,_{A^*}$ is monomorphism. On account of res $_{A^*}$ being epimorphism, we deduce from the diagram (3.2) that the homomorphism j is epimorphic. $\qquad\square$

Proposition 8. *For* $x \in K \setminus W$, *the class* $[\alpha_x^*]$ *of the cycle* α_x^* *in* $H_{n-1}(\mathbb{C}P^{n-1} \setminus A^*; X^* \setminus A^*, \mathbb{C})$ *is non-zero.*

Proof. Since $x \in K$, the plane $X^* = \{\zeta : (x, \zeta) = 0\}$ has no intersection with the image Γ^* of the cone $\Gamma \subset \mathbb{R}^n \subset \mathbb{C}^n$ (see Proposition 1) by the projection $\mathbb{C}^n \to \mathbb{C}P^{n-1}$. Thus, the index of intersection (α_x^*, Γ^*) is well-defined. Recall that the cycle α_x^* is obtained in the following way: The sphere $\{\xi \in \mathbb{R}^n, |\xi| = 1\}$ is shifted to the complex region by means of the vector field $-i\eta(\xi)$, where $(x, \eta(\xi)) = 0$, and if $P(\xi) = 0$, the expressions $(\eta(\xi), dP(\xi))$ and $(\eta, dP(\xi))$ are of the same sign. Then, the orientation is reversed on the part of the sphere where $(x, \xi) < 0$; and, finally, the resulting relative cycle is projected into $\mathbb{C}P^{n-1}$. Therefore, the index of intersection (α_x^*, Γ^*) is equal (to within sign) to the double index of the field $\eta(\xi)$ on the border of the $(n-1)$-dimensional disk $D = \Gamma \cap \{(x, \xi) = 1\}$. But $P(\xi) = 0$ on this border, and thus the vector field $\eta(\xi)$ is always directed to the inside of D; consequently, its index is equal to ± 1, i.e., $(\alpha_x^*, \Gamma^*) \neq 0$. Therefore $[\alpha_x^*] \neq 0$. $\qquad\square$

Proposition 9. *Assume that the set* A^* *is non-singular, C is a weak lacuna, and* $x \in C$. *Then*

$$
\partial[a_x^*] = 0 \quad \text{in} \quad H_{n-2}(X^* \setminus A^*, \mathbb{C}) .
$$

Proof. According to Proposition 6, in this case we have

$$[\beta_x^*] = 0 \quad \text{in} \quad H_{n-1}(\mathbb{C}P^{n-1} \setminus (X^* \cup A^*), \mathbb{C}) .$$

By definition, $\beta^* = \frac{1}{2}t_X \cdot \partial \alpha^*$, where t_{X^*} is the tube operation over X^*. The exact Leray sequence

$$H_n(\mathbb{C}P^{n-1} \setminus A^*, \mathbb{C}) \xrightarrow{\cap X^*} H_{n-2}(X^* \setminus A^*, \mathbb{C}) \xrightarrow{t_{X^*}}$$
$$\to H_{n-1}(\mathbb{C}P^{n-1} \setminus (X^* \cup A^*), \mathbb{C})$$

has zero as its first term, since $\mathbb{C}P^{n-1} \setminus A^*$ is a Stein manifold of dimension $(n-1)$. Consequently, t_{X^*} is a monomorphism, and therefore, $\partial \alpha^* = 0$. $\square$

Proposition 10. *There are no non-trivial stable lacunas for $m \geq n$.*

Proof. Let $C \subset K$ be a stable lacuna, and let $x \in C$. It then follows from (2.1) that

$$\int_{\alpha_x^*} \frac{\zeta^\nu}{P(\zeta)} \omega = 0 \quad \text{for} \quad |\nu| = m - n .$$

The same can be claimed, if $P(\zeta)$ is replaced by a close homogeneous polynomial of the form $P(\zeta) + \varepsilon Q(\zeta)$. Differentiating k times with respect to ε, and then setting $\varepsilon = 0$, we obtain

$$\int_{\alpha_x^*} \frac{\zeta^\nu Q(\zeta)^k}{P(\zeta)^{k+1}} \omega = 0 .$$

Since any homogeneous polynomial of degree mk can be represented as a linear combination of polynomials of type $Q(\zeta)^k$, we see that for any form

$$\frac{\zeta^\nu}{P(\zeta)^{k+1}} \omega \quad \text{with} \quad k \geq 0, \ |\nu| = m(k+1) - n , \tag{3.4}$$

its integral over α_x^* vanishes. Since the lacuna C is stable, the set A^* can be assumed non-singular (otherwise, we can replace P by a close polynomial with a non-singular set of complex roots, and repeat all our arguments). According to Proposition 9, we have $\partial[\alpha_x^*] = 0$. The exact sequence

$$H_{n-1}(\mathbb{C}P^{n-1}\setminus A^*, \mathbb{C}) \longrightarrow H_{n-1}(\mathbb{C}P^{n-1}\setminus A^*; X^*\setminus A^*, \mathbb{C}) \xrightarrow{\partial} H_{n-2}(X^*\setminus A^*, \mathbb{C})$$

implies that $[\alpha_x^*]$ is the image of a cycle in $H_{n-1}(\mathbb{C}P^{n-1}\setminus A^*, \mathbb{C})$. According to Proposition 8, we have $[\alpha_x^*] = 0$. Therefore, it suffices to show that the forms of type (3.4) generate the space $H^{n-1}(\mathbb{C}P^{n-1} \setminus A^*, \mathbb{C})$. Since $\mathbb{C}P^{n-1} \setminus A^*$ is a Stein manifold, the space $H^{n-1}(\mathbb{C}P^{n-1} \setminus A^*, \mathbb{C})$ is produced by holomorphic $(n-1)$-forms on $\mathbb{C}P^{n-1} \setminus A^*$. Each of these forms can be written as $f(\zeta)\omega$, where $f(\zeta)$ is a homogeneous function of order $-n$, analytic in $\mathbb{C}P^{n-1} \setminus A^*$. Every function of this kind can be expanded into a convergent series

$$\sum_k \sum_\nu \frac{c_\nu \zeta^\nu}{P(\zeta)^k}, \qquad \nu = (\nu_1, \ldots, \nu_n), \quad \nu_j \geq 0, \quad |\nu| = mk - n.$$

Therefore, holomorphic forms on $\mathbb{C}P^{n-1} \setminus A^*$ can be expanded into series with respect to the forms of type (3.4). It follows that the forms of type (3.4) generate the space $H^{n-1}(\mathbb{C}P^{n-1} \setminus A^*, \mathbb{C})$.   □

References

[1] Petrowsky, I.G. On the diffusion of waves and the lacunas for hyperbolic equations. *Mat. Sbornik* **17:3** (1945) 289–370.
[2] Atiyah, M.F.; Bott, R.; Gårding, L. Lacunas for hyperbolic differential operators with constant coefficients. I. *Acta Math.* **124** (1970) 109–189.
[3] Atiyah, M.F.; Bott, R.; Gårding, L. Lacunas for hyperbolic differential operators with constant coefficients. II. *Acta Math.* **131** (1973) 145–206.
[4] Malgrange, B. *Lectures on the Theory of Functions of Several Complex Variables.* Tata Institute of Fundamental Research, Bombay, 1958.

Topology of Real Algebraic Manifolds[*]

by V.M. Kharlamov

Introduction

1. Algebraic curves and surfaces have been an object of investigation for several centuries. The beginning of strictly topological study of real algebraic manifolds dates back to the last quarter of the 19th century, when topology took shape as a separate discipline.

Among the first objects of topological investigation were plane curves and surfaces in three-dimensional space. Almost immediately, more or less complete knowledge was gained about plane curves of degree 4 and 5, as well as surfaces of degree 3 in three-dimensional space (curves and surfaces of smaller degree had been studied previously). The first deep result of general character is due to Harnack: in 1876, he found the possible number of components a plane curve of a given degree may have.

In 1900, Hilbert included into his famous list of problems the topological investigation of plane curves and surfaces in three-dimensional space (the first part of the 16th Problem); in this connection, he laid stress on curves with the maximal number of components (the so-called M-curves), and indicated the first non-trivial cases: curves of degree 6 and surfaces of degree 4. Hilbert made the following conjecture: an M-curve of degree 6 cannot have all its components outside each other; among the components of an M-curve of degree 6, there must be one that contains within itself one component and has nine components outside, or, vice versa, nine components within and one outside. These hypotheses provided material for the works published in 1906–1913 by Ragsdale, Hilbert and his students, Rohn, et al. The major contribution should be credited to Rohn, who went far ahead of the other analysts in the matter of proving the first conjecture (however, there were some gaps in his proof, which were successfully filled

---

[*]A commentary on Articles 7 and 8 of this volume.

by Gudkov in the '60s; Gudkov also showed the second conjecture to be false; see Section 3.6 below).

In 1933–1938, Petrowsky discovered first fundamental general restrictions on the position occupied by the components of a plane curve of a given degree. These restrictions imply the validity of the first Hilbert's conjecture, and seem to provide its first complete proof.

There is a radical difference between Petrowsky's approach and the methods used by the authors mentioned above. The fundamental diversity lies in Petrowsky's transition to the complex region with the application of the well-known Euler-Jacobi's formula. This method was later elaborated by Petrowsky and Oleinik, and applied to surfaces in a space of arbitrary dimension. (As regards the surfaces of degree 4, the results of Petrowsky-Oleinik and those of Rohn have certain points in common).

The transition to the complex region for the purposes of studying real algebraic curves seems to have been initially performed by Klein in 1876. Klein made several important observations in this direction but could not go further in view of the lack of adequate topological tools. A remarkable achievement in this field is due to Comessatti, who, in his works of 1911–1932, considered some fairly general cases but, as far as we know, never applied his results to plane curves and surfaces in three-dimensional space. Nowadays such applications are known; some of them have been given in the works of Petrowsky and Petrowsky-Oleinik.

At present, the topology of real algebraic manifolds is a flourishing branch of mathematics. Its intensive development started with the works of Arnold and Rokhlin in 1971–1972. The achievements in this field can be explained by the application of new methods which appeared in algebraic geometry, number theory, and the theory of singularities. Of these methods, the following can be mentioned here: Smith's theory of periodic transformations, the index theorem of Atiyah-Singer, the special technique of the topology of four-dimensional manifolds, branched coverings, the Hodge theory, K3-surfaces, arithmetic of integer quadratic forms, toric manifolds and toric resolution of singularities, Levin-Eisenbud-Khimshiashvili's theorem. The efficiency of these methods is based on a systematic treatment of real algebraic manifolds as sets of fixed points on complex algebraic manifolds equipped with the involution of taking the complex conjugate.

Among previous studies underlying these new developments, an important place belongs to the works of Gudkov about curves of degree 6, as well as the works of Utkin about surfaces of degree 4, published in 1954–1969.

2. The aim of this commentary is to give a brief sketch of the history and the present state of the topology of real algebraic manifolds, emphasizing the aspects closely related to the works of Petrowsky. The results are described here without proof.

Because of its brevity, this review is far from being complete. The reader willing to acquire a more adequate impression of the subject may turn for general references to [1]–[9] (however, no comprehensive exposition of the subject based on a unified approach can be found). A vast bibliography given in the reviews [1]–[9] should be supplemented by the papers [10]–[15] and [80] (Khovansky, in [12], [80], considers topological characteristics of real algebraic manifolds from a new standpoint; these characteristics can be estimated from above in terms of the number of monomials entering the equations of the manifolds, however high their degree may be).

3. In what follows, a real plane algebraic curve of degree m is given by an equation of the form $F = 0$, where F is a homogeneous real polynomial of degree m and three arguments. A similar construction is adopted to specify a real algebraic surface of degree m in three-dimensional projective space, as well as a real algebraic hypersurface of degree m in projective space of any dimension.

1. Topology of Real Algebraic Manifolds before Petrowsky's Works

Generally, a historical description of the topology of real algebraic manifolds would start with the work of Harnack [16] published in 1876. And indeed, before that, only some very simple facts had been known. For instance, real branches of plane projective algebraic curves and spatial projective surfaces had been known to have either one-sided or two-sided position (on the plane, or in space, respectively). It had been known that non-singular curves of even degree consist of ovals only (two-sided components), and that every non-singular curve of odd degree has a unique one-sided component. It had also been known that non-singular surfaces of even degree consist only of orientable (two-sided) components, and that non-singular surfaces of odd degree have a unique non-orientable (one-sided) component. To conclude our list of previously known facts, we mention a relatively complete theory of real plane curves of degree ≤ 4, and real surfaces of degree ≤ 3 in three-dimensional space (the curves of degree 4 and surfaces of degree 3 were examined by Schläfli [17], Zeuthen [18], [19], and Klein [20]).

Harnack [16] was the first to raise the question: what is the maximal number of components a real plane algebraic projective curve of a given degree may have? And it was Harnack who gave an exhaustive answer to this question, at least for non-singular curves or those with the simplest type of singularities. He proved that for such curves the number of components cannot exceed $\frac{1}{2}(m - 1)(m - 2) + 1$, where m is the degree of the curve;

to be more exact, this number cannot exceed the genus of the curve plus 1 (these estimates coincide only for non-singular curves). Moreover, Harnack suggested a method for constructing curves with any number of components from $\frac{1}{2}(m-1)(m-2)+1$ to 0 for m even, and to 1 for m odd. At present, according to Petrowsky's terminology, curves with the maximal number of components are called M-curves.

In his proof, Harnack made use of the Bésout theorem, which states that a plane curve of degree q that occupies a generic position with respect to another plane curve of degree m cannot have more than mq points in common with the latter. Klein [21] immediately proposed another proof, based on a topological analysis of taking the complex conjugate in the set of complex points of a curve, and extended Harnack's inequality to arbitrary real algebraic curves (not only the plane ones).

Another question may also be asked: what possible position on the real projective plane can have the components of a non-singular curve of a given degree? This question is also considered in the works of Harnack and Klein. For curves of degree 5, it was actually solved by Harnack (see the list of possible positions in Section 3.6 below).

Klein [22] also considered a further problem: to describe non-singular real plane curves of a given degree m to within isotopies formed by non-singular curves of the same degree. Klein gave answer to this problem in the first non-trivial case, namely, for curves of degree 4, having reduced it to a similar, previously solved, problem about surfaces of degree 3 (see Sections 3.6 and 4.6; the reduction is based on the fact that the visible contour of a surface of degree 3 is a curve of degree 4, provided that the observer is on the surface itself).

Hilbert was also concerned with examining the relative positions of the components of a plane curve. In [23], he suggested a method for the construction of curves whose components have relative positions which cannot be obtained by the method of Harnack. He noticed that no M-curve of a degree $m > 3$ can have a nested structure deeper than $\frac{m}{2} - 1$ (i.e., it cannot have more than $\frac{m}{2} - 1$ two-sided components consecutively enclosed in one another), and constructed, by his own method, M-curves with a nested structure of depth $\left[\frac{m}{2} - 1\right]$ for any $m > 3$.

In 1906, Ragsdale [24] partially extended Hilbert's hypotheses from the curves of degree 6 to those of arbitrary even degree $m = 2k$ (with any number of components). According to Ragsdale's conjecture, a non-singular curve cannot have more than $\frac{3}{2}k(k-1)+1$ even components (i.e., those belonging to an even number of other components), and the number of its odd components (the remaining ones) is $\leq \frac{3}{2}k(k-1)$. She established these estimates for curves that can be constructed by the methods of Harnack and Hilbert. Moreover, Ragsdale formulated a generalization of Hilbert's hypotheses in weaker form, which can be interpreted as the following two-sided inequality

$$-\frac{3}{2}k(k-1) \le P - N \le \frac{3}{2}k(k-1) + 1 \qquad (1)$$

(here P and N are the numbers of even and odd components, respectively), and pointed out its relation to the characteristics considered by Kronecker. Finally, she constructed curves consisting of $\frac{3}{2}k(k-1)+1$ components positioned outside each other, and thereby showed that the estimate $P - N \le \frac{3}{2}k(k-1)+1$ cannot be improved.

In 1911–1913, Rohn [25], [26] examined curves of degree 6 and obtained results (although leaving some gaps in their proof) implying the correctness of Hilbert's first hypothesis and, partly, of the second one (namely, the part coinciding with Ragsdale's conjecture for $k = 3$; the gaps in his arguments were filled by Gudkov [27] in the '60s).

In 1909–1913, Hilbert [28] and Rohn [29], [26] studied real non-singular surfaces of degree 4 in three-dimensional projective space. Hilbert [28] indicated (without proof) that the results of Rohn [30], published as early as 1886, imply that the sum of Betti numbers for surfaces of degree 4 is bounded by 24; he also constructed surfaces of degree 4 with the sum of Betti numbers equal to 24. Rohn, in [29], [26], proved that for surfaces of degree 4 the maximal number of components homeomorphic to a sphere is equal to 10 (as shown by Utkin in the '60s, the flaws in his arguments can be set right owing to the results of Gudkov; see [27]).

In the works of Rohn [25], [26] and the preceding papers of Hilbert's students, the crucial point in the analysis of non-singular curves of degree 6 is the deformation of a non-singular curve into a curve with 9 singular points, and subsequent examination of such singular curves. A similar role in Rohn's investigation of surfaces of degree 4 is played by surfaces with 10 singular points.

In 1932, Comessatti [31], [32] introduced another method for the investigation of the topology of non-singular real algebraic surfaces, namely, the method based on the involution that is induced in the homologies of the set of complex points on the surface by taking the complex conjugate; thereby, he proved the inequalities:

$$-h^{1,1} + h_\alpha \le \chi - 2 - \operatorname{tr} \le h^{1,1} - h_\alpha , \qquad (2)$$

where χ is the Euler characteristic for the set of real points of the surface; $h^{1,1}$ is the Hodge number for the set of complex points of the surface; h_α is the dimension of the space generated by two-dimensional algebraic homology classes for the set of complex points; tr is the trace of the involution induced in this space by taking the complex conjugate. Comessatti [33]–[35] used a similar approach in relation to Abelian manifolds, Kummer's manifolds, etc.

2. The Works of Petrowsky and Petrowsky-Oleinik

In 1933-1938, Petrowsky [36], [37] established the estimates (1) and proved
the corresponding estimates for curves of odd order.

The inequalities (1) can be expressed in another form (more convenient
for what follows): for any non-singular real plane curve of degree $2k$, the
sets B_+ and B_-, defined in $\mathbb{R}P^2$ by $F > 0$ and $F < 0$, respectively, have
their Euler characteristics satisfying the inequalities

$$\chi(B_+) \le \frac{3}{2}k(k-1) + 1 \,, \qquad \chi(B_-) \le \frac{3}{2}k(k-1) + 1 \,. \tag{3}$$

Petrowsky's inequalities for odd order curves can be formulated as follows: if a non-singular real plane curve of degree $2k-1$ transversally crosses
the straight line $x_2 = 0$ $(x_0, x_1, x_2$ are homogeneous coordinates in $\mathbb{R}P^2)$,
then the Euler characteristics of the sets B_+ and B_- defined in $\mathbb{R}P^2$ by
$x_2 F(x_0, x_1, x_2) > 0$ and $x_2 F(x_0, x_1, x_2) < 0$ satisfy the inequalities

$$\chi(B_+) \le \frac{3}{2}k(k-1) + 1 - s \,, \quad \chi(B_-) \le \frac{3}{2}k(k-1) + 1 - s \,, \tag{4}$$

where s is the number of pairs formed by complex-conjugate solutions of
the system $x_2 = 0$, $F = 0$ $(2s = 2k - 1 - s'$, where s' is the number of real
solutions of the same system).

The inequalities (3), obviously, remain valid for curves of degree $2k$
with arbitrary singularities, and therefore, (4) can be regarded as a more
accurate version of (3) in the case of some special singular curves, namely,
the curves reducible to a straight line and a curve of odd degree.

Petrowsky, as evidenced by [36], [37], had not been conversant with
Ragsdale's paper [24]. Once again, he proved the upper bound in (1) to be
precise, noticing, as he did so, that the inequalities

$$P \le \frac{3}{2}k(k-1) + 1 \,, \quad N \le \frac{3}{2}k(k-1) + 1 \tag{5}$$

hold for all curves of degree $2k$ that can be constructed by the methods
of Harnack and Hilbert. (The inequality for N in (5) differs by 1 from the
inequality $N \le \frac{3}{2}k(k-1)$ in Ragsdale's conjecture).

In 1980, Viro [38] constructed curves with $N = \frac{3}{2}k(k-1) + 1$ for every
even $k \ge 4$ (for these curves, $P = \frac{1}{2}k(k-3) + 1$ and all odd ovals lie outside
each other and are contained in a single even oval), thereby proving false
this part of Ragsdale's conjecture. The question whether the inequalities (5)
are valid remains open. Moreover, Petrowsky made an observation which
can be interpreted as follows: supplemented by a straight line, any curve of
degree $2k-1$ that can be constructed by the methods of Harnack or Hilbert

satisfies inequalities stronger than (4) and obtained from (4) by replacing the Euler characteristic with the number of components.

The inequalities (1) imply the correctness of Hilbert's first hypothesis; the earliest complete proof of this hypothesis seems to have been given by Petrowsky in [37] (cf. Section 1).

In 1949, Petrowsky & Oleinik [39] extended the inequalities (3) and (4) from plane curves to hypersurfaces in real projective space $\mathbb{R}P^q$ of arbitrary dimension q. The principal results of this paper can be formulated as follows: the Euler characteristic for a non-singular hypersurface A of degree m in $\mathbb{R}P^q$, with q odd, satisfies the inequality

$$|\chi(A) - 1| \leq \Pi_{q+1}(m) , \tag{6}$$

where $\Pi_{q+1}(m)$ is the number of points that belong to the open cube $0 < x_i < m$ $(1 \leq i \leq q)$ in $\mathbb{R}^q$, have integer coordinates, and satisfy the two-sided inequality:

$$\frac{1}{2}(q - 1)m < x_1 + \cdots + x_q < \frac{1}{2}(q + 1)m ;$$

for a non-singular hypersurface of an even degree m in $\mathbb{R}P^q$, with q even, the sets B_+ and B_- defined in $\mathbb{R}P^q$ by $F > 0$ and $F < 0$, respectively, have their Euler characteristics satisfying the inequality

$$|\chi(B_+) - \chi(B_-)| \leq \Pi_{q+1}(m) ; \tag{7}$$

if a given hypersurface of an even degree m in $\mathbb{R}P^q$, with q even, reduces to a non-singular hypersurface and a hyperplane transversally crossing the hypersurface, then

$$|\chi(B_+) - \chi(B_-)| \leq \Pi_{q+1}(m - 1) . \tag{8}$$

The expression $\Pi_{q+1}(m)$ is a polynomial of degree q and argument m, if $\frac{1}{2}m(q + 1)$ is integer; otherwise, $\Pi_{q+1}(m)$ is a polynomial of degree q and arguments $[m/2]$ and m. For small q, this polynomial has the form:

$$(m - 1)^2 - 2C^2_{[\frac{m}{2}]} \quad (q = 2) ,$$

$$(m - 1)^3 - 2C^3_m \quad (q = 3) ,$$

$$(m - 1)^4 - 2C^4_{[\frac{3m}{2}]} + 8C^4_{[\frac{m}{2}]+1} \quad (q = 4) ,$$

$$(m - 1)^5 - 2C^5_{2m} + 10C^5_{m+1} \quad (q = 5) .$$

For curves, the inequality (7) yields (3) (and thereby, (1)), and the inequality (8) yields (4), since $\chi(B_+) + \chi(B_-) = 1$ in the case (3), and $\chi(B_+) + \chi(B_-) = 1 + s$ in the case (4).

For surfaces, the inequality (6) results in

$$-\frac{1}{3}(2m^3 - 6m^2 + 7m - 6) \le \chi(A) \le \frac{1}{3}(2m^3 - 6m^2 + 7m) . \qquad (9)$$

Petrowsky and Oleinik observed that the estimates (6) and (7) remain valid if the hypersurface has finitely many real singular points; the inequality (8) remains valid if the hypersurface (involved in this inequality) of odd degree has finitely many real singular points and the hyperplane (also entering this inequality) does not pass through any of these points and transversally crosses the hypersurface. It should be added that the inequality (8) itself can be considered as a stronger version of (7) valid in the case when the hypersurface has singularities resulting in a hyperplane being split off.

As mentioned in the Introduction, the results of Petrowsky and Oleinik have some common points with those of Comessatti: the inequalities (3), (4) and (9) can be derived by the application of Comessatti's inequality (2) to suitable surfaces; the inequality (9) is obtained by applying (2) directly to surfaces of degree m in $\mathbb{C}P^3$; the inequalities (3) and (4) are obtained by applying (2) to two-sheeted branched covering of the plane $\mathbb{C}P^2$ with branching along a curve of degree $2k$ (in the case of (4), the covering surface has singular points and, therefore, resolution of singularities should be used).

3. Topology of Real Plane Algebraic Curves after Petrowsky's Paper

3.1. On a Stronger Version of Petrowsky's Inequality

In 1971, Arnold [40], among other results (see Sections 3.3 and 3.5), gave a new proof of the estimates (1) and found a stronger version of these inequalities. This new version admits the following, more precise (see [41] or [6]), reading

$$P - N^- \le \frac{3}{2}k(k-1) + 1 , \quad N - P^- \le \frac{3}{2}k(k-1) , \qquad (10)$$

where P^- is the number of even hyperbolic ovals, and N^- is the number of odd hyperbolic ovals (an oval is said to be hyperbolic, if it limits from the outside a component forming part of the curve's complement and having a negative Euler characteristic).

The first estimate in (10) is precise for any k, which follows from the examples constructed by Ragsdale and Petrowsky (see Sections 1 and 2) with the purpose of showing the upper bound in (1) to be precise; the second estimate in (10) is precise for k even, as shown by the examples given by

Viro (see Section 2) in order to reject Ragsdale's hypothesis. For $k = 3$, the second estimate in (10) is not precise, and it is an open problem whether it is precise for odd $k > 3$. Every example demonstrating exactness of the second estimate in (10) automatically proves the hypothesis of Ragsdale, namely that $N \leq \frac{3}{2}k(k-1)$, to be false (otherwise, by the second inequality in (10), we should have $P^- = 0$, and therefore, the number of even ovals would not be less than the number of odd ovals, and their total number would be larger than that allowed by the Harnack theorem).

If k is odd and there is an oval enclosing all the others, then the second estimate in (10) can be improved by 1 (see Wilson [5]). In particular, for a curve consisting of a single encompassing oval and σ enclosed ones exterior to each other, we have

$$\sigma \leq \frac{3}{2}k(k-1) \, .$$

The examples constructed by Petrowsky in [37] show this upper bound to be precise.

Only a brief description can be given here of the methods used for the derivation of these estimates. The key role belongs to a two-sheeted branched covering of the plane $\mathbb{C}P^2$, with branching along the curve under consideration. On the covering manifold, two involutions of taking the complex conjugate are considered, the involutions being different by an automorphism of covering. To these two, the Lefschetz theorem about fixed points is applied, together with the signature theorem of Hirzebruch; the cycles belonging to the covering manifold over the domains in $\mathbb{R}P^2$ bounded by the ovals of the curve are taken into account (linear independence of these cycles is studied on the basis of Smith's theory).

3.2. Curves with Singularities

Two generalizations of the estimate (1), apparently having a close relation to each other, are known for curves with arbitrary isolated singularities (see Viro [42], Kharlamov [43]). In the case of the curve's singular set consisting of s simple double points and t cusps, both generalizations reduce to the inequality

$$|\chi_+ - \chi_-| \leq 3k^2 - 3k + 1 - s - 2t \, , \qquad (11)$$

where χ_+ and χ_- are Euler characteristics of the sets defined in $\mathbb{R}P^2$ by $F \geq 0$ and $F \leq 0$, respectively.

It follows from (10) that for curves with arbitrary singularities the following inequalities are valid:

$$\beta_0^-(B_-) \leq \chi(B_-) + \frac{3}{2}k(k-1) \, , \quad \beta_0^-(B_+) \leq \chi(B_+) + \frac{3}{2}k(k-1) \, , \qquad (12)$$

where β_0^- is the number of orientable components homeomorphic to a ring or a disk. However, the inequalities (12) cannot be regarded as a satisfactory generalization of (10), since, for a curve that reduces to a straight line and a curve of an odd degree, and therefore, satisfies (4), the estimate (12) happens to be weaker than (4). For curves with simple double points, an adequate generalization has been found by Zvonilov [44] and Viro [42]. This generalization can be formulated as follows.

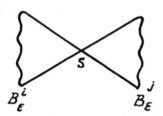

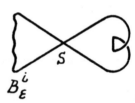

Fig. 1.   Fig. 2.

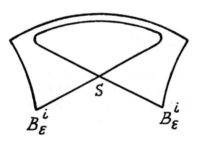

Fig. 3.

For k odd, denote by $B_\varepsilon^1, \ldots, B_\varepsilon^r$ (with $\varepsilon = +$ or $-$) the components of the set B_ε, and take $\delta_s = 1$ for any point s of the set Sing formed by all non-isolated singular points of the curve. For k even, we denote by $B_\varepsilon^1, \ldots, B_\varepsilon^r$ the orientable components of the set B_ε, arbitrarily choose orientation on the surface $B^\varepsilon = B_\varepsilon^1 \cup \ldots \cup B_\varepsilon^r$, and, for any point $s \in \text{Sing} \cap \text{Cl}\, B^\varepsilon$ (here Cl stands for 'closure' and Sing for 'singularities'), set $\delta_s = -1$, if the orientation of B^ε can be extended as the orientation of the surface obtained from $B^\varepsilon \cup \{s\}$ by the standard resolution of singularity at s; otherwise, set $\delta_s = 1$. Further, denote by f_s^ε the quadratic form such that $f_s^\varepsilon = 2^{-1}x_i^2 + 2^{-1}x_j^2 + \delta_s x_i x_j$, if s joins the surfaces B_ε^i and B_ε^j for $i \neq j$ (Fig. 1); $f_s^\varepsilon = 2^{-1}x_i^2$ if s joins a surface B_ε^i and a non-orientable portion of the surface B^ε (Fig. 2); $f_s^\varepsilon = (1 + \delta_s)x_i^2$, if one and the same component B_ε^i joins upon itself at s from both sides (Fig. 3). Finally, consider the quadratic form

$$f^\varepsilon = \sum_{s \in \text{Sing} \cap (\text{Cl}\, B^\varepsilon \backslash B^\varepsilon)} f_s^\varepsilon - 2 \sum_{i=1}^r \chi(B_\varepsilon^i)x_i^2 \; ;$$

denote by $\sigma_-(f^\varepsilon)$ and $\sigma_+(f^\varepsilon)$, respectively, the number of negative and positive squares in the diagonal expression of this form (on $\mathbb{R}$); let $\sigma_0(f^\varepsilon)$ be its defect. Using these notations, we can write the Viro-Zvonilov inequalities in the form

$$\sigma_-(f^\varepsilon) \leq \chi(B_\varepsilon) + \frac{s_0 - s_1}{2} + \frac{3}{2}\frac{(k-1)s}{2} \; ,$$

where s the number of all (real and imaginary) singular points of the curve; $s_0 + s_1$ is the number of real singular points; s_0 is the number of isolated (and therefore, singular) points of the curve. If k is even here, and B_ε has a non-orientable component, then

$$\sigma_-(f^\varepsilon) + \sigma_0(f^\varepsilon) - c + 1 + a \leq \chi(B_\varepsilon) + \frac{s_0 - s_1}{2} + \frac{3}{2}k(k-1) - \frac{s}{2} \; ,$$

where c is the number of irreducible components of the curve, and a is equal to 1 or 0, depending on whether the curve has irreducible components of odd degree or not. If the surface B_ε is orientable or k is odd, then either

$$\sigma_-(f^\varepsilon) + \sigma_0(f^\varepsilon) - c + 1 + a \leq \chi(B_\varepsilon) + \frac{s_0 - s_1}{2} + \frac{3}{2}k(k-1) - \frac{s}{2} \; ,$$

or the annihilator of the form f^ε contains a vector $(a_1, \ldots, a_r)$ with odd components (and therefore, every surface B_ε^i with $\text{Cl}\, B_\varepsilon^i \cap \text{Sing} = \emptyset$ is homeomorphic to a ring) and

$$\sigma_-(f^\varepsilon) + \sigma_0(f^\varepsilon) = \chi(B_\varepsilon) + \frac{s_0 - s_1}{2} + \frac{3}{2}k(k-1) - \frac{s}{2} + c - a \; .$$

3.3. New Inequalities

Together with the inequalities (10) and under the same assumptions, Arnold [40] obtained some other inequalities, which admit the following stronger version (see [41], [5], [6]):

$$N^- + N^0 \leq \frac{1}{2}(k-1)(k-2) , \tag{13}$$

$$P^- + P^0 \leq \begin{cases} \frac{1}{2}(k-1)(k-2) & \text{for } k \text{ odd,} \\ \frac{1}{2}(k-1)(k-2) + 1 & \text{for } k \text{ even,} \end{cases} \tag{14}$$

where $N^- + N^0$ is the number of non-empty (in the sense of containing at least one oval within itself) odd ovals; $P^- + P^0$ is the number of non-empty even ovals (P^- and N^- are defined in Section 3.1). Moreover, if $N^- + N^0 = (k-1)(k-2)$ and k is odd, then the curve has only one exterior oval, and each odd oval limits from the outside a component forming part of the complement to the curve and homeomorphic to a ring. If $P^- + P^0 = \frac{1}{2}(k-1)(k-2)$ and k is even, then every even oval bounds from the outside a component belonging to the complement to the curve and homeomorphic to a ring.

Zvonilov [44] and Viro [42] have extended these estimates to non-singular curves of odd degree and also to curves of even degree with simple double points. In notations of Section 3.2, their inequalities have the form

$$\sigma_+(f^\varepsilon) \leq \frac{1}{2}(k-1)(k-2) .$$

Moreover, if k is even and B_ε has a non-orientable component, then

$$\sigma_+(f^\varepsilon) + \sigma_0(f^\varepsilon) - c + 1 + a \leq \frac{1}{2}(k-1)(k-2) .$$

If B_ε is orientable or k is odd, then either

$$\sigma_+(f^\varepsilon) + \sigma_0(f^\varepsilon) - c + 1 + a \leq \frac{1}{2}(k-1)(k-2) ,$$

or the annihilator of the form f^ε contains a vector $(a_1, \ldots, a_r)$ with $a_1, \ldots, a_r$ being odd (and therefore, every surface B_ε^i with $\operatorname{Cl} B_\varepsilon^i \cap \operatorname{Sing} = \emptyset$ is homeomorphic to a ring) and

$$\sigma_+(f^\varepsilon) + \sigma_0(f^\varepsilon) = \frac{1}{2}(k-1)(k-2) + c - a .$$

3.4. Complex Orientations

As a basis for his proof of Harnack's inequality, Klein [21] used the following observation: the set of real points of a non-singular curve either causes no splitting of the set of its complex points (then the number of real components does not exceed the genus of the curve), or splits it into two connected halves which turn into one another on taking the complex conjugate and have as their common border the set of real points of the curve (then the number of components does not exceed the genus of the curve plus 1). The curves realizing the first case, Klein called *curves of type II*, and those corresponding to the second case – *curves of type I*. Klein's observation implies that all *M*-curves are of type I.

As shown by Klein in [21], [22], the type of a plane curve of degree ≤ 4 is determined by the scheme describing the position of its components in $\mathbb{R}P^2$ (the recent term is: the real scheme of the curve). As shown by Rokhlin in [6], for curves of degree ≥ 5, the type of the curve is no longer determined by its real scheme. The first example of this kind is provided by the curves of degree 5 with 4 ovals: such curves have all the 4 ovals exterior to each other; but, among such curves, there are curves of type I as well as those of type II.

Given a real scheme, there are three *a priori* possibilities for the type: any curve with this scheme belongs to type I; any curve with this scheme belongs to type II; there exist curves with this scheme belonging to type I as well as those belonging to type II. According to the terminology of Rokhlin, in the first case, the scheme is of type I; in the second case, the scheme is of type II; and in the third case, the scheme is of indefinite type. These three possibilities can be realized (see Section 3.6, where the types of all schemes for plane curves of small degree are given).

For a curve of type I, the two halves (mentioned in the beginning of this section) making up its complexification induce on the set of its real points, regarded as a border, two opposite orientations; these are called complex orientations. Petrowsky [45] introduced complex orientations in connection with lacunas for partial differential equations; into the studies on the topology of real algebraic curves, they were introduced by Rokhlin [41], who established the following formula relating the complex orientations of a curve's components to the position of the components on the plane.

First, we introduce some necessary definitions and notations. Two ovals are said to form an injective pair, if they limit a ring-shaped domain. An injective pair is called positive, if the orientation of the ovals is induced by that of the ring; otherwise, the pair is called negative. The number of positive (resp., negative) pairs corresponding to the complex orientation is denoted by Π^+ (resp., Π^-). Both complex orientations, being opposite to one another, result in the same values for Π^+ and Π^-. For a non-singular curve of degree $2k$ and type I, the formula for the complex orientations can be written as

$$2\left(\Pi^+ - \Pi^-\right) = l - k^2 ,\tag{15}$$

where l is the number of the curve's ovals. This formula implies that

$$\Pi^+ + \Pi^- \geq \frac{1}{2}(l - k^2) .$$

In particular, for $k = 3$, we obtain another (possibly the most elementary) proof of Hilbert's first hypothesis (see Sections 1 and 2).

For curves of odd degree, the use of complex orientations allows us to classify as positive or negative not only the injective pairs of ovals but the ovals themselves. Every oval divides the plane into two parts: the topological circle and the band of Möbius. An oval and a doubled one-sided component of the curve generate on the band homology classes which either coincide or differ by their sign; in the first case the oval is regarded as negative, and in the second case the oval is called positive. The number of positive (resp., negative) ovals is denoted by Λ^+ (resp., Λ^-). Mishachev showed in [46] that for a non-singular curve of degree $2k - 1$ and type I we have

$$\Lambda^+ - \Lambda^- 2(\Pi^+ - \Pi^-) = l - k(k - 1) ,$$

where l is again the total number of ovals of the curve. (In fact, Mishachev considered only M-curves. However, it is not very difficult to pass to arbitrary curves of type I).

The type and the complex orientations of a non-singular curve are invariant under its isotopies formed by the curves of the same degree. According to Rokhlin, such isotopies are called rigid. The study of rigid isotopic classes has as long a history as that of real schemes. The first non-trivial case – curves of degree 4 – was considered by Klein [22]. At present, it is known (see Section 3.6) that for curves of degree ≤ 6, the rigid isotopic class of a curve is determined by its real scheme and its type. For curves of degree ≥ 7, this cannot be claimed (see [6]). Moreover, Marin [47] and Fidler [48] found curves of type I and degree 7 having the same real power and the same complex orientations, but not rigidly isotopic.

There are still no general theorems to give sufficient conditions for rigid isotopy (cf. Section 3.6). It is only known that curves of the same degree with no real points are rigidly isotopic to each other (in the space of all curves of a given degree such curves form a convex subset). Moreover, the following, in some sense converse, case has been examined: all curves of degree m having $\left\lceil \frac{m}{2} \right\rceil$ ovals consecutively enclosed in one another are rigidly isotopic, as shown by Nuij [10] (such curves are called hyperbolic in the sense of Petrowsky; they have no other ovals).

3.5. Comparisons

In 1969–1972, it became clear that a characterization of real schemes for the curves of a given degree, in addition to the inequalities, should also involve the so-called comparisons. It turns out that for any M-curve of an even degree $m = 2k$ we have

$$P - N \equiv k^2 \bmod 8 . \tag{16}$$

This comparison had been guessed by Gudkov, and proved by him in 1969 (see [27]) for the curves of degree 6. In 1971, a weaker comparison

$$P - N \equiv k^2 \bmod 4 \tag{17}$$

was established by Arnold in [40] (actually, he proved (17) to be valid for curves of type I with any number of components). In its complete form, the comparison (16) was proved by Rokhlin [49].

Later, Gudkov & Krakhnov [50], and also Kharlamov [51], established a similar comparison for $(M - 1)$-curves (i.e., those consisting of $M - 1$ components, where $M = \frac{1}{2}(m - 1)(m - 2) + 1$): if an $(M - 1)$-curve of degree $m = 2k$ has no singularities, then

$$P - N \equiv k^2 \pm 1 \bmod 8 .$$

Some other comparisons have been proved by Nikulin [52] and Kharlamov [7] (see also: Marin [47], Fidler [53]).

Most of the comparisons are known to admit alternative proofs: some authors, following Arnold [40] and Rokhlin [49], use general results of differential and algebraic topology (such as the signature theorem of Hirzebruch, Smith's theory, etc.) and arithmetic of integer forms; others rely on some special techniques of four-dimensional topology indicated by Rokhlin [54] and Marin [47].

3.6. Curves of Small Degree

By now, the problem of describing the real schemes for non-singular curves of a given degree m has been completely solved for $m \leq 7$. For $m \leq 5$, its solution is based on almost elementary arguments (in fact, the case $m = 5$ was fully examined by Harnack in [16]). A considerable contribution to the study of real schemes for curves of degree 6 is due to Rohn and Petrowsky (see Sections 1 and 2). Complete solution of the problem for the curves of degree 6 was obtained in 1969 by Gudkov [27] (his method is the outcome of that used by Hilbert and Rohn; cf. Section 1); in particular, he proved false the second hypothesis of Hilbert (see the Introduction and Table 1).

For $m = 7$, final results were obtained in 1980 by Viro [38] (the principal point of this work is the new method found by Viro for the construction of curves; this method, in contrast to the classical ones, involves some intricate singularities).

In order to write down the schemes, we adopt the following notations (suggested by Viro [38]): a collection of α ovals exterior to one another we denote by $\langle \alpha \rangle$ (thus, $\langle 0 \rangle$ is an empty collection); if $\langle \sigma \rangle$ is a collection of ovals, then $\langle 1 \langle \sigma \rangle \rangle$ stands for another collection of ovals formed by adding to $\langle \sigma \rangle$ one new oval enclosing all the ovals of $\langle \sigma \rangle$; if two collections of ovals $\langle \sigma_1 \rangle$ and $\langle \sigma_2 \rangle$ belong to mutually disjoint domains, then their union is denoted by $\langle \sigma_1 \perp\!\!\!\perp \sigma_2 \rangle$; a connected one-sided curve is denoted by $\langle \mathcal{J} \rangle$; a collection of ovals $\langle \sigma \rangle$ supplemented with a connected one-sided curve that does not meet the ovals of this collection is denoted by $\langle \mathcal{J} \perp\!\!\!\perp \sigma \rangle$ (Fig. 4).

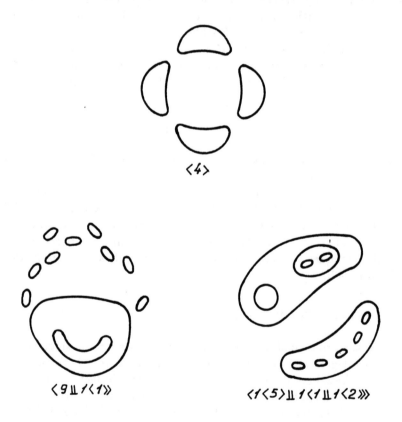

Fig. 4.

We start the list of schemes for non-singular curves of degree ≤ 7 with $m = 3$, omitting the trivial cases $m = 1$ and $m = 2$. For $m = 3$ we have two schemes $\langle \mathcal{J} \perp\!\!\!\perp 1 \rangle$ and $\langle \mathcal{J} \rangle$. For $m = 4$, there are 6 schemes:

$$\langle 4 \rangle \, , \, \langle 3 \rangle \, , \, \langle 2 \rangle \, , \, \langle 1 \langle 1 \rangle \rangle \, , \, \langle 1 \rangle \, , \, \langle 0 \rangle \ .$$

For $m = 5$, there are 8 schemes

$$\langle \mathcal{J} \perp\!\!\!\perp 6 \rangle \, , \, \langle \mathcal{J} \perp\!\!\!\perp 5 \rangle \, , \, \langle \mathcal{J} \perp\!\!\!\perp 4 \rangle \, , \, \langle \mathcal{J} \perp\!\!\!\perp 3 \rangle \, , \, \langle \mathcal{J} \perp\!\!\!\perp 2 \rangle \, ,$$

$$\langle \mathcal{J} \perp\!\!\!\perp 1 \langle 1 \rangle \rangle \, , \, \langle \mathcal{J} \perp\!\!\!\perp 1 \rangle \, , \, \langle \mathcal{J} \perp\!\!\!\perp 0 \rangle \, ,$$

For $m = 6$, the number of schemes is 56 and they are represented in Table 1. For $m = 7$, the number of schemes is 121 and they are as follows:

$$\langle \mathcal{J} \perp\!\!\!\perp a \perp\!\!\!\perp 1 \langle b \rangle \rangle \quad \text{for} \quad a + b \leq 14, \ 0 \leq a \leq 13, \ 1 \leq b \leq 13 \, ;$$

$$\langle \mathcal{J} \perp\!\!\!\perp a \rangle \quad \text{for} \quad 0 \leq a \leq 15; \quad \langle \mathcal{J} \perp\!\!\!\perp 1 \langle 1 \langle 1 \rangle \rangle \rangle \ .$$

The classification of curves with respect to rigid isotopies (see Section 3.4) has so far been brought up to degree 6. The first non-trivial case, $m = 4$, was examined by Klein [22] in 1876. The classification of curves of degrees 5 and 6 was completed in 1978–1981 by Rokhlin [6], Nikulin [55], and Kharlamov [56]. Rokhlin pointed out some rigidly isotopic invariants, and, in collaboration with his students, established which combinations of the invariants are realized for curves of degree ≤ 6. The fact that these invariants determine the curve to within rigid isotopies has been established by Kharlamov in the case of degree 5, and by Nikulin, in the case of degree 6. It should be observed that the classification of the curves of degree 4 has a close relation to the classification of the surfaces of degree 4 in $\mathbb{R}P^3$ (cf. Sections 1 and 4.6).

The rigid isotopic classification of non-singular curves of degree $m \leq 4$ coincides with the classification of their real schemes. For $m = 5$ and 6, the curves are rigidly isotopic, if they belong to the same type in the sense of Klein and have identical real schemes. For $m = 5$, there are 9 classes: the curves of type I with the schemes $\langle \mathcal{J} \perp\!\!\!\perp 1 \langle 1 \rangle \rangle$, $\langle \mathcal{J} \perp\!\!\!\perp 4 \rangle$, $\langle \mathcal{J} \perp\!\!\!\perp 6 \rangle$, and the curves of type II with the schemes $\langle \mathcal{J} \perp\!\!\!\perp 5 \rangle$, $\langle \mathcal{J} \perp\!\!\!\perp 4 \rangle$, $\langle \mathcal{J} \perp\!\!\!\perp 3 \rangle$, $\langle \mathcal{J} \perp\!\!\!\perp 2 \rangle$, $\langle \mathcal{J} \perp\!\!\!\perp 1 \rangle$, $\langle \mathcal{J} \rangle$ (the scheme $\langle \mathcal{J} \perp\!\!\!\perp 4 \rangle$ is of indefinite type; see Section 3.4). There are 64 classes for $m = 6$. A complete list of classes can be derived from Table 1, if we take into account the dotted lines there. Every scheme underlined by dots corresponds to two classes; the remaining schemes correspond to one class each. (To be more exact, the schemes underlined by dots are those of indefinite type; the schemes underlined by straight lines are those of type I; the remaining schemes are of type II.)

Table 1.

3.7. Inequalities for Curves of Type I

Arnold's comparison (17) shows that real schemes for curves of type I have certain special properties. The inequalities obtained by Rokhlin [57] imply some other properties, related, to a certain extent, to the inequalities of Petrowsky (1) and those of Arnold (10).

Rokhlin suggested the following two-fold division of the components forming the complement to a curve in $\mathbb{R}P^2$: a component is called *even*, if each oval limiting the component from the inside encircles an odd number of other ovals; otherwise, the component is said to be *odd*. Rokhlin denotes by π the number of non-empty even ovals limiting from the outside an even component of the complement to the curve; and he denotes by ν the number of non-empty odd ovals limiting from the outside an even component of the complement to the curve. The inequalities obtained by Rokhlin have the form: for a non-singular curve of degree $2k$ and type I

$$N - 2\nu \geq \frac{3k - 6}{2} - \frac{r}{2} \quad \text{for } k \text{ even,}$$

$$P - 2\pi \geq \frac{3k - 5}{2} - \frac{r}{2} \quad \text{for } k \text{ odd,}$$

where $r = 2k^2 - 3k + 2 - (P + N)$ (thus, r indicates by what amount is the number of the curve's components smaller than that of the M-curves). In particular, for M-curves we have

$$N - 2\nu \geq \frac{3k - 6}{2} \quad \text{for } k \text{ even,}$$

$$P - 2\pi \geq \frac{3k - 5}{2} \quad \text{for } k \text{ odd.}$$

The proof of these inequalities, as regards its technique, is similar to the proof of Petrowsky-Arnold's inequalities (10), but involves another branched covering.

4. Development of the Topology of Real Algebraic Surfaces after the Works of Petrowsky-Oleinik

4.1. Harnack's Inequalities for Surfaces

The problem of adequate generalization of Harnack's inequality from plane curves to arbitrary algebraic manifolds will be considered in Section 5.2. Here, we just mention that for non-singular surfaces A of degree m in $\mathbb{R}P^3$ a justified analogue of Harnack's inequality would be

$$\dim H_*(A, \mathbb{Z}_2) \le m^3 - 4m^2 + 6m \ . \tag{18}$$

For $m = 4$, this estimate was announced by Hilbert, who is also credited with the proof of its exactness in this case (see Section 1). The next step in this direction is due to Oleinik [78], who obtained an estimate with the same highest order term m^3 as in (18) but with a larger sum of inferior terms. The estimate (18) was proved by Thom [66]; and its exactness was established by Viro [67].

4.2. A Stronger Version of the Inequality (9) in the Case of Non-Singular Surfaces

The stronger version of Petrowsky's inequality (see Section 3.1), found by Arnold, can be extended to the case of non-singular surfaces and gives a more accurate value for the lower bound in (9) (see Kharlamov [58]):

$$k_+ + k_0 \le \frac{1}{2}\chi(A) + \frac{1}{6}(2m^3 - 6m^2 + 7m - 6) \ , \tag{19}$$

where k_0 (resp., k_+) is the number of the components of the surface A homeomorphic to a torus (resp., sphere). Viro [59] proved the estimate (19) to be precise for any m; for each m, he constructed a surface of degree m with $\frac{1}{6}(m^3 - 6m^2 + 11m)$ components, all of which, except one, homeomorphic to a sphere, and the remaining component, for m even, homeomorphic to a sphere with $\frac{1}{6}(2m^3 - 6m^2 + 7m)$ handles, and, for m odd, homeomorphic to the projective plane with $\frac{1}{6}(2m^3 - 6m^2 + 7m - 3)$ handles.

The inequality (19) can be extended to arbitrary (not only those belonging to $\mathbb{R}P^3$) non-singular real algebraic surfaces, and even to the case of A being the set of fixed points for an arbitrary anti-holomorphic involution acting on an arbitrary compact (without edge) non-singular complex surface X. In the latter, more general, case the inequality becomes

$$k_+ \le \frac{1}{2}\chi(A) + \frac{1}{2}(\sigma_-(X) - 1) \ ,$$

where σ_- is the number of negative squares in the diagonal expression for the manifold's quadratic form on $\mathbb{R}$; moreover, if A is non-orientable, then

$$k_+ + k_0 \le \frac{1}{2}\chi(A) + \frac{1}{2}(\sigma_-(X) - 1) + \dim H_3(X; \mathbb{Z}_2) - \frac{1}{2}\dim H_3(X; \mathbf{Q}) \ ;$$

if the surface A is orientable, then we either have

$$k_+ + k_0 \le \frac{1}{2}\chi(A) + \frac{1}{2}(\sigma_-(X) - 1) + \dim H_3(X; \mathbb{Z}_2) - \frac{1}{2}\dim H_3(X; \mathbf{Q}) \ ,$$

or

$$k_+ = 0 , \quad k_0 = \frac{1}{2}(\sigma_-(X) + 1) + \dim H_3(X; \mathbb{Z}_2) - \frac{1}{2}\dim H_3(X; \mathbb{Q}) ,$$

$$k_0 = \dim H_0(A; \mathbb{Z}_2) .$$

Essentially, these inequalities are a special case of the results obtained by Viro in connection with the study of singular curves; (cf. Sections 3.2 and 3.3).

4.3. Generalization

The inequalities of the preceding section can be extended to the case of the surface A having singular points. In the latter case, the inequalities involve, apart from the Euler characteristics of the surface's components, the properties characterizing the juncture of the components (cf. Section 3.2).

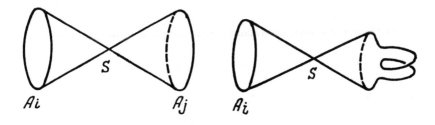

Fig. 5.   Fig. 6.

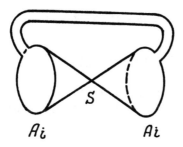

Fig. 7.

Let X be a compact complex surface whose singularities are exhausted by simple double points, and let $t : X \to X$ be an anti-holomorphic involution. Denote by $A_1, \ldots, A_r$ the orientable components of the surface $\text{Fix}\, t \setminus \text{Sing}$ (here $\text{Fix}\, t$ is the set of fixed points of the involution t), and choose an arbitrary orientation on the surface $A^0 = A_1 \cup \ldots \cup A_r$. For each singular point $s \in \text{Cl}\, A^0$, set $\varepsilon_s = 1$, if the orientation of A^0 can be extended to the orientation of the surface obtained from $A^0 \cup \{s\}$ by the standard resolution of singularity at the point s; otherwise, set $\varepsilon_s = -1$. Further, let us define a quadratic form f_s by: $f_s = \varepsilon_s x_i x_j - 2^{-1} x_i^2 - 2^{-1} x_j^2$, if s joins the surfaces A_i and A_j with $i \neq j$ (Fig. 5); $f_s = -2^{-1} x_i^2$, if s joins the surface A_i and a non-orientable portion of the surface $\text{Fix}\, t \setminus \text{Sing}$ (Fig. 6); $f_s = (\varepsilon_s - 1) x_i^2$, if one and the same component A_i joins upon itself at s from both sides (Fig. 7). Finally, set

$$f = \sum_{s \in \text{Cl}\, A^0 \setminus A^0} f_s - \sum_{i=1}^{r} \chi(A_i) x_i^2 \,,$$

and denote respectively by $\sigma_-(f)$ and $\sigma_+(f)$ the number of negative and positive squares in the diagonal expression (over $\mathbb{R}$) of this form; let $\sigma_0(f)$ denote its defect.

In these notations, the (generalized) inequalities have the form

$$\sigma_-(f) \leq \frac{1}{2}\chi(\text{Fix}\, t) + \frac{1}{2}(\sigma_-(X) - 1)\,.$$

Moreover, if the surface $\text{Fix}\, t$ is non-orientable, then

$$\sigma_-(f) + \sigma_0(f) - \dim H_3(X; \mathbb{Z}_2) + \frac{1}{2}\dim H_3(X; \mathbf{Q}) \leq$$

$$\leq \frac{1}{2}\chi(\text{Fix}\, t) + \frac{1}{2}(\sigma_-(X) - 1)\,.$$

If the surface $\text{Fix}\, t$ is orientable, we either have

$$\sigma_-(f) + \sigma_0(f) - \dim H_3(X; \mathbb{Z}_2) + \frac{1}{2}\dim H_3(X; \mathbf{Q}) \leq$$

$$\leq \frac{1}{2}\chi(\text{Fix}\, t) + \frac{1}{2}(\sigma_-(X) - 1)\,,$$

or the annihilator of f contains a vector $(a_1, \ldots, a_r)$ with odd components a_i (and therefore, every surface A_i with $\text{Cl}\, A_i \cap \text{Sing} = \emptyset$ is homeomorphic to a torus) and the following relations hold:

$$\sigma_0(f) - \dim H_3(X; \mathbb{Z}_2) + \frac{1}{2}\dim H_3(X; \mathbf{Q}) - 1 \leq$$

$$\leq \dim H_2(X; \mathbf{Q}) - \sigma_-(X)\,,$$

$$\sigma_-(f) + \sigma_0(f) = \frac{1}{2}\chi(\operatorname{Fix} t) + \frac{1}{2}(\sigma_-(X) - 1) + \dim H_3(X; \mathbb{Z}_2) -$$

$$- \frac{1}{2} \dim H_3(X; \mathbb{Q}) + 1 .$$

In the special case of X being a surface of degree m in $\mathbb{C}P^3$ whose singularities are exhausted by simple double singular points, the group $H_3(X; \mathbb{Q})$ is trivial, and the above estimates become as follows: for any m,

$$\sigma_-(f) \le \frac{1}{2}\chi(\operatorname{Fix} t) + \frac{1}{6}(2m^3 - 6m^2 + 7m - 6) - \frac{s}{2} ,$$

where s is the number of the surface's singular points; for m odd,

$$\sigma_-(f) + \sigma_0(f) - \dim H_3(X; \mathbb{Z}_2) \le$$

$$\le \frac{1}{2}\chi(\operatorname{Fix} t) + \frac{1}{6}(2m^3 - 6m^2 + 7m - 6) - \frac{s}{2} ;$$

for m even, we either have

$$\sigma_-(f) + \sigma_0(f) - \dim H_3(X; \mathbb{Z}_2) \le$$

$$\le \frac{1}{2}\chi(\operatorname{Fix} t) + \frac{1}{6}(2m^3 - 6m^2 + 7m - 6) - \frac{s}{2} ,$$

or the annihilator of the quadratic form f contains a vector $(a_1, \dots, a_r)$ with odd components (and therefore, every surface A_i with $\operatorname{Cl} A_i \cap \operatorname{Sing} = \emptyset$ is homeomorphic to a torus) and

$$\sigma_0(f) - \dim H_3(X; \mathbb{Z}_2) - 1 \le \frac{1}{3}(m^3 - 6m^2 + 11m - 3) ,$$

$$\sigma_-(f) + \sigma_0(f) - \dim H_3(X; \mathbb{Z}_2) =$$

$$= \frac{1}{2}\chi(\operatorname{Fix} t) + \frac{1}{6}(2m^3 - 6m^2 + 7m - 6) - \frac{s}{2} .$$

4.4. New Inequalities

The inequalities, established by Arnold for plane curves, which are supplementary to the inequalities found by Petrowsky (see Section 3.3), can be generalized to the case of surfaces; similar generalization is also possible for the stronger version of Petrowsky's inequalities (see Section 3.1). We limit ourselves to Arnold's inequalities as formulated for non-singular surfaces.

Let A be the set of fixed points for the anti-holomorphic involution on a compact non-singular complex surface X (with no boundary). Then

$$k_- \leq \frac{1}{2}(\dim H_2(X; \boldsymbol{Q}) - \sigma_-(X) - 1) \,,$$

where k_- is the number of the components of A with negative Euler characteristic. Moreover, if the surface A is non-orientable, then

$$k_0 + k_- \leq \frac{1}{2}(\dim H_2(X; \boldsymbol{Q}) - \sigma_-(X) - 1) + \dim H_3(X; \boldsymbol{Z}_2) - \frac{1}{2}\dim H_3(X; \boldsymbol{Q}) \,;$$

if A is orientable, then we either have

$$k_0 + k_- \leq \frac{1}{2}(\dim H_2(X; \boldsymbol{Q}) - \sigma_-(X) - 1) + \dim H_3(X; \boldsymbol{Z}_2) - \frac{1}{2}\dim H_3(X; \boldsymbol{Q}) \,,$$

or

$$k_- = 0 \,, \quad k_0 = \dim H_0(A; \boldsymbol{Z}_2)$$

$$k_0 = \frac{1}{2}(\dim H_2(X; \boldsymbol{Q}) - \sigma_-(X) + 1) + \dim H_3(X; \boldsymbol{Z}_2) -$$

$$-\frac{1}{2}\dim H_3(X; \boldsymbol{Q}) \,.$$

In the case of A being a non-singular surface of degree m in $\mathbb{R}P^3$, these estimates yield: either m is even, $k_- = 0$, $k_+ = 0$, $k_0 = \dim H_0(A; \boldsymbol{Z}_2)$ and

$$k_0 = \frac{1}{6}(m^3 - 6m^2 + 11m) \,,$$

or

$$k_0 + k_- \leq \frac{1}{6}(m^3 - 6m^2 + 11m - 6) \,.$$

4.5. Comparisons

Like M-curves, the surfaces with

$$\dim H_*(A; \boldsymbol{Z}_2) = \dim H_*(\mathbb{C}A; \boldsymbol{Z}_2)$$

(cf. Section 4.1; the latter are called M-surfaces) have some peculiar topological properties and are of special interest from the standpoint of topological analysis (cf. the 16th problem of Hilbert and his paper [23]). A notable illustration is the fact established by Rokhlin [49]: if A is an M-surface of degree m in $\mathbb{R}P^3$, then

$$\chi(A) \equiv \frac{1}{3}(4 - m^2) \bmod 16 \,.$$

This comparison is a special case of a more general theorem covering, in particular, arbitrary (not merely those in $\mathbb{R}P^3$) M-surfaces (see [49]). Recently, many kindred comparisons have been found to be valid under approximately the same conditions (see: Gudkov & Krakhnov [50], Nikulin [52], [55], Kharlamov [60], [51], [61]).

4.6. Surfaces of Small Degree

For surfaces in $\mathbb{R}P^3$, it is a non-trivial problem to find the maximal number of components a surface of a given degree m may have. The problem is open for $m \geq 5$; from (18) and the inequality of Petrowsky-Oleinik (9), it follows that this number is $\leq \frac{1}{12}(5m^3 - 18m^2 + 25m)$; however, this estimate seems far from being exact. All the more non-trivial becomes the problem of topological classification: to characterize the topological types of surfaces of a given degree.

At present, the topological classification of non-singular surfaces is known for $m \leq 4$. Moreover, for $m \leq 4$, the following classifications are known: isotopic (isotopies of the set of real points of a surface in $\mathbb{R}P^3$ are meant here), rigidly isotopic (in the sense of isotopies consisting of non-singular surfaces of the same degree), and rough projective classification (two surfaces are said to be projectively equivalent in rough sense, if one of them is rigidly isotopic to the projective image of the other).

For $m \leq 3$, the four classifications coincide. The first non-trivial case, i.e. $m = 3$, involves the following topological types: $S(7)$, $S(5)$, $S(3)$, $S(1) \perp\!\!\!\perp S$ and $S(1)$, where $\perp\!\!\!\perp$ denotes unconnected summation; $S(q)$ is a sphere with q handles; no knot formation can happen.

For $m = 4$, the four classifications are different from one another. The topological types are listed in Tables 2 and 3 (S_p denotes a sphere with p handles). Table 2 gives the topological types of surfaces that cannot be contracted in $\mathbb{R}P^3$ into a point; Table 3 gives the types of surfaces that contract into a point in $\mathbb{R}P^3$. For surfaces of the first kind, the isotopic classification is the same as the topological one: surfaces of type $S_1 \perp\!\!\!\perp S_1$ are isotopic to a union of hyperboloids; and surfaces of type $S_p \perp\!\!\!\perp aS$ ($a \geq 0$, $p \geq 1$) are isotopic to the union of a ellipsoids exterior to one another and a hyperboloid with standard handles such that one of the components of its complement contains all the ellipsoids. For surfaces of the second kind, except those of type $2S$, the isotopic classification is also identical to the topological one: to within isotopies, these surfaces consist of standard components in the affine portion of the space, and these components are exterior to each other; the surfaces of type $2S$ can have two types of position: one sphere inside another, and mutually exterior spheres. The rough projective classification differs from the isotopic one in that the surfaces of the same isotopic type can have unlike behavior in the complex region. A surface A is said to belong to type I, if A realizes zero element of the group $H_2(\mathbb{C}A; \mathbb{Z}_2)$; A belongs to type II_1, if it realizes in $H_2(\mathbb{C}A; \mathbb{Z}_2)$ the class of hyperplane section (recall that this class is non-zero); otherwise, A is said to be of type II_2. We obtain the entire list of rough projective classes from Tables 2 and 3, if we take into account the dotted lines: every item underlined by a single dotted line corresponds to two types, the other items correspond to one type

$$S_1$$

$$S_2 \qquad \underline{S_1 \amalg S_1}$$

$$S_3 \qquad S_2 \amalg S \qquad S_1 \amalg 2S$$

$$S_4 \qquad S_3 \amalg S \qquad S_2 \amalg 2S \qquad S_1 \amalg 3S$$

$$S_5 \qquad S_4 \amalg S \qquad \underline{S_3 \amalg 2S} \qquad S_2 \amalg 3S \qquad S_1 \amalg 4S$$

$$S_6 \qquad S_5 \amalg S \qquad S_4 \amalg 2S \qquad S_3 \amalg 3S \qquad S_2 \amalg 4S \qquad S_1 \amalg 5S$$

$$S_7 \qquad S_6 \amalg S \qquad S_5 \amalg 2S \qquad \underline{S_4 \amalg 3S} \qquad S_3 \amalg 4S \qquad S_2 \amalg 5S \qquad S_1 \amalg 6S$$

$$S_8 \qquad S_7 \amalg S \qquad S_6 \amalg 2S \qquad S_5 \amalg 3S \qquad S_4 \amalg 4S \qquad S_3 \amalg 5S \qquad S_2 \amalg 6S \qquad S_1 \amalg 7S$$

$$S_9 \qquad S_8 \amalg S \qquad \underline{S_7 \amalg 2S} \qquad S_6 \amalg 3S \qquad \underline{S_5 \amalg 4S} \qquad S_4 \amalg 5S \qquad \underline{S_3 \amalg 6S} \qquad S_2 \amalg 7S \qquad \underline{S_1 \amalg 8S}$$

$$S_{10} \qquad S_9 \amalg S \qquad S_8 \amalg 2S \qquad S_7 \amalg 3S \qquad S_6 \amalg 4S \qquad S_5 \amalg 5S \qquad S_4 \amalg 6S \qquad S_3 \amalg 7S \qquad S_2 \amalg 8S \qquad S_1 \amalg 9S$$

$$\underline{S_{10} \amalg S} \qquad \dots \qquad \underline{S_6 \amalg 5S} \qquad \dots \qquad \underline{S_2 \amalg 9S} \qquad S_1 \amalg 9S$$

Table 2.

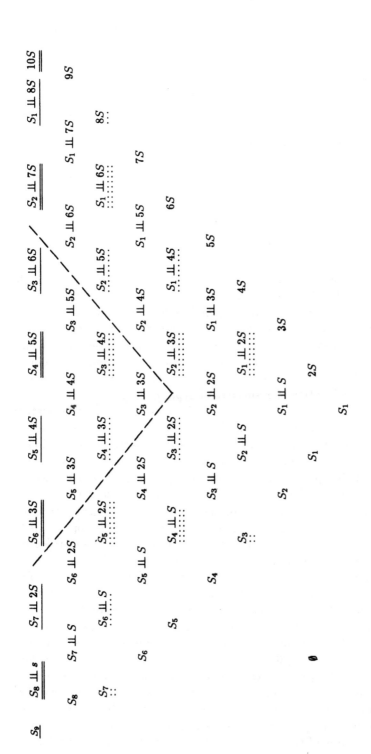

Table 3.

each. To be more exact, a single dotted line means that one class consists of surfaces of type I and the other class consists of surfaces of type II_2; a double dotted line means that one class consists of surfaces of type II_1 and the other is formed by surfaces of type II_2; a single straight line indicates that the corresponding class consists of surfaces of type I; double straight lines show the class to be formed by surfaces of type II_1; in the remaining cases, the class consists of surfaces of type II_2.

The rigid isotopic classification differs from the rough projective one in that the surface may or may not be rigidly isotopic to its mirror image. Every class of the rough projective classification in Tables 2 and 3 above the dash-line corresponds to two classes in the rigid isotopic classification; and every class below the dash-line corresponds to a single class in the rigid isotopic classification.

The above results pertaining to surfaces of degree 3 belong to Schläfli [17] and Klein [20]. The study of surfaces of degree 4 has been commenced by Rohn [30], [25], [26] and Hilbert [28] (see Section 1). Utkin [27], [62] developed the methods of Rohn and Hilbert and should be credited with a substantial contribution to the investigation of surfaces of degree 4. Kharlamov [58], [63], making use of a more elaborate technique of algebraic geometry (periods of K3-surfaces), brought to completion their topological and isotopic classification (for surfaces of type $S_1 \perp\!\!\!\perp S_1$, the examination of their possible positions was conducted by Kharlamov in collaboration with Nikulin; see [55], [63]). Then Nikulin, using some additional arithmetic and geometric tools, established the rough projective classification of these surfaces (this classification is part of his description obtained for the components of the space of modules of real algebraic polarized surfaces of type K3). Finally, Kharlamov [64] gave their rigid classification.

The reader is advised to compare the results just mentioned with those described in Section 3.5. In the case of plane curves, the isotopic classification is just the classification of the real schemes. The rough projective classification and the rigid isotopic classification are identical in this case, since the group of projective transformations of a plane is connected. The last circumstance is the reason why Nikulin's theorem about K3-surfaces yields the rigid isotopic classification for the curves of degree 6.

Regarding the surfaces, it should be added that for large degrees the problem of classification becomes much more difficult (as is the case with the curves). For degrees ≥ 6, rough projective equivalence of surfaces does not imply their isotopy even in the ordinary sense, since in this case one can easily construct surfaces forming knots with no mirror structure. No theorems of general nature have been established to give sufficient conditions for rigid isotopy or rough projective equivalence (cf. Section 3.4). It is only known that surfaces of the same degree without real points are rigidly isotopic to each other; and, in some sense, the opposite case has been analyzed:

all surfaces of degree m having $[m/2]$ spherical components consecutively enclosed in one another are rigidly isotopic to each other, as shown by Nuij [10]. Such surfaces are called hyperbolic in the sense of Petrowsky; they have no other orientable components; for m odd, there is also a component homeomorphic to $\mathbb{R}P^2$.

5. Generalizations

5.1. Principal Manifolds

Following Rokhlin [49], [65], we consider two cases. In the first case (the so-called even-dimensional case), we are supposed to be given a compact connected non-singular projective complex analytic manifold of even dimension n with an anti-holomorphic involution t. Set $A = \mathrm{Fix}\, t$. In the second case (odd-dimensional), the following objects are supposed to be given: a connected compact non-singular projective complex analytic manifold $\mathbb{C}B$ of even dimension $n + 1$; this manifold contains a complex analytic submanifold $\mathbb{C}A$ (possibly, unconnected) of odd dimension n; this submanifold realizes in $H_{2n}(\mathbb{C}B; \mathbb{Z})$ an element which is a multiple of 2; we are also given an anti-holomorphic involution $t : \mathbb{C}B \to \mathbb{C}B$ such that $t(\mathbb{C}A) = \mathbb{C}A$; a cohomology class $\omega \in H^1(\mathbb{C}B \setminus \mathbb{C}A; \mathbb{Z}_2)$ invariant with respect to t^* and mapped by the homomorphism

$$H^1(\mathbb{C}B \setminus \mathbb{C}A; \mathbb{Z}_2) \xrightarrow{\approx} H_{2n+1}(\mathbb{C}B, \mathbb{C}A; \mathbb{Z}_2) \to H_{2n}(\mathbb{C}A; \mathbb{Z}_2)$$

into the fundamental class of the manifold $\mathbb{C}A$. (If $H^1(\mathbb{C}A; \mathbb{Z}_2) = 0$, then there is exactly one class ω which satisfies the above conditions, and therefore, we do not need to specify ω beforehand in this case).

The manifold $\mathbb{C}A$ for n even and $A \neq \emptyset$, as well as the manifold $\mathbb{C}B$ for n odd and $B \neq \emptyset$, can be regarded as an invariant submanifold of some projective space $\mathbb{C}P^q$ with respect to the operation of taking the complex conjugate: $\mathrm{conj} : \mathbb{C}P^q \to \mathbb{C}P^q$; moreover, the involution t can be regarded as the restriction of the involution conj.

For n odd, the following (classical) subcase can be distinguished: a homogeneous real polynomial F of even degree m in $q + 1$ variables is given, for which the hypersurface $F = 0$ in $\mathbb{C}P^q$ has regular intersection with $\mathbb{C}B$ along $\mathbb{C}A$. In this situation, ω is taken to be the class induced by the unique non-zero element of the group $H^1(\mathbb{C}P^q \setminus \{F = 0\}; \mathbb{Z}_2)$.

For an arbitrary n, the simplest case (in particular, from the standpoint of specifying manifolds by equations) is $\mathbb{C}A$ being defined in $\mathbb{C}P^q$ by a system of equations

$$F_1(x_0, \ldots, x_q) = 0, \ldots, F_s(x_0, \ldots, x_q) = 0, \tag{20}$$

where $F_1, \ldots, F_s$ are homogeneous real polynomials, and $s = q - n$. Denote the respective degrees of these polynomials by $m_1, \ldots, m_s$. System (20) is assumed to have no singularities on $\mathbb{C}A$; moreover, if n is odd, it is assumed that: 1) $F_s = F$, so that $m_s = m$; 2) $\mathbb{C}B$ is defined in $\mathbb{C}P^q$ by the reduced system of equations

$$F_1(x_0, \ldots, x_q) = 0, \ldots, F_{s-1}(x_0, \ldots, x_q) = 0 ;$$

3) the latter system has no singularities in $\mathbb{C}B$. In this situation we are dealing with the so-called case of complete regular intersections.

Note that for $q - n = 1$, the manifold $\mathbb{C}A$ is always a non-singular hypersurface in $\mathbb{C}P^q$, i.e., a complete regular intersection with $s = 1$, while $\mathbb{C}B$ (for n odd) coincides with $\mathbb{C}P^q$. If $q = 2$, $s = 1$, then $\mathbb{C}A$ and A are non-singular algebraic curves in $\mathbb{C}P^2$ and $\mathbb{R}P^2$; if $q = 3$, $s = 1$, then $\mathbb{C}A$ and A are non-singular algebraic surfaces in $\mathbb{C}P^3$ and $\mathbb{R}P^3$. It should also be mentioned that for $q = 3$, $n = 1$ the manifold $\mathbb{C}B$ is always a non-singular algebraic surface in $\mathbb{C}P^3$.

Since in the case of n odd the manifold $\mathbb{C}A$ realizes in $H_{2n}(\mathbb{C}B)$ a class which is a multiple of 2, we see that in this case there exists a two-sheeted branched covering $\pi : Y \to \mathbb{C}B$ with branching along $\mathbb{C}A$, its characteristic class being ω. This branched covering is unique; let aut $: Y \to Y$ be its non-identical automorphism. Since $t^*\omega = \omega$, the involution $t : \mathbb{C}B \to \mathbb{C}B$ is covered by two involutions of the manifold Y. The manifolds formed by the fixed points of these involutions lie over B and, being projected by π, yield two manifolds $\overline{B}_+ \subset B$, $\overline{B}_- \subset B$ with common border A. (The manifolds $\overline{B}_+$ and $\overline{B}_-$ can be described without passing to a branched covering. In the classical case, when $\mathbb{C}A$ is cut out on $\mathbb{C}B$ by the hypersurface $F = 0$, these manifolds are defined by the inequalities $F \geq 0$ and $F \leq 0$. If $q = 2$, $n = 1$, then

$$\overline{B}_+ = \mathrm{Cl}\, B_+, \quad \overline{B}_- = \mathrm{Cl}\, B_- \quad \text{and} \quad \chi(\overline{B}_+) = \chi(B_+), \quad \chi(\overline{B}_-) = \chi(B_-),$$

where B_+, B_- are the sets defined in Section 2). To the manifold $\overline{B}_\varepsilon$, with $\varepsilon = +, -$, one can associate an integer a_ε equal to the dimension of the intersection of two kernels: that of the border homomorphism

$$\partial : H_*(\overline{B}_\varepsilon, A; \mathbb{Z}_2) \to H_*(A; \mathbb{Z}_2)$$

and that of the homomorphism $\cap\omega : H_*(\overline{B}_\varepsilon, A; \mathbb{Z}_2) \to H_*(A; \mathbb{Z}_2)$; let a be the largest of a_+ and a_-.

5.2. Generalization of Harnack's Inequality

The present state of the topology of real algebraic manifolds gives grounds to consider the following inequality as a justified generalization of Harnack's inequality (cf. Section 4.1):

$$\dim H_*(A; \mathbb{Z}_2) \leq \dim H_*(\mathbb{C}A; \mathbb{Z}_2) \quad \text{(even-dimensional case)} \qquad (21)$$

$$2a + \dim H_*(A; \mathbb{Z}_2) \leq \dim H_*(Y; \mathbb{Z}_2) \quad \text{(odd-dimensional case)}. \quad (22)$$

In the case of curves, the estimate (22) (as well as (21)) turns into Harnack's inequality. The principal characteristic features of M-curves can be regarded as special cases of general properties of M-manifolds (i.e., manifolds turning (21) (resp., (22)) into equality in the even-dimensional (resp., odd-dimensional) case); the same can be said about $(M-1)$-manifolds (i.e., manifolds for which the halved difference between the right-hand side and the left-hand side of the corresponding inequality is equal to 1; this difference always has an even value). The inequalities (21) and (22) are exact, at least in the case of complete regular intersections for arbitrary $m_1, \ldots, m_s$.

Both inequalities (21) and (22) are obtained from one and the same inequality appearing in the Smith theory and written out with respect to the involution of complex conjugate in $\mathbb{C}A$ for n even, and with respect to the involution of complex conjugate in Y for n odd. The inequality (21) was stated by Thom [66] in 1965. The inequality (22) and the basic properties of M-manifolds were established by Rokhlin [49] in 1972; later, similar properties were found for $(M-1)$-manifolds by Gudkov & Krakhnov [50], and also by Kharlamov [51] (see Section 5.3 for references to further publications). These inequalities were proved to be exact by Viro [67] in 1979 (their exactness, in an equivalent form, was established by Harnack for plane curves, and by Hilbert for curves belonging to surfaces of degree 2, as well as for surfaces of degree 4).

In the case of complete regular intersections, the right-hand sides of (21), (22) are polynomials of degree q and the variables $m_1, \ldots, m_s$. Explicit formulas for these polynomials are given, for instance, in [49]. The right-hand sides of (21) and (22) were represented in another form by Khovansky, who, apart from the sum of Betti numbers, considered Hodge numbers for complete regular intersections, not only in projective space but also on other toric manifolds. For hypersurfaces in $\mathbb{C}P^q$ ($s = 1$), the right-hand side of (21), according to Khovansky, can be calculated as follows. Consider hyperplanes passing through the vertices of the cube $]0, m[^q$ and orthogonal to its main diagonal. The sum of Betti numbers for a complex hypersurface is equal to q plus the number of integer points which belong to the interior of the cube and do not belong to any of the said hyperplanes.

The numbers of integer points belonging to the layers between the said hyperplanes are exactly the Hodge numbers of middle (equal to $q - 1$) dimension (for q odd, the number of points in the middle layer is reduced by 1).

In the case of hypersurfaces ($s = 1$), the highest order term in the right-hand side of (21) is equal to m^q. The same highest order term appears in

the estimate obtained in 1951 by Oleinik [78] for the sum of Betti numbers
of a non-singular hypersurface of degree m in $\mathbb{R}P^q$ (cf. Section 4.1).

In 1964, Milnor [79] gave estimates for the sum of Betti numbers for
the sets defined in $\mathbb{R}^q$ and $\mathbb{R}P^q$ by an arbitrary system of polynomial
equations and inequalities. His estimates are written in terms of polynomials
with respect to the maximal degree of these equations and inequalities: the
highest order terms in his estimates have the same order as those of the
estimates resulting from (21) and (22), but a larger coefficient.

5.3. Comparisons

Almost all comparisons mentioned in Sections 3.5 and 4.5 can be generalized
to the case of many dimensions (see [50], [51], [55], [61]). Here, we limit
ourselves to the comparisons established by Rokhlin [49], [65]: let n be even
and

$$\dim H_*(A; \mathbb{Z}_2) = \dim H_*(\mathbb{C}A; \mathbb{Z}_2) ,$$

then

$$\chi(A) \equiv \sigma(\mathbb{C}A) \bmod 16;$$

if n is odd and

$$2a_\varepsilon + \dim H_*(A; \mathbb{Z}_2) = \dim H_*(Y; \mathbb{Z}_2) ,$$

where $\varepsilon = +$ or $-$, then

$$2\chi(\overline{B}_\varepsilon) \equiv \sigma(Y) \bmod 16 .$$

5.4. Generalizations for the Inequalities of Comessatti, Petrowsky and Oleinik

In 1974, Kharlamov [68], analyzing the proof of Petrowsky's inequalities
suggested by Arnold [40], discovered the following inequalities: if n is even,
then

$$|\chi(A) - 1| \leq h^{\frac{n}{2}, \frac{n}{2}}(\mathbb{C}A) - 1 , \tag{23}$$

where $h^{a,b}$ is the dimension of the space $\mathcal{H}^{a,b}$ consisting of bihomogeneous
classes of cohomologies of degree (a, b) (see [69]); if n is odd, then

$$|\chi(\overline{B}_+) - \chi(\overline{B}_-)| \leq h_-^{\frac{n+1}{2}, \frac{n+1}{2}}(Y) , \tag{24}$$

where $h_-^{a,b} = \dim \operatorname{Ker} \{1 + \operatorname{aut}^* : \mathcal{H}^{a,b} \to \mathcal{H}^{a,b}\}$, and

$$|2\chi(\overline{B}_-) - 1| \leq h^{\frac{n+1}{2}, \frac{n+1}{2}}(Y) - 1 . \tag{25}$$

The right-hand sides of the inequalities $(23) - (25)$, in the case of complete regular intersections, have the form of polynomials in $m_1, \ldots, m_s$. As shown by Zvonilov [70], for hypersurfaces in $\mathbb{C}P^q$ ($s = 1$), the right-hand side of (23) and that of (24) coincide with the polynomial $\Pi_{q+1}(m)$ introduced by Petrowsky and Oleinik. Therefore, in this case, (23) and (24) yield the inequalities of Petrowsky-Oleinik (6) and (7).

For $q = 3$, $s = 2$, (24) implies the inequality obtained by Oleinik [71]:

$$|\chi(\overline{B}_+) - \chi(\overline{B}_-)| \leq \frac{2}{3}m_1^3 + \frac{3}{4}m_1 m_2^2 + \frac{1}{2}m_1^2 m_2 - 2m_1^2 - 2m_1 m_2 + \frac{7}{3}m_1 ; \quad (26)$$

the latter provides a generalization of Petrowsky's inequality (1) from plane curves to curves lying on surfaces in three-dimensional space.

The inequalities (23) and (24) can be improved as follows: for n even,

$$\left| \chi(A) - \sum_{k=0}^{n/2} (-1)^k T_k \right| \leq h^{\frac{n}{2}, \frac{n}{2}}(\mathbb{C}A) - \rho , \quad (27)$$

where T_k is the trace of the involution t^* on the subspace $\mathcal{P}_k^{\mathrm{alg}}$ of the space $\Omega^k \cup \mathcal{P}^{\frac{n}{2}-k, \frac{n}{2}-k}(\mathbb{C}A)$ (here $\mathcal{P}^{a,b}$ is the space of bihomogeneous primitive classes of cohomologies of degree (a, b), and Ω is a fundamental class induced by the inclusion $\mathbb{C}A \to \mathbb{C}P^q$; see [69]) generated by algebraic classes; $\rho = \sum_{k=0}^{n/2} \mathcal{P}_k^{\mathrm{alg}}$; if n is even, then

$$\left| \chi(\overline{B}_+) - \chi(\overline{B}_-) - \sum_{k=0}^{(n+1)/2} (-1)^k T_k^- \right| \leq h_-^{\frac{n+1}{2}, \frac{n+1}{2}}(Y) - \rho^- , \quad (28)$$

where T_k^- is the trace of the involution t^* on the subspace $\mathcal{P}_k^{\mathrm{alg}}$ of the space $\Omega^k \cup \mathcal{P}^{\frac{n+1}{2}-k, \frac{n+1}{2}-k}(Y)$ generated by algebraic classes ω with $\mathrm{aut}^* \omega = -\omega$; $\rho^- = \sum_{k=0}^{(n+1)/2} \dim \mathcal{P}_k^{\mathrm{alg},-}$.

For $n = 2$, (27) coincides with the inequality of Comessatti (2).

5.5. Manifolds with Singularities

This case has never been an object of systematic analysis, although in recent years its role has become considerably more prominent, and part of the results for non-singular manifolds has been extended to manifolds with singularities (see [72], [73]). These investigations started with the paper by Arnold [72], who linked the local theory of singularities to the examination of the topology of non-singular hypersurfaces in real projective space, and suggested methods for the application of modern achievements in the theory of singularities to the topology of real algebraic manifolds (see Sections 5.5–5.7). Another impetus to bring the theory of singularities into the topology of real algebraic manifolds was due to the new methods, discovered by Viro [38], for the construction of real algebraic manifolds; these methods involve the use of some complicated singularities.

In [39], Petrowsky and Oleinik also consider the question of whether the inequalities (6), (7) remain valid if the hypersurface has singular points. Moreover, they establish a stronger version of the inequality (7) in the form of (8), provided that the singularities of the hypersurface are of some special type, namely, if the hypersurface splits into a hyperplane and a non-singular hypersurface with a transversal intersection (see Section 2).

The remarks made by Petrowsky and Oleinik can be refined and extended to the situation described in Section 5.1; if we allow, in the odd-dimensional case (see Section 5.1), that $\mathbb{C}A$ has arbitrary singularities, then, for a fixed degree m, the extremal values of the difference $\chi(\overline{B}_+) - \chi(\overline{B}_-)$ remain the same as in the non-singular case; if, in the even-dimensional case, we assume that $\mathbb{C}A$ is cut out of a non-singular manifold $\mathbb{C}B$ by a hypersurface of even degree m, and allow $\mathbb{C}A$ to have arbitrary singularities, then, for m fixed, the extremal values of $\chi(A)$ will be the same as in the non-singular case. Actually, in the situations just described, one can prove even stronger estimates, obtained from (23) and (24) by replacing the Hodge numbers for the manifold in question by the Hodge-Steenbrink numbers for a close non-singular manifold; the latter numbers pertain to the root space corresponding to the eigenvalue 1 of the monodromy operator (cf. Section 5.7).

The stronger version (8) of the inequality (7) admits generalization to the case of many dimensions, in analogy to the generalizations of the inequalities (6) and (7) given in Section 5.3 (see [74], and also [75]).

5.6. The inequalities of Petrowsky-Oleinik in Terms of Vector Fields

As noticed by Arnold [72], the inequalities of Petrowsky-Oleinik can be expressed in the form of estimates for indexes of certain polynomial vector fields in a real affine space (for the inequality (1), an interpretation of this kind was given by Ragsdale [24]). On the basis of this interpretation and modern developments in the theory of singularities, Arnold [72] found new proofs for Petrowsky-Oleinik's inequalities (6), (7), and established generalized versions of these inequalities in the form of estimates for the index of a homogeneous polynomial vector field and the index of the gradient field. Subsequently, Khovansky [76] obtained further generalizations from the homogeneous to the arbitrary polynomial fields and proved his own estimates and those of Arnold (for the homogeneous fields) to be exact.

Arnold [72] introduced the gradient vector field $(\partial F/\partial x_0, \ldots, \partial F/\partial x_q)$ on $\mathbb{R}^{q+1}$ into the study of the topology of the manifolds A, $\overline{B}_+$ and $\overline{B}_-$ in the case of hypersurfaces in $\mathbb{R}P^q$, $s = 1$; (see Section 5.1 for the notations). This gradient field has a singularity only at the point $0 \in \mathbb{R}^{q+1}$, since the equation $F = 0$ is assumed to be free from singularities in $\mathbb{R}^{q+1} \setminus \{0\}$ (see Section 5.1). The index of the field at this point, as noticed by Arnold, coincides with $1 - \chi(A)$ for q odd, and with $\chi(\overline{B}_+) - \chi(\overline{B}_-)$ for q even.

Therefore, (6) and (7) can be expressed as a single inequality, namely

$$|\operatorname{ind}| \leq \Pi_{q+1}(m) . \tag{29}$$

The right-hand side of (29) is equal to the dimension (over $\mathbb{R}$) of the $\mathbb{R}$-algebra

$$\frac{\mathbb{R}[[x_0, \ldots, x_q]]}{\left(\dfrac{\partial F}{\partial x_0}, \ldots, \dfrac{\partial F}{\partial x_q} \right)}$$

reduced by the double dimension of the ideal generated in this algebra by the polynomials of degree $> \frac{1}{2}(q+1)(m-2)$ (see [72]).

Khovansky [76] passed from the polynomial F to its affine part

$$h(x_1, \ldots, x_q) = F(1, x_1, \ldots, x_q)$$

and then to the gradient vector field $\partial h/\partial x_1, \ldots, \partial h/\partial x_q$ in $\mathbb{R}^q$. Under the assumption that the field has no singularities at infinity (i.e., the system $\partial F/\partial x_1 = 0, \ldots, \partial F/\partial x_q = 0$, $x_0 = 0$, admits no real solutions) and all its singularities are isolated, the left hand sides of (6) and (7) happen to coincide with $|\operatorname{ind}_+ - \operatorname{ind}_-|$, where ind_+ is the sum of the indexes of singular points belonging to the region $h > 0$, and ind_- is the sum of the indexes in the region $h < 0$. Therefore, (6), (7) become once again expressed in terms of a single inequality

$$|\operatorname{ind}_+ - \operatorname{ind}_-| \leq \Pi_{q+1}(m) . \tag{30}$$

It should also be said that a unified expression for the inequalities (6), (7) dates back to the paper of Petrowsky and Oleinik (see [39], Lemma 4), where it has the form[1]

$$\left| \chi(M_0) - \frac{1}{2} \left(\chi(M_{c_1}) + \chi(M_{c_2}) \right) \right| \leq \frac{1}{2} \Pi_{q+1}(m) ; \tag{31}$$

here M_c is the closure of the subset in $\mathbb{R}P^q$ defined in $\mathbb{R}^q \subset \mathbb{R}P^q$ by the inequality $h(x_1, \ldots, x_q) \leq c$; and the interval $]c_1, c_2[$ contains all real critical

---

[1]The symbol $<$ is replaced by $\leq$ and, accordingly, the right-hand side is reduced by one.

points of the polynomial h. Formally, no indexes are involved in (31). On
the other hand, the proof (of Lemma 4 in [39]) suggested by Petrowsky
and Oleinik contains all the basic elements necessary for passing from (31)
to (30).

5.7. Generalization to Arbitrary Polynomial Fields

Arnold [72] proved the inequality (29) to remain valid for an arbitrary (not
only gradient) vector field in $\mathbb{R}^{q+1}$ which has no singularities in $\mathbb{R}^{q+1} \setminus \{0\}$
and is formed by homogeneous polynomials $F_0, \ldots, F_q$ of degree $m - 1$.

Further generalizations are due to Khovansky [76], who considered a
vector field in $\mathbb{R}^q$ formed by polynomials $h_1, \ldots, h_q$ of degrees $m_1, \ldots, m_q$,
respectively. It is assumed that all real singular points are isolated and the
field has no singularities at infinity. As shown by Khovansky, in this case
we have

$$|\mathrm{ind}| \leq \Pi(m_1, \ldots m_q) , \tag{32}$$

where ind is the sum of indexes of the field's singular points; $\Pi(m_1, \ldots, m_q)$
is the number of integer points of $\mathbb{R}^q$ belonging to the cross-section of the
parallelepiped $0 \leq x_1 \leq m_1 - 1, \ldots, 0 \leq x_q \leq m_q - 1$ by the hyperplane
$x_1 + \cdots + x_q = \frac{1}{2}(m_1 + \cdots + m_q - q)$. Moreover, Khovansky considered
the case with an additional given polynomial h_0 of degree m_0 such that the
hypersurface $h_0 = 0$ contains no singular points of the field. In this situation
he proved the inequality

$$|\mathrm{ind}_+ - \mathrm{ind}_-| \leq \Pi(m_1, \ldots, m_q; m_0) , \tag{33}$$

where ind_+ (resp., ind_-) is the sum of indexes of the field's singular points
in the region $h_0 > 0$ (resp., $h_0 < 0$), and $\Pi(m_1, \ldots, m_q; m_0)$ is the number
of integer points of the parallelepiped $0 \leq x_1 \leq m_1 - 1, \ldots, 0 \leq x_q \leq m_q - 1$
satisfying the inequalities

$$\frac{1}{2}(m_1 + \cdots + m_q - q - m_0) \leq x_1 + \cdots + x_q \leq \frac{1}{2}(m_1 + \cdots + m_q - q + m_0) .$$

Khovansky constructed some vector fields, which show the estimates (32),
(33) to be exact for any $m_1, \ldots, m_q, m_0$. Moreover, Khovansky gave an
exhaustive answer to the question about the pairs ind_+, ind_- that can
be realized if $q, m_1, \ldots, m_q, m_0$ are fixed and the aforesaid conditions of
non-degeneracy are satisfied.

Both inequalities (32) and (33) generalize the estimates of Petrowsky-
Oleinik, since (32) implies (29), and (33) implies (30). Finally, (32) coincides
with (33) with $m_0 = 0$. The inequality of Oleinik (26) can be incorporated
into approximately the same scheme, which can be easily seen from [71].

How the estimates (23), (24) for complete intersections with $s \geq 2$ and $q > 3$ can be obtained along these lines is still unknown.

In [76], apart form the estimates (32), (33), one can find estimates of the same type in the (degenerate) case of a field with singularities at infinity.

The main tool in Arnold's proof of the inequality (29) is the well-known Levin-Eisenbud-Khimshiashvili formula; in its stead, the Euler-Jacobi formula was used by Khovansky. Later, Khovansky [77] extended the Euler-Jacobi formula from systems consisting of polynomials of fixed degrees to systems consisting of polynomials with fixed Newton polyhedra. As shown in [77], the generalized Euler-Jacobi formula can be used to obtain the corresponding generalizations for the inequalities (32), (33), and therefore, the corresponding generalizations for the Petrowsky-Oleinik inequalities.

5.8. Generalization to Gradient Fields

Arnold [72] also suggested an alternative type of generalization, which makes the subject matter of the preceding section somewhat nearer to that of Section 5.3.

Consider a real analytic function of $2k$ variables with a critical point of finite multiplicity, and let ind be the index of the gradient of this function at this point. According to Arnold, we have

$$|\operatorname{ind}| \leq h_1^{k,k} , \tag{34}$$

where $h_1^{k,k}$ is the Hodge number of degree (k, k) for the root space corresponding to the eigenvalue 1 of the monodromy operator in the space of cohomologies of a local non-singular level-manifold of this function. If the function is a homogeneous polynomial and $2k = q+1$, then (34) implies (29), as shown by direct calculation. For a function of an odd number of independent variables, in order to estimate the index of its gradient, it suffices to consider the sum of this function and a squared new variable; in particular, this allows us to obtain (29) from (34) in the case of an odd $q+1$. Thereby, (34) becomes another generalization of the Petrowsky-Oleinik inequalities.

The proof of (34), as suggested by Arnold, relies on the Lefschetz theorem about fixed points, as well as on the major properties of the mixed Hodge structure (constructed by Steenbrink) in the cohomologies of a local non-singular level-manifold of a function.

Of immediate relevance to the topics considered in this commentary are the papers [81]–[83] and the review [84].

References

[1] Brusotti, L.; Galafassi, V.E. Topologia degli enti algebrici reali. In: *Atti del Quinto Congresso dell'Unione Matematica Italiana Tenuto a Pavia. Torino, 1955*. Roma, Edizione Cremonese, 1956, pp. 57–84.

[2] Brusotti, L. Su talune questioni di realità, nei loro metodi, risultati e problemi. In: *Colloq. sur les Questions de Réalité en Géométrie. Liege, 1955*. Paris, Masson, 1956, pp. 105–129.

[3] Galafassi, V.E. Classici e recenti sviluppi sulle superficie algebraiche reali. In: *Colloq. sur les Questions de Réalité en Géométrie. Liege, 1955*. Paris, Masson, 1956, pp. 130–147.

[4] Gudkov, D.A. Topology of real projective algebraic manifolds. *Uspekhi Mat. Nauk* **29**:4 (1974) 3–79.

[5] Wilson, G. Hilbert's sixteenth problem. *Topology* **17** (1978) 53–73.

[6] Rochlin, V.A. Complex topological characteristics of real algebraic curves. *Uspekhi Mat. Nauk* **33**:5 (1978) 77–89.

[7] Kharlamov, V.M. Real algebraic surfaces. In: *Proc. Intern. Congr. Math. Helsinki, 1978*, Vol. 1, pp. 421–428.

[8] Arnold, V.I.; Oleinik, O.A. Topology of real algebraic manifolds. *Vestnik Mosk. Univ. Mat. Mekh.* **6** (1979) 7–17.

[9] Viro, O.Ya. Advances in the topology of real algebraic manifolds during recent 5 years. In: *Proc. Intern. Congr. Math. Warszawa, 1983*, Vol. 1, pp. 595–611.

[10] Nuij, W. A note on hyperbolic polynomials. *Math. Scand.* **23** (1968) 69–72.

[11] Gudkov, D.A.; Shustin, E.I. Classification of non-singular curves of degree 8 on an ellipsoid. In: *Methods in Qualitative Theory of Differential Equations*. Gorky, 1980, pp. 104–107.

[12] Khovansky, A.G. On a class of systems of transcendental equations. *Dokl. Akad. Nauk SSSR* **255**:4 (1980) 804–807.

[13] Zvonilov, V.I. Complex orientations of real algebraic curves with singularities. *Dokl. Akad. Nauk SSSR* **268**:1 (1983) 22–26.

[14] Krasnov, V.A. Harnack-Thom's inequalities for mappings of real algebraic manifolds. *Izvest. Akad. Nauk SSSR, Ser. Mat.* **47**:2 (1983) 268–297.

[15] Shustin, E.I. Hilbert-Rohn's method and the elimination of singularities on real algebraic curves. *Dokl. Akad. Nauk SSSR* **281**:1 (1985) 33–36.

[16] Harnack, A. Über die Vieltheiligkeit der ebenen algebraischen Kurven. *Math. Ann.* **10** (1876) 189–198.

[17] Schläfli, L. On the distribution of surfaces of the third order into species, in reference to the absence or presence of singular points, and the reality of their lines. *Phil Trans. Roy. Soc. London* **153** (1863) 195–241.

[18] Zeuthen, H.G. Sur les différentes formes des courbes du quatrième ordre. *Math. Ann.* **7** (1874) 410–432.

[19] Zeuthen, H.G. Études des propriétés de situation des surfaces cubiques. *Math. Ann.* **8** (1875) 1–30.

[20] Klein, F. Über Flächen dritter Ordnung. *Math. Ann.* **6** (1873) 551–581.

[21] Klein, F. Über eine neue Art von Riemann'sche Flächen. *Math. Ann.* **10** (1976) 398–416.

[22] Klein, F. Über den Verlauf der Abel'schen Integrale bei den Kurven vierten Grades. *Math. Ann.* **10** (1876) 364–397.

[23] Hilbert, D. Über die reellen Zuge algebraischer Kurven. *Math. Ann.* **38** (1891) 115–138.

[24] Ragsdale, V. On the arrangement of the real branches of plane algebraic curves. *Amer. J. Math.* **28** (1906) 377–404.

[25] Rohn, K. Die ebene Kurve 6. Ordnung mit elf Ovalen. *Ber. Verhandl. Leipzig* **63** (1911) 540–555.

[26] Rohn, K. Maximalzahl und Anordnung der Ovale bei der ebenen Kurve 6. Ordnung und bei der Flache 4. Ordnung. *Math. Ann.* **73** (1913) 177–229.

[27] Gudkov, D.A.; Utkin, G.A. Topology of 6th order curves and 4th order surfaces. *Ucheniye Zapiski Gorkovsk. Univ.* **87** (1969) 3–213.

[28] Hilbert, D. Über die Gestalt einer Fläche vierter Ordnung. *Nachr. Kgl. Ges. Wiss. Göttingen. Math.-phys.* **K1** (1909) 308–313.

[29] Rohn, K. Die Maximalzahl von Ovalen bei einer Fläche 4.Ordnung. *Ber. Verhandl. Leipzig* **63** (1911) 423–440.

[30] Rohn, K. Die Flächen 4.Ordnung hinsichtlich ihrer Knotenpunkte und ihrer Gestaltung. *Preisschriften herausgegeben von der Fürstlich Jablonowskieschen Gesellschaft.* Leipzig, 1886.

[31] Comessatti, A. Sulla connessione delle superficie algebriche reali. *Verhandl. Intern. Math. Kongr. Zürich* **2** (1932) 169.

[32] Comessatti, A. Sulla connessione e sui numeri base delle superficie algebriche reali. *Rend. Semin. Mat. Univ. Padova* **3** (1932) 141–162.

[33] Comessatti, A. Sulle varietá abeliane reali. *Ann Mat.* (4), 1924–1925, Vol. 2, pp. 67–106.

[34] Comessatti, A. Sulle varietá abeliane reali. *Ann Mat.* (4), 1925–1926, Vol. 3, pp. 27–71.

[35] Comessatti, A. Problemi di realtá per le superficie e varietá algebriche. *Reale Accad. Ital. Atti del 9 Convegno, Roma,* 1943, Vol. 9, pp. 15–41.

[36] Petrowsky, I.G. Sur la topologie des courbes planes réelles et algebriques. *C.R. Acad. Sci. Paris* **197** (1933) 1270–1272.

[37] Petrowsky, I.G. On the topology of real plane algebraic curves. *Ann. Math.* **39** (1938) 189–209. (See Article 7 of this volume).

[38] Viro, O.Ya. Curves of degree 7, curves of degree 8 and Ragsdale's conjecture proved false. *Dokl. Akad. Nauk SSSR* **254:6** (1980) 1306–1310.

[39] Petrowsky, I.G.; Oleinik, O.A. On the topology of real algebraic surfaces. *Izvest. Akad. Nauk SSSR. Ser. Mat.* **13:5** (1949) 389–402.

[40] Arnold, V.I. On the position of the ovals of real plane algebraic curves, involutions of four-dimensional smooth manifolds, and arithmetic of integer valued quadratic forms. *Funkt. Anal. Prilozh.* **5:3** (1971) 1–9.

[41] Rochlin, V.A. Complex orientations of real algebraic curves. *Funkt. Anal. Prilozh.* **8:4** (1974) 71–75.

[42] Viro, O.Ya. Generalization of the inequalities of Petrowsky and Arnold to curves with singularities. *Uspekhi Mat. Nauk* **33:3** (1978) 145–146.

[43] Kharlamov, V.M. Petrowsky's inequalities for real plane curves. *Uspekhi Mat. Nauk* **33:3** (1978) 146.

[44] Zvonilov, V.I. A stronger version of the inequality of Petrowsky and Arnold for curves of odd degree. *Funkt. Anal. Prilozh.* **13:4** (1979) 31–39.

[45] Petrowsky, I.G. On the diffusion of waves and the lacunas for hyperbolic equations. *Matem. Sbornik* **17:3** (1945) 289–370. (See Article 6 of this volume).

[46] Mishachev, N.M. Complex orientations of plane M-curves of odd degree. *Funkt. Anal. Prilozh.* **9:4** (1975) 77–78.

[47] Marin, A. Quelques remarques sur les courbes algébriques planes réelles. Preprint no. 2205. Orsay, 1979.

[48] Fidler, T. Bundles of straight lines and the topology of real algebraic curves. *Izvest. Akad. Nauk SSSR. Ser. Mat.* **46:4** (1982) 853–863.

[49] Rokhlin, V.A. Comparisons mod 16 in the 16th problem of Hilbert. *Funkt. Anal. Prilozh.* **6:4** (1972) 58–64.

[50] Gudkov, D.A.; Krakhnov, A.D. On the periodicity of the Euler characteristic of real algebraic $(M-1)$-manifolds. *Funkt. Anal. Prilozh.* **7:2** (1973) 15–19.

[51] Kharlamov, V.M. New comparisons for the Euler characteristic of real algebraic manifolds. *Funkt. Anal. Prilozh.* **7:2** (1973) 74–78.

[52] Nikulin, B.B. Involutions of integer valued quadratic forms and their applications to real algebraic geometry. *Izvest. Akad. Nauk SSSR. Ser. Mat.* **47:1** (1983) 109–188.

[53] Fidler, T. New comparisons in the topolofy of real plane algebraic curves. *Dokl. Akad. Nauk SSSR* **270:1** (1983) 56–58.

[54] Rokhlin, V.A. A proof of Gudkov's hypothesis. *Funkt. Anal. Prilozh.* **6:2** (1972) 62–64.

[55] Nikulin, V.V. Integer valued symmetric bilinear forms and their applications in geometry. *Izvest. Akad. Nauk SSSR, Ser. Mat.* **43:1** (1979) 111–177.

[56] Kharlamov, V.M. Rigid isotopic classification of real non-singular plane curves of degree 5. *Funkt. Anal. Prilozh.* **15:1** (1981) 88–89.

[57] Rokhlin, V.A. New inequalities in the topology of real plane algebraic curves. *Funkt. Anal. Prilozh.* **14:1** (1980) 37–43.

[58] Kharlamov, V.M. Topological types of non-singular surfaces of degree 4 in $\mathbb{R}P^3$. *Funkt. Anal. Prilozh.* **10:4** (1976) 55–68.

[59] Viro, O.Ya. Construction of multi-component real algebraic surfaces. *Dokl. Akad. Nauk SSSR* **248:2** (1979) 279–282.

[60] Kharlamov, V.M. The maximal number of components for a surface of degree 4 in $\mathbb{R}P^3$. *Funkt. Anal. Prilozh.* **6:4** (1972) 101.

[61] Kharlamov, V.M. Additional comparisons for the Euler characteristic of even-dimensional real algebraic manifolds. *Funkt. Anal. Prilozh.* **9:2** (1975) 51–60.

[62] Utkin, G.A. Construction of an M-surface of degree 4 in $\mathbb{R}P^3$. *Funkt. Anal. Prilozh.* **8:2** (1974) 91–92.

[63] Kharlamov, V.M. Isotopic types of non-singular curves of degree 4 in $\mathbb{R}P^3$. *Funkt. Anal. Prilozh.* **12:1** (1978) 86–87.

[64] Kharlamov, V.M. On the classification of non-singular surfaces of degree 4 in $\mathbb{R}P^3$ with respect to rigid isotopies. *Funkt. Anal. Prilozh.* **7:2** (1984) 49–56.

[65] Rokhlin, V.A. Comparisons mod 16 in the 16th problem of Hilbert. *Funkt. Anal. Prilozh.* **7:2** (1973) 91–92.

[66] Thom, R. Sur l'homologie des variétés algébriques réelles. In: *Differential and Combinatorial Topology. Symposium in Honor of M.Morse.* Princeton Univ. Press, 1965, pp. 255–265.

[67] Viro, O.Ya. Construction of M-manifolds. *Funkt. Anal. Prilozh.* **13:3** (1979) 71–72.

[68] Kharlamov, V.M. Generalized Petrowsky's inequality. *Funkt. Anal. Prilozh.* **8:2** (1974) 50–56.

[69] Weil, A. *Introduction à l'Étude des Variétés Kählériennes.* Paris, Hermann, 1958.

[70] Zvonilov, V.I. The inequalities of Kharlamov and the inequalities of Petrowsky-Oleinik. *Funkt. Anal. Prilozh.* **9:2** (1975) 69–70.

[71] Oleinik, O.A. On the topology of real algebraic curves on an algebraic surface. *Mat. Sbornik* **29:1** (1951) 133–156.

[72] Arnold, V.I. The index of a vector field's singular point, Petrowsky-Oleinik's inequalities and mixed Hodge structures. *Funkt. Anal. Prilozh.* **12:1** (1978) 1–14.

[73] Viro, O.Ya.; Kharlamov, V.M. Comparisons for real algebraic curves with singularities. *Uspekhi Mat. Nauk* **35:4** (1980) 154.

[74] Kharlamov, V.M. Generalized Petrowsky's inequality. II. *Funkt. Anal. Prilozh.* **9:3** (1975) 93–94.

[75] Krasnov, V.A. Generalized Petrowsky's inequality in the case of odd degree. *Funkt. Anal. Prilozh.* **10:2** (1976) 41–48.

[76] Khovansky, A.G. Index of a polynomial vector field. *Funkt. Anal. Prilozh.* **13:1** (1979) 49–58.

[77] Khovansky, A.G. Newton's polyhedra and the formula of Euler-Jacobi. *Uspekhi Mat. Nauk* **33:6** (1978) 237–238.

[78] Oleinik. O.A. Estimates for the Betti numbers of real algebraic hypersurfaces. *Mat. Sbornik* **28:3** (1951) 635–640.

[79] Milnor, J. On the Betti numbers of real varieties. *Proc. Amer. Math. Soc.* **15** (1964) 275–280.

[80] Khovansky, A.G. Fewnomials and Pfaff manifolds. In: *Proc. Intern. Congr. Math. Warszawa, 1983,* Vol. 1, pp. 549–563.

[81] Gusein-Zade, S.M. Index of a gradient vector vield. *Funkt. Anal. Prilozh.* **18:7** (1984) 7–12.

[82] Varchenko, A.N. On the local residue and the form of intersections in vanishing cohomologies. *Izvest. Akad. Nauk SSSR. Ser. Mat.* **49:1** (1985) 32–54.

[83] Krasnov, V.A. Degenerations of real algebraic manifolds. *Izvest. Akad. Nauk SSSR. Ser. Mat.* **49:4** (1985) 798–827.

[84] Viro. O.Ya. Advances in the topology of real algebraic manifolds in recent 6 years. *Uspekhi Mat. Nauk* **41:3** (1986) 45–67.

List of Petrowsky's Scientific Publications[1]

1928 **Einige Bemerkungen zu den Arbeiten von Herren O. Perron und L.A. Lusternik über das Dirichlet'sche problem. *Mat. Sbornik* **35:1** (1928) 105–110.

1929 Sur les fonctions primitives par rapport à une fonction continue arbitraire. *C.R. Acad. Sci. Paris* **189** (1929) 1242–1244.

1933 Sur la topologie des courbes planes réelles et algébriques. *C.R. Acad. Sci. Paris* **197** (1933) 1270–1272.

1934 **On the reduction of the second variation to canonical form by triangular transformations. *Uchebniye Zapisky Mosk. Univ.* **2:2** (1934) 5–16. (In collaboration with L.A. Lyusternik).

Sur l'unicité de la fonction primitive par rapport à une fonction continue arbitraire. *Mat. Sbornik* **41:1 (1934) 48–59.

Über die Lösung der ersten Randwertaufgabe der Wärmeleitungsgleichung. *Uchebniye Zapiski Mosc. Univ.* **2:2** (1934) 55–60.

Über das Verhalten der Integralkurven eines Systems gewöhnlicher Differentialgleichungen in der Nähe eines singulären Punktes. *Mat. Sbornik* **41:1 (1934) 107–156.

Über das Irrfahrtproblem. *Math. Ann.* **109:3 (1934) 425–444.

1935 **Zur ersten Randwertaufgabe der Wärmeleitungsgleichung. *Compos. Math.* **1:3** (1935) 383–419.

Nachtrag zu meiner Arbeit "Über das Verhalten der Integralkurven eines Systems gewöhnlicher Differentialgleichungen in der Nähe eines singulären Punktes." *Mat. Sbornik* **42:3** (1935) 403.

1936 On the works of Academician J. Hadamard on partial differential equations. *Uspekhi Mat. Nauk* **2** (1936) 82–91. (In collaboration with S.L. Sobolev).

---

[1]The articles included in Parts I and II are marked by * and **, respectively.

551

Sur le problème de Cauchy pour un système d'équations aux dérivées partielles dans le domaine réel. *C.R. Acad. Sci. Paris* **202** (1936) 1010–1012.

Sur le problème de Cauchy pour un système linéaire d'équations aux dérivées partielles dans un domaine réel. *C.R. Acad. Sci. Paris* **202** (1936) 1246–1248.

1937 On systems of differential equations that can have only analytic solutions. *Dokl. Akad. Nauk SSSR* **17:7** (1937) 339–342.

*On the Cauchy problem in classes of non-analytic functions. *Uspekhi Mat. Nauk* **3** (1937) 234–238.

*Über das Cauchysche Problem für Systeme von partiellen Differentialgleichungen. *Mat. Sbornik* **2:5** (1937) 815–870.

Studies of the diffusion with the increasing quantity of the substance; its application to a biological problem. *Bull. Mosk. Univ. Mat. Mekh.* **1:6 (1937) 1–26. (In collaboration with A.N. Kolmogorov and N.S. Piskunov.)

1938 *On the topology of real plane algebraic curves. *Ann. Math.* **39:1** (1938) 189–209.

*On the Cauchy problem for systems of linear partial differential equations in classes of non-analytic functions. *Bull. Mosk. Univ. Mat. Mekh.* **1:7** (1938) 1–72.

1939 *Lectures on the Theory of Ordinary Differential Equations.* Moscow, ONTI, 1939.

*Sur l'analyticité des solutions des systèmes d'équations différentielles. *Mat. Sbornik* **5:1** (1939) 3–70.

1941 **On the first boundary value problem (the Dirichlet problem) for elliptic equations and some properties of functions satisfying these equations. *Uspekhi Mat. Nauk* **8** (1941) 8–31. (In collaboration with S.N. Bernstein.)

Perron's method for solving the Dirichlet problem. *Uspekhi Mat. Nauk* **8 (1941) 107–114.

A new proof of the existence of solutions for the Dirichlet problem by the method of finite differences. *Uspekhi Mat. Nauk* **8 (1941) 161–170.

1943 On the dependence of solutions of the Cauchy problem on the initial data. *Dokl. Akad. Nauk SSSR* **38:5-6** (1943) 163–165.

1944 *On the diffusion of waves and the lacunas for systems of hyperbolic equations. *Izvest. Akad. Nauk SSSR, Ser. Mat.* **8:3** (1944) 101–106.

1945 *On the diffusion of waves and the lacunas for hyperbolic equations. *Mat. Sbornik* **17:3** (1945) 289–370.

On propagation of discontinuities of displacement derivatives on the surface of a non-homogeneous elastic body of arbitrary shape. *Dokl. Akad. Nauk SSSR* **47:4 (1945) 258–261.

1946 **On some problems in the theory of partial differential equations. *Uspekhi Mat. Nauk* **1:3-4** (1946) 44–70.

1947 On the theory of partial differential equations. In: *A Collection of Articles Dedicated to the 30th Anniversary of the Great October Socialist Revolution.* Moscow, Izd.-vo Akad. Nauk SSSR, 1947, Vol. 1, pp. 214–230.

Lectures on the Theory of Ordinary Differential Equations. 2d Ed. Moscow, Gostekhizdat, 1947.

On the behavior of integral curves of a system of ordinary differential equations near a singular point. In: Poincaré, A. *On Curves Defined by Differential equations.* Moscow, Gostekhizdat, 1947, 336–347.

1948 *Lectures on the Theory of Integral Equations.* Moscow, Gostekhizdat, 1948.

1949 On the topology of real algebraic surfaces. *Dokl. Akad. Nauk SSSR* **67:1** (1949) 31–32. (In collaboration with O.A. Oleinik.)

*On the topology of real algebraic surfaces. *Izvest. Akad. Nauk SSSR, Ser. Mat.* **13:5** (1949) 389–402. (In collaboration with O.A. Oleinik.)

On topological properties of algebraic curves and surfaces. *Vestnik Mosk. Univ. Mat. Mekh.* **11** (1949) 23–27.

Lectures on the Theory of Ordinary Differential Equations. 3d Ed. Moscow, Gostekhizdat, 1949.

1950 *Lectures on Partial Differential equations.* Moscow, Gostekhizdat, 1950.

1951 On the topology of real algebraic surfaces. *Uspekhi Mat. Nauk* **6:4** (1951) 208–209. (In collaboration with O.A. Oleinik.)

Lectures on the Theory of Integral Equations. 2d Ed. Moscow, Gostekhizdat, 1951.

1952 *Lectures on the Theory of Ordinary Differential Equations.* 4th Ed. Moscow, Gostekhizdat, 1952.

1953 *Lectures on Partial Differential equations.* 2d Ed. Moscow, Gostekhizdat, 1953.

1954 On lines and surfaces of discontinuity of solutions of the wave equation. *Uspekhi Mat. Nauk* **9:3** (1954) 175–180. (In collaboration with L.A. Chudov.)

1955 **On the number of limit cycles for the equation $\frac{dy}{dx} = \frac{P(x,y)}{Q(x,y)}$, where P and Q are polynomials of degree 2. *Mat. Sbornik* **37:2** (1955) 209–250. (In collaboration with E.M. Landis.)

On the number of limit cycles for the equation $\frac{dy}{dx} = \frac{M(x,y)}{N(x,y)}$, where M and N are polynomials of degree 2. *Dokl. Akad. Nauk SSSR* **102:1** (1955) 29–32. (In collaboration with E.M. Landis.)

1956 Some remarks concerning my articles on the Cauchy problem. *Mat. Sbornik* **39:2** (1956) 267–272.

Ordinary differential equations. In: *Mathematics, Its Sibject, Methods, and Meaning.* Moscow, Izd.-vo Akad Nauk SSSR. Vol. 2, pp. 3–47.

1957 On the number of limit cycles for the equation $\frac{dy}{dx} = \frac{P(x,y)}{Q(x,y)}$, where P and Q are polynomials. *Mat. Sbornik* **43:2** (1957) 149–168. (In collaboration with E.M. Landis.)

On the number of limit cycles for the equation $\frac{dy}{dx} = \frac{P(x,y)}{Q(x,y)}$, where P and Q are polynomials of degree n. *Dokl. Akad. Nauk SSSR* **113:4** (1957) 748–751. (In collaboration with E.M. Landis.)

1958 *Theory of systems of partial differential equations. In: *Procceedings of the 3d All-Union Mathematical Congress.* Moscow, Izd.-vo Akad. Nauk SSSR, 1958, Vol. 3, pp. 65–72. (In collaboration with I.M. Gelfand and G.E. Shilov.)

1959 Corrections to the articles: "On the number of limit cycles for the equation $\frac{dy}{dx} = \frac{P(x,y)}{Q(x,y)}$, where P and Q are polynomials of degree 2" and "On the number of limit cycles for the equation $\frac{dy}{dx} = \frac{P(x,y)}{Q(x,y)}$, where P and Q are polynomials". *Mat. Sbornik* **49:2** (1959) 253–255. (In collaboration with E.M. Landis.)

1961 The contribution of S.N. Bernstein to the theory of partial differential equations. *Uspekhi Mat. Nauk* **16:2** (1961) 5–20. (In collaboration with N.I. Akhiezer.)

Lectures on Partial Differential equations. 3d Ed. Moscow, Izd.-vo Akad. Nauk SSSR, 1961.

1964 *Lectures on the Theory of Ordinary Differential Equations.* 5th Ed. Moscow, Nauka, 1964.

1965 *Lectures on the Theory of Integral Equations.* 3d Ed. Moscow, Nauka, 1965.

1967 Letter to the editor. *Mat. Sbornik* **73:1** (1967) 160.

1970 *Lectures on the Theory of Ordinary Differential Equations.* 6th Ed. Moscow, Nauka, 1970.

1984 *Lectures on the Theory of Ordinary Differential Equations.* 7th Ed. Moscow Univ. Press, 1984.

   Lectures on the Theory of Integral Equations. 4d Ed. Moscow Univ. Press, 1984.

Index

9 780367 449162